MICROBIOLOGY SERIES

Series Editor: Dr. Ajit Varma

MICROBES

FOR

HUMAN LIFE

Editors

Ashok Kumar Chauhan

Chairman, AKC Group of Companies
Founder President
Ritnand Balved Education Foundation
Amity University, Uttar Pradesh, India

Ajit Varma

Amity Institute of Herbal and
Microbial Studies
Noida, India

Harsha Kharkwal

Amity Institute of Herbal and
Microbial Studies
Noida, India

First published in the UK by

ANSHAN LTD
In 2007

6 Newlands Road,
Tunbridge Wells,
Kent,
TN4 9AT. UK

Tel: +44 (0) 1892 557767
Fax: +44 (0) 1892 530358

e-mail: info@anshan.co.uk
Web Site: www.anshan.co.uk

ISBN 978 1905740 765

British Library Cataloguing in Publication Data
A catalogue record for this book is available from the British Library

Printed in India

R A Mashelkar, FRS
CSIR Bhatnagar Fellow
President, Indian National Science Academy
President, Global Research Alliance

National Chemical Laboratory
Dr. Homi Bhabha Road, Pashan
Pune - 411 008, India

Foreword

The great scientific news that greeted this century was the campaign to decode the human genome. Much of the biological composition of our bodies consists of genomes other than that of humans. Several bacteria and viruses occupy our skin, mucous membranes, and intestinal tract. They are likely to play a major role in the development of disease.

Throughout the history, infectious diseases have been known to regulate our lives. It was in the 20th century, thanks to simple hygienic measures like washing hands regularly and separating drinking water from sewage runoff, that we have played a major role, for better or for worse, in controlling the microorganisms.

Microbes have a profound impact on every facet of human life. Pathogens harm us, yet other microbes protect us. Some microbes are pivotal in the growth of food crops, but others can kill the plants or spoil the produce. Technological advances and broad knowledge of the microbial world have brought about significant advances in one of the special areas of biology— microbiology. Microbes are the ancestors of all the complex and varied biological forms that now exist on earth.

This book begins with the historical development of microorganisms related to human health. In the next chapter, the therapeutic value of medicinal plants for various diseases is discussed. It also reflects on the present trend in health care i.e., the role and growing use of neutraceuticals. Continuing on the same subject, the role of microbes has been stressed by taking the fungus *Piriformospora indica* as an example, which not only increases the overall productivity, but also increases the content of secondary metabolites present in it. Food is another field, where microbes have a definite and defined role to play.

Some chapters in the book deal with the therapeutic value of fermented foods, like yogurt, and traditional native fermented foods of India, which are famous not only for their rich taste and delicacy, but also for their use against different diseases.

Infectious diseases like Tuberculosis, Malaria, AIDS, SARS, etc., with their clinical and socio-economic impact on Indian subcontinent and the entire world have been dealt with effectively.

A chapter describing biosorption as an economically feasible technology for efficient removal and recovery of metal(s) from aqueous solution is a welcome addition. The importance of genetically engineered micro-organisms, with an enhanced metal uptake efficiency, has been highlighted.

Strategies in the form of enzyme therapeutics have been discussed for controlling antimicrobial drug resistance. The last two chapters deal with proteomics and genomics as tools to study plant-microbe cross communication.

I am sure this book will prove to be very useful for the undergraduate and postgraduate students of biological sciences and technology, environmental science, soil science, etc.

March, 2007

R. A. Mashelkar

Preface

Microbes for Human Life describes the extraordinary impact of the microbial community in every sphere of our life.

The discovery of microorganisms and their effects on human body opened a new area of biology in which prevention and cure of devastating infectious diseases became possible. New insights into the beneficial aspects of microorganisms gave birth to the biotechnology industry, and introduced many useful products and processes. Studies on microorganisms have also revealed the secrets of genetic code—the blue print of all living things. These secrets have changed the direction of biology and led to new concepts and products, which will improve human health and welfare for many years to come.

The book begins with the historical development of microorganisms related to human health. The next chapter reflects on the therapeutic value of medicinal plants. It also describes present trends in health care i.e., role and growing use of neutraceuticals. Moving further on the same context, the role of microbes beneficial for medicinal plants has been stressed by taking the fungus *Piriformospora indica* as an example, which not only increases the overall productivity, but also increases the content of secondary metabolites present in it. Food is another aspect that has not remained untouched by the microbes. Some of the chapters deal with the therapeutic value of fermented foods, like yogurt, and traditional native fermented foods of India, which are not only famous for their rich taste and delicacy, but also for their use against different diseases.

One of the chapters describing *biosorption* as an economically feasible technology for efficient removal and recovery of metal(s) from aqueous solution has also been widely covered. The other chapter focuses on the potential of microorganisms to degrade the dyes present in the textile effluent, and factors affecting dye biodegradation in contaminated wastewaters.

Subsequent chapters deal with infectious diseases like Tuberculosis, Malaria, AIDS, Ebola hemorrhagic fever, and SARS, with their clinical and socio-economic impact on Indian subcontinent and the entire world, and the measures taken to counter these infections. A ray of hope is illustrated by one of the authors for completely curing Malaria by the combination of gene and protein predictions with micro-array and proteomic data, which will utilize only significant stage specific proteins or enzymes as components of multivalent vaccine, filtering out others that are not relevant. For Avian Flu, the author prepares us for the pandemic, emphasizing on the need to develop the vaccine that would prevent humans from catching the Flu.

One of the chapters illustrates the need for the search of a novel antibiotic.

Another chapter, taking *Candida albicans* as the theme, reveals the use of genomic information as an additional exploration tool in drug target screening for antifungal drug discovery.

Strategies in the form of enzyme therapeutics have been discussed in the following chapters for controlling anti-microbial drug resistance. Subsequent two chapters deal with proteomics and genomics as a tool to study plant-microbe cross-communication. The last chapter illustrates the benefits of polyunsaturated fatty acids, which some of the microorganisms are capable of producing in their lipid.

Microbes are the life support system of this planet, and a key to the life on other planets. They are sometimes our enemies, but more often our friends. They are the ultimate renewable resource, and a potential goldmine of new processes and products. Surely, we need to know as much about them as we can.

Our heartfelt gratitude to all the contributors/authors of the chapters, without whose efforts this volume would not have been possible.

Ashok Kumar Chauhan
Harsha Kharkwal
Ajit Varma

Contents

List of Contributors

A. C. Kharkwal, Amity Institute of Microbial Sciences, Amity University Uttar Pradesh, Noida- 201303, India E-mail: ackharkwal@amity.edu

A. Das, Amity Institute of Microbial Sciences, Amity University Uttar Pradesh, Noida-201303, India E-mail: adas@ amity.edu

A. K. Chauhan, AKC House, E-27, Defence Colony, New Delhi -110024, India E-mail: drchauhan@akcgroup.com

A. K. Dubey, School of Applied Sciences and Division of Biotechnology, Subhash Institute of Technology, Dwarka, New Delhi-11075, India E-mail: akdubey@nstic.ac.in

A. Madni, Department of Biotechnology, Faculty of Science, Hamdard University, Hamdard Nagar, New Delhi–110062, India E-mail: skjain@jamiahamdard.ac.in

A. P. Garg, Chaudhari Charan Singh University, Meerut-250005, Uttar Pradesh, India E-mail: amarprakashgarg@yahoo.com

A. Sharma, Special Centre for Molecular Medicine, Jawaharlal Nehru University, New Delhi-110067, India E-mail: skdhar2002@yahoo.co.in

A. Singh, Department of Biology, University of Waterloo, Waterloo, Ontario N2L 3G1, Canada E-mail: ajasingh@sciborg.uwterloo.ca

A. Singh, Petrozyme Techonologies, 7496 Wellington Road 34 R.R.#3 Guelph, Ontario, N1H6H9, Canada E-mail: asingh@petrozyme.com

A. Varma, Amity Institute of Microbial Sciences, Amity University Uttar Pradesh, Noida-201303, India E-mail: ajitvarma@aihmr.amity.edu

C. S. Nautiyal, Division of Plant-Microbe Interactions, National Botanical Research Institute, Rana Pratap Marg, Lucknow-226001, Uttar Pradesh, India E-mail: csnnbri@yahoo.com

D. Goyal, Department of Biotechnology & Environmental Sciences, Thapar Istitute of Engineering & Technology, PO Box No. 32, Bhadson Road, Patiala-147004 (Panjab), India E-mail: d_goyal_2000@yahoo.com

D. Srivastava, School of Applied Sciences and Division of Biotechnology, Subhash Institute of Technology, Dwarka, New Delhi-11075, India E-mail: adubey_nsit @yahoo.co.in

H. Kharkwal, Amity Institute of Microbial Sciences, Amity University Uttar Pradesh, Noida- 201303, India E-mail: hkharkwal@amity.edu

J. Isar, Department of Microbiology, Delhi University, South Campus, Benito Juarez Road, New Delhi-110021, India E-mail: rksmicro@hotmail.com

J. P. Tamang, Food Microbiology Laboratory, Department of Botany, Sikkim Government College, Gangtok-737102, Sikkim, India E-mail: jyoti_tamang@ hotmail.com

K. Bhatnagar, Amity Institute of Microbial Sciences, Amity University Uttar Pradesh, Noida-201303, India E-mail: kamyabhatnagar@gmail.com

K. Dutt, Department of Microbiology, Delhi University, South Campus, Benito Juarez Road, New Delhi-110021, India E-mail: rksmicro@ hotmail.com

L. Agarwal, Department of Microbiology, Delhi University, South Campus, Benito Juarez Road, New Delhi-110021, India E-mail: rksmicro@ hotmail.com

M. Bose, Department of Microbiology, Vallabhbhai Patel Chest Institute, University of Delhi, Delhi-110007, India E-mail: mandirav@ rediffmail. com

M. Fernado, Virology Group, International Centre of Genetics and Engineering Biotechnology, Aruna Asif Ali Marg, New Delhi - 110067, India E-mail: sunillal@icgeb.res.in

M. S. K. Jayadev, School of Applied Sciences and Division of Biotechnology, Subhash Institute of Technology, Dwarka, New Delhi-11075, India E-mail: adubey_nsit@yahoo.co.in

M. S. Kumar, Amity Institute of Microbial Sciences, Amity University Uttar Pradesh, Noida-201303, India E-mail: Sangeetanbri@yahoo.com

M.V. Basil, Department of Microbiology, Vallabhbhai Patel Chest Institute, University of Delhi, Delhi-110007, India E-mail: mandirav@ rediffmail.com

N. Hamid, Department of Biotechnology, Faculty of Science, Hamdard University, Hamdard Nagar, New Delhi-110062, India E-mail: skjain@jamiahamdard.ac.in

N. Satija, Virology Group, International Centre of Genetics and Engineering Biotechnology, Aruna Asif Ali Marg, New Delhi-110067, India E-mail: sunillal@icgeb.res.in.

N. Verma, Amity Institute of Microbial Sciences, Amity University Uttar Pradesh, Noida-201303, India E-mail: nverma@amity.edu.

O. P. Ward, Department of Biology, University of Waterloo, Waterloo, Ontario N2L 3G1, Canada E. mail: ajasingh@sciborg.uwterloo.ca

P. Pushpangadan, Director General, Amity Institute for Herbal and Biotech Products Development, Amity University Uttar Pradesh, Noida-201303 & Consultant, Rajiv Gandhi Centre for Biotechnology, Thycaud P.O, Poojappura Thiruvanthapuram-695 014, Kerala, India E-mail: palpuprakulam@yahoo.co.in

P. S. Bisen, Seedling Academy of Design, Technology and Management, Khorebariyan, Jagatpura, Jaipur-302004, India E-mail: psbisen@gmail.com.

P. S. Chauhan, Division of Plant-Microbe Interactions, National Botanical Research Institute, Rana Pratap Marg, Lucknow-226001, Uttar Pradesh, India E-mail: csnnbri@yahoo.com

R. Gautam, Chaudhari Charan Singh University, Meerut-250005, Uttar Pradesh, India E-mail: amarprakashgarg@yahoo.com

Rajani Malla, Tribhuvan University, Trichandra College, King's way, Kathmandu, Nepal E-mail: rajanimalla2000@yahoo.com

R. Prasad, Amity Institute of Microbial Sciences, Amity University Uttar Pradesh, Noida-201303, India E-mail: rprasad@amity.edu.

R. P. Tiwari, Seedling Academy of Design, Technology and Management, Khorebariyan, Jagatpura, Jaipur-302004, India E-mail: psbisen@gmail.com.

R. Sarin, Department of Biotechnology, All India Institute of medical Sciences, Ansari Nagar, New Delhi-110029, India E-mail : ydsharma_aims@yahoo.co.in

R. K. Saxena, Department of Microbiology, Delhi University, South Campus, Benito Juarez Road, New Delhi-110021, India E-mail: rksmicro@hotmail.com

S. Bhatnagar, School of Applied Sciences and Division of Biotechnology, Subhash Institute of Technology, Dwarka, New Delhi-11075, India E-mail: adubey_nsit@yahoo.co.in

S. Gupta, Lignocellulose Biotechnology Laboratory, Department of Microbiology, University of Delhi, South Campus, Benito Juarez Road, New Delhi-110021, India E-mail: kuhad@hotmail.com

S. K. Lal, International Centre of Genetics and Engineering Biotechnology, Aruna Asaf Ali Marg, New Delhi-11007, India E-mail: sunillal@icgeb.res.in

S. K. Dhar, Special Centre for Molecular Medicine, Jawaharlal Nehru University, New Delhi-110067, India E-mail: skdhar2002@yahoo.co.in

S. K. Jain, Department of Biotechnology, Faculty of Science, Hamdard University, Hamdard Nagar, New Delhi-110062, India E-mail: skjain@jamiahamdard.ac.in

S. M. Dasgupta, Amity Institute of Microbial Sciences, Amity University Uttar Pradesh, Noida-201303, India E-mail: sangeetanbri@yahoo.com

S. Saxena, Department of Biotechnology & Environmental Sciences, Thapar Istitute of Engineering & Technology, PO Box No. 32, Bhadson Road, Patiala-147004 (Punjab), India E-mail: sajaibiotech@yahoo.com.

T. Kumar, Amity Institute of Microbial Sciences, Amity University Uttar Pradesh, Noida-201303, India E-mail: ajitvarma@aihmr.amity.edu

U. Pokharel, Punjab University, Department of Microbiology, Chandigarh, India E-mail: rajanimalla2000@yahoo.com

V. P. Sharma, Centre for Rural Develoment and Technology, Indian Institute of Technology, Hauz Khas, New Delhi-110 016, India E-mail: vinodpsharma@gmail.com.

Y. D. Sharma, Department of Biotechnology, All India Institute of Medical Sciences, Ansari Nagar, New Delhi-110029, India E-mail: ydsharma_aims@yahoo.co.in

History of Medical Microbiology

A. C. KHARKWAL, H. KHARKWAL, K. BHATNAGAR and A. VARMA

1. INTRODUCTION

Microorganisms are present everywhere. They can colonize almost every natural surface. Some microorganisms can live in hot springs and others in frozen sea ice. Most microorganisms are harmless to humans; we swallow millions of microbes everyday with no ill effects. In fact, we are dependent on microbes to digest our food. Microbes also keep the biosphere in flow by carrying out essential functions such as decomposition of dead animals and plants. They make possible the cycles of carbon, oxygen, nitrogen and sulfur in terrestrial and aquatic systems. Besides some advantages, microorganisms have also harmed humans and disrupted society over the millennia.

Microorganisms, sometimes, cause diseases in humans, animals and plants. They are also involved in food spoilage. Infectious diseases have played major roles in shaping human history (decline of Roman Empire and conquest of the New World). The "Great Plague" reduced population of Western Europe by 25%. Smallpox and other infectious diseases introduced by European explorers to the Americas in 1500s were responsible for decimating native American populations.

Amity Institute of Microbial Sciences, Amity Campus, Sector 125, Noida-201303, India
E-mail: ajitvarma@aihmr.amity.edu

2. DISCOVERY OF MICROORGANISMS

Until late 1800s, no one had proved that specific microbes caused infectious diseases. Invisible creatures were thought to exist long before they were observed. **Antony van Leuwenhoek** (1632–1723) who invented the microscope (50–300x) was the first to accurately observe and describe microorganisms.

In the early 19th century, people believed in spontaneous generation, but many challenged it. **Louis Pasteur** (1822-1895) settled the conflict once for all with his experiment of swan necks. He developed vaccines for chickenpox, anthrax, rabies and also demonstrated that all fermentations were due to the activities of specific yeasts and bacteria. He also developed pasteurization to preserve wine during storage.

Robert Koch (1843 – 1910) established the relationship between *Bacillus anthracis* and anthrax and also isolated the *Tubercle* bacillus that causes tuberculosis. During Koch's studies, it became necessary to isolate suspected bacterial pathogens. He cultured bacteria on the sterile surfaces of cut, boiled potatoes but didn't get satisfactory results. Then, he tried to culture bacteria on regular liquid medium solidified by adding gelatin. Fannie Eilshemius suggested the use of agar. Koch gave his postulates for any recurring of the disease.

Koch's Postulates

Microorganisms must be present in every case of the disease but absent from healthy individuals. The suspected microorganism must be isolated and grown in pure cultures. The disease must result when the isolated microorganism is inoculated into a healthy host.

3. HISTORY OF MEDICAL MICROBIOLOGY—19TH CENTURY

During 430 BC	Thucydides (Historian of Peloponnesian War) observed that only those who survived bubonic plague were nurse patients.
Early 1700s	Lady Montague (wife of British Ambassador to Turkey) observed variolation of smallpox and reduced deaths overall. Lady Montague
	• Learned variolation and introduced it to England and also immunized her son in 1718.
Late 1700s	Edward Jenner (British Physician)
	• Observed that variolation is not effective on patients with cowpox and investigated it for 25 years.

	• Differentiated true cowpox from imitation diseases and hypothesized that vaccination is superior to variolation.
	• Devised clinical trial (1796) to 10-year old James Phipps.
Late 1800s	Louis Pasteur
	• Called attenuation and developed vaccine against Anthrax and Rabies
	• Devised Human trial (1885, Joseph Meister) and the boy recovered.
1888	George Nuttall (American bacteriologist)
	• Disovered that defibrinated blood is bactericidal.
	• Proposed that phagocytes engulfed killed bacteria and showed that immunity associated with blood.
1890	Von Behring (German Physician)
	• Discovered antibacterial serum properties of persons immune to diphtheria and scarlet fever.
	• Developed antiserum therapy.
1890-1896	Paul Ehrlich (German Physician)
	• Standardized toxins and antitoxins.
	• Found immunity could be transferred to fetus through placenta.
	• Developed concept of self vs. non-self immune system.
1896	Gruber and Durham
	• Discovered agglutination reactions.
	• Proposed that only specific antibodies adhere to specific bacterial cells.
1897	Rudolf Kraus discovered precipitation reactions, and soluble antigens and antibodies that are now called precipitin reactions.
1894	Alexander Yersin isolated *Yersinia* (*Pasteurella*) *pestis*, the organism that is responsible for bubonic plague. Shibasaburo Kitasato also observed the bacterium in cases of plague.
1894	Richard Pfeiffer observed that a heat stable toxic material bound to the membrane of *Vibrio cholerae* is released only after the cells are disintegrated. He calls the material endotoxin, to distinguish it from filterable material released by bacteria.
1894	Martinus Beijerinck isolated the first sulfate-reducing bacterium, *Spirillum desulfuricans* (*Desulfovibrio desulfuricans*).
1895	Sergei Winogradsky isolated the first free-living nitrogen-fixing organism, *Clostridum pasteurianum*.
1895	David Bruce described in great detail the Tsetse fly disease. He also described the parasite and demonstrated transmission by infected blood or fly bite.

1896	Max Gruber and Herbert Durham extended the 1889 observation of Charrin and Roger to show the agglutination of bacteria by serum is specific. This was recognized as a new disease diagnostic tool.
1896	Christan Eijkman, while searching for an infectious agent, discovered that beriberi is the result of a vitamin deficiency. He was awarded the Nobel Prize in 1929.
1897	Paul Ehrlich proposed "side-chain" theory of immunity and developed standards for toxin and antitoxin.
1897	Edward Buchner launched the field of enzymology by publishing the first evidence of a cell-free fermentation process using extracts isolated from yeast. This discovery refuted Pasteur's claim that fermentation requires the presence of live cells. Buchner was awarded the Nobel Prize in Chemistry in 1907.
1897	Waldemar Haffkine produced immunity against the plague with killed organisms.
1897	Almroth Wright and David Sample developed an effective vaccine with killed cells of *Salmonella typhi* to prevent typhoid fever.
1898	Friedrich Loeffler and Paul Frosch proved that foot-and-mouth disease in livestock is caused by organisms that are tiny enough to pass through bacteriological filters and too small to be seen through a light microscope.
1898	R. Schenck presented the first unequivocal case of sporotrichosis and included a description of the organism that was first isolated from the patient. This organism was later named *Sporotrichum schenckii*.
1899	Ronald Ross showed that the malarial parasite undergoes a cycle of development in mosquitoes and that the disease is transmitted by the bite of female mosquito. Ross was awarded the Nobel Prize in Medicine or Physiology in 1902.
1899	The organizing meeting of the Society of American Bacteriologists was held at Yale, December 28. The Society is the first independent organization devoted to the promotion and service of bacteriology in the United States. It was later named the American Society for Microbiology.

4. HISTORY OF MEDICAL MICROBIOLOGY—20TH CENTURY

- **1900** It was demonstrated that yellow fever is caused by a filterable virus transmitted by mosquitoes. The agent is similar to that reported in 1898 by Loffler and Frosch for foot-and-mouth disease of cattle. This is the first report of a viral agent known to cause human disease. Based on the findings of the Yellow Fever Commission, the mosquito was eradicated.

- **1901** Jules Bordet and Octave Gengou developed the complement fixation test. They showed that antigen-antibody reaction leads to the binding of complement to the target antigen.

- **1901** E. Wildiers published the first description of a microbial growth factor, opening the field of vitamin research. He found that a water-soluble extract of yeast has a compound that is required for the growth of yeast. The material is later found to be vitamin B.

- **1903** William Leishman observed *Leishmania donovani* in the spleen of a soldier who died from Dum-Dum fever. Charles Donovan helped to identify the protozoan causing the disease.

- **1905** Fritz R. Schaudinn and Erich Hoffman identified *Treponema pallidum*, the cause of syphilis. The bacterium was isolated from fluid leaking from a syphylitic chancre and is spiral in appearance.

- **1905** Shigetane Ishiwata discovered the cause of a disease outbreak in silkworms by bacteria, later called *Bacillus thuringiensis*, or Bt. Ishiwata called the organism "Sotto-Bacillen". ("Sotto" in Japanese signifies sudden collapse.)

- **1906** August von Wasserman described the "Wasserman reaction" for the diagnosis of syphilis in monkeys. The test uses complement fixation and becomes the basis for the general uses of complement tests as diagnostics.

- **1906** Samuel Darling, performed an autopsy on a patient with a disease resembling tuberculosis and an agent resembling *Leishmania* sp. He recognized significant differences between the etiologic agent and *Leishmania* sp., and named the organism *Histoplasma capsulatum*, believing that it is a protozoan, which is now known to be a fungus.

- **1909** Howard Ricketts showed that Rocky Mountain spotted fever is caused by an organism that is intermediate in size between a virus and bacterium. This organism, *Rickettsia*, is transmitted by ticks. Ricketts died from typhus, another rickettsial disease, in 1910.

- **1909** Sigurd Orla-Jensen proposed that physiological characteristics of bacteria are of primary importance in their classification.

- **1909** Carlos Chagas discovered the trypanosome, which he named *Trypanosoma cruzi*, and its mode of transmission, via reduviid bugs, as the cause of the human disease.

- **1910** Charles Henry Nicolle demonstrated that typhus fever is transmitted from person to person by the body louse. This information was used in both world wars to reduce the incidence of typhus. Nicolle was awarded the Noble Prize in Medicine or Physiology in 1928.

• **1910** Raimond Sabouraud summarized about twenty years of his systematic and scientific studies of dermatophytes and dermatophytoses in a classic treatise, Les Teinges. He introduced a medium for the growth of pathogenic fungi.

• **1911** Francis Peyton Rous discovered a virus that can cause cancer in chickens by injecting a cell free filtrate of tumors. This was the first experimental proof of an infectious etiologic agent of cancer. Rous was awarded the Nobel Prize in Medicine or Physiology in 1966.

• **1912** Paul Ehrlich announced the discovery of an effective cure (Salvarsan) for syphilis, the first specific chemotherapeutic agent for a bacterial disease. Ehrlich was a researcher in Koch's lab, where he worked on immunology. He sought an arsenic derivative. He brought news of the treatment to London, where Fleming became one of the few physicians to administer it.

• **1915** The first discovery of bacteriophage, by Frederick Twort. Twort's discovery was something of an accident. He had spent several years growing viruses and noticed that the bacteria infecting his plates became transparent.

• **1915** Chaim Weizmann, using the knowledge of Pasteur's discovery that yeast ferments sugar, used *Clostridium acetobutylicum* to produce acetone and butyl alcohol. These were essential to the British munitions program during World War.

• **1915** McCrady established a quantitative approach for analyzing water samples for coliforms using the most probable number, multiple-tube fermentation test. The test was based on the ability of coliforms to grow in selective broth at 35°C producing acid or gas within 24 to 48 hours. The number of coliforms and the 95% confidence limit can be determined using MPN tables for the volumes and number of fermentation tubes used.

• **1917** Felix d'Herrelle described bacterial viruses and coined the name "bacteriophage".

• **1918** Alice Evans established that members of the genus *Brucella* are responsible for the diseases of Malta fever, cattle abortion, and swine abortion. She reported that the bacteria are bacilli and not micrococci.

• **1918** An influenza pandemic of unprecedented virulence swept the globe, leaving some 40 million dead; in its wake research on Influenza virus started.

- **1919** Theobald Smith and M. S. Taylor described the microbe, *Vibrio fetus* responsible for fetal membrane disease in cattle.
- **1919** James Brown used blood agar as a medium to study the hemolytic reactions for the genus *Streptococcus* and divided it into three types, alpha, beta and gamma.
- **1920** The Society for American Bacteriologists committee presented a report on the Characterization and Classification of Bacterial Types that became the basis for the classic work of D. H. Bergey, later published in 1923. Even today Bergey's manual is an authentic document in bacteriology.
- **1923** Michael Heidelberger and O. A. Avery showed that carbohydrates from the *Pneumococcus* can serve as virulence antigens and are serologically specific. This overturned the current wisdom that only proteins or glycoproteins are antigenic.
- **1924** George and Gladys Dick described the "Dick test", a skin test for scarlet fever. They purified a soluble extoxin from hemolytic *Streptococccus pyogenes* and used it as a diagnostic. They used Koch's postulates to show that scarlet fever is caused by streptococci, recovered the bacteria from all cases of the disease and infected others with cultures of the bacterium.
- **1924** Albert Calmette and Camille Guerin introduced a living non-virulent strain of tuberculosis (BCG) to immunize against the disease. This was the result of work begun in 1906 on attenuating a strain of bovine tuberculosis bacillus. More than 200 subcultures were grown before the resulting strain was tested.
- **1924** Albert Jan Kluyver published an article "Unity and Diversity in the Metabolism of Micro-organisms" that demonstrated common metabolic events occurring in different microbes. The processes he referred to are oxidation, fermentation and biosynthesis. Klyuver also pointed out that life on earth without microbes would not be possible.
- **1926** Thomas Rivers distinguished between bacteria and viruses, establishing virology as a separate area of study. This paper was published after he presented it at an SAB meeting held in December 1926.
- **1926** Albert Jan Kluyver and Hendrick Jean Louis Donker proposed a universal model for metabolic events in cells based on a transfer of hydrogen atoms. The model applies to aerobic and anaerobic organisms.

- **1926** Everitt Murray isolated a bacterium from rabbits that is responsible for listeriosis in man. The organism can grow at low temperatures and is frequently found in food. He named it *Bacterium monocytogenes*. It was later renamed *Listeria monocytogenes*.
- **1928** Frederick Griffith discovered transformation in bacteria and established the foundation of molecular genetics.
- **1929** Alexander Fleming published the first paper describing penicillin and its effect on Gram-positive microorganisms, effectively launching the "Antibiotics Era,". With Florey and Chain, Fleming was awarded the Nobel prize in Medicine or Physiology in 1945.
- **1930** Henning Karstrom identified the phenomenon of enzyme adaptation and of constitutive synthesis, in which synthesis of an enzyme is either increased in response to the presence of the substrate of the environment or is independent of the growth medium. His work was based on studies of carbohydrate metabolism in Gram-negative enteric bacteria.
- **1931** Rene Dubos working with Oswald Avery discovered *Bacillus brevis*, an organism that breaks down the capsular polysaccharide of Type III *S. pneumocci* and protects mice against pneumonia.
- **1931** Margaret Pittman identified variation, such as encapsulated forms, and type specificity, such as type b, of the *Haemophilus influenzae* as determinants of pathogenicity.
- **1931** William Joseph Elford discovered that viruses range in size from large protein molecules to tiny bacteria.
- **1931** Alice Woodruff and Ernest Goodpasture devised a technique of cultivating viruses in eggs.
- **1934** Ladislaus Laszlo Marton was the first to examine biological specimens with the electron microscope, which achieved magnifications of 200-300,000x. Later in 1937, he published the first electron micrograph of bacteria.
- **1934** Alice Evans accomplished the first typing of a strain of bacteria with bacteriophage.
- **1934** William de Monbreun described the dimorphic nature of *Histoplasma capsulatum* after being surprised by the growth of a mold from patient tissues displaying yeasts.
- **1935** Gerhard J. Domagk used a chemically synthesized antimetabolite, Prontosil, to kill Streptococcus in mice. It was later shown that Prontosil is hydrolyzed *in vivo* to an active compound, sulfanilamide. One of the first patients to be treated with Protonsil was

Domagk's daughter who had a Streptococcal infection that was unresponsive to other treatments. When she was near death, she was injected with large quantities of Protonsil and she made a dramatic recovery. Domagk was awarded the Nobel Prize in Medicine or Physiology in 1939.

- **1935** Wendell Stanley crystallized tobacco mosaic virus and it remained infectious. However, he did not recognize that the infectious material is nucleic acid and not protein. Together with Northrop and Sumner, Stanley was awarded the Nobel Prize in Chemistry in 1946.

- **1938** Field tests of Max Theiler's vaccine against yellow fever proved successful. The vaccine is based on a mouse passage virus. The Rockefeller Foundation manufactured more than 28 million doses by 1947. Theiler was awarded the Nobel Prize in Medicine or Physiology in 1951.

- **1939** E. L. Ellis and Max Delbruck established the concept of one-step viral growth cycle for a bacteriophage active against *E. coli*.

- **1940** Pathologist Howard Florey and biochemist Ernest Chain produced an extract of penicillin, the first powerful antibiotic. They isolated the antibiotic from Fleming's mold cultures and demonstrated that it can cure infections in animals. With Fleming, Florey and Chain were awarded the Nobel Prize in Medicine or Physiology in 1945.

- **1940** Helmuth Ruska used an electron microscope to obtain the first picture of a virus.

- **1940** Charles E. Smith and his colleagues demonstrated the usefulness of a tuberculin-like preparation of *Coccidiodes immitis* in detecting prior exposure to the fungus. This preparation allowed for the delineation of the endemic area for the fungus.

- **1940** Donald O. Woods described the relation of para-aminobenzoic acid to the mechanism of action of sulfanilamide, which was used by Domagk to treat Streptococcal infections in mice.

- **1940** Selman Waksman and H. Boyd Woodruff discovered actinomycin, the first antibiotic obtained pure from an actinomycete, leading to the discovery of many other antibiotics from that group of microorganisms. After Renee Dubos discovered two antibacterial substances in soil, Waksman decided to focus on the medicinal uses of antibacterial soil microbes.

- **1941** George Beadle and Edward Tatum jointly published a paper on their experiments using the fungus *Neurospora crassa* to establish that particular genes are expressed through the action of correspondingly specific enzymes.

- **1941** Charles Fletcher first demonstrated that penicillin is non-toxic to human volunteers, by injecting a police officer suffering with a lethal infection.
- **1941** McFarlane Burnet proposed that descendents of antigen reacting cells produce antibodies specific to the antigen.
- **1941** George Hirst demonstrated that influenza virus agglutinates red blood cells. Since the cell attachment proteins of most viruses also agglutinate red blood cells, this property provides a rapid, accurate and quantitative method of counting virus particles.
- **1942** Selman Waksman suggested the word "antibiotic" (coined in 1889 by P. Vuillemin) after Dr. J. E. Flynn, the editor of Biological Abstracts asked him to suggest a term for chemical substances, including compounds and preparations that are produced by microbes and have antimicrobial properties.
- **1947** Waksman published a comprehensive definition "an antibiotic is a chemical substance produced by microbes that inhibits the growth of microbes and even destroys other microbes (and is active in dilute solutions)" was added later.
- **1942** Thomas Anderson and Salvador Luria photographed bacteriophages with the aid of an electron microscope, confirming earlier work by Ruska. They demonstrated that an E. coli T2 phage has a head and a tail.
- **1943** Salvador Luria and Max Delbruck provided a statistical demonstration that inheritance in bacteria follows Darwinian principles. The results, known as fluctuation analysis, showed that resistance occurs before exposure to the phage and argued against the adaptation hypothesis of mutations. With Delbruck and Hershey, Luria is awarded the Nobel Prize in Medicine or Physiology in 1969.
- **1944** Oswald Avery, Colin MacLeod and Maclyn McCarty showed that DNA is the transforming material in cells.
- **1944** Albert Schatz, E. Bugie and Selman Waksman discovered streptomycin, soon to be used against tuberculosis. Streptomycin has the same specific antibiotic effect against Gram-negative microorganisms as penicillin does on Gram-positive microorganisms. Waksman was awarded the Nobel Prize in Medicine or Physiology in 1952.
- **1944** Ira L. Baldwin, in his 1944 Presidential Address to the Society, provided an interesting mid-century perspective on the challenges facing microbiologists.

- **1945** Salvador Luria and Alfred Day Hershey demonstrated that bacteriophages mutate, thereby making it difficult to develop immunity to such things as flu and colds. They also introduced criteria for distinguishing mutations from other modifications. With Delbruck, Luria and Hershey were awarded the Nobel Prize in Medicine or Physiology in 1969.
- **1946** Joshua Lederberg and Edward L. Tatum published paper on conjugation in bacteria.
- **1949** Microbiologist, John Franklin Enders; virologist, Thomas H. Weller and physician, Frederick Chapman Robbins together developed a technique to grow polio virus in test tube cultures of human tissues. This approach gave virologists a practical tool for the isolation and study of viruses. Enders, Weller and Robbins were awarded the Nobel Prize in Medicine or Physiology in 1954.
- **1951** Andre Lwoff and Louis Siminovitch demonstrated that irradiation with ultra-violet light terminates the lysogenic state in bacteria and permits bacteriophage to replicate and then lyse the host cell. This opened the field of lysogeny to molecular analysis. With Jacob and Monod, Lwoff was awarded the Nobel Prize in Medicine or Physiology in 1965.
- **1952** Joshua Lederberg and Esther Lederberg published their replica plating method and provided firm evidence that mutations in bacteria yielding resistance to antibiotics and viruses are not induced by the presence of selective agents. With Beadle and Tatum, J. Lederberg was awarded the Noble Prize in Medicine or Physiology in 1958.
- **1952** Joshua Lederberg and Norton Zinder reported transduction, or transfer of genetic information by viruses. They showed that a phage of *Salmonella typhimurium* can carry DNA from one bacterium to another.
- **1952** Alfred Hershey and Martha Chase suggested that only DNA is needed for viral replication. They used radioactive isotopes ^{35}S to trace protein and ^{32}P to trace DNA and showed that progeny T2 bacteriophage isolated from lysed bacterial cells had only labeled nucleic acid.
- **1952** Salvador Luria and Mary Human; and Jean Weigle, described a non-genetic heritable variation in bacteriophage imposed on the host in which it was grown. They called this phenomenon host-controlled modification and noted that the incorrectly modified phages were "restricted" in the inappropriate host. This later led to the study of bacterial systems of restriction and modification, and eventually the discovery of restriction endonucleases.

- **1952** James T. Park and Jack L. Strominger concluded that penicillin acts by inhibiting murein synthesis in the cell wall. This is the first discovery of the mode of action of a natural antibiotic.

- **1953** Jonas Salk began preliminary testing of polio vaccine. The vaccine was composed of three types of killed virus.

- **1957** Alick Isaacs and Jean Lindemann discovered interferon, an antiviral protein produced by the body to fight viral infections. The first experiments took place with chick embryo tissue cultures infected with influenza virus.

- **1957** Joseph H. Burkhalter and Robert Seiwald make an essential contribution to the identification of antigens by developing the first antibody labeling agent, flourescein isothiocyanate (FITC). Widespread use of FITC catalyzed the generation of other labeling procedures such as the radio-immunoassay and enzyme-linked immuno-absorbant assay.

- **1959** Maxwell Finland, W. F. Jones, Jr. and M. W. Barnes commented on the development of antibiotic resistance, as a response to the introduction of antibacterial agents.

- **1959** The oral polio vaccine developed by Albert Sabin was approved for use in the U.S. after trials were conducted abroad on more than 100 million people.

- **1962** James Gowans determined that small lymphocytes can initiate both cellular and humoral immune responses to specific antigens. They are the units of selection in the Burnet theory of clonal selection.

- **1963** Baruch Blumberg described the "Australia Antigen" (hepatitis B antigen) that is found in the blood of viral hepatitis sufferers. Together with Irving Millman, Blumberg developed a vaccine against the virus. Some consider it to be the first vaccine against cancer because of the strong association of hepatitis B with liver cancer. With Gadjusek, Blumberg was awarded the Nobel Prize in Medicine or Physiology in 1976.

- **1966** Bruce Ames used auxotrophic strains of *Salmonella typhimurium* to screen for mutagens and potential carcinogens. There is a high correlation between mutagenicity and carcinogenicity in the "Ames test".

- **1966** William Kirby and Alfred Bauer established standards for antibiotic susceptibility testing based on a single disc diffusion procedure that distinguishes susceptible strains of bacteria from their resistant variants. This method permits clinical laboratories to provide physicians with accurate, reproducible and reliable information to chose antimicrobials.

- **1967** R. John Collier described the mechanism by which diphtheria toxin inhibits protein synthesis in a cell-free system from recticulocytes. This is the first definition at the molecular level of the function of a bacterial protein virulence factor.
- **1968** Levin and Bang discovered that the lysate of the amebocytes from the hemolymph of the horseshoe crab, Limulus polyphermus clots in the presence of the lipopolysaccharides in the cell walls of Gram-negative bacteria. This finding led to the development of an *in vitro* assay for the pyrogens that contaminated injectable products and replaced the rabbit pyrogen test.
- **1973** George Laver and Robert Webster demonstrated that the genomes of influenza virus strains responsible for pandemics possess genome fragments acquired by genome segment reassortment from influenza strains circulating in animals.
- **1973** Peter Doherty and Rolf Zinkernagl showed that the cellular immune system requires that lymphocytes recognize both the virus invader and major histocompatibility antigens in order to kill virus-infected cells. This establishesd the principle of simultaneous recognition, of both self and non-self molecules, as the basis of the specificity of the cellular immune system. Doherty and Zinkernagl were awarded the Nobel Prize in Medicine or Physiology in 1996.
- **1975** Georges Kohler and Cesar Milstein physically fused mouse lymphocytes with neoplastic mouse plasma cells to yield hybridomas that can produce specific antibodies and can survive indefinitely in tissue culture. This approach offers a limitless supply of monoclonal antibodies. Monoclonal antibodies permit the generation of diagnostic tests that are highly specific and also function as probes to study cell function. With Jerne, Kohler and Milstein were awarded the Nobel Prize in Medicine or Physiology in 1984.
- **1976** J. Michael Bishop and Harold Varmus identified oncogenes from the Rous sarcoma virus that can also be found in the cells of normal animals, including humans. Proto-oncogenes appear to be essential for normal development, but can become cancer genes when cellular regulators are damaged or modified. Bishop and Varmus were awarded the Nobel Prize in Medicine or Physiology in 1989.
- **1976** William Trager and Jim Jensen succeeded in cultivating the human malaria parasite *Plasmodium falciparum*, which allowed its study, in the laboratory, for the first time.

- **1977** Joseph McDade and Charles C. Shepard isolated and identified *Legionella pneumophilia* as the bacterial pathogen in a newly discovered pulmonary disease. There are now known to be more than 40 species which occur in water settings.
- **1982** U.S. Pharmaceutical manufacturer Eli Lilly marketed the first genetically-engineered human insulin.
- **1982** Stanley Prusiner found evidence that a class of infectious proteins which he called prions cause scrapie, a fatal neurodegenerative disease of sheep. Prusiner was awarded the Nobel Prize in Medicine or Physiology in 1997.
- **1982** Willy Burgdorfer and colleagues reported successful investigation and treatment of Lyme disease. They were able to isolate the organism, a treponema-like spirochete, and to show antibody formation in patients with clinically diagnosed Lyme disease.
- **1982** Luc Montagnier and Robert Gallo announced their discovery of the human immunodeficiency virus (HIV) believed to cause AIDS.
- **1984** Barry Marshall demonstrated that isolates from patients with ulcers contain a bacterium called *Campylobacter pylori*, later called *Helicobacter pylori*. Marshall proved that the bacterium is the etiologic agent by swallowing a dose and developing gastritis, the precursor to ulcer disease. The bacterium uses a novel urease to produce ammonia, allowing it to survive at low pH.
- **1985** U.S. Department of Agriculture granted first license to market a genetically-engineered living organism, a virus, to vaccinate against a herpetic disease in swine.
- **1988** Stanley Falkow proposed a molecular version of Koch's postulates which has applicability to the assessment of whether a gene or its products are required for virulence.
- **1990** D.A. Relman, J.S. Loutit, T.M. Schmidt, Stanley Falkow and Lucy Tompkins showed that cat scratch fever or bacillary angiomatosis is caused by a bacterium that can't be cultured. The authors used two molecular techniques, analysis of 16S rRNA and polymerase chain reaction amplification to identify the causative agent.
- **1992** The entire sequence of 315,000 units of one of the sixteen chromosomes of the yeast *Saccharomyces cerevisiae* was identified, representing a major advance towards the sequencing of all the chromosomes of yeast, a eucaryote. This breakthrough came 6000 years after the first known use of yeast by humans.

- **1995** Craig Venter, Hamilton Smith, Claire Fraser and colleagues at TIGR elucidated the first complete genome sequence of a microorganism - *Haemophilus influenza*.
- **1995** C. J. Peters, V. E. Chizhikov, S. F. Spiropoulou, S. P. Morzunov, and M. C. Monroe reported the complete genome of the hantavirus Sin Nobra NMH10, detected in autopsy tissue of a patient who died of hantavirus pulmonary syndrome.
- **1998** Fungus promotes the active ingredient in a large number of herbal plants tested. For details see Chapter No. 4.
- **2006** Bird Flu—A Potential Pandemic.

BirdLife seeks the complete removal of the H5N1 highly pathogenic avian influenza virus from the ecosystem—while recognising that the virus is so entrenched now in some regions that this cannot be achieved rapidly. BirdLife is greatly concerned and saddened by the human death toll from the ongoing infection, and by the massive economic loss suffered by those communities affected by the virus and dependent on poultry. We also recognise and share the real concerns about a potential human pandemic.

Outbreaks among wild birds in Europe and Iran during 2006 show that wild birds are capable of carrying the virus to new sites after infection. Many questions remain concerning the effects of the virus on wild birds and how efficiently they can spread it to other wild birds or to domestic poultry, especially over long distances. By contrast, outbreaks in Cameroon, Egypt, India, Israel, Jordan, Niger, Nigeria, Djibouti, Lao and Pakistan in 2006 originated within the poultry industry. Recent outbreaks in 2007 in Hungary, South Korea, Japan and Thailand, and ongoing outbreaks elsewhere, especially in Indonesia, are also associated with commercial poultry production.

Cumulative Number of Confirmed Human Cases of Avian Influenza A/(H5N1) Reported to WHO

12 January 2007

Country	2003		2004		2005		2006		2007		Total	
	cases	deaths	cases	deaths	cases	deaths	cases	deaths	cases	deaths	cases	deaths
Azerbaijan	0	0	0	0	0	0	8	5	0	0	8	5
Cambodia	0	0	0	0	4	4	2	2	0	0	6	6
China	1	1	0	0	8	5	13	8	0	0	22	14
Djibouti	0	0	0	0	0	0	1	0	0	0	1	0

Egypt	0	0	0	0	0	0	18	10	0	0	18	10
Indonesia	0	0	0	0	19	12	56	46	2	1	77	59
Iraq	0	0	0	0	0	0	3	2	0	0	3	2
Thailand	0	0	17	12	5	2	3	3	0	0	25	17
Turkey	0	0	0	0	0	0	12	4	0	0	12	4
Viet Nam	3	3	29	20	61	19	0	0	0	0	93	42
Total	4	4	46	32	97	42	116	80	2	1	265	159

Noble Prize Story of Two Australians

Dr Barry Marshall and research partner Dr Robin Warren, made the history by discovering that stomach ulcers are not caused by stress, spicy foods and too much acid. They were able to prove that instead the bacteria *Helicobacter pylori* were the cause of most stomach ulcers.

This theory was initially ridiculed by the establishment scientists and doctors, who did not believe that any bacteria could live in the acidic stomach. In order to force people to pay attention, Dr Marshall undertook that rather drastic action of drank a petri-dish of the bacteria and soon developed gastritis. The bacteria disappeared after two weeks and the illness resolved spontaneously with the aid of antibiotics.

Future Perspectives

Microbes are the most significant life forms sharing this planet with humans because of their pervasive presence and utilization of any available food source, including humans whose defenses may be breached. Microbial diseases are frequent and often severe, e.g. AIDS, cholera, tuberculosis, and rabies, etc. The ubiquitous presence of microbes and their large numbers give rise to the many mutants that account for rapid evolutionary adaptation and in part for emerging diseases such as AIDS, ebola and antibiotic-resistant tuberculosis. Depending on the food source, microbes may have either beneficial roles in maintaining life or undesirable roles in causing human, animals and plant disease.

Understanding and employing the principles of medical microbiology and the molecular mechanisms of pathogenesis will enable the physician and medical scientist to control an increasing number of infectious diseases.

SUGGESTED READING

Jawetz, Melnick, & Adelberg's Medical Microbiology, 23rd Edition (2004), Geo. F. Brooks, Janet S. Butel, Stephen A. Morse, Publisher: McGraw-Hill Medical

Medical Microbiology, 3rd Edition (2004), Cedric Mims, John Playfair, Ivan Roitt, Derek Wakelin, Rosamund Williams, Publisher : C.V. Mosby

Medical Microbiology, 4th Edition (2002), George S. Kobayashi, Patrick Murray, Ken S. Rosenthal, Michael A. Pfaller, Publisher : C.V. Mosby

Medicinal Plants for Human Life

H. KHARKWAL, A.C. KHARKWAL, N. VERMA, R. PRASAD, A. DAS, K. BHATNAGAR and A. VARMA

1. INTRODUCTION

Since antiquity, man has been using plants to treat common infectious diseases and some of these traditional medicines are still included as part of the habitual treatment of various maladies. Several diverse lines of evidences indicate that medicinal plants represent the oldest and most widespread form of medication. Until the last century, most medicines were derived directly from plant or animal sources. Despite major scientific and technological progress in combinatorial chemistry, drugs derived from natural products still make an enormous contribution to drug discovery today.

2. SOURCE OF ANTI-CANCER AGENTS

Plant derived compounds constitute several clinically useful anti-cancer agents. An analysis of the number of chemotherapeutic agents and their sources indicate that over 60% of the approved drugs are derived from natural compounds. The discovery and development of the Vinca alkaloids, vinblastine and vincristine, and isolation of the cytotoxic podophyllotoxins had led to search for anti-cancer agents from plant sources, late in 1950s.

Drug discovery from medicinal plants has played an important role in the treatment of cancer and, indeed, most new clinical applications of plant

Amity Institute of Microbial Sciences. Amity University, Uttar Pradesh, Noida-201303, India. E-mail: hkarkwal@amity.edu.

secondary metabolites and their derivatives over the last half century have been applied towards combating cancer (Newman et al. 2003). Of all the available anticancer drugs between 1940 and 2002, 40% were natural products per se or natural product-derived with another 8% considered natural product mimics (Newman et al. 2003). Anticancer agents from plants currently in clinical use can be categorized into four main classes of compounds: Vinca alkaloids, epipodophyllotoxins, taxanes and camptothecins. Vinblastine and vincristine were isolated from *Catharanthus roseus* (L.) G. Don, family Apocynaceae (formerly *Vinca rosea* L.) and have been used clinically for over 40 years (van Der Heijden et al. 2004), Table 1. The Vinca alkaloids and several of their semi-synthetic derivatives block mitosis with metaphase arrest by binding specifically to tubulin resulting in its depolymerization (Okouneva et al. 2003).

Table 1 Anticancer agents derived from Plants

Compound	Cancer use	Status
Vincristine	Leukemia, lymphoma, breast, lung, pediatric, solid cancers and others	Phase III/IV
Vinblastine	Breast, lymphoma, germ cell and renal cancer	Phase III/IV
Paclitaxel	Ovary, breast, lung, bladder, and head and neck cancer	Phase III/IV
Docetaxel	Breast and lung cancer	Phase III
Topotecan	Ovarian, lung and pediatric cancer	Phase II/III
Irinotecan	Colorectal and lung cancer	Phase II/III
Flavopridol	Experimental	Phase I/II
Acronycline	Experimental	Phase II/III
Bruceantin	Experimental	Preclinical/Phase I
Thalicarpin	Experimental	Preclinical/Phase I

Table 2 Anticancer agents derived from Microbes

Compound	Cancer use	Status
Actinomycin	Sarcoma and germ-cell tumors	Phase III/IV
Bleomycin	Germ-cell, cervix, and head and neck cancer	Phase III/IV
Daunomycin	Leukemia	Phase III/IV
Doxorubicin	Lymphoma, breast, ovary, lung and sarcomas	Phase III/IV
Epirubicin	Breast cancer	Phase III/IV
Idarubicin	Breast cancer and leukemia	Phase III/IV
Mitomycin C	Gastric and endocrine tumors	Phase III/IV
Streptozocin	Gastric and endocrine tumors	Phase III/IV
Wortmannin	Experimental	Preclinical
Rapamicin	Experimental	Preclinical
Geldanamycin	Experimental	Preclinical

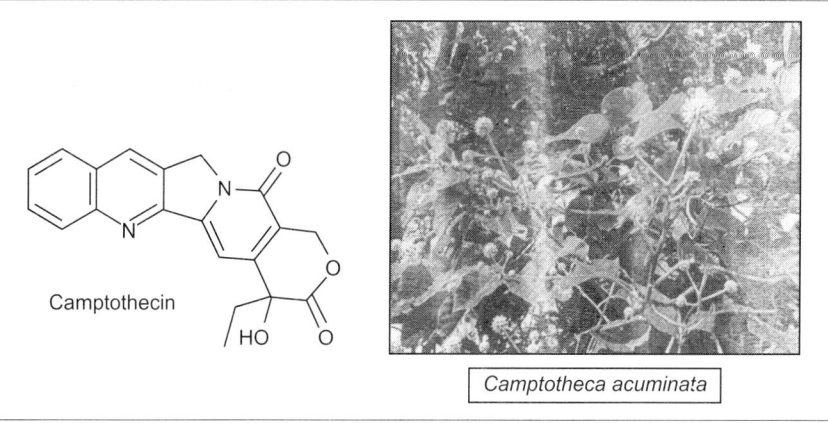

Camptothecin

Camptotheca acuminata

Fig. 1   Source of Anti-Cancer agent: *Camptotheca acuminata*

Table 3   Anticancer agents derived from Marine Organisms

Compound	Cancer use	Mechanism of action	Status
Citarabine	Leukemia, lymphoma	Inhibition of DNA synthesis	Phase III/IV
Bryostatin 1	Experimental	Activation of AKC	Phase III/IV
Dolastatin 10	Experimental	Inhibition of microtubules and pro-apoptotic effects	Phase III/IV
Ecteinascidin 743	Experimental	Alkylation of DNA	Phase III
Aplidine	Experimental	Inhibition of cell cycle progression	Phase I/II
Halicondrin B	Experimental	Interaction with tubulin	Preclinical
Discodermolide	Experimental	Stabilization of tubulin	Preclinical
Cryptophycin	Experimental	Hypephosphorylation of Bcl-2	Preclinical

3. EXTRACTION OF ALKALOIDS

The bioactive compound taken for study are mostly alkaloids. A schematic presentation for extraction of alkaloids is depicted in Fig. 6.

4. PLANTS USED FOR PROBLEMS RELATING TO CENTRAL NERVOUS SYSTEM (CNS)

During the history of mankind, drugs affecting the CNS have focussed essentially on those that bring relief to psychiatric disorders. Recently, a lot of focus has been made on those likely to Abring relief to those acting on Parkinsonism, epilepsy, more potent analgesics, etc. Drugs of plant origin

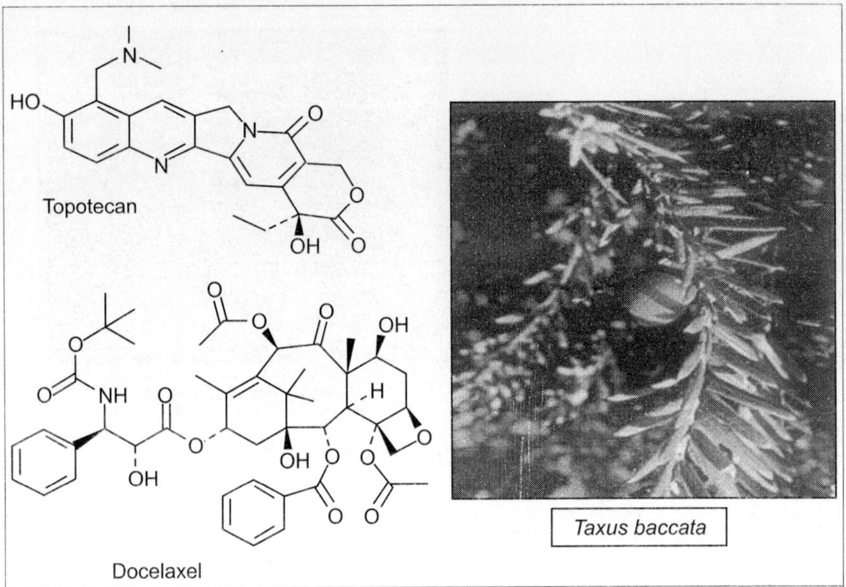

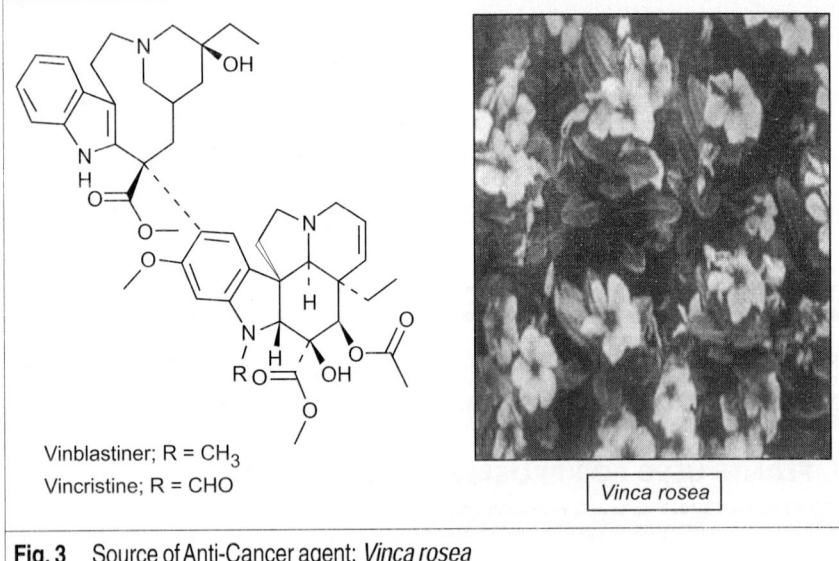

Topotecan

Docelaxel

Taxus baccata

Fig. 2 Source of Anti-Cancer agent: *Taxus baccata*

Vinblastiner; R = CH$_3$
Vincristine; R = CHO

Vinca rosea

Fig. 3 Source of Anti-Cancer agent: *Vinca rosea*

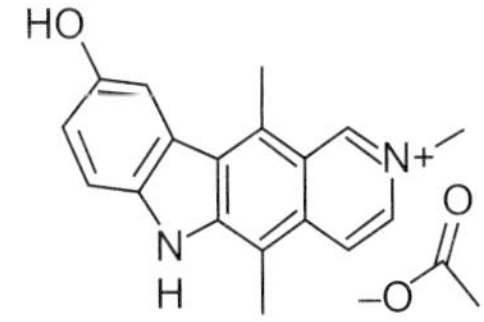

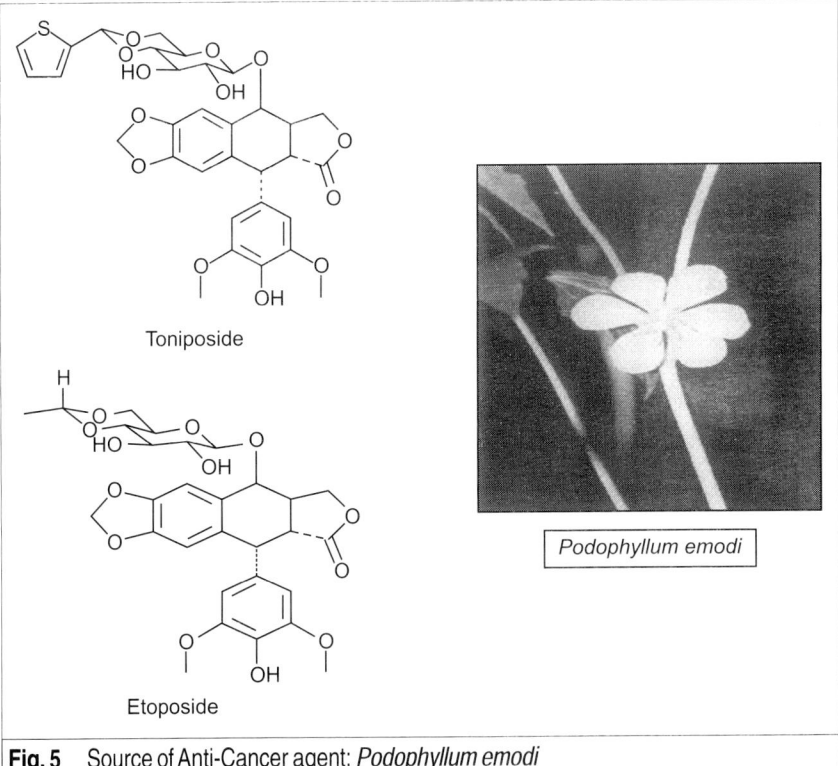

Elliptinium

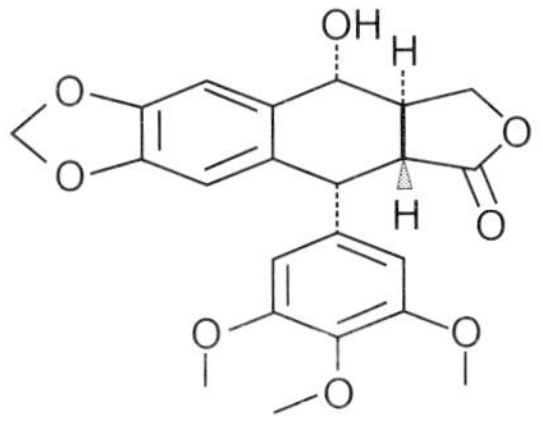

Podophyllotoxin

Podophyllum peltatum

Fig. 4   Source of Anti-Cancer agent: *Podophyllum peltatum*

Toniposide

Podophyllum emodi

Etoposide

Fig. 5   Source of Anti-Cancer agent: *Podophyllum emodi*

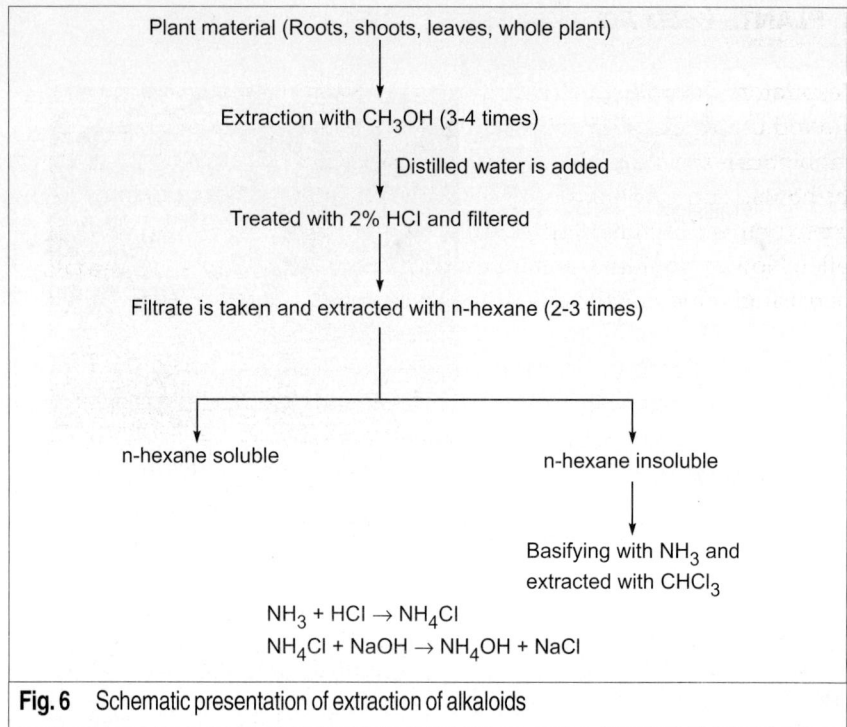

Fig. 6 Schematic presentation of extraction of alkaloids

(Fakim 2006) are important in all these areas although not usually for self-medication. Reserpine has been a classical example where this anti-psychotic drug has revolutionized the treatment of schizophrenia and has enabled patients to avoid hospitalization before the introduction of drugs, such as chlorpromazine, olanzines and risperidone. Reserpine in the meantime has shown some side effects in depleting the neurotransmitter levels in the brain, thus causing severe depression and has recently been involved in the development of breast cancer.

For milder psychiatric conditions, phytotherapy can still provide support when one takes into account the statistics, whereby depression and anxiety still affects one in six persons and 40% of the people having mental problems will also develop symptoms of anxiety and depression. The latter is more prevalent in women than in men with associated problems like sleep disturbances, etc. It is in this context again that phytotherapy is called upon to re-establish a regular pattern of sleep. Migraines, dementia, Alzheimer disease are many of the problems associated with CNS, which are being addressed by plant extracts.

5. PLANTS USED AGAINST THE RESPIRATORY SYSTEMS

Respiratory disorders such as cold, asthma and bronchitis can be treated by phytotherapy. For such ailments leading to infections, the recourse to antibiotics is inevitable. Nonetheless, throughout the duration of colds and flu-bouts, decongestants (Eucalyptus, Mint), broncholytics and expectorants (Thyme, Mint), demulcents (Mallow) all help in providing relief. Nowadays, immune system modulators (Echinaceae) are becoming increasingly popular and effective (Fakim 2006).

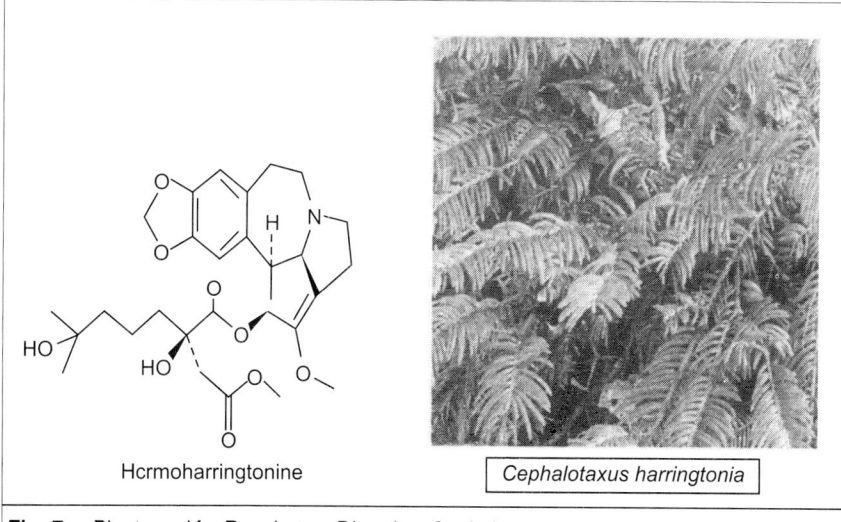

Hermoharringtonine

Cephalotaxus harringtonia

Fig. 7   Plant used for Respiratory Disorder: *Cephalotaxus harringtonia*

Asthma is also becoming prevalent in many countries. They are being treated more aggressively by steroids and bronchodilators, although the latter can also be of natural origins (Ephedrine and Theophylline). Although isolated components of Ephedra (e.g., pure ephedrine or pure Pseudoephedrine) are contraindicated in the event of asthma. It must be highlighted that Ephedra drugs have a long history of use without apparent side effects and this feature would be attributable to the presence of other components in the whole plant. This feature would appear to be a general feature in phyto-therapy where the synergistic properties of other molecules may affect the performance of the medication.

6. PLANTS EXHIBITING ANALGESIC ACTIVITY

Some of the Indian medicinal plants *Sida acuta* (whole plant), Malvaeae; *Stylosanthes fruticosa* (whole plant), Papilionaceae; *Toona ciliata* (heart wood), Meliaceao; *Bougainvilla spectabilis* (leaves), Nyctaginaceae; *Ficus glomerata* (bark, leaves), Moraceae; and *Polyalthia longifolia* (leaves), Annonaceae exhibited analgesic activity (Malairajan P et al. 2006).

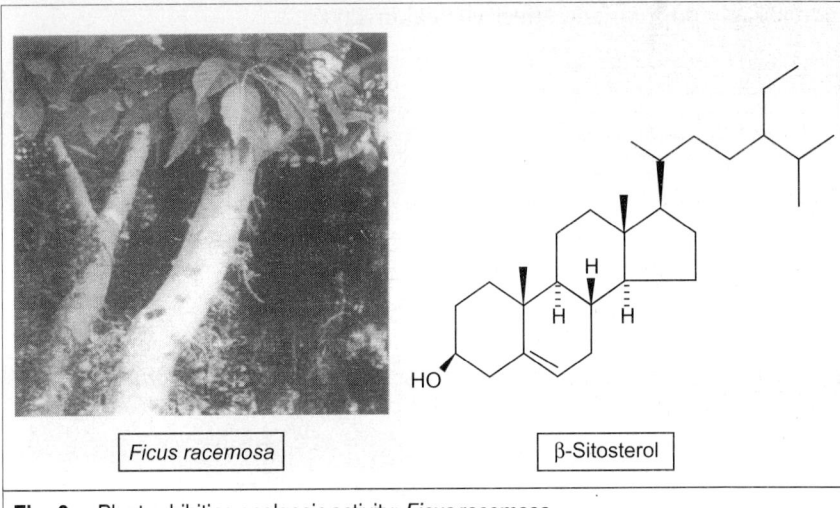

Ficus racemosa	β-Sitosterol

Fig. 8 Plant exhibiting analgesic activity: *Ficus racemosa*

The different plants were used in folklore medicine in the treatment of toothache and strengthening of gums, anthelmintic, kidney diseases, analgesic, anti-inflammatory, hepatoprotective, anti-hyperglycaemic and anticancer. Analgesic activity was significant with *Toona ciliata* (heart wood) ethanolic extract when compared with other extracts and its activity was confirmed by tail immersion method.

7. PLANTS EXHIBITING ANTIMICROBIAL ACTIVITY

The crude extracts of *Dorema ammoniacum, Sphaeranthus indicus, Dracaena cinnabari, Mallotus philippinensis, Jatropha gossypifolia, Aristolochia indica, Lantana camara, Nardostachys jatamansi, Randia dumetorum* and *Cassia fistula* exhibited significant antimicrobial activity and properties that support folkloric use in the treatment of some diseases as broad-spectrum antimicrobial agents (Prashanth Kumar V et al. 2006).

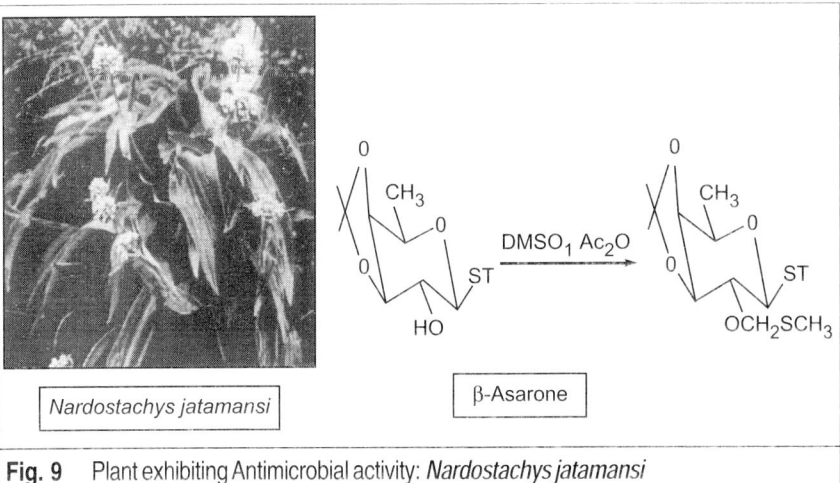

Nardostachys jatamansi

β-Asarone

Fig. 9   Plant exhibiting Antimicrobial activity: *Nardostachys jatamansi*

Hence, these plants are used by the indigenous people against a number of infections since generations.

8. PLANTS EXHIBITING ANTI-VIRAL ACTIVITY

Methanol and methanolic-water extracts of *Bergenia lingulata, Bombax cieba, Carea arborea, Coccinia cordifolia, Dypsacus mitis, Holoptelia integrefolia, Nerium indicum, Rhododendron anthopogon, Salvia coccinia* are reported to have shown significant activity against Herpes Simplex Virus (Khan MTH et al. 2005).

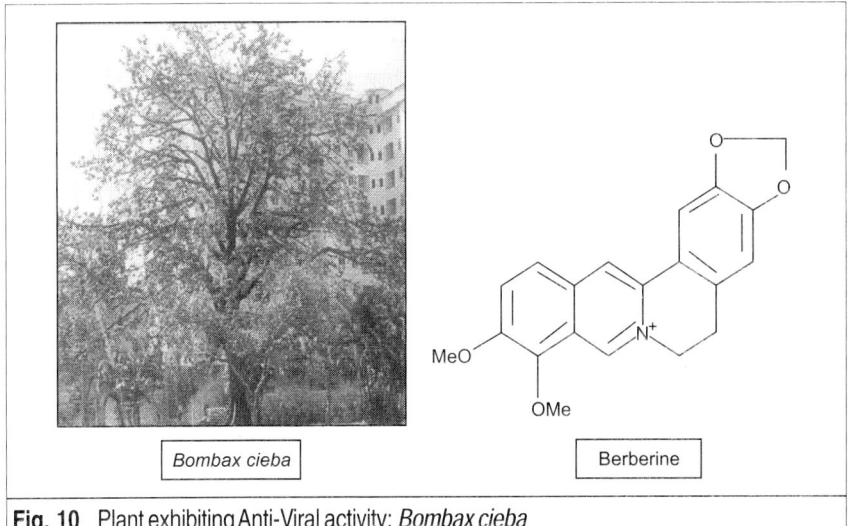

Bombax cieba

Berberine

Fig. 10   Plant exhibiting Anti-Viral activity: *Bombax cieba*

9. PLANTS EXHIBITING ANTIOXIDANT ACTIVITY

Eleven polysaccharides have been isolated from the leaves of *Arctium lappa* var. *herkules*, *Aloe barbadensis*, *Althaea officinalis* var. *robusta*, *Plantago lanceolata* var. *libor*, aerial parts and roots of *Rudbeckia fulgida* var. *sullivantii*, stems of *Mahonia aquifolium*, and peach-tree *Prunus persica* gum exudates. The polysaccharides were investigated for their ability to inhibit peroxidation of soyabean lecithin liposomes by OH radicals (Crescenzi et al. 1997; Fry 1998). The highest inhibition was found with glucuronoxylans of *A. officinalis* var. *robusta* and *P. lanceolata* var. *libor*, aerial parts. Their antioxidant activity accounted for ~69% of the activity of the reference compound α-tocopherol. The activity of eight polysaccharides ranged from 20 to 45%, while the fructofuranan from *P. lanceolata* var. *libor* roots was practically inactive.

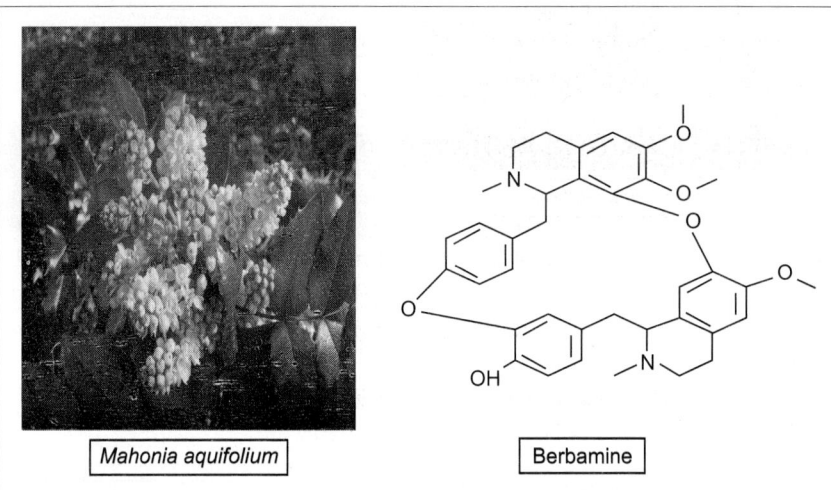

Mahonia aquifolium   Berbamine

Fig. 11 Plants exhibiting Antioxidant activity: *Mahonia aquifolium*

10. PLANTS WITH ANTI-DIABETIC POTENTIAL

Diabetes mellitus is one of the most common metabolic disease in the world and type II diabetes is the most common form of diabetes. It comprises a group of disorders that are associated with high blood sugar levels. It is a devastating disease that often leads to complications such as blindness, kidney failure, coronary heart disease, circulatory disease that may result in amputation, nerve problem and premature death.

There are many medicinal plants showing potent anti-diabetic activity (Alarcon-Aguilera et al. 1998). To name a few: *Azadirachta indica* (juice of leaves, bark, flower and seed oil), *Gymnema sylvestre* (leaves and extract of fruit, leaves and stems), *Eugenia jambolana* (seed and pulp), *Trigonella foenum-graecum* (aqueous extract of leaf and seed).

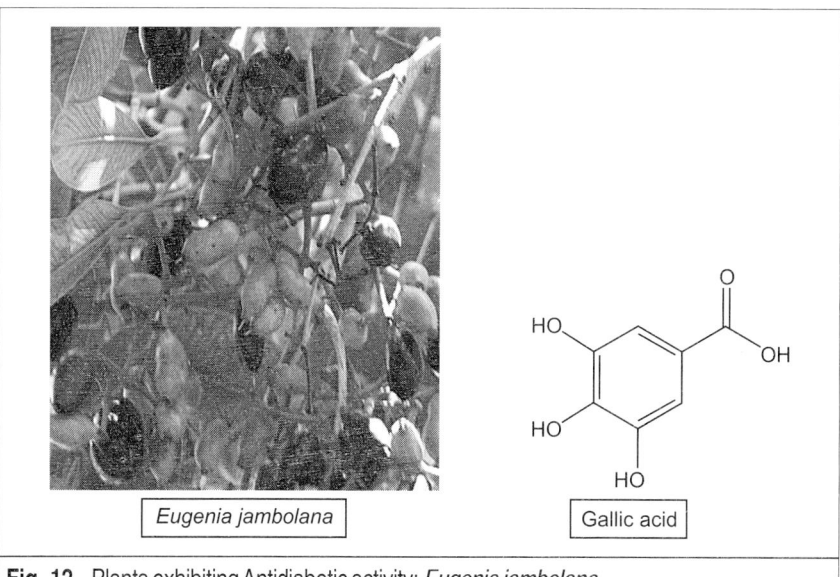

| Eugenia jambolana | Gallic acid |

Fig. 12   Plants exhibiting Antidiabetic activity: *Eugenia jambolana*

Modern therapies are too costly to be practical for the majority of diabetes referrers, so the ethanopharmcological use of herbal remedies for the treatment of diabetes in the area of study ripe with potential as a starting point in the development of alternative, inexpensive therapies for treating the disease.

11. FEW COMMERCIALLY USED INDIAN MEDICINAL PLANTS

As we know, India has 47,000 species of plants and is ranked 8th in the world biodiversity. Out of these, medicinal plants comprise 8000 species. Indian System of Medicine (ISM) uses 2500 plant species belonging to more than 1000 genera. About 800 species are used by industry of which approximately 25% are cultivated. The commercial importance of medicinal plants has trigerred the attention and following selection criteria was adopted for the purpose of short listing, such as, endemic nature of plants, high domestic as well as export demand, endangered nature of the plants, documented use in traditional system of medicines. Keeping the aforementioned criteria in view, a few of the medicinal plants of commercial importance are listed below:

Ammi majus Xanthotoxin (Coumarin) from the seeds of Ammi majus are used in the treatment of Leucoderma.

Aloe vera The gel and powder from A. vera are in great international demand. They act as a souce of nourishment for HIV patients and also stimulate immune response against cancer.

Artemesia annua Active principle (artemisinine) is used in treatmentof fever, malaria and cerebral malaria.

Aswagandha (Withania somnifera) Aswagandha is used for the treatment of rheumatism, gout, hypertension, cancer and as tonic and sex stimulant.

Bacopa monnieri (Brahmi) Bacosides A and B from the aquatic plant *Bacopa* have shown properties like : Faster performance time, improved acquisition and retention of learning, decreased forgetfulness.

Atropa accuminata Royle The leaves and roots of belladonna (*Atropa acuminata* Royle) are important crude drugs due to their anticholinergic, antipasmodic and mydriatic properties.

Catharanthus roseus A newly developed disease resistant variety called Nirmal provides good yield of root and leaves making the commercial cultivation a profitable venture. This plant yields a number of alkaloids, out of which Vincristine and Vinblastine are used in the treatment of leukemia.

Centella asiatica (Mandookaprni, Goty kola) It is called "Food for the brain"owing to its ability to increase mental and physical power. It has the ability to rebuild energy reserves.

Doscorea composita It is commonly known as medicinal yam or pill plant. Disogenin, which is used for the synthesis of number of steroidal drugs e.g., corticosteroids has been obtained from the rhizome of Dioscorea deltoidea.

Rauwolfia sepentina (Sarpagandha) Sarpagandha is a tropical evergreen climbing shrub. It was used for centuries to treat insanity as well as physical illness such as fevers and snakebites. The thick dry root yield akaloid such as reserpine.

Solanum khasianum A hardy plant that can be cultivated in a wide range of agro-climatic conditions. Berries of this plant are a rich source of splasodine used for commercial production of steroidal drugs.

CONCLUSION

India is rich in both biological resources i.e., medicinal plants and traditional knowledge; a combination of these can become the starting point for developing traditional medicine as well as chemical entities. Some of the medicinal plants have been discovered, and being used as herbal medicines by many Indian tribes, while the availability of so many varieties of medicinal plants in India indicates that the climate and soil condition are actually very supportive for the sustainable development of medicinal plant cultivation. The supportive environment will not only make cultivation to become cost-effective, but also preserve the genetically good quality of the plants as medicine.

There is an ever-increasing global market for medicinal plants, but at the same time, value is still small when compared to the global market. The main problem in India is collection of medicinal plants from wild by unskilled people, problems of inconsistent quality, adulteration, destructive methods of collection leading to extiction or endangered status of number of plants, are severe. As more and more people are becoming increasingly aware of the side effects of using allopathic medicines, there is an increasing demand for the herbal medicines.

REFERENCES

Alarcon-Aguilera FJ, Roman-Ramos R, Perez-Gutierrez S, Aguilar-Contreras A, Contreras-Weber CC, Flores-Saenz JL (1998). Study of the anti-hyperglycaemic effect of plants used as anti-diabetics. J. Ethnopharmacol. 61 (2), 101–110.

Crescenzi V, Belardinelli M, Rinaldi C. J. Carbohydr. Chem. 1997;16:561.

Fakim Ameenah Gurib (2006). Medicinal plants: Traditions of yesterday and drugs of tomorrow Molecular Aspects of Medicine 27: 1–93.

Fry SC (1998). Biochem J; 332:507.

Khan MTH, Ather Arjumand, Thompson KD, Gambari Roberto (2005). Extracts and molecules from medicinal plants against herpes simplex viruses. Antiviral Research 67: 107–119.

Malairajan P, Gopalakrishnan Geeta, Narasimhan S, Jessi Kala Veni K (2006). Analgesic activity of some Indian medicinal plants. Journal of Ethnopharmocology (available online).

Newman DJ, Cragg GM, Snader KM (2003). Natural products as sources of new drugs over the period 1981–2002. Journal of Natural Products 66: 1022–1037.

Okouneva T, Hill BT, Wilson L et al. (2003). The effects of vinflunine, vinorelbine and vinblastine on centromere dynamics. Mol Cancer Ther 2(5):427-436.

Prashanth Kumar V, Chauhan Neelam S, Padh Harish, Rajani M (2006). Search for antibacterial and antifungal agents from selected Indian medicinal plants. Journal of Ethnopharmocology (available online).

Herbal Drugs and Nutraceuticals in 21st Century Health Care

P. PUSHPANGADAN

1. INTRODUCTION

Food and medicine are indispensable companions of humans, since the very beginning of existence. The early man explored his surroundings mainly the biological materials to feed his hunger and selected many plants that nourished him. He continued his search in the plant world to broaden his food basket to heal his ailments/discomforts. The desire to attain vitality and longevity also prompted the early man to explore the natural surroundings. By a process of trial and error or even by empirical reasoning and experimentation, the early man made conscious selections of a variety of biological materials, predominantly plants from his surroundings, to meet his food and nutritional requirements, to enhance his health, to alleviate pain or to treat other physical and mental sufferings. The knowledge thus gathered by one generation was passed on to the next generation orally. Creative members of the succeeding generations incrementally improved and even added new knowledge to this body of traditional knowledge system. This effort has gone into history by the name traditional knowledge system or ethnic knowledge systems on food, medicine and other biological and non-biological materials of human use. This traditional knowledge system is also at times referred to as folklore knowledge or folklore traditions. All ancient

Amity Institute for Herbal and Biotech Products Development (An Institution of Ritnand Balved Education Foundation) Rajiv Gandhi Centre for Biotechnology Campus Thiruvananthapuram – 14, Kerala, India cultures of the world had thus evolved their own traditional food, nutrition and medicine from their ambient biological wealth. The earliest documented record of herbal medicine dates from Paleolithic age (5,000 BC) was found in the grave of Neanderthal man in southern part of Hakkari (present-day Turkey) (Baytop 1999).

A number of plant remedies were described on the Clay tablets that have survived from the Mesopotamian civilizations i.e., Sumarians, Assyrians and Akkadians and the papyrus documents from the ancient Egyptian civilization (Yesilada 2005). In eastern cultures such as India, China, Egypt and the middle east, medical wisdom and practices were systematically recorded, more particularly of those plants used for medicinal purposes, into organized regular systems of medicine that became the Materia Medica of the classical systems of medicine in these countries. The classical system of medicine was, however, practiced mainly by the elite members of the community who were mostly inhabiting the urban areas and constituted only 20-25 per cent of the population. The majority of the people living in the rural areas or the poor people living in urban areas had continued to depend on the traditional health care practices which are mostly oral in tradition. Such traditional food and medicine are continued to be practiced even today by majority of the world population particularly those in the third world countries.

2. IMPORTANCE OF ETHNIC FOOD AND MEDICINE

A significant aspect to ethnic food and medicine is that it is mostly location specific and is best suited for the local climate and environment. Such systems of food and practices are almost deep-rooted in the community's social life, tradition and cultural values. It remained in the mainstream for several centuries. In fact, this was the main reason for WHO to recognize intrinsic importance of such traditional food nutrition and medicine particularly in maintaining good primary health care. WHO also emphasized the strategic role of medicinal plants in ensuring the primary health care needs of the people particularly those in the rural sector. The introduction of modern medicine in 19th century began to exert a negative influence on the traditional health care sector. The advances made in biological science, chemistry and the technical tools and methods brought in quick and powerful healing devices. Such developments in the modern medicine and food processing caused a rapid decline in ethnic medicine and food. End of the 20th century, however, witnessed an unprecedented revival of interest in ethnic foods and medicine. In 21st century, once again, medicinal plants began to play a major role in the health care programme of mankind.

The revival of interest in plant based drugs and other herbal products is mainly because of the widespread belief that 'green medicine' is healthier than the synthetic products (Pushpangadan and Govindarajan 2005). This is mainly due to the increasing evidences of the health hazards associated with the harmful side effects of many synthetic drugs and the indiscriminate use of modern medicines such as antibiotics, steroids and other synthetic drugs. The increasing popularity of the plant based drugs is now realised all over the world leading to a fast growing market for plant based drugs, pharmaceuticals, nutraceuticals, functional foods and even cosmaceuticals. In 1980s, this led to the rapid spurt of demand for health products like herbal tea, ginseng and products of traditional medicine. The health promotive and disease precautionary strategies in treatment, prevalent in oriental system, especially the Indian ('Ayurveda', 'Siddha', 'Unani' and 'Amchi') and the Chinese systems of medicine are finding increasing popularity and acceptance all over the world. Because of this sweeping 'greenwave', a large number of herbal drugs and the plant derived herbal products are sold in the health food shops all over the developed countries. According to some health care experts, there will be more dieticians rather than physicians in coming years, as many diseases can be prevented and better health can be maintained if one takes the right kind of food and nutrition which contains certain plants with specific functional attributes.

2.1 Advent of Food Medicine and Nutraceuticals

The doctrine "Let food be thy medicine…." espoused by Hippocrates nearly 2500 years ago is receiving renewed interest. Hippocrates clearly recognized the essential relationship between food and health and emphasized that "…. difference of diseases depend on nutriment". The Ayurvedic masters of ancient India had a clear understanding of the delicate cellular mechanism of the body and the deterioration of the functional capacity of human being. To arrest such deterioration of the functional efficiency and to revive and revitalize the body system, the Ayurvedic masters developed an elaborative rejuvenation therapy known as 'Rasayana' therapy. 'Rasa' in Sanskrit means the essence of nutrients and 'ayana' means to circulate in the body without any obstruction. 'Rasayana' is one of the eight clinical specialities of Indian classical Ayurveda that is aimed for the rejuvenation and geriatric care. Rasayana is not a drug therapy but a specialized procedure practised to cleanse body from the toxic or other microbial substances and then with the help of special diet and nutritional agents, comprising mainly highly powerful antioxidant, rejuvenates the body by providing greater immunity, enhance vitality, longevity, improve all faculties and attain youthfulness of the whole body.

There has been an explosion of consumer interest in the health enhancing role or physiologically active food components, so called nutraceuticals or functional foods. It is essentially a direct outcome on the understanding of the health enhancing and physiologically active food and other nutritional agents. The term nutraceuticals was coined in 1989 by the foundation for Innovation in Medicine (New York, USA) to provide a name for this rapidly growing area of biomedical research. A nutraceutical was defined as "any substance that may be considered a food or a part of food that provides medical or health benefits including the prevention and treatment of diseases". Nutraceuticals may range from genetically engineered "designer foods", herbal products, processed foods, dietary supplements, extracts, beverages or other products, which can be used as preventive medicines. These nutraceuticals are likely to play a significant role in health promotion and disease prevention and will be our preferred prescription of tomorrow.

2.2 Nutraceuticals and Functional Foods

The role of food and nutrition is now fairly well understood. With the advancement in science, molecular biology and genetic engineering, our ability to understand and manage health at molecular level is manifold increased. It is now scientifically demonstrated that it is possible for one to achieve the high-level of health and well-being if one takes right food and nutrition that suits his genetic constitution. Molecular biologists are now busy in designing individualized food, customized food based on ones genetic make up called 'nutrigenomics'. Also, now it has become very clear that the human communities who developed the traditional food and nutritional recipes, now called ethnic food, are the best suited for the people inhabiting that particular locality or in similar agroclimatic conditions. This understanding towards the end of 20th century led the health scientists and nutritional experts to scientifically investigate the traditional foods, which has led to the discovery that the traditional food and other traditional nutritional recipes can be best for maintaining a healthy life. It has also led to the development of designer food that suited different groups and also for different categories of people suffering from what is now called lifestyle diseases like diabetes, obesity, cancer, arthritis, hypertension, etc. Functional foods or medicinal food or pharma food or nutraceuticals are the best treatment regime for curing or managing such diseases. It is now said that within another five years time, there will be more dieticians rather than physicians. One may first go to genomic expert who will make your genomic profile and based on the genomic profile the dietician will prescribe you a new diet regime or a 'Rasayana' therapy of Ayurveda or advise for a proteomic therapy or a gene therapy.

2.3 Market Size of Health Foods, Functional Foods and Nutraceuticals

The nutraceutical is a big business today, the hot market in terms of the priorities, optimum business potential and trend in world. Nutraceuticals include antioxidants, 'Ayurveda' based products, nutraceuctical beverages, omega 3 fatty acids and anti-obese preparations. One of the most important markets is the mushrooms and mushroom-based products. Since ancient times, mushrooms have also been consumed by humans not only as a part of normal diet, but also as a delicacy, because they have highly desirable taste and aroma. In addition to this, the medicinal properties and the nutritional value have also been recognised since very long time. Even our ancient literatures like the Veda have mentioned their medicinal importance with the Romans calling them as "Foods of the God's" and the Chinese referring them as "Elixir of life".

There are more than 250 companies both national and international selling more than 1000 products in the name of nutraceuticals. The global market of functional food is estimated upto 33 billion US $, the respective market estimation for Europe exceeds 2 billion US $, representing less than 1% of the European food market. USA have an estimated market of more than 50% (Pushpangadan 2000). In total, functional food have a market enhance of around 2% in the US food market. Another important market is Japan. In 1984, the concept of functional food was floated by the scientists studying the relationship between nutrition sensory satisfaction, fortification and modulation of physiological systems. FOSHU (Food for Specific Health Uses) was subsequently introduced by Japan Ministry of Health in 1991. There are over 1700 products that have been launched as functional foods between 1988-98 with an estimated turn-over of around 14 billion US$ in 1999. Health foods, functional foods and nutraceuticals constitute some of the fastest growing markets in the developed countries. The worldwide market for these products is estimated to be 100 billion US $ with an annual growth of 17%. The important areas with the market trends are given in Tables 1, 2 and 3.

Table 1 Major Category of Nutraceutical Market (million US $) in 2003

Disease	Risks
Cardiovascular diseases – 60	Obesity – 143
Arthritis – 43	High cholesterol lowering – 93
Diabetes – 16	Plague – 50-75
Cancer – 10	High blood pressure – 50
Atherosclerosis – 10	Stress – 42

Table 2 Health Food Market Indicators (in 2003)

Lifestyle	Treatment
Joint pain – 90	Immunity – 104
Mental – 61	Digestion – 70
Bone – 33	Aging – 49
Sleep – 45	Energy – 35
Vision – 33	Impotence – 30

Type	Growth % per year	1999 market US $ (billion)	Projected market US $ (billion)
Natural	15	25.4	29.4
Organic	24	4.2	6.6
Vitamins/minerals	6	13.2	16.7
Herbs	16	3.6	6.6
Functional foods	11	14.2	17.6

Table 3 Mega Markets (Consumer concerns on functional food) (in 2003)

• Healthy eyesight -	85%
• Cancer -	81%
• Fatigue/energy -	75%
• Heart disease -	73%
• Joint pain/arthritis -	73%
• High cholesterol -	73%
• High blood pressure -	66%

From the tables given above, it is very clear that there is a rapidly growing demand for health products and nutraceuticals. Since the last two decades, health promotive and preventive strategy and therapeutic potential of the eastern medical traditions particularly those of India and China are receiving increasing popularity in the world. Because of this increasing 'green trend', a large number of health foods and nutraceuticals are sold in the health food shops all over the world. Now it has been scientifically proved that many lifestyle diseases and metabolic disorders can be prevented, corrected or maintain good health by taking certain functional foods or nutritional supplements.

2.4 Traditional Diets and Nutraceuticals

The key to the development of health foods or nutraceuticals lies in the value addition in the traditional natural diets. India has over 5000 years of heritage in health science where food has been given an important role in maintaining healthy life. People living in different agro-climatic regions of the country had experimented and made a variety of food and diet, and health care practices, which are now termed as ethnic foods and ethnic nutritional diets. Ayurvedic medicine, as explained earlier, deals with an unique system of management called 'Rasayana' which is essentially a combination of food and medicinal herb recipes intended to rejuvenate the whole body system and make it fully healthy and functional.

Phytonutrients/ phytochemicals have tremendous impact on the health care system and may provide health benefits including prevention and treatment of diseases and physiological disorders. Polyphenols are one of the most widely distributed groups of phytochemicals that are responsible for the health promoting effects of nutraceuticals. They range from simple phenols to highly polymerised tannins. They protect plants from oxidation damage and the same in human protecting the tissue from oxidative decay thereby acting as antioxidants. The outstanding feature of these phyto-nutrients is their ability to block specific enzymes that cause inflammation. They also modify prostaglandin pathways and thereby protect platelets from clumping. There is also tremendous amount of research going on in the field of nutraceuticals with around 900 publications with nutraceuticals as the keyword.

Another class of nutraceuticals is represented by the Polyunsaturated Fatty Acids (PUFAs) especially of those n-3 and n-6 fatty acid (FA) families. Current interest is devoted to the so called fish oils containing a high share of n-3 FA (eicosapentanoic acid [EPA] and decosahexaenoic acid (DHA). It is claimed that these particular FA exert a positive effect on the development of cardiovascular and inflammatory diseases and the beneficial effects of fish oil supplementation in many other chronic diseases have been advocated.

Many recent observations suggest a potential role of fish oils in the treatment of atopic dermatitis and psoriasis. There is also indications that premature infants have limited dietary support of the n-3 FA required for normal composition of brain and retinal lipids.

3. NUTRACEUTICALS IN AYURVEDA

The ancient Indian codified systems of medicine, namely, Ayurveda and Siddha seem to have in depth knowledge and understanding about the delicate relationship between food, nutrition and health. They also had a clear understanding of the delicate cellular mechanisms of the body and the deterioration of the functional capacity of human beings. These ancient medical masters had developed certain dietary and therapeutic measures to arrest or delay ageing and rejuvenating whole functional dynamics of the body system. This revitalization and rejuvenation is known as the 'Rasayan Chikitsa' (Rejuvenation therapy) in Ayurveda. It is specifically adopted to increase the power of resistance to disease (enhance immunity) and improve the general vitation and efficiency of the human being. 'Rasayana' therapy is done for a particular period of time with strict regimen on diet and conduct. Rasayana drugs are very rich in powerful antioxidants and hepatoprotective and immunomodulators. Rasayana is not a drug therapy, but a specialized procedure practised in the form of rejuvenation recipes, dietary regimen (Achara Rasayana) and special health promoting conduct and behaviour i.e., 'Achara rasayana'. Sushruta while defining rasayana therapy says that it arrests ageing ('Vayasthapam'), increase lifespan ('Ayushkaram'), intelligence ('Medha') and strength ('Bala') and thereby enable one to prevent disease. There are over 30-35 medicinal plants mentioned in different treatise of Ayurveda and Siddha having rasayan properties. The important among them are *Sida cordifolia/ S. cordata, Abulition indicum, Tinospora cordifolia, Acorus calamus, Ocimum sanctum, Withania somnifera, Emblica officinalis, Asparagus racemosus, Piper longum, Commiphora mukul, Semicarpus anacardium, Centella asiatica, Curcuma longa, Chlorophytum borivilianum, Chlorophytum tuberosum and Dactylorhiza hatagirea*, etc.

In 'Ayurveda' the term 'Rasayana' therapy, thus, refers to the use of the plants or its extracts as rejuvenators or as an elixir to enhance longevity, improve memory intelligence, good health, promote youthfulness, improve the texture and luster of the skin, improve the complexion and voice, promote optimum strength of the body and sense organs and the personality becomes very attractive. Rasayan materials can be special foods or nutritional items, medicinal herbs or a combination of all these three. Thus, the use of the medicinal plants as a source of dietary supplement or as a nutraceutical is well documented for centuries.

Ayurveda considers that an individual with advancing age accumulates waste and toxic substances and declines in vitality and loss of resistance or immunity called,

- **'Dhatu Kshaya'** weakening of the functional dynamics of the cell or tissue system of the body.
- **'Ojas'** the state of excellent health expressed in general strength vitality and luster of the individual – with 'Bala' = immunity against diseases.
- **'Dhatuvridhi'** i.e., rejuvenation of the whole tissue system is done by 'Ojasvardhaka Dravyas'- the substance that improves the functional efficiency and immunity of the individual. This therapeutic process is known as 'Rasayana Chikitsa' – Rejuvenation therapy.

The ancient Ayurvedic physicians treated every individual as unique. According to them, normally, there cannot be two individuals with same constitutional nature what they referred as 'Prakriti', and therefore the treatment is prescribed only after diagnosing the constitutional nature of the individual. This constitutional nature of the individual is based on the 'Tridosha' philosophy…. The various permutation combination of the 'dosha' in conjunction with 'triguna'- the qualitative nature could offer countless variation in the constitutional nature of the individual and an experienced physician can very well diagnose it. Interestingly, the modern molecular geneticists also now speak a language similar to this i.e., genomic composition, i.e., DNA finger print is unique to an individual and we are now talking about gene profiling to understand the genetic predisposition and then suggest treatment to correct it either by proteomic therapy or using other substances that can alleviate the defects or even the genomic therapy- proteomics, metabolomics and genomic methods for correcting disorders or treating diseases and nutrigenomics genetically designed nutrition or food items. The ancient Ayurvedic masters had advised to consume specific food that suits to the constitutional nature of the individual whom they have categorized in seven major groups. They have insisted certain do's and don'ts with regard to the food and nutrition according to the constitutional nature of the individual (Prakriti). Modern molecular biology and genetic engineering is offering genetically modified nutrition or food that suits to the genomic background of the individual or design drug suited to the individual – known as nutrigenomics and pharmacogenomics, respectively, are the current fast developing research today. With the perfection of technology for mapping human genome, it is now possible to get the DNA profile of individuals and then develop customized nutrition and treatment regimen.

Pharmacogenomics is the study of the hereditary basis for differences in response of populations to a drug (Patwardhan et al. 2004). The same view was expressed by the ancient Ayurvedic master Charaka, some 4000 years ago. Charaka observed that 'Every individual is different from another and hence should be considered as a different entity'. As many variations are there in the universe, all are seen in human beings. Patwardhan (2000) referred it as Ayugenomics and explained that it has quite clear similarities with the pharmacogenomics that is expected to become the basis of designer medicine.

An "in-depth study and analysis" of the constitutional concept of Ayurveda, namely, 'Prakriti' and with that of modern genotype will yield highly valuable insight in understanding the functional dynamic of the human health and can lead to the development of a customized treatment regimen. Less than 20% of the plant species have been evaluated chemically or biologically (Cordell 2003). Approximately 21,200 alkaloids have been isolated and described out of which hardly 70% have been evaluated in a single bioassay (Cordell et al. 2005). About 5000 compounds which enter advanced pharmacological development, only one will become a drug and that from the prior screening of about millions of compounds (Cordell 2005). It is also now a well established fact that drug discovery for a single agent drug is an inefficient and extremely expensive process and the best choice is to develop phytomedicine or pharmacomedicine, which involves activity guided isolation of fraction on selected traditional polyherbal formulations and their various permutation combinations so that one could develop effective therapeutic remedies gaining increasing acceptance and popularity. Such an approach could lead to the development of evidence based herbal formulations. Automation and application of nanotechnology, proteomics and metabolomics, etc. may further fine tune ethnopharmacology research.

4. RESEARCH IN HERBAL DRUGS AND NUTRACEUTICALS: ETHNOPHARMACOLOGY

Ethnopharmacology research is the commonly accepted method by which herbal drugs, nutraceuticals and other traditional remedies are evaluated. Ethnopharmacology is defined as "a multidisciplinary area of research, concerned with the observation, description and experimental investigation of indigenous drugs and their biological activities"(Rivier and Bruhn 1979). This definition has been modified, time and again, and its contemporary definition addresses the "interdisciplinary study of the physiological actions of plant, animal and other substances used in indigenous medicine of past and present cultures" (International Society of Ethnopharmacology constitution, 2005; Soejarto et al. 2005). Journal of Ethnopharmacology volume 100, Nos. 1-2 carried authoritative articles on the trend of ethnopharmacology research in the world. Ethnopharmacological research in India was initiated at Regional Research Laboratory, Jammu (India) in 1985 under the initiative by Atal and Pushpangadan. Pushpangadan and his associates later established National Society of Ethnopharmacology in 1986. Combining the strengths of the knowledge base of traditional systems such as Ayurveda with the dramatic power of combinatorial sciences and HTS will help in the generation of structure activity libraries. Ayurvedic knowledge and experimental database can provide new functional leads to reduce time, money and toxicity, the three main hurdles in drug development. Herbal drug research initiative to develop scientifically validated and standardized medicine under the New Millennium Indian Technology Leadership Intiative (NMITLI) launched by Council of Scientific Research (CSIR) , India is worth mentioning in this context. Randomised controlled clinical trails of rheumatoid and osteoarthritis, hepatoprotectives, hypolipidemic agents, asthma, parkinsons disease, etc. have now thus reasonably established clinical efficacy. A review of some exemplary evidence based research and approaches now resulted in wider acceptance of Ayurvedic medicine (Vaidya 2000; Vaidya et al. 2003; Chopra et al. 2000). National Botanical Research Institute jointly with Deenadayal Research Institute , Chitrakoot organized a national workshop in 2003 that led to the development of a 'Golden Triangle' approach (Mashelker 2003). 'Golden Triangle' refers to the converging Ayurveda, modern medicine and modern sciences to form a real discovery engine (Fig. 1) that can result in newer, safer, cheaper and effective therapies.

New technologies are constantly being developed to isolate and identify the components responsible for the activity of these plants. But these technologies should consider and possibly use the fact that the biological activity of plant extracts often results from additive or synergistic effects of its components. Another possibility is the qualitative and quantitative variations in the content of bioactive phytochemicals, which are currently considered major detriments in its use as a medicine. Different stresses, locations, climates, microenvironments and physical and chemical stimuli,

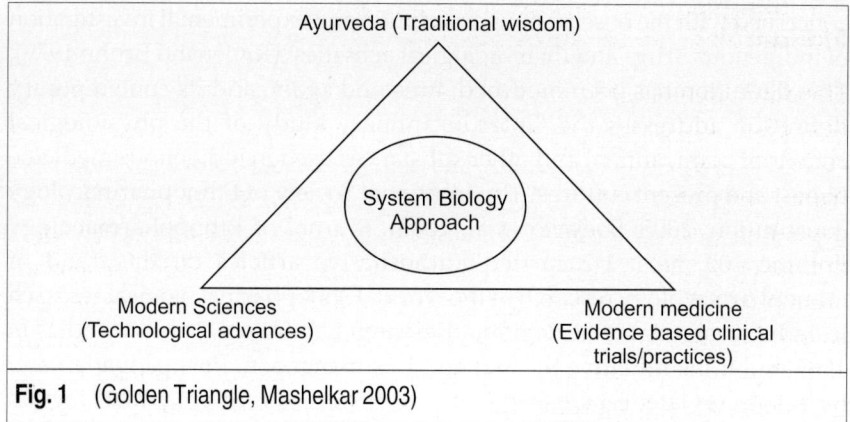

Ayurveda (Traditional wisdom)

System Biology
Approach

Modern Sciences
(Technological advances)

Modern medicine
(Evidence based clinical
trials/practices)

Fig. 1   (Golden Triangle, Mashelkar 2003)

often called elicitors, qualitatively and quantitatively alter the content of bioactive secondary metabolites. Enzymatic pathways leading to the synthesis of these phytochemicals are highly inducible (Ebel and Costa 1994). This is particularly true for phytochemicals that are well documented for their pharmacological activity, such as alkaloids (Facchini, 1994), phenylpropanoids (Dixon and Palva 1995) and terpenoids (Trapp and Croteau, 2001) whose levels often increase by two to three orders of magnitude following stress or elicitation (Darvill and Albershelm 1984). Thus, elicitation-induced, reproducible increases in bioactive molecules, which might otherwise be undetected in screens, should significantly improve reliability and efficiency of plant extracts in drug discovery while at the same time preserving wild species and their habitats. Molecular biologists and genetic engineers are currently engaged in designing food and medicinal plants with desired genetic make up so as to make custom made nutritional composition food or therapeutically desirable agents in plants–known as nutrigenomics and pharmacogenomics or proteomic approach to healthcare. Another emerging research area in medicinal plants is the metabolomics and system biology. Metabolomics is considered as a key technology in the system biology approach to study the mode of action in the therapeutic activity of traditional medicine and medicinal plants (Roos et al. 2004; Rao et al. 2004; Mei Wang et al. 2005). By measuring the activity of living organisms (which can be anything from a cell culture, animals to patients) for extracts with different composition, possibly one may identify a compound or a combination of compounds that correlate with the activity. This system biology approach is a major challenge for the coming years in studying medicinal plants (Verpoorte et al. 2005).

4.1 Regulation of Quality Control and Standardization of Nutraceuticals

Standardization of phytonutraceuticals raw drugs include passport data of raw plant drugs (crude drugs), correct taxonomic identification and authentication, study on the medicinal part (root, stem, bark, leaves, flowers, fruits, nuts, gum, resins, etc.), collection details (location, stage and development or growth of the plants, time, pre-processing storage, etc.), organoleptic examination of raw drug—evaluation by means of sensory organs (touch, odour taste), microscopic and molecular examination, chemical composition (TLC, GLC, HPLC, DNA fingerprinting), biological activity of the whole plant and shelf life of raw drugs. This is followed by well-defined Good Manufacturing Practices (GMP) and scientific validation including toxicity evaluation, chemical profiling, pharmacodynamics – effect of drug in the body, pharmacokinetics – absorption, distribution, metabolism, mechanism of action and execution, proper dosage form, proper presentation and packing and proper claim of therapeutic merits – Compared with other drugs. This should be followed by good survey of literature (Ancient & Modern), development and observation of norms of: Good Agricultural Practices (GAP), Good Collection/Harvesting and Post Harvest Handling Practices (GCP/ GHP & GPHP), Good Laboratory Practices (GLP), Good Clinical Practices (GCP), Good Manufacturing Practices (GMP) and Good Marketing Techniques (GMT).

Within the realm of alternative medicine, nutraceuticals represent one of the boldest challenges to government regulation. Historically, the Food and Drug Administration (FDA) took the position that a product that made health claims was a drug. Beginning in 1976, vitamins and minerals were exempted from regulation as drugs, but only so long as they did not make health claims. Then, in the early 1980s, studies began to show that certain food ingredients, such as fiber, provided specific benefits to health. Food manufacturers wanted to proclaim these benefits to consumers without having to obtain drug approval. The FDA wrestled with this problem until Congress passed the Nutrition Labelling and Education Act of 1990. This law authorized the FDA to issue regulations permitting certain health claims for foods, which led the agency to allow claims associating low levels of calcium with osteoporosis, dietary fats with cancer and cholesterol with heart disease.

After passage of the FDA Modernization Act of 1997, a food could make a health claim without FDA regulatory authority so long as the claim was based on an authoritative statement by a governmental or quasi-governmental scientific body, such as the National Institutes of Health (NIH) or National Academy of Sciences (NAS) and the agency was given advanced notice of the manufacturer's intent. This leeway did not apply to dietary supplements, however.

So, in 1994, Congress passed the Dietary Supplement Health and Education Act. Dietary supplements without drug approval can make so-called "structure/function" claims, such as "Vitamin A promotes good vision" or "St. Johns Wort maintains emotional well-being," so long as the label bears a disclaimer that the claim has not been evaluated by the FDA and that the product is not intended to diagnose, treat, cure or prevent disease. Until recently, dietary supplements could not make overt health claims (e.g., "Vitamin A prevents, cures or treats poor vision" or that "St. Johns Wort cures or treats depression") unless the FDA specifically permitted such a claim (as it has in the case of folic acid and the prevention of neural tube defects in babies).

Products that claim to offer various kinds of health benefits include prescription drugs, non-prescription drugs sold "over the counter" (OTC), and dietary supplements (also called "nutraceuticals"). The US Food and Drug Administration (USFDA) regulates these products under two different legal frameworks: (1) Prescription and non-prescription medicines are regulated as drugs and (2) Dietary supplements are considered a special class of foods. (www.alz.org/ResourceCenter/FactSheets/FSTherapy_claims.pdf).

The US Federal Food and Drug Administration (FDA) recently published guidance for standardized multifunctional and multicomponent plant extracts, referred to as botanical drugs, thus making it possible to market these products under the New Drug Application (NDA) Approval Process. (http://www.fda.gov/cder/guidance/ 1221dft.htm#P131_3293).

In response to the public demand for trustworthy and effective alternatives to NCE pharmaceuticals, the agency proposed abbreviated preclinical and clinical testing protocols for botanical drugs derived from plants with a safe history of human use. Botanical drugs are fully accepted and widely prescribed in China, Japan, India and other Asian and African countries. In addition, some countries in Europe, such as Germany, allow physicians to prescribe botanical drugs.

REFERENCES

Chopra, A. Chopra A, Lavin P, Patwardhan B, Chitre D (2000) Randomized double blind trial of an ayurvedic plant derived formulation for treatment of rheumatoid arthritis. J Rheumato 27:1365-1372.

Cordell.GA (2003) Discovery over Gifts from nature now and in the future Part II Revista de Quimica 17:13 – 15

Cordell.GA (2005) Some thoughts on the future of ethnopharmacology. Journal of Ethnopharmacology 100: 5-14

Darvill AG, Albersheim P (1984) Phytoalexins and their elicitors—a defense against microbial infection in plants. Ann Rev Plant Physiol 35: 245-277.

Dixon RA, Paiva NL (1995) Stress-induced phenylpropanoid metabolism. Plant cell 7: 1085-1097

Ebel J, Csio EG (1994) Elicitors of plant defense responses. Int Rev Cytol 148:1-36

Facchini PJ (2001) Alkaloid Biosynthesis in Plants: Biochemistry, Cell Biology, Molecular Regulation, and Metabolic Engineering Applications. Ann Rev Plant Physiol Plant Mol Biol 52: 29-36

Mashelkar RA (2003) Chitrakoot Declaration, National Botanical Research Institute & Arogyadham convention 2003, Chitrakoot.

Wang M, Lamers RJ, Korthout HA, van Nesselrooij JH, Witkamp RF, van der Heijden R, Voshol PJ, Havekes LM, Verpoorte R, van der Greef J (2005) Metabolomics in the context of system biology bridging traditional Chinese Medicine and Molecular Pharmacology Phytotherapy Research 19: 173-182.

Patwardhan.B, Ashok.DB, Vaidya, Mukund Chorghade (2004). Ayurveda and natural products drug discovery. Current science 86 (b): 789-799.

Pushpangadan P (2003) Biodiversity and IPR issues in medicinal plants in proceeding of Post-harvest Technology of Medicinal & Aromatic plants Eds. Tiwari K, P.Pushpangadan, NBRI, Lucknow, 134-142.

Pushpangadan P, Govindarajan R (2003) Role of medicinal plants in nutraceutical sector for nutritional security and export promotion in the Proceeding of the Second International Conference- Herbal Drugs & Health Tourism, AUUAHAN –Centre for Sustainable Development and Poverty Alleviation, Allahabad, pp. 234-242.

Pushpangadan P, Govindarajan R (2005) Need for Developing Protocol for Collection/ Cultivation and Quality parameters of Medicinal Plants for Effective Regulatory Quality Control of Herbal Drugs Proceedings – International conference on Botanicals, Kolkata, India. pp.63-69.

Rao Chv, Ojha SK, Radhakrishnan K, Govindarajan R, Rastogis, Mehrotra.S, Pushpangadan P (2004) Antiuleer activity of ulteria salicilofia rhizome extract. Journal of Ethnopharmacology 91: 243-249.

Rivier L, Bruhn JG (1979) Editorial, Journal of Ethnopharmacology.

Roos G, Roeseler C, Bueter KB, Simmen U (2004) Classification and correlation of ST.John's wort extracts by nuclear magnetic resonance spectroscopy, multivariate data analysis and pharmacological activity. Planta Medica 70:771-777

Soejarto DD , Fong HHS, Tan GT, Zhang HJ, Ma CY, Franzblau SG, Gyllenhaal C, Riley MC, Kadushin MR, Pezzuto JM, Xuan LT, Hiep NT, Hung NV, Vu BM, Loc PK, Dac LX, Binh LT, Chien NQ, Hai NV, Bich TQ, Cuong NM, Southavong B, Sydara K, Bouamanivong S, Ly HM, Thuy TV, Rose WC, Dietzman GR (USA, Vietnam, Laos) (2005) Seeking a transdisciplinary and culturally germane science: The future of ethnopharmacology. Journal of Ethnopharmacology 100: 15-22

Trapp S, Croteai R (2001) Defensive resin biosynthesis in conifers. Ann Rev Plant Physiol Plant Mol Biol 52:689-724.

Vaidya AD, Vaidya RA, Nagaral SI, Ayurveda and a different level of evidence: From Lord Macaulay to Lord Walton (1835-2001 AD) (2001) JO Assoc Physicians India 49: 534-537; Approach Paper, New Millennium Indian Technology Leadership Initiative Herbal Drug Development Programme, CSIR New Delhi, 2002.

Vaidya A Reverse pharmacology approach (2002) CSIR-NMITLI Herbal Drug Development Programme.

Vaidya RA, Vaidya AD, Patwardhan B, Tillu G, Rao Y Ayurvedic pharmacoepidemiology- a new discipline (2003) J Assoc Phys India 51:528.

Verpoorte YH Choi, Kim HK(2005) Ethnopharmacology and system biology: A perfect holistic match. Journal of Ethnopharmacology 100: 53-56.

Yesilada E (2005) Past and future contributions to traditional medicine in healthcare system of the Middle-East. Journal of Ethnopharmacology 100: 135-137.

Beneficial Microorganisms: Herbal and Medicinal Plants

R. PRASAD[1], R. MALLA[2], H. KHARKWAL[1], K. BHATNAGAR[1], A. DAS[1], N. VERMA[1], A. P. GARG[3] and A. VARMA[1*]

A mighty creature is the germ, Though smaller than the pachyderm,
His strange delight he often pleases, By giving people strange diseases.

~ Ogden Nash

1. INTRODUCTION

Do you know that there are more microorganisms in our body than there are people on earth? We spend millions of dollars each year on anti-bacterial soaps and antibiotics to fend off germs but, in fact, microorganisms play an essential role in human health, plant health and in the functioning of all ecosystems.

Microorganisms include viruses, bacteria, fungi, protozoa, algae and nematodes. They are the oldest form of life on earth and are found virtually everywhere, from boiling hot springs, deep in the earth and the oceans to the Arctic.

Microorganisms have a critical role in biogeochemical cycles, particularly the carbon, nitrogen and sulfur cycles. It is believed that it is the biological activity of microorganisms that is responsible for producing

1 Amity Institute of Microbial Sciences, Amity University, Uttar Pradesh, Sector 125, Noida, Uttar Pradesh, India
* Corresponding author: E-mail: ajitvarma@aihmr.amity.edu
2 Tribhuvan University, Trichandra College, Kingsway, Kathmandu Nepal
3 Department of Microbiology, Ch. Charan Singh University, Meerut, Uttar Pradesh, India

sufficient amount of oxygen in the earth's atmosphere more than two billion years ago to support life.

Microorganisms are the most important component of plant and soil ecosystems. There are one to ten million microorganisms present in each gram of soil. Similar numbers may occur on all plants and animals. Microorganisms break down dead plant and animal tissues and make their nutrients, including carbon and nitrogen available to support plant growth.

For the past century, the trend in University/Institute research and in modern farmer's practice is to focus on the physical and mechanical properties of soil. Therefore, in recent times the living component gained recognition for its central role in land productivity and plant health. Communities of bacteria, fungi, algae, protozoa and other microorganisms aerate the soil, make nutrients available to plants, create water and air channels, maintain soil structure and recycle nutrients and organic matter that allow vegetation to grow. Every chemical transformation that happens in soil involves microorganisms. Some of them excrete enzymes and other growth substances that stimulate plant feeding. Microorganisms provide a living reserve of nutrients like nitrogen and sulfur that would otherwise be easily leached. A healthy population of soil microorganisms can also maintain ecological balance, prevent the onset of major problems from the viruses or other pathogens that live in the soil.

The beneficial microorganisms assist in well functioning of ecosystem. They bring nutrient elements into the ecosystem from atmospheric or mineral reserves in soluble form; the roots take up the nutrients, break down the detritus and also protect roots from pathogens.

Microorganisms are vast frontier and a potential goldmine for the biotechnology industry because it offers countless new genes and biochemical pathways to probe for enzymes, antibiotics and other useful molecules.

Soil microorganisms are involved in various beneficial activities within the soil:

- Decomposition of crop residues
- Mineralisation of soil organic matter
- Immobilization of mineral nutrients
- Synthesis of soil organic matter
- Formation of organic substances which may be both stimulative and toxic to plant growth
- Nitrification
- Nitrogen-fixation

2. IMPROVING LAND PRODUCTIVITY WITH MICROORGANISMS

Often in dealing with land that has been cleared of natural vegetation and depleted of microlife, modern farmers have tried to improve productivity by adding soluble fertilizers. However, today and in future these microorganisms, may be implemented to focus on fostering and, when necessary, reintroducing healthy populations of microlife in the soil to enhance productivity. The use of mycorrhiza and rhizobium inoculants has become almost standard in forestry and agriculture. Now, various other ways are also being researched and practiced worldwide to increase productivity with various microlife. These techniques range from simply bringing in small amounts of healthy soil to inoculate new plantings in degraded areas to the manufacture and application of special biostimulants to feed and encourage microlife or introducing blends of selected beneficial microorganisms to be used in nursery or field production.

The farmers can help in maintaining and improving healthy populations of soil microorganisms, and therefore improving the productivity of the land. The use of mulch and organic matter is vital for healthy soil life, providing the moist, fertile conditions that allow microlife to thrive. Other activities that are highly beneficial include:

- Using no-till practices
- Using green manure or cover crops
- Maintaining species diversity of vegetation (encourages diversity of soil organisms)
- Reduce or eliminate the use of soluble fertilizers
- Reduce the use of fungicides, disinfectants and other chemicals that kill microlife unnecessarily
- Control erosion

3. THE RHIZOSPHERE

The rhizosphere is the zone of soil surrounding a plant root where the biology and chemistry of the soil are influenced by the root (Fig. 1). As roots grow through soil, they release water-soluble compounds such as amino acids, sugars and organic acids that supply food for the microorganisms. The food supply means microbiological activity in the rhizosphere is much greater than in soil away from plant roots. In return, the microorganisms provide nutrients for the plants. All these activities make the rhizosphere the most dynamic environment in the soil. Some microorganisms, including bacteria and mycorrhizal fungi form associations with roots that are mutually beneficial to both the plant and the microorganism.

The rhizosphere is a centre of intense biological activity due to the food supply provided by the root exudates. Bacteria, actinomycetes, fungi, protozoa, slime moulds, algae, nematodes, enchytraeid worms, earthworms, millipedes, centipedes, insects, mites, snails and small animals compete constantly for water, food and space.

Most soil microorganisms do not interact with plant roots, possibly due to the constant and diverse secretion of antimicrobial root exudates. However, there are some microorganisms that interact with specific plants. These interactions can be pathogenic (invade and kill roots and plants), symbiotic (benefit plant growth), harmful (reduce plant growth), saprophytic (live on plant debris) or neutral (no effect on plants). Interactions that are beneficial

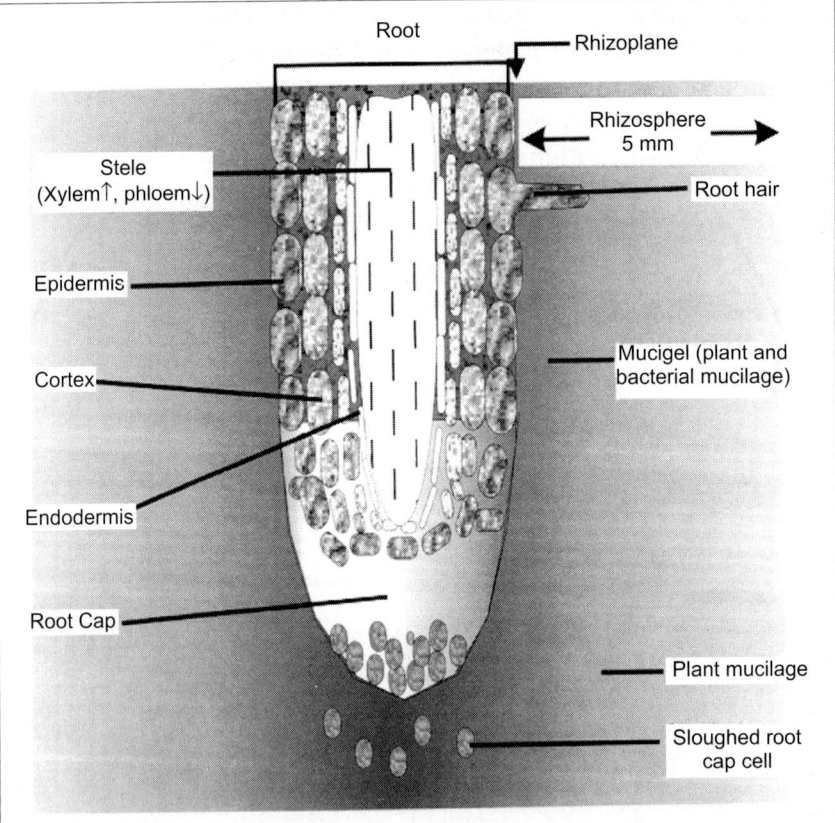

Fig. 1   Diagrammatic representation of the structure of the rhizosphere in and around live root (Raina Maier et al. 2000).

to agriculture include mycorrhizae (Fig. 2), legume nodulation and production of antimicrobial compounds that inhibit the growth of pathogens.

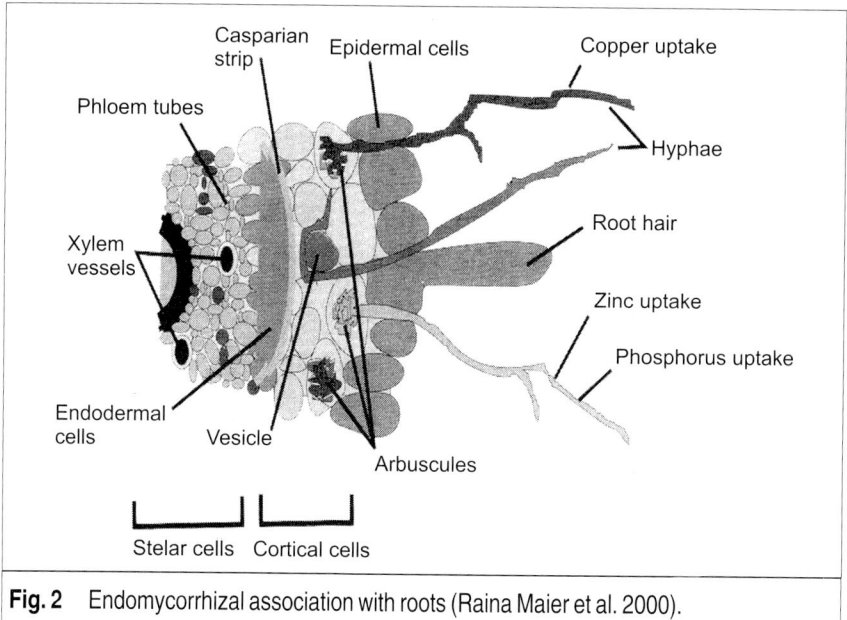

Fig. 2   Endomycorrhizal association with roots (Raina Maier et al. 2000).

Rhizosphere microorganisms produce vitamins, antibiotics, plant hormones and communication molecules that all encourage plant growth.

4. BENEFICIAL MICROORGANISMS FOR HERBAL AND MEDICINAL PLANTS

The most important beneficial microorganisms include mycorrhizae, plant growth promoting rhizobacteria (PGPRs), nitrogen fixers, decomposers, pathogen suppressive organisms, soil structure builders and protective surface crusts.

4.1 Mycorrhizae

Mycorrhizae (modern Latin of Greek words *mykes* + fungi, *rhiza* + root) are highly evolved, mutualistic associations between soil fungi and plant roots (Fig. 3). The history of mycorrhizae started with their description, interpretation and named by Frank (1885). Arbuscular mycorrhizal (AM) fungi are obligatory biotrophic organisms that live symbiotically with the roots of most plants (Fig. 4 and Fig. 5). AM fungi are found under all climates

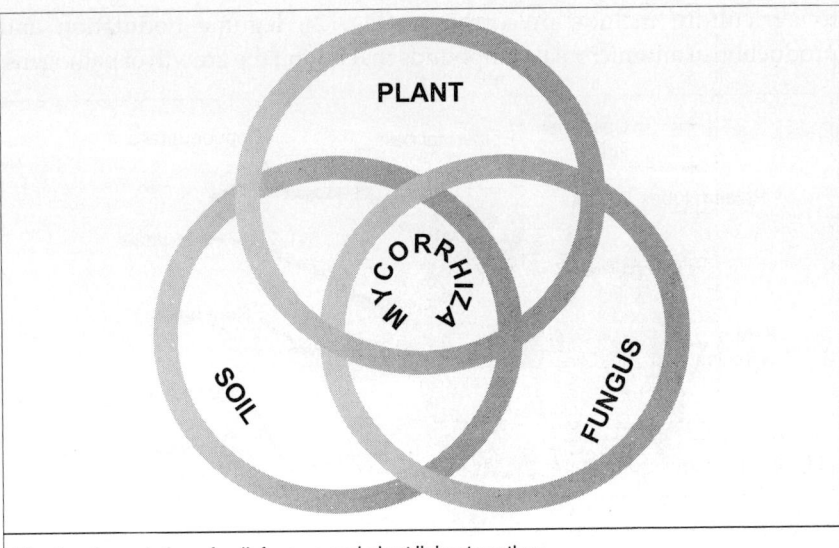

Fig. 3 Association of soil, fungus and plant living together.

and in all ecosystems, regardless of the type of soil, vegetation or growing conditions. The fungus delivers phosphate to the plant while receiving carbon and lipid (Bago et al. 2003). When phylogeny and ecology are considered, the arbuscular mycorrhizal symbiosis is probably the most important interaction between plants and microbes. About 80% land plants form AM, which improve their nutrient (especially phosphate) uptake. AM are an integral part of many ecosystems and are considered to be particularly advantageous to plants growing in tropical soils, in which nutrients are cycled rapidly and are present in low steady-state concentrations (Parniske 2004). Worldwide it has been estimated that 5 billion tons of carbon are transferred from plants to fungi via AM each year. AM fungi cannot complete their lifecycle in the absence of a host root.

In recent years, mycorrhizal studies have intensified to understand and improve plant fitness, to develop safe bioherbicides and to produce compounds for industrial and pharmaceutical applications (Shearer 2001). These fungi can be used as a tool for biological hardening of micropropagated plantlets and potent biological control agent against root pathogens. Not much progress has been made in genetic engineering and commercialization of AM fungi in forestry, agriculture and flori-horticulture due to the absence of axenic culture. To express foreign genes of importance to the host plants in a symbiosis specific manner, protoplast would serve as excellent biological systems. Viable protoplasts can also be used to exchange

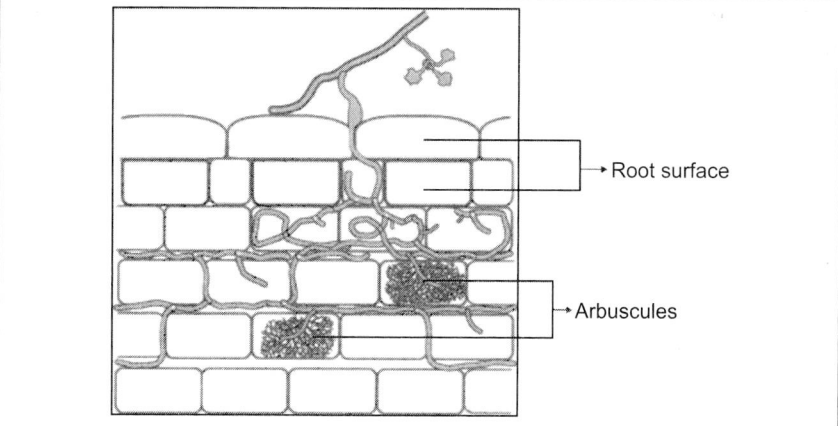

Root surface

Arbuscules

Fig. 4   Mycorrhizal structures produced by *Scutellospora* and *Gigaspora* species.

Fig. 5   Arbuscules- a unique feature of arbuscular mycorrhizae. This organ is responsible for bi-directional flux for the transport of photosynthates to the mycelium and vice versa nutrient and water to the plant.

metabolites, which are not accessible with intact hyphae, root tissues or fungi under functional symbiosis.

The development of detection and identification methods for the AM fungal matter within and outside the host plant root is of crucial importance since these symbiotic fungi cannot be raised axenically. The characterization of enzymes participating in symbiosis is an important aspect in mycorrhizal study. Till date there are no reports of isolation of pure enzymes from AM fungi as it suffers due to the non-cultivability of these fungi *in vitro* condition.

Heavy fertilization and incompatible fungicides commonly prevent mycorrhizal establishment. It takes a large-scale commitment to switch away from chemicals and towards beneficial organisms, but most of the problems now solved by chemicals are amenable to other solutions. Readily available phosphorus can be quite inhibitory but such insoluble forms as rock phosphate are not.

AM are also capable of inducing pathogen resistance in plants. The key functions of AM symbiosis can be summarized as follows: (1) plant establishment and improving rooting, (2) enhancing plant tolerance to biotic and abiotic stresses, (3) improving nutrient cycling, and (4) enhancing plant community diversity (Zhang et al. 2003). Use of AM fungi in the long-term would thus favor an agricultural system, that is, both production and protection oriented, thus enhancing and stabilizing agro-ecosystems (Leake 2001; Waceke et al. 2001).

4.1.1 How Mycorrhizal Fungi Benefit Plants

(1) **Better uptake of nutrients**   With the help of mycorrhizal fungi, a plant can take up many times more nutrients, particularly phosphorus than would be possible in the absence of the fungi. When a dependent plant lacks mycorrhizae, growers often have to load the soil with high levels of soluble nutrients. This heavy feeding is expensive and further damages the health of the soil and water.

(2) **Soil improvement**   Mycorrhizae enhance the soil by improving the structure of soil. This helps to increase water holding capacity and traps nutrients that otherwise could be leached by rains.

(3) **Faster rehabilitation of degraded sites**   By enhancing the plant's ability to take up nutrients and water, mycorrhizae fungi can help plants compensate for low nutrient availability, poor soil structure, and low water holding capacity that are often prevalent on harsh sites.

(4) **Healthier plants, less diseases and fewer pests**   Most experts in integrated pest management say that plant health is the most important aspect of pest management. Healthy plants have much fewer pest problems. Better nutrition and water uptake through mycorrhizae helps plants to stay healthy.

(5) **Biocontrol of certain pathogenic organisms**   By infecting the root system of a plant, mycorrhizae can interfere with pathogenic organisms, effectively protecting the host plant from diseases.

(6) **Tolerance for problem soils**   Mycorrhizal fungi may also help regulate the uptake of soil toxins, allowing plants to better tolerate salty or tough soil conditions.

4.1.2 Strategies for Improving Mycorrhizal Activity

1) Using green manures, mulch and plenty of organic matter will foster beneficial soil microorganisms.
2) Refrain from using soluble chemical fertilizers, especially those that have high levels of phosphorus. Use organic fertilizers when possible.
3) If the soils are degraded, consider adding a small amount of soil from a nearby healthy forest area to each planting hole to inoculate the soil with healthy microlife.
4) For badly degraded sites, plants can be inoculated with commercially available mycorrhizae prior to planting.

4.2 Plant Growth Promoting Rhizobacteria (PGPRs)

Microbes that exert beneficial effect on plant development are termed as "Plant Growth Promoting Rhizobacteria" (PGPRs). Chances for rhizospheric microorganism's interaction with mycorrhizal fungi are very high (Rovira et al. 1983). Many rhizospheric bacteria are plant growth promoters stimulating seedling growth and development, while mycorrhizal fungi provide vegetation with increased efficiency of nutrient uptake, increased productivity, drought stress and may contribute to plant diversity. These facts, among others are leading to a possible paradigm shift to a more microbially dominated, or at least highly reciprocal view of the relationship between plant and associated microbiota. PGPRs enhance plant growth either by producing plant hormones or by enhancing nutrient uptake.

Pseudomonas spp are typical PGPRs and their reaction with arbuscular mycorrhizal fungi has been studied by Barea et al. (1998). *Pseudomonas* spp had a positive effect on the spore germination and mycelial development of AMF in the soil as well as in root colonization. These bacteria are called Mycorrhization Helper Bacteria (MHBs) (Garbaye 1994). PGPRs have stimulatory effect on the arbuscular mycorrhizae formation and plant nutrition (Barea et al. 2004).

Siderophores, including salicylic acid, pyochelin and pyoverdine, which chelate iron and other metals, also contribute to disease suppression by conferring a competitive advantage to biocontrol agents for the limited supply of essential trace minerals in natural habitat (Loper and Henkels 1997). Siderophores may indirectly stimulate the biosynthesis of other antimicrobial compounds by increasing the availability of these minerals to the bacteria. Antibiotics and siderophores may further function as stress factors or signal inducing local and systemic host resistance (Leeman et al. 1996). Biosynthesis of antibiotics and other antifungal compounds are regulated by a cascade of endogenous signals (Corbell and Loper 1995).

PGPRs are of great potential importance and their culture and modes of action are the subject of active research (Burr and Caesar 1984; Linderman and Paulitz 1990; Okon and Hadar 1987). The possible mechanisms by which PGPRs aid plant growth include suppression of root pathogens through production of siderophores (compounds secreted by microorganisms that bind iron, making it less available to pathogens) or production of antibiotics (Burr and Caesar 1984; Kloepper et al. 1991; Weller 1985), nitrogen-fixation (Chanway and Holl 1991) and production of plant hormones (Holl et al. 1988). PGPRs are synergistic with mycorrhizae in stimulating plant growth (Chanway and Holl 1991, Meyer and Linderman 1986) and PGPRs may stimulate root colonization by mycorrhizal fungi (Meyer and Linderman 1986). There has been some success with PGPRs in agriculture and commercial preparations are likely to become available (Burr and Caesar 1984; Linderman and Paulitz 1990; McIntyre and Press 1991; Okon and Hadar 1987).

Colonization of the rhizosphere by some non-pathogenic microorganisms can protect the plant from a variety of bacterial, fungal and viral diseases. This is known as induced systemic resistance. Interaction between the plant and root-colonizing microorganisms trigger signaling pathways and the production of specific gene products that enhance the ability of the plant to resist pathogens. Secondary metabolites involved in these pathways include phenolics, flavonoids, alkanoids and terpenoids.

PGPRs and AMF can be used to enhance the growth of plants with natural health products. Pre-inoculation of hosts with PGPR and AMF can:

1) Induce or enhance specific human health promoting compounds in medicinal plants.
2) Enhance root health.
3) Increase resistance to environmental stress.
4) Increase yield and quality of medicinal products.

Although PGPRs and AMF have not been used specifically to increase the production of medicinal compounds in plants before, their ability to enhance plant growth and root health has been demonstrated with many crop species (Glick 1995; Maier et al. 1995; Van Loon et al. 1998).

The use of microbial associations for medicinal crops provides a sustainable approach to improve crop quality and yield and is suitable for use in organic agriculture. It provides the potential to increase production, value and export of human health-enhancing crops and products. It is expected that this will open new avenues products and markets for inoculant manufacturers.

4.3 Nitrogen-fixing Microorganisms

Biological nitrogen fixation is an important part of many agro-forestry, sustainable agriculture and land rehabilitation practices. Nitrogen is commonly the most limiting element in agricultural production. There is an abundant supply of nitrogen in the air (the air is 80% nitrogen gas, amounting to about 8000 pounds of nitrogen in the air over every acre of land, or 6400 Kg above every hectare). However, the nitrogen in the air is a stable gas, normally unavailable to plants. Many leguminous plants are able to utilize this atmospheric nitrogen through an association with rhizobia (bacteria), which are hosted by the root system of certain nitrogen-fixing plants.

Plants with nitrogen-fixing associations are important in some native vegetation types. Among our native nitrogen-fixing plants are alders, *Ceanothus, Acacia, Leucaena, Gliricidia, Erythrina, Sesbania, Dalbergia, Cajanus* and *Albizia*. Ground cover or annual crops include: *Crotalaria* species, *Mucuna pruriens* (velvet bean), *Dolichos lablab* (hyacinth bean), *Canavalia* species (jack or sword beans), *Arachis pintoi* (perennial peanut) and various other legumes. Some of the microorganisms are available commercially and some are probably available only from native soil. Some may be difficult to establish, few just require introduction of the microorganism, and others appear to take care of themselves.

Rhizobia are able to convert the nitrogen gas in the atmosphere into amino acids, which are the building blocks of proteins. The legume is then able to use this for its nitrogen needs. Rhizobia exchange nitrogen for carbohydrates from the plant. As the plants drop organic matter, or when the plants die, the nitrogen from their tissues is made available to other plants and organisms through soil. This process of accumulating atmospheric nitrogen in plants and recycling it through organic matter is the major source of nitrogen in tropical ecosystems (Fig. 6). Various agro-forestry practices such as alley cropping, improved fallow and green manure/cover

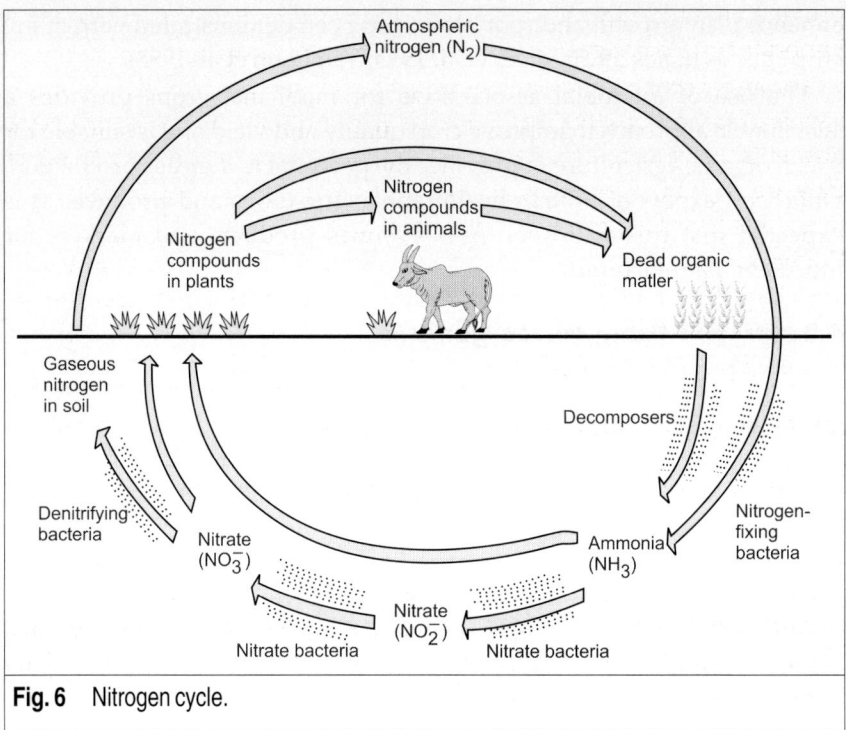

Fig. 6   Nitrogen cycle.

cropping exploit this natural fertility process by using nitrogen-fixing plants.

4.4 Decomposer Organisms

The degradation of plant remains is a critical step in nutrient cycling. Decomposition is carried out by a sequence of functional groups. The first on the scene are the opportunistic "sugar fungi" such as *Rhizopus stolonifer, Cunninghamella echinulata, Syncephalastrum racemosum* and fast-growing bacteria viz. *E. coli* and *Bacillus subtilis* which quickly metabolize readily available cell contents and leachable materials and complete their lifecycle within a few days. A second wave of organisms also arrives quickly and makes use of slightly more resistant plant compounds such as hemicellulose and cellulose (Alexander 1971).

The most resistant compounds are vulnerable only to specialized organisms, many of which are disperse and grow slowly. Lignin and other phenolics are quite resistant to biochemical breakdown. Their degradation may require an infusion of "starter glucose," which in nature often comes through fungal tissue from other regions of the soil. Soil dwelling invertebrates provide comminuting of plant materials (Witkamp 1971) and are also important in local dispersal of the primary decomposers.

The animal components of the detritus food web are the most important agents of mineralization. The primary decomposers tend to absorb plant nutrients into their own tissues and reproductive structures. Only when they are consumed by the next trophic level are the nutrients released in forms available to higher plants (Coleman et al. 1977).

The opportunistic organisms usually take care of themselves, since they are ubiquitous and quick to disperse (Alexander 1971). The process of decomposition might be hastened by adding invertebrates and specialized microorganisms in topsoil.

4.5 Pathogen-suppressive Microorganisms

Natural soil contains organisms that inhibit pathogens (Papavizas 1970). In the nursery or on a newly disturbed planting site, the complex balance of microbes has often been severely altered. There may be a nearly sterile environment, where the first organisms to arrive on the scene have a competitive advantage. All too often, the first ones to arrive are seed-borne plant pathogens (Leach and Garber 1970).

Fungi of the genus *Trichoderma* have repeatedly been implicated as suppressive factors in soils (Baker and Martinsen 1970, Leach and Garber 1970). Other pathogen-suppressive fungi have been reported from the genera *Gliocladium* (Sreenivasaprasad and Manibhushanrao 1990; Wu and Lu 1984) and *Penicillium* (Wu and Lu 1984). Certain fluorescent pseudomonad bacteria are able to suppress various root diseases (Weller 1985).

Kwok et al. (1987) reported that combinations of bacterial antagonists with *Trichoderma hamatum* were more effective than the fungal isolate alone. Hyphae of fungal antagonists may coil around hyphae of the pathogen. Antagonistic bacteria lysed the cell wall and plasmolysed conidia of the pathogen (Papavizas 1970; Wu and Lu 1984). *T. viride* produces antibiotics that can protect seedlings from damping off (Leach and Garber 1970).

Topsoil is a good source of organisms for field planting. In the nursery, a commercial preparation of *Trichoderma* or pathogen-suppressive bacteria would be safer than introduction of field soil. Certain ectomycorrhizal fungi are among the organisms that suppress pathogens (Zhao and Guo 1989) and could theoretically be dual-purpose symbionts. Even soil invertebrates may be of value in disease control. Certain Collembola, soil insects that eat fungi, may be able to help control pathogens (Curl and Gudauskas 1985).

Other Collembola eat AM fungi (Moore et al. 1985; Warnock et al. 1982), casting doubt on whether the good effects outweigh the damage.

4.6 Soil Structure-building Organisms

Soil is composed of weathered rock and organic matter, water and air. But the hidden "magic" in a healthy soil are the organisms—small animals, worms, insects and microbes-that flourish when the other soil elements are in balance.

Soil particles bind together to form aggregates, which help define the movement of air and water through the soil. Relatively large spaces between aggregates (macropores) allow rapid movement of water and air and permit ready penetration by roots. Graded or otherwise severely disturbed soil has no structure. The particles function as individuals and there are no distinct pore spaces (Kramer 1969). Since roots must grow into pores near their own diameter (Taylor 1974), it is very likely that many plant species cannot become established in non-aggregated soil.

Binding forces that hold soil particles into aggregates include chemical reactions and biological processes. Soil bacteria produce adhesive mucilages that cement small aggregates. The hyphae of soil fungi, particular mycorrhizal fungi, confer stability to mid-size aggregates (Miller and Jastrow 1990; Tisdall and Oades 1979). Macropores are short-lived unless the aggregates are water-stable (Kemper and Rosenau 1986).

4.7 Cryptogamic Crust Organisms

The surface layer of many natural soils is stabilized by a thin biological layer sometimes called a "cryptogamic crust." Cryptogams are plants such as mosses, lichens, algae and fungi that produce spores (Rushforth and Brotherson 1982).

There has been considerable interest in the intermountain west, where certain cyanobacteria (blue-green algae) form conspicuous layers on the soil surface (Skujins 1984). Some forms of cyanobacteria have also been used traditionally in rice agriculture to fix nitrogen for rice crops.

St. Clair et al. (1986) showed that a simple re-inoculation can greatly hasten the return of cyanobacteria after fire. However, a commercial cyanobacteria inoculant was apparently ineffective in a range of forest soils (Tiedemann et al. 1980). Quality native topsoil often contains propagules of the important cryptogamic crust organisms.

Crust forming cyanobacteria and green algae have filamentous growth forms that bind soil particles. These filaments exude sticky polysaccharide sheaths around their cells that aid in soil aggregation by cementing particles together with fungi (both free-living and as a part of lichens) contribute to soil stability. Apart from this, microorganisms and cryptogamic associations help the plants in the following ways:

- Lichens and mosses assist in soil stability by binding particles with rhizines/rhizoids, increasing resistance to wind and water action.
- The increased surface topography of some crusts, along with increased aggregate stability, further improves resistance to wind and water erosion.
- The microenvironment aids in the reduction of wind erosion and increases the probability for site stabilization through enhanced seed germination and plant establishment (St. Clair et al. 1984).
- Within this microenvironment, algal propagules are spread and dispersed via air currents.
- The microclimate provided by the irregular surface area aids in collecting organic matter for the seedbed and improved water availability to plants.

5. INTERACTION OF THE NOVEL SYMBIOTIC FUNGUS *PIRIFORMOSPORA INDICA* WITH HERBAL AND MEDICINAL PLANTS

Piriformospora indica (Hymenomycetes, Basidiomycota) is a newly described, cultivable, endophytic, root-colonizing fungus which has great potential for application in plant industry. This axenically culturable mycorrhiza-like-fungus has been recently described by Varma (Jawaharlal Nehru University, New Delhi, India) and his collaborators (Verma et al 1998; Pham et al. 2004a). The fungus was named *Piriformospora indica* based on its characteristic pear-shaped chlamydospores (Fig. 7). *P. indica* tremendously improves the growth and overall biomass production of diverse hosts, including legumes (Varma et al. 1999, 2001), medicinal and other economically important plants (Kumari et al. 2004; Pham et al. 2004b; Rai et al. 2001a; Peškan-Berghöfer et al. 2004; Shahollari et al. 2005). *P. indica* colonizes the roots of host plants as diverse group of plants belonging to monocots, dicots including orchids (Blechert et al. 1999; Singh and Varma 2001; Pham et al. 2004b; Prasad et al. 2005), herbs, shrubs and woody trees. The effect of *P. indica* interaction with various plants such as *Bacopa monieri*, *Azadirachta indica*, *Tridex procumbans* (Kumari 2006), *Abrus precatorius* (Kumari 2006), *Withania somnifera*, *Chlorophytum borivilianum* and *Spilanthes*

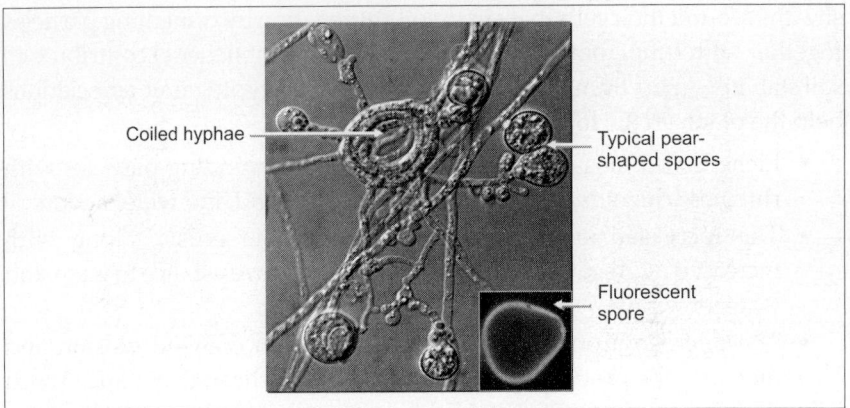

Fig. 7 *Piriformospora indica*, a model organism with hyphae and typical pear-shaped spore (in box).

calva (Rai et al. 2001a, 2004) (Figs. 8 and 9) have been tested in the laboratory conditions as well as in the extensive field trial.

Fig. 8 Interaction of *P. indica* (A) *Spilanthus calva* (B) *Withania somnifera* and (C) *Tridax procumbens*. The plant growth and size of flower increase when treated with the fungus.

Fig. 9 A case study for transfer of technology from lab to field. (A) Micropropagated Safed Musli (*Chlorophytum borivilianum*) treated with P. *indica* grown in single row, (B) grown in beds and (C) plants under flowering. An overall 90% plants survived well in the field.

Recent experiments on maize and cabbage have demonstrated the enhanced biomass production as a result of fertilisation with the fungus culture filtrate (Kumari 2006). The fungus also has the potential to act as bioprotectant against fungus root pathogens and soil insects. This fungus is documented to act as Biofertilizer (Singh et al. 2000a,b, 2001a,b, 2003; Malla et al. 2005), Biopromotor (Sahay et al. 2000; Das et al. 2005), Bioprotector (Rai et al. 2001a,b, 2004, 2005), Bioregulator (Kaldorf et al. 2005; Shahollari et al. 2005a,b; Peskan-Berghoefer et al. 2004) and Phytoremediator (Oelmüller et al. 2004, 2005; Kharkwal et al. 2006). Similar to arbuscular mycorrhizal fungi, *P. indica* stimulates nitrate assimilation in the roots and solubilizes insoluble phosphatic components in the soil. The interaction of *P. indica* with the model plant *Arabidopsis thaliana* is being used to understand the molecular basis of this beneficial plant-microbe interaction.

Recent experiments have amply demonstrated that *P. indica* provides resistance against heavy metal contamination in the soil. The soil treated with fungal bioinoculant, promoted the growth of the grass, turned greener, early flowering and seeding were set in, whereas, in the untreated soil plants failed to grow. Agro-forestry, flori-horticulture, arboriculture, viticulture and especially for better establishment of tissue culture raise-plants much needed for the application in plant industry (Singh et al. 2003). This would open up numerous opportunities for the optimization of plant productivity in both managed and natural ecosystems, while minimizing risks of environmental damage.

There is a renewed interest in the use of medicinal plants and medicinal plant products as an alternative to synthetically produced pharmaceuticals. The industry producing nutraceuticals and pharmaceuticals grew exponentially. However, the explosive growth of the plant-based medicine industry has been accompanied by issues of safety, quality, consistency and efficacy, and serious health problems have been reported. For commercial production of plant-based medicines, field grown or wild-harvested plant material has generally been used and these tissues may be contaminated with abiotic and biotic pollutants. Additionally, the natural resources of the medicinal plants are depleting rapidly, and concerns have arisen regarding the vulnerability of wild populations of medicinal plants due to over-harvesting and extinction. Current study represents a novel approach to medicinal plant research highlighting the optimized production conditions for biomass production and phytochemical contents in plant (Fig. 10).

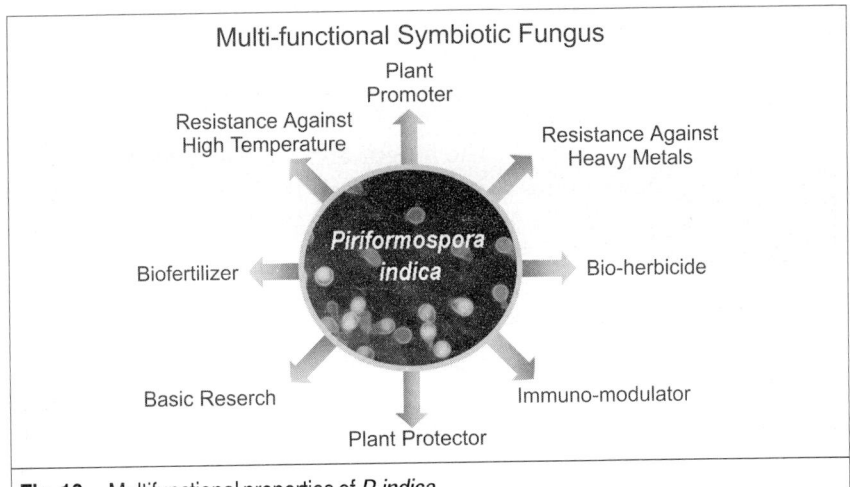

Fig. 10   Multifunctional properties of *P. indica*.

CONCLUSION

Soil microorganisms are very important for the plants as almost every chemical transformation taking place in soil involves active contribution from soil microorganisms. They play an active role in soil fertility as a result of their involvement in the cycle of nutrients like carbon and nitrogen, which are required for plant growth, nutrient uptake, decomposition of organic matter (e.g., plant litter) leading to recycling of the nutrients, protection of plants from pathogens, and formation of soil structure.

Certain soil microorganisms such as mycorrhizal fungi can also increase the availability of mineral nutrients (e.g., phosphorus) to plants. Other soil microorganisms can increase the amount of nutrients present in the soil e.g., nitrogen-fixing bacteria can transform nitrogen gas present in the soil into soluble nitrogenous compounds that can be utilized for plant growth. The microorganisms improving the fertility of the soil and contribute to plant growth are the 'biofertilizers' and are receiving tremendous attention to be used as microbial inoculants in agriculture. Some soil microorganisms have also been found to produce compounds (such as vitamins and plant hormones) that can improve plant health and contribute to higher crop yield. These phytostimulators are under study for their possible role as microbial inoculants to improve crop yield.

The survival and performance of many herbal and medicinal plants depends on beneficial microorganisms, especially after planting and cessation of artificial inputs. Nursery production of a diversity of plant species will depend upon effective use and management of beneficial microorganisms.

Piriformospora indica inoculated plants show better growth, early flowering than control plant. In contrast, the non-*piriformospora indica* plants were severely wilted and had reduced number of apical buds. It is postulated, that there were benefits of early rejuvenation of natural microflora on the sterile substratum which helped the plantlets in early acclimatization, fast establishment, enhanced growth and possibly improved root mycorrhization.

ACKNOWLEDGEMENTS

We express our gratitude to the Founder President, Dr. Ashok K Chauhan, Chancellor Amity University, Mr. Atul Chauhan, and Vice-Chancellor, Major General K Jai Singh for providing the necessary facilities in compiling this article.

REFERENCES

Alexander M (1971) Microbial ecology. John Wiley and Sons, New York.

Bago B, Pfeffer PE, Abubaker J, Jun J, Allen JW, Brouillette J, Douds DD, Lammers PJ, Shachar-Hill Y (2003). Carbon export from arbuscular mycorrhizal roots involves the translocation of carbohydrates as well as lipid. Plant physiol 131:1496-1507.

Baker R, Martinsen CA (1970). Epidemiology of diseases caused by *Rhizoctonia solani*. In: Parmeter JR (ed) *Rhizoctonia solani*, biology and pathology. University of California Press, Berkeley. pp 172-188.

Barea JM, Andrade G, Bianciotto V, Dowling D, Lokhrke S, Bonfante P, O'gara F. Azcon-Aguilar C (1998). Impact on arbuscular mycorrhiza formation of *Pseudomonas* strains used as inoculants for biocontrol of soil-borne fungal plant pathogens. Appl Environ Microbiol 64: 2304-2307.

Barea JM, Azcon R, Azcon-Aguilar C (2004). Mycorrhizal fungi and PGPR. In: A Varma, L Abbott, D Werner, R Hampp (eds) Plant Surface Microbiology. Springer-Verlag, Berlin pp 351-362.

Blechert O, Kost G, Hassel A, Rexer RH, Varma A (1999). First remarks on the symbiotic interactions between *Piriformospora indica* and terrestrial orchids. In: Varma A, Hock B (eds) Mycorrhizae. Springer-Verlag, Germany pp 683-688.

Burr TJ, Caesar A (1984). Beneficial plant bacteria. Critical Reviews in Plant Sciences 2:1-20.

Chanway CP, Holl FB (1991). Biomass increase and associative nitrogen-fixation of mycorrhizal *Pinus contorta* seedlings inoculated with a plant growth promoting *Bacillus* strain. Can J Bot 69:507-511.

Coleman DC, Cole CV, Anderson RV, Blaha M, Campion MK, Clarholm M, Elliott ET, Hunt HW, Shaefer B, Sinclair J (1977). An analysis of rhizosphere-saprophage interactions in terrestrial ecosystems. In: Rosswall T, Persson T (eds) Soil organisms as components of ecosystems. Ecological Bulletin (Stockholm) 25: 299-309.

Corbell N, Loper JE (1995). A global regulator of second metabolites production in *Pseudomonas fluorescence* Pf-5. J Bacteriol 177: 6230-6236.

Curl EA, Gudauskas RT (1985). Effects of soil insects on populations and germination of fungal propagules. In: Parker CA, Rovira AD, Moore KJ, Wong PTW (eds) Ecology and management of soil borne plant pathogens. Proceedings of section 5 of the fourth international congress of plant pathology. The American Phytopathological Society, St. Paul. pp 20-23.

Das A., Prasad R, Bhatnagar K, Lavekar GS, Varma A (2005). Synergism between Medicinal Plants and Microbes. In: Chauhan AK, Varma A (eds) Microbes: Health and Environment. Microbiology Series Vol. 3, IK International- India pp 13-64.

Frank AB (1885). Ueber die auf Wurzelsymbiose beruhende Ernaehrung gewisser Baeume durch unterirdische Pilze. Ber Dtsch Bot Ges 3: 128-145.

Garbaye J (1994). Helper bacteria: A new dimension to the mycorrhiza symbiosis. New Phytol 128: 197-200.

Glick BR (1995). The enhancement of plant growth by free-living bacteria. Can J Microbiol 41: 109-117.

Holl FB, CP Chanway, R Turkington, RA Radley (1988). Response of crested wheatgrass *Agropyron cristatum* L. perennial ryegrass *Lolium perenne* and white clover *Trifolium repens* L. to inoculation with *Bacillus polymyxa*. Soil Biol Biochem 20:19-24.

Kaldorf M., Koch B, Rexer KH, Kost G, Varma A (2005). Patterns of Interaction between *Populus* Esch5 and *Piriformospora indica*: A Transition from Mutualism to Antagonism. Plant Biology 7: 210-218.

Kemper WD, Rosenau RC (1986). Aggregate stability and size distribution. In: A Klute (ed) Methods of soil analysis, Part 1. Physical and mineralogical methods- Agronomy monograph no. 9. (2nd Edition). American society of agronomy, Soil science society of America, Madison WI. pp 425-442.

Kharkwal H, Kharkwal A, Bisaria VS, Varma A (2006). Symbiotic Fungi and medicinal plants: Their Nano-Biotechnological approach. ASSOCHAM, Nanotechnology and Biotechnology Meet the future nano billionaires 27-29 March 2006 pp 116-128.

Kramer PJ (1969). Plant and soil water relationships: a modern synthesis. McGraw-Hill Book Company, New York.

Kloepper JW, Zablotowicz RM, Tipping EM, Lifshitz R (1991). Plant growth promotion mediated by bacterial rhizosphere colonizers. In: Keister DL, Cregan P (eds) The rhizosphere and plant growth. Beltsville Symposia in Agricultural Research No. 14. pp 315-326.

Kumari R, Pham GH, Prasad R, Sachdev M, Srivastava A, Yadav V, Verma PK, Sharma S, Malla R, Singh A., Maurya AK, Prakash S, Pareek A, Rexer K-H, Kost G, Garg AP, Oelmueller R, Sharma MC and Varma A (2004). *Piriformospora indica*: Fungus of the Millenium. In: Basic Research and Applications: Mycorrhizae (eds Podila G and Varma A), Microbiology Series, IK International- India, New York, pp 259-295.

Kumari R (2006). Plant-microbe interaction in *in vitro* and *in vivo* conditions. Ph.D. Thesis, Meerut University, Uttar Pradesh, India.

Kwok OCH, Fahy PC, Hoitink HAJ, Kuter GA (1987). Interactions between bacteria and *Trichoderma hamatum* in suppression of *Rhizoctonia* damping-off in bark compost media. Phytopathology 77:1206-1212.

Leach L D, Garber RH (1970). Control of *Rhizoctonia*. In: Parmeter JR (ed) *Rhizoctonia solani*, biology and pathology. University of California Press, Berkeley pp. 189-199.

Leake JR (2001). Is diversity of ectomycorrhizal fungi important for ecosystem function? New Phytologist 152: 1-3.

Leeman M, Den Ouden FM, van Pelt JA, Dirkz FPM, Steijl H, Bakker PAHM, Schippers B (1996). Iron availability affects induction of systemic resistance to *Fusarium* wilt of radish in commercial greenhouse trials by seed treatment with *Pseudomonas fluorescence* WCS374. Phytopathol 85: 149-155.

Linderman RG, Paulitz TC (1990). Mycorrhizal rhizobacterial interactions. In: Hornby D (ed) Biological control of soil borne plant pathogens. Wallingford, Oxon, UK pp. 261-283.

Loper JE, Hankels MD (1997). Availability of Iron to *Pseudomonas fluorescence* in rhizosphere and bulk soil evaluated with an ice nucleation reported gene. Appl Environ Microbiol 63: 99-105.

Malla R, Prasad R, Kumari R, Giang PH, Pokharel U, Oelmüller R, Varma A (2005). Phosphorus solubilizing symbiotic fungus: *Piriformospora indica*. Endocytobiosis Cell Research 15: 579-600.

Maier W, Peipp H, Schmidt J, Wray V, Strack D (1995). Levels of a Terpenoid glycoside (Blumenin) and Cell Wall-Bound Phenolics in Some Cereal Mycorrhizas. Plant Physiol 109:465-470.

McIntyre JL, Press LS (1991). Formulation, delivery systems and marketing of biocontrol agents and plant growth promoting rhizobacteria (PGPRs). In: DL Keister, P Cregan (eds) The rhizosphere and plant growth. Beltsville Symposia in Agricultural Research No. 14 pp 289-295.

Meyer JR., RG Linderman (1986). Response of subterranean clover to dual inoculation with vesicular arbuscular mycorrhizal fungi and a plant growth promoting bacterium, *Pseudomonas putida*. Soil Biology and Biochemistry 18:185-190.

Miller RM, JD Jastrow (1990). Hierarchy of root and mycorrhizal fungal interactions with soil aggregation. Soil Biology and Biochemistry 22:579-584.

Moore JC, St. John TV, Coleman DC (1985). Ingestion of vesicular-arbuscular mycorrhizal hyphae and spores by soil microarthropods. Ecology 66: 1979-1981.

Oelmüller R, Peškan-Berghöfer T, Shahollari B, Trebicka A, Sherameti I, Varma A (2005). MATH-domain proteins represent a novel protein family in *Arabidopsis thaliana* and at least one member is modified in roots in the course of a plant/microbe interaction. Physiologia Plantarum 124:152-166.

Oelmüller R, Shahollari B, Peškan-Berghöfer T, Trebicka A, Giang PH, Sherameti I, Oudhoff M, Venus Y, Altschmied L, Varma A (2004). Molecular analyses of the interaction between *Arabidopsis* roots and the growth-promoting fungus *Piriformospora indica*. Endocytobiosis Cell Research 15: 504-517.

Okon Y, Y Hadar (1987). Microbial inoculants as crop yield enhancers. CRC Critical Reviews in Biotechnology 6: 61 85.

Papavizas GC (1970). Colonization and growth of *Rhizoctonia solani* in soil In: JR. Parmeter Jr. (ed.) *Rhizoctonia solani*, biology and pathology. University of California Press, Berkeley, pp. 108-124.

Parniske M (2004). Molecular genetics of the arbuscular mycorrhizal symbiosis. Current opinion in plant biology 7: 414-421.

Peškan-Berghöfer T, Shahollari B, Giang PH, Hehl S, Markent C, Blank V, Kost G, Varma A, Oelmüller R (2004). Association of *Piriformospora indica* with *Arabidopsis thaliana* roots represent a novel system to study beneficial plant-microbe interactions and involve in early plant protein modifications in the endocytoplasmic reticulum and in the plasma membrane. Plant Physiology 122: 465-471.

Pham GH, Kumari R, Singh An, Sachdev M, Prasad R, Kaldorf M, Buscot F, Oelmuller R, Tatjana P, Weiss M, Hampp R, Varma A (2004a). Axenic Cultures of *Piriformospora indica*. In: Varma A, Abbott L, Werner D, Hampp R (eds) Plant Surface Microbiology Springer-Verlag, Germany 593-616.

Pham GH, Singh An, Malla R, Kumari R, Prasad R, Sachdev M, Luis P, Kaldorf M, Tatjana P, Harrmann S, Hehl S, Declerck S, Buscot F, Oelmüller R, Rexer KH, Kost G, Varma A (2004b). Interaction of *P. indica* with other microorganisms and plants. In: Varma A, Abbott L, Werner D, Hampp R (eds) Plant Surface Microbiology. Springer-Verlag, Germany. pp 237-265.

Prasad R, Pham GH, Kumari R, Singh A, Yadav V, Sachdev M, Peskan T, Hehl, S, Oelmuller R, Varma A (2005). Sebacinaceae: culturable mycorrhiza–like endosymbiotic fungi and their interaction with non-transformed and transformed roots. In: Declerck S (ed) Root Organ Culture of Mycorrhizal Fungi Springer-Verlag, Germany pp. 291-312.

Rai M, Acharya D, Singh A, Varma A (2001a). Positive growth responses of the medicinal plants *Spilanthes calva* and *Withania somnifera* to inoculation by *Piriformospora indica* in a field trial. Mycorrhiza 11:123-128.

Rai MK, Varma A (2001b). New spectrum of fungal interactions in immunocompromised hosts. In *Innovative Approaches in Microbiology* (Ed: Maheshwari, D.K. and Dubey, R.C.) 299-320, B.S. Publishers, Dehradun, India.

Rai MK, Varma A, Pandey AK (2004). Enhancement of antifungal potential in *Spilanthes calva* after inoculation of *Piriformospora indica*, a new growth promoter. Mycoses 47: 479-481.

Rai M, Varma A (2005). Arbuscular mycorrhiza-like biotechnological potential of *Piriformospora indica*, which promotes the growth of *Adhatpda vasica* Nees. Electronic Journal of Biotechnology 8: 107-111.

Raina M M, Ian L Pepper, Charles P Gerba (2000). Environmental Microbiology.

Rovira AD, Bowen GD, Foster RC (1983). The significance of rhizosphere microflora and mycorrhiza in plant nutrition. In: Leuchli A, Bienleski RL (eds) Encyclopedia of plant physiology, New Series 15A: Mineral plant nutrition. Springer-Verlag, Berlin 61-68.

Rushforth SR, Brotherson JD (1982). Cryptogamic soil crusts in the deserts of North America. American Biology Teacher 44:472-475.

Sahay NS, Varma A (2000). Biological approach towards increasing the survival rates of micropropagated plants Current Science 78: 126-129.

Shahollari B, Varma A, Oelmüller R (2005a). Expression of a receptor kinase in *Arabidopsis* roots is stimulated by the basidiomycete *Piriformospora indica* and the protein accumulates in Triton X-100 insoluble plasma membrane microdomains. Journal of Plant Physiology 162: 945-958.

Shahollari B, Varma A, Oelmüller, R (2005b). Expression of receptor kinase in roots is stimulated by the basidiomycete *Piriformospora indica* and the protein accumulates in Triton-X-100 insoluble plasma membrane microdomains. Journal of Plant Physiology 162: 945-958.

Shearer JF (2001). Recovery of endophytic fungi from *Myriophyllum spicatum*. ERDC TN-APCRP-BC-03. pp 1-11.

Singh A, Varma A (2000a). Orchidaceous mycorrhizal fungi. In: Mukherji KG (ed) Mycorrhizal Fungi. Kluwer Academic Press, Amsterdam. 265-288.

Singh A, Rexer KH, Varma A (2000b). Plant productivity determinants beyond minerals, water and light. Current Science 79: 101-106.

Singh A, Varma A (2001a). AMF-like fungus, *Piriformospora indica*, shows promising applications. *International Symbiosis* 1: 1-4.

Singh An, Singh A, Rexer KH, Kost G, Varma A (2001b). Root endosymbiont: *Piriformospora indica*-a boon for orchids. The Journal of Orchid Society of India 15: 89-102.

Singh An, Singh Ar, Kumari M, Rai MK, Varma A (2003). Biotechnology importance of *Piriformospora indica*-A novel symbiotic mycorrhiza-like fungus: An Overview. Indian Journal of Biotechnology 2: 65-75.

Skujins J (1984). Microbial ecology of desert soils. Advances in Microbiology and Ecology 7: 49-91.

Sreenivasaprasad S, Manibhushanrao K (1990). Biocontrol potential of fungal antagonists *Gliocladium virens* and *Trichoderma longibrachiatum*. Zeitschrift fur flanzenkrankheiten und Pflanzenschutz 97: 570- 579.

St. Clair LL, Webb BL, Johansen JR and Nebeker GT (1984). Cryptogamic soil crusts: enhancement of seedling establishment in disturbed and undisturbed areas. Reclamation and Revegetation Research 3:129-136.

St. Clair LL, Johansen JR, Webb BL (1986). Rapid stabilization of fire-disturbed sites using a soil crust slurry: Inoculation studies. Reclamation and Revegetation Research 4: 261-269.

Tisdall JM, Oades JM (1979). Stabilization of soil aggregates by the root systems of ryegrass. Australian Journal of Soil Research 17: 429-441.

Taylor Howard M (1974). Root behavior as affected by soil structure and strength. In: E W Carson (ed) The plant root and its environment. University Press of Virginia, Charlottesville. P 691.

Van Loon LC, Bakker PAH, Pietesse CMJ (1998). Systemic resistance induced by rhizosphere bacteria. Annu Rev Phytopathol 36: 453-483.

Varma A, Singh A, Sudha, Sahay NS, Sharma J, Roy A, Kumari M, Rana D, Thakran S, Deka D, Bharti K, Franken P, Hurek T, Blechert O, Rexer KH, Kost G, Hahn A, Hock B, Maier W, Walter M, Strack D, Kranner I (2001). *Piriformospora indica*: A cultivable mycorrhiza-like endosymbiotic fungus. In: Hock B (ed) Mycota, Springer Verlag, Berlin, Heidelberg 9:125-150.

Varma A, Verma S, Sudha, Sahay N, Britta B, Franken P (1999). *Piriformospora indica*-a cultivable plant growth promoting root endophyte with similarities to arbuscular mycorrhizal fungi. Appl Environ Microbiol 65: 2741-2744.

Verma S, Varma A, Rexer Karl-Heinz, Hassel A, Kost G, Sarbhoy A, Bisen P, Buetehorn P and Franken P (1998) *Piriformospora indica* gen. nov., a new root-colonizing fungus. Mycologia-USA, 90: 895-909.

Warnock AJ, Fitter AH, Usher MB (1982). The influence of a springtail *Folsomia candida* (Insecta, Collembola) on the mycorrhizal association of leek *Allium porrum* and the vesicular-arbuscular mycorrhizal endophyte *Glomus fasciculatus*. New Phytol 90: 285-292.

Weller DM (1985). Application of fluorescent Pseudomonads to control root diseases. In: Parker CA, Rovira AD, Moore KJ, Wong PTW (eds.) Ecology and management of soil borne plant pathogens. Proceedings of section 5 of the fourth international congress of plant pathology. The American Phytopathological Society pp 137-140.

Witkamp M (1971). Soils as components of ecosystems. Annual Review of Ecology and Systematics 2: 85-110.

Wu WS, JH Lu (1984). Seed treatment with antagonists and chemicals to control *Alternaria brassicicola*. Seed Science and Technology 12: 851-862.

Zhao ZP, XZ Guo (1989). Study on hyphal hyperparasitic relationships between *Rhizoctonia solani* and ectomycorrhizal fungi. Acta Microbiologica Sinica 29:170-173.

Fermented Foods for Human Life

J. P. TAMANG

1. INTRODUCTION

A fermented food is prepared from the raw or cooked substrates of animal or plant origins by microorganism(s) either naturally or by adding mixed or pure starter culture(s) (Holzapfel 2002). The unique feature of food fermentations is the vital role of microorganisms that bring about essential biotransformations of the substrates during fermentation, contributing a number of desirable properties such as improved product shelf life, enriched diet with improved flavour and texture, increased safety, enriched nutritional supplements and probiotic functions in some foods. Indigenous technology of food fermentation represents a distillation of knowledge and wisdom of the people gained by trial-error basis, and also factors such as agro-climatic conditions, ethnic preference, socio-economic development status, religion and cultural practices of the region (Tamang 2005).

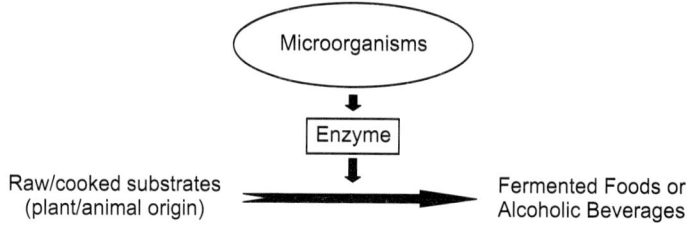

Food Microbiology Laboratory, Department of Botany, Sikkim Government College, Gangtok-737102, Sikkim, India. (jyoti_tamang@hotmail.com)

Bacteria, mostly lactic acid bacteria (LAB), yeasts and filamentous moulds constitute the microbiota of fermented foods and beverages, which are present in or on the ingredients, utensils, environment, and are selected through adaptation to the substrates (Hesseltine 1983; Tamang 1998). Table 1 shows the major functional microbial genera present in fermented foods and beverages of the world (Vandamme et al. 1996; Stiles and Holzapfel 1997; Kurtzman and Fell 1998; Pretorius 2000; Nout and Aidoo 2002; Tsuyoshi et al. 2005). Majority of the fermented foods involving filamentous moulds are produced in the East and South-east Asia. In Africa, Europe, America and Australia, fermented products are prepared exclusively using bacteria (mostly LAB) or mixed cultures of bacteria-yeasts; moulds seem to be little used (Tamang 1998). However, in the Indian subcontinent, mostly due to wide variation in agro-climatic conditions and diverse form of dietary culture of different ethnic groups of people, in most cases all three major types of microorganisms (bacteria-yeasts-filamentous moulds) are involved in traditional fermented foods and beverages (Tamang and Holzapfel 1999).

Table 1   Major functional microorganisms in fermented foods and beverages

Microbial group	Functional Genera
Filamentous moulds	*Actinomucor, Amylomyces, Aspergillus, Monascus, Mucor, Neurospora, Penicillium* and *Rhizopus*.
Yeasts	*Brettanomyces* (its perfect stage, *Dekkera*), *Candida, Cryptococcus, Debaryomyces, Galactomyces, Geotrichum, Hansenula, Hanseniaspora* (its asexual counterpart *Kloeckera*), *Hyphopichia, Kluyveromyces, Metschnikowia, Pichia, Rhodotorula, Saccharomyces, Saccharomycodes, Saccharomycopsis, Schizosaccharomyces, Torulopsis, Trichosporon, Zygosaccharomyces* and *Yarrowia*.
Bacteria	LAB: *Carnobacterium, Enterococcus, Lactobacillus, Lactococcus, Leuconostoc, Oenococcus, Pediococcus, Streptococcus, Tetragenococcus, Vagococcus* and *Weissella*.
	Non-lactic genera: *Acetobacter, Bacillus, Bifidobacterium, Brevibacterium, Citrobacter, Klebsiella, Micrococcus, Propionibacterium* and coagulase-negative *Staphylococcus*.

Fermented foods have immense functional and nutritional values, some of these are considered as potential source of medical therapy for human life. Ever since humans have started gathering and processing the foods for consumption, fermented foods have evolved as a part of requirement in daily diet. Extensive research has revealed that some of these age-old cultural foods, mainly fermented dairy products, fermented soybeans, etc. have therapeutic and functional properties. Today, some of these fermented foods such as *yoghurt, cheese* (by name *festivo), tempe (fote, tempe formula), natto, soy-sauce,* etc. are marketed as functional or therapeutic or nutraceuticals or health foods. This research paper deals with the review on role of functional microbes in fermented foods for human life concerning health.

2. ENRICHMENT OF DIET

Biotransformation of bland vegetable protein into meat-flavoured amino acids sauces and pastes by mould-fermentation (spp. of *Rhizopus, Aspergillus*) is common in Japanese *miso* and *shoyu*, Chinese *soy-sauce* and Indonesian *tauco* (Steinkraus 1996). In *ang-kak*, a traditional fermented rice food of South-east Asia, *Monascus purpureus* produces a purple-red water-soluble colour in the product, which is used in colouring meats and rice wine (Beuchat 1978). Halophilic microorganisms such as *Pediococcus halophilus*, contribute flavour and quality to fermented fish products of South-east Asia (Itoh et al. 1993). Fermentation improves the taste of otherwise bland foods, imparts typical flavour and texture to the fermented products such as *kinema* (Tamang 2001). In fermented milk products, LAB produce diacetyl and other desirable flavours (Kosikowski 1977). During *tempe* fermentation, mycelia of *Rhizopus oligosporus* knit the soybean cotyledons into a compact cake which when sliced resemble non-textured bacons (Steinkraus 1994). Similarly, in *ontjom*, an Indonesian fermented peanut product, *Neurospora sitophila* knits the particles into firm cakes imparting meat-like texture (Steinkraus 1996).

3. BIOPRESERVATION

Biological preservation refers to extended storage life and implies a significant approach to improve the microbiological safety of foods without refrigeration, by lactic acid fermentation. Species of LAB during fermentation produce lactic acid, the characteristic fermentative product, which reduce pH of the substrate to a level where growth of putrefactive, pathogenic and toxinogenic bacteria are inhibited, thus, the LAB can exert a 'biopreservative' (Holzapfel et al. 1995). During fermentation of *gundruk* and *sinki*, common traditional non-salted fermented vegetable products of the Himalayas, *Lactobacillus plantarum, Lb. brevis, Pediococcus pentosaceus, Leuconostoc fallax,* produce lactic acid and acetic acid and lower the pH of the substrates making the products more acidic in nature (Tamang and Sarkar 1993; Tamang et al. 2005). Due to low pH (3.3-3.8) and high acid content

(1.0-1.3%), *gundruk* and *sinki*, after sun-drying, can be preserved without refrigeration and addition of any synthetic preservative for more than two years. This can be cited as an example of biopreservation of perishable vegetables, which are plenty in the winter season in the Himalayan regions (Tamang and Holzapfel 2004). Common fermented vegetable products preserved by lactic acid fermentation are *kimchi* in Korea, *sauerkraut* in Germany and Switzerland, *sunki* in Japan, etc. Pickled vegetables, cucumbers, radishes, carrots, even some green fruits such as olives, papaya and mango are acid fermented in presence of salt (Steinkraus 1996).

4. BIO-ENRICHMENT OF NUTRITIONAL VALUE

Biological enrichment of food substrates by traditional fermentation with protein, essential amino acids, vitamins and bioactive compounds enhancing nutritive value of raw material, has high significance for developing countries where the majority of the people cannot afford to have commercially available expensive fortified nutritive foods. In *tempe*, traditional fermented soybean food of Indonesia, the levels of vitamins such as niacin, nicotinamide, riboflavin and pyridoxine are increased by *Rhizopus oligosporus*, whereas cyanocobalamine or vitamin B12 is synthesized by non-pathogenic strains of *Klebsiella pneumoniae* and *Citrobacter freundii* during fermentation (Liem et al. 1977; Keuth and Bisping 1994). Thiamine and riboflavin contents in *idli*, traditional fermented acid-leavened steamed cakes prepared from coarsely ground rice and dehulled black gram in South India, are increased during fermentation (Rajalakshmi and Vanaja 1967). Increase in methionine from 10.6 to 60.0% during *idli* batter fermentation has been observed (Rao 1961; Steinkraus et al. 1967). *Pulgue*, produced by lactic acid fermentation of juices of the cactus plant, is one of the oldest traditional fermented beverages of Mexico, which is rich in vitamins such as thiamine, riboflavin, niacin, pantothenic acid, pyridoxine and biotin, and serves as important diet of low-income children in Mexico (Steinkraus 1996). *Dawadawa*, a *Bacillus*-fermented locust bean food condiment is an important source of riboflavin in the local diet of West Africa (Campbell-Platt 1980). In yoghurt and cheese, LAB largely converts the lactose into more digestible lactate, proteins into free amino acids imparting digestibility to the product (Campbell-Platt 1994).

During *tempe* fermentation, isoflavone particularly Factor-II and aglycone contents were found to increase (Pawiroharsono 2002). During production of *kinema*, a traditional soybean food of the Himalayas, fermented by *Bacillus subtilis* (Tamang 2003), proteolytic enzymes produced by *Bacillus subtilis* into peptides and amino acids enhancing digestibility (Tamang and Nikkuni 1998) hydrolyze soya-proteins, which have been denatured by cooking process. Remarkable increase in water-soluble nitrogen and trichloroacetic acid (TCA)-soluble nitrogen contents were observed during *kinema* fermentation (Sarkar and Tamang 1995). Total amino acids, free amino acids and mineral contents are increased during *kinema* fermentation, and subsequently enriched the nutritional value of the product (Nikkuni et al. 1995). On a protein cost per kilogram basis, *kinema* is the cheapest source of high plant protein than animal and dairy products in the local diet of the people of Eastern Himalayan regions of India, Nepal and Bhutan (Tamang 2001).

5. DEGRADATION OF UNDESIRABLE COMPOUNDS

Functional microorganisms present in the fermented foods produce desirable amounts of enzymes which may degrade unsatisfactory or antinutritive compounds and thereby convert the substrates into consumable products with enhanced flavour and aroma. Bitter varieties of cassava (*Manihot esculenta*) tubers, the main staple crop in West Africa contain the cyanogenic glycoside linamarin, which can be detoxified by spp. of *Leuconostoc, Lactobacillus* and *Streptococcus* in *gari*, a fermented cassava product, are thereby rendered safe to eat (Westby and Twiddy 1991). Species of *Lactobacillus, Lacotococcus, Leuconostoc* and *Pediococcus*, isolated from fermented vegetables and tender bamboo shoots of the Eastern Himalayas, were found to degrade anti-nutritive factors such as phytic acids and flatulence causing oligosaccharides (Tamang and Holzapfel 2004).

During *tempe* fermentation, trypsin inhibitor is inactivated by *Rhizopus oligosporus* and eliminates the flatulence causing indigestible oligosaccharides, such as stachyose and verbascose into the absorbable monosaccharides and disaccharides (Hesseltine 1983). Microorganisms associated with *idli* batter fermentation reduce the phytic acid content of the substrates (Reddy and Salunkhe 1980).

6. IMPROVEMENT IN LACTOSE METABOLISM

Many people suffering from lactose intolerance or lactose mal-absorption, a condition in which lactose, the principal carbohydrate of milk, is not completely digested into glucose and galactose (Savaiano et al. 1984; Onwulata et al. 1989). Since lactose is cleaved into its constituent monosaccharides by the enzyme ß-D-galactosidase, lactose mal-absorption results from a deficiency of this enzyme (Shah and Jelen 1991). *Lb. delbrueckii* sub-sp. *bulgaricus* and *S. thermophilus*, the cultures used in making yoghurt, contain substantial quantities of ß-D-galactosidase, and it has been observed that the consumption of yoghurt may assist in alleviating the symptoms of lactose mal-absorption (Shah 1993).

Yoghurt or probiotic yoghurt is tolerated well by lactose-intolerant consumers, may be due to some lactose hydrolysed by lactics and bifids during fermentation, the bacterial enzyme autodigests lactose intracellularly before reaching the intestine, and it may be due to slower oral-caecal transit time (Shah and Jelen 1991). Yoghurt as viscous foods may delay gastric emptying and, thus, may be effective in alleviating lactose intolerant symptoms (Shah 1994). As a result, fermented acidophilus milk may be better tolerated than sweet acidophilus milk, as coagulated milk because of its viscous nature may pass more slowly through the gut than unfermented milk (Shah et al. 1992; Shah 2005).

7. PROBIOTIC FUNCTION

Probiotic foods are defined as foods containing live microorganisms, which actively enhance health of consumers by improving the balance of gut microflora when ingested live in sufficient numbers (Holzapfel et al. 1998). Recently, probiotics have been added to drinks and marketed as supplements including tablets, capsules and freeze dried preparations, and more than 70 bifidus and acidophilus containing products are produced worldwide, including sour cream, buttermilk, yoghurt, powdered milk and frozen desserts (Shah 2004). In Japan, more than 53 different types of probiotic-milk products are marketed, whereas in Europe their use is largely restricted to the yoghurt sector (Hilliam 2000). Some strains of LAB primarily species of *Lactobacillus* and *Enterococcus,* and species of *Bifidobacterium* predominant in fermented dairy products is not considered true lactics due to high (55-57 Mol %) G + C contents in DNA than true lactics] are used as probiotic adjuncts and as bio-therapeutic agents as protection against diarrhoea, stimulation of the immune system, alleviation of lactose intolerance symptoms, reduction of enzymes implicated in carcinogenesis, and reduction of serum cholesterol (Shah 2005). Probiotic cultures are considered to provide substantial health benefits by means of stabilising the gastrointestinal tract (Adachi 1992).

Species of LAB are normal residents of the complex ecosystem of gastrointestinal tract (GIT) (Holzapfel et al. 1997). The high degree of hydrophobicity by LAB isolated from traditional fermented cow-milk and yak-milk products of the Sikkim Himalayas probably indicates the potential of adhesion to gut epithelial cells of human intestine, suggesting a possible probiotic character, provided these strains are consumed in a viable state (Dewan 2002; Tamang et al. 2006).

Lactobacillus johnsonii (= *Lb. acidophilus*) which is used in production of acidophilus milk is now considered as important representative of probiotic bacteria which is used in production of novel yoghurt-like products with claims on their healthfulness in developed countries (Stiles and Holzapfel 1997). Some common probiotic cultures used in production of fermented functional foods are: *Lactobacillus acidophilus* La2, La5, Johnsonii; *Lb. bulgaricus* Lb12; *Lb. lactis* L1a; *Lb. plantarum* 299v, Lp01; *Lb. rhamnosus* GG, GR-1; *Lb. reuteri* MM2; *Lb. casei* Shirota; *Lb. paracasei* CRL 431; *Lb. fermentum* RC-14; *Lb. helveticus* B02; *Bifidobacterium adolescentis*; *B. longum* BB536; *B. breve* Yakult; *B. bifidus* Bb-11; *B. essensis* Danone; *B. lactis* Bb-12; *B. infantis* Shirota; *B. laterosporus* CRL 431; *B. lactis* DR-10; *B. longum* UCC35624 (Krishnakumar and Gordon 2001; Shah 2004). Table 2 shows some common functional foods marketed worldwide (Shah 2005).

Table 2 Gut active fermented functional dairy foods by country (adapted from Shah, 2005)

Brand	Manufacturer	Country
Actimel	Danone	Europe
Bioactivia	Danone	
Evolus	Valio	
LC1	Nestle	
Yakult	Yakult Honsha	
Actimel	Danone	USA
Danon	Danone	
Yoplait	General Mills	
All natural fruit blends	Stonyfields Farms	
Calpis	Calpis	Japan
Yakult	Yakult Honsha	
Danone Bio	Dairy Farmers	Australia
LC1	Nestle	
Vaalia	Pauls/Parmalat	
Yakult	Yakult Honsha	

Ingestion of probiotic yoghurt has been reported to stimulate cytokine production in blood cells and enhance the activities of macrophages (Solis and Lemonnier 1996). Translocation of small number of *Lb. acidophilus* and bifidobacteria via M cells to the Payer's patches of the gut associated lymphoid tissue in the small intestine is claimed to be responsible for enhancing immunity (Marteau et al. 1997).

8. PRODUCTION OF ENZYMES

Functional microorganisms, isolated from fermented foods showed a wide spectrum of enzymatic activities such as amylase, glucoamylase, protease, lipase, etc. Some of these strains produce a high amount of enzymes (Tamang and Nikkuni 1996; Thapa 2001; Tsuyoshi et al. 2005), which may be exploited for commercial production. Some enzymes may degrade unsatisfactory compounds and thereby convert the substrates into consumable products with enhanced flavour and aroma. *Bacillus subtilis* (*natto*) produces enzymes such as proteinase, amylase, mannase, cellulase and catalase during *natto* fermentation, a sticky fermented soybean food of Japan (Ueda 1989). *Bacillus subtilis* KK2:B10 (MTCC 2756) and GK2:B10 (MTCC 2757), isolated from *kinema*, produce desirable amounts of protease and α-amylase (Tamang and Nikkuni 1996).

9. ANTIMICROBIAL PROPERTIES

The widespread consumption of LAB in foods without any adverse health effects is taken as an indication that they can generally be regarded as safe (GRAS) and therefore, their bacteriocins might have potential as natural food preservatives (Adams 1999). Antimicrobial properties of the LAB are useful in traditional food fermentation, making foods safe to eat (Tamang, 1999). Some of the antimicrobial activities of the LAB are shown in Table 3. LAB compete with other microbes by screening antagonistic compounds and modifying the micro-environment by their metabolism (Lindgren and Dobrogosz 1990). Species of LAB isolated from fermented foods including fish, milk and vegetable products of the Eastern Himalayas showed antimicrobial properties against pathogenic bacteria such as *Listeria innocua* DSM 20649, *Listeria monocytogenes* DSM 20600, *Bacillus cereus* CCM 2010, *Staphylococcus aureus* S1 and *Klebsiella pneumoniae* subsp. *pneumoniae* BFE 147 (Dewan 2002; Thapa 2002; Thapa et al. 2006; Dewan and Tamang 2006). The inhibitory effect of LAB against *Bacillus cereus* was also observed in *dahi* samples collected from South Indian markets (Balasubramanyam and

Varadaraj 1994). The antimicrobial compounds produced by LAB are natural preservatives which could be used for safety of minimally processed foods (Niku-Paavola et al 1999).

Study of the effect of nisin-producing cultures has been shown to inhibit the growth of *Literia monocytogenes* in Camembert cheese (Maisnier-Patin et al. 1992). However, there have been several reports of work employing LAB-producing bacteriocins other than nisin to inhibit *Literia monocytogenes* in fermented sausages, cottage cheese and smoked salmon (McAuliffe et al. 1999; Nilsson et al. 1999).

Table 3 Antimicrobial properties of lactic acid bacteria

Product	Main target organisms
Lactic acid	Putrefactive and Gram-negative bacteria, some fungi
Acetic acid	Putrefactive bacteria, clostridia, some yeasts and fungi
Hydrogen peroxide	Pathogens and spoilage organisms, especially in protein-rich foods
Lactoperoxidase	Pathogens and spoilage organisms (milk and dairy products)
Diacetyl	Gram-negative bacteria
Bacteriocins → nisin	Spore-forming Gram-positive bacteria

10. ANTICARCINOGENIC PROPERTIES

Fermented products made by lactics and bifids have potential anticarcinogenic activity (Goldin and Gorbach 1977). Consumption of fermented foods containing viable cells of *Lb. acidophilus* decreased the enzymes, ß-glucuronidase, azoreductase, and nitroreductase (catalyse conversion of procarcinogens to carcinogens), thus possibly remove procarcinogens, and activate the immune system of consumers (Goldin and Gorbach 1984). Removal of procarcinogens by probiotic bacteria might involve a reduction in the rate at which nitrosamines are produced, due to the fact that certain species of *B. breve* have high absorbing properties for carcinogens, such as those produced upon charring of meat products (Mitsuoka 1989). Kimchi, fermented vegetables of Korea has large amounts of ascorbic acid, carotene and dietary fibre, which have anticarcinogenic effects (Cheigh and Park 1994).

11. REDUCTION IN SERUM CHOLESTEROL

Consumption of fermented milks containing very large populations of probiotic bacteria ($\sim 10^9$ bacteria/g) by hypercholesterolemic persons has resulted in lowering cholesterol level from 3.0 to 1.5 g/L (Homma 1988). Mann and Spoerry (1974) suggested that consumption of fermented dairy products could help lower serum cholesterol. They observed a decrease in serum cholesterol levels in men fed large quantities of milk fermented with lactobacilli, which may have been due to the production of hydroxymethyl-glutarate by LAB, which inhibit hydroxymethylglutaryl-CoA reductases required for the synthesis of cholesterol. Rao *et al.* (1981) reported that metabolites from orotic acid formed during fermentation of fermented milk may help lower cholesterol level.

12. THERAPEUTIC VALUES

Koumiss, a fizzy gray acidic-alcoholic beverage prepared from horse or donkey milk in Russia, contains *Lactococcus lactis, Lactobacillus bulgaricus* along with some yeasts *Candida kefyr* and *Torulopsis* spp. (Kosikowski 1977). *Koumiss* is not only regarded as a food high in nutritional quality, it is also considered therapeutic, particularly in the treatment of pulmonary tuberculosis (Auclair and Accolas 1974). Kosikowski (1977) reported that more than 50 Russian sanatoria offer *koumiss* treatment for tuberculosis. *Kvass*, a rye or wheat-based sour-alcoholic beverage of the Ukraine is fermented by *Saccharomyces cerevisiae* and *Lactobacillus* spp. and is suggested to provide protection to the digestive tract against cancer (Wood and Hodge 1985).

Consumption of *natto* in Japan prevents haemorrhage caused by vitamin K deficiency in infants (Ueda 1989). Consumption of *tempe* reduces the cholesterol level which is due to inhibition of hydroxymethylglutaryl-coenzyme A reductase, a key enzyme in cholesterol biosynthesis, by oleic acid and linoleic acid during fermentation (Hermosilla et al. 1993).

Kodo ko jaanr, a traditional mild-alcoholic beverage prepared from dry seeds of finger millet in Sikkim, the Darjeeling hills and North-east hills of India, Nepal, Bhutan, Tibet in China contributes to the mineral intake in daily diet of the local people (Thapa 2001). Because of high calorie, ailing persons and post-natal women consume the extract of *kodo ko jaanr* to regain the strength (Thapa and Tamang 2004).

CONCLUSION

Fermented foods are the distinct food culture of the community and the region. These have been prepared and consumed by the people for centuries, and they believe to have certain therapeutic values. The diversity of functional microorganisms in fermented foods has several novel properties.

Probiotic function of LAB strains associated with the gastrointestinal tract are selected for their health benefits such as protection against diarrhoea, stimulation of the immune system, anticarcinogenic, reduction of serum cholesterol, etc. and increasingly applied in functional foods.

REFERENCES

Adachi S (1992). Lactic acid bacteria and the control of tumours. In: The Lactic Acid Bacteria in Health and Disease, I (Ed. B.J.B. Wood), Elsevier Applied Science, London.

Adams MR (1999). Safety of industrial lactic acid bacteria. J Biotechnol 63: 17– 78.

Auclair J, Accolas JP (1974). Koumiss (koumiss, coomys). In: Encyclopedia of Food Technology (Eds: Johnson, A.H. and Peterson, M.S.), pp. 537-538. CT: AVI, Westport.

Balasubramanyam BV, Vardaraj MC (1994). *Dahi* as a potential source of Lactic Acid Bacteria active against foodborne pathogenic and spoilage bacteria. J Food Sci Technol 31(3): 241-243.

Beuchat LR (1978). Traditional fermented food products. In: Food and Beverage Mycology (Ed. L.R. Beuchat), Westport, CT:Avi, 224-253.

Campbell-Platt G (1980). African locust bean (*Parkia* species) and its West African fermented food product, dawadawa. Ecol. Food Nutr 9: 123-132.

Campbell-Platt G (1994). Fermented foods - a world perspective. Food Res Int 27: 253 257.

Cheigh HS, Park KY (1994). Biochemical, microbiological and nutritional aspects of Kimchi (Korean fermented vegetable products). Cr Rev Food Sci Nutri 34(2): 175-203.

Dewan S (2002). Microbiological evaluation of indigenous fermented milk products of the Sikkim Himalayas. Ph.D. Thesis, Food Microbiology Laboratory, Sikkim Government College, Gangtok, India.

Dewan S, Tamang J.P. (2006). Microbial and analytical characterization of Chhu, a traditional fermented milk product of the Sikkim Himalayas. J. Sci. Ind Res 65: 747-752.

Goldin BR, Gorbach SL (1977). Alterations in faecal microflora enzymes related to diet, age, lactobacillus supplement and dimethyl hydrazine. Cancer 40:2421-2426.

Goldin BR, Gorbach SL (1984). The effect of milk and lactobacillus feeding on human intestinal bacterial enzyme activity. Amer J Clin Nutr 39: 756-761.

Hermosilla J AG, Jha HC, Egge H, Mahmud M (1993). Isolation and characterization of hydroxymethylglutaryl coenzyme A reductase inhibitors from fermented soybean extracts. J. Clin Biochem. Nutr. 15: 163-174.

Hesseltine CW (1983). Microbiology of oriental fermented foods. Ann Rev Microbiol 37: 575-601.

Hilliam M (2000). Functional food: How big is the market? World Food Ingred 12:50-53.

Holzapfel WH (2002). Appropriate starter culture technologies for small-scale fermentation in developing countries. Int J Food Microbiol 75: 197-212.

Holzapfel WH, Geisen R, Schillinger U (1995). Biological preservation of foods with reference to protective cultures, bacteriocins and food-grade enzymes. Int. J Food Microbiol 24: 343-362.

Holzapfel WH, Schillinger U, Toit MD, Dicks L (1997). Systematics of probiotic lactic acid bacteria with reference to modern phenotypic and genomic methods. Microecol Therapy 26: 1-10.

Holzapfel WH, Harberer P, Snel J, Schillinger U, Huis in't Veld, JHJ (1998). Overview of gut flora and probiotics. Int J Food Microbiol 41: 85-101.

Homma N (1988). Bifidobacteria as a resistance factor in human beings. Bifidobact. Microflora 7: 35-43.

Itoh H, Tachi H, Nikkuni S (1993). Halophilic mechanism of isolated bacteria from fish sauce. In: Fish Fermentation Technology (Eds. C.H. Lee, K.H. Steinkraus and P.J. A. Reilly). The UNU Press, Tokyo, 249-258.

Keuth S, Bisping B (1994). Vitamin B12 production by *Citrobacter freundii* or *Klebsiella pneumoniae* during tempeh fermentation a proof of enterotoxin absence by PCR. Appl Environ Microbiol 60: 1495-1499.

Krishnakumar V, Gordon IR (2001). Probiotics: challenges and opportunities. Dairy Ind. Int 66(2): 38-40.

Kosikowski F (1977). Cheese and Fermented Milk Food, 2nd edn. Edwards Brothers, Ann Arbor, MI.

Kurtzman CP, Fell JW (1998). The Yeast, A Taxonomic Study, 4th Edition, Elsevier, Amsterdam.

Liem ITH, Steinkraus KH, Cronk TC (1977). Production of vitamin B12 in tempeh, a fermented soybean food. Appl. Environ. Microbiol 34: 773-776.

Lindgren S E, Dobrogosz W J (1990). Antagonistic activities of lactic acid bacteria in food and feed fermentations. FEMS Microbiol Rev 87:149-164.

Maisnier-Patin S, Deschamps N, Tatini SR, Richard J (1992). Inhibition of *Listeria monocytogenes* in Cambert cheese made with a nisin-producing starter. Lait 72: 249-263.

Mann G V, Spoerry A (1974). Studies of a surfactant and cholesterolemia in the Massai. Amer. J Clin Nutr 27: 484-469.

Marteau P, Vaerman JP, Bord JP, Brassart D, Pochart P, Desjeux, JF, Rambaud J C (1997). Effects of intrajejunal perfusion and chronic ingestion of *Lactobacillus johnsonii* strain LA1 on serum concentrations and jejunal secretions of immunoglobulins and serum proteins in healthy humans. Gastroenterol Clin Biol 21: 293-298.

McAuliffe O, Hill C, Ross RP (1999). Inhibition of *Listeria monocytogenes* in cottage cheese manufactured with a lacticin 3147-producing starter cultures. J Appl Microbiol 86: 251-256.

Mitsuoka T. (1989). Microbes in the intestine. Yakult Honsha Co. Ltd., Yokyo, Japan.

Nilsson L, Gram L, Huss HH (1999). Growth control *Listeria monocytogenes* on cold-smoked salmon using a competitive lactic acid flora. J Food Prot 62: 336-342.

Niku-Paavola, M.-L, Laitila A, Mattila-Sandholm T, Haikara A (1999). New types of antimicrobial compounds produced by Lactobacillus plantarum. J Appl Microbiol 86: 29-35.

Nikkuni S, Karki TB, Vilku K S, Suzuki T, Shindoh K, Suzuki C, Okada N (1995). Mineral and amino acid contents of Kinema, a fermented soybean food prepared in Nepal. Food Sc. Technol Int. 1: 107-111.

Nout, M.J.R, Aidoo, K.E. (2002). In: Mycota, A Comprehensive Treatise on Fungi as Experimental Systems and Applied Research, Industrial Applications, vol. X (ed. H.D. Osiewacz), Springer-Verlag Berlin Heidelberg New York, pp. 23-47.

Onwulata C I, Ramkishan Rao D, Vankineni P (1989). Relative efficiency of yoghurt, sweet acidophilus milk, hydrolysed-lactose milk, and a commercial lactose tablet in alleviating lactose maldigestion. Amer. J Clin Nutr. 49: 1233-1237.

Pawiroharsono S (2002). Tempe fermentation: products, technologies, and improvement of bioactive contents. In: The Proceedings of China and International Soybean Conference 2002, (Eds. K. Liu, J. Gui, R. Tschang, N. Zhuo and Y. Yu) CCOA and AOCS, Beijing. November 6-9, 2002. pp. 222-223.

Pretorius IS (2000). Tailoring wine yeast for the new millennium: novel approaches to the ancient art of winemaking. Yeast 16: 675-729.

Rajalakshmi R, Vanaja K (1967). Chemical and biological evaluation of the effects of fermentation on the nutritive value of foods prepared from rice and grams. Brit J Nutr 21: 467-473.

Rao MVR (1961). Some observations on fermented foods. In: Progress in Meeting Protein Needs of Infants and Preschool Children. National Academy of Sciences, Washington, DC, 291-293.

Rao DR, Chawan CB, Pulusani SR (1981). Influence of milk and thermophilus milk on plasma cholesterol levels and hepatic cholesterogenesis in rats. J Food Sci 46:1339-1341.

Reddy NR, Salunkhe DK (1980). Effect of fermentation on phytate phosphorus, and mineral content in black gram, rice, and black gram and rice blends. J Food Sc 45, 1708-1712.

Sarkar PK, Tamang JP (1995). Changes in the microbial profile and proximate composition during natural and controlled fermentation of soybeans to produce kinema. Food Microbiol 12: 317-325.

Savaiano DA, ElAnouar AA, Smith DJ, Levitt MD (1984). Lactose malabsorption from yoghurt, pasteurized yoghurt, sweet acidophilus milk, and cultured milk in lactase-deficient individuals. Ame. J Clin Nutr 40: 1219.

Shah NP (1993). Effectiveness of dairy products in alleviation of lactose intolerance. Food Australia 45: 268-271.

Shah NP (1994). Lactobacillus acidophilus and lactose intolerance: A review. ASEAN Food J 9: 47-54.

Shah NP (2004). Probiotics and prebiotics. Agro Food Industry Hi-tech 15 (1): 13-16.

Shah NP (2005). Fermented functional foods—an overview. In The Proceeding of the Second International Conference on "Fermented Foods, Health Status and Social Well-being", organized by Swedish South Asian Network on Fermented Foods and Anand Agricultural University, Anand, Gujarat, India, Dec. 17-18, 2005, p.1-6.

Shah NP, Jelen P (1991). Lactose absorption by post-weanling rats from yoghurt, quarg, and quarg whey. J Dairy Sci 74:1512-1520.

Shah NP, Fedorak RN, Jelen P (1992). Food consistency effects of quarg in lactose absorption by lactose intolerant individuals. Inert. Dairy J 2: 257-269.

Solis P, Lemonnier D (1996). Induction of human cytokines by bacteria used in dairy foods. Nutr. Res. 13:1127-1140.

Steinkraus KH (1994). Nutritional significance of fermented foods. Food Res Int 27: 259-267.

Steinkraus KH. (1996). Handbook of Indigenous Fermented Foods, second edition. Marcel Dekker, New York.

Steinkraus KH, van Veen AG, Thiebeau DB (1967). Studies on Idli - an Indian fermented black gram-rice food. Food Technol 21: 110-113.

Stiles ME, Holzapfel WH (1997). Lactic acid bacteria of foods and their current taxonomy. Int J Food Microbiol 36: 1-29.

Tamang J P (1998). Role of microorganisms in traditional fermented foods. Indian Food Indus 17 (3): 162-167.

Tamang JP (1999). Lactic acid bacteria in traditional food fermentation. In: Proceeding of the Symposium on Biotech Industry – a Challenge for 2005 A.D. with Special Reference to Fermentation, November 4-6, 1998, (Eds: Chand SC and Jain SC) pp. 102-108. All India Biotech Association and DBT, Delhi.

Tamang JP (2001). Kinema. Food Culture 3: 11-14.

Tamang JP (2003). Native microorganisms in fermentation of kinema. Indian J Microbiol 43(2): 127-130.

Tamang JP (2005). Food Culture of Sikkim, Sikkim Study Series volume IV, p. 120, Information and Public Relations Department, Government of Sikkim, Gangtok.

Tamang JP, Holzapfel WH (1999). Biochemical identification techniques-modern techniques: Microfloras of fermented foods. In: Encyclopedia of Food Microbiology (Eds: R.K. Robinson, C.A. Batt, C.A. and P.D. Patel), Academic Press: London. p. 249-252.

Tamang JP, Holzapfel WH (2004). Role of lactic acid bacteria in fermentation, safety and quality of traditional vegetable products in the Sikkim Himalayas. Final Project Report, Volkswagen Foundation, Karlsruhe, Germany, pp. 77.

Tamang JP, Sarkar PK (1993). Sinki: a traditional lactic acid fermented radish tap root product. J Gen Appl Microbiol 39: 395-408.

Tamang JP, Nikkuni S (1996). Selection of starter culture for production of kinema, a fermented soybean food of the Himalaya. World J. Microbiol Biotechnol. 12: 629-635.

Tamang JP, Nikkuni S (1998). Effect of temperatures during pure culture fermentation of kinema. World J Microbiol Biotechnol 14 (6): 847-850.

Tamang JP, Dewan S, Holzapfel WH (2006). Technological properties of predominant lactic acid bacteria isolated from indigenous fermented milk products of Sikkim in India. Int J Food Microbiol (in press).

Tamang JP, Tamang B, Schillinger U, Franz CMAP, Gores M, Holzapfel WH (2005). Identification of predominant lactic acid bacteria isolated from traditional fermented vegetable products of the Eastern Himalayas. Int J Food Microbiol 105 (3): 347-356.

Thapa S (2001). Microbiological and biochemical studies of indigenous fermented cereal-based beverages of the Sikkim Himalayas. Ph.D. Thesis, Food Microbiology Laboratory, Sikkim Government College, Gangtok, India.

Thapa N (2002). Studies on microbial diversity associated with some fish products of the Eastern Himalayas. Ph.D. Thesis, North Bengal University, Siliguri, India.

Thapa S, Tamang JP (2004). Product characterization of kodo ko jaanr: fermented finger millet beverage of the Himalayas. Food Microbiol 21: 617-622.

Thapa N, Pal J, Tamang JP (2006). Phenotypic identification and technological properties of lactic acid bacteria isolated from traditionally processed fish products of the Eastern Himalayas. Int. J Food Microbiol 107 (1): 33-38.

Tsuyoshi N, Fudou R, Yamanaka S, Kozaki M, Tamang N, Thapa S, Tamang JP (2005). Identification of yeast strains isolated from marcha in Sikkim, a microbial starter for amylolytic fermentation. Int J Food Microbiol 99 (2): 135-146.

Ueda S (1989). *Bacillus subtilis*. Molecular Biology and Industrial Applications, ed. B. Maruo and H. Yoshikawa. Elsevier and Kodansha Ltd. pp. 143-161.

Vandamme P, Pot B, Gillis M, de Vos P, Kersters K, Swings J (1996). Polyphasic taxonomy, a consensus approach to bacterial systematics. Microbiol Rev 60: 407-438.

Westby A, Twiddy DR (1991). Role of microorganisms in the reduction of cyanide during traditional processing of African cassava products. In: Proceeding of Workshop on Traditional African Foods - Quality and Nutrition (Eds. A. Westby and P.J.A. Reilly). I.F.S., Stockholm pp. 127-131.

Wood BJB, Hodge MM (1985). Yeast-lactic acid bacteria interactions and their contribution to fermented foodstuffs. In: Microbiology of Fermented Foods, vol I (Ed. B.J.B. Wood). Elsevier Applied Science Publication, London, pp. 263-293.

Intervention of Microbes in Promotion of Nutritional Attributes of Yogurt

A. DAS and A. VARMA

1. INTRODUCTION

Yogurt, less commonly known as yoghurt, joghurt or yogourt, is a fermented milk product produced by bacterial fermentation of any sort of milk. It is the fermentation of milk sugar (lactose) into lactic acid that gives yogurt its gel-like texture and characteristic tang. It is often sold in a fruit, vanilla, or chocolate flavors, but can also be unflavored.

Yogurt is produced by the controlled fermentation of milk by specific group of bacteria, e.g., *Lactobacillus* and *Streptococcus* species. The acid also restricts the growth of food poisoning bacteria and some spoilage bacteria. So, whereas milk is a potential source of food poisoning and only has a shelf life of a few days, yogurt is safer and can be kept for upto ten days, under proper storage conditions.

Yogurt can be made from any variety of mammal milk, but is most often made from cow, buffalo or goat milk. There are several types of yogurt with differing fat levels readily available in the market. Plain yogurt is generally made from cow milk and is unflavored and unsweetened. Flavored yogurt generally has fruit or flavoring added, along with plenty of sugar. The sugar is added not only for sweetness but to aid in preserving the fruit. Frozen yogurt is the yogurt version of soft ice cream. It is generally stabilized by the

Amity Institute of Microbial Sciences. Amity University Uttar Pradesh, Block A, Amity Campus, Sector 125, Noida, Uttar Pradesh, India E-mail: ajitvarma@aihmr.amity.edu addition of gelatin at home. All types of yogurt can be found in regular, low fat and non-fat.

Yogurt remained primarily a food of India, Central Asia, Western Asia, South Eastern Europe and Central Europe until the 1900s, when a Russian biologist named Ilya Ilyich Mechnikov theorized that large amount of consumption of yogurt was responsible for the unusually long lifespan of Bulgarian peasants. Believing *Lactobacillus* to be essential for good health, Mechnikov worked to popularize yogurt as a foodstuff throughout Europe. It fell to a Spanish entrepreneur named Isaac Carasso to industrialize the production of yogurt. In 1919 he started a commercial yogurt plant in Barcelona, naming the business Danone after his son, the group trades as Dannon in the US.

Yoghurt with added fruit marmalade was invented (and patented) in 1933 in dairy Radlická Mlékárna in Prague. The original intention of this combination was to protect yoghurt better against decay. Yoghurt was first commercially produced and sold in the United States in 1929 by Armenian immigrants, Rose and Sarkis Colombosian, whose family business later became Colombo Yogurt.

Yogurt has been a staple food in the Middle East for millennia, but it didn't become a mainstay in the United States until the later part of the 19th century. The Dannon Company achieved some minor success in 1947 when it first mixed strawberry preserves into yogurt and offered it as a sweetened dessert. However, it was health guru Gayelord Hauser who pushed yogurt into the limelight in United States when he proclaimed it a wonder food in his book "Look Younger, Live Longer," published in 1950. Sales of yogurt skyrocketed, increasing production 500% by 1968.

2. FERMENTATION

Fermentation is the energy yielding anaerobic metabolic breakdown of a nutrient molecule such as glucose, without net oxidation. Fermentation yields lactate, acetic acid, ethanol or some other simple product. The process is often used to produce or preserve food, typically wine or beer, but also includes the making of yogurt. The science of fermentation is known as zymology (zymosis). Pasteur coined the term zymosis. Buchner (1897) found that yeast extract could perform fermentation of sugary solution. The enzyme complex present in yeast, which could perform fermentation, was named as zymase. Fermentation is also used much more broadly to refer to the bulk growth of microorganisms on a growth medium.

Fermentation includes all metabolic processes such as:

(a) There is release of energy from a sugar or other organic compound.
(b) It does not require molecular oxygen.
(c) It does not require an electron transport system.
(d) There is a use of an organic compound as the final electron acceptor.

As in fermentation many energy rich bonds are not broken it tends to produce much less ATP than cellular respiration. Net gain is two molecules of ATP per glucose molecule. Fermentation also tends to produce waste products that can accumulate in the extracellular environment. Ethanol is a waste product that still contains a lot of energy. The energy in ethanol is unavailable in the absence of oxygen. Though fermentation of organic substrates is one way to make a living, but it is a poor choice as energy yields are low and microbes have to ferment large amounts of substrate to get enough energy for cellular processes. Anaerobes were probably the first microbes to evolve billions of years ago, but once oxygen became prevalent in the atmosphere, a better method of catabolism evolved which is respiration.

2.1 Microbes Used in Fermentation

2.1.1 Yeast

Yeasts constitute a group of single-celled (unicellular) fungi and are an important type of fermenting agent that are used in the production of bread and alcohol, as well as most other fermentations involving carbohydrates. The most commonly used yeast is *Saccharomyces cerevisiae*, which was domesticated for wine, bread and beer production thousands of years ago. Many types of yeast can be isolated from sugar-rich environmental samples like fruits and berries (such as grapes, apples or peaches), exudates from plants (such as plant saps or cacti), etc. Yeast species can have either obligate aerobic or facultative anaerobic physiology. There is no known obligate anaerobic yeast. In the absence of oxygen, fermentative yeasts produce their energy by converting sugars into carbon dioxide and ethanol (alcohol). In brewing, the ethanol is bottled, while in baking the carbon dioxide raises the bread and the ethanol evaporates.

A common medium used for the cultivation of yeasts is called potato dextrose agar (PDA) or potato dextrose broth. Yeast fermentations comprise the oldest and largest application of microbial technology. Baker's yeast is used for bread production, brewer's yeast is used for beer fermentation, and yeast is also used for wine fermentation.

2.1.2 Moulds

Moulds are also important organisms in the food industry, both as spoilers and preservers of foods. Moulds from the genus *Penicillium* are associated with the ripening and flavor of cheese. Moulds are aerobic and, therefore, require oxygen for growth. Other microbes involved in fermentation are *Aspergillus oryzae*, a fungus, which produces soy sauce, growing on soybeans.

2.1.3 Bacteria

Acetobacter and *Lactobacter* are the two main useful types of bacteria used in fermentation. The most important bacteria in desirable food fermentations are from the family lactobacillaceae, which have the ability to produce lactic acid from carbohydrates. *Lactobacter* are anaerobic which work on dairy products and produce cheese, yogurt, etc. They are naturally found in unpasteurised milk. *Acetobacter* are aerobic bacteria, which produce acetic acid or vinegar. *Erwinia dissolvens,* another type of bacteria, is essential for coffee bean production; it is used for softening and removal of the outer husk of beans.

2.2 Types of Bacterial Fermentation

Bacterial fermentation can be homofermentative or heterofermentative. In homofermentative fermentation or homolactic fermentation, a single by-product is produced that is lactic acid by glucose. In heterofermentative fermentation or heterolactic fermentation more than one by-product is produced and glucose produces lactic acid, carbon dioxide and ethanol.

2.2.1 Homolactic Fermentation or Homofermentative Fermentation

The fermentation of 1 mole of glucose yields two moles of lactic acid.

$$C_6H_{12}O_6 \text{------------} 2\,CH_3CHOHCOOH$$

Glucose   Lactic acid

Normal conditions required for this pathway are excess sugar and limited oxygen. The organisms performing homolactic fermentation are *Lactobacillus* spp. like *L. delbrueckii, L. leichmannii, L. plantarum* and *Streptococcus* spp. like *S. bovis, S. thermophilus,* etc.

2.2.2 Heterolactic Fermentation or Heterofermentative Fermentation

The fermentation of 1 mole of glucose yields 1 mole each of lactic acid, ethanol and carbon dioxide.

$$C_6H_{12}O_6 \dashrightarrow CH_3CHOHCOOH + C_2H_5OH + CO_2$$

Glucose   (lactic acid)   (ethanol)   (carbon dioxide)

The organisms undergoing heterolactic fermentation are *Lactobacillus* spp. like *L. brevis*, *L. fermentum*, *L. buchneri*, *Leuconostoc oenos*, etc. *L. brevis* is the most common, which is distributed in plants and animals and it partially reduces fructose to mannitol.

2.2.3 Bifidobacterium Fermentation

Bifidobacteria are non-motile, anaerobic, Gram-positive and ferment lactose and other sugars to acetic and lactic acids. They are residents of human intestinal tract and were discovered in 1906. These organisms are the initial residents of the colon of the breast fed infants, because human milk contains a disaccharide amino sugar required as a growth factor by these bacteria. With the ingestion of solid food, these bacteria are eventually replaced by other typical bacteria, which are composition of adult's microbiota (Prescot et al. 1996).

2.3 Overall Chemistry of Fermentation

The overall process of fermentation is to convert glucose sugar ($C_6H_{12}O_6$) to alcohol (CH_3CH_2OH) and carbon dioxide gas (CO_2). The reactions within the yeast to make this happen are very complex but the overall process is as follows:

$$C_6H_{12}O_6 \dashrightarrow 2(CH_3CH_2OH) + 2(CO_2)$$

Sugar   Alcohol   Carbon dioxide

(Glucose)   (Ethyl alcohol)

Fermentation products contain chemical energy (they are not fully oxidized); but are considered waste products, since they cannot be metabolized further without the use of oxygen (or other more highly-oxidized electron acceptors). A consequence is that the production of ATP by fermentation is less efficient than oxidative phosphorylation, where pyruvate is fully oxidized to carbon dioxide. Fermentation produces two ATP molecules per molecule of glucose compared to approximately 36 ATP molecules produced by aerobic respiration.

2.4 Types of Fermentation

Different methods of fermentation can be distinguished on the basis of formation of their products and/or on the basis of presence or absence of oxygen. Types of fermentation differentiated on the basis of formation of end products are:

(a) Lactic acid fermentation

(b) Alcohol fermentation

(c) Mixed acid fermentation

Type of fermentation that are determined by the presence or absence of air are:

(a) Aerobic fermentation

(b) Anaerobic fermentation

For details, please refer to chapter 7 (Fermented Health Foods: Boon to Mankind) of this book.

2.5 Importance of Fermentation

There are various kinds of fermentation carried out by different types of microorganisms. Many of these form highly useful end products. Such useful microbial activity is used on large industrial scale to obtain industrial products for the benefit of mankind. Some of the industrial products of the microbial fermentation activities are: (a) antibiotics, (b) vitamins, (c) industrial alcohol, (d) bakery and dairy products, (e) tanning of leather, (f) curing of tea and coffee, (g) lactic acid, (h) butric acid, (i) acetic acid, etc.

3. HISTORY OF YOGURT

Yogurt is traditionally believed to be an invention of the Bulgars from Central Asia, although there is evidence of other cultured milk products in other cultures 4500 years ago. The earliest yogurts were probably spontaneously fermented, perhaps by wild bacteria residing inside goatskin bags used for transportation. Milk stored in animal skins would acidify and coagulate. The acid helps to preserve the milk from further spoilage and from the growth of pathogens (disease-causing microorganisms).

The word derives from the Turkish yogurt deriving from the verb yogurmak, which means "to blend", a reference to how yogurt is made. Yogurt remained primarily a food of India, Central Asia, Western Asia, South-eastern Europe and Central Europe until the 1900s, when a Russian biologist named Ilya Ilyich Mechnikov theorized that heavy intake of yogurt by Bulgarian peasants was responsible for their long lifespan. Believing *Lactobacillus* to be essential for good health, Mechnikov popularized yogurt as a foodstuff throughout Europe.

The Russian-born scientist, Illya Metchnikoff can be credited for bringing yogurt to the Western world. He identified and isolated the bacilli that make yogurt. Metchnikoff was the co-winner with Paul Ehrlich for the Nobel Prize for Physiology and Medicine in 1908. In his studies on longevity in humans, he came to feel that the bacteria in yogurt help to purify the large intestine of poorly digested food that poisons the body. It was his belief that if people ate yogurt regularly, they could live to be well over 100 years old.

In Western countries, yogurt was virtually unheard of until the 1960s. Flavoring and sweetening yogurt has given it more appeal to sweet-toothed North Americans. With more interest in nutrition and natural foods, yogurt has become a favourite food for all.

4. MICROBES INVOLVED IN YOGURT PRODUCTION

The most commonly used microbes involved in yogurt production are *Streptococcus salivarius* and *Lactobacillus bulgaricus*, although sometimes another member of the *Lactobacillus* genus is also used, such as *L. acidophilus*.

4.1 Lactobacillus Bulgaricus

Lactobacillus bulgaricus (official name *Lactobacillus delbrueckii* subspecies *bulgaricus*, LBB) is one of the several bacteria used for the production of yogurt. It is named after Bulgaria, the country where it was first used (it thrives freely on the Balkan Peninsula). The bacterium feeds on milk and

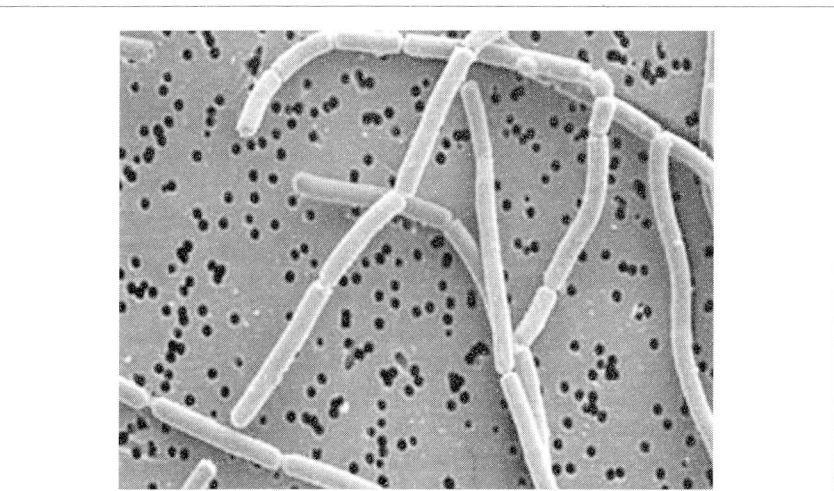

Fig. 1   *Lactobacillus bulgaricus*

produces lactic acid, which also helps to preserve the milk. It breaks down lactose and is often helpful to sufferers of lactose intolerance, whose digestive systems lack the enzymes, to break down lactose to simpler sugars. While fermenting milk, *Lactobacillus bulgaricus* produces acetaldehyde, which perfumes yogurt. One of the biggest importers of the bacterium is Japan. *L. bulgaricus* produces hydrogen peroxide, which can be used to inhibit growth of other organisms found on seafood.

4.2 *Streptococcus salivarius*

Streptococcus salivarius subspecies *thermophilus* is a streptococcus bacteria used to produce yogurt and some cheeses. It works symbiotically with *Lactobacillus delbrueckii* subspecies *bulgaricus*. Both are homofermentative; produce lactic acid quantitatively in addition to carbon dioxide and alcohol. The optimum temperature for their growth is about 42°C. *Streptococcus* ferments lactose to lactic and acetic acids.

    Streptococcus salivarius colonizes the mouth and upper respiratory tract of people just a few hours after birth, making further exposure to the bacteria harmless. The bacteria do have some negative effects, though, playing a major role in tooth decay and, rarely, finding its way into the bloodstream, where it has been implicated in septicemia cases in people with neutropenia.

Fig. 2   *Streptococcus salivarius*

5. TYPES OF YOGURT

5.1 Dahi Yogurt

Dahi yogurt, of the Indian subcontinent, is known for its characteristic taste and consistency.

5.2 Bulgarian Yogurt

Bulgarian yogurt is popular for its specific taste, aroma and quality and is commonly consumed plain. The qualities are specific to the particular culture strains used in Bulgaria, *Lactobacillus bulgaricus* and *Streptococcus thermophilus* bacteria.

Bulgarian yogurt is often strained by hanging in a cloth for a few hours to reduce water content. The resulting yogurt is creamier, richer and milder in taste because of increased fat content.

5.3 Greek Yogurt

Greek "full" yogurt is made from milk that has been blended with cream to a fat content of exactly ten percent. Standard (5%), low-fat (2%) and non-fat (0%) versions are also made. It is often served with honey, walnuts or fruit preserves as a dessert.

5.4 Yogurt based Beverage

5.4.1 Lassi

Lassi is a yogurt-based beverage, originally from India where two basic varieties are known: salty and sweet. Salty lassi is usually flavoured with ground-roasted cumin and chile peppers; the sweet variety with rosewater and/or lemon, mango, or other fruit juice.

5.4.2 Ayran

Ayran is a yogurt-based beverage, a salty drink, is quite popular in Turkey and Bulgaria. It is made by mixing yogurt with water and adding salt. The same drink is known as tan in Armenia.

5.4.3 Doogh

Doogh, is popular in the Middle East between Lebanon and Iran; it differs from ayran by the addition of herbs (usually mint) and being carbonated (carbonation is achieved by letting it ferment).

5.5 Kefir

Kefir is a fermented milk drink originating in the Caucasus, enjoyed in Turkey largely. A related Central Asian-Mongolian drink made from mare's milk is called kumis or, in Mongolia, airag.

5.6 Homemade Yogurt

Homemade yogurt is consumed by many people throughout the world, and is the norm in countries where yogurt has an important place in traditional cuisine, such as Bulgaria, Turkey and India. Yogurt can be made at home using a small amount of store-bought plain live active culture yogurt as the starter culture. One very simple recipe starts with a litre of low-fat milk, but requires some means to incubate the fermenting yogurt at a constant 43°C (109°F) for several hours. Yogurt-making machines are available for this purpose.

6. MILK FERMENTATION

Milk is an excellent food source for humans and is full of vitamins, fats, minerals, nutrients and carbohydrates. It is rich in the protein casein, which gives milk its characteristic white color. The most abundant carbohydrate is the disaccharide lactose, "milk sugar." At room temperature, milk undergoes natural souring caused by lactic acid produced from fermentation of lactose by fermentative lactic acid bacteria.

To ferment milk, concentrated cultures of bacteria are inoculated into pasteurized milk, which has been enriched in milk protein and is then incubated at 40-44°C for 4 to 5 hours.

6.1 Milk Fermentation of Yogurt Production

In the first step of milk fermentation, the milk is held at 46° to 48°C, which encourages the growth of *Streptococcus thermophilus*, a thermophile. This organism rapidly ferments lactose (milk sugar) to lactic acid. During the fermentation, lactic acid is produced from lactose by the yogurt bacteria, whose population increases 100 to 10000 fold to a final concentration of approximately 10^9/ml. Lactose is a disaccharide (2 covalently linked sugars) composed of glucose and galactose, which are connected by a glucosidic bond. Lactose is transported into the cell and the glucosidic bond is broken by the enzyme beta-galactosidase. After the lactose is broken into galactose and glucose, both sugars are converted to glucose-6-phosphate.

Then, *S. thermophilus* oxidizes glucose-6-phosphate to lactic acid using the Embden-Meyerhoff-Parnas (EMP) pathway or glycolysis.

Streptococcus thermophilus is called a homolactic acid bacterium, because they ferment sugars, mostly to lactic acid, which they excrete. All homolactic acid bacteria appear to ferment sugars using the EMP pathway, and most of the carbon atoms in glucose end up in the excreted lactic acid molecules; therefore, the EMP pathway is called the major pathway for the homolactic acid bacteria. This excreted acid increases the hydrogen-ion concentration, thus, lowering the pH.

Freshly drawn bovine milk has a pH of about 6.6. At this pH, casein (a milk protein) exists as a colloidal suspension of a calcium salt called calcium caseinate, which gives milk its white color and turbid appearance. When streptococci begin to excrete lactic acid, some of it reacts with the calcium caseinate to form calcium lactate and free (soluble) casein. The acidification is due to the production of lactic acid, causes a destabilization of the micellar casein at a pH of 5.1 - 5.2 and with a complete coagulation occurring around 4.6. As the pH approaches 4.6, casein begins to denature (coagulate), forming a smooth semi-solid curd and whey. Such an acid curd is typical of milk fermentation by the lactic acid bacteria.

In the second step, the temperature is lowered by refrigeration. As the temperature drops, the growth of the thermophile (*S. thermophilus*) slows, and a mesophilic bacterium called *L. bulgaricus* begins to ferment the remaining lactose in the milk. These lactobacilli are usually more acid resistant than other lactic acid bacteria and therefore are often responsible for the final stages of lactic acid fermentations. *L. bulgaricus* also carries out a homolactic fermentation using the EMP pathway to ferment lactose to lactic acid, but will use other minor pathways to produce volatile organic compounds that give yogurt its characteristic flavor and aroma.

This step lasts until the temperature falls below the point at which the active metabolism of *L. bulgaricus* stops; consequently, the excretion of both lactic acid and the minor fermentation products also stops. Thus, at the desired final pH, the coagulated milk is cooled quickly to 4-10°C to slow down the fermentation. Further production of acid not only increases sourness but also increases the compactness of the curd. Curds are large, white clumps of casein and other proteins. Whey is the yellow liquid that is left behind after the casein has formed curds. Thus, bacteria obtain nutrients from the milk, inadvertently curdle it and humans use it as the first step in making many dairy products.

Refrigeration is not really necessary; it just prolongs the shelf life of the product. The acid alone will prevent growth of other types of microorganisms that may cause spoilage, but refrigeration also prevents the live lactobacilli from continuing to make lactic acid and other fermentation products thus lowering the pH below what is palatable for most people. The fermentation of milk is among the oldest known methods of food preservation.

The microbes that are important for dairy products manufacturing can be divided into two groups, primary and secondary microflora. Products that are undergoing fermentation only by primary microflora are called unripened and those products processed by both primary and secondary microflora are called ripened. Primary microflora is fermentative lactic acid bacteria, which cause the milk to curdle. During dairy product production, milk is first pasteurized to kill bacteria that cause unwanted spoilage of the milk and of the downstream milk products. Primary microflora consists of certain kinds of *Lactococcus*, *Lactobacillus* and *Streptococcus* that are intentionally added to pasteurize milk and grown at 30°C or 37°C (temperature depends on the type of bacteria added to the pasteurized milk). Secondary microflora include several different types of bacteria (*Leuconostoc*, *Lactobacillus*, and *Propionibacterium*), yeasts and molds. These are only used for some types of surface ripened and mold ripened cheeses. The various combinations of microflora determine the type of milk products obtained at the end.

Yogurt normally is prepared using pure culture. The selection of species is based on their ability for rapid production of lactic acid and development of flavor. The inoculums are selected for their rapid growth so even if the heated milk has some bacteria remaining after heating, the inoculated bacteria quickly overtakes them. Use of pure starter culture ensures uniform results batch after batch giving consistent production of yogurt with same acidity, texture and flavor. The fermentation of milk to yogurt is dependent upon growth and acid production by *Streptococcus salivarius* var. *thermophilus* (commonly named *S. thermophilus*) and *Lactobacillus delbrueckii* var. *bulgaricus* (commonly named *S. bulgaricus*).

7. NUTRITIONAL VALUE OF YOGURT

Changes in the carbohydrates and proteins during fermentation not only alter the flavor and texture of the milk but may also alter the digestibility of these nutrients. The mineral content of the milk is quantitatively unchanged even though mineral bioavailability may be altered. The presence of a large number of live and active bacterial cells and/or metabolites formed during fermentation have a beneficial effect on human health.

7.1 Carbohydrates

Lactic acid bacteria ferment about 20-30% of the lactose present in milk through different pathways. Yogurt bacteria are homofermentative, producing one major end product, in this case lactic acid, which accounts for greater than 95% of the fermentation products. In other types of bacteria, other fermentation products are formed.

The final concentration of lactic acid is 0.7 to 1.2% in yogurt, which accounts for the mildly sour, refreshing taste. This lactic acid is a mixture of both the L(+) and D(-) isomers. Although the quantity of each isomer present depends on the specific culture, the L(+) isomer generally represents between 50-70% of the total lactic acid.

The reduction in the lactose concentration coupled with the presence of a high number of viable bacteria containing β-galactosidase explains why lactose maldigesting individuals can consume yogurt. The bacterial cells protect the β-galactosidase from denaturation by the acid in the stomach and deliver it to the intestine. The action of bile in the small intestine increases the permeability of the bacterial cells, facilitating the entry and subsequent hydrolysis of lactose (Gilliland 1991).

7.2 Proteins

The proteolytic activity of yogurt bacteria is low, resulting in a breakdown of only 1-2% of milk protein. This proteolytic activity is necessary to release small peptides and amino acids for the growth of these bacteria. *L. bulgaricus* is more proteolytic, but both yogurt bacteria contain peptidases, which are necessary to hydrolyze large peptides into smaller peptides for transport into the cell. The principal substrate for proteolysis is casein, but limited degradation of whey proteins may also occur (Chandan et al. 1982; Khalid et al. 1991). The net effect of this proteolytic activity is that fermented milks have a higher content of peptides and free amino acids, especially valine, histidine, serine and proline than that of milk (Tamime and Deeth 1980; Vaitheeswaran and Bhat 1988).

Even though the limited proteolytic action of yogurt bacteria does not significantly alter the nutritional value of milk proteins (Hewitt and Bancroft 1985), yogurt is more digestible than the milk mixture from which it was made (Breslaw and Kleyn 1973). A study with rats found that feeding yogurt compared to the milk from which it was prepared resulted in increased feed efficiency (McDonough et al. 1982). The increased digestibility of proteins in fermented milks may be related to the fine flocculation of caseins resulting from the joint action of proteolysis and acidification. This increased digestibility is especially important for people with gastric atrophy, gastrointestinal disturbances and protein malnutrition.

7.3 Lipids

Yogurt and fermented milks contain small amounts of dairy lipids (from 0 to 4g/100 g at room temperature). During fermentation and shelf life the lipids are hardly metabolized. All nutrition studies show that yogurt can be included in a healthy diet; daily consumption of yogurt seems unlikely to alter blood lipid parameters (McNamara et al. 1989).

7.4 Minerals and Vitamins

While the lactic acid bacteria require some minerals and vitamins for growth, the change of these constituents, compared to milk before fermentation is negligible, with the exception of some B vitamins. The pasteurization of milk before fermentation may destroy some vitamins such as B6, B12 and folic acid while the level of thermo-stable vitamins (niacin and pantothenic acid) remains unchanged. In general, *L. bulgaricus* uses folic acid, whereas *S. thermophilus* produces it (Kneifel et al. 1992). During cold storage after fermentation, the levels of some vitamins, especially B12 and folic acid decrease.

Yogurt, like milk, is an excellent source of calcium and phosphorus, which are essential for bones. Moreover, yogurt also contains relatively high amounts of potassium and can be considered a good source of these minerals. For lactose maldigesting individuals, yogurt provides a rich, easily digestible source of these minerals.

8. HEAT TREATED FERMENTED MILKS

Post-fermentation heat treatment significantly alters some properties of fermented milks. Heat treatment above 65°C appreciably reduces the level of some thermosensitive vitamins. In addition, enzymatic activity, notably ß-galactosidase, is markedly reduced. This reduced enzymatic activity dramatically lowers the ability of lactose maldigesters to tolerate the same amount of lactose that otherwise could be tolerated with live yogurt. This modification of properties may not be limited to lowering vitamin content and lactose absorption but may also affect some other probiotic properties.

9. EFFECTS OF REGULAR CONSUMPTION OF FERMENTED MILK

S. thermophilus and *L. bulgaricus* are not inhabitants of the intestinal tracts of humans and animals but are normally isolated from green plant material and milk, respectively. These bacteria are not highly acid and bile resistant. Only about 15% survive passage through the stomach and about 1% reaches the large intestine (Pochart 1989), but cannot colonize it (Rasic 1987). However, they may still exert an effect *in vivo* due to intracellular enzymes, cell surface antigenic receptors or metabolites produced during fermentation. Although yogurt bacteria and other lactic acid bacteria have been shown to inhibit pathogenic bacteria *in vitro* by the production of organic acids and antibiotic-like substances, this interaction has not been clearly demonstrated *in vivo*. It has been shown in studies with mice (Garvie et al. 1984; Kanbe 1992) that feeding yogurt resulted in an alteration in the intestinal flora, stimulating the growth of lactobacilli and bifidobacteria. Such changes in intestinal microflora are known to affect the intestinal transit time, which may have a significant impact on nutrient absorption.

10. CONDITIONS REQUIRED FOR BACTERIAL FERMENTATIONS

Control of fermentation process is achieved by controlling the conditions in which the medium is stored. All organisms have a particular range of conditions in which they are able to thrive; outside this range their action is suppressed.

10.1 Temperature

Incubation temperature affects all microbes and enzymes greatly. Different bacteria can tolerate different temperatures, which provide enormous scope for a range of fermentations. While most bacteria have a temperature optimum of between 20 to 30°C, there are some (thermophiles) which prefer higher temperatures (50 to 55°C) and others with colder temperature optima (15 to 20°C). Most of lactic acid bacteria work best between 18 to 22°C. Temperatures above 22°C, favors the growth of *Lactobacillus* species.

10.2 Water Activity

Water is essential for the growth and metabolism of all cells. There are two types of water in a medium: bound and available. Bound water (e.g., ice) is not available to the microbes. The presence of some available water is necessary for fermentation to occur. In general, bacteria require a fairly high water activity (0.9 or higher) to survive. There are a few species, which can tolerate water activities, lower than this, but usually the yeasts and fungi will predominate on foods with a lower water activity.

10.3 Hydrogen Ion Concentration (pH)

The optimum pH for most bacteria is near the neutral (pH 7.0). Certain bacteria are acid tolerant and will survive at reduced pH levels. Notable acid-tolerant bacteria include the *Lactobacillus* and *Streptococcus* species, which play a role in the fermentation of dairy and vegetable products. In general, yeast prefers acidic conditions. Fruits are naturally acidic, so the action of yeasts and fungi are encouraged.

10.4 Oxygen

Oxygen affects microbes and their end products formed. Some microbes require its presence and others do not. Some of the fermentative bacteria are anaerobes, while others require oxygen for their metabolic activities. Some, lactobacilli in particular, are microaerophilic. These organisms grow in the presence of reduced amounts of atmospheric oxygen. In aerobic fermentations, the amount of oxygen present is one of the limiting factors. It determines the type and amount of biological product obtained, the amount of substrate consumed and the energy released from the reaction. *Acetobacter* require oxygen for the oxidation of alcohol to acetic acid.

10.5 Nutrients

All bacteria require a source of nutrients for metabolism. The fermentative bacteria require carbohydrates either simple sugars such as glucose and fructose or complex carbohydrates such as starch or cellulose. The energy requirements of microorganisms are very high. Limiting the amount of substrate available can check their growth.

11. YOGURT AND ITS BENEFITS

11.1 Nutritional Value

Natural whole milk yogurt has a similar nutritional value to whole boiled milk, being rich in protein and minerals, especially calcium and phosphorus. Low fat and fat free yogurts are made from skimmed milk powder; they have a slightly higher carbohydrate and protein content than whole milk yogurts. Yogurt is rich in potassium, calcium, protein and B vitamins, including B-12. The bonus is that protein, calcium and phosphorus are more easily absorbed from yogurt than from milk as they are partially digested during the fermentation process. It is much more digestible than fresh milk and it is as low or high in fat as the milk it is made from.

11.1.1 Rich Source of Calcium

A 225g serving of most yogurts provides 450 mg of calcium, one-half of a child's Recommended Dietary Allowances (RDA) and 30 to 40 percent of the adult RDA for calcium. Live-active cultures in yogurt increase the absorption of calcium; a 225g serving of yogurt provides more calcium into the body than the same volume of milk.
(www.askdrsears.com).

The lactic acid of yogurt is a perfect medium to maximize calcium absorption. In yogurt the process of growth from milk into yogurt involves the conversion of lactose into lactic acid. Lactic acid helps to digest lactose. In other words, yogurt provides the enzyme needed to digest milk products. Calcium, which is found in dairy products, needs to enter the body in an acid matrix or your body will not absorb it. So the lactic acid of yogurt is a perfect medium to maximize calcium absorption. Eight ounces of yogurt will equal 400 mg of calcium, 25% more calcium than you would get out of a glass of milk.

11.1.2 Excellent Source of Protein

Plain yogurt contains around ten to fourteen grams of protein per eight ounces, which amounts to twenty percent of the daily protein requirement for most persons. In fact, eight ounces of yogurt that contains live and active cultures, contains 20 percent more protein than the same volume of milk (10 grams versus 8 grams). Besides being a rich source of proteins, the culturing of the milk proteins during fermentation makes these proteins easier to digest. For this reason, the proteins in yogurt are often called "predigested". (www.askdrsears.com).

Research shows that yogurt strengthens and stabilizes the immune system. The *Lactobacillus* in yogurt feeds the intestines, insures the digestive system stays healthy, and stabilizes the immune system. Culturing of yogurt increases the absorption of calcium and B-vitamin. The lactic acid in the yogurt aids in the digestion of the milk calcium, making it easier to absorb. (www.askdrsears.com).

"Yogurt has strong medicinal properties, including the ability to stimulate the immune system and kill bad "bugs" or bacteria in the human gut. Research at the University of California at Davis showed that eating live-culture yogurt was associated with higher-than-average levels of gamma interferon, a key component of the body's immune system".(Bell 1994).

Benefits of yogurt are thought to be lost when yogurt is frozen, which destroys most of the beneficial bacteria.

11.2 The Digestive System and Fighting Infection

Lactobacillus acidophilus is probably the most commonly used bacteria culture today and bacterium is thought to colonize the intestines with essential digestive micro-organisms, probably due to the fact that yogurt bacteria aid the synthesis of valuable vitamins which in turn stimulate the growth of beneficial intestinal bacteria, discouraging and destroying the harmful ones. Live yogurts are especially beneficial in this manner. Yogurts containing probiotic bacteria will improve the gut flora and aid in the digestion of food.

Yogurt can be helpful in restoring the digestive tract to its normal condition after a course of antibiotics, which are liable to indiscriminately destroy all intestinal bacteria, both good and bad. Yogurt can also be used in a similar way in the treatment of thrush where bacteria have reached a state of severe imbalance.

Yogurt aids healing after intestinal infections. Some viral and allergic gastrointestinal disorders injure the lining of the intestines, especially the cells that produce lactase. This results in temporary lactose malabsorption problems. This is why children often cannot tolerate milk for a month or two after an intestinal infection. Yogurt, however, because it contains less lactose and more lactase, is usually well tolerated by healing intestines and is a popular "healing food" for diarrhea. Many pediatricians recommend yogurt for children suffering from various forms of indigestion. Research shows that children recover faster from diarrhea when eating yogurt. It's good to eat yogurt while taking antibiotics. The yogurt will minimize the effects of the antibiotic on the friendly bacteria in the intestines. A 1999 study reported in Pediatrics showed that lactobacillus organisms can reduce antibiotic-associated diarrhea.
(www.askdrsears.com).

11.3 Lactose Intolerance

People who are unable to digest dairy products where the condition is due to the loss of the enzyme lactase during adulthood can tolerate yogurt. Up to 70 percent of the world's population have digestive systems that cannot tolerate lactose, so these people cannot eat most dairy products. Enzyme lactase converts lactose to lactic acid, without which any lactose ingested in milk products will sit undigested in the intestine, attracting water and causing bloating, abdominal cramps and diarrhea. Because live yogurt culture contains enzymes that break down lactose, some individuals who are otherwise lactose intolerant can easily digest and enjoy yogurt without ill effects. The low lactose contents of yogurt also make it suitable for some people with lactose intolerance.

11.4 Yogurt can be used as an Effective Douche

Douche is a stream of water, often containing medicinal or cleansing agents, that is applied to a body part or cavity for hygienic or therapeutic purposes. It is the irrigation with a jet of water or medicated solution into or around a body part (especially the vagina) to treat infections or cleanse from odorous contents. Research shows women who consume regular yogurt have less vaginal and bladder infections than others.

In women there may be yeast overgrowth in their vaginas. A *Candida* vaginal infection is caused by the condition, which allows overgrowth of the *Candida*. Antibiotics have the unwanted effect of wiping out normal flora in human bodies. *Candida* species will grow in abundance when other bacteria and yeasts are depleted. Yogurt can be used internally and externally, and it is prime good yeast replace for female's vaginal and intestinal flora. A good way to restore normal vaginal pH and flora without using commercial chemicals is to use a plain yogurt douche. A study at Long Island Jewish Medical Center showed that "women prone to vaginal yeast infections experienced a threefold decrease in infections when they ate a cup of *Lactobacillus acidophilus* yogurt daily for six months". (www.leaflady.org).

12. FOOD VALUE OF YOGURT

Food value of a particular food item is the nutritional value expressed in terms of its protein, carbohydrate, fat, fiber and energy content. Yogurt is one of the most versatile foods that can be used as a substitute for many high-fat foods e.g., yogurt can be used in place of mayonnaise or can be substituted milk, buttermilk, and sour creams in baking recipes. Yogurt shakes and smoothies are a low-fat alternative to ice cream. Food values for yogurt and different milk samples are tabulated in Tables 1 and 2.

Table 1 Food Value of Yogurt (Source: www.IndianGyan.com)

Food (Values per 100 g of edible portion)	Percent
Moisture	89
Protein	03
Fat	04
Minerals	0.8
Carbohydrates	03
Minerals and Vitamins	100
Calcium	149 mg
Phosphorus	93 mg
Iron	0.2 mg
Vitamin A	102 IU
Vitamin C	Small amount
Vitamin B Complex	Small amount
Calorific Value/ 100gms	60

Table 2 A typical comparative analysis of sheep, goat and cow's milk (www.sheepdairying.com)

Food value (%age)	Sheep	Goat	Cow
Total Solid	18.3	11.2	12.1
Fat	6.7	3.9	3.5
Protein	5.6	2.9	3.4
Lactose	4.8	4.1	4.5
Calorific value/100g	102	77	73
Vitamins (mg/l)			
Riboflavin B_2	4.3	1.4	2.2
Thiamine	1.2	0.5	0.5
Niacin B_1	5.4	2.5	1.0
Pantothenic acid	5.3	3.6	3.4
B_6	0.7	0.6	0.5
Folic acid (microgm/l)	0.5	0.06	0.5
B_{12}	0.09	0.007	0.03
Biotin	5	4.0	1.7
Minerals (mg/100g)			
Calcium	162-259	102-203	110
Phosphorus	82-183	86-118	90
Sodium	41-132	35-65	58
Magnesium	14-19	13-19	11
Zinc	0.5-1.2	0.19-0.5	0.3
Iron	0.03-0.1	0.01-0.1	0.04

13. YOGURT PRODUCTION

Yogurt production takes place by the controlled fermentation of milk by two species of bacteria, which are *Lactobacillus* and *Streptococcus*. The sugar in milk, which is called lactose, is fermented to lactic acid. The acid also restricts the growth of food poisoning bacteria and some spoilage bacteria. So, whereas milk is a potential source of food poisoning and only has a shelf life of a few days, yogurt is safer and can be kept for up to ten days, under proper storage conditions.

Yogurt can be easily produced in small-scale. The procedure is as follows:

- Collect the milk in cleaned and covered vessels.
- Bring the milk to 85°C (185°F) over a stove and boil it to kill any undesirable microbes.
- Pour the re-pasteurized milk into a tall, sterile container and allow it to cool to 43°C (110°F).
- Cool the milk to 40-45°C as quickly as possible and add a starter culture of the yogurt bacteria.
- After about six hours of incubation at precisely 43°C (110°F), the entire mixture will become very plain but edible yogurt with a loose consistency. If possible, then cool the yogurt in a refrigerator until it is eaten or sold.

13.1 Potential Problems in Yogurt Production

There are three potential problems in yogurt production on commercial scale:

13.1.1 Spoilage by bacteria or moulds

This is due to unclean equipment, contaminated milk or poor hygiene of the production staff. Ensure that all equipment is thoroughly scrubbed, sterilized with diluted bleach (two tablespoons of bleach per gallon of water) and thoroughly rinsed in clean water before production starts. Pasteurization should ensure that fresh milk is not contaminated, but do not use old milk. Make sure operators wash their hands before starting work and do not allow anyone with stomach complaints, coughs or skin infections (e.g., boils) to work with the milk.

13.1.2 Maintenance of Optimum Incubation Temperature

A commercial yogurt maker of kitchen may be used or something made locally from a shallow water bath with a small electrical element, keeping the water warm and the whole thing controlled by a simple variable temperature thermostat can also be used for this purpose. Another alternative way would be to fill the yogurt mix at 40-45°C into a large commercial thermos flask. Even a block of 4 inches polystyrene into which indentations are made of such size that those small cream containers can fit comfortably. The warm yogurt mixture is, thus, filled into the containers inside the block of polystyrene, and a polystyrene lid is placed on top. The insulating effect of the block will then prevent the loss of heat sufficiently to maintain the temperature of the product at the required 40-45 °C. A similar idea consists of a hollow polystyrene box approximately 0.75m^3 fitted with a 40W electric light bulb. The heat from the bulb maintains the temperature within the required range.

13.1.3 Yogurt Culture

The correct balance of the two *Lactobacillus* bacteria is important for good quality yogurt. In commercial practice, a dried culture can be obtained from most large towns/cities and this can be grown up on pasteurized milk and kept in a refrigerator. A part of this 'master culture' can then be used each day for a week and the last part re-inoculated into milk to form a new master culture. This method can be continued for several months, provided good hygiene is used, but eventually undesirable bacteria will contaminate the culture and it must be replaced.

CONCLUSION

Beyond their good taste, fermented milks like yogurts have unique nutritional attributes. Yogurt has been regarded as a wholesome food and has been an important part of the human diet. They supply high quality proteins, are an excellent source of calcium, phosphorus and potassium, and contain significant quantities of several vitamins. The carbohydrate content is easily absorbed even by lactose maldigestors. Fermented milks like yogurt are a valuable adjunct to any healthy diet. Domestic production of yogurt is being practiced all over India. However, quality of yogurt varies from house to house and these variations can also be recorded in day-to-day preparation of yogurt. Sometimes they are thick semi-solid and sometime they are diluted in slurry form. At times yogurt is sweet and at other time it is sour. It is high time to consider preparing a capsule of standard certified inocula, which on use and application can give consistent and uniform quality of yogurt. This aspect needs commercialization.

ACKNOWLEDGEMENTS

The authors acknowledge the support and encouragement received from Dr. Ashok K Chauhan, Founder President, Ritnand Balved Education Foundation (an umbrella organization of Amity Institutions), Mr. Atul Chauhan, Chancellor, Amity University, UP and Mr. Aseem Chauhan, Trustee and Board Member, Ritnand Balved Education Foundation. The technical support received from Mr. Anil Chandra Bahukhandi is highly appreciated.

REFERENCES

Bell B (1994). The hidden world of yoghurt. View Magazine pp 16 – 19.

Breslaw ES, Kleyn DH (1973). *In vitro* digestibility of protein in yogurt at various stages of processing. J Food Sci 38: 1016-1021.

Chandan RC, Argyle PJ, Mathison GE (1982). Action of *Lactobacillus bulgaricus* proteinase preparations on milk proteins. J Dairy Sci. 65: 1408-1413.

Garvie EI, Cole CB, Fuller R, Hewitt D (1984). The effect of yoghurt on some components of the gut microflora and on the metabolism of lactose in the rat. J. Appl. Bacteriol. 56:237-245.

Gilliland SE (1991). Properties of Yogurt. In: Robinson RK (ed) Therapeutic Properties of Fermented Milks. Elsevier Applied Science London pp 5.

Hewitt D, Bancroft HJ (1985). Nutritional value of yogurt. J. Dairy Res. 52: 197-207.

Kanbe M (1992). Cancer control and fermented milk. In: Nakazawa Y, Hosono A (eds) Functions of Fermented Milk Elsevier Applied Science, London pp 383.

Khalid NM, El-Soda M , Marth EH (1991). Peptide hydrolases of *Lactobacillus helveticus* and *Lactobacillus delbrueckii* sub-sp. *bulgaricus*. J. Dairy. Sci. 74: 29-45.

Kneifel W, Kaufmann M, Fleischer A, Ulberth F (1992). Screening of commercially available mesophilic dairy starter cultures: biochemical, sensory and microbiological properties. J. Dairy Sci. 75: 3158-3166.

McDonough FE, Hitchins AD, Wong NP (1982). Effect of yogurt and freeze-dried yogurt on growth stimulation rate. J. Food Sci. 47: 1463-1465.

McNamara DJ, Lowell A, Sabb JE (1989). Effect of yogurt intake on plasma lipid and lipoprotein levels in normolipidemic males. Atherosclerosis 79:167-171.

Pochart P, Dewit O, Desjeux JF, Bourlioux P (1989). Viable starter culture, ß-galactosidase activity and lactose in duodenum after yogurt ingestion in lactase-deficient humans. Amer. J. Clinic. Nutr. 49: 828-831.

Prescott LM, Harley JP, Klein DA (1996). The diversity of the microbial world. In: Prescott LM, Harley JP, Klein DA (eds) Microbiology. WCB Publishers, Dubuque, Iowa. Pp 557, 894.

Rasic JL (1987). Nutritive value of yogurt. Cultured Dairy Prod. J. 22: 6-9.

Tamime AY, Deeth HC (1980). Yogurt: Technology and biochemistry. J. Food Protect. 43: 939-977.

Vaitheeswaran NI, Bhat GS (1988). Influence of lactic cultures in denaturation of whey proteins during fermentation of milk. J. Dairy Res. 55: 443-448.

Fermented Health Foods: Boon to Mankind

A. DAS, K. BHATNAGAR, R. PRASAD, A. PAUL, T. KUMAR, M. S. KUMAR, H. KHARKWAL, N. VERMA and A. VARMA

1. INTRODUCTION

Fermentation of food had been considered a way to preserve food as well as to retain nutritional value since ancient times. Ancient humans used the process of natural fermentation without the knowledge that there were microorganisms involved in it.

Ancient Egyptians found when dough made from ground up wheat and rye was left for a period of time before cooking it unintentionally resulted in bread which were lighter and tastier than the normal ones cooked earlier. They found this process was not completely reproducible so, gradually it became the norm for producing leavened breads; a small portion of soft lump of one day's fermented dough was used to produce tastier and lighter bread. Although nowadays mostly commercial yeasts are used for bread-making, bread from a sour dough starter is also not uncommon and is more or less similar to what the ancient Egyptians would have been baking.

Fruits ferment naturally as ripe fruits when they get contaminated with yeasts, which are present on their skin surface causing conversion of sugars into ethanol. Alcoholic fermentation is a process that was known from time immemorial. Before 2000 B.C., the Egyptians apparently knew that crushed fruits stored in a warm place would produce a substance with a pleasant intoxicating power. By 1500 B.C.s the production of beer from germinating

Amity Institute of Microbial Sciences, Amity University Uttar Pradesh, Noida-201303, India E-mail: adas@amity.edu cereals (malt) and the preparation of wines from crushed grapes were established in most of the Middle East. Aristotle also believed that grape juice was an infantile form of wine and that fermentation was, therefore, the maturation of the grape extract. But the exact cause of fermentation was unknown until 1859 when Louis Pasteur discovered how yeast works (Mansi, et al. 2003). He revealed that only active living cells could cause fermentation. Fermentation or anaerobic respiration is the process that has been used since ages in the production of certain food and beverages. In the absence of oxygen, certain living things are capable of breaking down carbohydrates (starches and sugars) to form alcohol and carbon dioxide. This process is known as fermentation. In fermentation, the energy is released from a sugar or other organic molecule. This process does not require oxygen or an electron transport system, and use of organic molecule as the final electron acceptor. For a cell, fermentation is a way of getting energy without using oxygen.

The science of fermentation is known as zymology. Fermentation is caused by enzymes, catalysts in chemical reactions similar to the digestive enzymes in the human body. Certain enzymes act on starch to break down the long chain-like molecules into smaller units of sugar. Then other enzymes convert one kind of sugar molecule into another. Still other enzyme reactions break apart the sugar molecule (composed of carbon, hydrogen and oxygen) into ethyl alcohol and carbon dioxide gas. Discovery of fermentation revolutionized the modern food industry as for the first time the agent of fermentation was identified and commercially exploited. Thus, fermentation can be expressed as any process that produces alcoholic beverages or causes any spoilage of food by microorganisms (general use).

Biochemically, fermentation is a process that is important in anaerobic conditions when there is no oxidative phosphorylation to maintain the production of ATP (Adenosine Triphosphate) by glycolysis. During fermentation, pyruvate is metabolised to various different products. Homolactic fermentation is the production of lactic acid from pyruvate; alcoholic fermentation is the conversion of pyruvate into ethanol and carbon dioxide; and heterolactic fermentation is the production of lactic acid as well as other acids and alcohols. Typical examples of fermentation products are ethanol (drinkable alcohol), lactic acid and hydrogen. However, more exotic compounds can be produced by fermentation, such as butyric acid and acetone. The end products of various types of fermentations are depicted in the Figure 1.

Although the final step of fermentation (conversion of pyruvate to fermentation end products) does not produce energy, it is critical for an

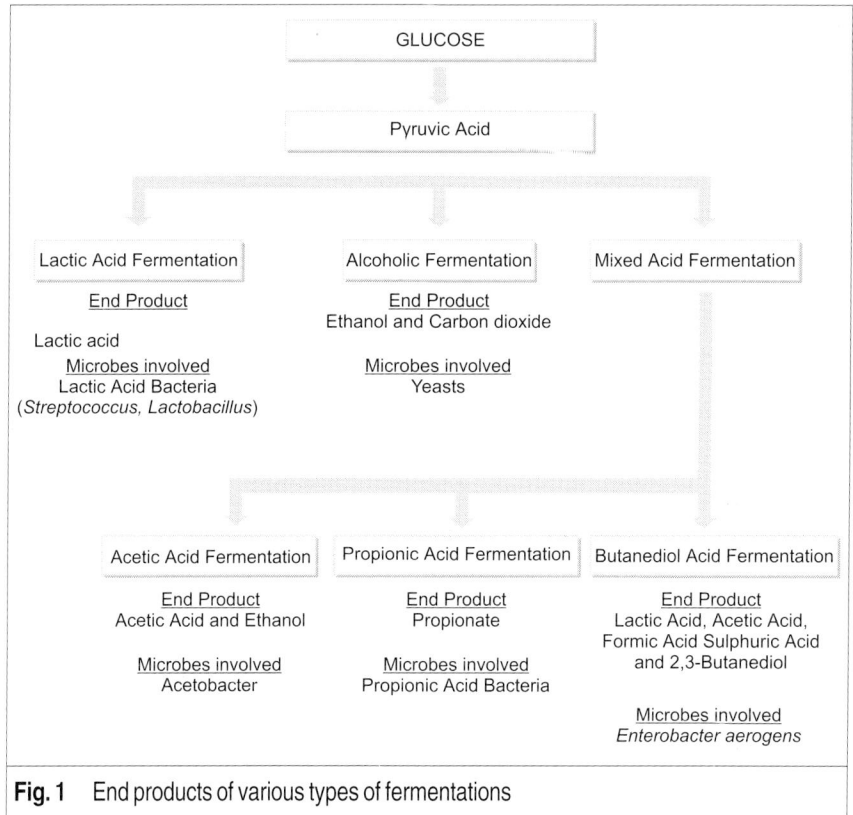

Fig. 1   End products of various types of fermentations

anaerobic cell since it regenerates nicotinamide adenine dinucleotide (NAD^+), which is required for glycolysis. This is important for normal cellular function, as glycolysis is the only source of ATP in anaerobic conditions. Fermentation products contain chemical energy (they are not fully oxidized) but are considered waste products since they cannot be metabolised further without the use of oxygen (or other more highly-oxidized electron acceptors). A consequence is that the production of ATP by fermentation is less efficient than oxidative phosphorylation, where pyruvate is fully oxidized to carbon dioxide. Fermentation produces two ATP molecules per molecule of glucose compared to approximately 36 ATP molecules produced by aerobic respiration. Fermentation end product of some microorganisms and comparison between aerobic respiration, anaerobic respiration and fermentation are given in Tables 1 and 2.

Table 1 Fermentation end product of some microorganisms (Source: Ingraham and Ingraham 2000)

Microbes involved	End products
Saccharomyces	Ethyl alcohol and carbon dioxide
Streptococcus and Lactobacillus	Lactic acid
Propionibacterium	Propionic acid, acetic acid and carbon dioxide
Escherichia coli	Acetic acid, lactic acid, succinic acid, ethyl alcohol, carbon dioxide and hydrogen
Clostridium	Butyric acid, butyl alcohol, acetone, isopropyl alcohol, carbon dioxide and hydrogen

Table 2 Comparison between Aerobic respiration, Anaerobic respiration and Fermentation

Parameters	Aerobic respiration	Anaerobic respiration	Fermentation
Growth conditions	Aerobic	Anaerobic	Aerobic and anaerobic
Final electron acceptor	Molecular oxygen	Usually an inorganic substance such as carbonates, sulphates, etc.	Organic molecule
Types of phosphorylation used to form ATP	Substrate level and oxidative	Substrate level and oxidative	Substrate level
ATPs produced per glucose molecule	38	Variable	2

2. DIVERSE FERMENTATION PROCESSES

Different methods of fermentation can be distinguished on the basis of formation of their products or on the basis of presence or absence of oxygen.

2.1 End Product Formation

Types of fermentation on the basis of formation of end products are:

(a) Lactic Acid fermentation
(b) Alcohol fermentation
(c) Mixed Acid fermentation

The products produced by fermentation are actually waste products produced during the reduction of pyruvate to regenerate NAD^+ in the absence of oxygen.

2.1.1 Lactic Acid Fermentation H_3

In lactic acid fermentation, pyruvic acid forms lactic acid (enzyme lactic dehydrogenase is involved). It is carried out by some bacteria (e.g., lactic acid bacteria that convert lactose into lactic acid in yogurt, giving it a sour taste), and also by animals (muscle glycolysis in animals, under oxygen deficiency, results in the formation of lactic acid which we experience in muscle cramps, or "Charley horse"). Thus, it occurs in the muscles of animals when they need energy faster than the blood can supply oxygen. Lactic acid fermentation produces no gas, which is unusual among fermentation pathways. Net gain is two ATP molecules per glucose molecule.

2.1.2 Alcohol Fermentation H_3

When yeast ferments, it breaks down the sugar ($C_6H_{12}O_6$) into exactly two molecules of ethanol (C_2H_6O) and two molecules of carbon dioxide (CO_2). In this type of fermentation, pyruvic acid is first decarboxylated to form acetaldehyde (enzyme pyruvic acid decarboxylase) and carbon dioxide. Acetaldehyde is then changed into ethyl alcohol with the help of NADH (enzyme alcohol dehydrogenase). Net gain is 2 ATP molecules per glucose molecule. The last two steps of alcohol fermentation accomplish the release of NAD^+ so it can be used again in glycolysis.

In ethanol fermentation (done by yeast and some types of bacteria) the pyruvate forms ethanol and carbon dioxide. It is important in bread-making, brewing and wine-making. When the ferment has a high concentration of pectin, minute quantities of methanol can be produced. Usually only one of the products is desired; in bread the alcohol is baked out, and in alcohol production the carbon dioxide is released into the atmosphere.

2.1.3 Mixed Acid Fermentation H_3

This type of fermentation is a characteristic of members of family Enterobacteriaceae. *Escherichia coli* performs this type of fermentation. Mixed acid fermentation results in the formation of (a) lactic acid, (b) ethanol, (c) formic acid and (d) succinic acid.

2.2 Air Fermentation

Types of fermentation that are determined by the presence or absence of air are:

(a) Aerobic fermentation
(b) Anaerobic fermentation

2.2.1 Aerobic Fermentation H_3

Aerobic fermentation is the fermentation that takes place in the presence of air. Aerobic fermentation usually occurs at the beginning of the fermentation process in wine-making process, before the wine is transferred to an air-locked vessel where anaerobic fermentation takes place. Aerobic fermentation is usually the shorter, but more intense fermentation.

2.2.2 Anaerobic Fermentation H_3

Anaerobic fermentation is the fermentation in the absence of air. It occurs in the fermentation vessel once an airlock has been affixed. Any air that was present in the fermentation vessel is expelled through the airlock (that is, through the process of bubbling) and replaced with carbon dioxide, a by-product of the fermentation process. Anaerobic fermentation is usually the long fermentation and the one in which almost all of the alcohol in mead is produced.

3. IMPORTANCE OF FERMENTATION

1. Fermented foods are cheap and energy efficient form of preservation.
2. Fermentation makes food products tasty and more interesting by intensifying the flavour through the conversion of sugars into acids. Examples of flavor producing microorganisms are tabulated in Table 3.
3. Fermenting makes grains more digestible. For example, porridge that has been fermented (as is common in Africa and even in Wales) hydrolyses starch into shorter chains of glucose and dextrose.
4. Fermented products often contain higher levels of vitamins (particularly thiamine, nicotinic acid, biotin and riboflavin) and proteins. Examples are the Mexican drink called *pulque*, the Indian fermented bread called *idli*, sorghum beer from South Africa and palm wine from West Africa.
5. Some fermented products have meat-like flavor and odour, which is important for cultures where meat is scarce (e.g., soy and fish sauces). South-east Asian countries had an excess of fish products and a paucity of meat. It made sense to make something that reminded the senses of meat.
6. Fermentation can reduce naturally occurring toxins in some foods which render them safe to eat. (e.g., cassava which is widely eaten in Africa has semi-dangerous levels of cyanide. By fermenting cassava to produce Kawal the cyanide is rendered harmless.)

7. Fermented foods often contain a higher level of convertible energy than non-fermented foods of the same weight.

8. There are medical advantages associated with a constant intake of some fermented foods (koumiss is used in Russia to treat tuberculosis).

Table 3 Examples of Flavor Producing Microorganisms

Organisms	Flavor	Microorganisms involved
Bacteria	Limburger	*Brevibacterium linens*
Bacteria	Swiss	*Propionibacterium freunenreichii*
Molds	Camembert	*Penicillium camemberti*
Molds	Roquefort	*Penicillium roqueforti*
Molds	Tempeh	*Rhizopus oligosporus*

Fermentation technology of the world which is responsible for the survival of man can be classified into six groups: alcoholic, lactic acid, leavened breads, meat substitutes, meat flavored sauces/pastes and protein/flavoring agents fermentation.

4. FERMENTED MILK PRODUCTS

The process of fermentation was used all over the world as a means of preserving a highly perishable product and to produce new flavours for an old food staple. Earlier in milk fermentation, milk was fermented by its normal microbiota in its natural course, but the actual process was not completely understood. But these days, commercially fermented milk products are produced. Cultures could be maintained by inoculating fresh milk with fermented milk (Kerr and McHale 2001). Currently, lactic acid producing milk products include sour cream, yogurt, cheese, etc. Microbes involved in different fermented milk products are listed in Table 4.

Table 4 Different microbes involved in different fermented milk products (Source: www.arches.uga.edu)

Microbe	Major Known function	Product
Propionibacterium shermanii	Flavor and eye formation	Swiss cheese family.
Lactobacillus bulgaricus	Acid and flavor	Bulgarian buttermilk, yogurt,
Lactobacillus lactis		kefir, koumiss, Swiss,
Lactobacillus helveticus		Emmental, and Italian cheeses.
Lactobacillus acidophilus	Acid	Acidophilus, buttermilk.
Streptococcus thermophilus	Acid	Emmental, Cheddar, and Italian cheeses, and yogurt.

Streptococcus diacetilactis	Acid	Sour cream, ripe cream, butter, cheese, buttermilk and starter cultures.
Streptococcus lactis *Streptococcus cremoris*	Acid	Cultured buttermilk, sour Cream, cottage cheese, all types of foreign and domestic cheeses, and starter cultures.
Streptococcus durans *Streptococcus faecalis*	Acid and flavor	Soft Italian, cheddar, and some Swiss cheeses.
Leuconostoc citrovorum *Leuconostoc dextranicum*	Flavor	Cultured buttermilk, sour cream, cottage cheese, ripened cream butter, and starter cultures.

4.1 Yogurt

Yogurt is produced from whole or skim milk. It is inoculated with a starter culture, which usually contains *Lactobacillus bulgaricus* and *Streptococcus thermophilus*. These bacteria ferment lactose in the milk to lactic acid, causing the milk to curd and form yogurt. Details about the yogurt are given in Chapter no. 6 of this book.

4.2 Cheese

The term, cheese comes from the Latin word *caseus*. There is evidence of cheese making as long as 8000 years ago. Residues of cheese were discovered in an Egyptian pot that is 2300 years old. People around the world discovered that milk carried in pouches made of animals' stomach coagulated into cheese. In its simplest form, the curds found in soured milk, when drained from the liquid, is cheese. It is a solid food made from the curdled milk of cow, goat, sheep, buffalo or other mammals. The milk is curdled using some combination of rennet (or rennet substitutes) and acidification. Bacteria acidify the milk and play a role in defining the texture and flavor of most cheeses. Some cheeses also feature molds, either on the outer rind or throughout.

There are hundreds types of cheese produced all over the world. Different styles and flavors of cheese are the result of using different species of bacteria and molds, different levels of milk fat, variations in length of aging, differing processing treatments (cheddaring, pulling, brining, mold wash) and different breeds of cow, sheep, or other mammals. Other factors include animal diet and the addition of flavoring agents such as herbs, spices or wood smoke.

4.2.1 Microbes Involved

Cheese is made by coagulating milk to give curds, which are then separated from the liquid, whey, after which they can be processed and matured to produce a wide variety of cheeses. Milk is separated into curds (solid lumps) and whey (liquid containing sugars and some protein) milk is first fermented a bit by lactic acid bacteria then rennet containing the enzyme rennin is added. Rennin causes the milk proteins to clump up forming the curds, the curds are then pressed into shape and more whey removed. Soft cheeses are high in water (80%), while hard cheeses are low in water (<40%). The active ingredient of rennet is the enzyme, chymosin (also known as rennin). The usual source of rennet is the stomach of slaughtered newly born calves. Vegetarian cheeses are manufactured using rennet from either fungal or bacterial sources. Widely available vegetarian cheeses are made by using rennet produced by fermentation of the fungus *Mucor miehei*. Vegetarian cheese may also be made using rennet from the bacteria *Bacillus subtilis* or *Bacillus prodigiosum*. Through advances in genetic engineering techniques, some vegetarian cheeses may be produced by using chymosin formed by genetically engineered microorganisms.

Unripened cheeses are fermented in a single step, whereas ripened cheeses are fermented more than once. The main players in cheese fermentation are *Streptococci* and *Lactobacillium*. Aged or ripened cheese comes from milk that is fermented (soured) by lactic acid bacteria and also coagulated by enzymatic action of rennin. The curd is pressed, salted, and then ripened (aged) for months. Cheddar, Parmesan, and Swiss are ripened cheeses. Very hard cheeses are aged up to a year, and the flavor usually gets stronger over time (www.chemistry.org). Examples of bacterial ripened cheese are Swiss cheese (in which holes are due to carbon dioxide produced by bacteria), Cheddar cheese, and Parmesan cheese. Examples of mold-ripened cheese are Blue cheese, Camembert, etc. (www.public.iastate.edu).

4.2.2 Nutritional Aspects

Cheese is a good source of protein, calcium, zinc, and vitamin B12. However, full fat cheese is a major source of saturated fat, which can lead to raise serum cholesterol levels. Also, it contains no carbohydrate or fibre, and is a very poor source of iron. Vegetarians, particularly new vegetarians, should be wary of too high consumption of cheese.

4.3 Sour Cream

Sour cream has long been a traditional ingredient in Eastern European cooking, and is an important ingredient in Hungarian cooking. It gives the pleasant tang to many dishes and thus has gained popularity in the rest of Europe, North America, and other parts of the world in the past 50 years. Traditionally made by letting fresh cream sour but these days, it is commercially produced. Commercially, sour cream is made by adding bacteria cultures to cream, allowing the bacteria grow until the cream is both soured and thick, then pasteurizing it to stop the process. By definition, sour cream must contain at least 18% milk fat by weight.

4.3.1 Microbes Involved

Sour cream is produced in the same way as yogurt. However, light cream is used instead of whole or skim milk. *Streptococcus* spp. and *Lactobacillus* spp. are used as starter cultures to form lactic acid from lactose. The lowered pH causes the cream to coagulate, resulting in sour cream.
http://www.public.iastate.edu/~cfford/101Milk.htm

4.3.2 Uses

Sour cream is widely used in dips, spreads, sauces, cakes, soufflés, in savoury dishes such as beef stroganoff, Hungarian goulash and top baked potatoes. It is rather rich in taste but is actually lower in calories than comparable amounts of salad oils and most salad dressings. Sour cream cannot be made at home with pasteurized cream due to the lack of natural bacteria, which will cause the cream to spoil instead of sour.

5. BEVERAGES

5.1 Kanji

In Northern India and Pakistan carrots, particularly a variety that is deep purple in color, is fermented to make a conventional ready to serve beverage known as *Kanji*. It is extremely admired and considered to have cooling and soothing properties and on the same side also acquire high nutritional value. It is usually made in terracotta (earthen ware) with a long narrow neck tied with folded cheesecloth on the aperture.

5.1.1 Microbes Involved

The fermentation involved is known as lactic fermentation, which must be carried out in the absence of air. Generally, lactic acid bacteria produce lactic acids, which lessen the pH i.e., augment the acidity to a level that prevents the growth of food poisoning organisms. The final product is slightly acidic in taste and has an attractive purple-red color. After fermentation the drink is strained through fine muslin and has to be consumed within 3 or 4 days after which it becomes stale. (Berry 1998, Shah 1986).

5.1.2 Uses

Kanji is a popular, spicy purple color nutritious beverage of northern India. The ingredients for its preparation comprise washed grated carrots, water, salt and pepper. Carrots are thoroughly washed and grated delicately. In certain regions, the carrots are sliced lengthwise and the fermented carrot sticks are consumed along with beverage. Each kilogram of grated carrot is mixed with 7 liters of water, 200 g of salt, 40 g of crushed mustard seed and 8 g of hot chilli powder. This is mixed and placed in a glazed earthenware vessel. It is almost entirely sealed, and only a tiny hole is left for escape of gases, released during fermentation. The mixture is then allowed to ferment for 7-10 days. It is like making wine at home.

5.2 Feni

Feni is a distilled beverage of about 40% alcohol content made from cashew wine distillation (www.uga.edu). Feni is written differently as fenny, fenim or fenn, but however, it should be pronounced as Fenim. The word feni is derived from the word fenn, which means froth. In fact, a good feni when poured in a glass produces a little froth, which is an indication of the superior quality of the product.

Two types of fenis are: Cashew feni and Coconut feni. Cashew feni is made out of the juice of cashew apples, whereas coconut feni is made from the toddy collected by toddy tappers. While coconut feni has its own charm, many people prefer cashew feni because of its flavour and between the two, cashew feni costs more. The liquor produced from cashew is of three grades Urrac, Cazulo and feni. The Urrac is the product of first distillation and is light. The Cazulo is the product of second distillation and is moderately strong. The product obtained after the process of third distillation is called feni. It has a long shelf life. When the Cazulo is not made, feni is produced after second distillation itself. The second or third hand feni is a product par excellence.

5.2.1 Microbes Involved

Yeast is the main microbe involved in production of feni. Yeast is a one-celled fungus that converts sugar and starch into carbon dioxide bubbles and alcohol. This has made it a useful ally in the production of bread, beer and wine. There are many varieties of yeast. Brewer's yeast (also known as brewers yeast or brewing yeast) are used in brewing industries.

5.2.2 Uses

Feni has distinct taste and strong smell. The smell of feni cannot be hidden once the bottle is opened and the drink is poured in the glass. All feni tends to stink up the area within a radius of a mile. If the glass is kept unwashed after consumption of feni, it does not lose the smell easily. Many people like to drink feni since it helps bowel movement and provides relief from constipation.

5.3 Toddy

Toddy or tari is an alcoholic drink made by the fermentation of the sap collected from all the trees that are from the palm tree family, including the date palm and coconut palm. It is white and sweet with a characteristic flavor. It has 4 to 6% alcohol and a shelf life of about 24 hours.

5.3.1 Microbe Involved

Yeast is the major microbe involved in the production of toddy.

5.3.2 Processing

The juice is sweet and refreshing only at the early morning. But if, after collection in the morning, it is stored for about two hours or more, it gradually ferments and becomes an intoxicating drink popularly known as *"tari or toddy"*. The fermentation starts as soon as the sap collects in the pots on the palms, particularly if a small amount of toddy or "tari" is left in the pots. The toddy is fully fermented in six to eight hours. The product is usually sold immediately due to its short shelf life. When the tree is tapped before dawn, and the fresh or slightly fermented juice is consumed, say, before 10 O'clock, it is a delightful drink. In its pure and natural form, toddy is supposed to be extremely good for the health. Toddy is prepared when the tree is tapped a little after sunrise, and then the juice is fermented and sour. The juice gets increasingly heady as the day advances. To the extent that a glass of undiluted toddy can even be considerably intoxicating to the uninitiated. Best sap for toddy making is obtained from those trees that are just flowering and are about 20 years old.

5.4 Beer

One of the world's oldest and most popular alcoholic beverage is beer. It is produced through the process of fermentation of starch-based materials, commonly barley, though cassava roots in Africa, potato in Brazil, and agave in Mexico, etc. are also used. As the ingredients and procedures used for making beer can differ, so characteristics such as taste and color may also vary accordingly.

Although beer had its origins in Mesopotamia, fermented beverages of some sort or another were produced in various forms around the world. For example, Chang (Tibetan beer) and Chicha (corn beer) and kumis is a drink produced from fermented camel milk. The word beer comes from the Latin word bibere i.e., "to drink", and the Spanish word cerveza originated from the Greek goddess of agriculture, Ceres. Apparently, beer was the first alcoholic beverage known to civilization.

5.4.1 Microbes Involved

Yeast, a single-celled microorganism is responsible for creating the alcohol and carbon dioxide found in beer. Different kinds of yeasts are used for making beer as different types of beer yeast help to give beer its various tastes.

There are two main categories of beer yeast: ale yeast and lager yeast. Ale yeast is top fermenting, meaning it rises near the surface of the beer during fermentation, and typically prefers to ferment at temperatures around 70°F (21°C). Lager yeasts are bottom fermenting. They ferment more slowly and prefer colder temperatures, around 50°F (10°C). Thus, essentially these two types of yeast are used viz. *Saccharomyces cervisiae* and *S. carlsbergensis*. *S. cerevisiae* is known as "top-fermenting" as it floats on the surface of the wort. It is used in brewing darker beers such as English "Bitter", whereas continental lager-type beers are made with *S. carlsbergensis*, which is bottom-fermenting yeast.

5.4.2 Production of Beer

Brewing is fundamentally a natural process. The art and science of brewing lies in converting natural food materials into a pure and pleasing beverage. Beer today is still a beverage brewed from natural products in a traditional way. Although the main ingredients of beer have remained constant (water, yeast, malt and hops), it is the precise recipe and timing of the brew that gives one a different taste from another. Apart from brewing company expenditures on research and quality control designed to achieve the highest standards of uniformity and purity in the product, the production of beer is also subject to regular inspection and review by federal and provincial Health Departments. On an average, a batch of beer takes about 30 days to produce. The beer contains mainly four ingredients, which are as follows:

5.4.2.1 Water

It is the main ingredient of a beer and the quantity and variety of dissolved salts in the water play an important role in the character of the final beer. The salts play a part in the extraction of fermentable sugars from the grain as well as affecting the way the yeast behaves during fermentation. The important salts are viz. (a) calcium increases the extract (efficiency of extracting sugars during mashing) and helps in making the beer clearer, (b) sulphates enhance the bitterness of the hops and (c) chlorides help in enhancing sweetness. Modern brewing processes involve modifying the content of these key ions to produce water that is best suited to the style of beer being produced.

5.4.2.2 Malt

Barley is used to make brewers' malt. When harvested grain starts to germinate, its sugar content increases. This is due to the conversion of starches into sugars as the seed began to germinate. The grain will contain some sugars plus enzymes to aid the extraction of fermentable sugars. Barley is soaked, germinated (sprouted), then dried and/or kilned/roasted to arrest further growth.

5.4.2.3 Hops

Winemakers to add aroma in beer generally put spices, fruits and mostly the flower of the hop vine. The herbs and spices once added to wine also act as a preservative. The hops add alpha and beta acids that provide bitterness and aroma to the final product. Hops are chosen for their content in these products as required by the beer being produced. They are also added at different stages in the process depending on whether they are being used to provide bitterness or aroma.

5.4.2.4 Yeast

These are yeasts that form foam on the top of the beer during fermentation. This foam is skimmed at a certain stage in the fermentation and used to start the next beer fermenting. These yeasts are used at higher temperatures. The sugar content of the liquid is monitored throughout the fermentation and the process is stopped when the desired alcohol strength is reached.

5.4.3 Types of Beer

5.4.3.1 Lagers

The word lager is derived from the German verb "lagern" meaning to store. Lager is brewed with bottom fermenting yeast at cooler temperature for a smooth flavour. It is pale golden color, carbonated and lightly hopped, tastes crisp and refreshing. Lagers typically take more time to brew and are aged longer than ales and are best enjoyed at cooler than room temperature.

5.4.3.2 Ales

It is brewed with top fermenting yeast at cellar temperature. Ales are fuller-bodied, with nuances of fruit or spice and a pleasantly hoppy finish. Ales are often darker than lagers, ranging from rich gold to reddish amber.

Top fermentation and the inclusion of more hops in the wort give these beers a distinctive fruitiness, acidity and a pleasantly bitter seasoning. All ales typically take less time to brew and age then lagers and have a more assertive, individual personality, though their alcoholic strength may be the same. Some of the more popular ales are as follows:

5.4.3.2.1 Barley wine

It is a very intense and complex beverage with alcohol content equal to most wines. It has a hearty, sweet malt flavor, which is offset by a strong and bitter flavoring from the hops for balance. The color ranges from copper to medium brown. Because of the preserving qualities of alcohol, this is the best beer for storing over a long period of time. The bitterness ranges from medium to the highest of all beer types.

5.4.3.2.2 English bitter

There are three classic types of English Bitters viz., Ordinary (mild), the Special (moderate strength) and the Extra Special (a strong bitter). They are typically characterized with traditional hops such as Kent Goldings, Fuggles, or Brewers Gold. They range from mild to strong, the color and alcohol percentage also follow.

5.4.3.2.3 Pale ale

There are various types of pale ales sharing a pronounced hop flavor and aroma with low to medium maltiness. There is also a good deal of fruity esters. The types of pale ales are the English, the India (IPA) and the American. English have a dry character usually due the high sulfate content of the water. The India Pale Ale is usually stronger and hoppier because the higher alcohol content and hop acids acted as a preservative. The American is usually amber in color and has more maltiness flavor than the other two. When brewing pale ales, fresh, quality hops is a necessity.

5.4.3.2.4 Scottish ale

Scottish ales are close to the English ales. They are usually darker, maltier and have less carbonation. They range in color, maltiness and strength in the order of Scottish Light (60 Shilling), Scottish Heavy (70 Shilling), Scottish Export (80 Shilling) and the Strong Scotch (wee heavy). The term 60-80 shilling dates back to when beer was taxed by gravity and strength. The Strong Scotch is usually dark brown, high in alcohol (6-8 percent) and can have a slightly smoky character.

5.4.3.2.5 Belgian strong dark ale

Though very diverse, they are usually medium to dark in color with high alcohol content. They are very malty, with a low hop flavor and aroma. The most important ingredient in this style of beer is the strain of yeast. The yeast and warm fermentations create a unique biscuity flavor with fruity and spicy overtones and a good deal of carbonation. These beers are usually very aromatic. These are often considered the champagne of beers.

5.4.3.3 Porter

The name comes from the Porters at London's Victoria Station. Arthur Guinness and Sons was the first brewer to offer a Porter commercially. Later on, they increased the alcohol content of the Porter and the new drink became known as the Stout Porter (which eventually became Stout).

5.4.3.4 Imperial stout

An Imperial Stout is dark copper to very black in color. It has a rich and complex maltiness with noticeable hop bitterness. The two main ingredients are the dark roasted barley and black malts. The Imperial Stout is like the espresso of beer styles, full flavored and intense.

5.4.3.5 Dry beers

The term "Dry" refers to the amount of residual sugar left in a beer following fermentation. This type of beer is fermented for longer than normal brews. These beers are crisp and clean, since most of the natural sugars are turned into alcohol during brewing resulting in a medium-golden beer that tastes less bitter.

5.4.3.6 Light beers

Light beers are highly carbonated with low bitterness and no aftertaste. These are extremely light in color and body, and mild in flavor. These have fewer calories and/or lower alcohol content.

5.4.3.7 Draught

Fresh tasting and easy to drink, draught beer in bottles or kegs has slightly lower carbonation levels so is less filling than other bottles or cans. Light or dark, in any style, draught beer is simply any beer served from a keg or cask.

5.4.3.8 Malt beers (Specialty Grains)

Malt beers are higher in alcohol (5.5% to 8.0%), boast a rich, full of flavour, heavier and sweeter than other beers. Their colour ranges from deep gold to amber to firelight red. These are of various types depending on the grains used.

5.4.3.8.1 Black patent malt (black malt)

It is very dark malted barley and gets its black color from very high roasting temperatures.

5.4.3.8.2 Chocolate malt

It is similar to black malt, except that it has not been roasted longer. It is also dark in color, but does not have the burnt flavor.

5.4.3.8.3 Crystal malt

It is produced by a special malting process that allows some of the starches to be converted to simpler sugars inside the intact grain. These simple sugars are fermentable. However, a significant percentage of more complex sugars remain intact and can add body, sweetness and mouth feel to a beer. It has a fairly mild flavor and will generally not overpower the final product.

5.4.3.8.4 Roasted barley

Unmalted barley is roasted in an oven at a fairly high temperature until it turns to the desired color. This specialty grain will not contribute to any of the final alcohol content of the beer and is used primarily for flavoring.

5.4.3.9 CaraPils

CaraPils or dextrine can give a range of characteristics to a beer. Roasted time determines what level of flavor it will impart to the final product. CaraPils has no enzyme by itself, so it should generally be used in conjunction with other types of malts that contain enzymes. There are three main types of CaraPils malts viz., Mild Malt, Vienna Malt and Munich Malt.

5.4.3.9.1 Ice beers

Ice beer owes its concentrated flavor and smooth finish to its unique brewing process. The beer is cooled until ice crystals form, and then filtered, resulting in a higher alcohol content. Ice beers are served well-chilled, with seafood or poultry dishes.

5.4.3.10 Stout beers

Deep, dark and flavorful, stout earns its character from brewing with highly roasted malts. Stout features intense malt and caramel flavours, and depending on the variety, ranges from sweet to dry and distinctively bitter. Serve as a unique compliment to shellfish, hearty stews and wild game.

Whether dry or sweet, flavored with roasted malt barley, oats or certain sugars, stouts and porters are characterized by darkness and depth. Both types of beer are delicious with hearty meat stews and surprisingly good with shellfish. The pairing of oysters and stout has long been acknowledged as one of the world's great gastronomic marriages.

5.4.3.11 Bock beer

The other bottom-fermented beer is bock, named for the famous medieval German brewing town of Einbeck. Heavier than lager and darkened by high-colored malts, bock is traditionally brewed in the winter for drinking during the spring.

5.4.4 Beer Types Based on Fermentation

There are three main types depending on the fermentation method.

5.4.4.1 Bottom-fermenting beer

It is the most recent type. The process dates from 1840 and produces a specific type of beer, Pils or lager. This type accounts for 90% of worldwide beer production. Pils is a light, clear, golden beer. It has a fresh, bitter and refined hoppy flavor.

5.4.4.2 Top-fermenting beer

It is a much older and more traditional variety. This process is involved in the production of many different types of beer, in particular Amber or "Special Belgian" beer. Originally, this type of beer had the same density and alcohol content as Pils. The amber color is obtained by using coloured or caramelised malt. The alcohol content is now slightly higher, and this beer is considered to be typical "sampling" beer.

5.4.4.3 Spontaneous fermentation

A process that is characteristic of the Brussels region, is used to produce Lambic. Lambic is a flat beer with no head. It is produced through spontaneous fermentation of the yeasts found specifically in the valley of the river Senne. It matures in barrels made of different types of wood and has a wide range of tastes.

Gueuze is obtained by the fermentation caused by mixing "old" Lambic (that has not completely fermented) and "young" Lambic. This new fermentation produces a sparkling, sharp beer, the "champagne of beers".

Fruity beers (bières fruitées) are prepared by mixing different fruits (cherries for Kriek) and Lambic. Traditional Kriek is a mixture of 50 kg of cherries and around 250 litres of Lambic. This mixture matures for 6 months in the barrel and produces a beer with a fruity but not sugary taste. The industrial process involves fruit juice and extract. The result is a sweet, fruity beer. Faro is a sharp Lambic to which candy sugar is added (or sometimes caramel or syrup) to give it a sweeter taste.

5.4.5 Health Effects

A beer may be non-alcoholic or alcoholic. Alcoholic beer has a number of health risks and benefits. However, it includes a number of other chemicals that are currently undergoing scientific evaluation.

A beer may contain magnesium, selenium, potassium, phosphorus, biotin, and B vitamins which are nutritionally helpful to a person. It is realized that the darker the brew, the more nutrient dense it is.

Non-alcoholic beer may possess strong anti-cancer properties and mirror the cardiovascular benefits associated with moderate consumption of alcoholic beverages. It is also considered that over-eating and lack of muscle tone is the main cause of a beer belly, rather than beer consumption.

5.5 Wine

Wine is an alcoholic beverage produced by the fermentation of fruits like grapes, plum, elderberry and blackcurrant. Most wines are made from grapes. Fruit juice contains simple carbohydrates, which the yeasts break down straightaway, without any other processing. Non-grape wines are called fruit wine or country wine. Other products made from starch-based materials, such as barley wine, rice wine, and sake, from other fermentable material such as honey (mead), or that is distilled, such as brandy, are not wines.

5.5.1 Microbes Involved

Louis Pasteur discovered that microbes were primarily responsible for fermentation. Yeasts i.e., *Saccharomyces cerevisiae* are concerned with wine fermentation via alcoholic fermentation. The sugars of the grape (glucose and fructose) produce ethyl alcohol, carbonic anhydride and other products. Gay-Lussac explained the chemistry of the reaction: -

$$C_6H_{12}O_6 + 2\,ADP + 2Pi \longrightarrow 2\,CH_3\text{-}CH_2OH + 2\,CO_2 + 2H_2O + 2ATP$$
$$2\,CH_3\text{-}C\,H_2OH \text{ (energy released 118 KJ}$$
$$\text{per mole)}$$

According to Neuberg, the wine fermentation is not a pure alcoholic fermentation because it continues with a glyceropyruvic fermentation, which produces glycerine and pyruvic acid.

Other secondary products appear in addition to the majority compounds in the alcoholic fermentation. Secondary products of the alcoholic fermentation can be obtained from glycerin and pyruvic acid that are listed in Table 5.

Table 5 Secondary products of the alcoholic fermentation (Source: www.ejeafche.uvigo.es)

From Glycerin	From Pyruvic acid
Glycerol	Acetic acid
Superior alcohols	Formic acid
Lactates	Succinic acid
Esters	Malic acid
Sorbite	Propionic acid
Arabitol	Acetylmetylcarbinol
Eritritol	Diacetyl
Polyhydric alcohols	Citramalic acid

During the primary fermentation of wine, the two grape sugars, glucose and fructose are converted to alcohol (ethanol) by the action of yeast. The by-products of primary fermentation are aromas and flavors, the gas carbon dioxide, and heat. The production of heat during fermentation (i.e., it is an exothermic process) implies that the temperature of the fermentation vessel will rise that require action on the part of the winemaker to cool it down. If temperature rises higher, then there can be a loss of desirable aroma and flavor compounds, and unattractive aroma characters in the spectrum of caramel, burnt or cooked characters can be produced. While fermentation is usually conducted in the range of 8-19°C, and red wine fermentations are allowed to run at between 25 and 32°C. There are many types of yeast, but two closely related types *Saccharomyces cerevisiae* and *S. bayanus* are accountable for fermentation. These species of yeast are encouraged to undergo fermentation because they are alcohol tolerant and more tolerant to sulphur dioxide than other yeast and bacteria. It can establish a viable population in environment of high sugar (190-270 gm/l) and high acidity. The yeast produces wine like aroma and flavor characters.

5.5.2 Brief Process of Wine Making

Fresh grapes may be crushed to take out the juice. Most wineries add sulphur dioxide either as a gas or as a solid salt to prevent the growth of other yeasts and bacteria, which cause spoilage. Sulphur dioxide is also produced by *Saccharomyces* during fermentation.

The grapes are crushed and the grape is allowed to ferment in vats, usually after the addition of sulphur dioxide. A selected strain of wine yeast may be added at this point. Heat is produced by the fermentation, and the temperature may have to be controlled within optimum limits. Air is excluded from the vats as much as possible to discourage the action of the vinegar-forming bacterium, and other harmful organisms. When fermentation is well advanced, the "free-run" wine is drawn off. Fermentation continues and is completed after several weeks. The wine is racked off to separate it from the lees, or sediment of yeast, acid potassium, and other substances. Classification of wines according to their sugar content is tabulated in Table 6.

Table 6 Classification of wines according to their sugar contents. (Source: www.glue.umd.edu)

Type	Specific Gravity	Sugar Content (wt.%)
Dry Wine	1.085-1.100	21-25
Medium Sweet Wine	1.120-1.140	29-33
Sweet Wine	1.140-1.160	33-37

5.5.3 Uses

5.5.3.1 Beverage

Wine is a popular and important beverage that accompanies and enhances a wide range of European and Mediterranean-style cuisines, from the simple and traditional to the most sophisticated and complex. Red, white and sparkling wines are the most popular, and are also known as light wines, because they only contain approximately 10-14% alcohol. The apéritif and dessert wines contain 14-20% alcohol, and are fortified to make them richer and sweeter than the light wines. Red wine is traditionally served with beef, lamb and game but there are no hard and fast rules and nowadays reds are served with fish, poultry and even with desserts. Cooking wine typically contains a significant quantity of salt, is unpalatable and intended for use only in cooking. Ten largest wine producing countries are given in Table 7.

Table 7 Ten largest Wine producing Countries in the year 2003. (Source: www.enwikipedia.org)

Rank	Country	Production (tones)	Rank	Country	Production (tones)
1	France	4,735,260	6	China	1,200,000
2	Spain	4,623,750	7	Australia	1,019,400
3	Italy	4,408,611	8	South Africa	885,300
4	USA	2,350,000	9	Germany	828,855
5	Argentina	1,322,500	10	Portugal	709,300

5.5.3.2 Religious use

Wine is also used in religious ceremonies in many cultures and the wine trade is of historical importance for many regions. Wine is an essential part of the Eucharistic rites in the Orthodox, Catholic, Lutheran and Anglican denominations of Christianity.

5.5.3.3 Wine based drinks

Well-known wine based drinks include Brandy, Calimocho, Mulled wine, Sangria, etc. Their various compositions are given in Table 8.

Table 8 Some popular wine based drinks

Name	Composition
Brandy	A general term for distilled wine, which has been aged for at least 2 years.
Calimocho	A cheap alcoholic drink, comprising 50% red wine and 50% cola drink.
Mulled wine (Glögg in Scandinavia and Glühwein in Germany)	A red wine, combined with spices, and usually served hot.
Sangria	A wine punch, comprising red wine, chopped fruits, sugar, and a small amount of brandy or other spirits.
Spritzer	A tall, chilled drink, usually made of white wine and soda water.
Wine cooler	An alcoholic beverage made from wine and fruit juice, often in combination with a carbonated beverage and sugar.

| Rebujito | A mixture of manzanilla wine, mixed with a soft drink like Sprite or 7 UP. |
| Zurracapote | A popular Spanish alcoholic drink comprised mainly of red wine, spirit, fruit juice, sugar and cinnamon. |

6. FERMENTED FOOD PRODUCTS

6.1 Dosa and Idli

6.1.1 Dosa

Dosa (Dose in Telugu, Dosay in Kannada, Dosai in Tamil, Dosha in Malayalam) is a very popular and common South Indian snack food (Gopalan et al. 1999) that comes in many varieties, flavors and with various side dishes like sauce and chutneys.

Dosa is prepared by spreading batter of rice and lentil flour mixed with water in a thin circular disc on a flat, preheated pan and frying it with a dash of edible oil or ghee till the dosa gets evenly fried in a golden brown color. The other side is partially fried next by turning the dosa over. The end product is neatly folded and served. The batter, which is made of rice and lentil flour mixed with water, is fermented overnight. The crispness, the color, the amount of batter used and the time for which the batter was allowed to ferment prior to cooking determine the variety of dosa made.

A stuffing inside a folded dosa is also common. Generally in masala dosa or Mysore masala dosa type freshly cooked, crisp dosa is stuffed with mashed potatoes lightly cooked with fried onions and spices. When the onions are mixed into the batter itself then it becomes an onion dosa. Cracking an egg onto a dosa as it is frying turns it into an 'egg dosa'.

6.1.2 Idli

Idli, also spelled Idly or Iddly, is a food native to southern India and consumed as popular snack item. The traditional idli is a small, round patty of batter made of rice and black lentils or urad dal and steamed. The 2"-3" diameter idli is usually served in pairs with chutney, sambar, or other condiments such as dry, crushed spice mixtures.

6.1.3 Idli Dosa Batter

Batter consists of rice, whole decorticated (without skin but not split) urad dal (*Vigna mungo*), fenugreek seeds and water. Batter is allowed to ferment

(Sharavathy et al. 2001) overnight by incubating it in a warm place (86° F to 90° F) till the batter has reached two and a half times the original volume.

6.1.4 Microbes Involved

The fermentation in dosa batter is caused by airborne wild yeast. Urad and Fenugreek seeds draw the wild yeast from air. Hence, urad dal or fenugreek seeds should not be over washed, as it will wash away the collected wild yeast. The fermentation can be retarded by yogurt, baking yeast, baking soda or baking powder. Only when the fermentation is over, yogurt or baking agents may be added as per the requirement of the recipe. Decorticated whole urad dal is preferred for idli dosa batter as decortications process involves introducing moisture to remove the skin. The split urad has a problem as the pulse is split mechanically which generates heat and destroys much of the wild yeast. To compensate for this fenugreek seeds may be added to aid in fermentation process.

6.1.5 How the Dosa Batter differs from Idli Batter?

The dosa batter is thinner than the idli batter. This consistency is required to make the dosas crepe. There is less Amylopectin in dosa than idli. The dosa crepe has to be crusty and crisp, while the idli is soft and fluffy. So, water can be added and idli batter is made thin to get crispy dosa. When yogurt is added to idli batter, then the lactic acid and the butterfat in the yogurt, tenderizes the gluten, and makes the idli soft.

6.1.6 Chemical Analysis

Nutritional composition and nutritionally important starch fractions of idli and dosa are given in Tables 9 and 10 (Paradkar et al. 2002).

6.2 Dhokla

Dhokla is steamed bread. It is popular fermented sweet-sour snack and savored a lot in West India, especially Gujarat. The main constituents of it are suji, besan (chana flour or flour of *Cicer aierentum*), curd/yogurt, grated ginger, chopped green chilli, salt, chopped coriander. These all are mixed and a thick batter is prepared. This mixture is left as such for 1-2 hours for household fermentation. Batter is placed in a greased serving dish and steamed for 15 minutes. It is slashed into any shape and fried in some oil containing mustard seed/sesame seeds. It is usually garnished with green or sweet sour ketchup or grated coconut.

Table 9 Nutrient composition of the cereal based food preparations

Food (g)	Protein (g) Nx 625	Fat (g)	Free sugars (g)	Starch (g)	Dietary fiber (g)	Dietary fiber (g)
					SDF	IDF
Dosa	12.6+0.3	13.6+0.3	2.1+0.1	46.1+2.4	1.4+0.2	5.3ı0.7
Idli	12.5+0.2	2.7+0.2	2.0+0.1	54.2+0.7	1.4+0.3	9.9+0.6

SDF- Soluble Dietary Fiber; **IDF** – Insoluble Dietary Fiber

Table 10 Nutritionally important starch fractions in cereal-based preparations

Foods (g)	Dry matter %	Rapidly digestible starch (g)	Slowly digestible starch (g)	Resistant starch (g)	Total starch (g)
Dosa	30	3.2+0.2	15.1+0.1	8.1+0.4	26.4+2
Idli	36	3.0+0.2	16.7+1.0	9.7+0.1	28.7+1.3

6.2.1 Microbes Involved

The batter of Dhokla ferments when it is left as such for 1-2 hours. During this time *Leuconostoc mesenteroides* and *Streptococcus faecalis*, which are naturally present on the grains/curd/utensils, grow up in logathmic phase. These organisms produce lactic acid and carbon dioxide that make the batter anaerobic and leavens the product. In the next few hours, the batter is steamed to prepared tasty dhoklas. (Ramakrishnan 1979a, 1979b; Steinkraus 1996).

6.3 Jalebi and Imarti

Jalebis are a striking dessert with their impressive shapes, brilliant colour and intriguing texture. These are deep-fried pretzel-shaped loops made from unbleached white flour, rice flour and small amount of yogurt in saffron syrup. Though the fried loops do not take long to cook, the batter must be started at least 18 hours before it is required. A simple flour-water mixture is allowed to ferment slightly thus when the batter falls from a spoon, it flows in a broad solid band, without breaking. The consistency appears gooey and somewhat gelatinous when whisked. The batter is forced through a plain nozzle of a pastry bag, the arm and hand rhythmically moving over the surface of the hot ghee, forming interlocking three-ring spirals or double figure eights, either individually or in a chain. After they are fried on both sides until golden and crisp, the jalebis are submerged for several seconds in a hot saffron-scented syrup, which saturates their hollow insides. Although skill is required to master the uniform shapes of jalebis, novice squiggles will taste just as good.

6.3.1 Microbes Involved

In batter preparation of jalebis, yogurt is involved. Microbes involved are *Lactobacillus bulgaricus* and *Streptococcus thermophilus* as in yogurt.

6.4 Chocolates

Fermentation is a vital step in the transformation of cacao beans to chocolate. Without fermentation cacao beans yield little or no chocolate flavor. The principal effect of fermentation is to eliminate or, at least, drastically reduce astringency and to develop the full cacao flavor. Cacao seeds must be fermented, dried, and roasted to produce the desired chocolate flavor. Fermentation and drying are done at the farm that grows the chocolate trees.

6.4.1 Microbes Involved

Fermentation involves removing the cacao seeds and the pulp that surrounds them from the protective pod shortly after the harvest. Yeast in the environment settles on the pulp and ferments natural sugars into alcohol. After production of alcohol by yeast, acetic acid (vinegar) producing bacteria and that produce lactic acid from residual sugar take over the fermentation process. As acids are produced during this procedure, the pH of the seeds decreases and the temperature rises as high as 140°F. The combined effects of the acidity and alcohol production in the fermentation process acts to kill the cacao seed.

The pulp contains mostly water with 10-15% sugars. The high sugar content in the pulp favors the growth of yeasts, which ferment sugars to ethyl alcohol in the anaerobic heap. Eleven different species of yeasts have been isolated with *Saccharomyces cerevisiae*, and of that group, *Candida rugosa*, and *Kluyveromyces marxianus* are the most plentiful. In addition to producing ethyl alcohol, the yeasts hydrolyze the pectin that covers the seeds. It has been shown that *S. cerevisiae* decreases the bitterness of the final product. Without pectin, the bitter alkaloids may leach out of the seed or be altered by alcohol that can now enter the seeds. (http://skylinecollege.net/case/chocolate.html).

The alcohol they produce kills the yeast and, as the temperature rises, lactic acid bacteria such as *Lactobacillus* and *Streptococcus* grow. The pulp is stirred and drained to aerate it. The presence of oxygen and the lower pH now favor the growth of acetic acid bacteria, Acetobacter and Gluconobacter.

After five days, the fermented mass contains upto 108 microbes per gram. The beans are then dried and as they dry, molds including *Geotrichium* grow. *Geotrichium* oxidizes the lactic acid to acetic acid and succinic acid.

If fermentation is allowed to continue beyond five days, microbes may start growing on the beans instead of on the pulp. Off-tastes result when *Bacillus* and filamentous fungi, including *Aspergillus*, *Penicillium*, and Mucor, hydrolyze lipids in the beans to produce short-chain fatty acids. As the pH approaches about neutral (7.0), *Pseudomonas*, *Enterobacter*, or *Escherichia* may grow and produce off-tastes and odours.

Dried beans are bagged for sale to chocolate manufacturing companies. Microorganisms are also used in the production of finished chocolate products. Alpha amylase obtained from *Aspergillus* is used for hydrolyzing starch for chocolate syrup and invertase from *Saccharomyces* is used for hydrolyzing sucrose in filling mixtures to make soft-centered chocolate-covered candies.

6.5 Pizza Base

The pizza dough is basically bread only. Pizza base is usually prepared having the amalgamation of yeast (fresh or dried), lukewarm water, sugar, olive oil, flour and a touch of salt. Bowl containing yeast, water and sugar together should be left in a warm place for some time so that froth forms on the top to facilitate the fermentation process. Then stir it in the olive oil. Sieve and mix the flour with right amount of salt. The yeast mixture is then poured at the center of the flour and smooth dough should be prepared. Dough is then wrapped in a greased plastic sheet and left in a warm place for at least an hour. The dough should be fermented and swollen to at least twice its size in this duration. The cause of puffy dough is that the carbon dioxide bubbles become trapped inside it and raise it. Recently in the European Journal reported that the more pizza eaten halves the risk of heart attack. The authors suggest that some of the ingredients of pizza have been shown to have a favorable influence on the risk of cardiovascular disease (Gallus et al. 2004).

6.5.1 Microbes Involved

Yeast is the main microbe responsible for formation of Pizza base. There are many varieties of yeast. Bread is made with baker's yeast, which creates lots of bubbles that become trapped in the dough, making the bread rise so it's light and airy when baked. A small amount of alcohol is also produced during baking, but this burns off and leaves as sweetness as the bread bakes.

6.6 Vinegar

Vinegar is a sour-tasting liquid made from the oxidation of ethanol in wine, cider, beer, fermented fruit juice, or nearly any other liquid containing alcohol. The word is derived from old French word vinaigre, meaning, "sour wine". Vinegar is one of the oldest products of fermentation used by man. It is the acetic acid produced by the fermentation of alcohol (ethanol), which gives the characteristic flavor and aroma to vinegar. Vinegar is also prepared by the action of certain bacteria operating on sugar-water solutions directly, without intermediary conversion to ethanol (acetic acid). To produce high quality vinegar, it is essential that the raw material used for its production is mature, clean and in good condition. Commercially available vinegar usually has a pH of about 2.4.

6.6.1 Microbes Involved

A wide range of raw materials can be made into vinegar with the help of strains of *Leuconostoc mesenteroides* and *Streptococcus faecalis* and by applying controlled fermentation. It requires the presence of alcoholic substrate, strains of acetic-acid forming bacteria (*Acetobacter*) and oxygen to enable the oxidation of alcohol for production of good quality vinegar. Range of raw materials used for manufacturing different varieties of vinegar from all over the world is given in Table 11.

Table 11 Different Types of Vinegar

Product	Raw Materials Involved	Location
Malt Vinegar	Barley	Australia, Britain, Canada
Wine Vinegar	Red or White Wine	Mediterranean countries, Germany
Fruit Vinegar	Fruit Wine (Black Currant, Raspberry, etc.)	Europe
Rice Vinegar	Rice	Japan, China
Coconut Vinegar	Sap of Coconut Palm	India, South-east Asia
Cane Vinegar	Sugarcane	Philippines, France, USA
Raisin Vinegar	Raisin	Turkey, Middle East
Beer Vinegar	Beer	Germany
Pineapple Peel Vinegar	Pineapple Peel	Latin America, Asia

6.6.2 Uses

Vinegar is commonly used in food preparations, particularly in pickling processes, vinaigrettes and other salad dressings. It is an ingredient in sauces such as mustard, ketchup and mayonnaise. It is also often used as a condiment. It is also used as cleaning agent.

7. FERMENTED VEGETABLES

7.1 Pickles

Pickles are the fermented food products, which are prepared by soaking and storing vegetables in a brine (salt) or vinegar solution. This process is used for preserving otherwise perishable foods for months and also allows relishing the out of season food.

If the food contains enough moisture, then pickling brine may be produced by addition of dry salt. In sauerkraut and Korean kimchi, vegetables are salted to draw out excess water, then natural fermentation takes place to create a vinegar-like solution containing lactic acid. Other pickles are prepared by adding the vegetable to vinegar. In pickling where fermentation is required, the food may not be completely sterile before it is sealed. The acidity or salinity of the solution, the temperature of fermentation, and the exclusion of oxygen determine which microorganisms will dominate and determine the flavor of the end product (McGee 2004).

7.1.1 Microbes Involved

Pickling in brine often results in anaerobic fermentation, by either lactic acid bacteria or by yeast. When the salt concentration and the temperature are low, *Leuconostoc mesenteroides* dominates, producing a mixture of acids, alcohol, and aroma compounds (http://en.wikipedia.org/wiki/Pickling). When the temperatures are higher, *Lactobacillus plantarum* dominates, which produces primarily lactic acid. Many pickles start with *Leuconostoc*, and change to *Lactobacillus* with higher acidity (McGee 2004).

7.1.2 Different Types of Pickles

Spices or sugar or both are added while pickles are being processed. Vegetables that are generally pickled include the beet, cabbage, cauliflower, cucumber, olive, onion, pepper and tomato. Mixed pickles include piccalilli, chowchow, mustard pickles and chutney. Dill pickles are cucumbers matured in a brine of dill leaves and seed heads. Sweet pickles are prepared from various fruits or vegetables e.g., tomatoes, cucumbers, peaches, or plums in which sugar is added. Different types of pickles used all over the world are given in Table 12.

7.1.3 Uses

Pickles have limited nutritional value and are often used as appetizers and sort of winter substitute for salads. There are hundreds of different recipes

Table 12 Different types of pickles from all over the world

Product	Raw materials	Location
Achar	Mango, lime, green chilli, lotus stem, radish, etc.	India
Khalpi (Pickled Cucumber)	Cucumber	Nepal
Pak-Gard-Dong (Pickled leafy vegetable)	Mustard leaf (*Brassica juncea*)	Thailand
Hum choy	Local leafy vegetable	South of China
Tempoyak (pickled durian)	Durian fruit (*Durio zibethinus*)	Malaysia
Lamoun makbous and Msir	Lemons	West Asia and North Africa
Kimchi (pickled cabbage)	Cabbage	Korea
Naw-mai-dong	Bamboo shoots (*Bambusa glaucescens*)	Thailand
Nukamiso-zuke	Vegetables fermented in rice bran, salt and water	Japan
Hom-dong	Red onions	Thailand
Gundruk	Leafy vegetable	Nepal
Kocho	False banana (*Ensete ventricosum*)	Ethiopia

utilizing locally available fruits and vegetables for pickling. For instance, the book "Pickles of Bangladesh" has recipes for mango sour pickle, sliced mango pickle, sweet olive pickle, hot olive pickle, sweet tamarind pickle, chalta pickle and green chilli pickle (Azami, 1994).

7.2 Sauerkraut

The Sauerkraut production is an effect of natural anaerobic fermentation by bacteria naturally present on cabbage in the presence of 2 to 3% salt. The fermentation yields lactic acid as the major product. Characteristic flavor and texture in sauerkraut is due to formation of lactic acid, along with other minor products as a result of fermentation. At the end of the fermentation period the pH should be around 2.0 and the sauerkraut should contain about 1% lactic acid. The word comes from the German Sauerkraut, which literally translates to sour cabbage. Sauerkraut is a prominent feature of cuisines from most of the cold regions of Europe, and it is eaten in many parts in the USA and Canada as well (http://en.wikipedia.org/wiki/Sauerkraut).

7.2.1 Microbes Involved

Sauerkraut is produced from the fermentation of cabbage by a group of bacteria that produce mostly lactic acid as an end product of the metabolism of the nutrients in the juices of the cabbage. The bacteria involved are *Leuconostoc* and *Lactobacillus* spp. Coliforms like *Klebsiella pneumoniae, K. oxytoca* and *Enterobacter cloacae* are initially involved in fermentation process. As acid is produced, an environment becomes more favorable for *Leuconostoc* and as a result coliform population declines with the increase in population of a strain of *Leuconostoc*. The pH continues to drop with the increased acid production and a strain of *Lactobacillus* succeeds the *Leuconostoc*.

7.2.2 Uses

Raw sauerkraut is an excellent source of Vitamin C, lactobacilli (even more than yogurt), and other nutrients. However, the low pH and overabundance of lactobacilli can easily upset the stomach of people who are not used to eating raw sauerkraut. Sauerkraut is also a source of biogenic amines such as tyramine, which in sensitive people can cause adverse reactions (http://en.wikipedia.org/wiki/Sauerkraut).

8. FERMENTED MEAT PRODUCTS

8.1 Sausages

The meat eating habit of western culture needed food preservation technology in order to keep their perishable meat and milk edible for longer period (Incze 1998). Meat products can be preserved through the process of fermentation. The process of fermentation and drying for meat preservation has been used for hundreds of years. Dry or semi-dry fermented sausages are prepared by mixing ground meat with various combinations of spices, flavorings, salt, sugar, additives and bacterial cultures. The mixtures, in bulk or after stuffing, are allowed to ferment at different temperatures for varying periods of time and, thus, these have a characteristic "tangy" flavor due to the accumulation of lactic acid produced from a microbial fermentation of added sugars (or in some cases by direct addition of encapsulated acids). Following fermentation, the product may be smoked and/or dried under controlled conditions of temperature and relative humidity. These sausages are dried to varying extents during processing. Semi-dry fermented sausages (slight drying) include summer sausage and snack sticks. Dry

Fig. 2 Photographs of some commonly fermented foods (a) Dhokla, (b) Dosa, (c) Idli, (d) Sauerkraut, (e) Jalebi and (f) Sausages

fermented sausage (extended drying) includes pepperoni, hard salami, and Genoa salami. The two factors primarily responsible for the keeping qualities of fermented meat products are low pH and low water activity. With the proper amount of acidification and drying, these sausages can be shelf stable (www.uwex.edu).

8.1.1 Microbes Involved

Microflora and its effect on product quality play a major part in sausages manufacturing. The fermentation process, in particular, results in the desired flavor characteristics and tang. In the traditional production process, fermentation is accomplished by natural flora. In order to achieve safe fermentation of the raw sausage, it is of importance to give the microflora the proper growth conditions as well as the appropriate type of the meat.

The fermentation process involves the growth of lactic acid bacteria in order to acidify the product and produce the desired flavor. The use of commercially prepared starter cultures is recommended to ensure that the correct organisms are added in sufficient numbers. Commercial cultures should be stored according to the culture manufacturer's directions. The inoculum shall be carefully handled and stored to avoid any contamination. The storage temperature for that inoculum shall be maintained at 4 °C or less and a pH of 5.3 or less. Lactobacilli and Pediococci are primarily responsible for converting sugars into lactic acid thereby lowering the pH of the meat product. The pH value of 4.8 or below influences the taste, but does not contribute to better binding properties of the final product. Low pH gives fermented sausages excellent keeping quality. Where nitrate salts are used for curing in slow cured sausages, Micrococci convert nitrate salts to nitrite salts. Bacterial starter cultures used in the production of dry sausages are lactic acid producing and belong mainly to the general *Lactobacillus, Pediococcus* and *Streptococcus*. The use of starter cultures in the field of meat fermentation has been reviewed several times (Hammes and Hertel 1998; Hugas and Monfort 1997; Lucke 2000).

Lactobacilli with or without micrococci are components of starter cultures available for use in slow fermentation (25 °C), whereas pediococci with or without micrococci are used in starter cultures for rapid fermentation at higher temperatures (25 to 37 °C). Pediococci do not occur in fresh meat products in numbers large enough to be a significant factor in traditional slow fermentation and therefore, are only important in meat product fermentation if they are added in starter cultures.

When fermented cured sausages are subjected to an extended drying period, lactobacilli act to significantly reduce the number of undesirable microorganisms including pathogens. Contamination by pathogenic organisms at the outset of the process may have a critical effect on finished product. *Staphylococcus aureus* and the production of its enterotoxin are of significant concern. Once the pH of the meat mixtures reaches 5.3 or lower, the environment for *Staphylococcus aureus* is effectively controlled. During the fermentation and the lowering of the pH to 5.3, it is necessary to limit the time during which the sausage meat is exposed to temperatures higher than 15.6°C.

8.1.2 Operations in Fermented Sausage Production

8.1.2.1 Choice of raw materials

The chilled meat of adult, well-fed animals with a low pH is most appropriate in raw sausage production. The pH of 5.4-5.5 (for beef) and 5.7 – 5.8 (for pork) is desirable. Spices contribute mainly to the development of flavor in dry sausages but it has been shown that they also possess an inhibitory and stimulatory activity, thus influencing the growth of certain bacteria. It has been found that *Lactobacillus plantarun* is the species that is most frequently stimulated by different spices with respect to growth and to acid production.

8.1.2.2 Grinding, chopping and mixing

Different types of sausages have varied methods of grinding and chopping. Finer the degree of grinding and chopping the more complete will be the extraction of protein while the spreading or slicing properties of the finished product will also be improved.

8.1.2.3 Stuffing

After remixing again in the mixer, the meat mix is packed in the stuffer as firmly as possible to exclude air pockets.

8.1.2.4 Treatment prior to smoking

The stuffed sausages are placed in a special place or room to enable the surface moisture to escape. The optimum temperature of this preliminary drying varies from 20 to 23 °C and the relative air humidity should not exceed 75–80 percent. Care has to be taken to dry the sausages properly. It should neither be dried too fast nor retain surface moisture and become slimy.

8.1.2.5 Smoking

Only hardwood sawdust smoke is applied. Uniform distribution of temperature and smoke throughout the smoke house is essential. There is no standardized procedure for smoking different raw sausages.

8.1.2.6 Drying

Drying or ageing is a key operation, especially in dry sausage production. The drying rate for dry sausages should be as low as possible. The most critical point in drying is to avoid the pronounced surface coagulation of proteins and the formation of sausage surface skin.

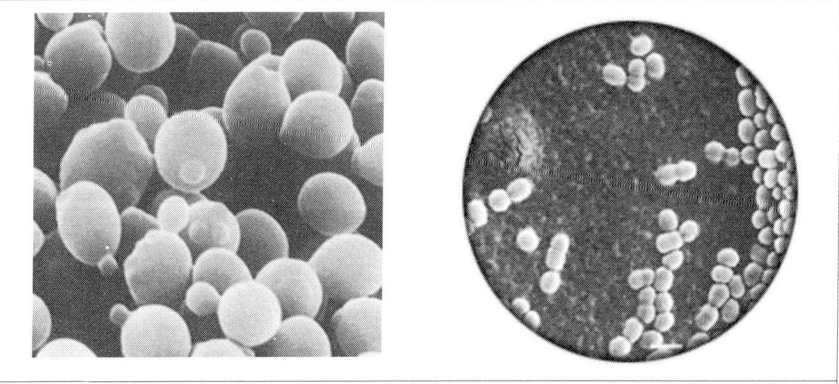

Fig. 3 (A) Yeast (B) *Leuconostoc mesenteroides*

The sausage should dry from the inside outwards. Thus, if the drying rate is adequately slow the sausage casing will enable gradual drying.

CONCLUSION

Fermentation has always been a significant part of our lives as foods can be spoiled by microbial fermentations and at the same time foods can be made by microbial fermentations. Hence, fermentation plays a vital role in our life because it gives us the basic food i.e., bread, yogurt, wine, etc.

Fermenting is a foremost method of food preservation. Not only does fermented food store better than fresh, but also it is safer, often tastier and subsequent cooking time is reduced. Without fermentation the world would be a much hungrier place. Since the invention of pasteurization and fermentation techniques the process have undergone radical changes and many more novel techniques need exploration. This area is the future microbial science.

ACKNOWLEDGEMENTS

The authors acknowledge the support and encouragement received from Dr. Ashok K Chauhan, Founder President, Ritnand Balved Education Foundation (an umbrella organization of Amity Institutions), Mr. Atul Chauhan, Chancellor, Amity University, UP and Mr. Aseem Chauhan, Trustee and Board Member, Ritnand Balved Education Foundation. The technical support received from Mr. Anil Chandra Bahukhandi is highly appreciated.

REFERENCES

Berry (1998). *Kanji,* Food Chain, 23, Intermediate Technology, UK.

Gallus S, Tavani A, Vecchia CL (2004). Pizza and risk of acute myocardial infarction. Eur J Clin Nutr 58:1543-6.

Gopalan C, Ramasastri BV, Balasubramanian SC (1999). Nutritive Value of Indian Foods. 8th Edition. National Institute of Nutrition, Indian Council of Medical Research, Hyderabad.

Hammes WP and Hertel. C (1998). New developments in meat starter culture. Meat Sci. 49:S125-S138.

Hugas M and Monfort JM (1997). Bacterial starter cultures for meat fermentation. Food Chem. 59:547-554.

Incze K. (1998). Dry fermented sausages. Meat Sci. 49:S169-S177.

Kerr TJ and McHale BB (2001). Applications in General Microbiology: A Laboratory Manual. Winston-Salem: Hunter Textbooks.

Lücke FK (2000). Utilization of microbes to process and preserve meat. Meat Sci. 56:105-115.

Mansi EMT, Bryce CFA and Hartley BS (2003). Fermentation Biotechnology: An Historical Perspective In: EL Mansi EMT and Bryce CFA (Eds) Fermentation Microbiology and Biotechnology, Taylor and Francis Group, London.

McGee H (2004). On Food and Cooking: The Science and Lore of the Kitchen. Simon & Schuster, New york.

Paradkar MM., Singhal RS, Kulkarni PR (2002). Quantification of blends of black gram and rice using pentosan as an indicator. 78: 47-51.

Ramakrishnan CV (1979a). Terminal report of American PL 480-project Nr GF-IN-491. Study of Indian Fermented foods from legumes and production of similar fermented foods from U.S. soybean. Biochemistry Department, Baroda University, Baroda, India.

Ramakrishnan CV (1979b). Studies on Indian fermented foods. Baroda J Nut 6:1-57.

Shah FH. (1986). *Fermented Foods of Pakistan,* in Traditional Foods: Some Products and Technologies, Central Food Technological Research Institute, India.

Sharavathy MK, Urooj A, Shashikala P (2001). Nutritionally important starch fractions in cereal based Indian food fractions. Food Chemistry. 75: 241-247.

Steinkraus KH (1996). Handbook of Indigenous Fermented Foods. 2nd Edition Revised and Enlarged. New York, NY: Marcel Dekker.

Emerging Infectious Diseases in Humans

M. S. K. JAYADEV, D. SRIVASTAVA, S. BHATNAGAR and A. K. DUBEY. *

1. INTRODUCTION

A dramatic global resurgence of infectious diseases has occurred during the past 25 years, such that infectious diseases have regained their position as the leading cause of morbidity and mortality worldwide. Movements of people, animals and insect vectors via jet travel have also increased the likelihood that infectious agents can spread swiftly and widely. Microbes that lead to infectious diseases are a major threat to human health. We live in an era of rapidly changing global landscapes and local environments. The microbes which are of major public health threat are rapidly evolving and making use of the so-called advancement to thrive better. There are many old identified microbes, which are re-emerging with more vigor and because of many factors some new and unidentified lethal infectious microbes are coming into limelight day after day. Scarcity of water, overwhelmed or nonexistence of sewage systems, rapid urbanization with global economy expansion, political conflicts and wars are compelling people from poor and developing countries to migrate to developed countries. Thus, infectious diseases affect the poor and developing nations and thereby becomes a source of global burden. A WHO report claims that infectious diseases are responsible for one in every two deaths in developing countries. Table 1 summarizes the leading infectious causes of death worldwide in accordance with

Division of Biotechnology, Netaji Subhash Institute of Technology, Sector 3, Dwarka, New Delhi-110075. Corresponding author. Email: adubey_nsit@yahoo.co.in; akdubey@nsit.ac.in. Phone: 011-25099028; Fax: 011-25099022

Table 1 Ranking of infectious diseases based on mortality rates according to WHO estimates of 2001.

Disease	Rank based on lethality
Respiratory infections	1
HIV/AIDS	2
Diarrheal diseases	3
Tuberculosis	4
Malaria	5

WHO estimates of deaths due to infectious diseases taken in 2001. Infectious diseases range from those occurring in tropical areas (such as malaria and dengue hemorrhagic fever, which are most common in developing countries) to diseases found worldwide (such as hepatitis and sexually transmitted diseases, including AIDS) and foodborne illnesses that affect large numbers of people in both the richer and the poorer nations.

Infectious diseases are illnesses caused to humans by viruses, bacteria, parasites, fungi and other microbes. They may be classified into different classes based on their mode of transmission as contact borne, foodborne, waterborne, vector borne and airborne. Once thought to be eliminated as a public health problem, infectious diseases remain the leading cause of death worldwide. Infectious diseases such as AIDS, Ebola hemorrhagic fever, Nipah virus tuberculosis, SARS, etc., are "emerging diseases". These are either new pathogens that have appeared and quickly spread through populations or already known pathogens that have become resurgent, once again threatening wider and wider populations.

Despite remarkable advances in medical research and treatments during the 20th century, infectious diseases remain among the leading causes of death worldwide for three reasons: (1) emergence of new infectious diseases; (2) re-emergence of old infectious diseases; and (3) persistence of intractable infectious diseases. Emerging diseases include outbreaks of previously unknown diseases or known diseases whose incidence in humans has significantly increased in the past two decades. Re-emerging diseases are known diseases that have reappeared after a significant decline in incidence. Within the past two decades, innovative research and improved diagnostic and detection methods have revealed a number of previously unknown human pathogens. For example, within the last decade, chronic gastric ulcers that were formerly thought to be due to stress or diet irregularities, were found to be the result of infection of Gram-negative bacterium *Helicobacter pylori* (Marshall and Warren 1984; Kelly et al. 1994).

New infectious diseases continue to evolve and "emerge". Changes in human demographics, behavior, land usage etc. are contributing to new disease emergence by changing transmission dynamics to bring people into closer and more frequent contact with pathogens. This may involve exposure to animal or arthropod carriers of diseases. Increasing trade in exotic animals for pets and as food sources has contributed to the rise in opportunity for pathogens to jump from animal reservoirs to humans. For example, close contact with exotic rodents imported to the US as pets was found to be the origin of the recent US outbreak of monkey pox, and use of exotic civet cats for meat in China was found to be the route by which the SARS corona virus made the transition from animal to human hosts (Gandey 2003). Table 2 summarizes the causes and mode of transmission of some emerging infectious diseases.

Table 2 Infectious diseases with their causative organism and mode of transmission.

Disease	Causative organism	Mode of transmission	References
Malaria	*Plasmodium*	Mosquito bite	
Tuberculosis	*Mycobacterium tuberculum*	Through droplets containing bacteria from infected persons.	Koch, 1932
Cholera	*Vibrio cholerae*	Contaminated food and water	Albert, 1994
Dengue hemorrhagic fever	*Dengue virus*	Mosquito bite	Alejandria, 2005
Ebola	*Ebola virus*	Viral particles through droplets from sneeze or cough of infected person	Bowen et al. 1977
SARS	*Coronavirus*	Inhalation or uptake of infected persons droplets released through sneeze or cough.	Rota et al. 2003
Hantavirus cardiopulmonary syndrome	*Sin nombre*	Inhalation of virus contaminated deer mouse excreta	Graziano and Tempest, 2002

This chapter begins by reviewing the causes and factors that influence emergence and re-emergence of infectious diseases and then limelighting the global burden and efforts that are putting into counter some of the most devastating infectious diseases like Cholera, Malaria, Tuberculosis, Viral hemorrhagic fevers, Viral respiratory infections and AIDS. We conclude this chapter with the measures to be taken to overcome the prevalence and spread of these infectious diseases.

2. FACTORS FOR EMERGENCE OF INFECTIOUS DISEASES

Many factors are making it easier for infectious diseases to become an even bigger problem in the future.

- The genetic makeup and reproductive ability of many infectious agents allows them to mutate or evolve into more deadly strains against which humans have little resistance.
- Mass migrations of refugees bring infectious diseases into new areas.
- Global travelers visiting exotic areas bring new diseases home with them.
- Growth of congested urban slums, lacking sanitation and clean water, result in large outbreaks of infections spread by food, water and environmental factors.
- Population shifts and urbanization disturb natural habitats and increase human contact with remote environments and poorly understood ecosystems that hide many unknown and dangerous microorganisms.
- Over time, animal infections can become transmissible to humans (zoo noses).
- The globalization of world commerce brings potential contaminants across our borders daily by way of food, plants, insects, and other products.
- Misuse and overuse have eroded the ability of once-dependable antibiotics to fight common infections. Many microorganisms have gained resistance against most powerful modern drugs. Likewise, disease-carrying insects are becoming resistant to pesticides.
- Human sexual behavior and substance abuse expedite the spread of infectious agents.
- Institutional settings, such as child care centers and hospitals; provide an ideal environment for transmission of infectious diseases because they bring susceptible individuals into close daily contact.
- Ecological changes (agriculture, reforestation and deforestation, famine, global warming, flood, drought).

- Human Demographics and Behavior (population growth and migration, urbanization, war, sexual behavior, injection drugs, use of high-density facilities).
- International Travel and Commerce (air travel, worldwide movement of goods, people, and vectors)
- Technology and Industry (globalization of food supplies, organ transplantation, antibiotic misuse).
- Microbial Adaptation and Change (microbial evolution, response to selection pressures).
- Breakdown in Public Health Infrastructure (reduction in prevention programs including immunization, inadequate sanitation and vector control measures, lack of trained personnel).

Throughout the world, immunosuppression, most notably that associated with HIV infection, has contributed to the recent increase in tuberculosis outbreaks. As the HIV endemic moves through Asia, we can expect to see significant increases in the incidence of tuberculosis in this population as well.

Human behavior continues to be an important part of the infectious disease picture, as illustrated by the HIV pandemic. Urbanization in regions of Africa and Asia has resulted in changes in sexual behavior, leading to accelerated spread of HIV infection. Sexual behavior and use of intravenous drugs continue to be primary modes of HIV transmission, and public health efforts over the last 10 years have shown how difficult such behaviors are to change.

3. IMPORTANT EMERGING INFECTIOUS DISEASES

3.1 Tuberculosis

Tuberculosis (TB) as an infectious disease has almost certainly caused more suffering and death than any other infection. The dreadful disease had been scourging humanity since time immemorial. One-third of the world population is infected with *Mycobacterium tuberculosis* (*M. tuberculosis*), the causative agent of the leading killer disease (Koch 1932). The usual site of infection is lung, but other organs like lymph nodes, intestine, bones, meninges, skin, etc. may also be involved. TB is an ancient disease. Humankind irrespective of age, gender, ethnicity and economic status are at risk from this debilitating, disabling and potentially fatal disease. In March 1882, Robert Koch discovered the causative agent *M. tuberculosis*, the most insidious and morbid pathogenic bacteria (Koch 1932). The lungs are the most common sites for TB; approximately 85% of TB cases are pulmonary. Patients with pulmonary TB usually have a cough, fever and an abnormal chest radiograph. However, TB is a systemic disease and may also occur in the central nervous system, lymph nodes, bones and joints or as disseminated disease (miliary TB). Extrapulmonary TB is more common in persons with HIV infection than in persons without HIV infection (Hopewell 1994).

3.2 The *M. tuberculosis* Complex Bacterium

This acid-fast, non-motile and rod-shaped aerobic bacteria belongs to mycobacteriaceae family, order actinomycetales. TB results from infection by any of the *M. tuberculosis* complex mycobacteria: *M. tuberculosis, M. bovis, M. africanum, M. microti,* and *M. canetti, M. bovis* and *M. africanum* causes a disease clinically indistinguishable from classical TB. *M. microti* causes a disease resembling TB in voles but rarely causes TB in humans (Niemann et al. 2000; Wayne and Kubica 1996; Pattyn et al. 1970).

3.3 Evolution

Mycobacteria are abundant in soil and water so the *M. tuberculosis* complex probably arose as the result of an ecological niche change that culminated in pathogenicity for mammals. It is generally believed that TB was acquired from cattle following the domestication of livestock at the beginning of the Neolithic period when the hunter–gatherer lifestyle was replaced by agriculture. Consequently, it is widely accepted that *M. bovis* was the ancestor of *M. tuberculosis* (Haas and Haas 1996). In seminal works by Kapur et al. 1994; Sreevatsan et al. 1997, who examined the population genetics of the *M. tuberculosis* complex by multi-locus sequence typing and found remarkably high conservation of gene sequences with little evidence for synonymous substitutions. This study concluded that the spread of TB was young in evolutionary terms and even suggested that *M. tuberculosis* emerged as a human pathogen as recently as 10000–15000 years ago, possibly coinciding with the Paleolithic–Neolithic transition (Kapur et al. 1994; Sreevatsan et al. 1997).

3.4 The Genome

Entire genome of drug-susceptible *M. tuberculosis* H37Rv strain was sequenced in 1998 (Cole et al. 1998). The genome comprises 4,411,529 base pairs, contains around 4,000 genes and has a high G + C content (65%). The genome is ridden with insertion elements and repetitive sequences, many new multigene families and duplicated housekeeping genes, but only one rRNA operon. There are 3924 open reading frames, of which approximately 40% encode proteins that have been identified by database comparisons. Large portion of the genome is devoted to code enzymes involved in lipogenesis and lipolysis. About 10% of the coding capacity of the genome is devoted to two large families of acidic, glycine-rich proteins.

3.5 Transmission

Koch demonstrated the aerosol transmission of *M. tuberculosis* to guinea pigs in 1884 (Wells 1955). Modern concepts of TB began with the work of William Wells in the 1930s. Microscopic particle, 1 to 5μ size called droplet nuclei generated by patient with pulmonary TB, consisting of not more than two or three viable bacilli surrounded by a layer of moisture is the central concept of airborne transmission (Wells 1955).

Factors that determine the likelihood of infection are the number, viability and virulence of the organism and how effectively droplet nuclei are produced. Environmental factors affect the concentration of droplet nuclei in the air, ventilation, humidity and exposure to sunlight and UV (Riley et al. 1957, 1989). Host factors include time spent with transmitters, resistance to infection, genetic susceptibility and immune status. All these key determinants in these transitions make the cycle of transmission inefficient, which makes it theoretically possible to consider disease elimination a reasonable objective (Enarson and Rouillon 1993).

Twenty-five to 50% of the household contacts of patients with active pulmonary TB become infected (Grzybowski et al. 1975). In persons not infected with HIV, 5-10% of those with *M. tuberculosis* infection manifest active TB in their lifetime (Rieder 1989; Comstock 1975).

WHO has published recommendations for practical interventions to prevent transmission (Control of tuberculosis transmission in health care settings (WHO 1993; Harries et al. 1997). Recommendations are focused on prompt diagnosis and treatment rather than generally unavailable technologies such as negative pressure isolation rooms, ventilation, UV radiation and personal respirators. The recent reports by WHO demonstrated extensive TB transmission especially MDR-TB in several countries. For known infectious cases, effective treatment remains the most important transmission-control intervention in high prevalence countries.

3.6 Pathogenesis

Much of the understanding about the pathogenesis of TB comes from Lurie and Dannenberg's animal model studies and other workers over the past 7 decades (Dannenberg 1989, 1991; Lurie 1964; Youmans 1979; Bloom et al. 1994). Dannenberg described the four stages of infection. In the first stage, scavenging non-activated alveolar macrophages (Mφ) ingest the bacilli. The bacilli then multiply or are inhibited or destroyed depending on the pathogenesis, virulence and the Mφ innate mycobactericidal activity. Infected Mφ release chemotactic factors such as complement component C5a that attract additional Mφ and circulating monocytes. In the second stage, the number of organisms increases logarithmically. Monocytes continue to migrate to the site of infection and this occurs from day 7 to day 21.

Third stage occurs after 3 weeks and is characterized by the onset of cell-mediated immunity and delayed-type hypersensitivity (Dannenberg 1991). Mφ and Th1 subset (CD4+ and CD8+) demonstrate an increased ability to destroy intracellular bacilli. Patients at this stage or later stages of disease demonstrate Tuberculin Skin Test (TST) positive. In most of the patients, infection is controlled. The stage ends with granuloma formation, often with caseation. Some caseous foci become replaced or encapsulated by fibrous tissue (Ghon's complex). Caseation inhibits the bacillary growth leading the pathogen to dormancy although it is viable, but the immune response does not sterilize the lesion. The bacilli may later reactivate (Smith and Weigeshaus 1989). The biological mechanisms leading to latency in *M. tuberculosis* are poorly understood (Parrish et al. 1998). In the fourth stage, hydrolytic enzymes liquefy and DTH continues to the point of cavity formation and destruction of bronchial structures and the liquefied material is infectious and is easily expectorated (Dannenberg and Sugimoto, 1976).

A single, dominant autosomal gene, the BCG gene, controls resistance to mycobacterial disease in mice. This gene is located on the centromeric portion of chromosome 1 (shows structural homology with chromosome 2 in human) regulates T-cell independent Mφ activation for antimycobacterial activity.

Stead and co-workers (1990) demonstrated the racial difference in susceptibility to infection with *M. tuberculosis* (black residents 13.2% versus white residents 7.2% at Arkansas nursing homes). However, among those infected there were no differences between rates in active TB cases.

3.7 Epidemiology of TB

3.7.1 Global Scenario

According to WHO, global prevalence (TB prevalence is the number of cases of TB per 100,000 people) of *M. tuberculosis* is 32%. By the end of 2003, 202 (96%) of 210 countries and territories reported globally showed 1.1% increase per year in the TB incidence cases and 2.4% increase per year in the number of cases. The total number of TB cases at any given time is 16-20 millions of whom 8-10 million are smear positive and highly infectious. New cases of TB increased from 7.96 million during 1997 to 8.8 million in the year 2002, whereby new smear-positive cases increased from 3.5 million to 3.9 million (WHO 1997). Eighty percent of all incident TB cases were found in 22 countries, with more than half the cases occurring in five South-east Asian countries.

In an estimate, TB has been featured to be the seventh leading specific cause of global Disability-Adjusted Life Years (DALYs) for both sexes in 1990 and it is expected to remain at the same position even by 2020 (Murray and Lopez 1997).

3.7.2 Indian Scenario

Grzybowski estimated that the current TB epidemic in India started probably in the mid 19th century, reached its peak some 50 years later at the beginning of the 20th century and has been naturally declining ever since. He considers India to be a low incidence but high prevalence country. Every year, approximately 2 million persons develop and 0.3-0.5 million die of TB.

The Revised National Tuberculosis Control Programme (RNTCP), based on the DOTS strategy, began as a pilot project in 1993 and gradually expanded to cover a population of 20 million by mid-1998. Rapid RNTCP expansion began in late 1998. By the end of 2000, 30% of the country's population was covered under the RNTCP. The present coverage is 829 million (around 76% of the population). All districts have been urged to prepare for implementation and it is planned that the RNTCP coverage will be extended to the entire country by 2005. Treatment success rates tripled from 25% in the earlier programme to 86%. There are about 14 million (accounting for one-fourth of the world's new TB cases) estimated TB cases, including 3.5 million smear-positive cases in India (www.tbcindia.org).

3.7.3 HIV – The Sinister Partner

TB is the most common opportunistic disease in people infected with HIV. As the HIV suppresses the immune system, HIV infected people are at increased risk of TB. Without HIV, the lifetime risk of developing TB in *M. tuberculosis* infected people is 10%, compared to at least 50% in HIV-infected, *M. tuberculosis* infected people (Girardi et al. 2000). HIV is also the most powerful risk factor for progression from *M. tuberculosis* infection to TB. TB, in turn, accelerates the progression of HIV to AIDS and shortens the survival of patients with HIV infection. In India, there are an estimated 4.58 million HIV-infected persons. It is estimated that 50-60% of HIV-infected persons in India will develop TB (NACO 2003). With an estimated 4.58 million HIV-positive individuals in India, it is likely that HIV will worsen the TB epidemic.

3.7.4 Drug-resistance

Microbiologically, resistance and susceptibility are relative terms, as any mycobacterial strain is resistant to a certain concentration of a given drug. This may not be clinically important, as this concentration may be too low to affect the outcome when treated with optimal dosages of the drug. Moreover, practically all populations of drug-susceptible tubercle bacilli will contain a certain proportion of resistant mutants. Mitchison defined drug-resistance as a decrease in the sensitivity to a drug of a sufficient degree to be reasonably certain that the strains concerned are different from a sample of wild strains of human type that have never come into contact with the drug. The proportions of these drug-resistant mutants are very small in wild strains that are not exposed to drugs.

The international group of specialists meeting held by the WHO in 1969 adopted the definition of drug-resistance of *M. tuberculosis*. A wild strain is defined as a strain of *M. tuberculosis* that has never been exposed to any antimycobacterial agent. Naturally resistant strains are wild strains with species-specific constitutional resistance to a specific drug, such as resistance of *M. bovis* to PZA or pencillin resistance in *M. tuberculosis*. Susceptible strains are those that have not been exposed to drug in a uniform manner. Resistant strains differ from susceptible strains in their capacity to grow in the presence of higher concentration of a drug. The drug-resistance definition was arrived at, by testing a large number of wild strains against three drugs available in 1969. The MIC of these drugs was established in starch-free LJ medium. It was suggested that a strain should be considered resistant if one per cent or more of the bacterial population was resistant to a designated concentration of drug (Canetti et al. 1969).

Drug-resistance in TB may be broadly classified as primary and acquired. Drug-resistance in a patient who has never received anti-TB treatment previously is termed as primary resistance. Acquired resistance is that which occurs as a result of previous treatment. The term initial resistance is used to indicate primary resistance and resistance among patients whose history of previous chemotherapy is not known. The WHO and the IUATLD have replaced the term primary resistance by the term "drug-resistance among new cases" and acquired resistance by the term "drug-resistance among previously treated cases.

4. EPIDEMIOLOGY OF DRUG-RESISTANCE

4.1 Global Scenario

WHO recently reported drug-resistance data from 77 settings, collected between 1999 and 2002. The primary resistance to at least one antiTB drug (any resistance) ranged from 0% in some Western European countries to 57.1% in Kazakhstan. Median prevalence of resistance to specific drugs: SM, 6.3%; INH, 5.9%; RIF, 1.4%; and EMB, 0.8%. Primary MDR-TB prevalence ranged from 0% in eight countries to 14.2% in Kazakhstan and Israel. The highest prevalence of MDR-TB was observed in Tomsk Oblast (Russian Federation; 13.7%), Karakalpakstan (Uzbekistan; 13.2%), Estonia (12.2%), Liaoning Province (China; 10.4%), Lithuania (9.4%), Latvia (9.3%), Henan Province (China; 7.8%), and Ecuador (6.6%).

The median prevalence of acquired resistance to at least one drug (any resistance) was 18.4%, with the highest prevalence 82.1%, in Kazakhstan. Median prevalence of resistance to specific drugs: INH, 14.4%; SM, 11.4%; RIF, 8.7%; and EMB, 3.5%. The median prevalence of MDR was 7.0%. The highest prevalence of acquired MDR-TB was reported in Oman (58.3%) and Kazakhstan (56.4%). Among countries of the former Soviet Union the median prevalence of resistance to the four drugs was 30%, compared with a median of 1.3% in all other settings. A significant increase in the prevalence of any resistance was observed in Botswana. Cuba, Switzerland and the USA showed significant decrease. The prevalence of MDR significantly increased in Estonia, Lithuania, and Tomsk Oblast (Russian Federation). Decreasing trends were significant in Slovakia and the USA.

4.2 Indian Scenario

Drug-resistance studies in India, with the exception of those from the Tuberculosis Research Center (TRC), Chennai and the National Tuberculosis Institute, Bangalore are deficient in varying technical aspects

(Paramasivan 1998). During the 1980s, though the levels of primary resistance to INH and SM were similar to those in the earlier studies from the 1960s, RIF-resistance was observed in all the centers studied, except Gujarat. The reason for this was the introduction of RIF-containing short course chemotherapy (SCC) regimens during this period. However, MDR-TB was <5% in all centres (Paramasivan 1998).

Studies undertaken by the TRC in the late 80s in Tamil Nadu, during 1997-99 and 1999-2002, revealed primary resistance in the range of 1.0-4.4%, with MDR-TB prevalence between 1-3%. Primary resistance in the case of INH ranged from 6.0 - 13.0%, in SM from 1.0-5.8% and in RIF from 0-1.9%. Acquired resistance in the case of INH was ranged between 34.5–67%, SM from 26-26.9% and RIF from 2.8-37.3%. In the case of MDR, primary resistance was found around 3.2% and acquired resistance was found to be equal to or less than 30% (Paramasivan 1998). A study carried out in Jodhpur district of Rajasthan showed there were 3.3% primary MDR-TB cases and 38.2% acquired MDR-TB cases (Mathur et al. 2000). The WHO/IUALTD Global Drug-resistance Survey (1997) has brought out the combined prevalence of MDR-TB in Delhi at 13.3% to be equated to that of the Baltics. The resistance to INH and RIF was 4.2%. In a study from Gujarat, it was observed that in patients with treatment failure and relapse resistance to RIF increased from 3% in 1980 to 37% in 1986, and to INH, from 35 to 56% over the same period (Trivedi and Desai 1988).

4.3 Diagnosis of TB

Control of TB relies on identifying and treating active cases and latent infections in order to interrupt transmission to uninfected hosts. Accurate and rapid diagnosis of active TB cases is a necessity for this strategy to succeed.

4.3.1 Tuberculin Skin Test (TST)

TST detects the presence of *M. tuberculosis* infection and not that of active TB. Robert Koch first prepared tuberculin test material known as old tuberculin (later termed as protein purified derivative [PPD]), from cultures of *M. tuberculosis* (Koch 1890). Later Mantoux introduced the intradermal technique. It contains substances belonging to the bacillus (antigens) to which an infected person has been sensitized. 5 tuberculin units, which equal 0.0001 mg of PPD, are intracutaneously injected in the forearm. The test is read within 48-72 hours. The lesion is characterized by erythema (redness) and swelling and induration (raised and hard). People who have a lesion of 15 mm or greater are currently infected with *M. tuberculosis*.

Specificity of TST is very low and sensitivity is only 70-75%. The test serves as an epidemiological tool (Diel et al. 2004).

4.3.2 Chest Radiography

On a chest X-ray, evidence of the disease appears as numerous white, irregular areas against a dark background, or as enlarged lymph nodes. The upper parts of the lungs are most often affected. To verify the test results, the clinician obtains a sample of sputum for smear microscopy and culture.

4.3.3 Smear Microscopy

Unlike many other types of bacteria, mycobacteria can retain certain dyes even when exposed to acid. This so-called acid-fast property is characteristic of the tubercle bacillus and was identified by Ehrlich in 1882. The present method of acid fast staining was developed by Ziehl, in 1992, and subsequently modified by Neelsen and hence the name Ziehl-Neelsen (ZN) staining (Bishop and Neuman 1970; Smithwick 1976). In developing countries, ZN-staining on smear microscopy forms the basis of the primary diagnosis of TB. Smear microscopy is inexpensive, rapid and easy to perform but suffers from low sensitivity (10^4 bacilli/ml). Other staining technique includes Auramine O fluorescence (Smithwick 1976).

4.3.4 Culture

Only culturing clinical specimens and testing the isolates further after preliminary identification can obtain a definitive diagnosis of TB. Traditionally two types of solid media are used for culture: egg-based media e.g., Lowenstein Jensen, Ogawa and American Trudeau Society and agar based media e.g. Middlebrook 7H10 and 7H11 (Middlebrook et al. 1960; Cohn et al. 1968). Cultures are incubated at 37 °C in an atmosphere of 5 to 10% CO_2. All cultures should be examined weekly for 8 weeks. The major advantage of culture on solid media is that it allows visualization of colony morphology and pigmentation, which is useful diagnostically for distinguishing colonies of *M. tuberculosis* from those of some non-tuberculous mycobacteria.

The recently developed MODS is a simple adaptation of the microagar 7H11, which uses microscopic observation to detect *M. tuberculosis* (Caviedes et al. 2000). Results obtained by Cavidedes et al. (2000) showed the advantage of using this liquid medium. Both the liquid culture methods, MODS and MGIT were performed. MODS was at least 14% more sensitive and 4 to 5 days faster than either the microagar 7H11 or LJ culture.

4.3.5 Molecular Diagnostic Tools

Several new methods based on molecular biology techniques have been developed and are increasingly being used in diagnostic TB bacteriology. The PCR, which is based on DNA amplification methods, has been proposed for rapid detection of mycobacterial DNA in clinical specimens as a replacement for culture and identification of the *M. tuberculosis* complex (Kirschner et al. 1993). This technique has been extensively tested because it has the potential to shorten the time required for diagnosis from 6 weeks to 24 hours. So far it has not lived up to expectations, because in the present format it is not as sensitive or as specific as was originally thought (Noordhok et al. 1994; Grosset and Mouton 1995).

Nucleic acid probes with non-radioactive detection systems have gained increased acceptance in the clinical TB laboratory as a replacement for fastidious, time-consuming mycobacterial identification tests (Eisenach et al. 1988). These probes can be used to identify isolates growing on conventional or radiometric media and are available for the detection of the *M. tuberculosis* complex. In this regard, they compare favorably with the BACTEC (Bergmann and Woods 1999).

New genotypic approaches for detection of drug-resistance based on the sequencing of DNA have either been developed or under development following the recent identification of mutations in the drug target genes (Caws and Drobniewski 2001; Fluit et al. 2001; Van Rie et al. 2001; Ramaswamy and Musser 1998). Current genetic tests typically detect only 40% to 95% of mutations, depending on the drug (Garcia et al. 2001; Hajj et al. 2001; Bobadilla et al. 2001; Mokrousov et al. 2004).

4.3.6 Antituberculosis Treatment

The chemotherapy era for TB began with the introduction of SM in 1964. ATT aims at (i) achieving very early sputum conversion; (ii) improving cure rates and (iii) preventing the selection of drug resistant mutants (Grosset 1990). The most notable standardized regimens were developed by the British Medical Research Council in East Africa, India, Hong Kong and Singapore.

TB is currently being treated with an initial intensive 2-month regimen comprising of multiple drugs, RIF, INH, EMB and PZA in order to ensure that mutants resistant to a single drug do not emerge. In the next four months, only RIF and INH are administered to eliminate any persisting tubercle bacilli (WHO 2003). This forms the basis of Short Course Chemotherapy (SCC).

To improve cure rates and preventing emergence of drug-resistance TB, the WHO has actively promoted the DOTS strategy (Maher 1997). Since its introduction in 1991, more than 13 million patients have received treatment under the DOTS strategy. By the end of 2002, all 22 high-burden countries had adopted the DOTS strategy. In total, 180 countries were implementing the DOTS strategy, and 69% of the global population was living in parts of countries where the DOTS strategy was in place (www.who.int/gtb/). In India alone, 740 million people (almost 70% of the total population) were living in parts of the country where the strategy had been implemented by August 2003 (www.tbcindia.org).

4.3.7 Basis for Chemotherapy of TB (M. tuberculosis Populations in TB Lesions (Theoretical Model for TB Infection)

According to theoretical model for TB infection, four different populations of *M. tuberculosis* are known to exist in the lesions of human TB (Grosset 1980). They are: intracellular actively growing tubercle bacilli (10^7–10^9 organisms; acted upon by INH, RIF and SM), extracellular bacteria that multiply slowly and intermittently (acted upon by INH and RIF), slow growing intracellular tubercle bacilli (10^5 organisms; acted upon by PZA, RIF and INH), and dormant bacilli ($<10^5$ organisms), which do not multiply at all against which no anti-TB agent works (Mitchison 1992).

4.3.8 Theoretical Basis for ATT

On the basis of animal experiments and clinical trials Mitchison, (1985) classified and graded antiTB agents into three categories: (i) *prevention of drug-resistance* is the ability of the drug to suppress the growth of tubercle bacilli, to prevent the emergence of mutants resistant to another drug, (ii) *early bactericidal activity* is the ability of a drug to reduce number of bacilli during the initial part of therapy (Jindani et al. 1980), (iii) *sterilizing activity* is the ability of a drug to kill semi-dormant bacilli (Grosset 1980).

4.3.9 Antimicrobial Resistance in M. tuberculosis

Drug-resistant organisms result from random mutations in drug target genes, which occur spontaneously in wild-type strains of *M. tuberculosis.* These mutations occur at different rates (probability of a mutation) for the anti-TB agents. It is 10^{-3} for less effective anti-TB drugs such as thioacetazone, EMB, capreomycin, cycloserine; 10^{-7} for EMB; 10^{-8} for INH and SM; and 10^{-9} for RIF and cycloserine. Mutants with resistance to two drugs occur at a rate determined by multiplying the rates for the individual drugs. Thus, in a population of tubercle bacilli the ratio of resistant to susceptible bacilli would be about $1:10^8$ for INH, $1:10^9$ for RIF and $1:10^{14}$ for both INH and RIF. Cavities in pulmonary TB contain about 10^7-10^9 bacilli, likely to contain a few mutants resistant to one anti-TB drug (Grange 1990). These naturally occurring mutants are selected during inadequate or inappropriate therapy. If TB is treated with a single drug, the strains sensitive to that drug are killed and drug-resistant mutants continue to multiply and become dominant. In such a large population of resistant bacilli, further mutants may occur, and the patient may eventually have bacilli resistant to two or more drugs. Thus, the MDR phenotype is caused by sequential accumulation of mutations in different genes involved in individual drug-resistance due to inappropriate treatment or poor adherence to treatment. Patients with drug-resistant strains constitute a pool of chronic infections, which propagate primary resistance.

5. CATEGORIES OF ANTI-TB DRUGS

Based on the drug target, resistance mechanisms to these drugs can be broadly classified into:

(I) Resistance to inhibitors of nucleic acid synthesis – RIF, FQs.
(II) Resistance to inhibitors of protein synthesis – SM.
(III) Resistance to inhibitors of cell wall synthesis – INH, EMB, ETH.
(IV) Resistance to other anti-TB drugs – PZA.

5.1 Drug Susceptibility Testing (DST)

5.1.1 Importance of DST

The determination of drug susceptibility of clinical isolates is crucial, as this serves three purposes. Firstly, it confirms the efficacy of an empirical treatment regimen in a given patient, or permits optimization of treatment in the case of drug-resistance. Secondly, it enables medical and public health officials to plan appropriate empirical regimens to address the problems of emerging drug-resistance at various settings. Thirdly, it serves as performance indicator of the efficacy of control programmes.

5.1.2 RIF-resistance as Surrogate Marker for MDR-TB

Analysing the differing roles of INH and RIF in short-course treatment, Ellard and Mitchison concluded that failures during treatment are much higher among patients with initially INH-resistant organisms, but relapses after termination of chemotherapy are unrelated to initial INH sensitivity.

Although the importance of detecting initial INH monoresistance for two or more drugs, especially RIF, on the outcome of therapy is beyond any doubt, especially in this MDR-TB era.

According to the Fastrack concept employed in UK, RIF drug susceptibility is done immediately following collection of specimens, as it is predictive that around 90% of RIF-resistant isolates are also INH-resistant (Vareldzis et al. 1994; Drobniewski and Wilson 1998). Moreover, the possibility of targeting a single region of the genome for diagnostic and drug susceptibility purposes, rather than, multiple genes as is the case with INH and other drugs, makes RIF a surrogate marker for MDR-TB. It was further confirmed in recent studies conducted in Estonia, Ethiopia and Latvia, where, of the total number of RIF-resistant strains, 100%, 100% and 96% respectively of the strains were MDR-TB (WHO 1997).

5.1.3 Conventional DST

Conventional DST methods require that pathogens are first isolated from human specimens by culture methods. In separate assays, isolated microorganisms are then exposed to various concentrations of antimicrobial agents under specified growth conditions, and the ability of these antimicrobials to inhibit growth is determined. The conventional DST of mycobacteria differs considerably from the testing of other organisms.

The conventional DST for *M. tuberculosis* strains can be performed as direct or indirect tests in liquid media, such as Middlebrook 7H9, or solid media, such as Lowenstein-Jensen slopes or Middlebrook 7H10 agar. In the indirect test a set of drug-containing and drug-free media inoculated directly with concentrated smear-positive specimen. The specimen may be diluted according to the number of AFB seen in the smear. The advantage of the direct method is that the results are available sooner; better represent the true bacterial population and it may be difficult to ensure a uniform distribution of bacilli in the inoculum. In the indirect test, primary culture serves as inoculum. Though preparation of a uniform inoculum is easier, diluting entire growth of the susceptible can avoid selection of predominant population of resistant or susceptible organisms.

There are three widely employed conventional DST methods. They are (i) the proportion method, (ii) the absolute concentration method or the MIC, (iii) the resistance ratio method.

The proportion method takes a defined proportion of the inoculum as the end point, while the MIC takes no growth as the end point. The RR method considers the ratio between the MIC of the test strain and the MIC of a reference strain (Canetti 1969).

In proportion method the ratio between the numbers of colonies growing on drug-free medium indicates the proportion of drug-resistant bacilli present in the bacterial population. High and low dilutions of the inoculum (several 10-fold dilutions) are plated on the media so that countable colonies can be obtained with at least one of the dilutions.

In the MIC method, media containing several serial two-fold dilutions of each drug are used, as resistance is indicated by the lowest concentration of the drug that will inhibit growth i.e., MIC (Canetti 1969).

The RR method is the variant of the MIC. The RR is defined as the MIC of the test strain divided by the MIC of the standard susceptible strain H37Rv in same set of tests. The RR of 2 or less defines susceptibility, that 8 or more attests resistance.

5.1.4 Rapid Methods for the Detection of Drug-resistance

Conventional methods that rely on mycobacterial growth for the assessment of drug activity are at major disadvantage, both in the time-scale and particularly if the compound under investigation is unstable on prolonged incubation. Thus, rapidity of the newer tests is an important and essential consideration. The following are the main requirements for a new, rapid test:

- High intra- and inter-laboratory reproducibility in differentiating resistant from susceptible clinical isolates, which is usually measured by the rates of discrepancies in repeated tests, and the rates of false-resistant and false-susceptible reports.
- The shortest possible turnaround time between the receipt of the specimen and the availability of result.
- Applicability for implementation in the clinical laboratory environment, with minimal training of laboratory technicians.
- Applicability of simple bio-safety measures.
- Minimal investment in special technology beyond the standard laboratory equipments.
- Minimal labour time and the ability to perform tests simultaneously with a large number of specimens.
- Minimal cost of additional supplies.

The major challenge for a new drug-resistance detection assay would lie in minimizing the cost and complexity. Although diverse in their scientific underpinnings, new methods can be divided into two major categories, viz., Phenotypic and Genotypic.

Phenotypic methods measure physiological responses of the pathogens to antibiotic challenge. They detect bacterial metabolic activities that are inhibited by the antimicrobial agents in drug-susceptible but not in drug-resistant bacteria. However, these methods are time-consuming and they require additional testing to differentiate *M. tuberculosis* from other pathogens. Some of these methods still need pre-grown pure and viable cultures but the effects of the drugs in subsequent culture are measured by means other than conventional growth inhibition. These methods detect drug-resistance regardless of its genetic basis, and the newer phenotypic technologies seek to employ novel (usually molecular) indicators that provide information on bacterial physiology more rapidly than do existing indicators. The best approaches will have some capacity for species-specificity, so that they can be applied directly to raw samples without bacteriological purification of the pathogens (direct testing).

Genotypic approaches detect the genetic basis of resistance rather than the resistance phenotype. Most of these approaches have two basic steps. First, a nucleic acid amplification method (PCR) is used to amplify segments of the *M. tuberculosis* genome known to be altered in resistant strains. Ideally, this step bypasses the need for time-consuming cultivation of the pathogen. Then in the second step, amplification products are analyzed for specific mutations correlated with resistance. As with phenotypic approaches, genotypic methods should ideally have some capacity for species-specificity (Drobniewski and Wilson 1998).

6. MALARIA

Malaria remains a devastating global health problem. Worldwide, an estimated 300–500 million people contract malaria each year, resulting in 1.5–2.7 million deaths annually. Malaria is one of the oldest known diseases. Romans associated the disease with foul air of the marshes and swamps, they called it malaria (mal= bad; aria= air). Obligate intraerythrocytic protozoa of the genus Plasmodium cause malaria. Humans can be infected with one (or more) of the following four species: *P. falciparum*, *P. vivax*, *P. ovale*, and *P. malariae*. Plasmodia are primarily transmitted by the bite of an infected female *Anopheles* mosquito, but infections can also occur through exposure to infected blood products (transfusion malaria) and by congenital transmission. Because of the increase in global travel to and immigration of people from areas endemic for malaria, the incidence of imported cases of malaria in developed countries has risen. Approximately 10,000–30,000 travelers from industrialized countries are expected to contract malaria each year (Kain and Keystone 1998). In addition, drug-resistant *Plasmodium falciparum* malaria continues to spread and at present involves almost all areas of the world. An increasing number of travelers are exposed to drug-resistant plasmodia.

In industrialized countries, most cases of malaria occur among travelers, immigrants, or military personnel returning from areas endemic for malaria (imported malaria). Exceptionally, local transmission through mosquitoes occurs (indigenous malaria). Female *Anopheles* mosquito is the vector host for the causative organism of this deadly infectious disease. Infection can also be transmitted by blood transfusion from infected persons or by congenital transmission. Malaria should be included in differential diagnosis for patients with unexplained fever or clinical deterioration who went to endemic areas in the past few years. Appropriate diagnosis and treatment in the right time period are must in malaria to control morbidity and mortality rates (Kain et al. 1998). After the failure of global eradication program of the 1960s and the immense burden it is imposing on poor countries in the tropical region, malaria had regained its importance as an infectious disease.

6.1 Life Cycle of Plasmodium

When the infected anopheline mosquito takes a blood meal, sporozoites are inoculated into the bloodstream. Within an hour, sporozoites enter hepatocytes and begin to divide into exoerythrocytic merozoites (tissue schizogony). For *P. vivax* and *P. ovale*, dormant forms called hypnozoites typically remain quiescent in the liver until a later time; *P. falciparum* does not produce hypnozoites. Once merozoites leave the liver, they invade erythrocytes and develop into early trophozoites, which are ring shaped, vacuolated and uninucleated. Once the parasite begins to divide, the trophozoites are called schizonts, consisting of many daughter merozoites (blood schizogony). Within the liver cells, the trophozoites start their intracellular asexual division. At the completion of this phase, thousands of extra erythrocytic merozoites are released from each liver cell. The time taken for the completion of the tissue phase is variable, depending on the infecting species; (8 - 25 days for *P. falciparum*, 8 - 27 days for *P. vivax*, 9 - 17 days for *P. ovale*, 15 - 30 days for *P. malariae*) and this interval is called as *pre-patent period*.

Eventually, the infected erythrocytes are lysed by the merozoites, which subsequently invade other erythrocytes, starting a new cycle of schizogony.

This repetitive cycle of invasion - multiplication - release - invasion continues. The intra erythrocytic cycle takes about 48 hours in *P. vivax, P. ovale* and *P. falciparum* infections and 72 hours in case of *P. malariae* infection. In non-immune humans, the infection is amplified about 20-fold each cycle. After several cycles, some of the merozoites develop into gametocytes, the sexual stage of malaria, which cause no symptoms, but are infective for mosquitoes.

6.2 Clinical Symptoms and Diagnosis of Malaria

In malaria the symptoms are due to rupture and destruction of the erythrocytes by the schizont stage of plasmodium in its life cycle. It may lead to delay in diagnosis due to non-specific symptoms and presenting itself to common viral infections. The majority of patients experience fever (>92% of cases), chills (79%), headaches (70%), and diaphoresis (64%). Other common symptoms include dizziness, malaise, myalgia, abdominal pain, nausea, vomiting, mild diarrhea, and dry cough. Physical signs include fever, tachycardia, jaundice, pallor, orthostatic hypotension, hepatomegaly, and splenomegaly. Clinical examination in non-immune persons may be completely unremarkable, even without fever. *P. falciparum* is the major cause for almost all forms of severe malaria and its associated deaths, but often *P. vivax* or *P. ovale* produce serious complications and even lead to death of the patient. In the year 2000 WHO has revised the criteria which they have set for future clinical and epidemiological studies to include other clinical manifestations and laboratory records which help in poor prognosis of the disease. The major complications of severe malaria include cerebral malaria, pulmonary edema, acute renal failure, severe anemia, and/or bleeding. Acidosis and hypoglycemia are the most common metabolic complications. These complications can develop individually or can develop together rapidly and progress to death within hours or days. Laboratory findings like thrombocytopenia the most common laboratory abnormality, followed by hyperbilirubinemia, anemia, and elevated hepatic aminotransferase levels (D'Acremont et al. 2002). The leukocyte count is usually normal or low, but neutrophilia with a marked increase in band forms (left shift) is present in the majority of cases. The erythrocyte sedimentation rate, C-reactive protein, and procalcitonin are almost invariably elevated. The severity of malaria corresponds to the degree of the laboratory abnormalities. In one study of travelers who returned from the tropics, thrombocytopenia and hyperbilirubinemia had a positive predictive value of 95% for malaria (Doherty et al. 1995). Other complications in severe malaria cases include neurological complications, pulmonary complications, renal complications, hypoglycemia, hypotension, shock and hematological abnormalities.

6.3 Drug Resistance in Malaria

Resistance has emerged to all classes of anti-malarial drugs except the artemisinins and is responsible for a recent increase in malaria-related mortality, particularly in Africa. The de novo emergence of resistance can be prevented by the use of anti-malarial drug combinations. Artemisinin-derivative combinations are particularly effective, since they act rapidly and are well tolerated and highly effective. Widespread use of these drugs could roll back malaria. The theory underlying combination drug treatment of tuberculosis, leprosy, and HIV infection is well known and is now generally accepted for malaria (Peters 1969; Peters 1990; Chawira et al. 1987; Curtis and Otoo 1986). If two drugs are used with different modes of action, and therefore different resistance mechanisms, then the per parasite probability of developing resistance to both drugs is the product of their individual per parasite probabilities. The cause of chloroquine resistance in *P. vivax* has not been found. Resistance to mefloquine and other structurally related arylaminoalcohols in *P. falciparum* results from amplifications (i.e., duplications, not mutations) in *Pfmdr*, which encodes an energy-demanding p-glycoprotein pump (Pgh) (Cowman et al. 1994; Reed et al. 2000; Trigilia et al. 1991). Resistance to antifols *P. falciparum* and *P. vivax* is due to sequential acquisition of mutations in *dhfr* (plowe 2003). Resistance to atovaquone results from point mutations in the gene cytB, coding for cytochrome b.

7. EMERGING VIRAL INFECTIOUS DISEASES

Viruses with RNA as their genetic material can quickly adapt to and exploit the varying conditions because of the high error rate of viral polymerase enzymes that replicate their genome. It comes as no surprise that several prominent recent examples of emerging and re-emerging diseases are caused by RNA viruses. This section of the chapter reviews some prominent emerging infectious diseases caused by RNA viruses like HIV, Ebola virus, Dengue virus, SARS virus, etc.

7.1 AIDS

Acquired Immuno Deficiency Syndrome (AIDS) caused by Human Immuno-deficiency Virus (HIV) ranks as one of the most important infectious diseases facing humankind in the 21st century. Over the past two decades, the HIV pandemic has entered our existence as an incomprehensible calamity and has laid claim to millions of lives, causing fear, pain and grief and threatening economic devastation. The impact of the epidemic on the economy is already being felt in most countries for life expectancy has been reduced significantly as many people in the 15-49 year age group are now dying of AIDS. HIV/AIDS has stricken on different regions of the world at different levels but the Sub-Saharan Africa is the worst affected region hosting 70% of people living with HIV/AIDS in the world. India stands second in the world after South Africa with high number of HIV infected individuals. HIV has been spreading around the world infecting millions of people. Already, more than twenty million people around the world have died of AIDS-related diseases. In 2004, 3.1 million men, women and children have died. Around twice the amount that have died until now - almost 40 million - are now living with HIV, and most of these are likely to die over the next decade or so. The most recent UNAIDS/WHO estimates show that, in 2004 alone, 4.9 million people were newly infected with HIV.

AIDS was first recognized in the United States in the summer of 1981, when the U.S. Centers for Disease Control and Prevention (CDC) reported the unexplained occurrence of *Pneumocystis carinii* pneumonia in five previously healthy homosexual men in Los Angeles and of Kaposi's sarcoma (KS) in 26 previously healthy homosexual men in New York and Los Angeles. In 1983, Human Immunodeficiency Virus (HIV), a lentivirus, which belongs to the family of human retroviruses (Retroviridae) and the subfamily lentivirinae was isolated from a patient with lymphadenopathy, and by 1984 it was demonstrated clearly to be the causative agent of AIDS. There are two types of HIV: HIV-1 and HIV-2. Worldwide, the predominant virus is HIV-1 and relatively uncommon HIV-2 type is concentrated in West Africa and is rarely found elsewhere. HIV-1 has been divided into three genetic groups, based on phylogenetic analysis (Simon et al. 1998). Retroviruses constitute a large and diverse family of enveloped RNA viruses that use as a replication strategy the transcription of virion RNA into linear double-stranded DNA with subsequent integration into the host genome. The characteristic enzyme used for this process, an RNA-dependent DNA polymerase that reverses the flow of genetic information, known as Reverse Transcriptase (RT). The unique lifestyle of the retrovirus involves two forms, a DNA provirus and an RNA-containing infectious virion. The basic structure, genetic organization, and life cycle of HIV-1 are similar to those of most retroviruses, but with some additional features.

As RNA viruses, retroviruses have the survival advantage of great genetic diversity. As viruses with a DNA intermediate in their replication cycle, they also have the advantage of latency, as do many DNA viruses, but even more so because the DNA provirus is integrated into the chromosomal DNA. As a CD4+ T-cell and macrophage-tropic virus, HIV has the advantage of reducing the effectiveness of host immune attacks.

Retroviruses are typically 100 nm in diameter and contain two single strands of RNA (diploid genome), which permits recombination between the strands. The typical genome is ~9.7 kilobases (Kb) in size and contains three major structural genes, namely, *gag, pol* and *env*. HIV-1 also contains several additional genes that are essential to viral replication. These complicated genomes are characteristic of human retroviruses. The four known human retroviruses- HIV-1, HIV-2, HTLV-I and HTLV-II and related animal viruses belong to two different groups.

7.2 HIV Genomic Structure

The HIV-1 proviral DNA is ~9.7 kbp in length that gets integrated into the host cell and follows the basic genomic structure common to most retroviruses: *gag-pol-env* genes flanked by two complete viral Long Terminal Repeats (LTRs). These LTRs contain transcriptional regulatory sequences, RNA processing signals, packaging sites and the integration sites. The 5′ end begins with the *gag* gene, which encodes core and matrix proteins; the *pol* gene, which begins in an overlapping frame encoding viral protease (Pr) (Erickson et al. 1999; Miller et al. 2001), Reverse Transcriptase (RT) (Huang et al. 1998; Kohlstaedt et al. 1992), and integrase (IN) (Bushman et al. 1990); and then the *env* gene, which encodes the outer and transmembrane envelope proteins (Capon and Ward 1991), gp120 and gp 41, respectively. One group of accessory genes is located between the *pol* and *env* sequences and a second group at the 3′ end of the virus. The different genes are usually represented on three levels corresponding to the three alternate reading frames of the nucleic acid. The corresponding proteins which are encoded by three structural genes *gag* (Bryant et al. 1990), *pol, env*, two regulatory genes *tat* (Ruben et al. 1989) and *rev* (Zapp and Green 1989), four accessory genes *vpu, vpr, vif*, and *nef* are divided into three classes, the major structural proteins (Gag, Pol, Env); the regulatory proteins (Tat and Rev) and the accessory proteins (Vpu, Vpr, Vif, and Nef), respectively. The genomes of HIV-2, SIVsm and SIVmac contain a new gene called vpx.

7.3 HIV Transmission

Sexual transmission of HIV is primarily a major route for worldwide expansion of the AIDS pandemic. Sexual route of transmission was first suggested by the appearance of AIDS (Padian et al. 1987) in homosexual men in New York, San Francisco and Los Angeles (Goedert et al. 1985). In Africa, the recognition of approximately equal number of AIDS cases among men and women pointed to heterosexual mode of transmission. HIV transmission can also take place by contaminated blood or blood products (Curran et al. 1984) or by the use of contaminated needles or surgical instruments. Vertical transmission of HIV is another major route of HIV transmission (Pizzo and Butler 1991). It is estimated that approximately 60% of vertical infections are due to perinatal and only 8-10% result from prenatal in utero infections (Palasanthiran et al. 1993).

7.4 Implications of Molecular Heterogeneity of HIV-1

For the detection of divergent strains subtype specific serological techniques such as synthetic peptide enzyme immunoassays (EIAs) were developed for accurate diagnosis and improved sensitivity and specificity.

Information on the presence of HIV multiple infections caused by distinct viral subtypes circulating within a population, and the long-term observation of patients dually infected with distinct HIV strains may contribute to a better understanding of pathogenesis of HIV infections. To study the magnitude of HIV-1 epidemic in India, more extensive molecular epidemiological analysis will be required from high-risk population from different geographical regions of the country, to substantiate the breadth of subtype representation in individual regions.

7.5 Antiretroviral Drugs and Antiretroviral Therapy (ART)

HIV prevention campaigns have proven widely successful at reducing the unsafe behavior and lowering infection rate (Coates and Collins 1998). Treatments using powerful antiretroviral drugs have greatly minimized the mortality rate of HIV infected individuals all over the world. Treatment goals mainly focus on maximal and durable suppression of viral load, restoration and preservation of immunologic function, improvement of quality of life and reduction of HIV-related morbidity and mortality. There has been reduction in number of new AIDS cases in the developed countries with the advent of Highly Active Anti-Retroviral Therapy (HAART). HAART, the name given to treatment regimens is to aggressively suppress the viral replication and progression of HIV disease. The usual HAART regimen combines three or more different antiretroviral drugs. These treatment regimens have been shown to reduce the amount of viral RNA to almost undetectable levels in patient's plasma.

Antiretroviral drugs that are currently used for the treatment of HIV infections belong to the following four classes: (i) Nucleoside/Nucleotide Reverse Transcriptase Inhibitors (NRTIs) are faulty versions of nucleic acid. Reproduction of the virus is stalled when HIV uses an NRTI instead of a normal nucleotide. The NRTIs are Zidovudine (AZT), Didanosine (ddI), Zalcitabine (ddC), Stavudine (d4T), Lamivudine (3TC), Abacavir (ABC), Emtricitabine [(-) FTC], Tenofovir (PMPA) disoproxil fumarate; (ii) Non-Nucleoside Reverse Transcriptase Inhibitors (NNRTIs) which bind to and disable reverse transcriptase enzyme. The NNRTIs are Nevirapine (NVP), Delavirdine (DLV), Efavirenz (EFV), (iii) Protease Inhibitors (PIs) block the protease enzyme. The PIs are Saquinavir, Ritonavir, Indinavir, Nelfinavir, Amprenavir, Atazanavir, Fosamprenavir, Lopinavir and (iv) Fusion Inhibitor (FIs), Enfuvirtide (Fuzeon, T-20) prevents HIV entry into cells. Since the introduction of protease inhibitors (PIs) three types of combination regimens may be employed as initial therapy. These include: (1) NNRTI-based regimens that are PI and FI-sparing, (2) PI-based regimens that are NNRTI and FI-sparing, and (3) Triple NRTI regimens that are PI, NNRTI, and FI.

In addition various other events in the HIV replicative cycle are potential targets for chemotherapeutic intervention: (i) Viral adsorption, through binding to the viral envelope glycoprotein gp120 (polysulfates, polysulfonates, polyoxometalates, zintevir, negatively charged albumins, cosalane analogues); (ii) Viral entry, through blockage of the viral coreceptors CXCR4 and CCR5 [bicyclams (i.e., AMD3100), polyphemusins (T22), TAK-779, MIP-1 alpha LD78 beta isoform]; (iii) Virus-cell fusion, through binding to the viral glycoprotein gp41 [T-20 (DP-178), T-1249 (DP-107), siamycins, betulinic acid derivatives]; (iv) viral assembly and disassembly, through NCp7 zinc finger-targeted agents [2,2'-dithiobisbenzamides (DIBAs), azadicarbonamide (ADA) and NCp7 peptide mimics]; (v) Proviral DNA integration, through integrase inhibitors such as L-chicoric acid and diketo acids (i.e., L-731,988); (vi) Viral mRNA transcription, through inhibitors of the transcription (transactivation) process (fluoroquinolone K-12, Streptomyces product EM2487, temacrazine, CGP64222). Also, in recent years new NRTIs, NNRTIs and PIs have been developed that possess, respectively, improved metabolic characteristics.

Although, a multitude of anti-HIV agents are being pursued actively, it has not been possible to eradicate HIV completely in an infected individual. Over a period of time-expansion of resistant variants takes place and these viral populations overtake the immune system leading eventually to AIDS (Mbewu et al. 2001). The use of existing therapies is limited in the developing world, where more than two-thirds of the total HIV infection prevails, owing to their high cost. Apart from the high cost and emergence of resistant mutants, another limiting factor is low patient compliance owing to the cumbersome drug regimens and side effects.

Based on the pathogenesis of HIV infection, therapy can be initiated as follows:

(1) *In asymptomatic patients*: CD4 count < 200 cells/mm^3- Treatment recommended (Cameron et al. 1998; Hammer et al. 1997) CD4 count 200-300 cells/mm^3- monitor CD4 count; If count decreases to 20-80 cells/mm^3- commence therapy; CD4 > 350 cells /mm^3 - Defer treatment.

(2) *In symptomatic patient* (excluding TB): Presence of HIV-related symptoms, current or previous HIV-associated disease (includes AIDS-defining illness, unexplained weight loss >10%, unexplained diarrhoea > 1 month, oral candidiasis or oral hairy leukopakia)- Treatment recommended.

(3) *Patient with TB*: CD4 count < 200 cells/mm^3- Delay treatment until after 2 month intensive phase of TB therapy, unless the patient has other serious HIV-related illness or a CD4 count of < 50 cells/mm^3, ART should be introduced as soon as the patient is stabilized on TB therapy. CD4 count > 200 cells/mm^3 (and no other HIV related symptoms) assess the need for ART after completing TB therapy, using CD4 and clinical criteria into consideration.

Potential benefits of early therapy include: (1) Earlier suppression of viral replication; (2) Preservation of immune function; (3) Prolongation of disease-free survival; (4) Lower risk of resistance with complete viral suppression; and (5) Possible decrease in the risk for viral transmission.

Potential risks of early therapy include: (1) The adverse effects of the drugs on quality of life; (2) The inconvenience of some of the available regimens, leading to reduced adherence; (3) Development of drug resistance because of suboptimal suppression of viral replication; (4) Limitation of future treatment options as a result of premature cycling of available drugs; (5) The risk of transmission of virus resistant to antiretroviral drugs;

(6) Serious toxicities associated with certain antiretroviral drugs; and (7) The unknown durability of effect of available therapies.

Potential benefits of delayed therapy include: (1) Avoidance of treatment-related negative effects on quality of life and drug-related toxicities; (2) Preservation of treatment options; and (3) Delay in the development of drug resistance.

Potential risks of delayed therapy include: (1) The possibility that damage to the immune system, which might otherwise be salvaged by earlier therapy, is irreversible; (2) The possibility that suppression of viral replication might be more difficult at a later stage of disease; and (3) The increased risk for HIV transmission to others during a longer untreated period.

However, among patients with advanced HIV disease, antiretroviral therapy should be initiated after the first viral load measurement is obtained to prevent a potentially deleterious delay in treatment. Treatment should be offered to persons plasma HIV ribonucleic acid (RNA) levels of >55,000 copies/ml (by b-deoxyribonucleic acid [bDNA] or reverse transcriptase-polymerase chain reaction [RT-PCR] assays). Therapy should be evaluated through measuring plasma HIV RNA levels, which are expected to indicate a 1.0 log10 decrease at 2–8 weeks and no detectable virus (<50 copies/mL) at 4–6 months after treatment initiation. CD4+ T-cell counts should be measured at the time of diagnosis and every 3–6 months.

7.6 HIV Drug Resistance

HIV-1 drug resistance is a result of mutations occurring within the wild-type viral genome that produce variants capable of efficiently replicating in the presence of antiretroviral agents. The evolution of HIV-1 drug resistance within an individual depends on the generation of genetic variation in the virus and on the selection of drug-resistant variants during therapy. HIV-1 genetic variability is a result of the inability of HIV-1 RT to proofread nucleotide sequences during replication (Mansky and Temin 1998). It is exacerbated by the high rate of HIV-1 replication *in vivo*, the accumulation of proviral variants during the course of HIV-1 infection, and genetic recombination when viruses with different sequences infect the same cell. As a result, innumerable genetically distinct variants (quasispecies) evolve in individuals in the months following primary infection (Coffin 1995).

The first report of HIV-1 drug resistance was to zidovudine (ZDV) in 1989 (Larder et al. 1989). Subsequently, drug resistance to all therapeutic antiretroviral agents has been observed (Hertogs et al. 1999). Although highly active antiretroviral therapy for the treatment of HIV-1 infection has produced substantial decreases in morbidity and mortality in recent years (Kaplan et al. 2000), as many as 50% of patients fail therapy within 1 year of initiation (Montaner et al. 1998). As drug-resistant HIV-1 variants are often selected during the course of antiretroviral therapy (Hertogs et al. 1998), drug resistance is considered a major contributor to treatment failure. Furthermore, cross-resistance has a considerable negative impact on future treatment options. Current consensus guidelines recommend the use of HIV-1 drug resistance testing as a clinical management tool to help guide the choice of new regimens, particularly after treatment failure and for guiding therapy for pregnant women (Hirsch et al. 2000; Vandamme et al. 2001). The high rate of treatment failure due to the evolution of drug resistance mutations makes the development of new drugs that are active against resistant variants of paramount importance.

Another resistance issue that has recently received considerable attention is the transmission of drug-resistant HIV-1 in primary infection. Wide ranging prevalence estimates from 1 to 26% (Boden et al. 1999; Little et al.1999; Salomon et al. 2000; Yerly et al. 1999) have led to considerable debate as to the seriousness of this problem. A re-evaluation of previously published data using uniform criteria found that the prevalence of HIV-1 resistance among treatment-naive subjects ranged from 1 to 11% (Little et al. 2000). In an effort made by Harrigan et al. (2001) to further clarify this issue, they examined over 1000 HIV-1 isolates in plasma samples collected from treatment-naive individuals in the United States, Germany, Canada, and South Africa. Based on the results obtained from these samples, in the large majority of cases (approximately 97.5%) it was fond that the ranges of phenotypic drug susceptibility were < 2.5-fold to 4.0-fold, < 3.0-fold to 4.5-fold, and < 5-fold to 10-fold decreases in susceptibility to five protease inhibitors (PI), six nucleoside reverse transcriptase (RT) inhibitors, and three non-nucleoside reverse transcriptase inhibitors (NNRTI), respectively. Given the fact that individuals are being treated for longer and longer, there is likely to be continuing evolution of virus and increases in transmission of drug resistant virus. In addition, as the use of antiretrovirals increases in the less developed countries, drug resistance issues are likely to become even more acute.

7.7 Genetic Basis of Drug Resistance

A comprehensive list of documented mutations that have been associated with HIV-1 drug resistance is available and periodically updated (Johnson et al. 2005). It is important to appreciate that not all of these mutations have been verified, via site-directed mutagenesis, to confer resistance to a specific drug or drugs. Such information is critical to build an accurate picture of those mutations that are linked to resistance, rather than genetic polymorphisms that do not play a role. It is generally accepted that while resistance to some drugs is conferred by a single point mutation e.g., lamivudine (3TC) and NNRTI (Joly and Yeni 2000; Tisdale et al. 1993) this resistance can be exacerbated by additional mutations and the causation of resistance to other inhibitors is highly complex. Given the ever-expanding list of resistance mutations, predicting phenotypic resistance from mutational patterns is far from straightforward.

8. EVOLUTION OF VIRAL MUTATIONS

HIV replication is a highly dynamic process, whereby large numbers of virions are created and destroyed by the immune system each day (Perelson et al. 1996). Recent studies have calculated that the half-life of an HIV virion is about 30 minutes, and the production of virus can amount to 10^9 to 10^{10} virions per day (Markowitz et al. 2003). The plasma viral load thus reflects the balance between the production and clearance of viral particles. Most HIV-infected cells are short-lived T-cells, which have a half-life of about 2 days. The pool of virus-producing cells is, therefore, maintained through the constant infection of new cells.

Mutations in the HIV genome are primarily generated during one of the initial steps in the HIV replication cycle. The genomic RNA carried by HIV is copied into DNA early in the replication cycle. An HIV enzyme called reverse transcriptase (RT) is responsible for the RNA to DNA transcription. RT makes spontaneous errors when copying the RNA, placing the incorrect nucleotide in the growing DNA strand about once in every 10,000 to 30,000 nucleotides (Mansky and Temin 1995). Because the HIV genome is about 10,000 bases long, 1 error (mutation) may be made each time a viral genome is replicated, on an average. These nucleotide errors may cause changes in the amino acid coding of the HIV proteins made from the HIV DNA, potentially altering the structure and/or function of these proteins and affecting the replication competence of the viral strain.

It is unclear whether the evolution of resistance is completely deterministic (Rouzine and Coffin 1999) or strongly influenced by chance (stochastic) effects (Frost et al. 2000). It has been suggested that all drug-resistant strains may exist before the initiation of therapy. However, it is likely that both deterministic and stochastic effects influence HIV evolution.

Genetic recombination also contributes to the development of resistance. During viral replication, RT can randomly jump from one RNA template to the other, exchanging segments of genetic information between viral genomes (Zhuang et al. 2002). This template switching can facilitate the accumulation of mutations in a viral genome. Recombination may be particularly important in the development of multidrug-resistant HIV strains (Moutouh et al. 1996).

8.1 Viral Hemorrhagic Fevers

Viral hemorrhagic fevers (VHFs) refer to a group of illnesses that are caused by several distinct families of viruses. In general, the term "viral hemorrhagic fever" is used to describe a severe multisystem syndrome (multisystem in that multiple organ systems of the body are affected). Characteristically, the overall vascular system is damaged, and the body's ability to regulate itself is impaired. These symptoms are often accompanied by hemorrhage (bleeding); however, the bleeding is itself rarely life-threatening. While some types of hemorrhagic fever viruses can cause relatively mild illnesses, many of these viruses cause severe, life-threatening diseases. VHFs are caused by viruses of four distinct families; arenaviruses, filoviruses, bunyaviruses, and flaviviruses. Each of these families share a number of features:

- They are all RNA viruses, and all are covered, or enveloped, in a fatty (lipid) coating.
- Their survival is dependent on an animal or insect host, called the natural reservoir.
- The viruses are geographically restricted to the areas where their host species live.
- Humans are not the natural reservoir for any of these viruses. Humans are infected when they come into contact with infected hosts. However, with some viruses, after the accidental transmission from the host, humans can transmit the virus to one another.
- Human cases or outbreaks of hemorrhagic fevers caused by these viruses occur sporadically and irregularly. The occurrence of outbreaks cannot be easily predicted.
- With a few noteworthy exceptions, there is no cure or established drug treatment for VHFs.

In rare cases, other viral and bacterial infections can cause a hemorrhagic fever; scrub typhus is a good example. In this section we are reviewing viral hemorrhagic fevers caused by Ebola virus a member of Filovirus family and Dengue virus a member of Flavivirus family.

8.2 Ebola Hemorrhagic Fever

Ebola virus is an aggressive pathogen that causes a highly lethal hemorrhagic fever syndrome in humans and non-human primates. First recognized near the Ebola river valley in 1976 in an outbreak in Zaire. The natural host for Ebola virus is unknown. There is currently no antiviral therapy or vaccine that is effective against Ebola virus infection in humans. Ebola virus is a non-segmented RNA virus, which together with Marburg virus makes up the filovirus family. This notorious group of viruses was discovered in 1967 when Marburg virus was identified as the etiologic agent of a hemorrhagic fever outbreak in research facilities in Europe, which handled tissues from African green monkeys imported from Uganda. Subsequently, Ebola viruses were shown to be the cause of simultaneously occurring hemorrhagic fever outbreaks in 1976 in the Democratic Republic of Congo (DRC, formerly Zaire) and Sudan (Bowen et al. 1977). These outbreaks were shown caused by two different subtypes of Ebola virus, which became known as the Zaire and Sudan subtypes. There are four known species of Ebola virus; EBO-Z Zaire, EBO-R Reston, Sudan, and cote-d'Ivoire.

Mortality rates of upto 80% were recorded in these and more recent outbreaks in DRC and Gabon in 1995–1996. Epidemiologic data from recent outbreaks indicate that close contact is necessary for efficient transmission of Ebola virus from one individual to another, and little evidence can be found for aerosol transmission of the virus. Despite considerable efforts to identify the natural reservoir for Ebola and Marburg viruses, the host species remains an enigma. Although non-human primates have been implicated as the source of introduction of the virus into humans during several of the identified outbreaks, they are not considered likely to represent reservoir species because of their susceptibility to high mortality hemorrhagic disease similar to that seen in humans. Little genetic difference has been detected between Ebola-Zaire viruses isolated 20 years apart and from locations over 1,000 km from one another, suggesting that ecological rather than genetic factors may play the dominant role in initiation of Ebola hemorrhagic fever outbreaks.

8.3 Symptoms and Diagnosis of Ebola Hemorrhagic Fever

Ebola virus infection runs its course within 14 to 21 days. Infection initially starts with non-specific flu like symptoms like fever, myalgia, and malaise. As the infection progress, patients exhibit severe bleeding and coagulation abnormalities, including gastrointestinal bleeding, rash and a range of hematological irregularities, such as lymphopenia and neutrophilia. Massive liver damage and viremia, leads to intravascular coagulopathy. Final stages of ebola virus infection usually include diffuse bleeding and hypotensive shocks that account for many Ebola virus fatalities (colebunders and Borchert 2000; Sanchez et al. 1996). Antigen-capture enzyme-linked immunosorbent assay (ELISA) testing (Saijo et al. 2001), IgM ELISA, polymerase chain reaction (PCR) (towner et al. 2004), and virus isolation can be used to diagnose a case of Ebola HF within a few days of the onset of symptoms. Persons tested later in the course of the disease or after recovery can be tested for IgM and IgG antibodies; the disease can also be diagnosed retrospectively in deceased patients by using immunohistochemistry testing, virus isolation, or PCR. There is no specific antiviral drug or vaccine available for Ebola hemorrhagic fever. There is no standard treatment for Ebola HF. Patients receive supportive therapy. This consists of balancing the patient's fluids and electrolytes, maintaining their oxygen status and blood pressure, and treating them for any complicating infections.

8.4 Genome Structure of Ebola Virus

The Ebola virus genome is 19 kb long with seven open reading frames that encode for four structural proteins and three non-structural proteins. The structural proteins include the virion envelope glycoprotein (GP) (Sanchez et al. 1996), nucleoprotein (NP) and matrix proteins VP24 and VP40. Non-structural proteins include VP30 and VP35 and the viral polymerase.

8.5 SARS Virus

Severe Acute Respiratory Syndrome (SARS) caused a fatal lung infection and created alarm across the globe in March 2003, when World Health Organisation scientist; Dr. Carlo Urbani died in Bangkok with the infection of new virus that he was trying to stop spreading throughout the world. SARS causes a severe lung infection that starts with symptoms similar to influenza (flu). It is caused by a new form of a *Coronavirus* that probably mutated inside a domesticated animal, such as a cat or chicken. After spreading it passed on to humans and the epidemic began. The outbreak originated in Southern China and within the space of eight months the virus had spread to over 24 countries including the USA, South America, Canada and Europe. The first infections were seen in China in November 2002. By July 2003, when the last case was reported, the SARS had infected over 8,000 people and killed 813. Then arise a question, where does this new virus come from? There are many speculations regarding the appearance of this virus. They are: (1) This Coronaviruses with 99% sequence similarity to the surface spike protein of human SARS isolates have been isolated in Guangdong province of china, from healthy masked palm civet cats. It is a regarded as delicacy in this province and people became infected as they raised and slaughtered the animals. (2) There might be a possibility that SARS coronavirus recombine with other human coronaviruses to produce an even more deadly virus. Fortunately, the coronaviruses of which we are aware indicate that recombination has not occurred between viruses of different groups, only within a group, so recombination does not seem likely given the distance between the SARS virus and HCoV.

8.6 Transmission of Disease

The SARS virus is believed to spread by droplets produced by coughing and sneezing, but other routes of infection may also be involved, such as fecal contamination. There is still a lot to learn about the SARS virus, but researchers think that it can survive outside of the body for only a short time. Cleaning infected surfaces or clothes with soaps and bleaches will destroy the virus immediately. Doctors in Hong Kong first noticed an increase in the number of people entering hospital with severe lung infections and medical investigators realized that many patients lived in the same apartment blocks. Health workers who had come into contact with infected patients also caught the disease. They realized that a new disease was spreading through the city and that action needed to be taken. Researchers looking at the Hong Kong outbreak realized that the virus could only be spread from person to person if there was close contact. The SARS virus was found in droplets of fluid from an infected persons cough or sneeze. The infection was spread if these droplets were inhaled or touched and transferred to a person's mouth or eyes.

8.7 Characteristic Symptoms and Diagnosis of SARS

The most common reported symptom is fever, with patients reporting general influenza-like symptoms, chills, malaise, loss of appetite, and myalgia. Gastrointestinal symptoms are less common at presentation, including diarrhoea, vomiting and abdominal pain. The mean incubation period of SARS is estimated to be 4-6 days. The estimated case fatality rate is 13·2% for patients younger than 60 years and 43·3% for patients aged 60 years or older.

8.8 Treatment and Prophylaxis for SARS

Unfortunately, there are no approved antiviral drugs that are highly effective against coronaviruses. However, many steps unique to coronavirus replication could be targeted for development of antiviral drugs. Coronavirus infection begins with binding of the spike protein (S) on the viral envelope to a specific receptor on the cell membrane. Conformational changes are induced in S that probably lead to fusion of the viral envelope with the host cell membrane (de Arriba et al. 2002; Rota et al. 2003; Marra et al. 2003). Molecules that block binding to the receptor or inhibit the receptor-induced conformational change in S might block SARS-CoV infection (matsuyama and taguchi 2002; Zelus et al. 2003; Lewicki and Gallagher 2002). Inhibitors of HIV-1 entry and membrane fusion are good models for new drugs that target this first step in coronavirus infection. The large polyprotein encoded by the polymerase gene of coronaviruses must be proteolytically cleaved at specific sites by several virus-encoded proteases in order to have RNA polymerase activity (Zeibuhr and Siddell 1999; Denison et al. 1999; Hegyi and Zeibuhr 2002). Protease inhibitors developed to treat other viral diseases as well as new protease inhibitors are being tested for the ability to inhibit cleavage of the SARS-CoV polymerase protein and viral RNA synthesis.

There is considerable experience of development of coronavirus vaccines for veterinary purposes – though not all of it is encouraging. On the whole, inactivated coronavirus vaccines induce poor protection. The spike protein alone can induce immunity, but the internal nucleoprotein has also been reported to induce protective immunity. The WHO has recommended that SARS vaccines be developed. The quickest and probably safest to develop would be an inactivated or subunit vaccine. Even if such a vaccine was not fully protective against SARS infection, it might still provide some protection against life-threatening SARS pneumonia.

9. APPROACHES FOR PREVENTION AND CONTROL OF INFECTIOUS DISEASES

There are basically two approaches in overcoming infectious diseases. They are:

1. Cure and prevention of the spread of the disease, and
2. Prevention of the emergence of the disease.

9.1 Cure and Prevention of the Spread of the Disease

For cure and prevention of spread of disease new specifically targeted drugs for the causative organism are to be introduced and proper diagnosis for the causative organism are must. Many of the infectious diseases as mentioned in this chapter are gaining resistance to the existing drugs. So as to overcome this, combination therapy is introduced. Novel drug derivatives of the existing drugs are being introduced. Educating the people about the causative organism and the mode of transmission of disease to help them to take care about themselves. The two basic public health policies existing for controlling the spread of the infectious disease in the absence of effective vaccine or treatment are (*i*) effective isolation of symptomatic individuals and (*ii*) tracing and quarantining of the contacts of symptomatic cases. Both measures rely on rapid dissemination of information to facilitate accurate diagnosis of the symptoms of the disease based on a clear and precise case definition. For SARS, the timing of the onset of symptoms relative to peak infectivity is likely to have been a crucial factor in the success of simple public health interventions aimed at reducing transmission.

9.2 Prevention of Emergence

For the prevention of emergence of infectious diseases there are many factors which need to be taken care of, like public health care infrastructure had to be improved, vaccine development and repeated immunization programs had to be initiated to eliminate the pathogen as the limited or non-availability of host may eliminate the pathogen, reservoir hosts which anchor the pathogens during their latent periods are to be treated or eliminated to get rid of the pathogen, vector hosts which are responsible for the transmission of the disease needs to checked, like proper measures are to be taken to eliminate mosquito breeding in slums to eliminate disease carried by mosquitoes like malaria, dengue, etc.

People are to be educated to control diseases which are transmitted by the abnormal human behavior like sexually transmitted diseases and diseases like AIDS and to control zoonoses, hygienic conditions are to be maintained in animal husbandry practices. World economy and travel are also one of the major factors in the spread of infectious diseases so strict norms and proper health checks are recommended for people traveling through susceptible places.

CONCLUSION

We would conclude this chapter with a brief description of different cause for the emergence infectious diseases with their mode of transmission and the different diseases caused by these. Most of them dealt in this chapter are in the view of Indian scenario. Diseases like tuberculosis, malaria and AIDS are more in India than many other developed nations. According to some estimates the number of cases of AIDS in India are growing rapidly and it overtook the lead from African countries that were believed to be the origin of the disease. There are no reports of Ebola hemorrhagic fever and SARS in the Indian subcontinent. The rate of infection and the mode of transmission are of major concern in the view of Indian public health care system and normal Indian behavior. It will be uncontrollable to prevent these lethal infectious agents from spreading in such a huge population with many remote areas that are still with scarce medical facilities. This chapter could be of some help in over viewing these diseases and help in the guidance of preventive measures to be taken in these diseased conditions.

ACKNOWLEDGEMENTS

We would like to thank all the researchers working in the field of infectious diseases for their timely contributions for the public health care. We would like to thank Council for Scientific and Industrial Research (CSIR) India, for providing fellowship to M.S.K. Jayadev.

REFERENCES

Albert MJ (1994). Vibrio cholerae O139 Bengal. J Clin Microbiol 32: 2345-2349.

Alejandria M (2005). Dengue fever. Clin Evid 13: 887-895.

Bergmann JS, Woods GL (1999). Enhanced Amplified Mycobacterium tuberculosis Direct Test for detection of *Mycobacterium tuberculosis* complex in positive BACTEC 12B broth cultures of respiratory specimens. J Clin Microbiol 37: 2099-2101.

Bishop PJ and Neuman G (1970). The history of the Ziehl-Neelsen stain. Tubercle 51: 196-206.

Bloom BR, Flynn J, McDonough K, Kress Y, Chan J (1994). Experimental approaches to mechanisms of protection and pathogenesis in M. tuberculosis infection. Immunobiology 191: 526-536.

Boden D, Hurley A, Zhang L, Cao Y, Guo Y, Jones E, Tsay J, Ip J, Farthing C, Limoli K, Parkin N, Markowitz M (1999). HIV-1 drug resistance in newly infected individuals. JAMA 282: 1135-1141.

Bowen ET, Lloyd G, Harris WJ, Platt GS, Baskerville A, Vella EE (1977). Viral haemorrhagic fever in southern Sudan and northern Zaire. Preliminary studies on the aetiological agent. Lancet 12: 571-573 .

Bryant M, Ratner L (1990). Myristoylation-dependent replication and assembly of human immunodeficiency virus 1. Proc Natl Acad Sci U S A. 87: 523-527 .

Bushman FD, Fujiwara T, Craigie R (1990). Retroviral DNA integration directed by HIV integration protein *in vitro*. Science 249:1555-1558.

Cameron DW, Heath-Chiozzi M, Danner S, Cohen C, Kravcik S, Maurath C, Sun E, Henry D, Rode R, Potthoff A, Leonard J (1998). Randomised placebo- controlled trial of ritonavir in advanced HIV-1 disease. The Advanced HIV Disease. Ritonavir Study Group. Lancet 351: 543-549.

Canetti G, Fox W, Khomenko A, Mahler HT, Menon NK, Mitchison DA, Rist N, Smelev NA (1969). Advances in techniques of testing mycobacterial drug sensitivity, and the use of sensitivity tests in tuberculosis control programmes. Bull World Health Organ 41: 21-43.

Capon DJ, Ward RH (1991). The CD4-gp120 interaction and AIDS pathogenesis. Annu Rev Immunol 9: 649-678.

Caviedes L, Lee TS, Gilman RH, Sheen P, Spellman E, Lee EH, Berg DE, Montenegro-James S (2000). Rapid, efficient detection and drug susceptibility testing of Mycobacterium tuberculosis in sputum by microscopic observation of broth cultures. The Tuberculosis Working Group in Peru. J Clin Microbiol 38: 1203-1208.

Caws M, Drobniewski FA (2001). Molecular techniques in the diagnosis of Mycobacterium tuberculosis and the detection of drug resistance. Ann N Y Acad Sci 953:138-145.

Chawira AN, Warhurst DC, Robinson BL, Peters W (1987). The effect of combinations of qinghaosu (artemisinin) with standard antimalarial drugs in the suppressive treatment of malaria in mice. Trans R Soc Trop Med Hyg 81: 554-558 .

Coates TJ, Collins C (1998). Preventing HIV infection. Sci Am 279: 96-97.

Coffin JM (1995). HIV population dynamics in vivo: implications for genetic variation, pathogenesis, and therapy. Science 267: 483-489.

Cohn ML, Waggomer RF, Mc Clatchy JK (1968). The 7H11 medium for the culture of mycobacteria. Am Rev Respir Dis 98: 295-296.

Cole ST, Brosch R, Parkhill J, Garnier T, Churcher C, Harris D, Gordon SV, Eiglmeier K, Gas S, Barry CE 3rd, Tekaia F, Badcock K, Basham D, Brown D, Chillingworth T, Connor R, Davies R, Devlin K, Feltwell T, Gentles S, Hamlin N, Holroyd S, Hornsby T, Jagels K, Krogh A, McLean J, Moule S, Murphy L, Oliver K, Osborne J, Quail MA, Rajandrean MA, Rogers J, Rutter S, Seeger K, Skelton J, Squares R, Squares S, Suloton JE, Taylor K, Whitehead S and Barrell BG, (1998). Deciphering the biology of Mycobacterium tuberculosis from the complete genome sequence. Nature 393: 537-544.

Colebunders R, Borchert M (2000). Ebola haemorrhagic fever: a review. Journal of nfection 40: 16-20

Comstock GW (1975). Frost Revisited. The modern epidemiology of Tuberculosis American J of Epid 101: 303-382.

Cowman AF, Galatis D, Thompson JK (1994). Selection for mefloquine resistance in *Plasmodium falciparum* is linked to amplification of the pfmdr1 gene and cross-resistance to halofantrine and quinine. Proc Natl Acad Sci U S A. 91:1143-7.

Curran JW, Lawrence DN, Jaffe H, Kaplan JE, Zyla LD, Chamberland M, Weinstein R, Lui KJ, Schonberger LB, Spira TJ, Alexander WJ, Swinger G, Ammann A, Solomon S, Auerbach D, Mildvan D, Stoneburner R, Jason JM, Haverkos HW and Evatt BL (1984).

Acquired immunodeficiency syndrome (AIDS) associated with transfusions. N Engl J Med 310: 69-75.

Curtis CF, Otoo LN (1986). A simple model of the build-up of resistance to mixtures of anti-malarial drugs.Trans R Soc Trop Med Hyg 80: 889-892.

D'Acremont V, Landry P, Mueller I, Pecoud A, Genton B (2002). Clinical and laboratory predictors of imported malaria in an outpatient setting: an aid to medical decision making in returning travelers with fever. Am J Trop Med Hyg 66: 481-486.

Dannenberg AM Jr (1989). Immune mechanisms in the pathogenesis of pulmonary tuberculosis. Rev Infect Dis 2: S369-S378.

Dannenberg AM Jr (1991). Delayed-type hypersensitivity and cell-mediated immunity in the pathogenesis of tuberculosis. Immunol Today 12: 228-233.

Dannenberg AM Jr, Sugimoto M (1976). Liquefaction of caseous foci in tuberculosis. Am Rev Respir Dis 113: 257-259.

David HL (1970). Probability distribution of drug-resistant mutants in unselected populations of Mycobacterium tuberculosis. Appl Microbiol 20: 810-814.

De Arriba ML, Carvajal A, Pozo J, Rubio P (2002). Mucosal and systemic isotype-specific antibody responses and protection in conventional pigs exposed to virulent or attenuated porcine epidemic diarrhoea virus. Vet Immunol Immunopathol 85: 85-97.

Denison MR, Spaan WJ, van der Meer Y, Gibson CA, Sims AC, Prentice E, Lu XT (1999). The putative helicase of the coronavirus mouse hepatitis virus is processed from the replicase gene polyprotein and localizes in complexes that are active in viral RNA synthesis. J Virol 73: 6862-6871.

Diel R, Rusch-Gerdes S, Niemann S (2004). Molecular epidemiology of tuberculosis among immigrants in Hamburg, Germany. J Clin Microbiol 42: 2952-2960.

Doherty JF, Grant AD, Bryceson AD (1995). Fever as the presenting complaint of travellers returning from the tropics.QJM 88: 277-281.

Drobniewski FA and Wilson SM (1998). New biomolecular assays must be tested by direct study in the developing world. British Med J 316: 940.

Drobniewski FA and Wilson SM (1998). The rapid diagnosis of isoniazid and rifampin resistance in Mycobacterium tuberculosis: a molecular story. J. Med. Microbiol 47: 189-196.

Eisenach KD, Crawford JT, Bates JH (1988). Repetitive DNA sequences as probes for Mycobacterium tuberculosis. J Clin Microbio l26: 2240-2245.

Enarson DA, Rouillon A (1993). The epidemiological basis of tuberculosis control. In: Davies PDO, ed. Clinical Tuberculosis ,Chapman Hall Medical. London,pp 19-32.

Erickson JW, Gulnik SV, Markowitz M (1999). Protease inhibitors: resistance, cross-resistance, fitness and the choice of initial and salvage therapies. AIDS 13 Suppl A: S189-204.

Fluit AC, Visser MR, Schmitz FJ (2001). Molecular detection of antimicrobial resistance. Clin Microbiol Rev 14: 836-871.

Frost SD, Nijhuis M, Schuurman R, Boucher CA, Brown AJ (2000). Evolution of lamivudine resistance in human immunodeficiency virus type 1-infected individuals: the relative roles of drift and selection. J Virol 74: 6262-6268

Gandey A (2003). Is it flu or SARS? MDs gear up for a difficult winter. CMAJ 14: 821.

Garcia L, Alonso-Sanz M, Rebollo MJ, Tercero JC, Chaves F (2001). Mutations in the rpoB gene of rifampin-resistant Mycobacterium tuberculosis isolates in Spain and their rapid detection by PCR-enzyme-linked immunosorbent assay. J Clin Microbiol 39:1813-1818.

Girardi E, Raviglione MC, Antonucci G, Godfrey-Faussett P, Ippolito G (2000). Impact of the HIV epidemic on the spread of other diseases: the case of tuberculosis. AIDS 14: S47-S56.

Goedert JJ, Biggar RJ, Winn DM, Mann DL, Byar DP, Strong DM, DiGioia RA, Grossman RJ, Sanchez WC, Kase RG, Greene MH, Hoover RN and Blattner WA (1985) Decreased helper T lymphocytes in homosexual men. I. Sexual contact in high-incidence areas for the acquired immunodeficiency syndrome.Am J Epidemiol 121: 629-636.

Grange JM Drug-resistance and tuberculosis elimination. Bull Int Union Tuberc Lung Dis 1990; 65: 57-79

Graziano KL, Tempest B (2002). Hantavirus pulmonary syndrome: a zebra worth knowing. Am Fam Physician 66: 1015-1020.

Grosset J (1980). Bacteriologic basis of short-course chemotherapy for tuberculosis. Clin Chest Med 1: 231-241.

Grosset J, Mouton Y (1995). Is PCR a useful tool for the diagnosis of tuberculosis in 1995? Tuber Lung Dis 76:183-184.

Grzybowski S, Barnett GD, Styblo K (1975). Contacts of cases of active pulmonary tuberculosis. Bull Int Union Tuberc 50: 90-106.

Haas F & Haas, SS (1996). The origins of Mycobacterium tuberculosis and the notion of its contagiousness. In: W. N. Rom & S. Garay.(eds) Little, Brown and Company Tuberculosis, Boston, pp 4-19.

Hammer SM, Squires KE, Hughes MD, Grimes JM, Demeter LM, Currier JS, Eron JJ Jr, Feinberg JE, Balfour HH Jr, Deyton LR, Chodakewitz JA, Fischl MA (1997). A controlled trial of two nucleoside analogues plus indinavir in persons with human immunodeficiency virus infection and CD4 cell counts of 200 per cubic millimetre or less. AIDS Clinical Trials Group 320 Study Team.N Engl J Med 337: 725-733.

Harries AD, Maher D, Nunn P (1997). Practical and affordable measures for the protection of health care workers from tuberculosis in low-income countries. Bull World Health Organ 75: 477-489.

Harrigan PR, Montaner JS, Wegner SA, Verbiest W, Miller V, Wood R, Larder BA (2001). World-wide variation in HIV-1 phenotypic susceptibility in untreated individuals: biologically relevant values for resistance testing. AIDS 15:1671-1677.

Hegyi A, Ziebuhr J (2002). Conservation of substrate specificities among coronavirus main proteases. J Gen Virol 83: 595-599.

Hertogs K, de Bethune MP, Miller V, Ivens T, Schel P, Van Cauwenberge A, Van Den Eynde C, Van Gerwen V, Azijn H, Van Houtte M, Peeters F, Staszewski S, Conant M, Bloor S, Kemp S, Larder B, Pauwels R (1998). A rapid method for simultaneous detection of phenotypic resistance to inhibitors of protease and reverse transcriptase in recombinant human immunodeficiency virus type 1 isolates from patients treated with antiretroviral drugs. Antimicrob Agents Chemother 42: 269-276.

Hertogs, K, Van Houtte, M, Larder, B., Miller, V(1999). Testing for HIV-1 drug resistance: new developments and clinical implications. Recent Res Dev Antimicrob Agents Chemother 3: 83-104

Hirsch MS, Brun-Vezinet F, D'Aquila RT, Hammer SM, Johnson VA, Kuritzkes DR, Loveday C, Mellors JW, Clotet B, Conway B, Demeter LM, Vella S, Jacobsen DM, Richman DD (2000). Antiretroviral drug resistance testing in adult HIV-1 infection: recommendations of an International AIDS Society-USA Panel. JAMA 283 :2417-2426.

Hopewell PC (1994). Tuberculosis: pathogenesis, protection, and control. In B. R. Bloom (ed.), American Society for Microbiology. Overview of clinical tuberculosis, Washington, D.C, pp 25-46.

Huang H, Chopra R, Verdine GL, Harrison SC (1998). Structure of a covalently trapped catalytic complex of HIV-1 reverse transcriptase: implications for drug resistance. Science 282: 1669-1675.

Jindani A, Aber VR, Edwards EA, Mitchison DA (1980). The early bactericidal activity of drugs in patients with pulmonary tuberculosis. Am Rev Respir Dis 121: 939-949.

Johnson VA, Brun-Vézinet F, Clotet B, Conway B, Kuritzkes DR, Pillay D, Schapiro J., Telenti A, and Richman DD (2005). Update of the Drug Resistance Mutations in HIV-1. International AIDS Society–USA, Topics in HIV Medicine.Volume 13 Issue 1 March/April 2005.

Joly V, Yeni P (2000). Non-nucleoside reverse transcriptase inhibitors. Ann Med Interne (Paris). 151: 260-267.

Kain KC, Harrington MA, Tennyson S, Keystone JS (1998). Imported malaria: prospective analysis of problems in diagnosis and management. Clin Infect Dis 27: 142-149.

Kain KC, Keystone JS (1998). Malaria in travelers. Epidemiology, disease, and prevention. Infect Dis Clin North Am 12 : 267-284.

Kaplan JE, Hanson D, Dworkin MS, Frederick T, Bertolli J, Lindegren ML, Holmberg S, Jones JL (2000). Epidemiology of human immunodeficiency virus-associated opportunistic infections in the United States in the era of highly active antiretroviral therapy. Clin Infect Dis 30: S5-S14.

Kapur V, Li LL, Iordanescu S, Hamrick MR, Wanger A, Kreiswirth BN, Musser JM (1994). Characterization by automated DNA sequencing of mutations in the gene (rpoB) encoding the RNA polymerase beta subunit in rifampin-resistant Mycobacterium tuberculosis strains from New York City and Texas. J Clin Microbiol 32: 1095-1098.

Kelly SM, Geraghty JM, Neale G (1994). H pylori, gastric carcinoma, and MALT lymphoma. Lancet 343:418.

Kirschner P, Springer B, Vogel U, Meier A, Wrede A, Kiekenbeck M (1993). Detection and identification of mycobacteria by nucleic acid sequence determination: report of a 2-year experience in a clinical laboratory. J Clin Microbiol 31: 2882-2889.

Koch R (1890). An address on bacteriological research. Br Med J 2: 380-383.

Koch R (1932). The etiology of tuberculosis. Am Rev Tuberc 25: 284-323.

Kohlstaedt LA, Wang J, Friedman JM, Rice PA, Steitz TA (1992). Crystal structure at 3.5 A resolution of HIV-1 reverse transcriptase complexed with an inhibitor. Science 256:1783-1790.

Larder BA, Darby G, Richman DD (1989). HIV with reduced sensitivity to zidovudine (AZT) isolated during prolonged therapy. Science 243:1731-1734.

Lewicki DN, Gallagher TM (2002). Quaternary structure of coronavirus spikes in complex with carcinoembryonic antigen-related cell adhesion molecule cellular receptors. J Biol Chem 277: 19727-19734.

Little SJ(2000). Transmission and prevalence of HIV resistance among treatment-naive subjects. Antivir Ther 5: 33-40.

Little SJ, Daar ES, D'Aquila RT, Keiser PH, Connick E, Whitcomb JM, Hellmann NS, Petropoulos CJ, Sutton L, Pitt JA, Rosenberg ES, Koup RA, Walker BD, Richman DD (1999). Reduced antiretroviral drug susceptibility among patients with primary HIV infection. JAMA 282:1142-1149.

Lurie MB (1964). Resistance to Tuberculosis: Experimental Studies in Native and Acquired Defensive Mechanisms. Cambridge, MA: Harvard University Press.

Maher D, Chaulet P, Spinaci S and Harries A. Treatment of tuberculosis: guidelines for national programmes, 2nd ed. World Health Organization publication 1997; WHO/TB/97/220. World Health Organization, Geneva, Switzerland. pp 77.

Mansky LM (1998). Retrovirus mutation rates and their role in genetic variation.J Gen Virol 79 : 1337-1345.

Mansky LM, Temin HM (1995). Lower in vivo mutation rate of human immunodeficiency virus type 1 than that predicted from the fidelity of purified reverse transcriptase. J Virol 69: 5087-5094.

Markowitz M, Louie M, Hurley A, Sun E, Di Mascio M, Perelson AS, Ho DD (2003). A novel antiviral intervention results in more accurate assessment of human immunodeficiency virus type 1 replication dynamics and T-cell decay in vivo. J Virol 77: 5037-5038.

Marra MA, Jones SJ, Astell CR, Holt RA, Brooks-Wilson A, Butterfield YS, Khattra J, Asano JK, Barber SA, Chan SY, Cloutier A, Coughlin SM, Freeman D, Girn N, Griffith OL, Leach SR, Mayo M, McDonald H,Montgomery SB, Pandoh PK, Petrescu AS, Robertson AG, Schein JE, Siddiqui A, Smailus DE, Stott JM, Yang GS, Plummer F, Andonov A, Artsob H, Bastien N, Bernard K, Booth TF, Bowness D, Czub M, Drebot M, Fernando L, Flick R, Garbutt M, Gray M, Grolla A, Jones S, Feldmann H, Meyers A, Kabani A, Li Y, Normand S, Stroher U, Tipples GA, Tyler S, Vogrig R, Ward D, Watson B, Brunham RC, Krajden M, Petric M, Skowronski DM, Upton C, Roper RL (2003). The Genome sequence of the SARS-associated coronavirus. Science 300:1399-1404.

Marshall BJ, Warren JR. (1984). Unidentified curved bacilli in the stomach of patients with gastritis and peptic ulceration. Lancet 1:1311-1315.

Mathur ML, Khatri PK, Base CS (2000). Drug resistance in tuberculosis patients in Jodhpur district. Indian J Med Sci 54: 55-58.

Matsuyama S, Taguchi F (2002). Receptor-induced conformational changes of murine coronavirus spike protein.J Virol 76: 11819-11826.

Mbewu A (2001). Antiretroviral therapy is only part of it. Bull World Health Organ. 79: 1152.

Middlebrook G, Cohn ML, Eye WE, Russell WF, Levy D (1960). Microbiologic procedures of value in tuberculosis. Acta Tuberc Scan 38: 66-81.

Miller MD, Margot NA, Hertogs K, Larder B, Miller V (2001). Antiviral activity of tenofovir (PMPA) against nucleoside-resistant clinical HIV samples. Nucleosides Nucleotides Nucleic Acids 20: 1025-1028.

Mitchison DA (1985). The action of antituberculosis drugs in short-course chemotherapy. Tubercle 66: 219-225.

Mitchison DA (1992). The Garrod lecture. Understanding the chemotherapy of tuberculosis current problems. J Antimicrob Chemother 29: 477-493.

Mokrousov I, Bhanu NV, Suffys PN, Kadival GV, Yap SF, Cho SN, Jordaan AM, Narvskaya O, Singh UB, Gomes HM, Lee H, Kulkarni SP, Lim KC, Khan BK, van Soolingen D, Victor TC, Schouls LM (2004). Multicenter evaluation of reverse line blot assay for detection of drug resistance in Mycobacterium tuberculosis clinical isolates. J Microbiol Methods 57: 323-335.

Montaner JS, Hogg R, Raboud J, Harrigan R, O'Shaughnessy M (1998). Antiretroviral treatment in 1998. Lancet 352:1919-1922.

Moutouh L, Corbeil J, Richman DD (1996). Recombination leads to the rapid emergence of HIV-1 dually resistant mutants under selective drug pressure. Proc Natl Acad Sci U S A. 93:6106-6111.

Murray CJ, Lopez AD (1997). Global mortality, disability and the contribution to risk factors: Global Burden of Diseases Study. Lancet 349:1436-1442.

NACO National AIDS Control Organization, 2003, HIV/AIDS. Indian Scenario. http://www.naco.nic.in.

Niemann S, Richter E, Dalugge-Tamm H, Schlesinger H, Graupner D, Konigstein B, Gurath G, Greinert U, Rusch-Gerdes S (2000). Two cases of Mycobacterium microti derived tuberculosis in HIV-negative immunocompetent patients. Emerg Infect Dis 6: 539-542.

Noordhoek GT, Kolk HJ, Bjunene G, Catty D, Dale JW, Fine PE, et al (1994). Sensitivity and specificity of PCR for detection of *Mycobacterium tuberculosis*: a blind comparison study among seven laboratories. J Clin Microbiol 32: 277-284.

Padian NS (1987). Heterosexual transmission of acquired immunodeficiency syndrome: international perspectives and national projections. Rev Infect Dis 9:947-960.

Palasanthiran P, Ziegler JB, Stewart GJ, Stuckey M, Armstrong JA, Cooper DA, Penny R, Gold J (1993). Breast-feeding during primary maternal human immunodeficiency virus infection and risk of transmission from mother to infant. J Infect Dis 167: 441-444.

Paramasivan CN (1998). An overview of drug resistant tuberculosis in India. Indian J Tuberc 45: 73-81.

Parrish NM, Dick JD, Bishai WR (1998). Mechanisms of latency in Mycobacterium tuberculosis. Trends Microbiol 6: 107-112.

Pattyn SR, Portaels FA, Spanogne L and Magos J (1970). Further studies on African strains of Mycobacterium tuberculosis. Comparison with M. bovis and M. microti. Ann. Soc. Belge Med. Trop 50: 211-228.

Perelson AS, Neumann AU, Markowitz M, Leonard JM, Ho DD (1996). HIV-1 dynamics in vivo: virion clearance rate, infected cell life-span, and viral generation time. Science 271:1582-1586.

Peters W (1969). Drug resistance in malaria–a perspective. Trans R Soc Trop Med Hyg 63: 25-45.

Peters W (1990). The prevention of antimalarial drug resistance. Pharmacol Ther 47: 499-508.

Pizzo PA, Butler KM (1991). In the vertical transmission of HIV, timing may be everything. N Engl J Med 325: 652-654.

Plowe CV (2003). Monitoring antimalarial drug resistance: making the most of the tools at hand. J Exp Biol 206: 3745-3752.

Ramaswamy S, Musser JM (1998). Molecular genetic basis of antimicrobial agent resistance in Mycobacterium tuberculosis: 1998 update. Tuber Lung Dis 79: 3-29.

Reed MB, Saliba KJ, Caruana SR, Kirk K, Cowman AF (2000). Pgh1 modulates sensitivity and resistance to multiple antimalarials in Plasmodium falciparum. Nature 403: 906-909.

Rieder HL (1989). Epidemiology of tuberculosis in the United States. Epidemiology Rev 11: 79-98.

Riley RL, Nardell EA (1989). Clearing the air. The theory and application of ultraviolet air disinfection. Am Rev Respir Dis 139: 1286-1294.

Riley RL, Wills WF, Mill CM (1957). Air Hygiene in Tuberculosis: Quantitative studies of Infectivity and Control in a pilot ward. Am Rev Tub Pulm 75: 420-431.

Rota PA, Oberste MS, Monroe SS, Nix WA, Campagnoli R, Icenogle JP, Penaranda S, Bankamp B, Maher K, Chen MH, Tong S, Tamin A, Lowe L, Frace M, DeRisi JL, Chen Q, Wang D, Erdman DD, Peret TC, Burns C, Ksiazek TG, Rollin PE, Sanchez A, Liffick S, Holloway B, Limor J, McCaustland K, Olsen-Rasmussen M, Fouchier R,Gunther S, Osterhaus AD, Drosten C, Pallansch MA, Anderson LJ, Bellini WJ (2003). Characterization of a novel coronavirus associated with severe acute respiratory syndrome. Science 300:1394-1399.

Rouzine IM, Coffin JM (1999). Linkage disequilibrium test implies a large effective population number for HIV in vivo. Proc Natl Acad Sci U S A. 96:10758-10763.

Ruben S, Perkins A, Purcell R, Joung K, Sia R, Burghoff R, Haseltine WA, Rosen CA (1989) Structural and functional characterization of human immunodeficiency virus tat protein. J Virol 63:1-8.

Saijo M, Niikura M, Morikawa S, Ksiazek TG, Meyer RF, Peters CJ, Kurane I (2001). Enzyme-linked immunosorbent assays for detection of antibodies to Ebola and Marburg viruses using recombinant nucleoproteins.J Clin Microbiol 39 :1-7.

Salomon H, Wainberg MA, Brenner B, Quan Y, Rouleau D, Cote P, LeBlanc R, Lefebvre E, Spira B, Tsoukas C, Sekaly RP, Conway B, Mayers D, Routy JP (2000). Prevalence of HIV-1 resistant to antiretroviral drugs in 81 individuals newly infected by sexual contact or injecting drug use. Investigators of the Quebec Primary Infection Study. AIDS 14: F17-23.

Sanchez A, Trappier SG, Mahy BW, Peters CJ, Nichol ST (1996). The virion glycoproteins of Ebola viruses are encoded in two reading frames and are expressed through transcriptional editing. Proc. Natl. Acad. Sci. USA 93:3602-3607.

Simon F, Mauclere P, Roques P, Loussert-Ajaka I, Muller-Trutwin MC, Saragosti S, Georges-Courbot MC, Barre-Sinoussi F, Brun-Vezinet F(1998). Identification of a new human immunodeficiency virus type 1 distinct from group M and group O. Nat Med 4:1032-1037.

Smith DW, Wiegeshaus EH (1989). What animal models can teach us about the pathogenesis of tuberculosis in humans. Rev Infect Dis 11:S385-S393.

Smithwick RW (1976). Laboratory manual for acid-fast microscopy. 2nd ed. Atlanta: Centers for Disease Control.

Sreevatsan, S, Pan X, Stockbauer KE, Connell ND, Kreiswirth BN, Whittam TS & Musser JM (1997). Restricted structural gene polymorphism in the Mycobacterium tuberculosis complex indicates evolutionarily recent global dissemination. Proc Natl Acad Sci USA 94: 9869-9874.

Stead WW, Senner JW, Reddick WT, Lofgren JP (1990). Racial differences in susceptibility to infection by Mycobacterium tuberculosis. N Engl J Med 322:422-427.

Tisdale M, Kemp SD, Parry NR, Larder BA (1993). Rapid in vitro selection of human immunodeficiency virus type 1 resistant to 3'-thiacytidine inhibitors due to a mutation in the YMDD region of reverse transcriptase. Proc Natl Acad Sci U S A. 90: 5653-5656.

Towner JS, Rollin PE, Bausch DG, Sanchez A, Crary SM, Vincent M, Lee WF, Spiropoulou CF, Ksiazek TG, Lukwiya M, Kaducu F, Downing R, Nichol S (2004). Rapid diagnosis of Ebola hemorrhagic fever by reverse transcription-PCR in an outbreak setting and assessment of patient viral load as a predictor of outcome.J Virol 78:4330-4341.

Triglia T, Foote SJ, Kemp DJ, Cowman AF (1991). Amplification of the multidrug resistance gene pfmdr1 in Plasmodium falciparum has arisen as multiple independent events. Mol Cell Biol 11: 5244-5250.

Trivedi SS and Desai SC (1988). Primary antituberculosis drug resistance and acquired rifampicin resistance in Gujarat, India. Tubercle 69: 37-42.

Van Rie A, Warren R, Mshanga I, Jordaan AM, van der Spuy GD, Richardson M, Simpson J, Gie RP, Enarson DA, Beyers N, van Helden PD, Victor TC (2001). Analysis for a limited number of gene codons can predict drug resistance of Mycobacterium tuberculosis in a high-incidence. J Clin Microbiol 39: 636-641.

Vandamme AM, Houyez F, Banhegyi D, Clotet B, De Schrijver G, De Smet KA, Hall WW, Harrigan R, Hellmann N, Hertogs K, Holtzer C, Larder B, Pillay D, Race E, Schmit JC, Schuurman R, Schulse E, Sonnerborg A, Miller V (2001). Laboratory guidelines for the practical use of HIV drug resistance tests in patient follow-up. Antivir Ther 6:21-39.

Vareldzis BP, Grosset J, de Kantor I, Crofton J, Laszlo A, Felten M, Raviglione MC, Kochi A (1994). Drug-resistant tuberculosis: laboratory issues. World Health Organization. recommendations. Tuber Lung Dis 75: 1-7.

Wayne LG and Kubica GP (1986). The mycobacteria, *In* P. H. A. Sneath, and J. G. Holt (ed.), vol. 2. Md. Bergey's manual of systematic bacteriology The Williams and Wilkins Co., Baltimore, pp 1435-1457.

Wells WF (1955). Airborne Contagion and Air Hygiene. Harvard University Press; Cambridge, pp 104-141.

Yerly S, Kaiser L, Race E, Bru JP, Clavel F, Perrin L (1999). Transmission of antiretroviral-drug-resistant HIV-1 variants. Lancet 354: 729-733.

Youmans GP (1979). Tuberculosis. W.B. Saunders Co., Philadelphia, pp322.

Zapp ML, Green MR (1989). Sequence-specific RNA binding by the HIV-1 Rev protein. Nature 342: 714-716.

Zelus BD, Schickli JH, Blau DM, Weiss SR, Holmes KV (2003). Conformational changes in the spike glycoprotein of murine coronavirus are induced at 37 degrees C either by soluble murine CEACAM1 receptors or by pH 8.J Virol 77: 830-840.

Zhuang J, Jetzt AE, Sun G, Yu H, Klarmann G, Ron Y, Preston BD, Dougherty JP (2002). Human immunodeficiency virus type 1 recombination: rate, fidelity, and putative hot spots. J Virol 76:11273-11282.

Ziebuhr J, Siddell SG (1999). Processing of the human coronavirus 229E replicase polyproteins by the virus-encoded 3C-like proteinase: identification of proteolytic products and cleavage sites common to pp1a and pp1ab. J Virol 73: 177-185.

Microbial Diversity in Heavy Metal Remediation

S. S. AHLUWALIA and D. GOYAL

1. INTRODUCTION

Microbial metal accumulation has received more attention in recent years because of its potential application in environmental protection and recovery of metals. (Tobin et al. 1994; Ledin 2000) There has been a gradual shift towards exploring biological means involving use of live or dead organisms to sequester heavy metals from aqueous solution (Volesky 1989; Vieira and Volesky 2000). Various microorganisms can detoxify a range of toxic heavy metals, (a) by forming less toxic compounds that persist in the environment, (b) by removing them from affected environment by making them insoluble, and (c) by volatilising them. Biosorption of heavy metals by certain types of inactive, non-living biomass of microbial origin has been found to be highly effective for metal removal from dilute aqueous solution. The process has been successfully employed for harvesting valuable heavy metals, nuclear fuel or radioactive elements (de Rome and Gadd 1991).

Metal removal by microbes may take place either by metabolism dependent accumulation referred to as bioaccumulation or through metabolism independent binding called biosorption (Gadd 1990, Gadd and White 1993). Uptake of metallic species by means of metabolism dependent processes, involves both transporting into the cell and partitioning into

Department of Biotechnology & Environmental Sciences, Thapar Institute of Engineering & Technology (Deemed University), Patiala-147004, Punjab (India). *Email: d_goyal_2000@yahoo.com* intracellular components. Metal sequestration mainly involves ion exchange, chelation or adsorption. Mostly, metals bind to extra cellular polymers such as the cell wall or cell envelope and to some extent with various components present in the cell cytoplasm. It is particularly the cell wall structure of certain algae, fungi and bacteria, which is responsible for this phenomenon (Volesky 1990). Biosorption capacity of dead cells may be greater, equivalent or less than that of living cells (Kapoor and Viraraghavan 1995). Biomass can be produced by growing organisms using unsophisticated fermentation technique and inexpensive growth media or dead biomass can be procured as a by-product of fermentation from pharmaceutical, distillery or other fermentation based industries (Kapoor and Viraraghavan 1995; Vieira and Volesky 2000). Such dead biomass types have advantages, such as, (a) toxicity of the metals will have no effect on the metal binding capacity, (b) biomass can be grown under optimum conditions prior to use, (c) biomass can be immobilized to provide maximum physical stability during operation, and (d) dead biomass does not require nutrient supply for the system and is cost-effective w.r.t living cells. Due to inherent advantage in the specificity and binding efficiency, biosorption offers excellent tool for removal of metals from wastewater.

1.1 Metal-microbe Interaction

Microorganisms are generally the first to be affected by the presence of heavy metals in the environment either directly or indirectly. Interaction of metals with specific groups on cell surface may enhance or inhibit metal transport into the cell affecting metal transformation and biomineralization (Barkay and Schaefer 2001). Sorption of metals to cells plays a critical role in all kinds of metal-microbe interaction processes. The capacity of a biomass to remove metals from a waste stream depends upon its physico-chemical and biological properties (Unz and Shuttleworth 1996). Metal sorption or uptake by microbial cells can be enhanced through selection of efficient strains from autochthonous microbial flora, genetic manipulation and direct physical/chemical tailoring of the cells. The microbial interaction with metal can be exploited for (a) metal removal to purify water and soil remediation and (b) metal recovery from natural deposits (bio-leaching) and process streams (biosorption) having very low (Ledin 2000). The basic principle in both the cases is essentially the same but the practical requirements are quite different as for water purification, microorganisms may be promiscuous in binding the metals but for metal recovery they have to be specific.

2. BIOREMEDIATION OF HEAVY METALS

2.1 Choice of Metal

Selection of a metal for its removal depends on aim of the exercise and the extent of their impact on the environment. There may be (a) toxic heavy metals, (b) strategic metals, (c) precious metals, and (d) radio nuclides. The metal of interest is extremely important for a coherent and relevant study of especially biosorptive phenomenon (Ahalya et al., 2003). Removal of metals from aqueous solution is, mainly for, (a) toxicity removal—an environmental aspect, and (b) recovery of metals of commercial value—a technological aspect.

In terms of environmental threats, it is mainly the toxic heavy metals and radionuclides that warrant sequestering at the source of effluent (Volesky 1990). Heavy metals pose a threat not only to aquatic life but also endanger to the whole food chain consequently influencing the human health. The second category includes metals of technological importance, strategic significance and of high value. Technologically important heavy metals, in many cases, also act as pollutants because of their increasing concentration beyond safe limits in the environment caused by anthropological activities.

2.2 Removal of Heavy Metals by Living Systems of Microbial Origin

Metal accumulation by microorganisms has received much attention because of their small size and high surface area to volume ratio for interaction with the metals in the surrounding environment. Bioaccumulation of heavy metals by different living systems of microbial origin is summarized in Table 1. Attempts have been made to remove heavy metals from the environment by biological methods using phytoremediation techniques involving higher plants (Lytle et al. 1998; Zayed et al. 1998; Ksheminska et al. 2003), biosorption, bioaccumulation and bioprecipitation (Gadd and White 1993; Kratochvil and Volesky 1989; Bhide et al. 1996) involving fungi (Tobin et al. 1984; Pillichshammer et al. 1995; Volesky and May-Phillips 1995; Kapoor and Viraraghavan 1997; Zhang et al. 1998; Prakasham et al. 1999) algae (Kratochvil and Volesky, 1989; Donmez et al. 1999; Lee et al. 2000), weeds (Chen et al. 1996; Gardea-Torresday et al. 1997; Upatham et al. 2002; Horsfall et al. 2003), aquatic plants (Rahmani et al. 1999; Osman et al. 1996), bacteria (Ishibashi et al. 1990; Philip et al. 1998; Hussein et al. 2004) and activated sludge (Rossin et al. 1982; Wang and Shen 1995).

Table 1 Bioremoval of heavy metals by living microbial systems.

Microorganism/Plant	Metal	Reference
Fungi		
Penicillium sp.	Pb	Siegel et al., 1986
Rhizopus arrhizus	Cr, Pb, Ni	Tobin et al., 1984; Volesky, 1986; Fourest and Roux, 1989
Polyporous versicolor	Pb	Yetis et al., 1998
Phanerochaete chrysosporium	Pb,	Yetis et al., 1998; Ahluwalia and Goyal, 2003,
Saccharomyces cerevisiae	Pb, Cu	Suh et al., 1998; Huang et al., 1990
Aureobasidium pullulans	Pb, Cu	Suh et al., 1998; Ahluwalia and Goyal, 2003; Gadd and de Rome, 1988
Rhizopus nigricans	Cr, Cd, Ni, Pb	Bai and Abraham, 2003; Holan and Volesky, 1995
Volveriella volvacea	Pb	Osman et al., 1996
Aspergillus niger	Pb, Ni, Cu	Ahluwalia and Goyal, 2003; Townsley et al., 1986; Venkobacher, 1990
Aspergillus foetidus	Pb	Ahluwalia and Goyal, 2003,
Aspergillus terricola	Pb	Ahluwalia and Goyal, 2003,
Cladosporium resinae	Pb, Cu	Ahluwalia and Goyal, 2003; Gadd et al., 1988; Gadd and de Rome, 1988
Acremonium strictum	Pb	Ahluwalia and Goyal, 2003
Paecilomyces variotii	Pb	Ahluwalia and Goyal, 2003
Bacteria		
Bacillus circulans	Cr	Srinath et al., 2002
Bacillus megaterium	Cr, Pb	Srinath et al., 2002; Roane, 1999
Bacillus coagulans	Cr	Srinath et al., 2002; Philip et al., 1998
Bacillus firmus	Pb	Salehizadeh and Shojaosadati, 2003
Thiobacillus ferroxidans	Zn, Cu	Baillet et al., 1998; Boyer et al., 1998
Pseudomonas putida PRS 2000	Cr	Ishibashi et al., 1990
Streptomyces	Cr	Desjardin et al., 2002;
Thermocarboxydus	Pb, Ni	Hartmann et al., 2003
Streptomyces noursei	Pb	Mattuschka and Straube, 1993
Zoogloea ramigera	CN	Sag and Kutsal, 1995
Cladosporium cladosporiodes	Pb, Ni, Cu,	Patil and Paknikar, 1999
Arthobacter sp.	Mn	Veglio et al., 1997
Algae		
Microcystis sp.	Zn	Ahuja et al., 1999
Oscillatoria anguistissima	Zn	- do-

Sargassum asperifolium	Cr	Hamdy, 2000
Cystoseira trinode	Cr	- do-
Tubinaria decurrens	Cr	- do-
Laurencia obtusa	Cr	- do-
Dunaliella tertiolecta	Pb	Santana-Casiano *et al.*, 1995
Aphanothece halophytica	Zn	Incharoensakdi and Kitjaharn, 2002
Anabaena variabilis	Cr	Garnham and Green 1994
Spirogyra sp.,	Cr	Gupta *et al.*, 2001
Ascophyllum nodosum	Pb, Ni,	Holan *et al.*, 1993

2.3 Removal of Metals by Fungi

Three strains of *Saccharomyces cerevisiae* and one of *Candida* spp. were screened for uptake of Ag and Cu. Maximum uptake by *Saccharomyces cerevisiae* was attributed to the difference in the cell wall composition (Simmons et al. 1995). Lead accumulation by live cells of *Saccharomyces cerevisiae* was higher than that of dead cells. While in *Aureobasidium pullulans,* due to the existence of extra-cellular polymeric substance, both the capacity and initial rate of lead accumulation in live cells was higher than that of dead cells (Suh et al. 1998). Biosorption studies carried out with white-rot fungi, *Polyporous versicolor* and *Phanerochaete chrysosporium* for removal of Cu (II), Cr (III), Cd (II), Ni (II) and Pb (II), showed that both the fungi were effective in removing Pb (II) from aqueous solution with maximum biosorption capacity of 57.5 and 110 mg Pb (II) g^{-1} dry biomass, respectively (Yetis et al. 1998). *Volveriella volvacea,* a mushroom removed 95% Cd from Cd spiked distilled water within 15 min of contact, whereas increasing ionic strength and presence of complexing agents such as EDTA, decreased Cd sorption. Presence of other divalent cations (Ca, Mg) impeded Cd uptake, while the presence of Cl⁻ had no significant effect (Osman et al. 1996). Biosorption of heavy metals Cd (II), Pb (II) and Cu (II) from synthetic wastewater onto dry fungal biomass of *Phanaerochaete chrysosporium* was in the range of 5-500 mg/l (Say et al. 2001).

2.4 Removal of Metals by Bacteria

Reduction of hexavalent chromium (chromate) to less toxic trivalent form was studied by using cell suspensions and cell free supernatant fluids from *Pseudomonas putida* PRS 2000 (Ishibashi et al. 1990). Chromate reductase activity was associated with soluble protein and not with membrane fraction. The crude enzyme activity was heat labile and showed a K_m of 40 µm for substrate, CrO_4^{2-}. Neither sulfate nor nitrate affected chromate reduction either *in vitro* or with intact cells. Feasibility of biotransformation of hexavalent chromium was studied by immobilized *Bacillus coagulans* in a reactor (Philip et al. 1998). The results showed that almost complete removal of Cr (VI) was achieved with an influent hexavalent chromium concentration of 26 mg l^{-1} and retention time of 24 hrs.

Biosorption of cadmium and lead ions by purple non-sulfur bacteria, *Rhodobacter sphaeroides* and hydrogen bacteria *Alcaligenes eutrophus* H16, inactivated by steam sterilization was studied (Seki et al. 1998). Removal of metal ions by whole cell bodies of the bacteria was due to monodentate binding to carboxylic and phosphatic sites and the number of metal binding sites 204 times greater in *A. eutrophus* than *R. sphaeroides*.

Thiobacillus ferroxidans could tolerate upto 75 mM Cr^{3+} without affecting its bacterial activity (53 µg protein mM^{-1} Fe^{2+} oxidized) (Baillet et al. 1998). *Micrococcus* sp. isolated from activated sludge process showed the biosorption capacity (q$_{max}$) of 36.5 mg of Cu^{2+}/g of dry cell at pH 5.0 and 52.1 mg of Cu^{2+} g^{-1} of dry cell at pH 6.0 (Wong et al. 2001). Cells harvested at exponential and stationary growth phase showed the similar biosorption characteristics for copper. The copper capacity of *Micrococcus* sp. remained unchanged after five successive sorption and desorption cycles when sulphuric acid (0.05M) was used the desorption medium. Immobilization of *Micrococcus* sp. in 2% calcium alginate and 10% polyacrylamide gel beads increased copper uptake by 61%. Biomass of *Micrococcus* sp. may be applicable to the development of potentially cost-effective biosorbent for removing and recovering copper from effluent.

Bacillus circulans and *Bacillus megaterium* isolated from treated tannery effluent were able to bioaccumulate 34.5 and 32.0 mg Cr g^{-1} dry weight, respectively and bring down residual concentration to permissible limits within 24 hrs from 50 mg Cr(VI) l^{-1} (Srinath et al. 2002). Living and dead cells of *B. megaterium* adsorbed 23.8 and 39.9 mg Cr g^{-1} dry weight, respectively, whereas *B. coagulans* adsorbed 15.7 and 30.7 mg Cr g^{-1} of dry weight showing that biosorption by dead cells was higher than the living cells. Adsorption of Pb, Cu and Zn on the polysaccharide produced by *Bacillus firmus*, was significantly affected by the initial pH of the solution, initial metal ions and polysaccharide concentration and presence of the other ions (Salechizadeh and Shojaosadati 2003). Metal ion removal was lower at neutral pH and initial adsorption rate was rapid reaching equilibrium after 10 min. The uptake of Pb, Cu and Zn was 98.3, 74.9 and 61.8%, respectively.

2.5 Removal of Metals by Algae

Metal biosorption was carried out with capsulated and decapsulated cells of field and laboratory grown *Microcystis*, a unicellular cyanobacterium.

Field-grown capsulated *Microcystis* showed higher efficiency than laboratory-grown decapsulated *Microcystis*, which was 2.16 and 1.98 for Fe^{3+}, 1.77 and 1.57 for Cu^{2+} respectively (Singh et al. 1998).

Oscillatoria anguistissima showed higher biosorption capacity from the solution containing Zn^{2+} ions and compared well with commercial ion-exchange resin IRA-400C (Ahuja et al. 1999). Adsorption of copper and zinc in a single component system and multicomponent mixture by *Cymodocea nodosa*, a brown algae was reported by Sanchez et al. 1999. Three brown algae *Sargassum asperifolium, Cystoseira trinode, Tubinaria decurrens* and one red alga *Laurencia obtuse* were studied for removal of $Cr^{3+}, Co^{2+}, Ni^{2+}, Cu^{2+}$ and Cd^{2+} from solution. Biosorption capacity of each alga was different towards different metals. Further, HCl treatment of biomass of *L. obtuse* decreased biosorption of Cr^{3+} (Hamdy 2000). Adsorption capacity of Cr (VI) by the biomass of filamentous algae *Spirogyra* sp., was strongly dependent upon equilibrium pH (Gupta et al. 2001). Equilibrium isotherms showed 14.7×10^3 mg Cr (VI) removal per kg of dry weight biomass at a pH of 2.0 in 120 min from 5 mg l^{-1} of Cr (VI) containing solution. Zinc biosorption by *Aphanothece halophytica* was low at pH 4-6 and increased progressively at pH 6.5-7.0 (Incharoensakdi and Kitjaharn 2002). EDTA treatment and energy dependent studies revealed that process of zinc biosorption by this alga was due to binding of zinc to the cell surface. The maximum binding capacity of zinc was 133 mg/ g and apparent binding constant of zinc was 28 mg l^{-1}. The biomass of *Ulothrix zonata* was found to be suitable biosorbent for both the removal and recovery of copper from wastewater (Nuhoglu et al. 2002).

Removal of cadmium and copper from aqueous solution by *Scenedesmus abundans*, a common green alga was studied to determine competitive effects between metals in a multicomponent system (Terry and Stone 2002). At low algal concentration, competitive effect of copper and cadmium was observed above 4mg l^{-1}, whereas at highest algal concentration, no competitive effects were observed. Removal of copper and cadmium from water was higher by living *S. abundans* than non-living biomass. They found that biological treatment of heavy metal contaminated water is possible and that at adequately high algae concentrations, multicomponent metal systems can be remediated to the same level as individual metals.

CONCLUSION

Among all mechanisms of biological uptake heavy metal cations, biosorption offers an economically feasible technology for efficient removal and recovery of metal(s) from aqueous solution. The process of biosorption has many attractive features including the selective removal of metals over a broad range of pH and temperature, its rapid kinetics of adsorption and desorption and low capital and operation cost. Biomass can easily be produced using inexpensive growth media or obtained as a by-product from industry. It is desirable to develop promiscuous biosorbent with wide range of metal affinity that can remove a variety of metal cations. These will be particularly useful for industrial effluent, which carry more than one type of metals. Alternatively, a mixture of non-living biomass consisting of more than one type of microorganisms can be employed as biosorbent. Such "Combo" biosorbent have to be tested for commercial applications. The use of immobilized biomass rather than native biomass has been recommended for large-scale application, but various immobilization techniques have yet to be thoroughly investigated for ease, efficacy and cost-effectivity. The biomass-based technologies can be employed as adjutants to the conventional water treatment routes. At present, information on different biomass types to be used as biosorbent is insufficient to accurately define various parameters essential for process scale up and design perfection. Biological processes are efficiently applicable to effluents containing low concentration of heavy metals for an extended period. This aspect makes it even more attractive for treatment of dilute effluent that originates either from an industrial plant or from the primary wastewater treatment facility. Isolation of stress compatible microbes from contaminated sites with higher bioaccumulation potential will provide ideal material for removal of heavy metal from wastewater. Tailored organisms developed through genetic engineering may further enhance their metal uptake efficiency.

REFERENCES

Ahalya N, Ramachandra TV, Kanamadi RD (2003). Biosorption of heavy metals. Res J Chem Environ 7(4): 71-79.

Ahluwalia SS, Goyal D (2003). Removal of lead from aqueous solution by different fungi. Indian J of Microbiol 43(4): 237-241.

Ahuja P, Gupta R, Saxena RK (1999). Zn^{2+} biosorption by *Oscillatoria anguistissima*. Process Biochem 34: 77-85.

Aksu Z, Sag Y, Kutsal T (1990). A comparative study of the adsorption of chromium(VI) ions to *C. vulgaris* and *Z. ramigera*. Environ Technol 11: 33-40.

Bai SR, Abraham TE (2003). Studies on chromium(VI) adsorption-desorption using immobilized fungal biomass. Biores Technol 87: 17-26.

Baillet F, Magnin JP, Cheruy A, Ozil P (1998). Chromium precipitation by the acidophilic bacterium *Thiobacillus ferrooxidans*. Biotech Letts 20(1): 95-99.

Barkay T, Schaefer J (2001). Metal and radionuclide bioremediation: issues, consideration and potentials. Curr Opin Microbiol 4: 318-323.

Bhide JV, Dhakephalkar PK, Paknikar KM (1996). Microbiological processes for the removal of Cr(VI) from chrome-bearing cooling tower effluent. Biotechnol Letts 18: 667-672.

Boyer A, Magnin JP, Ozil P (1998). Copper ion removal by *Thiobacillus ferrooxidans* biomass. Biotechnol Lett 20(2): 187-190.

Chen JP, Chen WR, Hsu RC (1996). Biosorption of copper from aqueous solutions by plant root tissue. J Ferment Bioengg 81(5): 458-463.

de Rome L, Gadd GM (1991). Use of pelleted and immobilized yeast and fungal biomass for heavy metal and radionuclide recovery. J of Indl Microbiol 7: 97-104.

Desjardin V, Bayard R, Huck N, Manceau A, Gourdon R (2002). Effect of microbial activity on the mobility of chromium in soils. Waste Manag 22: 195-200.

Donmez GC, Aksu Z, Ozturk A, Kutsal T (1999). A comparative study on heavy metal biosorption characteristics of some algae. Process Biochem 34: 885-892.

Fourest, E, Roux JC (1992). Heavy metal biosorption by fungal myceliel by-products: Mechanism and influence of pH. Appl Microbiol Biotechnol 37: 399-403.

Gadd GM, de Rome L (1988). Biosorption of copper by fungal melanin. Appl Microbiol Biotechnol 29: 610-617.

Gadd GM (1990). Heavy metal accumulation by bacteria and other microorganism Experentia 46: 834-840.

Gadd GM, White C (1993). Microbial treatment of metal pollution—a working biotechnology? Trends Biotechnol 11: 353-359.

Gadd GM, White C, de Rome L (1988). Heavy metals and radionuclide uptake by fungi and yeast. In: Norris PR, Kelly DP (eds) Biohydrometallurgy, Sci Technol Letts pp. 421-435.

Gardea-Torresdey JL, Bibb J, Hernandez A (1997). Uptake of toxic heavy metal ions by inactivated cells of *Larrea tridentata* (Creosote bush).
http://www.engg.ksu.edu/HSRC/97abstracts/p22.html

Garnham GW, Green M (1994). Chromate (VI) uptake by and interaction with cyanobacteria. J of Indl Microbiol 247-251.

Gupta VK, Shrivastava AK, Jain N (2001). Biosorption of chromium (VI) from aqueous solutions by green algae *Spirogyra* species. Water Res 35 (17): 4079-4085.

Hamdy AA (2000). Biosorption of heavy metals by marine algae. Curr Microbiol 41: 232-238.

Hartmann T, Venus J, Desjardin V, Bayard R, Gourdon R, Spyra W (2003). Microbial reduction of hexavalent chromium. Forum der Forschung 15: 99-102.

Holan ZR, Volesky B (1995). Accumulation of cadmium, lead and nickel by fungal and wood biosorbents. Appl Biochem Biotechnol 53: 133-146.

Holan ZR, Volesky B, Prasetyo I (1993). Biosorption of cadmium by biomass of marine algae. Biotechnol Bioengg 41: 819-825.

Horsfall MJ, Abia AA (2003). Sorption of cadmium (II) and zinc (II) ions from aqueous solutions by cassava waste biomass (*Manihot esculenta* Cranz). Water Res 37(20): 4913-4923.

Huang CP, Huang CP, Morehart AL (1990). The removal of Cu (II) from dilute aqueous solutions by *Saccharomyces cerevisiae*. Water Res 24 (4): 433-439.

Hussein H, Ibrahim SF, Kandeel K, Moawad H (2004). Biosorption of heavy metals from wastewater using *Pseudomonas* sp. Electronic J Biotechnol 7(1): 30-37.

Incharoensakdi A, Kitjaharn P (2002). Zinc biosorption from aqueous solution by a halotolerant cyanobacterium *Aphanothece halophytica*. Curr Microbiol 45: 261-264.

Ishibashi Y, Cervantes C, Silver S (1990). Chromium reduction in *Pseudomonas putida*. Appl Environ Microbiol 56 (7): 2268-2270.

Kapoor A, Viraraghavan T (1995). Fungal biosorption—an alternative treatment option for heavy metal bearing wastewater: a review. Biores Technol 53: 195-206.

Kapoor A, Viraraghavan T (1997). Heavy metal biosorption sites in *Aspergillus niger*. Biores Technol 61: 221-227.

Kratochvil D, Volesky B (1998). Advances in the biosorption of heavy metals. Tibtech 16: 291-299.

Ksheminska H, Fedorovych D, Babyak L, Yanovych D, Kaszycki P, Koloczek H (2003). Bioremediation of chromium by the yeast *Pichia guilliermondii*: toxicity and accumulation of Cr (III) and Cr (VI) and the influence of riboflavin on Cr tolerance. Microbiol Res 158: 59-67.

Kuyucak N, Volesky B (1989). Accumulation of cobalt by marine algae. Biotechnol Bioengg 33(7): 809-814.

Ledin M (2000). Accumulation of metals by microorganisms-process and importance for soil systems. Earth-Sciences Reviews 5: 1-31.

Lee DC, Park CJ, Yung JE, Jeong H (2000). Screening of hexavalent chromium biosorbent from marine algae. Appl Microbiol Biotechnol 54: 445-448.

Lytle DR, Coates JD (1998). Bioremediation of metal contamination. Curr Opin Biotechnol 8: 285-289.

Macaskie LE, Empson RM, Cheetham AK, Grey CP, Skarnulis AJ (1992). Uranium bioaccumulation by a *Citrobacter* sp. as a result of enzymically-mediated growth of polycrystalline HUO_2PO_4. Science 257: 782-784.

Mattuschka B, Straube G (1993). Biosorption of metals by waste biomass. J Chem Tech Biotechnol 58: 57-63.

Muraleedharan TR, Venkobachar C (1990). Mechanism of biosorption of copper (II) by *Ganoderma lucidium*. Biotechnol Bioengg 35: 320-325.

Nuhoglu Y, Malkoc E, Gurses A, Campolat N (2002). The removal of Cu (II) from aqueous solutions by *Ulothrix zonata*. Biores Technol 85(3): 331-333.

Osman MS, Bandyopadhyay M (1996). Cadmium removal from water environment by a fungus *Volveriella volvacea*. Bioprocess Engg 14(5): 249-254.

Patil YB, Paknikar KM (1999). Removal and recovery of metal cyanides using a combination of biosorption and biodegradation processes. Biotechnol Letts 21(10): 913-919.

Philip L, Venkobachar C, Lyengar L (1998). Immobilized microbial reactor for the biotransformation of hexavalent chromium. Intl J Environ Pollut 11(2): 202-210.

Pillichshammer M, Pumpel T, Poder R, Eller K, Klima J, Schinner F (1995). Biosorption of chromium to fungi. Biometals 8: 117-121.

Prakasham RS, Sheno Merrie J, Sheela R, Saswathi N, Ramakrishna SV (1999). Biosorption of chromium (VI) by free and immobilized *Rhizopus arrhizus*. Environ Pollut 104(3): 421-427.

Rahmani GNH, Sternberg SPK (1999). Bioremoval of lead from water using *Lemna minor*. Biores Technol 70: 225-230.

Roane TM (1999). Lead Resistance in Two Bacterial Isolates from Heavy Metal–Contaminated Soils. Microbiol Ecol 37: 218-224.

Rossin AC, Sterritt RM, Lester JN (1982). The influence of process parameters on the removal of heavy metals in activated sludge. Water Air Soil Pollut 17: 185-198.

Sag Y, Kutsal T (1995). Biosorption of heavy metals by Zoogloea ramigera: use of adsorption isotherms and a comparison of biosorption characteristics. Chem Eng J (Lausanne) 60(1-3): 181-188.

Salechizadeh H, Shojaosadati SA (2003). Removal of metal ions from aqueous solution by polysaccharide produced from Bacillus firmus. Water Res 37(17): 4231-4235.

Sanchez A, Ballester A, Blazuez ML, Gonzalez F, Munoz J, Hammaini A (1999). Biosorption of copper and zinc by Cymodocea nodosa. FEMS Microbiol Rev 23(5): 527-536.

Santana-casiano JM, Gonzalez-Davila M, Perez-Pena J, Milleron FJ (1995). Pb^{2+} interaction with the marine phytoplankton Dunaliella tertiolecta. Mar Chem 48: 115-129.

Say R, Denizil A, Arica MY (2001). Biosorption of cadmium (II), lead (II) and copper (II) with the filamentous fungus Phanerochaete chrysosporium. Biores Technol 76(1): 67-70.

Seki H, Suzuki A, Mitsueda SI (1998). Biosorption of heavy metal ions on Rhodobacter sphaeroides and Alcaligenes eutrophus H16. J Colloid Interface Sci 192(2): 185-190.

Siegel S, Keller BZ, Galun M, Lehr H, Siegel B, Galun E (1986). Biosorption of lead and chromium by Penicillium preparation. Water Air & Soil Pollut 27: 69-75.

Siegel SM, Galun M, Keller P, Siegel BZ, Galun E (1987). Metal Speciation, Separation and Recovery. In: Patterson JW, Passino R (eds), Lewis Publishers, Chelsea, Michigan pp. 339-361.

Simmons S, Tobin JM, Singleton I (1995). Consideration on the use of commercially available yeast biomass for the treatment of metal–containing effluents. J Indl Microbiol 14:240-246.

Singh S, Pradhan S, Rai LC (1998). Comparative assessment of Fe^{3+} and Cu^{2+} biosorption by field and laboratory-grown Microcystis. Process Biochem 33(5): 495-504.

Srinath T, Verma T, Ramteke PW, Garg SK (2002). Chromium (VI) biosorption and bioaccumulation by chromate resistant bacteria. Chemosphere 48: 427-435.

Suh JH, Yun JW, Kim DS (1998). Comparison of Pb^{2+} accumulation characteristics between live and dead cells of Saccharomyces cerevisiae and Aureobasidium pullulans. Biotechnol Letts 20(3): 247-251.

Terry PA, Stone W (2002). Biosorption of cadmium and copper contaminated water by Scenedesmus abundans. Chemosphere 47(3): 249-55.

Tobin JM, Cooper DG, Neufeld RJ (1984). Uptake of metal ions by Rhizopus arrhizus biomass. Appl Environ Microbiol 47 (4): 821-825.

Tobin JM, White C, Gadd GM (1994). Metal accumulation by fungi: applications in environmental biotechnology. J of Indl Microbiol 13: 126-130.

Townsley CC, Ross IS (1986). Copper uptake in Aspergillus niger during batch growth and in non-growing mycelial suspensions. Exper Mycol 10: 281-288.

Unz RF, Shuttleworth KL (1996). Microbial mobilization and immobilization of heavy metals. Curr Opin Biotechnol 7: 307-310.

Upatham ES, Boonyapookana B, Kruatrachue M, Pokethitiyook P, Parkpoomkamol K (2002). Biosorption of cadmium and chromium in duckweed Wolffia globosa. Intl J Phytoremediation 4 (2): 73-86.

Veglio F, Beolchini F, Gasbarro A (1997). Biosorption of toxic metals: an equilibrium study using free cells of Arthrobacter sp. Process Biochem 32(2): 99-105.

Venkobacher C (1990). Metal removal by waste biomass to upgrade wastewater treatment plants. Water Sci Technol 6: 319-320.

Vieira RHSF, Volesky B (2000). Biosorption: a solution to pollution? Intl Microbiol 3: 17-24.

Volesky B (1986). Biosorbent materials. Biotechnol Bioeng Symp Ser 16: 121-126.

Volesky B (1989). Evaluation of sorption performance.

http.//www.mcgill.ca/biosorption/publication/book/book.html

Volesky B (1990). in Biosorption of heavy metals, ed CRC press Boca Raton, FL .

Volesky B, May-Phillips HA (1995). Biosorption of heavy metals by *Saccharomyces cerevisiae*. Appl Microbiol Biotechnol 42: 797-806.

Volesky B (1988). Biosorbents for metal recovery. Trends Biotechnol 5: 96-101.

Wainwright M, Grayston SJ, Dejong PM (1986). Adsorption of insoluble compounds by the fungus *Mucor flavus*. Enzyme Microb Technol 8: 597-600.

Wang YT, Shen H (1995). Bacterial reduction of hexavalent chromium. J Indl Microbiol 14: 159-163.

Wong MF, Chua H, Lo W, Leung CK, Yu PH (2001). Removal and Recovery of Copper (II) ions by bacterial biosorption. Appl Biochem Biotechnol 91: 447-457.

Yetis U, Ozcengiz G, Dilek FB, Ergen N, Erbay A, Dolek A (1998). Heavy metal biosorption by white-rot fungi. Water Sci Technol 38(4-5): 323-330.

Zayed A, Gowthaman S, Terry N (1998). Phytoaccumulation of trace elements by wetland plants: I Duckweed. J Environ Qual 27: 715-772.

Zhang L, Zhao L, Yu Y, Chen C (1998). Removal of lead from aqueous solution by non-living *Rhizopus nigricans*. Water Res 32: 1437- 1444.

Microbial Management of Pollutants from Textile Industry

S. GUPTA[1], DEEPTI[1], S. KUHAR[1], K. K. SHARMA[1], A. SINGH[2] and R. C. KUHAD*

1. INTRODUCTION

The textile industry is one of the oldest and technologically most complex of all industries. Since ages human beings have been using colorants for painting and dyeing of their surroundings, skin and clothes. Until the middle of the 19th century, all colorants applied were from natural origin. These dyes are all aromatic compounds, originating from plants (e.g., the red dye alizarin from madder and indigo from wood), insects (e.g., the scarlet dye kermes from shield-louse *Kermes vermilio*), fungi (anthraquinone from *Aspergillus cristatus* and cynodontin from *Curvularia lunata*) and lichens (Archil from *Roccella tinctoria* and cud bear from *Ochrol echia*).

Synthetic dye manufacturing started in 1856, when an English chemist W. H. Perkin, in an attempt to synthesize quinine, instead obtained a bluish substance with excellent dying properties that was later known as aniline purple, tyrian purple or mauveine. In the beginning of the 20th century, synthetic dyestuffs had almost completely supplanted natural dyes

1 Lignocellulose Biotechnology Laboratory, Department of Microbiology, University of Delhi, South Campus, Benito Juarez Road, New Delhi-110021, India, e-mail: kuhad@hotmail.com

2 Petrozyme Techonologies, 7496 Wellington Road 34, R.R.#3, Guelph, Ontario, N1H6H9, Canada, e-mail: asingh@petrozyme.com

* Corresponding Author: *Dr. R.C. Kuhad*, Department of Microbiology University of Delhi, South Campus, Benito Juarez Road, New Delhi-110021, India. Tel.: +91-11-24112062; Fax: +91-11-24115270

(Welham 2000). At present over 100,000 dyes are commercially available with 7×10^5 tons of dyestuff being produced annually (Meyer 1981; Zollinger 1987; Banat et al. 1996; Robinson et al. 2001). Improper handling practices in dyeing process results in 10-15% of all dyestuffs being directly lost to wastewater, which ultimately find its way into the environment (Zollinger 1987; O'Neill et al. 2000). Colored industrial effluent is the obvious indicator of water pollution, that are not removed from industrial effluents by conventional wastewater treatments (Cripps et al. 1990; Moran et al. 1997; Willmott et al. 1998). Most of the dyes are difficult to treat because of their synthetic origin and mainly complex aromatic molecular structures. Such structures are often constructed to resist fading on exposure to sweat, soap, water, sunlight and oxidizing agents (Poots et al. 1976; McKay 1979; Seshadri et al. 1994; Sorek et al. 1995; Tsatsaroni 1995; Koprivanac et al. 1997; Gabelica et al. 2001; Ramalho et al. 2004; Khehra et al. 2005) and this renders them more stable and less amenable to biodegradation. This necessitates developing novel processes for treatment of effluent from the textile industry to ensure safety of the environment from these potential pollutants.

Textile industry effluents are characterized as having the high visible color (3000-4500 units), chemical oxygen demand (800-1600 mg/L), alkaline pH range of 9-11 and total suspended solids (6000-7000 mg /L) (Manu and Chaudhari 2002). A variety of effective physical and chemical treatment methods are available for treating textile effluents (Nigam et al. 2000; Robinson et al. 2001). There has been considerable interest in development of biological methods because these methods are considered attractive due to their potential low cost, environmental compatibility and public acceptability (Dubin and Wright 1975; Paszczynski and Crawford 1991). A wide variety of microorganisms including bacteria, fungi and algae are capable of decolorizing a diverse range of dyes (McMullan et al. 2001; Kuhad et al. 2004). Biological dye removal techniques are based on microbial bio-transformation of dyes e.g., activated sludge, anaerobic digestion, trickling filter, etc. However, majority of these studies are of little help in the development of biological processes for treatment of textile effluent as they were carried out under conditions unlikely to occur in industrial situations. Therefore, the isolation of potent microbial species and its usage in degradation are of main interest in biological effluent treatment (Banat et al. 1996; Aksu et al. 1997; Addour et al. 1999; Conneely et al. 1999; Kapdan et al. 2000). Among microorganisms, ligninolytic fungi have been used extensively to remove color from textile wastewater and also in partial mineralization of azo dyes (Paszczynski et al. 1992; Spadaro et al. 1992). Several potential white rot fungi, namely, *Bjerkandera adusta, Irpex lacteus, Phanerochaete chrysosporium, Phlebia radiata, Pleurotus ostreatus, Pycnoporous cinnabarinus* and *Trametes versicolor,* have been studied for decolorization of several groups of recalcitrant dyes and dye effluents (Chivukula and Renganathan 1995; Paszcynski and Crawford 1995; Heinfling et al. 1998; Schliephake et al. 2000; Novotony et al. 2001; Jorosz-Wilkolazka et al. 2002; Murugesan 2002; Kashinath et al. 2003). This chapter focuses on chemical constituents of textile industrial waste with special attention to the dyes, their possible health hazards to human beings and animal and various microbial techniques for treatment of textile industrial effluents.

2. EFFLUENT FROM TEXTILE INDUSTRIES AND ITS TOXIC EFFECT

Textile industries consume substantial volumes of water and chemicals for wet processing of textiles. These chemicals range from inorganic compounds and elements to polymers and organic products. There are more than 8000 chemical products associated with the dyeing process listed in the color index (Society of Dyes and colourists, 1976).

A great variety of synthetic dyes are used for textile dyeing and other industrial applications. The structural diversity of dyes derived from the use of different chromophoric groups like azo, anthraquinone, triarylmethane, phthalocyanines, etc. and from the use of different application technologies such as reactive, direct, disperse, vat dyeing, etc.

The total world colorant production is estimated to be on the order of 800,000 tons/year and only azo dyes containing one or more azo group (–N=N–) accounts for 60-70% of all textile dyestuffs used in textile industries (Easton 1995). Reactive azo textile dyes like orange 7, reactive black 5, food yellow 3, etc., due to their poor fixation to fabrics, occurs in wastewater in a concentration ranging from 5 to 1500 mg/L (Pierce 1994).

Besides different dyes, the materials that may be expected to predominate in textile wastewater include:

- Metals
- Non-degradable surfactants including detergents, emulsifiers and dispersants used frequently in textile wet processing.
- Toxic organics such as phenols, aromatic solvents methylene, chloride, perchloroethylene and oxalic acid.

Metals may be present in varying amounts in different dye classes (Table 1). Metal based complex dyes can lead to release of chromium, a carcinogenic metal, into water supplies (Anliker et al. 1981; Baughman and Perenich 1988). Another example is Food red # 5, a dye known to produce degradation products that are carcinogenic and mutagenic in nature.

Textile industry wastewater is heavily charged with unconsumed dyes, surfactants and traces of metals. These effluents cause a lot of damage to the environment. Soluble organic substances usually contribute the major fraction of the biochemical oxygen demand (BOD). Discharge off these effluents in water bodies affect their pH, color and gas solubility, which are not suitable for aquatic life and soil, and ultimately for human being.

Table 1   Average metal content of selected Dyes (ppm)

Metal	Acid	Basic	Direct	Disperse	Reactive	Vat
Arsenic	<1	<1	<1	<1	1.4	<1
Cadmium	<1	<1	<1	<1	<1	<1
Chromium	9	2.5	3.0	3.0	24	83
Cobalt	3.2	<1	<1	<1	<1	<1
Copper	79	33	35	45	71	110
Lead	37	6	28	37	52	6
Mercury	<1	0.5	0.5	<1	0.5	1.0
Zinc	<13	32	8	3	4	4

DYE CLASS

3. TEXTILE EFFLUENT AND ENVIRONMENTAL CONCERN

The major environmental problem of colorants is the removal of dyes from effluents. Many dyes are visible in water at concentrations as low as 1 ppm. Textile-processing wastewaters, typically with a dye content in the range 10-200 mg/L (O'Neill et al. 2000), are therefore usually highly colored and when discharged in open water presents an aesthetic problem. Interest in the pollution potential of textile dyes has been primarily prompted by concern over their possible toxicity and carcinogenicity. This is mainly due to the fact that many dyes are made from known carcinogens, such as benzidene and other aromatic compounds (Clarke and Alinker 1980). It has been shown that azo and nitro compounds are reduced in sediments (Weber and Wolfe 1987) and animal intestinal environment resulting in the regeneration of the parent toxic amines. Chronic effects of dyestuffs, especially of azo dyes, have been studied for several decades e.g., toxicity to fish and mammalian life, possible chronic risks of colorants and their intermediates, they are carcinogenic and to a lesser extent sensitizing and allergic. Intestinal cancer are common in highly industrialized societies and cerebral abnormality and skeletal abnormalities in fetuses. Azo dyes in purified form are seldom directly mutagenic or carcinogenic, except for some azo dyes with free amino groups (Brown and DeVito 1993). However, reduction of azo dyes, leads to the formation of aromatic amines which are known mutagens and carcinogens. Exposure to aromatic amines may cause methemoglobinemia, the amine oxidize the heme iron of hemoglobin from Fe (II) to Fe (III), blocking the oxygen binding. In mammals, metabolic activation (= reduction) of azo dyes is mainly due to bacterial activity in the anaerobic parts of the lower gastrointestinal tract. However, various other organs, especially the liver and the kidneys, can also reduce azo dyes. After azo dye reduction in the intestinal tract, the released aromatic amines are absorbed by the intestine and excreted in the urine. The acute toxic hazard of aromatic amines is carcinogenesis, especially bladder cancer. The carcinogenicity mechanism probably includes the formation of acyloxy amines through N-hydroxylation and N-acetylation of the aromatic amines followed by O-acylation. These acyloxy amines can be converted to nitremium and carbonium ions that bind to DNA and RNA, which induce mutations and tumor formation. The mutagenic activity of aromatic amines is strongly related to molecular structure. In 1975 and 1982, the International Agency for Research on Cancer (IARC) summarized the literature on suspected azo dyes, mainly amino-substituted azo dyes, fat-soluble azo dyes, and benzidine azo dyes. Most of the dyes on the IARC list were taken out of production (Brown and DeVito 1993). Generally stated, genotoxicity is associated with all aromatic amines with benzidine moieties, as well as with some other aromatic amines with toluene, aniline and naphthalene moieties. Dermal, immunological effects have been reported in workers exposed to benzidine and cancer is the documented toxic effect of benzidine in both human and animals. Metabolic transformations of benzidine result in the formation of reactive intermediate that are thought to produce DNA adduct, and initiate carcinogenesis by producing mutation.

Direct brown 95, Direct black 38, Direct blue 6, are generally used in textile industries and worker exposed to these compounds excrete high level of benzidine into their urine. Consequently, there exists a cancer risk for the workers.

The toxicity of aromatic amines depends strongly on the spatial structure of the molecule or in other words, the location of the amino-group(s). For instance, there is strong evidence that 2-naphthylamine is a carcinogen, 1-naphthylamine is much less toxic (Cartwright 1983). The toxicity of aromatic amines depends furthermore on the nature and location of other substituents. As an example, the substitution with nitro, methyl or methoxy groups or halogen atoms may increase the toxicity, whereas substitution with carboxyl or sulphonate groups generally lowers the toxicity (Chung and Cerniglia 1992).

Sulfonated azo compounds are widely used as dyes for textile and cosmetics. Both aromatic sulfonic acid and azo groups are rare among natural products and, thus, confer a xenobiotic character to sulfonated azo dyes. Several amino substituted azo dyes are mutagenic as well as carcinogenic and their toxicity has been extensively studied along with risk of occupational cancer associated with their use (Joachim et al. 1985; Gonazoles et al. 1988).

It is known that 90% of reactive textile dyes entering activated sludge sewage treatment plants will pass through unchanged and will be discharged into rivers. In river problems caused by dyes are (a) Sunlight penetration of the streams would be reduced, which is essential for photosynthesis and consequently, the ecosystem of the stream will be seriously affected (Kuo 1992) (b) toxicity to fish and mammals life (c) inhibits activity and growth of microorganisms particularly in high concentrations (d) some cationic species (mostly triphenylmethanes) affect the flora and fauna even at lesser concentrations (Meyer 1993).

Other compounds like formaldehyde, carbon disulphide, some phenolic and sulphur containing compounds are also used in textile industries. Formaldehyde is generally used in a textile industry to attain a permanent press finish. It is a well-known cause of ocular and airway irritation and it can cause certain skin reactions including contact dermatitis via either allergic or irritant mechanism. Carbon disulphide (CS_2) is prominent in the production of the rayon. It is used in the conversion process of cellulose to rayon fiber and cellophane and high level of CS_2 resulted in the severe central nervous system (CNS) disorder including acute mania and narcosis (Cassitto et al. 1993).

4. CLASSIFICATION OF DYES

Aromatic compounds that absorb light with wavelength in the visible range (~350-700) are colored and known as dye. Dyes contain two main groups, which are chromophore and auxochrome. Auxochrome group intensifies the color of the chromophore. Usual chromophore groups are –C=C–, –C=N–, –C=O–, –N=N–, –NO$_2$, and quinoid rings and auxochromes are –NH$_3$, -COOH, -SO$_3$H and –OH. Based on the dyeing mode and main structure moieties, chromophore group, dyes can be divided into 20-30 different groups like azo (monoazo, diazo, triazo, polyazo), anthraquinone, phthalocyanine and triarylmethane dyes are the most important groups.

According to color index, being edited since 1924 (and revised every three months) there are 15 different classes of dyes - Acid dyes, Reactive dyes, Basic dyes, Direct dyes, Metal complex dyes, Mordant dyes, Disperse dyes, Pigment dyes, Vat dyes, Sulphur dyes, Solvent dyes, Fluorescent brighteners and classes which include Food dyes and Natural dyes. Food dyes are not used in textile but the natural dyes like anthraquinone, indigoid, flavanol or chromone compounds can be used (Table 2) (Hachem et al. 2001).

4.1 Effluent Treatment Techniques

Various physical, chemical and biological pretreatment techniques have been employed to remove color from dye containing wastewaters (Judkins 1984; Grau 1991; Cooper 1993; Tunay et al. 1996; Vandevivere et al. 1998; Hao and Chang 2000; Robinson et al. 2001). In the past, municipal treatment systems were mainly used for the purification of textile mill wastewaters. These systems, however, depended mainly on biological activity and were mostly found inefficient in removal of the more resistant synthetic dyes. Less sensitive, yet more effective methods, therefore, were developed and tested for dye removal (Groff 1993; Wilking and Frahne 1993; Mishra and Tripathy 1993). Methods of effluent treatment for dyes may be classified into three main categories: physical, chemical and biological (Table 3).

Several factors determine the technical and economic feasibility of each single dye removal technique:

(1) Dye type.
(2) Wastewater composition.
(3) Dose and cost of required chemical.
(4) Operation costs.
(5) Environmental fate and handling cost of generated waste products.

In general, each technique has its limitations. The use of one individual process may often not be sufficient to achieve complete decolorization. Therefore, an effective and low cost treatment system would be of great value. Among the low cost viable alternatives available for effluent treatment and decolorization, the biological systems seem to be the best one. Biological systems are recognized by their capacity to reduce BOD and COD by conventional and aerobic biodegradation. But there is a problem with its inability to remove color (O'Neill et al. 2000). However, potential microbial decolorization systems have been developed with total color removal, in

Table 2 Important classes of dyes and their characteristics

Dye class	Colour index	Fiber	Type of interaction/ Reaction with fiber	Examples
Acid dyes	Largest class ~2300 different dyes listed ~40% of them are in current production	Mainly wool, Nylon, Polyamide, Modified acryl	Ionic interaction, anionic compound of dye bind with the cationic NH_4 ions of those fibers.	Azo dyes like orange II orange G and Carbonyl like Indigo carmine, anthraquinone, etc.
Basic dyes	Represent ~5% of all dyes listed in colour index	Mainly synthetic fibers like modified polyacryl	Ionic interaction	Triphenyl like Crystal violet and Malachite green, and anthraquinone, triarylmethane, etc.
Reactive dyes	Second largest dye class	Cotton, wool, silk and nylon, etc.	Covalent interaction with fiber	Reactive orange II, Reactive Black 5, Reactive Red198, and quinine type Remazol blue
Direct dyes	~1600 are listed and about 30% are in current production	Mainly cellulose fibers like cotton	Van der Waal forces	Mostly are azo Congo Red, Direct Blue86 Phthalocyanine, etc.
Disperse dyes	Third largest group of dye	Synthetic fibers like cellulose acetate, polyester, Polyamide acryl, etc.	These dyes penetrate synthetic fibers	Azo compounds like Methyl yellow, Disperse Yellow1 nitro, etc.
Vat dyes		Cellulose fibers mainly cotton and wool	Dyeing method based on the solubility of vat dyes in their reduced (Leuco) form.	All vat dyes are anthraquinone or indigoid like indigo, etc.
Mordant dyes	~600 Mordant dyes are listed in the colour index	Wool, Leather, Silk, Paper and modified cellulose fibers	Mordant dyes fixed to fabric by the addition of a mordant.	These are azo, Oxazine or Triarylmethane but usually these are dichromates or chromium complexes.
Pigment dyes	About 25% of all commercial dye name listed in the colour index	Used for printing diverse fibers.	To interact with fiber required the help of dispersing agent.	Mostly are azo (Yellow to Red) or metal complex and quinacridone like acridine

Table 3   Methods of effluent Treatment

Physical	Chemical	Biological
Adsorption	Neutralization	Stabilization
Sedimentation	Reduction	Aerated lagoons
Floatation	Oxidation	Trickling filters
Flocculation	Electrolysis	Activated sludge
Coagulation	Ion exchange	Anaerobic digestion
Foam fraction	Wet air oxidation	Bioaugmentation
Reverse osmosis		
Ultrafiltration		
Ionization		
Radiation		

some cases within a few hours (Shin et al. 1997; Balan 1999; Balan and Moteiro 2000; Nyanhongo et al. 2002; Neamtu et al. 2004; Nilsson et al. 2006).

5. BIODEGRADATION OF DYES

Biological methods are currently viewed as effective, specific, less energy intensive and environmentally benign since they have potential of resulting in partial to complete bioconversion of organic pollutants to non-toxic end products (Baker and Herson 1994). Biological dye removal techniques are based on biotransformation of dyes (Poots et al. 1976; Mc Kay 1979; Seshadri et al. 1994; Sorek et al. 1995; Tsatsaroni 1996; Koprivanac et al. 1997; Gabelica et al. 2001; Ramalho et al. 2004; Khehra et al. 2005). Microbes have been isolated which can catalyze anaerobic reductive fission of the azo linkage resulting in formation of colorless highly toxic and carcinogenic aromatic amines. Nevertheless, many researchers have demonstrated partial or complete biodegradation of dyes by pure and mixed cultures of bacteria, fungi and algae.

Table 4   Recent reports on bacterial cultures capable of dye decolorization

Culture	Dye	Mechanism	Reference
Aeromonas hydrophila var 24B	Various azo dyes	Azoreductase	Yatome et al. (1987)
Aeromonas hydrophila var 24B	Various azo dyes	Azoreductase	Idaka & Ogawa (1978)
Bacillus subtilis IFO 13719	2-carboxy 4 dimethy-leamino benzene	Azoreductase	Yatome et al. (1991)

Bacillus subtilis IFO 3002	p-aminobenzene	Azoreductase	Horitsu et al. (1977)
Klebsiella pneumoniae RS-13	Methyl Red	Azoreductase	Wong & Yuen (1996)
Pseudomonas cepacia 13NA	C.I. Acid Orange 12 C..I. Acid Orange 20 C.I. Acid Red 88	Azoreductase Azoreductase	Ogawa et al. (1986)
Pseudomonas cepacia 13 NA	Orange I	Azoreductase	Yatome et al. (1991)
Pseudomonas cepacia 13 NA	p-aminobenzene	Azoreductase	Idaka et al. (1987)
Pseudomonas luteola	Red G RBB	Azoreductase Azoreductase	Hu et al. (1994)
Pseudomonas stutzeri	Azo dyes	Azoreductase	Yatome et al. (1991)
Pseudomonas stutzeri IAM 12097	Orange I Orange II	Azoreductase Azoreductase	Yatome et al. (1991)
Streptomyces BW 130	Azo-reactive Red 147 Azo- copper Red 171 Anthraquinone Blue 114 Formazon Blue 209	Adsorption Adsorption Adsorption Adsorption	Zhou & Zimmermann (1993)
Mixed Bacterial culture	Mordant Yellow	Azoreductase	Haug et al. (1991)
Mixed Bacterial culture	Acid red 88 Acid red 119 Acid red 97 Acid red 113	Azoreductase Azoreductase Azoreductase Azoreductase	Khehra et al. (2005)

5.1 Bacterial Biodegradation

Numerous bacteria capable of dye decolorization have been reported. Efforts to isolate bacterial cultures capable of degrading azo dyes started in 1970s with reports of *Bacillus subtilis* (Horitsu et al. 1977), then *Aeromonas hydrophila* (Idaka and Ogawa 1978), followed by *Bacillus cereus* (Wuhrmann et al. 1980) (Table 4). Investigations to bacterial dye bio-transformation have so far mainly been focused to the most abundant chemical class, azo dyes. The electron withdrawing nature of the azo linkages obstructs the susceptibility of azo dye molecules to oxidative reactions (Fewson 1988). Therefore, azo dyes generally resist aerobic bacterial biodegradation (Pagga and Brown 1986; Jimenez et al.1988; Shaul et al. 1991; Pagga and Taeger 1994; Ganesh et al. 1994). Only bacteria with specialized azo dye reducing enzymes were found to degrade azo dyes under fully aerobic conditions. In contrast, breakdown of azo linkages by reduction under anaerobic conditions is much less specific. This anaerobic reduction implies decolorization as the azo dyes are converted to usually colorless but potentially harmful aromatic amines. Aromatic amines are generally not further degraded under anaerobic conditions. Anaerobic treatment must, therefore, be considered merely as the first stage of the complete degradation of azo dyes.

The second stage involves conversion of the produced aromatic amines. For several aromatic amines, this can be achieved by biodegradation under aerobic conditions. Haug et al. (1991) described a bacterial consortium capable of mineralizing the sulphonated azo dye mordant yellow. An alternation, however, from anaerobic to aerobic conditions was required to achieve complete degradation (Sponza and Isik 2005). This was necessary as different members of the consortium needed different conditions for optimum reactions and the main azo bond cleavage the reductase enzymes, mainly functional under anaerobic conditions (Haug et al. 1991). In a review, Groff and Kim (1989) described a host of bacterial cultures with capabilities to carry out decolorization, including *Rhodococcus* species, *Bacillus cereus*, a *Plesiomonas* species and *Achromobacter* species. Georgiou et al. (2004) demonstrated anaerobic digestion technique for decolorization of cotton textile wastewater. Complete decolorization of dyes was succeeded in 4-5 days. Lee and Pavlostathis (2004) reported reductive decolorization of anthraquinone reactive dyes under methanogenic conditions using mixed methanogenic culture. A part of decolorization (13-17%) was due to biosorption as confirmed by inactivated microbial cells (Aksu 2005).

Coughlin et al. (1999) isolated a *Sphingomonas* strain from a wastewater treatment plant capable of aerobically degrading a suite of azo dyes, as a sole source of carbon and nitrogen. After an analysis of the structures of dyes they suggested that there were certain positions and types of substituents on the azo dye which determined the degradation of the dye. Their strain decolorized dye with either 1-amino-2-napthol or 2-amino-1-napthol in their structure and the decolorization appeared to be through reductive cleavage of the azo bond. On the other hand, *Proteus mirabilis* strain decolorized RRBN by a combination of biodegradative and biosorptive processes. This organism displayed good growth on the contaminant in shake culture but color removal was best in anoxic static culture (Chen et al. 1999). An et al. (2002) recently reported optimum decolorization of several recalcitrant triphenylmethane and azo dyes by *Citrobacter* sp. at pH 7-9 and temperature 35-40°C. Color removal by *Citrobacter* sp. was both by adsorption to cells and enzymatic as evidenced by the experiments with extracellular culture filtrate. Sarnaik and Kanekar (1999) described the aerobic mineralization of the triphenyl methane dye and methyl violet by a strain of *Pseudomonas mendocina* MCMB 402. *P. mendocina* degraded the dye via a number of unidentified metabolites to phenol which then entered the β-ketoadipic acid pathway.

P. luteola cells, growing under shaking conditions for 24h, were capable of removing 59-99% of the color of 7 azo dyes in static conditions (Hu 2001). There are other reports on the "aerobic" metabolism of azo dyes, where the bacterial strain (e.g., *Aeromonas sp.*, *Bacillus subtilis*, *Proteus mirabilis*, *P. pseudomalli BNA*, *P. luteola*) were grown aerobically with complex media or sugars, then incubated (often using high cell densities) without shaking in the presence of different azo dyes (Horitsu et al. 1977; Chen et al. 1999; Chang and Lin 2000; Hayase et al. 2000).

In order to develop novel decolorization processes for practical use, Chang et al. (2001) attempted immobilized-cell system of *P. luteola* with a view to enhance the stability, mechanical strength and reusability of the biocatalyst. Cell immobilization by entrapment within natural or synthetic matrices is particularly suitable for bacterial dye decolorization, since it creates a local anaerobic environment favorable to dye metabolism (Stolz 2001; Pazarlioglu et al. 2005). *P. luteola* cells entrapped in natural and synthetic polymeric matrices efficiently decolorized azo dyes enzymatically. Immobilized cells were less sensitive to dissolved oxygen levels and pH as compared to suspended and immobilized cells. After four repeated experiments, the decolorization rate of the free cells decreased nearly by 45%, while immobilized cells retained 75-85% of their original activity in different matrices.

Chen et al. (2005) investigated fixed-bed bioreactors using gel-entrapped cells of *Pseudomonas luteola* for utilization of azo dye decolorization in continuous mode. The performance of the fixed-bed decolorizers was examined to investigate the effect of bed length, volumetric flow rate, dye loading rate, dye concentration in the feed, as well as the characteristics of two matrix types—calcium alginate and polyacrylamide for cell immobilization. With a constant feeding dye concentration of 50 mg/l, the beds with calcium alginate-immobilized cells had an optimal volumetric decolorization rate of 30.6 mg/h/L and a specific decolorization rate of 2.61 mg/g cell/h when hydraulic retention time (HRT) and dye loading rate was 1.12 hours and 2.25 mg/h, respectively. At approximately same biomass loading, the beds with calcium alginate cells seem to be more economically feasible than those with polyacrylamide cells due to significantly less mass transfer resistance and higher volumetric decolorization rates.

The decolorization of Reactive Black 5 (RB5) by immobilized *Funalia trogii* was investigated by Mazmanci and Unyayar (2005). Cultures of *F. trogii* immobilised on *Luffa cylindrica* sponge could effectively decolorize the dye. Decolorization rate of a 3 days old culture was higher (8.22 mg dye/g dry wt/day) than that of 0 and 6 days old cultures (6.86 and 7.80 mg dye/g dry wt/day). Macroscopic and microscopic examinations showed that dye was not adsorbed to the fungal mycelium.

The performance of aerobic and anaerobic-SBR in treating Orange II containing wastewater was investigated by Ong et al. (2005). The result from a specific oxygen uptake rate (SOUR) study indicated that the average fraction of Orange II removed was 15 and 80% in aerobic and anaerobic-SBR, respectively, demonstrating that the anaerobic microbes may be five times more effective in removing Orange II as compared to aerobic microbes. The anaerobic/aerobic sequential process provides simultaneous color, COD and carcinogenic amine removal in an anaerobic upflow anaerobic sludge blanket reactor and an aerobic completely stirred tank reactor system during decolorization of CI Direct Black 38 (DB 38) dye (Sponza and Isik 2005).

Khehra and co-workers (2005) reported a bacterial consortium consisting of *Bacillus cereus, Pseudomonas putida, Pseudomonas fluorescence* and *Stenotrophomonas acidaminiphila* capable of completely decolorizing Acid Red 119. It was also able to decolorize 99% Acid Red 119, 94% Acid Red 97, 99% Acid Red 113 and 82% of Reactive Red 120 dyes at an initial concentration of 60 mg/L of mineral salts medium in 24 hours.

Bacterial biodegradation of non-azo dyes has received little attention so far. Anthraquinone dyes may possibly be aerobically degraded (Meulenberg et al. 1997). At least it has been demonstrated that three bacterial strains could grow with anthraquinone dye Acid Blue 277:1 as sole source of energy (Walker and Weatherly 2000). Under anaerobic conditions, the transformations of anthraquinone dyes are presumably limited to reduction of quinone to hydroquinone, a reaction that reverses once the molecule is again exposed to oxygen (Pasti-Grigsby and Crawford 1990; Seignez et al. 1996; Fontenot et al. 2001). Some anthraquinone dyes have been observed to be removed from the water phase by formation of an 'insoluble pigment' under anaerobic conditions (Brown and Laboureur 1983). This is in line with the observation that electrochemical reduction of an anthraquinone dye increased its adsorptive properties (DeFazio and Lemley 1999). Aerobic decolourization of triphenylmethane dyes has been demonstrated repeatedly (Yatome et al. 1991; Yatome et al. 1993; Azmi et al. 1998; Sani and Banerjee 1999; Sarnaik and Kanekar 1999) but it also has been stated that these dyes resist degradation in activated sludge systems (Jank et al. 1998). Under anaerobic conditions, the transformation of triphenylmethane dyes is presumably limited to reversible reactions like the reduction of malachite green (Basic Green 4) to leucomalachite green (Henderson et al.1997). Phtalocyanine dyes are probably non-biodegradable, but reversible reduction and decolorization occurs under anaerobic conditions (Nigam et al. 1996; Beydilli et al. 1998).

It is generally recognized that azoreductases play an important role in bacterial dye decolorization. However, only a limited number of studies have attempted molecular characterization of dye decolorization. The genes encoding for azoreductase and other possible proteins involved in decolorization have not been clearly identified. Recently, the gene encoding for an aerobic azoreductase was cloned from *Xenophilus azovorans* KF46F (formerly *Pseudomonas* sp. KF46F), a strain able to grow with carboxylated azo compound 1-(4'-carboxyphenylazo)-2-napthol (Carboxy Orange II) as the sole carbon and energy source (Blumel et al. 2002). The enzyme was heterologously expressed in *E. coli*. A presumed NAD(P)H-binding site was identified in the amino-terminal region of the azoreductase. While the cells extracts from the recombinant strain demonstrated the turnover of several industrially relevant azo dyes, the whole cells of the recombinant *E. coli* were unable to take up sulfonated azo dyes and did not show *in vivo* azoreductase activity.

A recombinant *E. coli* strain NO_3 containing genomic DNA fragments from azoreducing wild type *P. luteola* effectively decolorized an azo dye Reactive Red 22 at the rate of about 17mg/g cell/h (Chang and Lin 2000). In another study from the same group (Chang and Lin 2001), a 6.3 Kb fragment from genome DNA of *Rhodococcus* sp. containing genes responsible for azo dye decolorization was cloned and expressed in *E. coli*. The recombinant strain *E. coli* CY1 decolorized Reactive Red 22 at the rate of 8.2 mg/g cells/h with performance of excellent stability during repeated batch operations.

5.2 Degradation by Actinomycetes

Actinomycetes are known to produce extracellular peroxidases that participate in the initial oxidation of lignin to produce various water-soluble polymeric compounds and have also been shown to catalyze hydroxylation, oxidation and dealkylation reactions against various xenobiotic compounds (Ball et al. 1989; Goszczynski et al. 1994; McMullan et al. 2001). Species of *Streptomyces* and *Thermomonospora* are good examples of actinomycetes capable of effective dye decolorization. Pasti-Grigsby and Crawford

(1990) investigated the ability of ligninolytic microbes (white rot fungi and Streptomyces) to mineralize and decolorize textile dyes and found strong correlation between dye decolorization ability and ligninolytic ability. Decolorization of mono-sulphonated monoazo dye derivatives of azo benzene by the *Streptomyces* sp. was observed with five azo dyes having the common structural pattern of a hydroxy group in the para position relative to the azo linkage and at least one methoxy and/or one alkyl group in an ortho position relative to the hydroxyl group. Paszczynski et al. (1992) compared the efficiency of a soil actinomycete culture, *Streptomyces chromofuscus*, to that of the soil fungus *Phanerochaete chrysosporium* and concluded that the soil bacterium could carry out decolorization, but to a lesser extent than the white-rot fungus. Several other actinomycete strains have been reported with a capability to decolorize reactive dyes including anthraquinone, pthalocyanine and azo dyes, through adsorption of dyes to the cellular biomass without any degradation (Zhou and Zimmermann 1993; Aksu 2005). Other copper based dyes, such as formazan-copper complex dyes, were completely decolorized through degradation by the same actinomycete strains (Zhou and Zimmermann 1993). Ball and Cotton (1996) have also studied three well-characterized lignocellulose degrading actinomycetes, *Streptomyces viridosporus*, *Streptomyces badius* and *Thermomonospora mesophila* and showed that they decolorize the polymeric dye, Poly-R with a maximum decolorization rate of 0.1 unit/day. The potential of different *Nocardia* species, such as *N. corallina* and *N. globeurulla* for their ability to degrade triphenylmethane dyes has also been demonstrated (Yatome et al. 1991).

5.3 Fungal Biodegradation

White-rot fungi are the most widely studied microorganisms for dye decolorization. This property is mainly due to the relatively non-specific activity of their ligninolytic enzymes, such as lignin peroxidase, manganese peroxidase and laccase. The reactions catalyzed by these extracellular enzymes are oxidation reactions, e.g. lignin peroxidase catalyzes the oxidation of non-phenolic aromatics, whereas manganese peroxidase and laccase catalyze the oxidation of phenolic compounds. The same unique non-specific mechanism that gives these fungi the ability to degrade lignin also allow them to degrade a wide range of pollutants and they possess a number of advantages not associated with other bioremediation systems (Thurston 1994; Yaropolov et al. 1994). Virtually all dyes from all chemically distinct groups are prone to fungal oxidation, but there are large differences between fungal species with respect to their catalyzing power and dye selectivity. A clear relationship between dye structure and fungal dye biodegradability has not been established so far (Fu and Viraraghvan 2001). Fungal degradation of aromatic structures is a secondary metabolite event that starts when nutrients (C, N and S) become limiting (Kirk and Farrell 1987). Therefore, while the enzymes are optimally expressed under starving conditions, supplementation of energy substrates and nutrients are necessary for propagation of the cultures. Other important factors for cultivation of white-rot fungi and expression of ligninolytic activity are the availability of cofactors and the pH of the environment.

The decolorization of three polymeric dyes e.g., Polymeric B-411, Polymeric R-481 and Polymeric Y-606 by *Phanerochaete chrysosporium* was first confirmed by Glen and Gold (1983). Their results suggested that decolorization was a secondary metabolite activity linked to the fungus ligninolytic activity. The process was, however, slow; and optimum decolorization took 8 days. Decolorization of the azo dyes such as orange II, tropeolin O, congo red, acid red 114, acid red 88, biebrich scarlet, direct blue 15, chrysophenin, tetrazine and yellow 9 (Cripps et al. 1990; Rafii et al. 1990; Paszczynski et al. 1991), triphenyl methane dyes, basic green 4, crystal violet, brilliant green, cresol red, bromophenol blue and para rose aniline (Bumpus and Brock 1988) by various white-rot fungi has been reported. The degradation of azo, anthraquinone, heterocyclic dyes, triphenyl methane and polymeric dyes by *Phanerochaete chrysosporium* was most extensively studied (Bumpus and Brock 1988; Cripps et al. 1990; Kling and Neto 1991; Ollika et al. 1993). *Phanerochaete chrysosporium* was also shown to biodegrade the azo and heterocyclic dyes, orange II, Tropaeofin O, congo red and azure G (Cripps et al. 1990). *Phanerochaete chrysosporium* removed 87-90% of these dyes and the mycelial mats were visibly colored after 5 days of incubation. The extent of color removal varied depending upon the dye complexity, nitrogen availability in the media and ligninolytic activity in the culture. At low nitrogen concentration, 90% of the color was removed within initial 6 hours, while when excess nitrogen was provided upto 5 days were required to achieve 63-93% decolorization of the dyes. A series of other dyes were also tested and decolorized by *Phanerochaete chrysosporium* at various concentrations, particularly when veratryl alcohol was present in the medium. Veratryl alcohol is believed to stimulate the ligninase activity, which seems to be linked to decolorization (Paszcynski and Crawford 1991; Caplash and Sharma 1992) tested the biodegradation of 18 azo dyes using *Phanerochaete chrysosporium* and only eight were degraded with 40-70%

color removal. This degradation was mainly through the lignin degrading enzyme system or adsorption to cell mass.

Substitution with sulfo groups on the aromatic component of some azo dyes did not seem to significantly affect the biodegradability of the azo dyes (Paszcynski et al. 1992; Pasti-Grigsby et al. 1992). Spadara et al. (1992), in contrast, showed that when aromatic rings of dyes had substituted hydroxyl, amino, acetamide, or nitro functional groups, mineralization was greater than those with unsubstituted rings. Bumpus and Brock (1988) described the biodegradation of crystal violet and six other triphenylmethane dyes in ligninolytic cultures of the *Phanerochaete chrysosporium*. Unexpectedly degradation of crystal violet also occurs in non-ligninolytic cultures of *Phanerochaete chrysosporium* suggesting that in addition to the lignin degrading system, other mechanism also exist in this fungus which is also able to degrade crystal violet. However, in the majority of the above cases it is generally observed that nitrogen limitation increased ligninolytic activities through increased lignin peroxidase and manganese-dependant peroxidase, and therefore enhanced decolorization (Perie and Gold 1991).

In addition to *Phanerochaete chrysosporium*, several other basidiomycetes such as *Cyathus*, *Trametes versicolor*, *Bjerkandera adusta*, *Pleurotus*, *Phelbia* and *Thelephora* species have also been observed for their dye decolorization activity (Vasdev and Kuhad 1994; Heinfling et al. 1998; Conneely et al. 1999; Swamy and Ramsay 1999; Kirby et al. 2000; Pointing et al. 2000; Selvam et al. 2003). Vasdev et al. (1995) observed effective decolorization of three triphenylmethane dyes by *Cyathus bulleri* and *C. stercoreus*. Abadulla et al. (2000) reported the use of purified laccase from *Trametes hirsuta* to degrade triarylmethane, indigoid, azo, and anthraquinonic dyes. Heinfling et al. (1998) described the transformation of six azo dyes and phthalocyanine dyes by ligninolytic peroxidases from *Bjerkandera adusta* and *Pleurotus erngii*. Wesenberg et al. (2002) reported that the white-rot fungus *Clitocybula dusenii* partially decolorized the dye wastewater and the fungus produced higher manganese peroxidase and laccase activities when grown in effluent. Novotny et al. (2000) have reported that the white-rot fungus *Irpex lacteus* and *Pleurotus ostreatus* are the potential candidates for the removal of dye remazol brilliant blue red present in the soil under *in vitro* conditions. Another white-rot fungus, *Pycnoporous cinnabarinus*, was also found to tolerate and decolorize high concentrations of dyes in a packed-bed reactor within 48-72h incubation. An extracellular oxidase activity similar to the veratryl alcohol, which acts as an inducer for ligninolytic activity in

Phanerochaete chrysosporium, was observed in *P. cinnabarinus*, which confirmed its suitability for the degradation of dye effluent (Schliephake et al. 1993). Nyanhongo et al. (2002) reported the decolorization of synthetic dyes by *Trametes modesta* under acidic conditions. The decolorization efficiency of *Trametes modesta* laccase was improved in the presence of mediators 1-hydroxybenzotriazole and 2-methoxyphenothiazine. Peralta-Zamosa (2003) reported the use of immobilized laccase for decolorization of Remazoll Brilliant Blue R, Remazoll Brilliant Black B, Reactive Orange 122 and Reactive Red 251. Blanquez et al. (2004) reported 90% decolorization of grey Lanaset G, which consists of metal complexed dyes by *Trametes versicolor*. D'Souza et al. (2006) reported the decolorization of textile effluent by marine fungus NIOCC≠2a to various degrees within 6 hours of incubation.

Besides white-rot fungi, other filamentous fungi like *A. niger, Trichoderme viride* and *A. foetidus* have been shown to be capable of decolorizing textile dyes such as scarlet direct red, fast greenish blue, and brilliant direct violet (Kousar et al. 2000; Sumathi and Manju 2000; Fu and Viraraghavan 2001). *Neurospora crassa* was reported to decolorize diazo dyes by Corso and co-workers (1981) (Table 5). They achieved 89-91% color removal within 24 hours incubation when the concentration of the dye Vermelho Reanil P8B ranged between 16.0 and 20.0 mg/L, which resemble the range of concentrations found in industrial dye effluents. De Angelis and Rodrigues (1987) tested color removal of textile dyes using a *Candida* sp. and demonstrated 93-98% decolorization for several Procyon dyes at 100 mg/L. A strain of *Aspergillus sojae* B-10 was also shown to be able to decolorize the azo dyes Amaranth, Congo red, and Sudan III in nitrogen-poor media after 3-5 days incubation (Ryu and Weon 1992). Recently, other facultative anaerobic fungi capable of growth on dyes as sole carbon sources have been reported. However, they do not seem to be able to carry out decolorization (Marchant et al. 1994; Nigam et al. 1995a;b). They appear to cleave some of the bonds in these dyes to use as carbon sources, yet do not affect the chromophore centre of the dyes. This capability might be of significance when a consortium of microorganisms is employed in degrading dye-containing effluents when other decolorizers are present. Both types, the degrading and the decolorizing microorganisms, would ultimately benefit from each other's activities to achieve more complete or faster biodegradation. Ramalho et al. (2004) reported the reduction of various azo dyes by growing cultures of an ascomycete yeast species, *Issatchenkia occidentalis*. Under microaerophilic conditions 80% of the dyes were

removed within 15 hours. Decolorization of Poly R-478 and Poly S-119 by
Penicillium appeared to involve initial adsorption and followed by
biodegradation (Zheng et al. 1999). *Geotrichum candidum* Dec 1 was found to
decolorize 21 kinds of synthetic dyes; an extracellular enzyme dye-
decolorizing peroxidase (DyP) was responsible for the decolorization of the
dyes (Kim and Shoda, 1999). In order to produce large amounts of DyP with
dye-decolorizing activity, Sugano et al. (2000) achieved efficient
heterologous expression of DyP from *Geotrichum candidum* Dec 1 in *A. oryzae*,
and by fusing mature cDNA encoding DyP with *A. oryzae* α-amylase
promoter (amyB). *A. oryzae* is a safe host with higher growth rate and with
the capacity to secrete gram per litre quantities of heterologous proteins.

Table 5   Recent Reports on various fungi capable of dye decolorization

Culture	Dye	Mechanism	Reference
Aspergillus sojae B-10	Amaranth	NR	Ryu & Weon (1992)
	Sudan III	NR	
	Congo Red	NR	
Myrothecium verrucaria	Orange II	Adsorption	Brahimi-Horn et al. (1992)
	10 B (Blue)	Adsorption	
	RS (Red)	Adsorption	
Myrothecum sp.	Orange II	Adsorption	Mou et al. (1991)
	10 B (Blue)	Adsorption	
	RS (Red)	Adsorption	
Neurospora crassa	Vermelho Reanil P8B	Adsorption	Corso et al. (1981)
Pycnoporus cinnabarinus	Pigment plant effluent	Extracellular oxidases	Schliephake et al. (1993)
Candida sp.	Procyon Black SPL	Adsorption	De Angelis & Rodrigues (1987)
	Procyon Blue MX2G	Adsorption	
	Procyon Red HE7B	Adsorption	
	Procyon Orange HER	Adsorption	
Trametes versicolor	Reactive Red 2	Ligninolytic enzymes	Nilsson et al. (2006)
	Reactive Blue 4		
Phanerochaete chrysosporium	Amaranth	Lignin Peroxidase	Chao and lee (1994)

5.4 Algal Biodegradation

Degradation of a number of azo dyes by algae has also been reported in a few
studies (Jinqi and Houtian 1992; Semple et al. 1999). The degradation

pathway is thought to involve reductive cleavage of the azo linkage followed by further degradation (mineralization) of the formed aromatic amines. Hence, algae have been demonstrated to degrade several aromatic amines, even sulphonated ones (Luther and Soeder 1987; Soeder et al. 1987; Luther and Soeder 1991). Recently, Marungrueng and Pavasant (2005) had reported the adsorption of astrazo blue FGRL by *Caulerpa lentillifera*. In open wastewater treatment systems, especially in (Shallow) stabilization ponds, algae may, therefore, contribute to the removal of azo dyes and aromatic amines from the water phase.

6. MECHANISM OF DYE DECOLORIZATION

From various studies conducted so far, it becomes clear that a two-step mechanism *viz.* the physical adsorption and enzymatic degradation are involved in dye decolorization by white-rot fungus. Knapp and Newby observed that in many cases adsorption of dye is in the microbial cell surface with primary mechanism of decolorization. Young and Yu (1997) suggested that the binding of dyes to the fungal hyphae and physical adsorption and enzymatic degradation by extracellular and intracellular enzymes as reasons for the color removal. The dye saturated mycelium can be regenerated and used for repeated dye adsorption. They have further stated that the dyes were not decolorized by manganese peroxidase while above 80% color was removed by ligninase-catalyzed oxidation. Dyes with different structures are decolorized at different intrinsic enzymatic rates and high dye concentration results in slower decolorization rates (Abadulla et al. 2000). The dye was adsorbed to the mycelial pellets in both ligninolytic and non-ligninolytic cultures (Akkaya and Ozer 2005). Wang and Yu (1998) reported the adsorption of acid green 27, acid violet 7 and indigo carmine dyes on living and dead mycelium of *Trametes versicolor*. The physical desorption and enzymatic degradation of the adsorbed dye molecules was the major mechanism in which the regeneration of dye adsorption capacity of the mycelium was achieved.

Zheng et al. (2005) studied the effects of pH, suspended solids, ionic strength and solution temperature on dye adsorption and thermodynamics by aerobic granules. Results show that pH was an important factor governing the adsorption with optimal pH around 7.0. However, at suspended solids concentrations >1 g/l, the dye adsorption density remained almost unchanged. An increase in ionic strength and solution temperature decreased the dye adsorption density. Isotherm model analysis indicated that the dye adsorption by the aerobic granules could be described by Langmuir equation very well.

The involvement of LME in the dye decolorization process has been confirmed in several studies using purified cell free enzymes. Enzymes such as lignin peroxidase, manganese peroxidase and laccase are involved in lignin degradation, which participates in the decolorization of the dyes (Vyas and Molitoris 1995). Kim et al. (1996) reported the presence of H_2O_2 dependent Remazol Brilliant blue R decolorizing enzymatic activity in the culture filtrate of *Pleurotus ostreatus* in chemically defined medium. Lignin peroxidase of *Phanerochaete chrysosporium* has been shown to decolorize azo, triphenylmethane and heterocyclic dyes in the presence of veratryl alcohol and H_2O_2 (Cripps et al. 1990; Ollika et al. 1993). It is interesting that the ability of enzyme preparation to decolorize identical dyes varied between the two studies. Extracellular fluid from cultures of *Phanerochaete chrysosporium* and purified lignin peroxidase were able to degrade crystal violet and six other triphenylmethane dyes by sequential N-demethylation (Bumpus and Brock 1988). The role of purified lignin peroxidase in the decolorization of azo dyes has been clearly demonstrated (Paszczynski and Crawford 1991). The different isoenzymes of lignin peroxidase produced by *Phanerochaete chrysosporium* are able to decolorize several dyes with different chemical structures including azo, triphenylmethane, hetero cyclic and polymeric dyes (Ollika et al. 1993). The chemical steps involved in the degradation of azo dyes by LiP and MnP have been elucidated (Pasti-Grigsby et al. 1992; Goszcaynski et al. 1994). The mechanism of azo dye oxidation by peroxidases such as lignin peroxidase probably involves the oxidation of the phenolic groups to reduce a radical at the carbon bearing the azo linkage. Then, water attacks this phenolic carbon to cleave the molecule producing phenyl diazine. The phenyl diazine can be oxidized by one-electron reaction generating N_2 (Spadaro et al. 1994; Chivukula et al. 1995). The early stages of azo dye decolorization is the breaking of the azo bond, the ease of which has been found to be depending on the identity, number and position of functional groups in the aromatic region and the resulting interactions with the azo bond. Further degradation of azo dyes involves aromatic cleavage which has also been found to be dependent on the identity of the ring substituents with the presence of phenolic, amino acetamide, 2-methoxyphenol, or other easily biodegradable functional groups resulting in a greater extent of degradation (Paszczynski and Crawford 1991; Paszczynski et al. 1992;). Manganese peroxidase from *Phanerochaete chrysosporium* was also able to decolorize several azo dyes *in vitro*, and the decolorization rate was dependent on the chemical structure of the dye (Pasti-Grigsby et al. 1992).

The large number of azo dyes can also be reduced by many different bacteria indicating that azo dye reduction is a non-specific reaction and that the capability of reducing azo dye can be considered a universal property of anaerobically growing bacteria. Anaerobic azo dye reduction is the reductive cleavage of azo linkages, i.e., the transfer of reducing equivalents resulting in the formation of aromatic amines. As aromatic amines are generally colorless, azo dye reduction is also referred to as azo dye decolorization. According to the mechanism of biological azo dye reduction by bacteria, enzymes transfer the reducing equivalents originating from the oxidation of organic substrates to the azo dyes. Enzymes that catalyze azo dye reduction may either be specialized enzymes (catalyzing only the reduction of azo dyes) or non-specialized enzymes (non-specific enzymes that catalyze the reduction of a wide range of compounds, including azo dyes). Evidence for the existence of specialized azo dye reducing enzymes, so-called azoreductases, has so far only been found in studies with some aerobic and facultative aerobic bacteria that could grow with mostly simple azo compounds as sole source of carbon and energy. These strains grew under strict aerobic conditions by using a metabolism that started with reductive cleavage of the azo linkage (Kulla et al. 1983 a,b). These intracellular azoreductases showed high specificity to dye structures. There is no clear evidence for the existence of specific azoreductases in anaerobically grown bacteria. However, also under anaerobic conditions, non-specific enzymes may be responsible for the almost ubiquitous capacity of many strains of anaerobic, facultative anaerobic and even aerobic bacteria to reduce azo dyes.

CONCLUSION

Colored dye wastewater treatment and decolorization presents a difficult task. Among the most economically viable choices available for effluent treatment, appear to be the biological system. At present biological systems are known to be capable of dealing with BOD and COD reduction or removal through conventional aerobic biodegradation. However, they have an inherent problem in their inability to remove color. Although decolorization is a challenging process to both the textile industry and the wastewater treatment facilities that treat them. Critical analysis of the literature shows that decolorization of dyes by white-rot fungi offers several advantages over the conventional treatment system due to its potential ligninolytic enzyme system. It is hoped that enzyme based treatment systems will be the technologies of the future. Genetically engineered strains, suitable for degradation process would play an important role in this. Though the white-rot fungi are efficient decolorizers, fungal-based color removal processes in contaminated industrial wastewater are still limited. White-rot fungal bioremediation treatments may be particularly appropriate for *in situ* remediation of soils, where recalcitrant compounds and bio-availability are problematic. A further application may lie in the operation of bioreactors for synthetic dyes in liquid waste, where near 100% degradation efficiencies have been achieved using white-rot fungi. Concerted efforts are still required to establish biological decolorization systems. The techniques by which decolorization occurs vary and among them adsorption seems of great significance for future development in bio-removal or bio-recovery of dye substances. The role of molecular biology has yet to feature prominently in this vital area of environmental protection.

ACKNOWLEDGEMENTS

We thankfully acknowledge Department of Biotechnology (DBT), Government of India, for financial support to carry out the research work on decolorization of textile industrial waste.

REFERENCES

Abadulla E, Tzanov T, Costa S, Robra KH, Cavaco-Paula A, Gubitz GM (2000). Decolourization and detoxification of textile dyes with laccase from *Trametes hirsutus*. Appl Environ Microbiol 66: 3357-3362.

Addour L, Belhocine D, Boudries N, Comeau Y, Pauss A, Mameri N (1999). Zinc uptake by *Streptomyces rimosus* biomass using a packed bed column. J Chem Technol Biotechnol 74: 1089-1095.

Akkaya G, Ozer A (2005). Biosorption of Acid Red 274 (AR 274) on *Dicranella varia*: Determination of equilibrium and kinetic model parameters. Proc Biochem 40:3559-3568.

Aksu Z (2005). Application of biosorption for the removal of organic pollutants: a review. Proc Biochem 40: 997-1026.

Aksu Z, Acikel U, Kutsal T, (1997). Application of multicomponent process adsorption isotherm to simultaneous biosorption of iron (III) and chromium (VI) by *Chlorella vulgaris*. J Chem Technol Biotechnol 70: 368-375.

An SY, Min SK, Cha IH, Choi YS, Kim CH, Lee YC (2002). Decolorization of triphenymethane and azo dyes by *Citrobacter* sp. Biotechnol Lett 24: 1037-1040.

Anliker R, Clarke EA, Moser P (1981). Use of the partition coefficient as an indicator of bioaccumulation tendency of dye stuffs in fish. Chemosphere 10: 263-274.

Azmi W, Sani RK, Banerjee UC (1998). Biodegradation of triphenylmethane dyes. Enzyme Microbial Technol 22: 185-191.

Baker KH, Herson DS (1994). Bioremediation of surface and subsurface soils, in bioremediation (Baker, K.H. and Herson, D.S.) In Handbook Introduction and overview of bioremediation p. 1, McGraw Hill, New York (1994).

Balan DSL (1999). Biodegradabilidade e toxicidade de effluents texties. Revista Brasileira de Quimica Textil XXII. 54: 26.

Balan DSL, Monteiro RTR (2000). Indigo dye effluent decolourization by lignolytic fungi. Symp. on biotechnology in the textile industry, Portugal 3-7 May 33.

Ball AS, Betts WB, McCarthy AJ (1989). Degradation of lignin-related compounds by actinomycetes. Appl Environ Microbiol 55: 1642-1644.

Ball AS, Cotton J (1996). Decolourization of two polymeric dye Poly R by *Streptomyces viridosporus* T7A. J Basic Microbiol 36: 13-18.

Banat IM, Nigam P, Singh D, Marchant R (1996). Microbial decolorization of textile-dye-containing effluents: a review. Biores Technol 58: 217-227.

Beughman GL, Perenich TA (1998). Fate of dyes in aquatic systems:1 solubility and partitioning of some hydrophobic dyes and related compounds. Environ Toxicol Chem 7:183-199.

Beydilli MI, Pavlostathis SG, Tincher WC (1998). Decolourization and toxicity screening of selected reactive azo dyes under methanogenic conditions. Water Sci Technol 38: 225-232.

Blanquez P, Casas N, Font X, Gabarrell X, Sarra M (2004). Mechanism of textile metal dye biotransformation by *Trametes versicolor*. Water Res 38:2166-2172.

Blumel S, Knackmuss HJ, Stolz A (2002). Molecular cloning and characterization of the Gene coding for the Aerobic Azoreductase from *Xenophilus azovorans* KF46F. Appl Envir Microbiol 68:3948-3955.

Brahimi-Horn MC, Lim KK, Liang SL, Mou DG (1992). Binding of textile azo dyes by *Myrothecium verrucaria*, J Ind Microbiol 10:31-36.

Brown D, Laboureur P (1983). The degradation of dye stuffs: Part 1- Primary biodegradation under anaerobic conditions. Chemosphere 12: 397-404.

Brown MA, DeVito SC (1993). Predicting azo dye toxicity. Crit Rev Env Sci Technol 23: 249-324.

Bumpus JA, Brock BJ (1988). Biodegradation of crystal violet by the white-rot fungus *Phanerochate chrysosporium*. Appl Environ Microbiol 56:1143-1147.

Capalash N, Sharma P (1992). Biodegradation of textile azo-dyes by *Phanerochaete chrysosporium* World J Microbiol Biotechnol 8:309-312.

Cartwright RA (1983). Historical and modern epidemiological studies on populations exposed to N-substituted aryl compounds. Environ Health Persp 4913-4919.

Cassitto MG (1993). Carbon disulphide and the central nervous system: a 15 year neurobehavioural survillence of an exposed population.Environ Res 63:252-257.

Chang JS, and Lin YC (2001). Decolorization kinetics of a recombinant *Escherichia coli* strain harbouring azo-dye-decolorizing determinants from *Rhodococcus* sp. Biotechnol Lett 23: 631-636.

Chang JS, Chou C, Chen SY (2001). Decolorization of azo dyes with immobilized *Pseudomonas luteola*. Proc Biochem 36:757-763.

Chang JS, Lin YC (2000). Fed-batch bioreactor strategies for microbial decolorization of azo dye using a *Pseudomonas luteola* strain. Biotechnol Prog 16:979-985.

Chen BY, Chen SY, Chang JS (2005). Immobilized cell fixed-bed bioreactor for wastewater decolorization. Proc Biochem 40:3434–3440.

Chen KC, Huang WT, Wu JY, Houng JY (1999). Microbial decolorization of azo dyes by *Proteus mirabilis*. J Ind Microbiol Biotechnol 23:686-690.

Chivukula M, Renganthan V (1995). Phenolic azo dye oxidation by laccase from *Pyricularia oryzae*. Appl Environ Microbiol 60:4374-4377.

Chung KT, Cerniglia CE (1992). Mutagenicity of azo dyes: structure-activity relationships. Mutat Res 277:201-220.

Clarke AE and Anliker R (1980). Organic dye and pigments, in: The Handbook of Environmental chemistry, Hutzinger, editor. Springer-Verlag, Berlin. p 178.

Conneely A, Smyth WF, McMullan G (1999). Metabolism of the pthallocyanin textile dye Remazol turquoise blue by *Phanerochate chrysosporium* FEMS Microbiol Lett 179:333-338.

Cooper P (1993). Removing colour from dye house wastewaters-a critical review of technology available. J Soc Dyers Col 109:97-100.

Corso CR, Angelis de DF, Oliveira de JE, Kiyan C (1981). Interaction between the diazo dye, "Vermelho Reanil" P8B, and *Neurospora crassa* strain 74A. Appl Microbiol Biotechnol 13(1):64-66.

Corso CR, De Angelis DF, De Oliveira JE, Kiyan C (1992). Interaction between the diazo dye Eur. J. Appl. Microbiol 240.

Coughlin MF, Kinkle BK, Bishop PL (1999). Degradation of azo dyes containing sacinonapthol by *Sphingomonas* sp. Strain 1Cx. J Ind Microbiol Biotechnol 23:341-346.

Cripps C, Bumpus JA, Aust, SD (1990). Biodegradation of azo and heterocyclic dyes by *Phanerochaete chrysosporium*. Appl Environ Microbiol 56:1114-1118.

D'Souza DT, Tiwari R, Sah AK, Raghukumar C (2006). Enhanced production of laccase by a marine fungus during treatment of colored effluents and synthetic dyes. Enzyme Microbial Technol 6:504-511.

DeAngelis FE and Rodrigues GS (1987). Azo dyes removal from industrial effluents using Yeast biomass. Arquiros De Biologia e Technologia 30: 301-309.

DeFazio SAK, Lemley AT (1999). Electrochemical treatment of acid dye systems: Sodium meta-bisulfite addition to the acidco system. J. Environ. Sci. Health 34: 217-240.

Dubin P, Wright KL (1975). Reduction of azo food dyes in culture of *Proteus Vulgaris*. Xenobiotica 5:563-571.

Easton JR (1995). The dye maker's view, in colour in dye house effluent, Cooper P, editor. UK: Society of Dyers and Colourist: Bradford, England. p 9-21

Fewson CA (1988). Biodegradation of xenobiotics and other persistent compounds: the causes of recalcitrance. Trends Biotechnol 6: 148-153.

Fontenot EJ, Beydilli MI, Lee YH Pavlostathis SG (2001). Kinetics and inhibition during the decolorization of reactive anthraquinone dyes under methanogenic conditions. In: 9th World Congress Anaerobic Digestion 2001 – Anaerobic conversion for sustainability. A.F.M. Van Valsen and Verstraete, W.H.(eds.), Antwerpen, Belgium, 2-6 Sept. Technologisch Instituut vzw, p.215-220.

Fu Y, and Viraraghavan T (2001). Fungal decolourization of dye waste water: a review. Biores Technol 79: 251262.

Gabelica Z, Valange S, Shibata M, Hotta H, Suzuki T (2001). Stability against color fading of azo dyes encapsulated in Ca-aluminosilicate mesoporous substrates. Microporous and Mesoporous Materials 44:645-652.

Ganesh R, Boardman GD, Michelsen D (1994). Fate of azo dyes in sludges. Water Res 28: 1367–1376.

Georgiou D, Metallinou C, Aivasidis A, Voudrias E, Gimouhopoulos K (2004). Decolorization of azo-reactive dyes and cotton-textile wastewater using anaerobic digestion and acetate-consuming bacteria. Biochem Engg J 19:75-79.

Glenn JK, Gold MH (1983). Mn (II) Decolorization of several polymeric dyes by the lignin degrading basidiomycete *Phanerochaete chrysosporium*. Appl Environ Microbiol 45:1741-1747.

Gonazoles CA, Elio A, Gonzalo LB (1988). Bladder cancer among workers in the textile industry; results of a Spanish case control study. Am J Indust Medic 14:637-641.

Goszczynski S, Paszczynski A, Pasti-Grigsby MB, Crawford RL, Crawford DL (1994). New pathway for degradation of sulfonated azo dyes by microbial peroxidases of *Phanerochaete chrysosporium* and *Streptomyces chromofuscus*. J Bacteriol 176:1339-1347.

Grau P (1991). Textile industry wastewaters treatment. Water Sci Technol 24: 97-103.

Groff KA (1993). Textile waste-textile industry wastewater waste disposal; a review. Water Environ. Res 65: 421-423.

Groff KA, Kim BR (1989). Textile wastes. J Water Pollut Control Fed 63: 872-876.

Hachem C, Bocquillon F, Zahraa O, Bouchy M (2001). Decolourization of textile industry wastewater by the photocatalytic degradation process. Dyes and Pigments 49:117-125

Hao OJ, Chang PC (2000). Decolourization of wastewater. Crit Rev Env Sci Technol 30: 449-505.

Haug W, Schmidt A, Nortemann B, Hempel DC, Stolz A, Knackmuss HJ (1991). Mineralization of the sulfonated azo dye Mordant Yellow 3 by a 6-aminonaphthalene-2-sulfonate-degrading bacteria consortium. Appl Environ Microbiol 57: 3144-3149.

Hayase N, Kouno K, Ushio K (2000). Isolation and characterization of *Aeromonas* sp. B-5 capable of decolorizing various dyes. J Biosci Bioengg 90:570-573.

Heinfling A, Martinez MJ, Martinez AT, Berbauer M, Szewzyk U (1998). Purification and characterization of peroxidases from the dye-decolorizing fungus *Bjerkandera adusta*. Appl Environ Microbiol 165:43-50.

Henderson AL, Schmitt TC, Heinze TM, Cerniglia CE (1997). Reduction of malachite green to leucomalachite green by intestinal bacteria. Appl Environ Microbiol 63: 4099-4101.

Horitsu H, Takada M, Idaka E, Tomoyeda M, Ogawa T (1977). Degradation of p-aminoazobenzene by *Bacillus subtilis*. Eur J Appl Microbiol 4: 217-224.

Hu TL (1994). Decolourization of reactive azo dyes by transformation with *Pseudomonas luteola*. Biores Technol 49:47-51.

Hu TL (2001). Kinetics of azoreductases and assessment of toxicity of anaerobic products from azo dyes by *Pseudomonas luteola*. Water Sci Technol 43:261-269.

Idaka E, Ogawa T, Horitsu H (1987). Oxidative pathway after reductions of p-aminobenzene by *Pseudomonas cepacia*. Bull Environ Contam Toxicol 39:108-113.

Idaka E, Ogawa Y (1978). Degradation of azo compounds by *Aeromonas hydrophila* var. 2413. J Soc Dyers Colorists 94: 91-94.

Jank M, Koser H, Lucking F, Martienssen M, Wittchen S (1998). Decolourization and degradation of Erioglaucine (acid blue 9) dye in wastewater. Environ Technol 19: 741-747.

Jimenez B, Noyola A, Capdeville B (1988). Selected dyes for residence time distribution evaluation in bioreactors. Biotechnol Techniques 2: 77-82.

Jinqi L, Houtian L (1992). Degradation of azo dyes by algae. Environ Pollut 75: 273-278.

Joachim F, burrel A, Anderson (1985). Mutagenicity of azo dyes in the *Salmonella* microsome assay *in vitro* and *in vivo* activation. J Muta Res 156:131-137.

Jorosz-Wilkolazka AJ, Rdest J, Malarczy E, EdwardsM, Leonowicz A (2002). Fungi and their ability to decolourize azo and anthraquinonic dyes. Enz Microb Technol 566-572.

Judkins Jr JF (1984). Textile wastewater. J. Water Pollut Control Fed 56: 642.

Kapdan IK, Kargi F, McMullanG, Marchant R (2000). Effect of environmental condition on biological decolorization of textile dye stuff by *Corilous versicolor.* Enz Microb Technol 26:381-387.

Kashinath A, Novotny CK, Kamalesh KS, Patel C, Vaclava (2003). Decolorization of synthetic dyes by *Irpex lacteus* in liquid culture and packed bed bioreactor. Enz Microb Technol 32:167-171.

Khehra MS, Saini HS, Sharma DK ,Chadha BS, Chimni SS (2005). Decolorization of various azo dyes by bacterial consortium. Dyes and Pigments 67:55-61.

Kim BS, Ryu SJ, Shin KS (1996). Effect of culture parameters on the decolourization of remazol brilliant blue R by *Pleurotus ostreatus.* J Microbiol 34:101-104.

Kim SJ, Shoda P (1999). Purification and characterization of novel peroxidase from *Geotrichum candidum* dec 1 involoved in decolorization of dyes. Appl Environ Microbiol 65:1029-1035.

Kirby N, Marchant R, McMullan G (2000). Decolorization of synthetic textile dyes by *Phelbia tremellosa.* FEMS Microbiol Lett 188:93-96.

Kirk TK, Farrell RL (1987). Enzymatic 'combustion'. The microbial degradation of lignin. Annu Rev Microbiol 41: 465-505.

Kling SH, Neto JSA (1991). Oxidation of methylene blue by crude lignin peroxidase from *Phanerochaete chrysosporium* J Biotechnol 21:295-300.

Knapp JS, Newby PS (1995). The Microbial Decolorization of an Industrial Effluent Containing Diazo Linked Chromophore. Water Res 29:1807-1809.

Koprivanac N, Papi S, Hergold-Brundi A, Nagl A, Parac-Osterman D, Grabari Z (1997). Constitution and dyeing properties of a 2:2 copper complex azomethine dye. Dyes and Pigments 35: 57-68.

Kousar N, Seshikala D, Singara C (2000). Decolorization of textile dyes by fungi. Ind J Microbiol 40:191-197.

Kuhad RC, Sood N, Tripathi KK, Singh A, Ward, OP (2004). Developments in microbial methods for the treatment of dye effluents. Adv Appl Microbiol 50:185-213.

Kulla HG (1981). Aerobic bacterial degradation of azodyes; 12th FEMS Symposium on Microbial Degradation of Xenobiotics and Recalcitrant Compounds, Academic Press, London.

Kulla HG, Klausener FK, Meyer U, Ludeke B, Leisinger T (1983). Interference of aromatic sulfo groups in the microbial degradation of azo dyes orange I and orange II. Arch Microbiol 135: 1- 7.

Kuo WG (1992). Decolourizing dye waste water with Fenton's reagent. Water Res 26: 881-886.

Lee YH, Pavlostathis SG (2004). Decolorization and toxicity of reactive anthraquinone textile dyes under methanogenic conditions. Water Res 38:1838-1852.

Luther M, and Soeder CJ (1987). Some naphthalenesulfonic acids as sulfur sources for the green microalga, *Scenedesmus obliquus*. Chemosphere 16: 1565 -1578. Luther M, and Soeder CJ (1991). 1-naphthalenesulfonic acid and sulfate as sulfur souces for the green alga, *Scenedesmus obliquus*. Water Res 25: 299-308.

Luther M, Soeder CJ (1991). 1-Naphthalenesulfonic acid and sulfate as sulfur sources for the green alga *Scenedesmus obliquus*. Water Res 25:299-307.

Manu B, Chaudhary S (2002). Anaerobic decolourization of simulated textile wastewater containing azo dyes. Biores Technol 82: 225-231.

Marchant R, Nigam P, Banat IM (1994). An unusual facultatively anaerobic fungus isolated from prolonged enrichment culture conditions. Mycol Res 98: 757-760.

Marungrueng K, Pavasant P (2005). Removal of basic dye (Astrazon Blue FGRL) using macroalga *Caulerpa lentillifera*. J Environ Managmt (in press).

Mazmanci MA, Unyayar A (2005). Decolourisation of Reactive Black 5 by *Funalia trogii* immobilised on *Luffa cylindrica* sponge. Proc Biochem 40:337-342.

McKay G (1979). Water colour removal from textile effluents. Am Dyestuff Reporter 68: 29-36.

McMullan G, Meehan C, Conneely A, Nirby N, Robinson T, Nigam P, Banat IM, Meulenberg R, Rijnaarts HHM, Doddema HJ, Field JA (2001). Mini review: Microbial decolourization and degradation of textile dyes. FEMS Microbiol.Lett 152: 45-49.

Meulenberg R, Rijnaarts HHM, Doddema HJ, Field JA (1997). Partially oxidized polycyclic aromatic hydrocarbons show an increased bioavailability and biodegradability. FEMS Microbiol Lett 152:45-49.

Meyer U (1981). Biodegradation of synthetic organic colorants. 12th FEMS Symposium on Microbial Degradation of Xenobiotic and Recalcitrant Compounds, Academic Press, London.

Meyer U (1993). Biodegradation of synthetic organic colourants. FEMS Symp 12:371-390.

Michaels GB, Lewis DL (1985). Sorption and toxicity of azo and triphenylmethane dyes to aquatic microbial populations. Environ Toxicol Chem 4: 45-50.

Mishra G, Tripathy M (1993). A critical review of the treatments for decolorization of textile effluents. Colourage 40:35-38.

Moran C, Hall ME, Howell RC (1997). Effect of sewage treatment on textile effluents. J Soc Dyers Colour 113:272-274.

Mou DG, Lin KK, Shen HP (1991). Microbial agents for decolorization of dye wastewater. Biotechnol Adv 9: 613-622.

Murugesan K (2002). Studies on production, purification, characterization and crystallization of laccase from a white-rot fungus *Pleurotus sajor-caju* and its application in bioremediation of textile dye effluent and dye contaminated soils. Madras University, Ph.D thesis.

Neamtu M, Yediler A, Siminiceanu I, Macoveanu M, Kettrup A (2004). Decolorization of disperse red 354 azo dye in water by several oxidation processes—a comparative study. Dyes and Pigments 60:61-68.

Nigam P, Armour G, Banat IM, Singh D, Marchant R (2000). Physical removal of textile effluents and solid-state fermentation of dye-adsorbed agricultural residues. Biores Technol 72: 219-226.

Nigam P, Banat IM, Oxspring D, Marchant R, Singh D, Smyth WF (1995a). A new facultative anaerobic filamentous fungus capable of growth on recalcitrant textile dyes as sole carbon source. Microbios 84:171-185.

Nigam P, Banat IM, Singh D, Marchant R (1996). Microbial process for the decolourization of textile effluent containing azo, diazo and reactive dyes. Process Biochem 31: 435 442.

Nigam P, Singh D, Marchant R (1995b). In: M.Moo-Young (ed) Environmental Biotechnology: Principles and Applications (Ed.), p.278, Kluwer Academic, The Netherlands, pp 278-292.

Nilsson I, Moller A, Mattiasson B, Rubindamayugi MST, Welander U (2006). Decolorization of synthetic and real textile wastewater by the use of white-rot fungi. Enzyme and Microbial Technol 38:94-100.

Novotny C, Erbanova P, Cajthami T, Rothschild N, Dosoretz D, Sasek V (2000). *Irpex lacteus*, a white-rot fungus applicable to water and soil bioremediation. Appl Microbiol Biotechnol 54:850-856.

Novotny C, Rawal B, Bhatt M, Patel M, Sasek V, Molitoris HP (2001). Capacity of *Irpex lacteus* and *Pleurotus ostreatus* for decolorization of chemically different dyes. J Biotech 89:113-122.

Nyanhongo GS, Gomes J, Gübitz GM, Zvauya R, Read J, Steiner W (2002). Decolorization of textile dyes by laccases from a newly isolated strain of *Trametes modesta* . Water res 84:1449-1456.

O'Neill C, Hawkes FR, Hawkes DW, Esteves S, Wilcox SJ (2000). Anaerobic-aerobic biotreatment of simulated textile effluent containing varied ratios of starch and azo dye. Water Res 34: 2355-2361.

Ogawa T, Idake E, Yatome C, Kamiya H (1986). Biodegradation of azo acid dyes by continuous cultivation of *Pseudomonas cepacia* 13NA. J Soc Dyers Colourists 102:12-14.

Ollikka P, Alhonmaki K, Leppanen VM, Glumoff T, Raijola T, Souminen I (1993). Decolorization of azo and triphenylmethane heterocyclic and polymeric dyes by lignin peroxidase isoenzymes from *Phanerochaete chrysosporium*. Appl Environ Microbiol 59: 4010-4018.

Ong SA, Toorisaka E, Hirata M, Hano T (2005). Treatment of azo dye Orange II in aerobic and anaerobic-SBR systems. Proc Biochem 40:2907–2914.

Pagga U, Brown D (1986). The degradation of dyestuffs: Part II. Behaviour of dyestuffs in aerobic biodegradation tests. Chemosphere 15: 479-491 Pagga U, Taeger K (1994) Development of a method for adsorption of dyestuffs on activated sludge. Water Res 28: 1051-1057.

Pagga U, Taeger K (1994). Development of a method for adsorption of dyestuffs on activated sludge. Water Res 28:1051-1057.

Pasti-Grigsby MB, Crawford DL (1990). Isolation of microorganisms able to reductively transform aromatic compounds and their relevance to coal liquefaction. Can J Microbiol 37: 902-907.

Pasti-Grigsby MB, Paszczynski A, Goszczynski S, Crawford DL, Crawford RL (1992). Influence of aromatic substitution patterns on azo dye degradability by *Streptomyces* sp. and *Phanerochaete chrysosporium*. Appl Environ Microbiol 58: 3605-3613.

Paszczynski A, Crawford RL (1991). Degradation of azo compounds by ligninase from *Phanerochaete chrysosporium:* Involvement of veratryl alcohol. Biochem Biophys Res Comm 178:1056-1063.

Paszczynski A, Crawford RL (1995). Potential for bioremediation of genobiotic compounds by the white-rot fungus *Phanerochaete chrysosporium*. Biotechnol Prog 11: 368-374.

Paszczynski A, Pasti-Grigsby MB, Goszczyanski S, Crawford RL, Crawford DL (1992). Mineralization of sulphonated azo dyes and sulfanilic acid by *Phanerochaete chrysosporium* and *Streptomyces chromofuscus*. Appl Environ Microbiol 58:3598-3604.

Pazarlioglu NK, Urek RO, Ergun F (2005). Biodecolourization of Direct Blue 15 by immobilized *Phanerochaete chrysosporium*. Proc Biochem 40:1923-1929.

Peralta-Zamora P, Pereira CM, Tiburtius RL, Moraes SG, Rosa MA Minussi RC, Duran N (2003). Decolorization of reactive dyes by immobilized laccase. Appl Catalysis B: Environmental 42:131-144.

Périé FH, Gold MH (1991). Manganese regulation of manganese peroxidase expression and lignin degradation by the white-rot fungus *Dichomitus squalens*. Appl Envir Microbiol 57:2240-2245.

Pierce J (1994). Colour in textile effluents: the origin of problem. J Soc Dyers Colourists 110:131-133.

Pointing SB, Bucher VVC, Vrijmoed LLP (2000). Dye decolorization by subtropical basidiomycetous fungi and the effect of metals on decolorizing ability. World J Microbiol Technol 16:199-205.

Poots VJ, McKay G, Heakt JJ (1976). The removal of acid dye from effluent using natural adsorbents—I peat. Water Res 10:1061-1066.

Radha KV, Regupathi I, Arunagiri A, Murugesan T (2005). Decolorization studies of synthetic dyes using *Phanerochaete chrysosporium* and their kinetics. Proc Biochem 40: 3337-3345.

Rafii F, Franklin W, Cerniglia CE (1990). Azoreductase activity of anaerobic bacteria isolated from human intestinal microflora. Appl Environ Microbiol 56: 2146-2151.

Ramalho PA, Cardoso MH, Cavaco-Paulo A, Ramalho TM (2004). Characterization of Azo Reduction Activity in a Novel Ascomycete Yeast Strain. Appl Envir Microbiol 70:2279-2288.

Robinson T, McMullan G, Marchant R, Nigam P (2001). Remediation of dyes in textile effluents: a critical review on current treatment technologies with a proposed alternative. Biores Technol 77: 247-255.

Ryu BH, Weon YD (1992). Decolorization of azo dyes by *Aspergillus sojae* B-10. J Microbiol Biotechnol 2: 215-219.

Sani RK, Banerjee UC (1999). Decolorization of triphenylmethane dyes and textile and dyestuff effluent by *Kruthia* sp. Enzyme Microbial Technol 24: 433-437.

Sarnaik S, Kanekar P (1999). Biodegradation of methyl violet by *Pseudomonas mendocina* MCM B-402. Appl Microbiol Biotechnol 52: 251-254.

Schliephake K, Lonergan GT, Jones CL, Mainwaring DE (1993). Decolorization of a pigment plant effluent by *Pycnoporus cinnabarinus* in a packed-bed bioreactor Biotechnol Lett 15: 1185-1188.

Schliephake K, Mainwaring DE, Lonergan GT, Jones IK, Baker WL (2000). Transformation and degradation of the diazo dye Chicago Sky Blue by a purified laccase from *Pycnoporus cinnabarinus*. Enz Microb Technol 27:100-107.

Seignez C, Adler N, Suard JC, Peringer P (1996). Aerobic and anaerobic biodegradability of 1-anthraquinone sulphonate. Appl Microbiol Biotechnol 45:719-722.

Selvam K, Swaminathan K, Chae KS (2003). Decolorization of azodyes and a dye industry effluent by a white-rot fungus *Thelephora* sp. Biores Technol 88:115-119.

Semple KT, Cain RB, Schmidt S (1999). Biodegradation of aromatic compounds by microalgae. FEMS Microbiol Lett 170: 291-300.

Seshadri S, Bishop PL, Agha AM (1994). Anaerobic/aerobic treatment of selected azo dyes in wastewater. Waste Mgmnt 14:127-133.

Shaul GM, Holdsworth TJ, Dempsey CR, Dostall KA (1991). Fate of water soluble azo dyes in the activated sludge process. Chemosphere 22:107-112.

Shin KS, Oh IK, Kim CJ (1997). Production and purification of Remazol brilliant blue R decolorizing peroxidase from the culture of *Pleurotus ostreatus*. Appl Environ Microbiol 63:1744-1749.

Soeder CJ, Hegewald E, Kneifel H (1987). Green microalgae can use napthalenesulfonic acids as sources of sulfur. Arch Microbiol 29: 260-263.

Sorek Y, Reisfeld R, Weiss AM (1995). Effect of composition and morphology on the spectral properties and stability of dyes doped in a sol-gel glass waveguide. Chem Phy Lett 244:371-378.

Spadaro JT, Gold MH, Renganathan V(1992). Degradation of azo dyes by the lignin-degrading fungus *Phanerochaete chrysosporium*. Appl Environ Microbiol 58:2397-2401.

Spadaro JT, Tenganathan V (1994). Peroxidase catalysed oxidation of azo dyes: mechanism of disperse Yellow 3 degradation. Arch Biochem Biophys 312:301-307.

Sponza DT, Isik M (2005). Reactor performances and fate of aromatic amines through decolorization of Direct Black 38 dye under anaerobic/aerobic sequentials. Proc Biochem 40:35–44.

Stolz A (2001). Basic and applied aspects in the microbial degradation of azo dyes. Appl Microbiol Technol 56:69-80.

Sugano Y, Nakano R, Saski K, Shoda M (2000). Efficient heterologous expression in *Aspergillus oryzae* of a unique dye-decolorizing peroxidase, of *Geotrichum candidum* Dec1. Appl Environ Microbiol 66:1754-1758.

Sumathi S, Manju BS (2000). Uptake of reactive dyes by *Aspergillus foetidus*. Enz Microb Technol 27:347-355.

Swamy J, Ramsay JA (1999). The evalution of white-rot fungi in the decolorization of textile dyes. Enz Microb Technol 24:130-137.

Thurston F (1994). The structure and function of fungal laccases. Microbiology 140:19-26.

Tsatsaroni EG (1996). Structure-stability relationships in some azo disperse dyes. Dyes and Pigments 31:301-307.

Tunay O, Kabdasli I, Eremektar G (1996). Color removal from textile wastewaters. Water Sci Technol 34: 9-16.

Vandevivere PC, Bianchi R, Weaver VJ (1998). Treatment and reuse of wastewater from the textile wet-processing industry: Review of emerging technologies. Chem Technol Biotechnol 72: 289-302.

Vasdev K, Kuhad RC (1994a). Decolourization of Poly R-478 (Polyvinylamine Sulfonate Anthrapyridone) by *Cyathus bulleri*. Folia microbiology 39: 61-64.

Vasdev K, Kuhad RC, Saxena RK, (1995). Decolorization of triphenylmethane dyes by the bird's nest fungus *Cyathus bulleri*. Current Microbiol 30: 269-272.

Vyas BRM, Molitoris HP (1995). Involvement of extracellular H2O2 dependent ligninolytic activity of the white-rot fungus *Pleurotus ostreatus* in the decolorization of Remazol brilliant blue R. Appl Environ Microbiol 61:3919–3925.

Walker GM, Weatherley LR (2000). Biodegradation and biosorption of acid anthraquinone dye. Environ Pollut 108: 219-223.

Wang Y, Yu J (1998). Adsorption and degradation of synthetic dyes on the mycelium of *Trametes versicolor*. Water Sci Technol 38: 233-238.

Weber EJ, Wolfe NL (1987). Kinetic studies of reduction of aromatic azo compounds in anaerobic sediment/water systems. Environ Toxicol Chem 6: 911-920.

Welham A (2000). The theory of dyeing (and the secret of life). J Soc Dyers Colour 116: 140-143.

Wesenberg D, Buchon F, Agathos SN (2002). Degradation of dye-containing textile effluent by the agaric white-rot fungus *Clitocybula dusenii*. Biotechnol Lett 24: 989-993.

Wilking A, Frahne D (1993). Textile effluent treatment methods of the 90s. Melliand English 74: E325-E328.

Willmott N, Guthrie J, Nelson (1998). The biotechnology approach to colour removal from textile effluents. J Soc Dyer Colour 114: 38-41.

Wong PK, Yuen PY (1996). Decolorization and Biodegradation of Methyl Red by *K. Pneumoniae* RS13. Water Res 30:1736-1744.

Wuhrmann K, Mechsner KI, Kappeler TH (1980). Investigation on rate determining factors in the microbial reduction of azo dyes. Eur J Appl Microbiol 9: 325-338.

Yaropolov AI, Skorobogat'ko OV, Vartanov SS, Varfolomeyev SD (1994). Laccase: properties, catalytic mechanism and applicability. Appl Biochem Biotechnol 47:257-261.

Yatome C, Ogawa T, Itoh K, Sugiyama A, Idaka E (1987). Degradation of azo dyes by cell-free extract from *Aeromonas hydrophila* var.24B. JSDC 103:395-398.

Yatome C, Ogawa T, Matsui M (1991). Degradation of crystal violet by *Bacillus subtilis*. J Environ Sci Eng 26:75-88.

Yatome C, Yamada S, Ogawa T, Matsui M (1993). Degradation of crystal violet dye by *Nocardia corallina*. J Environ Sci Health 38:565-569.

Young L, Yu J (1997). Ligninase catalysed decolorization of synthetic dyes. Water Res 31:1187-1191.

Zheng YM, Zhao QB, Yu HQ (2005). Adsorption of a cationic dye onto aerobic granules. Proc Biochem 40:3777-3782.

Zheng Z, Levin RE, Pinkham JL, Shetty K (1999). Decolorization of polymeric dyes by a novel *Penicillium* isolate. Proc Biochem 34:31-37.

Zhou W, Zimmermann W (1993). Decolorization of industrial effluents containing reactive dyes by actinomycetes. FEMS Microbiol Lett 107:157-162.

Zollinger H (1987). Colour chemistry –synthesis, properties of organic dyes and pigments. VCH Publishers, New York.

The Odyssey of Human Malaria Parasite: Past, Present and Future

R. SARIN and Y. D. SHARMA*

1. INTRODUCTION

Nature has probably witnessed the presence of malaria parasites since the outset of time. Thirty million years old fossilized remains of mosquitoes show that the vector for malarial parasite was present well before their documentation in the literature. It is believed that the disease co-evolved with human species and spread itself all over the tropics, subtropics, and temperate regions across the globe due to population movements. Malaria parasites have continued to dwell in man with a high hand. Malaria is correctly referred to as poor man's disease. Every year, close to hundreds of million cases occur due to malaria with 1 to 3 million deaths, mainly in children particularly in economically burdened countries and thus merit antimalarial therapy. The emergence of drug resistant parasite species and insecticide resistant vectors in the environment has contributed to increase in malaria incidence. Interactions between malaria and AIDS / tuberculosis have added new chapter in our fight against malaria. The present situation is grim and there is dire need to tackle the disease on a war footing.

Amongst over the 170 species of malaria parasites belonging to the genus *Plasmodium*, that are known to infect birds, reptiles and mammals, four species i.e., *P. falciparum*, *P. vivax*, *P. malariae* and *P. ovale* cause malaria in

*   Department of Biotechnology, All India Institute of Medical Sciences, Ansari Nagar, New Delhi-110029, India. Phone: 91-11-26588145; Fax: 91-11-26589286. Email: ydsharma_aiims@yahoo.com

humans (Garnham 1966). Out of these four species, *P. vivax* and *P. falciparum* scramble for the greatest prevalence in the world today, with *P. falciparum* being the most lethal parasite accounting for 40-60% malaria cases and >95% deaths worldwide. *P. vivax* is responsible for 30-40% malaria cases around the world, whereas *P. ovale* and *P. malariae*, the two uncommon parasites show distribution mostly in West Africa (5% cases) and around the world (1-5% cases), respectively.

The bipartite cycle of the parasite is completed in two hosts: vertebrate host to carry out the asexual part and the vector anopheles mosquito to complete the sexual phase of life cycle. The *Plasmodium* parasites display extreme specificity pointing towards a long and adaptive relationship with the humans.

Despite more than a century since the discovery of malaria parasite, our understanding on the molecular basis of the parasite and host relationship has increased very little until recently. A landmark in our futuristic battle to defeat malaria was reached with the sequencing of *P. falciparum*, one of the best model murine parasites *P. yoelii* and the vector *Anopheles gambiae* genomes (Gardner et al. 2002; Carlton et al. 2002; Holt et al. 2002). Thanks to the spin-offs from the human genome project (Lander et al. 2001) we now have the genomic infrastructure for a better understanding of the complex interactions within the malaria trio.

2. HISTORY

The earliest records on malaria like symptoms belong to those of Hippocrates in 500 B. C. An accurate description of clinical symptoms of the disease has been documented in Chinese, Assyrian and Indian writings. The word "malaria" is derived from two Italian words, mala (bad) and aria (air) meaning "bad air".

Malaria in those days was described as miasmatic emanation from the swamps (Harrison 1978). The malarial parasite (*Plasmodium*) was first seen in the blood of malarial patient by French army surgeon Charles Alphonse Laveran in 1880. In 1892, Richard Pfeiffer, a German doctor suggested the possibility of human blood infection by *Plasmodium* through a blood sucking insect. Later in 1894, Sir Patrick Manson, a Scottish doctor intuitively suggested that mosquito acted as a vector and persuaded Sir Ronald Ross, an Indian army doctor to carry out the future study on this. Ross, in 1897, discovered the oocysts of *P. falciparum* in the stomach of anopheline mosquito. A year later, he found out all the stages of bird malaria parasite, *P. relictum* in the culicine mosquito (*Culex fatigans*) (Ross 1898). In the 1898

Italian malariologists, Grassi and his co-workers worked out the complete life cycles of *P. falciparum*, *P. vivax* and *P. malariae* in the female anopheline mosquito. The mystery regarding the liver developmental stage in humans was solved 50 yrs later by Henry Shortt and Cyril Garnham in 1947 (Shortt and Garnham 1948). The dormant stages of some *P. vivax* strains in liver were shown by Krotoski in collaboration with Garnham's team (Krotoski et al. 1982).

3. LIFE CYCLE OF MALARIA PARASITE

Malarial infection sets in as a result of infectious blood sucking bite by mosquito, whereby sporozoites are injected into the bloodstream of humans. Within an hour sporozoites invade hepatocytes (Fig. 1.). Parasitic invasion

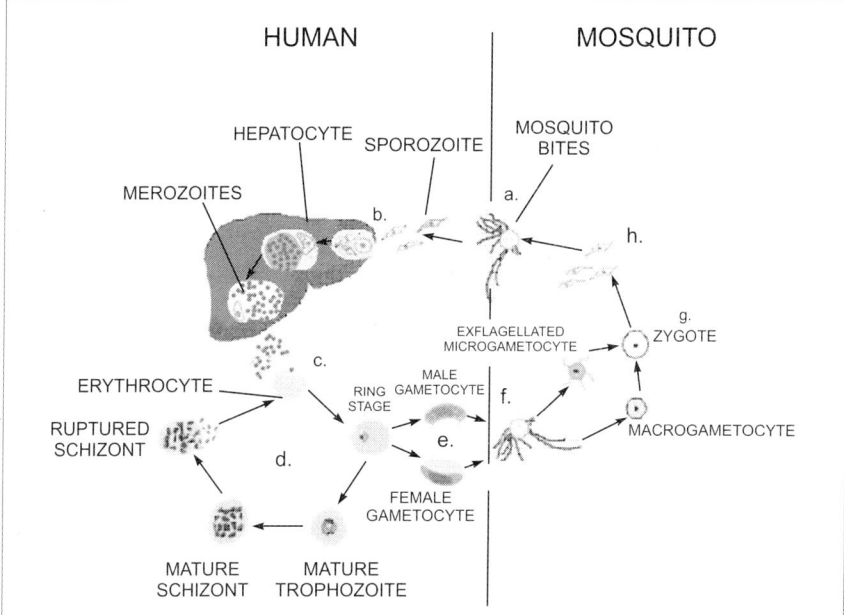

Fig. 1 **Life cycle of malaria parasite: a.** Blood sucking bite of a female anopheles injects infectious *Plasmodium* sporozoites into the bloodstream of humans. **b.** Sporozoites undergo exoerythrocytic schizogony within the liver cells and leads to the production of merozoites. **c.** Merozoites invade red blood cells **d.** Erythrocytic schizogony sets in with multiple rounds of infection in RBCs. **e.** Formation of male and female gametocytes. **f.** Infectious blood meal of mosquito takes up gametocytes with it. **g.** Zygote forms following fusion of micro- and macrogametocytes within the mosquito **h.** Encysted zygote enters sporogony to produce sporozoites which are taken up by mosquito to initiate the life cycle again.

of liver cells is facilitated by apical organelles. Once within the hepatocyte the parasite undergoes asexual replication. This stage is also called as **exoerythrocytic (pre-erythrocytic) schizogony** and leads to the production of progeny merozoites which result due to repetitive nuclear divisions followed by cytoplasmic segmentation or budding. Merozoites are released into circulation following hepatocyte rupture. The number of merozoites released as a result of hepatocyte rupture varies with the species (12,000 merozoites in *P. vivax*, 40,000 in *P. falciparum* and 15,000 in both *P. ovale* and *P. malariae*). *P. vivax* and *P. ovale* exhibit uniqueness in their lifecycle due to the formation of hypnozoites, a dormant stage where sporozoites don't form merozoites but can reactivate later giving rise to relapse. It has been observed that *P. vivax* strains isolated from temperate regions tend to exhibit longer latent period between the primary infection and first relapse than strains from tropical region with continuous transmission (Contacos et al. 1972).

Erythrocytic schizogony or **erythrocytic** cycle begins when merozoites invade red blood cells. Each merozoite within the RBC assumes a ring-like shape with a single nucleus. As the parasite grows, the ring morphology disappears and is called *trophozoite*. During the trophic phase, the parasite ingests a large portion of matrix or cytoplasm of red blood corpuscle and secretes a large number of digestive enzymes into the vacuole and vesicles. These break down hemoglobin into amino acids with the formation of hemozoin, malarial pigment or the by-product of hemoglobin digestion. Nuclear division brings trophozoite stage to an end and leads to the formation of schizonts. Depending on the type of species, 3-5 rounds of multiple fission take place. As a result of nuclear division, daughter nuclei move to the periphery while hemozoin granules collect at the centre. Finally, following cytoplasmic division, many small, uninucleate cells also termed as erythrocytic merozoites are formed. These are released as a result of RBC membrane rupture. The number of merozoites formed in each erythrocytic cycle varies with the species (12-24 merozoites /cycle in *P. vivax*, 18-24 merozoites /cycle in *P. falciparum* and 6 -12 merozoites /cycle in both *P. ovale* and *P. malariae*). These merozoites invade fresh corpuscles and initiate a new round of schizogony. The simultaneous rupture of the infected erythrocytes and the concomittent release of antigens and waste products accounts for the intermittent fever paroxysms associated with malaria. In contrast to other species during this stage, *P. falciparum* trophozoite and schizont-infected erythrocytes get sequestered by capillary endothelial cells and are not found in the peripheral circulation leading to cerebral malaria, the major cause of lethality (Sharma 1991).

Based on the period of recurrence of fever, four types of malaria caused by the four species of human malaria parasite exist:

1. **Benign tertian** or vivax malaria is caused by *P. vivax*. Fever recurs after every 48 hours.
2. **Malignant tertian** malaria is caused by lethal parasite *P. falciparum*. Fever recurs after every 36 to 48 hours.
3. **Quartan** malaria is caused by *P. malariae* and the fever repeats after every fourth day i.e., after 72 hours.
4. **Ovale** or **mild tertian** malaria is caused by *P. ovale*. The fever recurs third day or after 48 hours.

Multiple rounds of erythrocytic schizogony ultimately cause anaemia and eventually may cause the death of host. Hence, to perpetuate themselves, alternatively parasites will undergo a sexual cycle and terminally differentiate into either **micro- or macrogametocytes**. The gametocytes do not develop further in the vertebrate host, until they reach the alimentary canal of proper vector species through another blood sucking meal. Three rounds of nuclear replication in the microgametocyte within the mosquito form eight nuclei and become associated with flagella that emerge from the body of the microgametocyte. This process is called **exflagellation**. The macrogametocyte undergoes maturation into macrogametes.

The motile microgametes then seek out a macrogamete and undergo fusion. The resulting **zygote** develops into an **ookinete** within 12-24 hours. The ookinete is a motile invasive stage which transverses both the peritrophic matrix and the midgut epithelium of the mosquito. Transversing the midgut epithelium involves invading and exiting several epithelial cells before emerging on the basal side of the epithelium. As it rests against the basement membrane of the stomach it stops moving and becomes rounded and begins the process of encysting. The diploid encysted zygote (also called **oocyst**) now enters upon asexual phase of multiplication termed **sporogony** which culminates in the production of several thousand slender and sickle-shaped bodies, called **sporozoites**. Mature oocyst ruptures and releases the sporozoites which cross the basal lamina into the hemocoel (body cavity) of the mosquito. These invade and transverse the salivary gland epithelial cells and come to lie within its lumen. Some of these sporozoites will be expelled into the vertebrate host as the mosquito takes a blood meal, and thus reinitiate the infection in the vertebrate host.

4. INSIGHTS FROM THE GENOME

The malaria parasite comprises three genomes: the 23-Mb nuclear genome has 14 chromosomes, the 35-kb apicoplast circular DNA and the 6-kb mitochondrial element. Most of the information on genetic composition, metabolism, transport, repair, protein trafficking, DNA damage and immune evading systems of *Plasmodia* has been gathered from *P. falciparum*, owing to pervasive interest in devising strategies to combat this lethal parasite and from murine *P. yoelii*, a model parasite. However, genome sequencing projects of other species are also underway, which are likely to get completed within the next five years (Carucci 2004).

(http://www.plasmodb.org, http://www.tigr.org/tdb/euk/, http://www.sanger.ac.uk/Projects/Protozoa/)

The *P. falciparum* nuclear genome is 81% A+T rich and encodes ~5300 genes (Gardner et al. 2002). The genomes of *P. vivax* and *P. cynomolgi* have relatively lower A+T content. It also encodes for 43 tRNAs including an unusual selenocysteinyl tRNA . Instead of repeated rRNA units which are a characteristic of other eukaryotes, *Plasmodium* contains single rRNA units distributed on different chromosomes. The apicoplast genome, is also highly A+T rich with inverted repeats encoding a small and large subunit rRNA flanked by tRNA genes (Gardner et al. 1991). The mitochondrial element of *P. falciparum* (Feagin 1992; Sharma et al. 2001), the rodent malaria parasite *P. yoelii* (Vaidya et al. 1989) and *P. vivax* in Indian study (Sharma et al. 1998) as well as *P. vivax* Salvador strain have been completely sequenced. The *Plasmodial* mitochondrial genome lacks any tRNA but contains rRNA, cytochrome c, oxidase I, III, and cytochrome b genes.

The predicted 5268 nuclear encoded proteins of *P. falciparum* upon analysis for similarity to other proteins banked in currently available public databases didn't show much homology indicative of uniqueness of parasitic proteins (Gardner et al. 2002). In contrast to other eukaryotes, a large fraction of total proteins (60%) is exclusive to the parasite. Using the Gene Ontology (GO) database, 40% proteins could be manually assigned functions. Interestingly in *P. falciparum*, whereas 1.3% genes are involved in cellular attachment and host cell invasion, a larger percentage (3.9%) gene products participate in evasion of host immune response. The apicoplast, an essential organelle for parasite survival encodes for 23 proteins, but over 500 nuclear-encoded proteins is also estimated to be imported in it (Gardner et al. 2002; Ralph et al. 2004; He et al. 2001; Fichera and Roos 1997). The known functions of apicoplast include the synthesis of fatty acid (Gardner et al. 1998; Waller et al. 1998; Surolia and Surolia 2001), isoprenoids (Jomaa et al. 1999) and heme (Sato and Wilson 2002; van Dooren et al. 2002). By developing neural network-based computer software (*PlasMit*), 381 *P. falciparum* proteins likely to be located in the mitochondrion have been identified (Bender et al. 2003). Approximately, 5% nuclear encoded proteins are targeted to mitochondria (Gardner et al. 2002).

Despite the A+T richness of *P. falciparum*, a complete picture of the most of the biochemical pathways of the parasite could be drawn from the genome sequence. Out of the total predicted proteins of *P. falciparum*, only 14 % were identified as enzymes. Only 8 % out of these could be numbered according to Enzyme Commission nomenclature. Unlike other free-living eukaryotes, *P. falciparum* has relatively less genetic information for encoding enzymes. This may be attributed to unique and complex life cycle of the parasite or to the existence of genes in small non-sequenced portions.

The erythrocytic stages of the parasite mainly utilize glycolytic pathway and all the enzymes including phosphofructokinase could be found in the genome. Enzymes necessary for gluconeogenesis, trehalose, glycogen, or other storage carbohydrate synthesis are missing. All the enzymes except transaldolase of hexose monophosphate shunt could be identified in the genome (Gardner et al. 2002).

Further, the complete set of Kreb's cycle enzymes including pyruvate dehydrogenase complex are also present in *P. falciparum* genome, however, the precise role of the cycle still remains elusive. The TCA cycle could play a significant role in haem biosynthesis by producing succinyl-CoA in erythrocytic stages. As mentioned above, the main energy generating pathway in blood stages is glycolysis, it is apparent that oxidative phosphorylation pathway in mitochondria in these stages may not be significant (Gardner et al. 2002). The mitochondrion in the parasite is responsible for membrane potential maintenance, oxidative phosphorylation, haem and coenzyme Q biosynthesis. In addition, it also coordinates pyrimidine biosynthesis, the electron transport system and oxygen utilization via dihydroorotate dehydrogenase and coenzyme Q (Krungkrai 2004).

Malaria parasite obtains all of its amino acids either by salvage from the host or by globin digestion and enzymes required for the synthesis of other amino acids are absent except those required for glycine-serine, cysteine-alanine, aspartate-asparagine, proline-ornithine and glutamine-glutamate conversions (Gardner et al. 2002). Nucleotide biosynthetic pathways in *P. falciparum* include only *de novo* pyrimidine synthesis for which all the enzymes are present, whereas *de novo* purine biosynthetic pathway is absent from all the stages.

The gamut of membrane transporters in *P. falciparum* initially seemed very limited with the presence of only glucose/H^+ symporter; Na^+/H^+, Ca^{2+}/H^+ exchangers; water/glycerol channel; sugar, carboxylate, metal cation transporters. No amino acid transporter was identified in the genome of parasite (Gardner et al. 2002). Using a computer program that searches a genome database on the basis of the hydropathy plots of the corresponding proteins, 55 additional putative transport proteins including previously unidentified amino acid transporters could be identified, thus expanding the total number of putative/proven *P. falciparum* encoded transporters to 109 (Martin et al. 2005).

To prevent its DNA against damage, *P. falciparum* also possesses repair and recombination machinery. While most of eukaryotic nucleotide excision repair proteins including XPB/Rad25, XPG/Rad2, XPF/Rad1, XPD/Rad3 and ERCC1 could be identified, other highly conserved XPA/Rad4 and XPC could not be located. Core proteins of homologous recombination repair (MRE11, DMC1, Rad50 and Rad51) were found but not accessory less conserved proteins (NBS1 and XRS2). Whereas homologues of MutL and MutS proteins implicated in post replication mismatch repair exist in *P.falciparum* those required for non-homologous end joining are absent.

The genes involved in protein translocation and secretion could also be identified in the *P. falciparum* genome. *P. falciparum* contains homologues of signal recognition particle, the translocon, the signal peptidase complex and many components that allow vesicle assembly, docking and fusion, however, the organelles associated with the classical secretory pathway haven't been identified. Recent data suggest the role of Maurer's clefts, parasite-derived membranous structures in the cytosol of the host cell in protein sorting and trafficking (Przyborski et al. 2003).

A comparatively large number of genes are dedicated to immune evading mechanisms in *Plasmodium*. These genes undergo antigenic variation, thus enabling malaria parasites to establish chronic infections that persist for long periods of time, despite the presence of strong host immune response (Miller et al. 1994). First conclusive evidence in support of antigenic variation in malarial parasite was also presented long time back (Brown and Brown 1965). These genes are clustered towards telomeres and encode for highly variant cell surface proteins. This gene clustering towards subtelomeric regions enables high recombination frequencies which allow these genes to form gene families simultaneously creating novel antigenic variant phenotypes. In *P. falciparum* these include multigene families termed *var* (*P. falciparum* erythrocyte membrane binding protein 1, PfEMP1), *rif* (repetitive interspersed family, rifin), *stevor* (sub-telomere variable open reading frame) and *clag* (cytoadherance linked asexual protein). The *var* genes mediate cell-to-cell attachment and hence help in sequestering the infected RBCs in various organs. In *P. vivax vir* gene family products have been hypothesized to play a role in spleen-specific cytoadherance and macrophage-clearance escape (del Portillo et al. 2004). The functions of *rifin* and *stevor* gene products remain unknown, however, both undergo antigenic variation similar to *var* genes, thus, accounting for immune evading mechanisms.

Although PfEMP1 proteins also evoke protective humoral response within the host, but due to antigenic variation, effects of malarial pathogenesis can be seen. The presence of a putative homologue of human cytokine Macrophage Inhibitory Factor (MIF) in *P. falciparum* supports previous reports on modulation of dendritic cell function (Urban et al. 1999) as another means of parasites immune evasion mechanism.

5. UNVEILING THE POST-GENOMIC ERA: CLINICAL APPLICATIONS

With the genetic blueprint of annotated genomes of *P. falciparum* and model rodent parasite *P. yoelii* in hand along with the completed genomes of mosquito and human host, hope of vanquishing malaria never seemed so clear before as it is today. However, this clearly doesn't herald the doom of parasite but marks only the commencement of worldwide efforts to do so. With rapid advances in technology and bioinformatics, it has now become possible to apply global analytical tools such as transcriptome and proteome analysis for understanding and devising novel interventions to combat malaria.

5.1 Comparative and Functional Genomics

One of the offshoots of genome sequencing projects has been comparative genomics which not only compares the synteny between two completed genomes but also reveals significant information regarding the coding potential of related genomes. Thus, comparative genomics aim to discriminate conserved from divergent and functional from non-functional DNA, this approach is also contributing to identify the general functional class of certain DNA segments, such as coding exons, non-coding RNAs and some gene regulatory regions.

Comparative genomics may shed light on parasite-host interactions and glean information on pathways unique to the parasite. Combining comparative genomics with gene expression studies may identify new drug targets to overcome malaria. Comparison of *P. falciparum* genome with

P. yoelii yoelii revealed greater than 60% genes mostly constitutive had possible orthologues in rodent parasite (Carlton et al. 2002). However, the genes families present near telomeric or subtelomeric region of *P. falciparum* had no homologues in *P. yoelii yoelii*. These genes are implicated in host immune response evading mechanisms. The differences are quite intriguing keeping in mind the fact that *P. yoelii yoelii* is often used as a model of *P. falciparum* and telomeric *yir* family of *P. yoelii* is more close to *vir* family in *P. vivax*. In fact, it renders *P. yoelii yoelii* a good model for studying antigenic variation in less virulent parasite *P. vivax*. Comparisons have also indicated significant synteny between the human parasite *P. falciparum* and the rodent malaria parasite *P. berghei*. Members belonging to P48/45 gene family including the newly identified Pf36p and Pfs38 were also found to be conserved among the two parasite species (Thompson et al. 2001a).

The various *Plasmodium* genome initiatives have now provided the opportunity to perform comparative genomics within different species of malaria parasites in more detail, allowing the discovery of orthologues and paralogues of less well-conserved genes and addressing questions of conservation, evolution and structure of multi-gene families. Further, genes whose functions are elucidated in other organisms and which have homologues in *Plasmodia* can be investigated in a systematic manner in the latter, thereby accelerating the elucidation of specific gene functions in the parasite. Genome-wide experimental techniques, namely, DNA microarray and proteome analysis are providing global data on expression profiles of malaria parasite, host and its vector (http://www.plasmodb.org). A full length cDNA database such as full malaria offers advantages of comparative genomics and also enables the determination of transcription start sites (http://fullmal.ims.u-tokyo.ac.jp).

5.2 Gene Chips/DNA Microarrays

Long oligonucleotide based DNA microarrays are proving to be invaluable tool to identify and annotate functional genes in *Plasmodium*. One software program called ArrayOligoSelector offers the choice of gene-specific long oligonucleotide probes minimizing the possibility of cross-hybridization for determining temporal and stage specific expression profile in *P. falciparum* (Bozdech et al. 2003). Expression profiling results from throughout the intraerythrocytic cycle (oligonucleotide-based microarrays in both glass-slides and Affymetrix formats) are also available at PlasmoDB 4.0 (http://www.plasmodb.org).

Proteomics: Limitation imposed by difficulties in obtaining microgram quantities of RNA often makes microarray approach a daunting task in expression profiling. This is overcome by proteomic global data mining. The *P. falciparum* proteome represented by various stages was elucidated using Multidimensional Protein Identification Technology (MudPIT) which is a combination of in-line high resolution liquid chromatography and tandem mass spectroscopy (Washburn et al. 2002). It was published simultaneously with the *P. falciparum* genome (Florens et al. 2002; Lasonder et al. 2002). Differential patterns of protein expression have enabled us to identify vaccine candidate antigens. For example, proteomics has been used to identify vaccine candidate antigens in the sporozoite stage, by identifying the antigens that generate protective immune responses in volunteers challenged against infection by an attenuated vaccine (Doolan et al. 2003a, Doolan et al. 2003b).

6. POTENTIAL DRUG TARGETS IN MALARIA PARASITE

Malarial drug discovery has been by and large fortuitous with mechanism of actions of many still incompletely understood. Presently many drugs are available in the market (Tables 1 and 2), however, emerging resistance to many of these either demands development of new molecules, therapy with combination drugs or rotating the presently available drugs reviewed in detail (Baird 2005; Plowe 2003; Winstanley et al. 2004). Deciphering of the genetic information of malarial parasite has led to the identification of many new drug targets.

6.1 Food Vacuole

The 4-aminoquinolines interfere with the hemoglobin digestion process. Further, enzymes like plasmepsin aspartic proteinases and falcipain cysteine proteinase are also possible drug targets in food vacuole, reviewed in (Rosenthal 1998). Chloroquine, a 4-aminoquinoline has been the primary support for antimalarial therapy for decades, but due to widespread resistance of both *P.falciparum* and *P. vivax* to this drug, its use has been limited (Olliaro et al. 1996, Murphy et al. 1993). Mutation leading to K76T change in *pfcrt* gene has been implicated in chloroquine resistance phenotype in *P. falciparum* strains (Djimde et al. 2001; Vinayak et al. 2003; Ranjit et al. 2004). The advantages of using chloroquine include its safety, low cost and long half-life. Amodiaquine, analogue of chloroquine is more potent than chloroquine. However, resistance to this drug also develops. It is also toxic to bone marrow and liver.

Table 1 Malarial chemotherapy in use[1,2,3,4,5,6]

Drug	Application	Dosage†	Cost	Resistance	Geographical region for prophylaxis	Side effects/ Contraindications
Chloroquine (Aralen™)	Blood Stage Schizonticide; gametocytocidal in P. vivax but not in P. falciparum	600 mg base, followed by 600 mg in 24 h and 300 mg 2 days later	Very cheap	Widespread	Central America, Carribean, Middle East, N. Africa	Nausea, vomiting, headache, dizziness, blurred vision and itching. Serious effect includes death from overdose. Not recommended in epilepsy.
Mefloquine (Lariam™)	Blood Stage Schizonticide	15-25 mg/kg in a divided dose over 12 h	Moderately cheap	Resistance in South-East Asia	South America, Sub-Saharan Africa, Indian Subcontinent, South-East Asia	Headache, nausea, dizziness. Serious effects include seizures, depression and pyschosis. Contraindications include epilepsy, neuro-pyschiatric and cardiac conduction disorders.
Primaquine*	Tissue Stage Schizonticide, gametocytocide	30 mg base daily for 14 days	Moderately cheap	Increasing. Resistant P. vivax strains reported in South-East Asia and Oceania	South America, Sub-Saharan Africa, Indian Subcontinent, South-East Asia	Stomach cramps, nausea, vomiting Severe adverse effects include hemolysis in people with G-6-PD enzyme deficiency.

| Quinine[§] | Blood Stage Schizonticide | 600 mg salt every 8 h for 5-7 days | Cheap | Increasing | Sub-Saharan Africa or Indian subcontinent only | Tinnitus, vertigo, headache, fever, syncope, delirium, nausea. Severe adverse effects reported (hemolytic anaemia, coma). |
| Sulfadoxine-pyrimethamine (Fansidar™) | Blood Stage Schizonticide | Sulfadoxine: 1500 mg & Pyrimethamine: 75 mg (single dose) | Very cheap | Reported in South-East Asia and South America | Some parts of Africa | Serious adverse allergic effects. Not recommended in pregnancy and renal disease. |

1: http://www.nevdgp.org.au/travel/dis/mal2p96.htm.

2. J. K. Baird; N. Engl. J. Med., 352, 1565 (2005).

3. P. Winstanley, S. Ward, R. Snow and A. Breckenridge; Clin. Microbiol. Rev., 17, 612 (2004).

4. P. J. Rosenthal; Emerg. Infect. Dis., 4, 49 (1998).

5. http://www.cdc.gov/malaria/diagnosis_treatment/clinicians2.htm.

6. http://www.travmed.com/thg/thg_pdf_2001/06-Malaria-01.pdf.

†: Recommended adult dosage for oral treatment of uncomplicated malaria.

*: Administered to prevent relapses after chloroquine therapy in *P. vivax* and *P. ovale* malaria.

§: Administered along with tetracycline (250 mg every 6 h for 7 days) or doxycycline (100 mg every 12 h for 7 days).

Table 2 Combination therapy in use for treatment in resistant malaria cases[1,2,3,4]

Combination Drugs	Dosage	Cost	Disadvantages
a) Atovaquone/ Proguanil (Malarone™)	20 mg/kg atovaquone and 8 mg/kg proguanil once a day for 3 days	Very high	Stomach pain, nausea, vomiting.
b) SP based			
SP + Amodiaquine	SP single dose* + Amodiaquine 10 mg/ kg daily for 3 days	Cheap	Adverse effects of amodiaquine and SP.
SP + Chloroquine (CQ)	SP single dose as above + CQ -25 mg/ kg over 3 days	Cheap	Serious adverse effects to SP and drug resistance.
SP+ Mefloquine (Fansimef™)	SP single dose as above +Mefloquine 15 mg/kg	Expensive	Risk of serious adverse effects to combination, resistance reported.
Chlorproguanil + Dapsone (Lapdap™)	Chlorproguanil 2 mg/kg + Dapsone 2.5 mg/kg once daily for 3 days	Cheap	Methemoglobinaemia induced by dapsone, hemolysis in G-6-PD enzyme deficiency. Allergic reactions to dapsone include rashes and fever.
c) Artemisnin based			
Artesunate + Amodiaquine	Artesunate-4 mg/kg + Amodiaquine base 10 mg/kg once daily for 3 days	Moderate	Neutropenia and Pharmacokinetic mismatch.
Artesunate + Mefloquine	Artesunate-4 mg/kg once daily for 3 days + Mefloquine base split dose (15 mg/kg on day 2 and 10 mg/kg on day 3)	Expensive	Mefloquine induced neuropsychiatric effects, cardiotoxic effects though combination produces less adverse effects.
Artesunate + SP	Artesunate- 4 mg/kg once daily for 3 days + SP single dose as above	Moderate	Pharmacokinetic mismatch; adverse effects to SP.
Artemether + Lumefantrine (Coartem™, Riamet™)	Artemether-1.5 mg/kg + Lumefantrine-9 mg/kg at 0,8,24,36, 48 and 60 h	Expensive	Dizziness, palpitation. No serious adverse effect known. Irreversible hearing impairment? Not recommended for pregnant or lactating women.

1. Antimalarial drug combination therapy at http://mosquito.who.int/cmc-upload/0/000/015/082
2. P.G. Kremsner & S. Krishna; Lancet: 364, 285(2004)
3. http://www.paho.org/English/AD/DPC/CD/maladet.htm
4. P. Winstanley, S. Ward, R. Snow and A. Breckenridge: elin Microbiol. Rev. 17.612 (2004): Sulfadoxime 25 mg/kg and Pyrimethamine 1.25 mg/kg as single dose.

6.2 Folate Pathway

A large group of drugs exist that interfere with DNA synthesis by draining an important cofactor, tetrahydrofolate. Amongst these are competitive inhibitors of the enzyme dihydropteroate synthetase (DHPS) which is absent from mammals and those of the enzyme dihydrofolate reductase (DHFR). Sulfonamides such as sulfadoxine and sulfalene as well as sulfones (dapsone) inhibit DHPS, whereas pyrimethamine and biguanides, proguanil and chlorproguanil act against DHFR. Resistance against these drugs has also been reported and attributed to S108N, N51I, C59R and I164L point mutations in *dhfr* gene (Plowe et al. 1997; Ahmed et al. 2004). Double mutations at Ser 436 and Ala 613 or a single mutation Ala 581 in *dhps* gene have been associated with resistance to sulfonamides. Whereas sulphadoxine-pyrimethamine is frequently used to treat chloroquine resistant falciparum malaria, Dapsone in combination with pyrimethamine (Maloprim) is occasionally used for prophylaxis.

6.3 Alkylating Agents

Artemisinins products of plant *Artemisia annua* generate free radicals that rapidly alkylate parasite membrane (Meshnick 2002), the process being catalyzed by hemozoin. Artemisinins also have gametocytocidal effects on *P. falciparum* (Targett et al. 2001). Though extensive resistance to artemisinins hasn't been demonstrated, they form part of Artemisinin Combination Therapy (ACT) in which commonly used antimalarial drugs are combined with artemisinins for malaria treatment. Adverse effects include embryotoxic effects and long-bone shortening.

6.4 Mitochondria

Newly identified dihydroorotate dehydrogenase and cytochrome *c* reductase are drug targets in mitochondrion. Atovaquone inhibits cytochrome *c* reductase, however resistance to atovaquone develops fast. Atovaquone-proguanil is used for treatment and prophylaxis of multidrug resistant falciparum malaria. Cost factor limits its usage in poor nations.

6.5 Apicoplast

The apicoplast contains the machinery for mevalonate-independent isoprenoid synthesis. Because of its absence from mammals, the biosynthesis of isopentyl diphosphate from pyruvate and glyceraldehyde-3-phosphate provides several attractive targets for anti-malarial drugs.

Biosynthesis of type 2 fatty acids along with replication and transcription of 35-kb circle are also perceived as attractive drug targets (Vernick and Waters 2004). The known antibiotics like tetracyclines and doxycyclines work in the apicoplast, by interfering with protein synthesis.

Recent studies have indicated that protein kinases of parasite exhibit considerable structural and functional differences from the host protein kinases. Particularly, there exists divergence in the otherwise conserved signal transduction pathway. This suggests the possibility that protein kinases can act as effective targets for chemotherapy (Doerig 2004). Glucose transport, nutrient uptake mechanisms and Na^+/H^+ antiport are newly identified targets in parasite plasma membrane (Vernick and Waters 2004).

7. FROM GENOMICS TO VACCINES

The multiplying drug resistance in malaria parasite clearly compels researchers to look for effective vaccine to reduce the burgeoning burden of malaria. The fact that sterile protective immunity can be attained by immunization with irradiated sporozoites, and anti-malaria immunity can be induced in malaria endemic area residents suggests that malaria vaccine is feasible. With genomic and proteome picture in hand we can better understand the molecular basis of vector-human and host-parasite interactions and hence devise new strategies for designing new vaccines. DNA vaccines and recombinant viral vector vaccines have entered pre-clinical and clinical test stages and is not the subject of review here but is reviewed elsewhere in detail (Webster and Hill 2003, Mahanty et al. 2003). Though the present generation of DNA vaccines is insufficiently immunogenic to offer protective immunity, immunogenicity and malarial protein expression can be enhanced by various approaches like administration with plasmids expressing cytokines such as GM-CSF (Weiss et al. 1998; Sedegah et al. 2000) and IL-12 (Freidag et al. 2000) and optimizing gene codons for mammalian expression (Narum et al. 2001). Using the existing bioinformatic tools like CTLPred (http://www.imtech.res.in/raghava/ctlpred/algo.html) to identify protective T-cell epitopes and other platform technologies like recombinatorial cloning (Gateway™), transcriptionally active PCR (TAP), protein arrays and ImmunoSense epitope-based techniques reviewed elsewhere (Doolan et al. 2003), subunit vaccines representing subset of antigens expressed either by irradiated sporozoites within liver cells or by malarial blood stages can be created. There are basically three stages of parasitic development against which vaccines are being developed or can be developed (Table 3).

Table 3   Candidate antigen genes for vaccine development[1,2,3]

Plasmodial stage	Antigens	Characteristic features
1. Pre-erythrocytic (sporozoite/liver stage)	Circumsporozoite protein (CSP); Sporozoite-threonine-asparagine-rich protein (STARP); Liver Stage Antigen-1, 3 (LSA-1), (LSA-3); Thrombospondin related anonymous protein 2 (TRAP2); Sporozoite and liver stage antigen (SALSA); Exp-1; Sporozoite surface protein 2 (SSP- 2).	Stage and species specificity; Liver infection prevented; Requires adjuvants and delivery systems that maintain strong immune response.
2. Blood Stages	Rhoptry associated protein-1 & 2 (RAP-1, RAP-2); Serine Repeat Antigen-1 (SERA-1); Merozoite Surface Protein (MSP-1, MSP-2, MSP-3, MSP-4, MSP-5); Apical Membrane Antigen (AMA-1); Erythrocyte Binding Antigen (EBA-175); Glutamate rich protein (GLURP); Pf35; Pf55; Ring-infected erythrocyte surface antigen (RESA)	Stage/species specific. Prevents erythrocytic invasion, thus, reducing severity of disease. Repeated infection provides boosting.
3. Sexual Stages	Pfs25, Pfs230, Pfg27, Pfs45/48, Pfs16, Pfs27, Pfs28, Pvs25	Prevents infection of mosquitoes but does not provide protection from the disease; antibody blocks parasite transmission cycles, suitable for endemic areas. Requires adjuvants and delivery systems to maintain a high immunoglobulin titer.

1. Vaccine challenges at http://www.malariavaccine.org/mal-vac2-challenge.htm
2. Malaria vaccine at http://www.malariasite.com/malaria/malaria_vaccine.htm
3. Malaria vaccine development at http://www.niaid.nih.gov/dmid/malaria/malariavac.htm

7.1 Pre-erythrocytic Vaccine

These include vaccines directed against sporozoites and/or liver stages. The major antigens falling under this group are circumsporozoite protein (CSP) and thrombospondin related anonymous protein 2 (TRAP2), LSA1 and LSA3. DNA based immunizations with these antigens have been shown to elicit protective immunity (Sedegah et al. 1994; Schneider et al. 1998; Schneider et al. 2001; Sauzet et al. 2001).

7.2 Blood-stage Vaccine

These include vaccines directed against the asexual stages. The most commonly characterized vaccine candidate antigens include merozoite surface protein-1 and -2 (MSP-1 and MSP-2) (Holder and Freeman 1981; Blackman et al. 1993; Chappel and Holder 1993; O'Donnell et al. 2001; Triglia et al. 2001a). Other antigens included in this category are Apical Membrane antigen-1 (AMA-1) (Marshall et al. 1989; Narum and Thomas 1994), Erythrocyte binding antigen-175 (EBA-175) (Triglia et al. 2001b, Thompson et al. 2001b), Serine rich antigen (SERA) (Jakobsen et al. 1994), Rhoptries associate protein-1, -2 and -3 (RAP-1, RAP-2 and RAP-3) (Ridley et al. 1990a; Ridley et al. 1990b; Howard and Reese 1990; Howard et al. 1998a; Howard et al. 1998b; Baldi et al. 2002), and *P. falciparum* Glutamine rich protein (GLURP) (Oeuvray et al. 2000) and MSP-3 (Oeuvray et al. 1994; Hisaeda et al. 2002).

7.3 Transmission Blocking Vaccine

These include vaccines targeted against the sexual stages of the parasite that would prevent the parasite from infecting the mosquito vector. Target antigens being studied under this category include the surface antigens of male and female gametes (Pfs48/45, Pfs230 and Pf11.1 molecules) or ookinete stages (Pfs25 and Pfs28 of *P. falciparum*; Pvs25 and Pvs28 of *P. vivax*) (Kaushal et al. 1983a; Kaushal et al. 1983b; Grotendorst et al. 1984; Hisaeda et al. 2000; Tomas et al. 2001).

7.4 Multistage or Multivalent Vaccine

Though vaccines targeted against above stages seem sufficient, it is speculated that due to multistagic life style completed in two hosts, targeting multiple independent foci will produce more effective control and, hence, multistagic/multivalent vaccine may be needed. NYVAC-Pf7 is a genetically engineered, attenuated vaccinia virus, multistage, multicomponent *P. falciparum* vaccine. It comprises Pfs25, CSP, TRAP, LSA-1, MSP-1, AMA-1 and SERA. Recently, a project named MuStDO (Multi-Stage malaria DNA based vaccine Operation) was initiated to develop multivalent vaccine (Doolan and Hoffman 2001). The goal of MuStDO vaccine is to elicit protective response against 5 parasite antigens (CSP, SSP2, EXP1, LSA1 and LSA3) from the sporozoite, liver stages of *P. falciparum* and four encoding proteins (AMA1, EBA175 and two alleles of MSP1) from the blood stage. This two-tiered vaccine design will ensure that those parasites that escape the pre-erythrocytic stage defence will be suppressed by a second line of immune response directed at blood stages, thus decreasing the likelihood of serious illness in vaccinated individuals with breakthrough blood-stage infections.

The NIAID Malaria Vaccine Development Unit (MVDU) is focusing on recombinant proteins derived from blood stages (MSP3, MSP4, MSP5, and AMA1) and sexual stages (Pfs25 and Pvs25) of parasite development. FALVAC-1, an ~41 kd protein contains 21 B-and T-cell epitopes from a variety of pre-erythrocytic, erythrocytic and sexual stage antigens: CS, LSA1, MSP1, SSP2, MSP2, AMA1, RAP1, EBA-175, and Pfg27. FALVAC-1 has been expressed in a baculovirus expression system in collaboration with National Institute of Immunology, New Delhi, India, and Protein Sciences Corporation, Connecticut; USA. Mouse, rabbit and monkey immunization studies of FALVAC-1 with various adjuvants demonstrated induction of immune responses that recognizes different stages of the parasite. Research conducted at the Centre for Disease Control, Atlanta suggests that by administering FALVAC-1 with a suitable adjuvant such as QS-21, it is possible to elicit immune responses to most of the epitopes included in it (Rafi-Janajreh et al. 2002). A second candidate, FALVAC-2, containing the 19 kd fragment of MSP1, the third epidermal growth factor domain of Pfs25, Region II of EBA-175, as well as 30 B-cell epitopes and 25 T cell epitopes from a total of 13 stage-specific antigens, is also being developed. Various vaccine clinical trials conducted on humans show promising results (Table 4) and may lead us towards the development of an effective, multistagic, multisubunit vaccine in future.

Table 4 Malaria vaccine trials

Vaccine Candidate	Formulation	Clinical Trial	Type of response on conclusions from human trials	Reference
AMA1-C1	Equal mixture of recombinant proteins based on sequences from the FVO and 3D7 clones of *Plasmodium falciparum*. adsorbed on Alhydrogel.	Phase I	Humoral immunity	(Malkin et al. 2005)
ICC-1132 (Malarivax)	Modified hepatitis B virus coreprotein containing minimal epitopes of PfCSP.	Phase I	Cell mediated	(Nardin et al. 2004)
ME-TRAP/MVA	Multiepitope (ME) string and the TRAP sporozoite antigen	Phase I	Cell mediated	(Moorthy et al.

	modified vaccinia virus Ankara (MVA).			2004)
RTS, S/AS02	Circumsporozoite surface protein of *Plasmodium falciparum* fused to T-cell eiptopes and HBsAg, incorporating a new adjuvant (AS02) - circumsporozoite surface protein of *Plasmodium falciparum* fused to HBsAg, incorporating a new adjuvant (AS02)	Phase I/IIb	Cell mediated	(Alonso et al. 2004; Pinder et al. 2004)
NYVAC-Pf7	Sporozoite (circumsporozoite protein and sporozoite surface protein 2), liver (liver stage antigen 1), blood (merozoite surface protein 1, serine repeat antigen, and apical membrane antigen 1), and sexual (25-kDa sexual-stage antigen) stages of the parasite life cycle were inserted into a single NYVAC genome to generate NYVAC-Pf7.	Phase I/IIa	Cell mediated	(Ockenhouse et al. 1998)
Combination B	MSP1, MSP2 and RESA in oil based adjuvant.	Phase I/IIb	Humoral primarily; IFN-γ response to MSP1 only	(Genton et al. 2003)

8. DIAGNOSTICS

The development of diagnostic tools for malaria that are cheap, widely accepted, offer sensitivity and specificity is another challenging field. Multiple factors such as population movements, endemic malaria, changing morphological patterns in the species mounting due to drug pressure create encumbrances in the diagnosis of malaria in laboratories which lack skilled personnel (Moody 2002). Malaria is generally regarded as the disease of poorer nations and, hence, in those countries cost-effectiveness is the major concern. Further, since many of the field workers may not be as skilled as researchers there is need for procedures that are easy to perform. Various methods of diagnosis are presently available (Table 5) and are reviewed extensively by Moody (2002). Since many decades microscopy has remained the gold standard for detection of malaria parasites, however, this method

Table 5 Comparison of various diagnostic methods for malaria[1,2,3]

	Microscopy	Fluorescence	RDTs[§]	PCR[†]
REQUISITES				
Equipment	Microscope	Fluorescent microscope/ centrifuge/ quartz halogen source	None	PCR apparatus
Supplies	Blood collection, staining reagent, water, slides	Coated capillaries, Fluorescent dyes, blood	Blood collection, kits	Blood collection, molecular biology reagents
Training	Skilled technical staff	Moderately skilled	Only minimal training needed	Expert needed
EFFICACY				
Test duration	30-60 min	30-60 min	15-20 min	6 h
Labour intensiveness	High	Moderate	Low	High
Sensitivity (parasites/μl)	5-10	50	100-500 for *P. falciparum* and higher for others	*P. falciparum* = 0.38-1.35 *P. vivax* = 0.12
Species detection	Yes	Good for *P. falciparum*, others difficult	*P. falciparum*, others Some RDTs	Yes
Differentiation between P.vivax, P. ovale and P. malariae	Possible	Not possible	Not possible	Yes
Differentiation between stages	Possible	Difficult	Not possible	Not possible
Quantification	Possible	Not possible	Crude estimation	Possible only by real time PCR
COST	Low	Moderate/Low	Moderate	High

1. A. Moody clin Microbiol. Rev. 15, 66 (2002)
2. www.who.int/tdr/publications/pdf/malaria.diagnosis.pdf
3. www.malariasite.com/malaria/Diagnosis of Malaria.htm
§. rapid diagnostic tests
†. polymerase chain reaction

requires trained staff and is labor intensive. Fluorescence based techniques are a step above these as they are more sensitive and require moderate skill level; however, artifacts are a common problem with them. Newer **Rapid Diagnostic Tests** (RDTs) now employ dipstick based on immunochromatographic methods which involve the capture of the parasite antigens from the peripheral blood using either monoclonal or polyclonal antibodies against the parasite antigen targets. Presently, these tests can target the histidine-rich protein-2 of *P. falciparum*, a pan-malarial *Plasmodium* aldolase, and the parasite specific lactate dehydrogenase. RDTs are very easy to use, yield fast results and don't require skilled people. However, the major disadvantage associated with these is the cost and that they offer very crude estimation regarding the parasite density. These are also incapable in detecting cases of imported malaria. During the past decade, the diagnosis of infectious agents has also begun to include the use of nucleic acid-based technologies. Amongst these, polymerase chain reaction (PCR) offers sensitivity, specificity and is independent of immunocompetence or previous clinical history of the patient. However, its high cost, detection of false positives as result of carryover contaminations or false negatives due to PCR inhibitors makes it less preferred diagnostic tool. Other methods include serological detection of antibodies which only measures prior exposure and not specifically current infection.

9. FUTURE AHEAD

The *Plasmodium*, host and vector genome projects have clearly enabled us to identify potential new drug targets at an accelerating rate and this information may allow a drug already in use for some other purpose to be developed as an antimalarial quickly and relatively cheaply. Not only this, combining gene and protein predictions with microarray and proteomic data will also enable us to consider only significant stage specific proteins or enzymes as components of multivalent vaccine candidates while filtering out any which are not relevant. It is, thus, believed that the findings of the basic research will contribute to the development of novel intervention strategies for the control of malaria (Fig. 2). Despite the enormity of challenges, and disappointments we may have faced in the past, there still lies a gleam of hope and optimism to eradicate the malaria from its roots.

ACKNOWLEDGEMENTS

Prof Y.D. Sharma is thankful to DBT, Government of India for financial support.

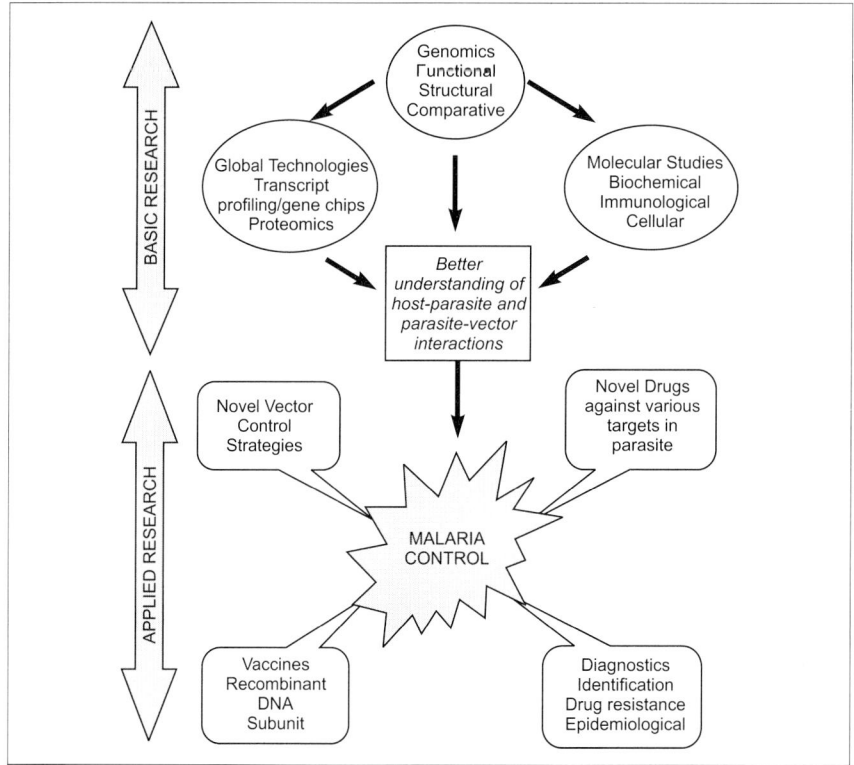

Fig. 2   Achieving malaria control through the combination of basic and applied research.

REFERENCES

Ahmed A, Bararia D, Vinayak S, et al. (2004). *Plasmodium falciparum* isolates in India exhibit a progressive increase in mutations associated with sulfadoxine-pyrimethamine resistance. Antimicrob Agents Chemother 48:879-889.

Alonso PL, Sacarlal J, Aponte JJ et al. (2004). Efficacy of the RTS,S/AS02A vaccine against *Plasmodium falciparum* infection and disease in young African children: randomised controlled trial. Lancet 364:1411-1420.

Baird JK (2005). Effectiveness of antimalarial drugs. N Engl J Med 352:1565-1577.

Baldi DL, Good R, Duraisingh MT, Crabb BS, Cowman AF (2002). Identification and disruption of the gene encoding the third member of the low-molecular-mass rhoptry complex in *Plasmodium falciparum*. Infect Immun 70:5236-5245.

Bender A, van Dooren GG, Ralph SA, McFadden GI, Schneider G (2003). Properties and prediction of mitochondrial transit peptides from *Plasmodium falciparum*. Mol Biochem Parasitol 132:59-66.

Blackman MJ, Chappel JA, Shai S, Holder AA (1993). A conserved parasite serine protease processes the *Plasmodium falciparum* merozoite surface protein-1. Mol Biochem Parasitol 62:103-114.

Bozdech Z, Zhu J, Joachimiak MP et al. (2003). Expression profiling of the schizont and trophozoite stages of *Plasmodium falciparum* with a long-oligonucleotide microarray. Genome Biol 4:R9.

Brown KN, Brown IN (1965). Immunity to malaria: antigenic variation in chronic infections of *Plasmodium knowlesi*. Nature 208:1286-1288.

Carlton JM, Angiuoli SV, Suh BB et al. (2002). Genome sequence and comparative analysis of the model rodent malaria parasite *Plasmodium yoelii*. Nature 419:512-519

Carucci D (2004). Know thine enemy. Nature 430:944-945.

Chappel JA, Holder AA (1993). Monoclonal antibodies that inhibit *Plasmodium falciparum* invasion *in vitro* recognise the first growth factor-like domain of merozoite surface protein-1. Mol Biochem Parasitol 60:303-311.

Contacos PG, Collins WE, Jeffery GM, Krotoski WA, Howard WA (1972). Studies on the characterization of *Plasmodium vivax* strains from Central America. Am J Trop Med Hyg 21:707-712.

del Portillo HA, Lanzer M, Rodriguez-Malaga S, Zavala F, Fernandez-Becerra C (2004). Variant genes and the spleen in *Plasmodium vivax* malaria. Int J Parasitol 34:1547-1554.

Djimde A, Doumbo OK, Cortese JF, et al. (2001). A molecular marker for chloroquine-resistant falciparum malaria. N Engl J Med 344:257-263.

Doerig C (2004). Protein kinases as targets for anti-parasitic chemotherapy. Biochim Biophys Acta 1697:155-168.

Doolan DL, Aguiar JC, Weiss WR, et al. (2003a). Utilization of genomic sequence information to develop malaria vaccines. J Exp Biol 206:3789-3802.

Doolan DL, Hoffman SL (2001). DNA-based vaccines against malaria: Status and promise of the Multi-stage Malaria DNA Vaccine Operation. Int J Parasitol 31:753-762.

Doolan DL, Southwood S, Freilich DA, et al. (2003b). Identification of *Plasmodium falciparum* antigens by antigenic analysis of genomic and proteomic data. Proc Natl Acad Sci. USA 100:9952-9957.

Feagin JE (1992). The 6-kb element of *Plasmodium falciparum* encodes mitochondrial cytochrome genes. Mol Biochem Parasitol 52:145-148.

Fichera ME, Roos DS (1997). A plastid organelle as a drug target in apicomplexan parasites. Nature 390:407-409.

Florens L, Washburn MP, Raine JD, et al. (2002). A proteomic view of the *Plasmodium falciparum* life cycle. Nature 419:520-526.

Freidag BL, Melton GB, Collins F, et al. (2000). CpG oligodeoxynucleotides and interleukin-12 improve the efficacy of *Mycobacterium bovis* BCG vaccination in mice challenged with *M. tuberculosis*. Infect Immun 68:2948-2953.

Gardner MJ, Feagin JE, Moore DJ, et al. (1991). Organisation and expression of small subunit ribosomal RNA genes encoded by a 35-kilobase circular DNA in *Plasmodium falciparum*. Mol Biochem Parasitol 48:77-88.

Gardner MJ, Hall N, Fung E, et al. (2002). Genome sequence of the human malaria parasite *Plasmodium falciparum*. Nature 419:498-511.

Gardner MJ, Tettelin H, Carucci DJ, et al. (1998). Chromosome 2 sequence of the human malaria parasite *Plasmodium falciparum*. Science 282:1126-1132.

Garnham PCC (1966). Malaria parasites and other haemosporidia. Oxford: Blackwell Scientific Publications.

Genton B, Al-Yaman F, Betuela I, et al. (2003). Safety and immunogenicity of a three-component blood-stage malaria vaccine (MSP1, MSP2, RESA) against *Plasmodium falciparum* in Papua New Guinean children. Vaccine 22:30-41.

Grotendorst CA, Kumar N, Carter R, Kaushal DC (1984). A surface protein expressed during the transformation of zygotes of *Plasmodium gallinaceum* is a target of transmission-blocking antibodies. Infect Immun 45:775-777.

Harrison G (1978). Mosquitoes, Malaria and Man: A History of the Hostilities since 1880, London: John Murray.

He CY, Striepen B, Pletcher CH, Murray JM, Roos DS (2001). Targeting and processing of nuclear-encoded apicoplast proteins in plastid segregation mutants of *Toxoplasma gondii*. J Biol Chem 276:28436-28442.

Hisaeda H, Saul A, Reece JJ, et al. (2002). Merozoite surface protein 3 and protection against malaria in Aotus nancymai monkeys. J Infect Dis 185:657-664.

Hisaeda H, Stowers AW, Tsuboi T, et al. (2000). Antibodies to malaria vaccine candidates Pvs25 and Pvs28 completely block the ability of *Plasmodium vivax* to infect mosquitoes. Infect Immun 68:6618-6623.

Holder AA, Freeman RR (1981). Immunization against blood-stage rodent malaria using purified parasite antigens. Nature 294:361-364.

Holt RA, Subramanian GM, Halpern A, et al. (2002). The genome sequence of the malaria mosquito *Anopheles gambiae*. Science 298:129-149

Howard RF, Jacobson KC, Rickel E, Thurman J (1998a). Analysis of inhibitory epitopes in the *Plasmodium falciparum* rhoptry protein RAP-1 including identification of a second inhibitory epitope. Infect Immun 66:380-386.

Howard RF, Narum DL, Blackman M, Thurman J (1998b). Analysis of the processing of *Plasmodium falciparum* rhoptry-associated protein 1 and localization of Pr86 to schizont rhoptries and p67 to free merozoites. Mol Biochem Parasitol 92:111-122.

Howard RF, Reese RT (1990). *Plasmodium falciparum*: Hetero-oligomeric complexes of rhoptry polypeptides. Exp Parasitol 71:330-342.

Jakobsen PH, Hundt E, Hansen MB, Knapp B (1994). Serine-stretch protein (SERP) of *Plasmodium falciparum* corresponds to the exoantigen Ag2, a target of antibodies associated with protection against malaria. Apmis 102:53-58.

Jomaa H, Wiesner J, Sanderbrand S, et al. (1999). Inhibitors of the non-mevalonate pathway of isoprenoid biosynthesis as antimalarial drugs. Science 285:1573-1576.

Kaushal DC, Carter R, Howard RJ, McAuliffe FM (1983a). Characterization of antigens on mosquito midgut stages of *Plasmodium gallinaceum*. I. Zygote surface antigens. Mol Biochem Parasitol 8:53-69.

Kaushal DC, Carter R, Rener J, et al. (1983b). Monoclonal antibodies against surface determinants on gametes of *Plasmodium gallinaceum* block transmission of malaria parasites to mosquitoes. J Immunol 131:2557-2562.

Kremsner PG, Krishna S (2004). Antimalarial combinations. Lancet 364:285-294.

Krotoski WA, Collins WE, Bray RS, et al. (1982). Demonstration of hypnozoites in sporozoite-transmitted *Plasmodium vivax* infection. Am J Trop Med Hyg 31:1291-1293.

Krungkrai J (2004). The multiple roles of the mitochondrion of the malarial parasite. Parasitology 129:511-524.

Lander ES, Linton LM, Birren B, et al. (2001). Initial sequencing and analysis of the human genome. Nature 409: 860-921.

Lasonder E, Ishihama Y, Andersen JS, et al. (2002). Analysis of the *Plasmodium falciparum* proteome by high-accuracy mass spectrometry. Nature 419:537-542.

Mahanty S, Saul A, Miller LH (2003). Progress in the development of recombinant and synthetic blood-stage malaria vaccines. J Exp Biol 206:3781-3788.

Malkin EM, Diemert DJ, McArthur JH, et al. (2005). Phase 1 clinical trial of apical membrane antigen 1: an asexual blood-stage vaccine for *Plasmodium falciparum* malaria. Infect Immun 73:3677-3685.

Marshall VM, Peterson MG, Lew AM, Kemp DJ (1989). Structure of the apical membrane antigen I (AMA-1) of *Plasmodium chabaudi*. Mol Biochem Parasitol 37:281-283.

Martin RE, Henry RI, Abbey JL, Clements JD, Kirk K (2005). The 'permeome' of the malaria parasite: An overview of the membrane transport proteins of *Plasmodium falciparum*. Genome Biol 6:R26.

Meshnick SR (2002). Artemisinin: mechanisms of action, resistance and toxicity. Int J Parasitol 32:1655-1660.

Miller LH, Good MF, Milon G (1994). Malaria pathogenesis. Science 264:1878-1883.

Moody A (2002). Rapid diagnostic tests for malaria parasites. Clin Microbiol Rev 15:66-78.

Moorthy VS, Imoukhuede EB, Milligan P, et al. (2004). A randomised, double-blind, controlled vaccine efficacy trial of DNA/MVA ME-TRAP against malaria infection in Gambian adults. PLoS Med 1:e33.

Murphy GS, Basri H, Purnomo, et al. (1993). Vivax malaria resistant to treatment and prophylaxis with chloroquine. Lancet 341:96-100.

Nardin EH, Oliveira GA, Calvo-Calle JM, et al. (2004). Phase I testing of a malaria vaccine composed of hepatitis B virus core particles expressing *Plasmodium falciparum* circumsporozoite epitopes. Infect Immun 72:6519-6527.

Narum DL, Kumar S, Rogers WO, et al. (2001). Codon optimization of gene fragments encoding *Plasmodium falciparum* merozoite proteins enhances DNA vaccine protein expression and immunogenicity in mice. Infect Immun 69:7250-7253.

Narum DL, Thomas AW (1994). Differential localization of full-length and processed forms of PF83/AMA-1 an apical membrane antigen of *Plasmodium falciparum* merozoites. Mol Biochem Parasitol 67:59-68.

Ockenhouse CF, Sun PF, Lanar DE, et al. (1998). Phase I/IIa safety, immunogenicity, and efficacy trial of NYVAC-Pf7, a pox-vectored, multiantigen, multistage vaccine candidate for *Plasmodium falciparum* malaria. J Infect Dis 177: 1664-1673.

O'Donnell RA, de Koning-Ward TF, Burt RA, et al. (2001). Antibodies against merozoite surface protein (MSP)-1(19) are a major component of the invasion-inhibitory response in individuals immune to malaria. J Exp Med 193: 1403-1412.

Oeuvray C, Bouharoun-Tayoun H, Gras-Masse H, et al. (1994). Merozoite surface protein-3: A malaria protein inducing antibodies that promote *Plasmodium falciparum* killing by cooperation with blood monocytes. Blood 84:1594-1602.

Oeuvray C, Theisen M, Rogier C, et al. (2000). Cytophilic immunoglobulin responses to *Plasmodium falciparum* glutamate-rich protein are correlated with protection against clinical malaria in Dielmo, Senegal. Infect Immun 68:2617-2620.

Olliaro P, Cattani J, Wirth D (1996). Malaria, the submerged disease. Jama 275:230-233.

Pinder M, Reece WH, Plebanski M, et al. (2004). Cellular immunity induced by the recombinant *Plasmodium falciparum* malaria vaccine, RTS,S/AS02, in semi-immune adults in The Gambia. Clin Exp Immunol 135:286-293.

Plowe CV (2003). Monitoring antimalarial drug resistance: Making the most of the tools at hand. J Exp Biol 206:3745-3752.

Plowe CV, Cortese JF, Djimde A, et al. (1997). Mutations in *Plasmodium falciparum* dihydrofolate reductase and dihydropteroate synthase and epidemiologic patterns of pyrimethamine-sulfadoxine use and resistance. J Infect Dis 176:1590-1596.

Przyborski JM, Wickert H, Krohne G, Lanzer M (2003). Maurer's clefts—a novel secretory organelle? Mol Biochem Parasitol 132:17-26.

Rafi-Janajreh A, Tongren JE, Kensil C, et al. (2002). Influence of adjuvants in inducing immune responses to different epitopes included in a multiepitope, multivalent, multistage *Plasmodium falciparum* candidate vaccine (FALVAC-1) in outbred mice. Exp Parasitol 101:3-12.

Ralph SA, van Dooren GG, Waller RF, et al. (2004). Tropical infectious diseases: metabolic maps and functions of the *Plasmodium falciparum* apicoplast. Nat Rev Microbiol 2:203-216.

Ranjit MR, Das A, Chhotray GP, et al. (2004). The PfCRT (K76T) point mutation favours clone multiplicity and disease severity in *Plasmodium falciparum* infection. Trop Med Int Health 9:857-861.

Ridley RG, Takacs B, Etlinger H, Scaife JG (1990a). A rhoptry antigen of *Plasmodium falciparum* is protective in Saimiri monkeys. Parasitology 101 Pt 2:187-192.

Ridley RG, Takacs B, Lahm HW, et al. (1990b). Characterisation and sequence of a protective rhoptry antigen from *Plasmodium falciparum*. Mol Biochem Parasitol 41:125-134.

Rosenthal PJ (1998). Proteases of malaria parasites: new targets for chemotherapy. Emerg Infect Dis 4:49-57.

Ross, R (1898). The role of the mosquito in the evolution of the malarial parasite. Lancet 2: 488-489.

Sato S, Wilson RJ (2002). The genome of *Plasmodium falciparum* encodes an active delta-aminolevulinic acid dehydratase. Curr Genet 40:391-398.

Sauzet JP, Perlaza BL, Brahimi K, Daubersies P, Druilhe P (2001). DNA immunization by *Plasmodium falciparum* liver-stage antigen 3 induces protection against *Plasmodium yoelii* sporozoite challenge. Infect Immun 69:1202-1206.

Schneider J, Gilbert SC, Blanchard TJ, et al. (1998). Enhanced immunogenicity for CD8+ T cell induction and complete protective efficacy of malaria DNA vaccination by boosting with modified vaccinia virus Ankara. Nat Med 4:397-402.

Schneider J, Langermans JA, Gilbert SC, et al. (2001). A prime-boost immunisation regimen using DNA followed by recombinant modified vaccinia virus Ankara induces strong cellular immune responses against the *Plasmodium falciparum* TRAP antigen in chimpanzees. Vaccine 19:4595-4602.

Sedegah M, Hedstrom R, Hobart P, Hoffman SL (1994). Protection against malaria by immunization with plasmid DNA encoding circumsporozoite protein. Proc Natl Acad Sci U S A 91:9866-9870.

Sedegah M, Weiss W, Sacci JB, Jr., et al. (2000). Improving protective immunity induced by DNA-based immunization: priming with antigen and GM-CSF-encoding plasmid DNA and boosting with antigen-expressing recombinant poxvirus. J Immunol 164:5905-5912.

Sharma I, Pasha ST, Sharma YD (1998). Complete nucleotide sequence of the *Plasmodium vivax* 6 kb element. Mol Biochem Parasitol 97:259-263.

Sharma I, Rawat DS, Pasha ST, Biswas S, Sharma YD (2001). Complete nucleotide sequence of the 6 kb element and conserved cytochrome b gene sequences among Indian isolates of *Plasmodium falciparum*. Int J Parasitol 31:1107-1113.

Sharma YD (1991). Knobs, knob proteins and cytoadherence in falciparum malaria. Int J Biochem 23:775-789.

Shortt HE and Garnham PCC (1948). Pre-erythrocytic stage in mammalian malaria parasites. Nature 161:126.

Surolia N, Surolia A (2001). Triclosan offers protection against blood stages of malaria by inhibiting enoyl-ACP reductase of *Plasmodium falciparum*. Nat Med 7:167-173.

Targett G, Drakeley C, Jawara M, et al. (2001). Artesunate reduces but does not prevent post-treatment transmission of *Plasmodium falciparum* to *Anopheles gambiae*. J Infect Dis 183:1254-1259.

Thompson J, Janse CJ, Waters AP (2001a). Comparative genomics in *Plasmodium*: a tool for the identification of genes and functional analysis. Mol Biochem Parasitol 118:147-154.

Thompson JK, Triglia T, Reed MB, Cowman AF (2001b). A novel ligand from *Plasmodium falciparum* that binds to a sialic acid-containing receptor on the surface of human erythrocytes. Mol Microbiol 41:47-58.

Tomas AM, Margos G, Dimopoulos G, et al. (2001). P25 and P28 proteins of the malaria ookinete surface have multiple and partially redundant functions. Embo J 20:3975-3983.

Triglia T, Thompson J, Caruana SR, et al. (2001a). Identification of proteins from *Plasmodium falciparum* that are homologous to reticulocyte binding proteins in *Plasmodium vivax*. Infect Immun 69:1084-1092.

Triglia T, Thompson JK, Cowman AF (2001b). An EBA175 homologue which is transcribed but not translated in erythrocytic stages of *Plasmodium falciparum*. Mol Biochem Parasitol 116:55-63.

Urban BC, Ferguson DJ, Pain A, et al. (1999). *Plasmodium falciparum*-infected erythrocytes modulate the maturation of dendritic cells. Nature 400:73-77.

Vaidya AB, Akella R, Suplick K (1989). Sequences similar to genes for two mitochondrial proteins and portions of ribosomal RNA in tandemly arrayed 6-kilobase-pair DNA of a malarial parasite. Mol Biochem Parasitol 35:97-107.

van Dooren GG, Su V, D'Ombrain MC, McFadden GI (2002). Processing of an apicoplast leader sequence in *Plasmodium falciparum* and the identification of a putative leader cleavage enzyme. J Biol Chem 277:23612-23619.

Vernick KD, Waters AP (2004). Genomics and malaria control. N Engl J Med 351:1901-1904.

Vinayak S, Biswas S, Dev V, et al. (2003). Prevalence of the K76T mutation in the *pfcrt* gene of *Plasmodium falciparum* among chloroquine responders in India. Acta Trop 87:287-293.

Waller RF, Keeling PJ, Donald RG, et al. (1998). Nuclear-encoded proteins target to the plastid in *Toxoplasma gondii* and *Plasmodium falciparum*. Proc Natl Acad Sci U S A 95:12352-12357.

Washburn MP, Ulaszek R, Deciu C, Schieltz DM, Yates JR, 3rd (2002). Analysis of quantitative proteomic data generated via multidimensional protein identification technology. Anal Chem 74:1650-1657.

Webster D, Hill AV (2003). Progress with new malaria vaccines. Bull World Health Organ 81:902-909.

Weiss WR, Ishii KJ, Hedstrom RC, et al. (1998). A plasmid encoding murine granulocyte-macrophage colony-stimulating factor increases protection conferred by a malaria DNA vaccine. J Immunol 161:2325-2332.

Winstanley P, Ward S, Snow R, Breckenridge A (2004). Therapy of falciparum malaria in sub-saharan Africa: from molecule to policy. Clin Microbiol Rev 17:612-637.

Global Review on Malaria: Impact on Human Life

V. P. SHARMA

1. INTRODUCTION

Malaria is one of the major scourges of humankind in the developing world and a disease of antiquity. From its origin in Africa, malaria has spread world-over with human migration. The disease was known in China for almost 5,000 years ago. Sumerian and Egyptian texts dating from 3,500 to 4,000 years ago mention fevers and splenomegaly suggestive of malaria. The ancient Hindu text Charaka Samhita (300BC) and the Susruta Samhita (100BC) refer to diseases with intermittent fever as the main symptom. Malaria had reached India 3,000 years ago, the Mediterranean region 2,500-2,000 years ago and northern Europe 1,000-500 years ago. The spread of malaria was accelerated by international trade. As for example, *P. vivax* and *P. malariae* were possibly brought to the New World from South-East Asia by early trans-Pacific voyages; and *P. falciparum* entered Americas through the African slaves. Over the next 100 years, malaria spread across America and Canada. By 19th century, malaria had reached its global limits and probably accounted for 10% deaths worldwide, and in India for over half of all deaths (Carter and Mendis 2002). By mid 20th century malaria controlled interventions, as well as improved living conditions resulted in the spontaneous decline in man-mosquito contact so that by early 1950s, malaria almost disappeared from North America and Europe. In the 21st century, malaria is a

Meghnad Saha Distinguished Fellow Centre for Rural Development and Technology, Indian Institute of Technology, New Delhi-110016, India major public health problem in the tropical and sub-tropical countries; and it is entering receptive and vulnerable areas in Europe and America. Return of malaria is a formidable challenge and its re-emergence is characterized by the resistance in the vectors, parasite, and man; and the environmental change that supports transmission (Sharma 1996, 1999).

The name *malaria* (from the Italian mala "bad" and aria "air") originated in Italy from the strong links of the disease with stagnant water and swamps. It was believed that exposure to marsh air causes intermittent fevers. Malaria is caused by four species of the genus Plasmodium viz., *Plasmodium vivax*, *Plasmodium falciparum*, *Plasmodium malariae* and *Plasmodium ovale*. Man is the only reservoir of human malaria parasites. These parasites share the characteristic of periodic paroxysm, chills, rigors, and sweating. They also cause body aches, headache, nausea, general weakness and prostration. Malaria causes anemia and spleen enlargement. Clinical malaria diagnosis may at times be misleading as malaria may mimic symptoms of many other diseases. Malaria diagnosis should rely on microscopic examination of blood smears or immunological tests e.g., dipstick. On biting by an infective anopheles mosquito, malaria parasites (sporozoites) enter the blood capillaries of the human body. From the blood circulation sporozoites enter the liver in a short time, say half an hour. The incubation period in liver, generally, lasts for about a week. In the liver sporozoites multiply and produce merozoites that are capable of invading the red blood cells. The pre-erythrocytic merozoites from the liver are released into the blood. Once inside the bloodstream, merozoites invade the red blood cells in less than a minute and start to grow within the RBC. They consume intracellular proteins especially the haemoglobin and complete schizogony within 24 to 36 hours depending on the parasite species. Eventually the infected RBCs rupture and release merozoites (asexual form) and gametocytes (sexual forms). Merozoites restart the erythrocytic cycle, and the circulating gametocytes enter the mosquito gut in the process of feeding by the mosquito on humans. In the gut gametocytes differentiate into microgametocytes (males) and macrogametocytes (females), fuse to form a zygote. Developing zygotes transform into spindle-shaped ookinete, cross the inner wall of the abdomen and finally transform into sporozoites. Thousands of sporozoites lodge into the salivary glands, ready to be injected into the humans through the bite of female anopheles mosquito. It is notable to mention that in the order Diptera Anopheles is the only genus capable of transmitting human malaria. Of the four malaria parasites, *P. falciparum* may cause severe anemia and life-threatening disease; damage lungs, kidneys, liver and brain i.e., "cerebral malaria." Almost all deaths due to malaria are

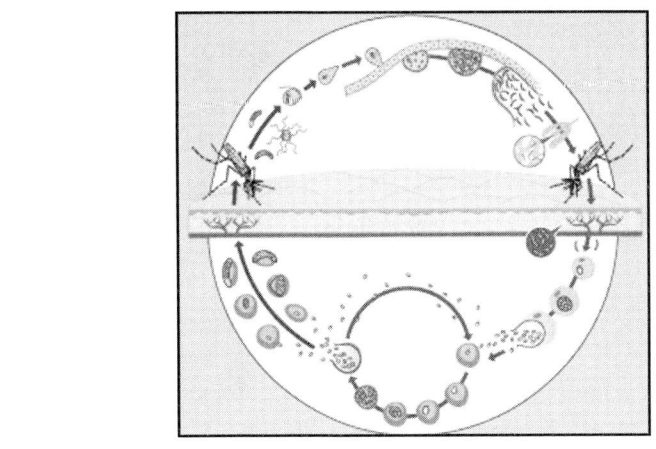

Fig. 1   Life cycle of a Plasmodium (Copyright: TDR/Welcome Trust)

caused by *P. falciparum*. *P. malariae* can cause nephrotic syndrome. *P. vivax* and *P. ovale* cause benign infections causing high morbidity and repeated relapses. *P. vivax* infection although rare, may cause rupture of the spleen or acute respiratory syndrome (ARDS).

2. IMPACT OF MALARIA ON HUMAN LIFE

Malaria affects almost all aspects of human life. The impact is profound and deadly. Human history and human evolution are the testimony to the ravages of malaria and acquisition of immunity against malaria. In the following section, an attempt has been made to highlight how malaria affects the health and well-being of men, women and children, the impact of malaria in the evolution of genetic traits as a means of survival at a high cost to the societies for centuries together, and the destruction of household and national economy. Freedom from malaria is, therefore, a prerequisite for poverty alleviation and human welfare.

2.1 Malaria Burden

Malaria is the leading cause of disease and death in the world today. Over 40% of the world's population lives in malaria endemic countries. At the end of 2004, 3.2 billion people in 107 countries and territories were living in areas at risk of malaria (Fig. 2). Ten new cases of malaria are detected every second and malaria alone kills 3,000 people each day in Africa. Malaria

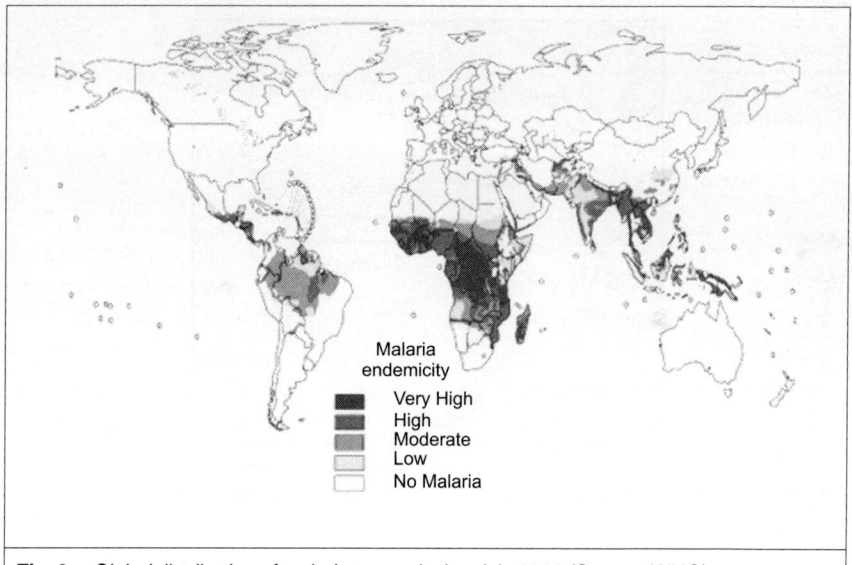

Fig. 2   Global distribution of malaria transmission risk, 2003 (Source: WHO)

causes 42,280,000 DALYs and 1,124,000 deaths each year (Hay et al. 2004; WHO 2005). Table 1 gives the malaria profile in various regions of the world. In Africa 10.8% population of the world contributes 87.9% DALYs, and rest is South-East Asia where 25.6% world's population contributes 6.0% DALYs. In the eastern Mediterranean region 8.1% world's population contributes 4.8% DALYs. America and Western Pacific region have <1% DALY and Europe has no malaria. Worst affected regions are South Africa of Sahara and South-East Asian countries.

Table 1   Malaria at a glance

Region	Percent of World Population	Estimated deaths from malaria 2002	Malaria deaths as a % of all deaths	Share of global burden burden of malaria*%	Percent of disease burden attributable to malaria*
Africa	10.80%	1,136,000	10.70%	87.9	11.3
South-East Asia	25.60%	65,000	<1.0%	6.0	<1.0%
America	13.70%	1,000	<1.0%	<1.0%	1.0
Western Pacific	27.60%	11,000	<1.0%	1.0	<1.0%
Eastern Mediterranean	8.10%	59,000	1.40%	4.8	1.6

| Europe | 14.10% | 0 | 0 | 0 | <1.0% |
| Global | 100% | 1,272,000 | 2.20% | 100 | 3.1 |

*As measured in DALYs. (DALYs = Disability Adjusted Life Years: The sum of years of potential life lost due to premature mortality and the years of productive life lost due to disability) Source: Global Health Reporting.org http://www.globalhealthreporting.org/malaria.asp?id=62

WHO estimates 350-500 million clinical disease episodes worldwide, of which 270-400 million are falciparum malaria. Snow et al. (2005) estimated around 515 (range 300-660) million clinical cases of malaria in 2002. These global estimates were 50% higher than those reported by WHO and 200% higher for areas outside Africa. WHO (2005) reports that around 60% cases of clinical malaria and over 80% deaths takes place in South Africa of Sahara. The case fatality rate among serious patients may exceed 30%. Epidemics are becoming more frequent in countries of South Africa of Sahara and South East Asia (Lopez et al. 2006).

The malaria mortality rates have increased during the 1990s, primarily in Sub-Saharan Africa from 15% to 22% in children younger than 5 years old. Malaria situation is grossly under-reported from other regions e.g., in India 1.5-2.0 million cases and 500-800 deaths are being reported annually. WHO SEA region reports 10 million cases and 19,500 deaths due to malaria. Based on environmental determinants WHO, Geneva estimates 70 million cases in India. This is also indicated by drug consumption. India consumed 61 mt chloroquine phosphate to treat 6.5 million cases in 1976 and in 2005 cases have reduced by 70% but anti-malarial usage has increased 10-fold (Sharma 2005).

Malaria epidemiology is highly complex and there are several factors for variation in malaria profile related to man, vector, parasite, and environment. Malaria burden is highly uneven within the countries and regions. *Inter alia* the leading factors are the fact that there are four human malaria parasites with highly variable distribution in space and time, and diseases they cause from mild to severe and death. *P. falciparum* is the killer parasite which is most common and dominant parasite in Africa, and its proportion is increasing in other malarious regions of the world. There are at least 40 malaria vectors, which differ greatly in their transmission potential. *Anopheles gambiae* is the most serious of all vectors. Besides climate is highly variable and constitute an important determinant of vector breeding, biting, survival and thus influencing transmission. *Inter alia* malaria determinants include the poverty, housing, and access to health care, education, hemoglobinopathies, environment and the quality of national malaria control programme. Sociological factors often ignored, profoundly affect the burden of malaria, as rightly said by Sinton (1956) "Malaria perpetuates ignorance, and ignorance malaria".

To add to the already complex malaria problem, the problem of malaria control failures is associated with resistance in the parasites and malaria vectors. Initially, DDT was very effective but gradually resistance to DDT resulted in control failures and new insecticides were pressed into service (Sharma 2003). Vectors developed multiple resistance e.g., in India replacement of insecticide policy resulted in DDT, BHC, Malathion and Synthetic Pyrethroid resistant *Anopheles culicifacies* (Sharma 1996,1999). Selective spraying by the malaria control programmes resulted in malaria outbreaks leading to chaotic situations. Concurrently *P. falciparum* developed resistance to chloroquine resulting in treatment failures. Introduction of new drugs such as sulfadoxine pyrimethamine, Mefloquine, Quinine resulted in multiple resistant strains. This is a global problem and wherever *P. falciparum* is present with history of the use of chloroquine drug resistant strains may be encountered (Sharma 2006a; Dua 2003; 2004). The problem of chloroquine resistance is now spreading to *P. vivax*. Vivax malaria was not a killer disease but recently *P. vivax* has been associated with clinical picture of *P. falciparum* causing severe malaria and deaths in confirmed cases of *P. vivax* (Kocher, 2005). Drug resistance enhances malaria morbidity, deaths by 5 to 11-fold and cost of treatment may become unaffordable for large sections of society (Sharma 2000).

In Africa, south of Sahara pregnant women from low or unstable malaria areas have two-or three-fold higher risk of developing severe disease as a result of malaria infection than are non-pregnant adults living in the same area. Malaria infection may cause severe anemia, abortion, neonatal death, and Low Birth Weight (LBW) babies. In moderate to high transmission areas malaria infection may results in anemia, the presence of parasites in the placenta, impaired fetal nutrition contributing to low birth weight, poor infant survival and development. In stable malaria transmission areas, *P. falciparum* infection during pregnancy is estimated to cause 10,000 maternal deaths each year, 8% to 14% of all low birth weight babies, and 3% to 8% of all infant deaths. In Africa 30 million women become pregnant each year in malaria endemic regions, of which 200,000 new born deaths as a result of malaria in pregnancy (WHO 2003).

In South-East Asia Region, the burden of malaria in pregnant women and their unborn children is consistently high since the early records. In the malaria epidemic in Punjab in 1908, out of 4,600 pregnant women, malaria caused 300 premature births, 1,100 still births, miscarriages, abortions, etc. amounting to 30% interrupted pregnancies (Christopher, 1911). In 1934-35 malaria epidemic in Sri Lanka Wickaramsuria (1937) reported that the case fatality rate in pregnant women was 13% (47/358) and fetal loss or neonatal death rate was 67% (178/266). Sinton (1956) reported that "there is much evidence to prove that malaria, both in the endemic and epidemic forms, has a very marked influence in lowering the birth rate of any population in which it is prevalent. The disease acts on the birth rate by (a) reducing the number of conceptions, and by (b) causing interruptions in pregnancy which results in abortions and still births. Malaria is, therefore, an important direct factor in checking the natural increase of any population which is afflicted by it." In Thailand, severe malaria was three times more common in pregnant women than in non-pregnant women (Luxemburger et al. 1997). In Myanmar Naing et al. (1988) reported that in 52 patients with *P. falciparum* infection, the prevalence of malaria was highest among primigravidae (40%), and in 958 asymptomatic pregnant women attending the antenatal clinic, overall parasite prevalence was 12.3% (*P. falciparum* 63%, *P. vivax* 34% and remaining were mixed infections). Besides, maternal death (36%); miscarriages (4%); intrauterine fetus death and still birth (3.4%); neonatal malaria (1%); low birth weight (25%); neonatal parasitaemia (3.2%); cord blood parasitaemia (4.2%), and placental malaria (7.3%) were common complications. A direct relationship between maternal malaria, anemia and mortality was also observed, e.g., 98 deaths/ thousand live births with moderate anemia (7-10g/dL) and 467 deaths/ thousand live births with severe anemia (<7g/dL). The major adverse effect of malaria in pregnancy on the mother was anemia. In an epidemic in Mandla, India 55% pregnant women were infected with malaria, of which 88% were due to *P. falciparum*; 7% due to *P. vivax*, and the remaining were mixed infections (Singh et al. 2001). All primigravidae (42%), secundigravidae (68%) or multigravidae (54%) were at great risk of developing severe malaria especially in the second trimester. Moderate anemia was recorded in 100% pregnant women (7.5-11g/dl). Of the women found infected with *P. falciparum*, 3% had abortions, 4% still births and 2% neonatal deaths. Of the babies born, 85% were of low birth weight (Singh et al. 1999, 2005).

About 40% children in the world live in malaria-endemic countries (Snow et al. 1999). Each year 0.75 million African children under 5 die of malaria with *Plasmodium falciparum*. Low birth weight babies with untreated hypoglycaemia may suffer brain damage. Approximately 7% children who survive cerebral malaria may live with permanent neurological problems e.g., weakness, spasticity, blindness, speech problems and epilepsy.

Children who recover from neurological problems following cerebral malaria, may eventually develop significant cognitive problems that adversely affect school performance (Holding et al. 1999). It has been estimated that severe malarial anemia causes between 190,000 and 974,000 deaths each year among children < 5 years (Murphy and Breman, 2001). African children suffer 1.6-5.4 malarial fever episodes each year from about 4 months of age, and in some regions upto 70% of one-year-olds may have malaria parasites in their blood. Fever reduces appetite, exacerbates malnutrition, and increases school absenteeism (Murphy and Breman 2001). In Sri Lanka, multiple attacks of uncomplicated malaria have a deleterious effect on school performance, irrespective of school absenteeism and socioeconomic circumstances (Fernando 2003).

2.2 Genetic Disorders and Malaria

High intensity of malaria transmission produced human genetic polymorphism especially those affecting red blood cells (RBCs). Genetic traits were eventually selected in high frequencies as innate mechanisms to protect human populations from the lethal effects of malaria. J.B.S. Haldane (1948) was the first to propose "malaria hypothesis" linking genetic disorders with both *P. falciparum* and *P. vivax* malaria or for that matter for human malaria. Genetic analysis of the *Plasmodium falciparum* genome sequence has revealed a recent expansion of *Plasmodium* species within the last 6,000 years, coinciding with the expansion of both human and mosquito populations, and establishing link of malaria with the advent of agriculture. In that time frame agriculture was crucial in sustaining *P. falciparum* transmission. The intensity of transmission resulted in mutations that provided malaria-protective polymorphisms. Origin of other *Plasmodium* species is rather uncertain. These parasites are considerably older than deadly *P. falciparum* (Hume 2003). A brief description of the natural selection of genetic disorders that provide immunity against malaria is given below.

2.2.1 Sickle Cell Haemoglobin

Sickle cell mutation is widely distributed across tropical Africa, non-African Mediterranean populations, and throughout the middle east and central India (Weatherall et al. 1988). The distribution of malaria and the sickle cell haemoglobin is shown in Fig. 3. The disease Sickle cell AS (HbAS) is inherited as a Mendelian trait. This is an example of best studied protective mutation in the red blood cell (RBC). Sickle cell haemoglobin is the product of a single point mutation in the β globin gene resulting in the replacement of glutamic acid by valine in the sixth position of β globin chain. Chromosome 11 where glutamic amino acid is replaced with valine amino acid in the sixth position. While in sickle cell trait (HbAS) the abnormal gene is inherited from one of the parents, in sickle cell anaemia (HbSS) abnormal genes are inherited from both the parents. Red blood cells carrying HbSS become elongated and sickle-shaped under conditions of hypoxia, acidosis, infection, etc. resulting in blockage of capillaries leading to tissue hypoxia. Repeated such episodes lead to multiple infarctions of the organs. The result is the production of both HbA and HbS. Red blood cells of the mutant hemoglobin (HbS) become elongated and sickle-shaped. Such distorted RBC entangles and blocks small vessels and capillaries depriving vital organs of oxygen. Sickled RBCs have a short life of 10-20 days instead of 120 days. In individuals with sickle cell anaemia (HbSS) repeated sickling episodes may lead to severe illness and a significant proportion of patients die young. In homozygous condition HbS gene causes severe illness and patient die before attaining puberty. In West Africa it is lethal for at least 1 birth in 10. The heterozygotes (HbAS) are generally healthy. Sickled RBCs produce higher levels of the superoxide anion (O_2^-) and hydrogen peroxide (H_2O_2), both toxic to malaria parasites. Sickle cell trait (HbAS) give at least 90% protection against malaria mortality and heterozygotes enjoy almost 80% protection from severe complications of malaria, especially cerebral malaria (Hill et al. 1991). Arese (2006) summarizing the protection from sickle-cell trait state 90% protection against cerebral malaria and severe anemia, and preventing hospitalization in 60% cases. Hemoglobin S clearly fulfils the conditions of J.B.S. Haldane's malaria hypothesis.

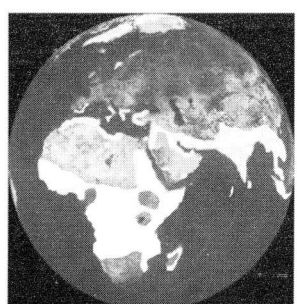 

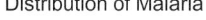

Distribution of Malaria   Distribution of the Sickle Cell Mutation

Fig. 3 Distribution of malaria and sickle cell mutation overlap each other and provide 90% protection from severe anemia and cerebral malaria in heterozygous (HbA and HbS) condition. http://sickle.bwh.harvard.edu/menu_sickle.html

2.2.2 Thalassemias

J.B.S. Haldane in 1949 suggested that heterozygote advantage to malaria might be the mechanism for very high gene frequency of thalassaemias in the tropical world reflecting a state of balanced polymorphism. Decreased production of one of the globin chains leads to thalassemias; decreased production of α globin chain leads to α-thalassemias and decreased synthesis of β globin chain leads to β-thalassemias. While all α-thalassemias are caused due to deletion in the α-globin gene, β-thalassemias may be deletional or non-deletional (e.g., intron splicing error, chain termination error, etc). Thalassemias are a class of anaemias of hemoglobin production involving the effective loss of alpha (alpha thalassemias) or beta (beta thalassemias) alpha or beta-globin chains. High frequencies of Thalassemias occur almost throughout the malaria endemic world upto malaria distribution limits about 2,000 years ago (Carter and Mendis 2002). Several studies show association of thalassemias with the long-standing exposure of a population to the presence of malaria. Allen et al. (1997) found approximately 50% reduction in patients hospitalized for malaria for both heterozygotes and homozygotes for certain alpha thalassemias, and later confirmed for beta thalassemias in Africa (Willcox et al. 1983). Among the thalassemias Hemoglobin C gene is allelic with Hb S but codes for lysine-to-glutamate substitution at position 6. Thalassemia HbC is found in West African populations. The frequency of the allele varies between 10-20% or more (Livingstone 1967,1971). HbC in homozygous condition provides 80-90% protection against severe malaria and risk of death by *P. falciparum* (Agarwal et al. 2000). Protection by the hemoglobin C in a heterozygous combination is about 29% reduction in risk of clinical malaria (Modiano et al. 2001). Hemoglobin E (Hb E) is the result of a single point gene mutation of the beta-chain with a glutamate-to-lysine substitution at position 26. HbE gene is most prevalent affecting 30 million population in the Indian subcontinent and Southeast Asian countries, its occurrence consistent with extremely high malaria burden e.g., in Thailand it is present in the range of 19-25% (Arese 2006). Hemoglobin E provides protection against *P. vivax*, and reduces severe complications of *P. falciparum*, thus, preventing hospitalization in some cases. Those treated with artemisinin drugs (but not chloroquine) clear the parasites from their blood more rapidly than those lacking this trait (Hutagalung et al. 2000).

2.2.3 Glucose-6-Phosphate Dehydrogenase (G-6-PD)

G-6-PD is an important housekeeping enzyme in the pentose phosphate pathway glycolytic pathways of the glucose metabolism. G-6-PD deficiency follows as similar distribution pattern to the hemoglobin variants. It occurs in high frequency throughout tropical Africa, the Mediterranean region, the middle east and the south east Asia (Livingstone 1985). G-6-PD deficiency results from the mutation within the G-6-PD gene located on the X-chromosome. G-6-PD enzyme deficiency affects nearly 400 million human population worldwide (Vulliamy et al. 1992). High disease burden of *P. falciparum* and *P. vivax* could have provided the force for the selection of G-6-PD deficiency in African region. The distribution and frequency of the G-6-PD deficiency is positively correlated with the regions where malaria was or is endemic (Siniscalco et al. 1961). The individuals with G-6-PD deficiency have relatively low *P. falciparum* load (Gilles et al. 1967). In heterozygous and hemozygous combinations, G-6-PD deficiency provides about 50% protection against severe *P. falciparum* malaria (Gilles,1967; Ringwald, 2001). The RBCs deficient in G-6-PD are under oxidative stress, which may create toxic environment for the parasite (Miller 1994). In subjects having abnormal haemoglobin variants or G-6-PD deficiency parasitaemia was limited compared to subjects with normal RBC and in turn disease severity. In populations where these variants were present under the influence of malaria got selected because non-immune subjects with normal variants were killed due to severity of disease, thus, reducing the gene frequency .On the other side, non-immune subjects with abnormal variants could cope up better by limiting parasitaemia and survived, thus in the population in due course of time frequency of abnormal variants became very high and got fixed. In Africa G-6-PD deficiency confers 46-58% reduction in risk of severe malaria (Ruwende et al. 1995). G-6-PD deficiency is hazardous in malaria endemic regions as it may induce hemolysis with certain antimalarial drugs such as Primaquine.

2.2.4 Ovalocytosis

Ovalocytosis is a dominant disorder found sporadically in every racial group. It occurs in high frequency in a milder form in Papua New Guinea. Ovalocytic RBCs resist invasion by *P. falciparum,* and this characteristic may be related to their relative rigidity due to abnormality of their cytoskeleton (Kidson et al. 1981). Ovalocytosis is caused in main structural proteins of the RBC, by the band-3 gene deletion mutation which is the main structural protein of the RBC. This results in abnormal shape of the affected RBCs in the heterozygote condition. The homozygotes for band-3 gene deletion are 100% lethal *in utero*. Neither the sickle cell gene nor hemoglobin E appears to be present among Melanesian populations. Other than ovalocytosis itself, only the moderately protective G-6-PD deficiency gene and some thalassemias are found (Livingstone 1967). Ovalocytosis occurs upto 20% in high malaria endemic lowlands of New Guinea but virtually disappears in the malaria-free highlands. Carriers of ovalocytosis trait are at reduced risk of malaria, especially *P. vivax* malaria (Cattani 1987).

2.2.5 RBC Duffy Negativity

Miller et al. (1975) noticed that human RBCs lacking the Duffy blood group (FyFy) are refractory to invasion by the merozoites of *P. knowlesi*, a monkey parasite similar *to P. vivax*. In West and Central African populations are almost entirely Duffy negative, and therefore, completely refractory to *P. vivax* malaria as they carry homozygous RBC Duffy negativity (Garnham, 1996). The frequency of FY^*B^{null} allele is close to fixation in this region and exceeds 97% leading to almost universal RBC Duffy negativity (Mourant,1976; Cavalli-Sforza,1994). *P. vivax* malaria must have been the selective agent for the near fixation of the FY^*B^{null} allele in the African populations, to the point that *P. vivax* was itself virtually eliminated (Carter and Mendis 2002).

The impact of malaria has undoubtedly protected the human population by selection of protective genetic traits but our past experience of malaria (and other poorly understood sweep of infectious diseases) has left us with a huge burden of genetic diseases. A recent report by the March of Dimes (Anonymous 2006) estimated that in the world, 8 million children or 6 percent of total children are born each year with serious birth defect of genetic or partially genetic origin. Of these 3.3 million children die of serious birth defects and 3.2 million may be mentally and physically disabled for life. In 2001 five birth defects of genetic origin accounted for 26 percent of all births: congenital heart defects (1,040,865 births); neural tube defects (323,904 births); the hemoglobin disorders thalassemia and sickle cell disease (307,897 births); Down syndrome (217,293 births); and G-6-PD deficiency (177,032 births).

2.3 Economic Impact of Malaria

Populations exposed to the perennial unremitting malaria lived and died in abject poverty and destitution. No economic enterprise, big or small, was possible in the presence of malaria. Malaria alone was enough for the stagnant economy preventing all developments and the tapping of natural resources. As, for example, resolution of the Government of India, Department of Education, 1914 states "The most important tropical disease in India is malaria. After allowance is made for the tendency to attribute to fever deaths from other causes, malaria stands out universally prevalent in India and in many tracts, it is a scourge greater than either plague or cholera. It maims as well as kills, and causes more sickness, misery and death than any other single disease." Throughout human history malaria has always been one of the severest of impediments in social and economic development. In words of Sinton (1956) "malaria begets poverty and poverty malaria".

Sinton (1935-36) estimated that in British India malaria caused 1,000,000 deaths and 100,000,000 people suffered from the disease in normal years. During regional epidemics these figures increased by 25-75 million and deaths doubled. Sinton provides a graphic account of what malaria costs India giving details of cost involved in almost every aspect of life from individual expenditure to cost involved in the treatment, loss in working efficiency, agriculture, industry, burial of the dead and indirect costs, etc. and I quote "Malaria gives rise to the greatest economic problem with which India is faced. The financial loss to the individual and the family alone have been calculated at not less than Rs. 100,000 lakhs annually, or about £80 million sterling per annum. This is apart from the effects of the disease upon all aspects of the labor problem, and thus upon the fullest exploitation of the natural resources of the country and the successful development of her manufacturing and other industries. Whilst it is not possible to evaluate with any degree of accuracy the immensity of these direct and indirect losses, there is little reason to doubt that they must run in to unbelievable millions of pounds sterling each year." Thus, Sinton summarized the malaria situation in British India as "the problem of existence in very many parts of India is the problem of malaria. There is no aspect of life in this country which is not affected, either directly or indirectly, by this disease. It constitutes one of the most important causes of economic misfortune, engendering poverty, diminishing the quantity and quality of the food supply, lowering the physical and intellectual standards of the nation, and hampering increased prosperity and economic progress in every way" (GOI 1960).

Malaria and poverty are two sides of the same coin. Malaria is disease of the poverty and the cause of the poverty. Fig. 4 shows that intense malaria transmission is confined to the tropical and sub-tropical regions; whereas poverty as measured by low GDP also predominate the same region as malaria. Almost all rich countries are outside the intensive malaria transmission belts (Sachs and Malaney 2002).

Malaria affects the poor and hapless severely. World Bank (2001) reports that world's poorest 20% contribute 60% malaria deaths (Table 2).

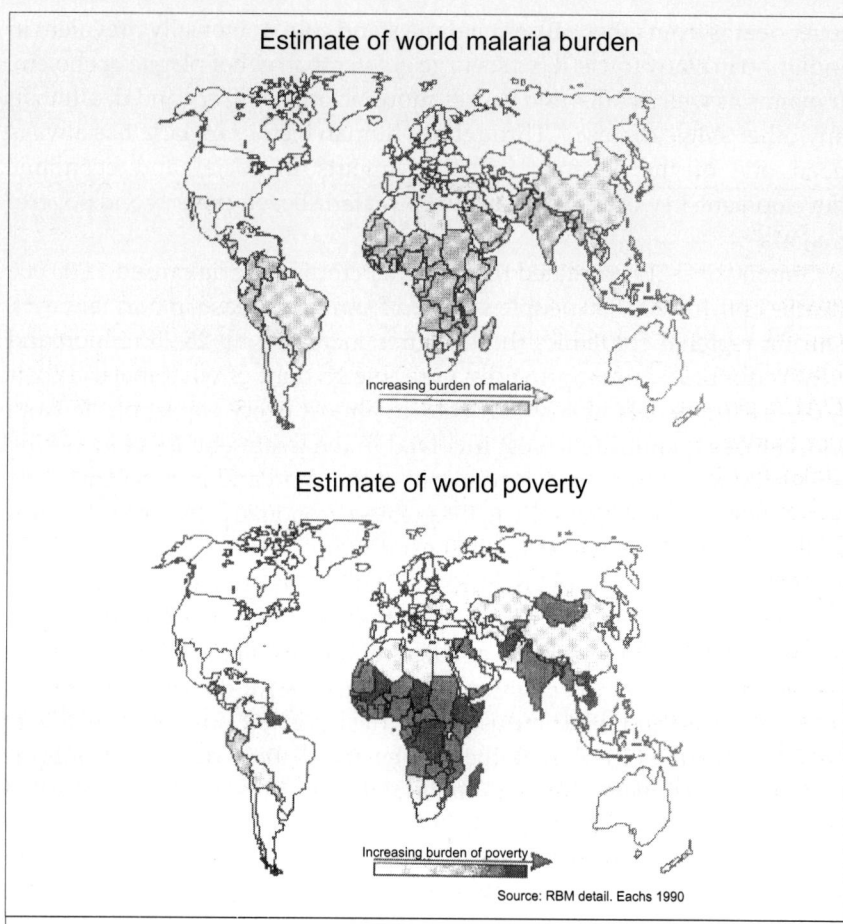

Estimate of world malaria burden

Increasing burden of malaria

Estimate of world poverty

Increasing burden of poverty

Source: RBM detail. Eachs 1990

Fig. 4   There is a clear overlap in the countries with the intensity of malaria and the level of the Gross National Product (GDP). http://www.rbm.who.int/cmc_upload/0/000/015/363/RBMInfosheet_10.pdf

Table 2   Malaria burden on the poor population

Disease	Percentage of death from disease that occur among the poorest 20% of the total global population
Malaria	57.9%
Childhood Diseases	55.0%
Diarrheal Diseases	53.2%
Perinatal Conditions	45.0%

Tuberculosis	44.4%
Maternal Conditions	43.2%
Respiratory Infections	42.6%
HIV/AIDS	41.8%
Weighted Average	48.6%

Furthermore, a recent study based on literature review and surveys worldwide has revealed that 24% of the global diseases are caused by environmental exposures, over 33 per cent of diseases in children under 5 are caused by environmental factors which can be averted by well-targeted interventions (WHO 2005). As, for example, malaria alone causes 19 million DALYs per year; largely as a result of poor water resource, housing and land use management which fails to curb vector populations effectively (WHO, 2006). In Africa majority of malaria deaths occur in areas where malaria is a major cause of poverty and hinders social and economic development. Malaria is responsible for the growth penalty of 1.3% per year in some African countries, and when compounded over the years this becomes a substantial difference between the countries with and without malaria (Gallup and Sachs 2001; Sachs and Malaney 2002). Total costs (direct and indirect) of malaria as a proportion of household income have been estimated to range between 5 and 18% for infected households in four African nations. In India national developments under 5-year plans resulted in the national average of population below poverty line (BPL) from 50% in 1965 to 26.1% in 2000. Malaria was rampant in all states when poverty levels were high but states performing better have low malaria incidence. Malaria remains a major problem only in states with population exceeding 26.1% BPL i.e., of the countries total malaria cases 69.57% positive for malaria, 49.28% *P. vivax* and 88.37% *P. falciparum* (Sharma 2003a). In 1995 malaria costs India upto 1 billion US$ annually, and this cost is rising steeply (Sharma 1996a). Malaria costs sub-Saharan Africa $12 billion in lost GDP per year or $4.9 billion per year estimated over the 1980-1995 period for 31 African nations (Sachs and Malaney 2002). Malaria constitutes 10% of the continent's overall disease burden, and in some countries with very high malaria burden, malaria may account for as much as 40% of public health expenditure, 30-35% of inpatient admissions and upto 50% of the outpatient visits. Malaria endemic countries in Africa or outside had income levels in 1995 of only 33% that of countries without malaria. A 10% reduction in malaria was associated with 0.3% higher growth (Gallup and Sacks 2001).

2.4 Global Efforts in Malaria Control

Malaria known as the "King of Diseases" is one of the world's most prevalent killer diseases. The discovery of DDT's residual insecticidal properties paved the way for malaria eradication. In 1950s World Health Organization (WHO) launched the malaria eradication campaign in endemic countries of the world. For logistic reasons malaria control in Africa was considered beyond the scope and never attempted. During the Global Malaria Eradication Programme (1957-1972), DDT spraying and treatment reduced/eliminated malaria transmission from America, Asia, Europe, and Transcaucasia. The success in malaria control in the tropical and sub-tropical countries was spectacular but shortlived. Epidemics frequently broke out in Central America and South East Asia. Malaria was at the point of extinction in Sri Lanka but a massive epidemic in 1968 and reports of resurgence from India and other countries finally led to the scrapping of malaria eradication programme in favor of malaria control. Malaria remained high on WHO's agenda, supporting and guiding member countries in strengthening the primary health care system, improving delivery of malaria control interventions and providing training. Since, malaria's return was associated with increasing drug resistance and control failures causing high morbidity and mortality, it was obvious that attack on malaria should be uniformly applied in all endemic countries primarily to provide relief to people living in endemic countries from the ill affects of malaria. Therefore, at the instance of WHO, in 1992 ministers and representatives of more than 100 countries attended a Ministerial Conference on Malaria in Amsterdam, Netherlands. Backed by WHO, this conference adopted the Global Malaria Control Strategy (GMCS). The four main objectives of the GMCS were: (i) to provide early diagnosis and prompt treatment of malaria; (ii) to plan and implement selective and sustainable preventive measures, including vector control; (iii) to detect early, contain and prevent epidemics; and (iv) to strengthen local capabilities in basic and applied research to permit and promote the regular assessment of a country's malaria situation, in particular the ecological, social and economic determinants of the disease (Anonymous,1993). In 1998 another landmark decision was taken by Dr. Gro H. Brundtland, the newly appointed Director General of the WHO. She called upon member countries to Roll Back Malaria (RBM), a scourge in the tropics and subtropics. In pursuance of this decision, a Global Partnership was launched in October 1998 to Roll Back Malaria in partnership with the UNICEF, UNDP and the World Bank. RBM envisaged reduction in malaria mortality to half by 2010 as compared to 2000 and to half again by 2015 (GCMS/RBM, 1992). This will be achieved by increasing speedy access for people to effective treatment and means of protection from mosquito bites enabling national authorities and non-governmental organizations to combat malaria intensifying efforts developing new products for the prevention and treatment of malaria. This objective was to be achieved by much broader partnership *inter alia* involving the private sector and civil societies worldwide; and through an effective delivery of anti-malaria interventions. RBM emphasized on the evidence based action, political mobilization, involvement of non-health sectors, advocacy and partnership. RBM strategy was consistent with the Global Malaria Control Strategy. RBM outlined six key elements for malaria control viz., early case detection, rapid treatment of cases within 24 hrs of fever, multiple prevention emphasizing on the insecticide treated mosquito nets, well-coordinated action, dynamic global movement, and focused research. The role of RBM was to strengthen the national capacities of health systems so that technical interventions are implemented to the required level of effectiveness. It is notable to mention that the years 2001-2010 were declared as the United Nations Decade to Roll Back Malaria. Several other important initiatives supported the global efforts to control malaria world wide.The Asian Collaborative Training Network for Malaria (ACTMalaria) is an inter-country initiative between nine countries to collect, develop and disseminate training materials and to implement practical training courses to meet the needs of malaria control programs in South East Asia and the Mekong valley, and to improve communication between these countries on malaria control problems affecting their common borders. The *P. falciparum* Genome Sequencing Consortium is a collaboration among The Sanger Centre (U.K.), The Institute for Genomic Research/Naval Medical Research Institute (U.S.A.), and the Stanford University (U.S.A.). The mapping malaria risk in Africa (The MARA/ARMA collaboration) was initiated to provide an atlas of malaria for Africa. These maps are based on remote sensing. The WHO's Medicines for Malaria Venture (MMV) is a joint public-private sector initiative which aims to develop anti-malarial drugs and drug combinations for distribution in poor countries. MMV is a non-profit organization bringing private sector management methods. MMV is committed to develop 1 new anti-malarial every 5 years. Special Program for Research and Training in Tropical Diseases (TDR) provides support and guidance on research and training in the control of tropical diseases through a task force approach in three areas viz., Strategic and Discovery Research (SDR) Product Development and Evaluation (PDE) and Research Capability Strengthening (RCS) (TDR/http://www.searo.who.int/en). RITAM was established during 1999 as a network of researchers and other people who are active or interested in the study and use of traditional, plant-based anti-malarials. Research Initiative on Traditional Anti-Malarials (RITAM) is a partnership between Global Initiative for Traditional Systems (GIFTS) of Health, University of Oxford and the Tropical Disease Research (TDR) Programme of WHO. The Multilateral Initiative on Malaria (MIM) was launched in 1999 to strengthen and sustain, through collaborative research and training, the capability of malaria endemic countries in Africa to carry out research required to develop or improve tools for malaria control (http://www.mim.su.se/). The Regional Malaria Control Programme in Cambodia, Laos and Vietnam is a European Community Regional Malaria Control Programme. The World Bank is boosting malaria control programme entitled "Rolling Back Malaria: The Global Strategy and Booster Program", which will make funding available to countries to enhance existing programs to combat the disease. The booster programme is building on the global knowledge base for malaria control and the World Bank's operational experience (http://web.worldbank.org/WBSITE).

Malaria Vaccine Initiative was created in 1999 through a grant from the Bill & Melinda Gates Foundation. MVI works with Global Alliance for Vaccines and Immunization (GAVI) to ensure availability of vaccines in countries most affected by malaria (http://www.gavialliance.org/). The MVI is expected to lead to field trials of one or more vaccine candidates; coordinate its efforts with malaria vaccine programs at various organizations and agencies; and identify gaps in current research efforts and apply resources to advance promising malaria vaccine candidates (http://www.malariavaccine.org/malaria-and-vaccines.htm).

WHO Initiative for Vaccine Research (IVR) was established in 2001 to coordinate and streamline the vaccine research and provide leadership particularly for diseases endemic in developing countries. The African Malaria Vaccine Testing Network (AMVTN) is a non-profit network established in 1995 in Arusha, Tanzania. Its objective is to provide a forum for scientists and policy makers involved in the planning, coordination, and execution of malaria vaccination trials in Africa. Malaria Research and Reference Reagent Resource Center (MR4) sponsored by NIAID parasite, vector, and host cell reagents (http://www.mr4.org/). The Global Fund on AIDS, Tuberculosis and Malaria is set up in Geneva to fund countries to control the three diseases. Total disbursement for malaria alone over a 2-year period exceeds US $ 2.0 billion (Source: WHO/UNICEF World Malaria Report 2005). Hopefully all these efforts would result in achieving the Millennium Development Goal to halt the rising incidence of malaria by 2015.

ACKNOWLEDGEMENTS

I am grateful to Dr. B.S. Das, Emeritus Medical Scientist (ILS, Bhubaneswar, Orissa), Dr. Hema Joshi, Dy Director NIMR and Dr. Neeru Singh, Sr Dy Director, RMRC Jabalpur for review and constructive suggestions.

REFERENCES

Agarwal A, Guindo A, Cissoko Y et al. (2000). Hemoglobin C associated with protection from severe malaria in the Dogon of Mali, a West African population with a low prevalence of hemoglobin S. Blood 96:2358-2363.

Allen SJ, Donnell AO, Alexander NDE, Alpers MP, Peto TE, Clegg JB, Weatherall DJ (1997). α - Thalassemia protects children against disease caused by other infections as well as malaria. Proc Natl Acad Sci USA 94:14736-14741.

Allen SJ, Donnell AO, Alexander ND, Mgone CS, Peto TE, Clegg JB, Alpers MP, Weatherall DJ (1999). Prevention of cerebral malaria in children in Papua New Guinea by South East Asian ovalocytosis band 3. Amer Jour Trop Med Hyg 60:1056-1060.

Allison AC (1964). Polymorphism and natural selection in human population. Cold Spring Harbor Symposium. Quant Biol 29:137-149.

Anonymous (1993). Implementation of the global malaria control strategy. Report of a WHO Study Group on the Implementation of the Global Plan of Action for Malaria Control 1993-2000. World Health Organ Tech Rep Ser 839:1-57.

Anonymous (2006). March of Dimes Birth Defects Foundation: Birth defects: 8 million annually worldwide. Hidden toll of dying and disabled children; A comprehensive global analysis identifies trends and interventions. Public release date: 30-Jan-2006.

Arese P, Ayi K, Skorokhod A, Turrini F (2006). Removal of early parasite forms from circulation as a mechanism of resistance against malaria in widespread red blood cell mutations. In Emerging Infectious Diseases of the 21st Century. Malaria: Genetic and evolutionary aspects. Eds. K.R. Dronamraju and Paolo Arese. Springer New York 2006 pp.25-53.

Carter R, Mendis KN (2002). Evolutionary and Historical Aspects of the Burden of Malaria. Clini Microbiol Rev 15: 564-594.

Cattani JA, Gibson FD, Alpers MP, Crane GG (1987). Hereditary ovalocytosis and reduced susceptibility to malaria in Papua New Guinea. Trans R Soc Trop Med Hyg 81:705-709.

Cavalli-Sforza LL, Menozzi P, Piazza A (1994). The history and geography of human genes. Princeton University Press, Princeton, N.J

Christopher SR (1911). Sci. Mem. Off Med. Sanit Depts. Govt. India (n.s.) 46:197.

Dua VK, Vas Dev, Phookan S, Gupta NC, Sharma VP, Subbarao SK (2003). Multi-drug resistant Plasmodium falciparum malaria in Assam, India. Timing of recurrence and antimalarial drug concentration in whole blood. Amer J Trop Med Hyg 69:555-557.

Dua VK, Kar PK, Sharma VP, Edwards G (2004). Monitoring chloroquine resistant malaria. Curr Sci 87(6):726.

Fernando SD, Gunawardena DM, Bandara MRSS, De Silva D, Carter R, Mendis KN, Wickremasinghe AR (2003). The Impact of repeated malaria attacks on the school performance of children. Am J Trop Med Hyg 69: 582-588.

Garnham PCC (1996). Malaria parasites and other haemosporidia. Blackwell, London, United Kingdom.

Gallup JL, Sachs JD (2001). The economic burden of malaria. Am J Trop Med Hyg 64 (Suppl):85-96.

GMCS/RBM (1992) http://www.searo.who.int/en/Section10/Section21/Section1368 9865.htm

Gilles HM, Fletcher KA, Hendrickse RG, Lindner R, Reddy S, Allan N (1967). Glucose 6 phosphate dehydrogenase deficiency, sickling and malaria in African children in South Western Nigeria. Lancet i:138-140.

GOI (1960). Ministry of Health. GOIGovernment of India Manual of the Malaria Eradication Operation. Issued by Office of the Directorate, National Malaria Eradication Programme, 22, Alipur Road, Delhi.

Haldane JBS 1948. The rate of mutation of human genes. Hereditas 35(Suppl.):267-273.

Haldane JBS (1949). In symposium on ecological and genetic factors in animal speciation. Pallanza, 1948, pp 68-76. Supplemento a la rocerca scientifica 19.

Hay SI, Rogers DJ Toomer JF Snow RW (2004). "Annual *Plasmodium falciparum* inoculation rates (EIR) across Africa: Literature survey. Internet access and review" Trans R Soc Trop Med Hyg 94:113-127.

Hill AVS, Catherine EM, Kwiatkowski, D, Anstey NTM, Twumasi P, Rowe, PA Bennett S, Brewster D, McMichael AJ, Greenwood BM (1991). Common West African HLA antigens are associated with protection against severe malaria. Nature 352:595-600.

Holding PA, Stevenson J, Peshu N, Marsh K (1999). Cognitive sequelae of severe malaria with impaired consciousness. Trans R Soc Trop Med Hyg 93:529-34.

Hume JCC (2003). Malaria in antiquity: a genetics perspective. World Archaeology, 35(2): 180-192.

Hutagalung R, Wilairatana P, Looareesuwan S, Brittenham GM, Gordeuk VR (2000). Influence of hemoglobin E trait on the antimalarial effect of artemisinin derivatives. J Infect Dis 181:1513-1516.

Kidson C, Lamont G, Saul A, Nurse GT (1981). Ovalocytic erythrocytes from Melanesians are resistant to invasion by malaria parasites in culture. Proc Natl Acad Sci USA 78:5829-5832.

Kochar DK, Thavi K, Joshi I, Subhakaran A, Aseri S, Kumawat BL (1998). Falciparum malaria and pregnancy. Indian J Malariol 35(3):123-130.

Kochar DK, Saxena V, Singh N, Kochar SK, Kumar SV, Das A (2005). *Plasmodium vivax* malaria. Emer Infec Dis 11(1):132-134.

Lopez AD, Mathers CD, Ezzati M, Jamison D, Murray CJL (2006). Global and regional burden of disease and risk factors, 2001: Systematic analysis of population health data. Lancet 367:1747-1757.

Livingstone FB (1967). Abnormal hemoglobins in human populations. Aldine, Chicago, Ill.

Livingstone FB (1971). Malaria and human polymorphisms. Annu. Rev. Genet. 5:33-64.

Livingstone FB (1985). Frequencies of hemoglobin variants. New York. Oxford University Press.

Luxemburger C, Ricci F, Nosten F, Raimond D, Bathet S, White NJ (1997). The epidemiology of severe malaria in an area of low transmission in Thailand. Trans R Soc Trop Med Hyg 91: 256-62.

Miller LH, Mason SJ, Dvorak JA, McGinniss MH, Rothman IK (1975). Erythrocyte receptors for (*Plasmodium knowlesi*) malaria: Duffy blood group determinants Science. 189:561-563.

Miller LH (1994). Impact of malaria on genetic polymorphism and genetic diseases in Africans and Africam Americans PNAS 91:2415-2419.

Medicines for Malaria Venture (MMV) /http://www.mmv.org/rubrique.php3? id_rubrique=15

Modiano D, Luoni G, Sirima BS, Simpores HV, Konate A, Rastrelli E, Olivieri A, Calissano C, Paganotti GM, D'Urbano L, Sanou I, Sawdogo A, Modiano G, Coluzzi M (2001). Haemoglobin C protects against clinical *Plasmodium falciparum* malaria. Nature 414:305-308.

Mourant AE, Kopec AC, Domaniewska-Sobczak K (1976). The distribution of the human blood groups and other polymorphisms. Oxford University Press, London, United Kingdom.

Murphy SC, Breman JG (2001). Gaps in the childhood malaria burden in Africa: cerebral malaria, neurological sequelae, anemia, respiratory distress, hypoglycemia, and complications of pregnancy. Amer J Trop Med Hyg 64(1-2 Suppl):57-67.

Naing T, Win H, New YY (1988). Falciparum malaria and pregnancy. Relationship and treatment response SEA J Trop Med Publ Hlth 19:253-258.

Ringwald P, Williams R (2001). Drug resistance in malaria. Pharmacol. News 8:15-20.

Ruwende C, Khoo SC, Snow RW, Yates SN, Kwiatkowski D, Gupta S, Warn P, Allsopp CE, Gilbert SC, Peschu N, Newbold CI, Greenwood BM, Marsh K, Hill AVS (1995). Natural selection of hemi- and heterozygotes for G6PD deficiency in Africa by resistance to severe malaria. Nature 376:246-249.

Sachs Jeffrey, Malaney P (2002).The economic and social burden of malaria. *Nature* 415, 680-685.

Sharma VP (1996). Re-emergence of malaria in India. Ind J Med Res 103:26-45.

Sharma VP (1996a). Malaria: Cost to India and future trends. SEA J Trop Med Publ Hlth 27: 4-14.

Sharma VP (1999). Current scenario of malaria in India. Parassitologia 41:349-353.

Sharma VP (2000). Status of drug resistance in malaria in India. In:Multi-drug resistance in emerging and re-emerging diseases, (eds.) R.C. Mahajan and Amu Therwath. Indian National Science Academy, New Delhi. Narosa Publishing House, New Delhi. pp 191-202.

Sharma VP (2003). DDT: The fallen angel. Curr Sci 85: 1532-1537.

Sharma VP (2003a). Malaria and poverty in India. Curr Sci 84: 513-515.

Sharma VP (2005). Getting to grips with malaria: A view from India. SciDev.net 7th November 2005.

Sharma VP (2006a). Artemisinin drugs in the treatment of *Plasmodium falciparum* malaria in India. Curr Sci 90(10): 1323-1324.

Singh Neeru, Shukla MM Sharma VP (1999). Epidemiology of malaria in Pregnancy. Bull Wrld Hlth Org 77: 567-571.

Singh N, Mehra RK Srivastava N (2001). Malaria during pregnancy and infancy in an area of intense malaria transmission in central India. Ann Trop Med Parasitol 95:19-29.

Singh N, Awadhia SB, Dash AP, Shrivastava R (2005). Malaria During Pregnancy: A Priority Area for Malaria Research and Control in South-East Asia Regional Health Forum WHO South-East Asia Region Volume 9 Number 1, 2005.

Siniscalco M, Bernini L, Litte B, Motulsky AG (1961). Fasism and thalassemia in Sardinia and their relationship to malaria. Nature 190: 1175-1180.

Sinton JA (1935-1936). What malaria costs India, nationally, socially and economically? Rec Mal Surv India 5: 223-264, 5:413-489, 6:92-169.

Sinton JA (1956). What malaria costs India. Health Bulletin No. 26, Malaria Bureau No. 13, Malaria Institute of India, 22-Alipore Road, Delhi.

Snow RW, Craig M, Deichmann U, Marsh K (1999). Estimating mortality, morbidity and disability due to malaria among Africa's non-pregnant population. Bull. Wrld Hlth Org 77:624-640.

Snow RW Guerra CA Noor AM Myint HY Hay SI (2005). The global distribution of clinical episodes of Plasmodium falciparum malaria. Nature 434:214-217.

Vuilliamy T, Mason P, Luzzatto L (1992). The molecular basis of glucose-6-phosphate dehydrogenase deficiency. Trends Gene 8:138-146.

Weatherall DW, Bell JI, Clegg JB, Flint J, Higgs DR, Hill AVS, Pasvol G, Thein SL, Blackwell CC (1988). Genetic Factors as Determinants of Infectious Disease Transmission in Human Communities. Phil Trans R Soc Lond Series B Biological Sciences 321: 327-348.

World Bank/RBM (March 2001) http://www.rbm.who.int/cmc_upload/0/000/014/813/Malaria_at_a_glance1.htm

WHO (2005). The World Malaria Report 2005. Roll Back Malaria, Geneva, WHO (UNICEF/RBM/WHO) http://rbm.who.int/wmr2005.

World Health Organization (2006). Preventing disease through healthy environment-towards an estimate of the environmental burden of disease.

WHO (2003). Lives at risk: malaria in pregnancy. http://www.who.int/features/2003/04b/en/

Wickaramsuria GAW (1937). Clinical features of malaria in pregnancy. Malaria and Ankylosto-miasis in pregnant women. London Oxford University Press UK 5-90.

Willcox M, Bjorkman A, Brohult J (1983). A case control study in northern Liberia of Plasmodium falciparum malaria in hemoglobin S and β-thalassaemia trait. Ann Trop Med Parasit 77:239-246.

The Bird-Flu Virus H5N1: A Pandemic in the Waiting

M. FERNANDO and S. K. LAL**

1. INTRODUCTION

Avian Flu is spreading at an alarming rate. Already there have been 244 confirmed human cases of the H5N1 infection and more than half the cases have proven fatal. The World Health Organization (WHO) has sounded a pandemic alarm. Most influenza virologists are beginning to believe that a new subtype of influenza will mutate to become stable in human populations and therefore spread easily between humans thus taking the form of a pandemic since humans have no pre existing immunity to the H5N1 viruses. It could be the subtype H5N1 or any of the 16 other H5 subtypes. There is growing concern for the front line workers who are in close contact with dead and sick animals. In June 2006, WHO reported evidence of human to human spread in Indonesia. In this incident, 8 members of a family were infected by the H5N1 strain. The first family member is thought to have come in contact with infected poultry. This person then infected 6 of the family members. One of the 6 people, a child, then infected another family. Apart from the first member, the other members did not come into contact with infected poultry leaving concern if it spread

Virology Group, International Centre for Genetic Engineering and Biotechnology, New Delhi 110067, India
** Corresponding author. Dr. Sunil K. Lal, Senior Research Scientist, Virology Group. International Centre for Genetic Engg. & Biotechnology, New Delhi 110067, India Phone: 91-11-26189360/61; Fax: 91-11-26162316; e-mail: sunilklal@gmail.com; sunillal@icgeb.res.in

due to the close and prolonged exposure of the members while tending to the initial victim. Human to human transmission of H5N1 probably occurred at least once before in Thailand in the year 2004 when the mother and an aunt of an infected girl fell ill after caring for her. But no instances of sustained transmission have been documented.

2. H5N1 AND ITS GEOGRAPHICAL SPREAD

Bird flu is a lethal and disastrous infection caused by the avian influenza (AI) virus and today scientists are faced with a challenge in controlling the spread of this virus which, in some cases is deadly. At present the threatening strain of Influenza A—H5N1 virus is generally carried in the intestines of wild birds. These wild birds world-wide carry the infection and are usually unaffected although for reasons unknown, occasionally do cause disease and death in them. However it is contagious among domesticated birds which include chickens, ducks and turkeys. The virus H5N1 was first isolated from birds in South Africa in 1961 and ever since out breaks among poultry have occurred in many countries in Asia which include Cambodia, China, Indonesia, Japan, Laos, South Korea, Thailand and Vietnam during late 2003 and early 2004. Over 100 million birds in the affected countries either died from the disease or were slaughtered to control the outbreak. The first human infection of influenza A – H5N1 through poultry was reported in Hong Kong in 1997 and ever since it has been reported in Cambodia, Indonesia, Thailand and Vietnam.

The influenza virus belongs to the family Orthomyxoviridae and it has 3 different types- A, B and C out of which only A infects humans, animals and birds while types B and C infect only humans. This is a pleomorphic enveloped virus approximately around 80-120 nm and has 8 negative RNA segments. The subtypes occur due to antigenic variation in the glycoproteins hemagglutinin (HA) and neuraminidase (NA) .There are 16 different HA subtypes and 9 different NA subtypes of Influenza A virus. Each combination is a different subtype. HA help virion enter into epithelial cells of humans by attachment to sialic acid receptors containing the alpha 2,6 galactose linkages while the NA enzyme prevents aggregation of the virions within the host cells and facilitates cell to cell spread by cleavage of the glycosidic linkages to sialic acid. The seventh segment or the M segment, codes for matrix 1 (M1) and matrix 2 (M2) proteins. The M1 protein located within the envelope facilitates virion assembly while the M2 protein forms an ion channel and serves as a link between the interior and the outer environment. It also provides a low pH which is essential for HA synthesis.

The RNA segments coding for NP, polymerases PB1, PB2 and PA are involved in viral transcription.

3. H5N1 PATHOGENESIS

Influenza enters the host through inhalation and causes infection by binding to host respiratory columnar epithelial cells, which contain target receptors for HA and other surface proteins. Replication occurs via the host cell's enzymatic processes, and infectious virions are cleaved from the host cell by NA proteins. Influenza causes a range of symptoms including fever, cough, sore throat, myalgia, nasal congestion, weakness, loss of appetite and headache. Respiratory symptoms are caused by local cellular damage and apoptosis. Other symptoms result from host immune responses. Cytokines produced by immune and epithelial cells in response to infection cause fever and other systemic manifestations. Fever usually resolves after a few days but myalgia may last for two weeks. Children may exhibit lower respiratory symptoms and gastrointestinal complaints, and febrile fits are common (Chiu SS et al. 2001).

4. DIAGNOSIS

Influenza infection causes non-specific signs and symptoms which are difficult to differentiate from similar viral respiratory illnesses, labeled as influenza-like illnesses. The two most predictive symptoms are fever and cough, especially if present for more than 36 hours. No single clinical definition provides a sensitive yet specific identification of the infection (Govaert TM et al. 1998). Viral culture of nasopharyngeal and/or throat samples remains the gold standard for diagnosis, but takes three to ten days, limiting its use in prevention strategies. However, viral culture is important to provide information on circulating subtypes for outbreak confirmation and vaccine formulation. Direct immunofluorescent tests offer quicker alternatives but are difficult and labour-intensive, with sensitivities much lower than viral culture. Other tests include reverse-transcriptase polymerase chain reaction, viral serology, and enzyme immunosorbent assays. There are many rapid tests for influenza currently available, providing results in less than 30 minutes, with good sensitivities. Disadvantages include high costs, dependency on specimen adequacy and stage of disease, and decreased sensitivity for H5N1 influenza, although there are new products being constantly developed.

5. VACCINE APPROACHES FOR H5N1

Currently, there is no commercially available vaccine to protect humans against H5N1 in Asia and Europe, but efforts of developing a vaccine are under way and many different approaches have made it to clinical trials. The development of a vaccine for the flu virus has become a major challenge for 2 main reasons. Firstly due to the phenomena of antigenic drift, through which the proteins HA and NA undergo frequent point-mutations due to the lack of proof-reading enzymes. Amino-acid substitutions occur commonly in the 5' antigenic regions which bind with virus neutralizing antibodies (Carton et al. 1982), the virus keeps mutating into different strains. Therefore the vaccine manufacturing companies face a challenge in reformulating vaccines according to the circulating strains. Secondly, unlike monogenic viruses the fragmented genome of influenza virus is more vulnerable to genetic variation. This phenomenon is called antigenic shift and it occurs commonly with influenza A virus. This has resulted in historically important pandemics, such as the strain H1N1 which caused the Spanish flu in 1918, H2N2 which caused Asian flu in 1957 and recently the appearance of H5N1 in Hong Kong in 1997. These general biological characteristics help set the stage for this virus as a potential candidate capable of causing a human pandemic. Understanding the interfaces between host susceptibility, strain diversity, and animal reservoirs is critical for developing vaccines for the control and prevention of this potentially hazardous virus (Hileman 2002; Nicholson 2003).

Inactivated influenza virus vaccine is a trivalent influenza vaccine (TIV) and is formulated with 15 microgram HA from A/H1N1, A/H3N2, and virus B. The seed strain is prepared by infecting the allantoic sac of the chick embryo with phenotype H1N1 (A/PR/8/34) along with the epidemic strain. Then viral replication and genetic re-assortment is allowed to occur followed by screening the new genotype for the absence of genes encoding PR8 or PR8 –like surface glycoproteins. By doing so, it ensures that the genes which carry infection are not included. The seed strain is distributed to vaccine manufacturers after harvesting and purification by zonal centrifugation. The selected seed strain containing HA and NA components of the epidemic strain is further mass propagated in fresh eggs to obtain sufficient vaccine virus. Finally it is chemically treated to obtain the inactivated vaccine. Treatment with formaldehyde leaves the whole inactivated vaccine while treating with sodium dodecyl sulphate leaves one with the split vaccine. These vaccines are at a disadvantage though of having low levels of cell mediated immunity and only effective against antigenically matched strains. In addition the risk in this whole approach remains that the inactivated virus can revert back to virulent strains by recombination inside the human body (Hannoun et al. 2004). Today, the conventional use of chick embryo technique is fast being replaced by the use of mammalian cells as the substrate since it is easier to find, the risk of microbial contamination is minimized and a lot lesser manpower is required in the latter technique compared to the former.

Virosomes are reconstituted virus like particles which are unilamellar liposomes carrying the spike proteins of influenza virus on their surface but lack any viral genetic material. Using virosomes as vaccines, they elicit high titers of influenza specific antibodies for HA and NA and exhibit strong immunogenicity. Virosomes support antibody formation and induction of T-helper cell responses. A virosome derived influenza vaccine has is currently been developed and has made it to clinical trials (Plante et al. 2001; Hezog et al. 2002).

Immunopotentiators increase the immunogenicity of the TIV by increasing the yield of available doses without an increase in the chick eggs used. An aluminium adjuvanted whole virus vaccine with 1.9- 7.5 micrograms of HA is similar to a 'whole virus' vaccine with 15 micrograms HA without alum. This method is at a disadvantage since it requires complex production protocol and cause local pain, temperature and mylagia (Podda 2001).

Although the vaccine approach described above provides better immune protection and increased specificity against the virus, the fear that it might revert back to the virulent phenotype is a major obstacle in its worldwide deployment. Live attenuated vaccines can be produced by denaturing of virus under elevated temperatures, passage of the virus through an unusual host, and by attenuating the master strain by serial passage in embryonated chick eggs at low temperatures ranging from $33 - 25\,^{0}$C. These processes are unable to cause illness in humans but are able to donate the genes other than those of HA and NA which are co-infected with the epidemic strain (Massab and DeBorde 1985).

Vaccines are the only principle medical intervention for protecting individuals in a pandemic influenza situation. If available readily and in sufficient quantities, they can reduce the mobility and mortality of the traditional pandemics which took many lives. Recent research on live but weakened vaccine development showed success. It had several strains of the H5N1 bird flu virus tested on ferrets instead of mice. Ferrets proved to be an ideal model since they develop fever and sneeze and cough like humans, on the other hand mice are unable to show these symptoms (Suguitan et al. 2006).

Recently recombinant live attenuated vaccines have been produced by deleting the virulent portion, mainly NS1 and M2 genes. These are still in the early stages of development but show great promise for the future (Watanbe et al. 2002). Since H5N1 began spreading through Asian poultry in 2003 the WHO has alerted the world to take precautions. Things can get more serious as the virus mutates and stabilizes in the population, causing spread between human to human. Thus advances in both vaccine quality and production are actively investigated by governments of many countries.

DNA vaccines have offered a new opportunity to immunize with materials that are entirely gene based, expressed by the recipient's own cells. This provides a greater control over the immunization process, because the investigator determines which antigens and co-stimulants to use, and where to elicit the response. The duration of the response can be controlled by repeated exposures to the genes which are expressed transiently by a variety of delivery mechanisms like direct injection and electroporation. This new technology helps in bypassing years of development for the production of efficacious vaccines. DNA vaccines are an ideal mimic of intracellular antigens. A plasmid DNA vaccine can be produced by using a non-infective and non-replicative plasmid that encodes only proteins of interest. Through homologous recombination the variable antigenic determinant domains are cloned into the vector to produce the desired plasmid DNA vaccine (Bright et al. 2003).

A new technology that's looking increasingly promising in the fight against the flu is plasmid-based reverse genetics (Hoffmann et al. 2002). This technique shows significant promise to speed up the process by which a vaccine can be created. With reverse genetics, scientists can customize a flu vaccine by assembling genes that code for the desired features. Two genes representing the HA and NA antigens are selected from the target virus, while the remaining six genes come from a virus that's time-tested for its ability to grow inside an egg. The DNA that codes for the viral genes is then inserted into animal cells, which make new copies of the virus. Researchers recover the resulting virus that will be used to manufacture the vaccine. Although the vaccine still needs to be grown inside eggs for large-scale production, animal cells could also be used as that technology advances. One of the major advantages of this new technology is that if portions of a targeted virus, such as the H5 and H7 antigens, are too toxic to grow inside eggs, the segments of the genes coding for these antigens that make them so dangerous can be clipped and removed. If successful, scientists will try to determine if reverse genetics technology can provide a quick and effective way for developing a suitable vaccine in the event of a pandemic.

6. ANTI-VIRALS

Currently many nations worldwide have stockpiled the anti-viral drug Tamiflu in case human to human transmission of H5N1 is reported. Tamiflu is a drug which is taken to control the spread of the virus within the body. It is a neuraminidase inhibitor and it works by inhibiting the critical neuraminidase protein on the surface of the virus. As the flu virus multiplies, the neuraminidase helps the virus to spread to other host cells but Tamiflu prevents spread by inhibiting the neuraminidase so that it doesn't leave it's host cell. Infected people with H5N1 are administered with the anti-viral drug Tamiflu, and currently oseltamavir and zanamavir are also used, but additional studies are needed to show their effectiveness. The H5N1 has shown resistant to the commonly used anti influenza drugs such as amantadine and rimantadine. However, Tamiflu will not be a solution in fighting a pandemic since it can just prevent the spread of the virus and does not help in eliminating the virus from a human body. Also, the virus is constantly evolving and the reassortant that causes the human pandemic may or may not respond to Tamiflu.

REFERENCES

Bright RA, Ross TM, Subbarao K, Robinson HL, Katz JM. (2003) Impact of glycosylation on the immunogenicity of a DNA-based influenza H5 HA vaccine. *Virology* 308:270-8.

Caton A, Brownlee G, Yewdell J, Gerhard W (1982) The antigenic structure of the influenza virus A/PR/8/34 hemagglutinin (H1 subtype). *Cell* 31:417-27.

Chiu SS, Tse CY, Lau YL, Peiris M. In.uenza A infection is an important cause of febrile seizures. Pediatrics 2001; 108:63.Comment in: *Pediatrics* 2001; 108:1004-5.

Govaert TM, Dinant GJ, Aretz K, Knottnerus JA. The predictive value of in.uenza symptomatology in elderly people. *Fam Pract* 1998; 15:16-22.

Hannoun C, Megas F., Piercy J (2004) Immunogenicity and protective efficacy of Influenza vaccination. *Virus Res.* **103**:133-8.

Hezog C, Metcalfe IC, Schaad UB (2002) Virosome influenza vaccine in children. *Vaccine* **20**:B24-8.

Hilleman MR (2002) Realities and enigmas of human viral influenza. *Vaccine* **20**:3068-87.

Hoffmann E, Krauss S , Perez D, Webby R , Webster RG (2002)Eight plasmid system for rapid generation of Influenza virus vaccines. *Vaccine.* **20**:3165-70.

Massab HF, DeBorde DC (1985) Development and characterization of cold adapted viruses for use as live virus vaccines. *Vaccine* 3:355-69.

Morein B,SundquistB,Hoglund S, Dalsgaard K (1984) A novel structure for antigen presentation of membrane proteins of enveloped viruses. *Nature* **308:**457-60.

Nicholson KG, Wood JM, Zambon M (2003) Influenza. *Lancet* **362:**1733-1745

Plante M, Jones T, Allard F, Torossian K, Gauthier J, St-Felix N, White GL, Lowell GH, Burt DS. (2001) Nasal immunization with submit proteosome influenza vaccines induces serum HAI, mucosal IgA, and protection against influenza challenge. *Vaccine* **20:**218-25.

Podda A (2001) The adjuvanted influenza vaccines with novel adjuvants: experience with the MF59-adjuvanted vaccine. *Vaccine* 19:2673-80.

Suguitan AL, McAuliffe J, Mills KL, Jin H, Duke G, Lu B, Luke CJ, Murphy B, Swayne DE, Kemble G, Subbarao K. (2006). Live, Attenuated Influenza A H5N1 Candidate Vaccines Provide Broad Cross-Protection in Mice and Ferrets. *PLoS Med.*3:9-12

Watanbe T , Watanbe S, Kida H, Kawaoka Y (2002) Influenza A virus with defective M2 ion channel activity as a live vaccine. *Virology* **299:**266-70.

The Severe Acute Respiratory Syndrome Coronavirus (SARS-CoV) Outbreak: Lessons Learnt and Future Perspective

N. SATIJA and S. K. LAL*

1. INTRODUCTION

First recognized in mid-March 2003, Severe Acute Respiratory Syndrome (SARS) is a newly emerged infectious disease that appeared in the Guangdong Province, mainland China, in November 2002. By March 2003, the disease had spread globally and by July there were 8447 probable SARS cases including 811 deaths reported to the World Health Organization (WHO), from 32 countries or regions worldwide. In response to this outbreak the WHO issued a global alert and warned against unnecessary travel to affected areas. WHO also coordinated an international collaboration led by thirteen principle laboratories in ten countries that included clinical, epidemiological and laboratory investigations and initiated worldwide efforts to control the spread of SARS.

Several possible causal agents were initially identified in patients, including influenza A subtype H5N1 (Drosten et al. 2003), virus-like particles resembling paramyxoviruses (World Health Organization Multicentre Collaborative Network for Severe Acute Respiratory Syndrome

Virology Group, International Centre for Genetic Engineering & Biotechnology, Aruna Asaf Ali Road, New Delhi-110067, India
* Address for correspondence: ICGEB Campus, P.O. Box: 10504, Aruna Asaf Ali Road, New Delhi-110067. Tel: 26189360 Fax: 91-11-26162316; e-mail: sunillal@icgeb.res.in

Diagnosis, 2003), metapneumovirus (Chan et al. 2004c; Poutanen et al. 2003), rhinovirus and chlamydia (Drosten et al. 2003; Hong et al. 2003; Ksiazek et al. 2003). However, since none of these pathogens were consistently detected in other suspected SARS patients and hence were quickly eliminated as the aetiological agent. Attempts to identify the etiological agent of the SARS outbreak were successful during the third week of March 2003, when laboratories in the United States, Canada, Germany, and Hong Kong isolated a novel coronavirus (SARS-CoV) from SARS patients. (Chan et al. 2004a; Drosten et al. 2003; Ksiazek et al. 2003; Kuiken et al. 2003). The complete genome sequence of SARS-CoV Tor2 (Toronto) isolate was published in April 2003 which established it as a novel member of the family (Rota et al. 2003; Marra et al. 2003). This novel member was shown to fulfil Koch's postulates, as modified by Rivers for viral diseases, in cynomolgus macaques and confirmed SARS-CoV as the causative agent of SARS (Fouchier et al. 2003).

The SARS epidemic brought to the forefront, benefits and dangers of the impact of globalization on infectious diseases. On one hand, the ease and frequency of international travel lead to a swift spread of the SARS coronavirus infection to six continents within no time and on the other hand worldwide telecommunication networks facilitated collaborative research among eleven geographically distinct labs, which led to the expedited identification of the causative agent within a short span of one month.

The spread of SARS-CoV worldwide, brought along with it sociological and economic repercussions. SARS exerted extreme pressure on the health care system both international and local. Today, although the chain of spread of disease has been broken successfully, there remains a frightening prospect of a future outbreak which may be more contagious or virulent than ever encountered before.

The most dramatic and immediate effects of SARS occurred in Asia. Experts estimated that the short-term global losses in economic activity due to SARS approximate to $80 billion. But it is believed that the true economic consequences remain to be determined. The way in which this epidemic was curbed, brought to light the continuing need for investments in a robust response system that has the capacity to tackle the next emerging disease. In case of SARS, WHO initiated and cocoordinated much of this response through its Global Outbreak Alert and Response Network (GOARN), as also the partner organizations comprising a total of 115 national health services, academic institutions, technical institutions and individuals. It is only through this multinational collaborative and coordinated surveillance, research and containment that the spread of SARS was limited.

1.1 Coronaviridae

Coronaviruses are a family of large, enveloped, positive-stranded RNA viruses belonging to order Nidovirales, family *Coronaviridae* and order Coronaviruses. The virions are spherical particles ranging between 100-120nm in diameter. Inside the virions the positive sense RNA genome ranges from 27-32 kb in size. The size range is the largest in all RNA virus genomes.

Coronaviruses have been divided into three subgroups, which differ with respect to their genome. The first group consists of viruses such as the human coronavirus 229E (HCoV-229E), porcine respiratory coronavirus (PRCV), porcine transmissible gastroenteritis virus (TGEV), feline infectious peritonitis virus (FIPV) and feline enteritis virus (FEV) or the canine coronavirus (CCoV). The second group comprises human coronavirus OC43 (HCoVOC43), bovine coronavirus (BCoV), and mouse hepatitis virus (MHV), and the third group mainly consists of avian species such as the chicken infectious bronchitis virus (IBV). Although the SARS-CoV has been shown to cross-react with some group I coronavirus antibodies, its genetic sequence does not belong to this group. Irrespective of which SARS-CoV RNA region is used for analysis, within the nucleic acid or protein sequence phylogenetic trees of the coronavirus family, the SARS-CoV has first been located at an equal distance from the second and third group. Therefore, the SARS-CoV currently represents the first member of a new group of coronaviruses. However, the taxonomy is still not clear, and recent studies that focused on the N-terminal domain of the spike protein and on poorly conserved proteins such as nsp1 (non-structural protein 1), matrix or nucleocapsid protein (N), have suggested a relation to group II viruses. A similar conclusion can be drawn if the polymerase gene is examined, pointing to an early split-off from the coronavirus group II lineage. Despite the fact that this new virus, most likely, jumped to humans from wild animal species, it has adapted remarkably well to the human organism as observed by its high person-to-person transmissibility.

1.2 Natural Host for SARS-CoV

It is generally believed that SARS-CoV evolved from an animal coronavirus that recently crossed the species barrier. This hypothesis is supported by data published by Guan and co-workers (2003), who isolated SARS-CoV-like viruses from Himalayan palm civets and a raccoon dog, which were sold in a live-animal market in Guangdong, China. Such live animal markets are common in China where a variety of live animals from chickens to cats, turtles and badgers are sold for food. Viruses in such markets have highly propitious conditions for jumping from one species to another, including to humans. Interestingly, an increased seroprevalence of the SARS-CoV was observed in people trading with these animals. Geographically, the Pearl River delta, which encompasses Hong Kong, Macau and much of Guangdong, has a history of producing many of the world's biggest pandemics, including some of the most virulent strains of influenza that swept the world in the 20th century. SARS, or severe acute respiratory syndrome, is believed to have started in Guangdong as well. The delta, with 57 million people living in overcrowded conditions, has long played a special role in the transmission of virulent diseases from animals to people. It is common for ducks, pigs, rodents and other carriers of disease to be in very close contact with people, sometimes under the same roof. Fine cuisine in Guangdong has also meant the consumption of a very wide range of wild animals, which may also carry diseases. Considering the living conditions of these people and their close proximity with animals sets the stage for viruses to easily jump species. It is believed that SARS originated as an animal virus which mutated and jumped species to humans, in the unhygienic and overcrowded markets of Guangdong. The potential for interspecies transmission of the SARS-CoV is also illustrated by the fact that a whole range of animals, including cats, ferrets, mice and macaques can be infected with SARS-CoV. The animal reservoir, however, of the SARS-CoV in nature yet remains to be identified. And although the chain of spread of the disease has been broken, one can't predict with certainty when this deadly virus might strike back. The virus today might be lying dormant in its natural host or it may still be circulating among animal reservoirs. Whatever may be the case, the world must prepare itself for an unwarranted attack.

2. TISSUE TROPISM

SARS mainly shows pneumonia like symptoms and lung is pathologically the most affected organ. But several investigations suggest that SARS is a systemic disease with widespread extrapulmonary dissemination, resulting in viral shedding in respiratory secretions, stools, urine and possibly even in sweat. SARS-CoV was found not only in the lung and intestine, but also in liver, distal convoluted renal tubules, sweat glands, parathyroid, pituitary, pancreas, adrenal gland and cerebrum when immunohistochemistry and *in situ* hybridization was used to examine organs from four SARS patients who died (Ding et al. 2004). SARS-CoV RNA was detected in lung, small and large bowels, lymph nodes, spleen, liver, heart, kidney and skeletal muscle, in descending order of viral load per gram of tissue using reverse-transcriptase polymerase chain reaction (Farcas et al. 2005).

3. TRANSMISSION

Majority of new infections in case of SARS-CoV occurred in close contacts of patients, such as household members, health care workers, or other patients who were not protected with contact or respiratory precautions. This indicates that the virus is predominantly spread by droplets or by direct and indirect contact (Seto et al. 2003). Oral-fecal transmission seems to be another possibility as the virus is present in the stools of the patients (Drosten 2003; Peiris et al. 2003b).

The airborne spread of SARS does not seem to be a major route of transmission. This route, however, cannot be ruled out as certain cases of spread of SARS indicate that although a rare event, it cannot be ruled out and is a matter of concern. Cases among health care workers especially seem to confirm airborne transmission via a contaminated environment (i.e., re-aerosolization when removing protective equipment, etc.).

4. CELLULAR RECEPTOR

A metallopeptidase, angiotensin converting enzyme 2 (ACE2), the C-type lectin CD209L (also known as L-SIGN) and DC-SIGN bind SARS-CoV, but ACE2 appears to be the key functional receptor for the virus. The virus seems to gain entry into the target cells by direct membrane fusion at the target cell surface. Apart from this it has been postulated that the SARS-CoV may gain cell entry via pH-dependent endocytosis, which is also mediated by the S protein (Yang, 2004). The ACE2 protein is reportedly present in type 1 and type 2 pneumocytes, enterocytes of all parts of the small intestine, the brush border of the proximal tubular cells of the kidney as well as the endothelial cells of small and large arteries and veins of all tissues studied, and arterial smooth muscle cells (Hamming, 2004). This localization of ACE2 explains the tissue tropism of SARS-CoV for the lung, small intestine and kidney; however, notable discrepancies include virus replication in colonic epithelium, which has no ACE2, and no virus infection in endothelial cells, which have ACE2. Other receptors or co-receptors such as L-SIGN may explain such discrepancies.

5. GENOME ORGANISATION AND PROTEIN PROCESSING

The SARS-CoV genome has been predicted to contain 14 functional open reading frames (ORFs). The comparison of different SARS-CoV ORFs with those of other coronaviruses revealed a similar pattern of structural gene arrangement with the replicase and protease genes (gene 1a-1b) and the spike (S), envelope (E), membrane (M) and nucleocapsid (N) genes in a 5' to 3' order of appearance. Interspersed between these well-characterized genes are a series of ORFs many of whose functions are yet unknown. There are two ORFs situated between the S and the E genes and three to five ORFs between the M and N genes. Such gene organization most closely resembles that of group III coronaviruses. Also, the SARS-CoV genomic sequence does not contain a gene for the hemagglutinin-esterase (HE) protein, which is present in most group II coronaviruses.

Two-thirds of the SARS-CoV RNA is organized in the genes 1a-1b. The sequence of these genes is highly conserved among all coronaviruses with the exception that SARS-CoV lacks the sequence coding for PL1pro, one of the two papain-like proteinases operating on cleavage sites at the N-terminus of the polyproteins. ORFs 1a and 1b encode two polyproteins, pp1a and pp1ab, respectively, the latter is predicted to be translated through a −1 ribosomal frameshifting mechanism. These two polyproteins are processed by virus-encoded proteinases, to yield 16 individual non-structural proteins (nsp). Many of the postulated nsp proteins to date have undesignated functions, but it is suggested that they participate primarily in viral RNA replication. Most potential gene 1a-1b products are fairly well conserved between SARS-CoV and other coronaviruses. PL2pro of the SARS virus is responsible for the cleavage of all the N-terminal proteins of gene 1a. The main proteinase (Mpro), also called the 3C-like protease (3CLpro), is responsible for the cleavage of all the remaining proteins encoded by gene 1a-1b. Spike(S), nucleocapsid (N), matrix (M) and envelope (E) form the major structural proteins of SARS-CoV.

The S protein forms typical petal-shaped spikes on the surface of the virion and is highly glycosylated. Entry of coronaviruses into target cells is initiated by binding of the viral S protein to receptor molecules with the minimal receptor-binding domain is located between the residues 318-510 (Wong *et al.* 2004). The C-type lectin CD209L (also called L-SIGN) has been discovered to be a human cellular glycoprotein that can serve as an alternative receptor for SARS-CoV.

Nucleocapsid or the N protein of SARS-CoV is a 46 kDa phosphoprotein which shows homology with other members of coronavirus family. It gets abundantly expressed during infection and antibodies against N have been detected in SARS patients. This makes N an excellent candidate for diagnostic purposes.

The N protein associates with the viral genomic RNA, and together they form the ribonucleoprotein. It has been shown that assembly of SARS-CoV

RNA packaging signal into VLPs is nucleocapsid dependent (Hsieh et al. 2005). The N protein can self-associate to form dimers (Surjit et al. 2004) using an interaction domain present in the C terminal 209 amino acid residues. In 2005, another group reported the same finding and the domain responsible for self-association was further narrowed down to 138 amino acid residues (between 285-422) at the C terminus of the protein. Apart from interactions amongst viral proteins, interactions with the proteins of the host play a major role in establishment of viral infection. The human cyclophilin A (cypA) (Luo et al. 2004) and human cellular heterogenous nuclear ribonucleoprotein A1 (hnRNP A1) (Luo et al. 2005) binds tightly to N protein of SARS-CoV. A report by R. He in 2003 showed that N protein activates AP1 signal transduction pathway. However, the mechanism by which N does so is not known yet (He et al. 2003). SARS-CoV N protein has also been shown to induce apoptosis in Cos 1 cells and induce actin reorganization (Surjit et al. 2004; Surjit et. al. 2006).

Research indicates that N protein of several coronaviruses exhibit cytoplasmic as well as nucleolar localization at some point of time during the infection cycle (Hiscox et al. 2001; Wurm et al. 2001; Rowland et al. 2005). Surjit *et al.* in their recent publication have shown that N protein of SARS-CoV shuttles between nucleus and cytoplasm and localizes in the cytoplasm. It has been shown that phosphorylated N is translocated to cytoplasm with the help of 14-3-3 (tyrosine 3-monooxygenase/tryptophan 5-monooxygenase activation protein) (Surjit et al. 2005). Moreover, binding of N protein to cyclin-CDK complex leads to downregulation of S phase gene products which in turn leads to inhibition of S phase progression (Surjit et al. 2006). Coronavirus Matrix (M) protein is the most abundant structural protein at the surface of virus particles and has been shown to interact with nucleocapsid through its 12 amino acid domain (194-205). The envelope (E) protein of the SARS-CoV is, as the name suggests, the main component of the virus envelope. Topology prediction suggested that the E protein is a type II membrane protein. Although E is only found at low levels in coronavirus envelope, it has a pivotal role in virus assembly.

6. CLINICAL PICTURE AND DIAGNOSIS

The clinical symptoms of the syndrome include fever, chills, rigors, cough, myalgia, malaise, shortness of breath, diarrhea and headache and the pathological aspects include leukopenia, lymphopenia, thrombocytopenia, elevated lactate dehydrogenase and creatine kinase levels. Unless specific laboratory tests confirm the initial suspicion of SARS infection, the diagnosis of SARS is based on the clinical findings of an atypical pneumonia not attributed to any other cause, as well as a history of exposure to a suspect or probable case of SARS, or to their respiratory secretions or other body fluids.

All tests for SARS-CoV available so far have limitations. Extreme caution is therefore necessary when management decisions are to be based on virological test results. The diagnostic tests currently available for SARS-CoV use detection methods like virus isolation on cell cultures, polymerase chain reaction, immunofluorescence assay (IFA), Enzyme-linked immunosorbent assay (ELISA) and Neutralization test (NT) on patients serum. SARS-CoV specific RNA can be detected in various clinical specimens such as blood, stool, respiratory secretions or body tissues by the polymerase chain reaction (PCR). RT-PCR based detection targets the polymerase genes or the nucleocapsid gene. For a specimen to be declared positive, two out of the three target genes are required to be amplified. The advantages of this method include an increased sensitivity and an increased speed of detection. Also the results can be quantified in a real time format.

But when the detection has been done using this method, a positive test result should be considered provisional until confirmed by an independent testing. Also, a negative result does not rule out SARS. Various methods provide a means for the detection of antibodies produced in response to infection with SARS-CoV. Different types of antibodies (IgM and IgG) appear and change in level during the course of infection Serology appears to be quite specific, as barring a few cases, specimen do not cross react with other coronavirus infection like OC43 and 229E. Detection using this method can ascribe a sample positive in as few as 8 to 10 days after onset of symptoms but a sample cannot be considered negative until >28 days after onset of symptoms. Serology utilizes tests like immunofluorescence assay (IFA), ELISA, neutralizing antibody test, western blot or immunodot. Sensitivity ranges between 90-100% and both IgG and IgM assays may be utilized. However, IgM antibodies may be detected earlier in the course of infection but is a transient response. IgG usually remains detectable after resolution of the illness.

The presence of the infectious virus can be detected by inoculating suitable cell cultures with patient specimens (such as respiratory secretions, blood or stool) and propagating the virus *in vitro*. Once isolated, the virus must be identified as SARS-CoV using further tests. This diagnostic method requires Biosafety Level III containment. Also culture range of SARS-CoV serves as another limitation. SARS-CoV does not grow well in usual cell lines used for culturing respiratory viruses like Hep-2, RMK, etc. but grows well and produces cytopathic effect in Vero E6 cells. The diagnostic assays used for detection of SARS are sensitive and specific but may not provide definitive diagnosis especially early in course of illness.

7. VACCINES

The genome of SARS-CoV was sequenced in 2003. But till date there is no successful vaccine available to combat the agent. Most of the work in this area relates to vaccines using either inactivated SARS-CoV, recombinant DNA or viral vector based on SARS-CoV S protein. The inactivated virus vaccine elicits humoral responses and does not provide long lasting protection. Also there is always a fear that the whole organism inactivated vaccine may induce immunopathology in the host.

On the other hand, using only a part of virus for eliciting immune response do not run the risk of inducing pathogenicity in the host. Both cell mediated as well as humoral protection can be elicited by presentation of multiple epitopes of immunogenic proteins. Therefore, this offers one of the best methods to address SARS. One of the biggest limitation in development of vaccines has been non-availability of a suitable animal model to test efficacy of vaccines and the probability that SARS-CoV may mutate in the future.

8. PREVENTION AND CONTAINMENT

Till date there is no vaccine available against SARS. Hence, breaking the chain of transmission from infected to healthy persons is the most effective way to control a disease like SARS.

For SARS, three activities—case detection, patient isolation and contact tracing—were used to reduce the number of people exposed to each infectious case and eventually break the chain of transmission (WHO, WER 20/2003):

1. Case detection aims to identify SARS cases as soon after the onset of illness as possible.
2. Once cases are identified, the next step is to ensure their prompt isolation in a properly equipped facility, and management according to strict infection control procedures.
3. The third activity - the detective work - involves the identification of all close contacts of each case and assurance of their careful follow-up, including daily health checks and possible voluntary home isolation.

Together, these activities limit the daily number of contacts possible for each potentially infectious case. They also work to shorten the amount of time that lapses between the onset of illness and isolation of the patient, thus reducing the opportunities for the virus to spread to other patients (WHO WER 20/2003).

Isolation refers to the separation and restricted movement of ill persons who have a contagious disease in order to prevent its transmission to others. It typically occurs in a hospital setting, but can be done at home or in a special facility. Usually individuals are isolated, but the practice may be applied in larger groups.

Quarantine refers to the restriction of movement or separation of well persons who have been exposed to a contagious disease, before it is known whether they will become ill. It is an ancient tool used to prevent the spread of disease which has been practiced for centuries, when properly applied and practiced can be a highly effective tool in prevention of contagious disease. It can take place either at individual or community level. During the SARS epidemic this approach was used effectively with the airline passengers arriving from areas of high transmission during the SARS epidemic.

Containment strategies employed during the recent SARS epidemic included case and contact management, infection control in hospitals and other facilities, community wide temperature screening, mask use, isolation and quarantine and monitoring of travelers and response at national borders. Combination(s) of these strategies were applied depending on factors such as magnitude and scope of the outbreak and availability of resources.

REFERENCES

Chan KH, Poon LL, Cheng VC, Guan Y, Hung IF, Kong J, Yam LY, Seto WH, Yuen KY, Peiris JS (2004a). Emerg Infect Dis 10 :294-299.

Chan PK, To KF, Wu A, Tse GM, Chan KF, Lui SF, Sung JJ, Tam JS, Tomlinson B (2004c). Emerg Infect Dis 10:497-500.

Ding Y, He L, Zhang Q, Huang Z, Che X, Hou J, Wang H, Shen H, Qiu L, Li Z, Geng J, Cai J, Han H, Li X, Kang W, Weng D, Liang P, Jiang S (2004). J Pathol 203(2):622-30.

Drosten C, Gunther S, Preiser W, van der Werf S, Brodt HR, Becker S, Rabenau H, Panning M, Kolesnikova L, Fouchier RA, Berger A, Burguiere AM, Cinatl J, Eickmann M, Escriou N, Grywna K, Kramme S, Manuguerra JC, Muller S, Rickerts V, Sturmer M, Vieth S, Klenk HD, Osterhaus AD, Schmitz H, Doerr HW (2003). N Engl J Med, 348:1967-1976.

Farcas GA, Poutanen SM, Mazzulli T, Willey BM, Butany J, Asa SL, Faure P, Akhavan P, Low DE, Kain KC (2005). J Infect Dis. 191:193-197.

Fouchier RA, Kuiken T, Schutten M, van Amerongen G, van Doornum GJ, van den Hoogen BG, Peiris M, Lim W, Stohr K, Osterhaus AD (2003). Nature 423 (6937):240.

Guan Y, Zheng BJ, He YQ, Liu XL, Zhuang ZX, Cheung CL, Luo SW, Li PH, Zhang LJ, Guan YJ, Butt KM, Wong KL, Chan KW, Lim W, Shortridge KF, Yuen KY, Peiris JS, Poon LL (2003). Science 302(5643):276-8.

Hamming I, Timens W, Bulthuis ML, Lely AT, Navis GJ, van Goor H (2004). J Pathol 203:631-637.

He R, Leeson A, Andonov A, Li Y, Bastien N, Cao J, Osiowy C, Dobie F, Cutts T, Ballantine M, Li X (2003). Biochem Biophys Res Commun. 28;311(4):870-6.

Hiscox JA, Wurm T, Wilson L, Britton P, Cavanagh D, Brooks G (2001). J Virol 75(1):506-12.

Hong T, Wang JW, Sun YL, Duan SM, Chen LB, Qu JG, Ni AP, Liang GD, Ren LL, Yang RQ, Guo L, Zhou WM., Chen J, Li DX, Xu WB, Xu H, Guo YJ, Dai SL, Bi SL, Dong XP, Ruan L (2003). Zhonghua Yi Xue Za Zhi. 83: 632-636.

Hsieh PK, Chang SC, Huang CC, Lee TT, Hsiao CW, Kou YH, Chen IY, Chang CK, Huang TH, Chang MF. (2005). J Virol. 79(22):13848-55.

Ksiazek TG, Erdman D, Goldsmith CS, Zaki SR, Peret T, Emery S, Tong S, Urbani C, Comer JA, Lim W, Rollin PE, Dowell SF, Ling AE, Humphrey CD, Shieh WJ, Guarner J, Paddock CD, Rota P, Fields B, DeRisi J, Yang JY, Cox N, Hughes JM, LeDuc JW, Bellini WJ, Anderson LJ; SARS Working Group (2003). N Engl J Med 348:1953-1966.

Kuiken T, Fouchier RA, Schutten M, Rimmelzwaan GF, van Amerongen G, van Riel D, Laman JD, de Jong T, van Doornum G, Lim W, Ling AE, Chan PK, Tam JS, Zambon MC, Gopal R, Drosten C, van der Werf S, Escriou N, Manuguerra JC, Stohr K, Peiris JS, Osterhaus AD Lancet (2003). 362:263-270.

Luo C, Luo H, Zheng S, Gui C, Yue L, Yu C, Sun T, He P, Chen J, Shen J, Luo X, Li Y, Liu H, Bai D, Shen J, Yang Y, Li F, Zuo J, Hilgenfeld R, Pei G, Chen K, Shen X, Jiang H (2004). Biochem Biophys Res Commun.;321(3):557-65.

Luo H, Chen Q, Chen J, Chen K, Shen X, Jiang H (2005). FEBS Lett, 579(12):2623-8.

Marra MA, Jones SJ, Astell CR, Holt RA, Brooks-Wilson A, Butterfield YS, Khattra J, Asano JK, Barber SA, Chan SY, Cloutier A, Coughlin SM, Freeman D, Girn N, Griffith OL, Leach SR, Mayo M, McDonald H, Montgomery SB, Pandoh PK, Petrescu AS, Robertson AG, Schein JE, Siddiqui A, Smailus DE, Stott JM, Yang GS, Plummer F, Andonov A, Artsob H, Bastien N, Bernard K, Booth TF, Bowness D, Czub M, Drebot M, Fernando L, Flick R, Garbutt M, Gray M, Grolla A, Jones S, Feldmann H, Meyers A, Kabani A, Li Y, Normand S, Stroher U, Tipples GA, Tyler S, Vogrig R, Ward D, Watson B, Brunham RC, Krajden M, Petric M, Skowronski DM, Upton C, Roper RL (2003) Science. 300:1399-1404.

Peiris JS, Chu CM, Cheng VC, Chan KS, Hung IF, Poon LL, Law KI, Tang BS, Hon TY, Chan CS, Chan KH, Ng JS, Zheng BJ, Ng WL, Lai RW, Guan Y, Yuen KY; HKU/UCH SARS Study Group. (Lancet 2003b) 361:1767-72.

Poutanen SM, Low DE, Henry B, Finkelstein S, Rose D, Green K, Tellier R, Draker R, Adachi D, Ayers M, Chan AK, Skowronski DM, Salit I, Simor AE, Slutsky AS, Doyle PW, Krajden M, Petric M, Brunham RC, McGeer AJ; National Microbiology Laboratory, Canada; Canadian Severe Acute Respiratory Syndrome Study Team (2003). N Engl J Med 348:1995-2005.

Rota PA, Oberste MS, Monroe SS, Nix WA, Campagnoli R, Icenogle JP, Penaranda S, Bankamp B, Maher K, Chen MH, Tong S, Tamin A, Lowe L, Frace M, DeRisi JL, Chen Q, Wang D, Erdman DD, Peret TC, Burns C, Ksiazek TG, Rollin PE, Sanchez A, Liffick S, Holloway B, Limor J, McCaustland K, Olsen-Rasmussen M, Fouchier R, Gunther S, Osterhaus AD, Drosten C, Pallansch MA, Anderson LJ, Bellini WJ (2003). Science 300:1394-1399.

Rowland RR, Chauhan V, Fang Y, Pekosz A, Kerrigan M, Burton MD (2005). J Virol 79(17):11507-12.

Seto WH, Tsang D, Yung RW, Ching TY, Ng TK, Ho M, Ho LM, Peiris JS; Advisors of Expert SARS group of Hospital Authority (2003). Lancet, 361: 1519–20.

Surjit M, Kumar R, Mishra RN, Reddy MK, Chow VT, Lal SK (2005). J Virol; 79(17): 11476-86.

Surjit M, Liu B, Chow VT, Lal SK (2006). J Biol Chem [Epub ahead of print]

Surjit M, Liu B, Jameel S, Chow VT, Lal SK (2004). Biochem J, 383 :13-8.

Surjit M, Liu B, Kumar P, Chow VT, Lal SK (2004). Biochem Biophys Res Commun. 317(4):1030-6.

WHO Update 49: SARS case fatality ratio, incubation period. May 7. http://www.who.int/csr/sarsarchive/2003_05_07a/en/Fisher DA.

Wong SK, Li W, Moore MJ, Choe H, Farzan M (2004). J Biol Chem 279: 3197-3201.

World Health Organization Multicentre Collaborative Network for Severe Acute Respiratory Syndrome Diagnosis. (2003). Lancet 361:1730-1733.

Wurm T, Chen H, Hodgson T, Britton P, Brooks G, Hiscox JA (2001). J Virol. 75(19):9345-56.

Yang ZY, Huang Y, Ganesh L, Leung K, Kong WP, Schwartz O, Subbarao K, Nabel GJ. (2004). J Virol 78:5642-5650.

Search for an ever Elusive "Guardian-angel" Novel Antibiotic: Myth or Reality

S. M. DASGUPTA[1], P. S. CHAUHAN[2] and C. S. NAUTIYAL[2]*

Nearly all the great discoveries have been made as a result of a false hypothesis or due to a so-called chance observation.

- Alexander Kohn

1. INTRODUCTION

Prior to the introduction of effective antimicrobial therapy, mortality due to bacterial infections was a leading cause of death worldwide (Lederberg 2000). Few would dispute the profound impact of antibiotics on human health. Indeed, much of the eight-year increase in average human lifespan between 1944 and 1972 has been attributed to their global introduction into medical practice. What is now considered the "antibiotic era" was ushered in when the industrial-scale fermentation of penicillin in 1943 enabled wide use of the antibiotic in the clinics? The discovery and deployment of the "miracle drugs", antibiotics, had a profound impact on clinical medicine. With the rapid discovery of new structural classes in the 1940s and 1950s, it seemed that any serious bacterial infection was quickly alleviated. No doubt antibiotic use has substantially improved the quality of life as well as life expectancy itself. Nevertheless, the clinical effectiveness of most antibiotics

1   Amity Institute of Microbial Sciences, Amity University Uttar Pradesh Sector 125, New Super Express Highway, Noida-201303

2   Division of Plant-Microbe Interactions, National Botanical Research Institute, Rana Pratap Marg, Lucknow-226001, India.

*   **Corresponding author:** Tele-fax: +91 0522-2206651; E.mail: csnnbri@yahoo.com.

has declined radically as resistant organisms have become dominant to a point where the use of antibiotics is seriously compromised. The potential emergence of untreatable "super-bugs" has alarmed scientists and clinicians alike, raising the specter of a post-antibiotic era. Such gloomy forecasts have provided added impetus to efforts to develop new strategies for managing infectious diseases.

Until recently, research and development (R&D) efforts have provided new drugs in time to treat bacteria that became resistant to older antibiotics. That is no longer the case. Unfortunately, both the public and private sectors appear to have been lulled into a false sense of security based on past successes. The potential crisis at hand is the result of a marked decrease in industry R&D, government inaction, and the increasing prevalence of resistant bacteria. Infectious diseases physicians are alarmed by the prospect that effective antibiotics may not be available to treat seriously ill patients in the near future. The systematic screening of natural product libraries from soil samples or marine environments identified most of the classes of antibacterial agents in use today. When these approaches began to yield diminishing returns in the 1960s, companies turned instead to semi-synthetic modification of existing antibiotics to produce second and third generation compounds with broadened antimicrobial activity, enhanced oral bioavailability, and improved toxicological and pharmacokinetic properties. These compounds, largely natural products isolated from organisms found in soil samples, were found by simple whole bacterial cell activity screens. Yet, even in the earliest days of antibiotic research, it was clear that bacteria could become resistant to their effects (Barber and Rozwadaska-Downzenko 1948). The discovery of transmissible resistance on plasmids was the harbinger of an alarming trend in which antibiotic resistance has grown to encompass virtually all the serious pathogens and classes of antimicrobial compounds (Davies 1994).

Growing public health problem has spurred renewed efforts to discover novel types of antibacterial agents with mechanisms radically different from existing compounds. With pharmaceutical companies banking on bacterial genomics to deliver novel targets for future antibiotics, several smaller companies are testing completely new classes of antibacterial agents that target bacterial cell membranes, adherence mechanisms, or gene expression. Although the clinical efficacy of many of these compounds remains unproven, they offer some promise of slowing the emergence of resistant strains. Measures have been attempted to control resistance by cycling antibiotics in the clinic, with somewhat mixed success, due to practical problems and the durability of resistance genes (Monroe and Polk 2000;

Gould 1999). The substitution of one antibiotic class for another has been termed "squeezing the balloon", in the sense that the antibiotic pressure moves the resistance problem from one class to another as usage is modified and new selection pressure applied (Burke 1998). At the same time that resistance was growing, efforts to identify novel classes of antimicrobial agents declined. Only within the last decade or so has the recognition of the serious nature of resistance been widely appreciated (Neu 1992; Levy 1998). The consensus has emerged that as part of the overall strategy to control resistant pathogens, novel antibiotic classes will be essential in the not too distant future (Moir et al. 1999). With the identification and availability of essential gene products through the use of microbial genomics, it will not only be possible to establish novel *in vitro* screens of increased sensitivity, it will also be possible to perform structural analyses on these proteins with the goal of designing specific inhibitors of function. These novel targets can also be exposed to the array of new compounds that have been generated by the advances in the field of combinatorial chemistry. The combination of past knowledge of biochemistry and physiology of microorganisms and new insights into biological functions derived from genome and functional genomic studies can guide more specific discovery strategies (Fig. 1).

2. DEVELOPMENT OF NEW ANTIMICROBIAL AGENTS

In 1929, Alexander Fleming observed the inhibition of multiplication of bacteria on the surface of a solid agar plate around a contaminating fungus, which eventually led to the clinical introduction of penicillin in the 1940s. From point of view of human benefit, never was a Nobel Prize so justified as Selman Waksman for discovery of streptomycin and other antibiotics produced from *Streptomyces spp.* He along with his team developed the concept of systematic screening of microbial culture products for biological activity, a technology which has provided the foundation of antibiotic industry. Since then, the inhibition of actively multiplying bacteria has been the assay with which potential antimicrobial agents are selected. Screening of large numbers of natural and synthetic compounds results in the selection of a lead compound, a process that we call the 'classic screening approach'. Early antibiotics, which were developed by screening in the 1940s and 1950s and beyond, were all natural products, which have been a vitally important source of new antimicrobial agents. Recent interest has been rekindled in the screening approach by the discovery of naturally occurring antimicrobial peptides. These include peptides known as cathelicidins, which are found in mammalian skin and other tissues (Nizet et al. 2001),

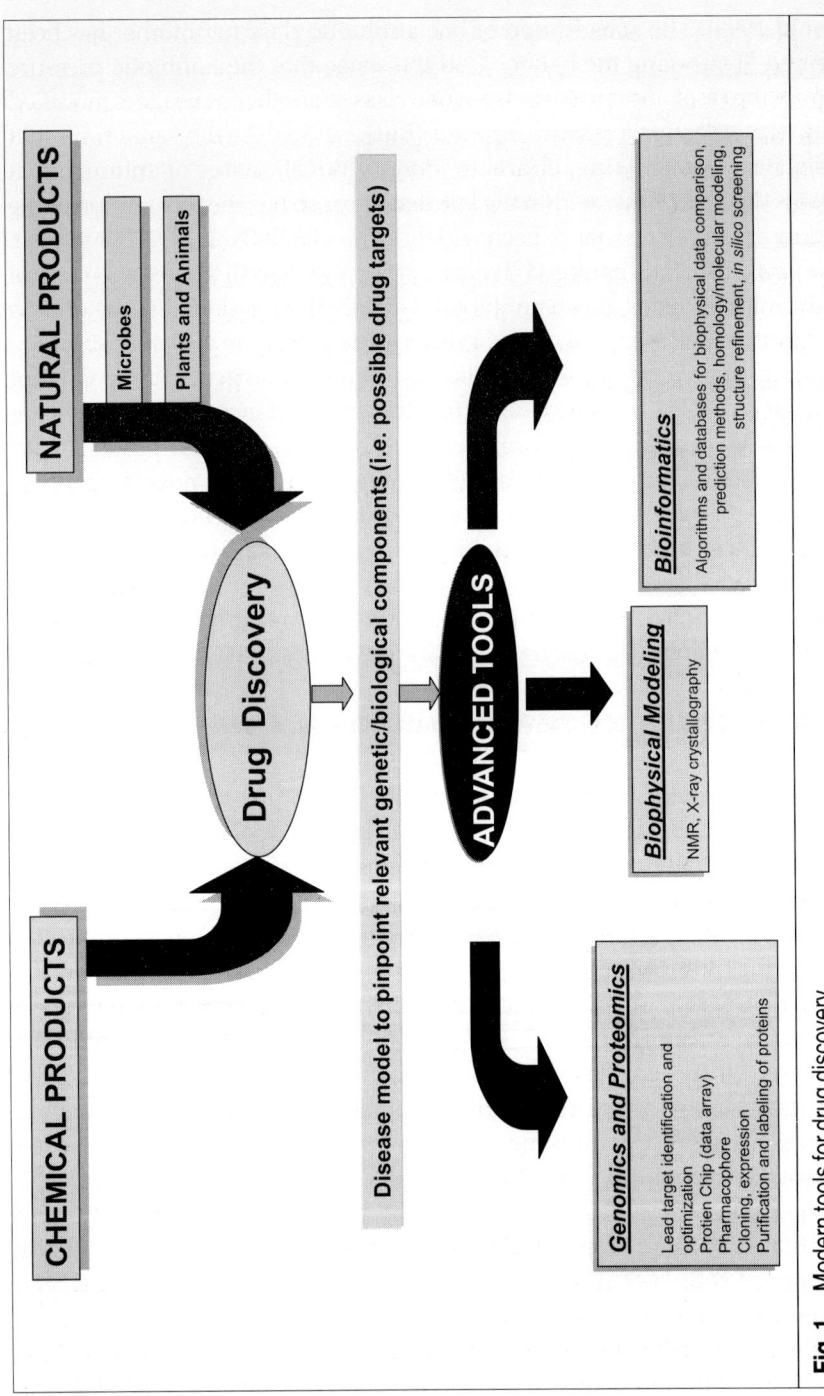

Fig. 1 Modern tools for drug discovery.

and piscidins, which are found in the mast cells of fish. Another new route for the development of natural products from soil has been proposed (MacNeil et al. 2001). It is thought that 99% of soil bacteria are non-cultivable. So, researchers are now extracting bacterial DNA from soil samples, cloning large fragments into, for example, bacterial artificial chromosomes, expressing in a host bacterium and screening the library for new antibiotics.

This could open up the exciting possibility of a large untapped pool from which new natural products could be discovered. Chemical synthesis can be used to make fundamentally new structures that might act at different bacterial targets to those already identified. Drugs arising from such approaches include the oxazolidinones and ketolides. Potentially, new antimicrobial agents could result from combinatorial chemistry (Bax and Mullan 1999; Bax et al. 2000; Chopra et al. 1996). However, the compounds identified so far have been shown to be either too toxic or too insoluble to allow them to be developed as drugs. One drawback is that although inhibitors of enzymatic reactions *in vitro* can be found, the compounds cannot be converted into effective inhibitors of bacterial cells. The overall problem with classic screening coupled with assays that use actively multiplying bacteria is that bacterial resistance arises soon after the new antimicrobial agent is widely used in the community. This means that it is necessary to produce new antimicrobial agents that have different mechanisms of action at regular intervals (Fig. 1).

Another reason which calls for such a need for speedy discovery of novel antibiotics is the dosage rationality of the antibiotics. In 1994, for example, penicillin appeared to be a magic bullet against staphylococcal infections. It killed most of the infecting agents but those left slowly began to multiply. The result, by 1950s most staph infections had become resistant to penicillin. The same fate met penicillin's successors, erythromycin and methicillin; now it appears to be vancomycin's turn. For this reason it is not enough to come up with new drugs; it is also imperative for us to try our utmost to extend their useful lifetime. This means we must stop misusing them. For decade, it has been prescribed by many physicians for sniffle and sneeze, even when source of problem was virus. To make the matter worse, antimicrobials have been introduced into hand creams, household cleansers and livestock feed.

The emergence of antibiotic resistance led to the synthesis of many derivatives of existing antibiotics in the hope of discovering drugs that would be effective against resistant strains—a process that has been dubbed 'molecular roulette'. Whole families of drugs have been made that are based on, for example, penicillin (the-cillins), erythromycin (macrolides), imipenem (carbapenems) and nalidixic acid (quinolones) (Knowles 1997; Gootz 1990). Cefuroxime is an early example of a second generation cephalosporin that was launched by Glaxo in the United Kingdom in 1978. New compounds in the β-lactam, macrolide, tetracycline and quinolone families were produced that have advantages in spectrum, potency (Neu 1983) and/or pharmacology (Allan 1987), such as improved oral absorption or longer half-life. An interesting twist is the genetic manipulation of bacteria, such as *Streptomyces spp.* (Hopwood 1985; Omura 1986) and *Escherichia coli* (Pfeifer 2001), in such a way that they make a hybrid antimicrobial agent based on, for example, erythromycin. This technology is based on the principles of heterologous gene expression, molecular biology and the promiscuous nature of enzymes that are involved in the production of secondary metabolites, such as antibiotics (Bentley and Bennett 1999). The advantage of this method is that numerous changes can be made to the antimicrobial agent by relatively cheap genetic means. Another intriguing approach is to merge natural product biosynthesis with combinatorial solid-phase chemistry. Recently, this approach has been used to produce novel cyclic peptides related to the antibiotic tyrocidine that have improved properties as potential drugs (Kohli 2002). The main disadvantage of synthesizing derivatives of existing agents is that, because the parent molecule led to bacterial resistance, the new one will too, with a high likelihood of cross resistance, which further reduces the effectiveness of related antimicrobial agents. Molecular roulette alleviates the problem of bacterial resistance to a point, but sooner or later, the bacteria win.

So, the strategy for attack against resistant strains is multipronged. Some microbiologists are trying to reengineer the older generation of miracle drugs to get around the mechanisms of resistance. Tetracycline, which kills bacteria by disabling ribosome, is target of one such effort. Bacteria become resistant to tetracycline by developing one protein that serves to shield the ribosome and another that acts as a molecular pump, forcibly ejecting the antibiotics from the cell. These insights have led to the discovery of a line of tetracycline analogs, against which neither shield nor pump is effective (Fig. 1).

Also some novel compounds are being discovered. One such compound is daptomycin isolated from filamentous bacteria. It disrupts the conformation of the bacterium's cell membrane. Another novel drug discovered is Linezolid which is totally synthetic and approved by FDA and has a great advantage as this compound has never been seen by the bacteria and it blocks protein synthesis much earlier in the cellular cycle. Both these novel drugs daptomycin and linezolid (branded under the trade name

Cidecin and Zyvox) are aimed at drug resistant enterococci and streptococci, which are a big problem for nursing homes and hospitals.

3. ROLE OF NATURAL PRODUCTS IN DRUG DISCOVERY

Chemical substances derived from animals, plants and microbes have been used to treat human disease since the dawn of medicine. The investigation of natural products as source of novel human therapeutics reached its peak in the Western pharmaceutical industry in the period 1970–1980, which resulted in a pharmaceutical landscape heavily influenced by non-synthetic molecules. Of the 877 small-molecule New Chemical Entities (NCEs) introduced between 1981 and 2002, roughly half (49%) were natural products, semi-synthetic natural product analogues or synthetic compounds based on natural product pharmacophores (Newman et al. 2003). Despite this success, pharmaceutical research into natural products has experienced a slow decline during the past two decades. Although these trends seemed obvious to investigators working in the field, their downstream effects are somewhat difficult to measure precisely, given the long product-development cycles encountered in the pharmaceutical industry. The lengthy delay—usually ten years or more—between the initial discovery of a potential therapeutic agent and subsequent market launch of a new molecular entity (NME) means that agents reaching the market today are typically the products of discovery research programmes initiated at least a decade ago. However, it is possible to gain a reasonable picture of natural product pharmaceutical discovery research by examining the general trends found in patent statistics.

Current thinking in the generation of drug leads embodies the concept of achieving high molecular diversity within the boundaries of reasonable drug-like properties (Ajay and Murcko 1998; Sadowski and Kubinyi 1998). It has long been recognized that natural product structures have the characteristics of high chemical diversity, biochemical specificity and other molecular properties that make them favorable as lead structures for drug discovery, and which serve to differentiate them from libraries of synthetic and combinatorial compounds (Newman 2003). Various investigators have worked to measure by means of computational chemistry those desirable chemical features that distinguish natural products from other sources of drug leads. Feher and Schmid (2003) examined representative combinatorial, synthetic and natural product compound libraries on the basis of molecular diversity and 'drug-likeness' properties such as molecular mass, number of chiral centers, molecular flexibility as measured by number of rotatable bonds and ring topology, distribution of heavy atoms, and LIPINSKI-type descriptors (Lipinski 1997). Other investigators have differentiated natural products, trade drugs or other synthetic molecular libraries on the basis of scaffold architecture and pharmacophoric properties (Lee and Schneider 2003), or other molecular descriptors (Stahura 2000).

These studies reveal that natural products typically have a greater number of chiral centers and increased steric complexity than either synthetic drugs or combinatorial libraries (Henkel 1999; Feher and Schmid 2003). Although drug and combinatorial molecules tend to contain a significantly higher number of nitrogen, sulphur, and halogen containing groups; natural products bear a higher number of oxygen atoms (Feher and Schmid 2003; Stahura 2000). Multivariate statistical analysis of molecular descriptors shows that natural products differ significantly from synthetic drugs and combinatorial libraries in the ratio of aromatic ring atoms to total heavy atoms (lower in natural products), number of solvated hydrogen-bond donors and acceptors (higher in natural products) and by greater molecular rigidity. Natural-product libraries also have a broader distribution of molecular properties such as molecular mass, octanol–water partition coefficient and diversity of ring systems compared with synthetic and combinatorial counterparts (Feher and Schmid 2003; Stahura 2000; Lee and Schneider 2003).

Indeed, less than one-fifth of the ring systems found in natural products are represented in current trade drugs. A perhaps unexpected finding is that of Schneider and Lee who revealed that the fraction of natural product structures with two or more 'rule-of-five'(Lipinski 1997) violations is quite low (approximately 10%) and equal to that of trade drugs (Stahura 2000). Recent advances in genomics and structural biology during the past 5–10 years are painting a clearer picture of the diversity of proteins targeted by natural product molecules. It has been shown that the number of unique protein architectures (or folds) in nature is much smaller than the number of protein families predicted by sequence similarity (Chothia 1992; Zhang and DeLisi 1998). Structural motifs of domains, the functional building blocks of proteins, are frequently conserved even when there is a low degree of sequence similarity. These units are combined in modular fashion to fine-tune the function of full proteins (Salem 1999; Holm and Sander 1996). This indicates that proteins populate the total fold space in a highly non-uniform fashion, such that localized concentrations of receptive binding 'space' are distributed onto clusters of super folds (Holm and Sander 1996). Furthermore, the same protein-binding target or fold can serve differing functional roles in a number of higher organisms (Anantharaman 2003). On the basis of this concept, a guiding principle has emerged that natural products, by virtue of their molecular evolution to preferentially bind to these folds, are validated starting points for screening-library design (Breinbauer 2002).

3.1 Natural Product Libraries and High Throughput Screening

The exercise of identifying natural product (and also synthetic) molecules as potential drug leads most often involves automated testing of large collections (libraries) of compounds for activity as inhibitors or activators of a specific biological target, such as a cell-surface receptor or enzyme. This process is commonly known as high throughput screening (HTS). The separation, identification, characterization, scale-up, and purification of natural products for large-scale libraries suitable for these high throughput screens are daunting. Many of these screening systems are not sufficiently robust to handle complex mixtures of natural products from ill-defined biological systems and may be inhibited by interaction with uncontrolled physiochemical conditions, simple toxic chemicals and known bioactive compounds (Fig. 1).

The natural product library itself might be composed of samples that are either themselves mixtures—such as crude extracts (10-100s of components), semi-pure mixtures (5-10 compounds) or, alternatively, single purified natural products. In the case of pure libraries, the hit-detection process is the same as that for synthetic pure libraries. In the first case, however, heterogeneity of the natural product library samples adds two additional levels of complexity to the screening process. The first of these is that once a response for the sample is detected by HTS, one or more rounds of chemical purification and biological assay might be necessary for identifying and isolating the active component in the mixture. This is described in greater detail later in this article. The second additional hurdle is that the complexity of crude or semi-pure natural product libraries, and the chemical nature of many of the components found in them, often challenges the robustness of HTS technology. Before the advent of biotechnology, the difficulty of obtaining purified protein targets directly from tissues necessitated that much of the compound screening was performed using cellular *in vitro* or even whole-animal systems. When applied to crude or otherwise partially purified natural product libraries, these platforms have a relatively good capacity to detect active components. Examples can be found in the antimicrobial *in vitro* and *in vivo* assays used in 1970s and 1980s, which effectively served to detect many novel antibiotics. Advances in genomics and molecular biology have now made it possible to obtain relatively large quantities of protein targets for high throughput, cell-free assay systems for directly detecting the catalytic inhibition or target binding (Silverman 1998). These developments have greatly expanded the number of targets amenable to HTS, but at the same time have introduced technical complexities into the screening of natural product libraries. These complications reduced the success rates for lead discovery from natural product libraries in the early and mid-1990s, but technical improvements during the past few years have circumvented the problems in many HTS platforms.

The decreased emphasis in the pharmaceutical industry on the discovery of natural products during the past decade can be attributed to a number of factors: first, the introduction of HTS against defined molecular targets, which prompted many companies to move from natural products extract libraries towards so-called 'screen friendly' synthetic chemical libraries; second, the development of combinatorial chemistry, which at first offered the prospect of simpler, more drug-like screening libraries of wide chemical diversity; third, advances in molecular biology, cellular biology and genomics, which increased the number of molecular targets and prompted shorter drug discovery timelines; fourth, a declining emphasis among major pharmaceutical companies on infectious disease therapy, a traditional area of strength for natural products (Projan 2003) and last, possible uncertainties with regard to collection of biomaterials as a result of the 1992 Rio Convention on Biological Diversity (Kirsop 1996).

The underlying reasons for these industry trends are as much commercial as they are scientific, particularly in the case of research into infectious disease (Projan 2003). As a result of these factors, today's drug discovery environment calls for rapid screening, hit identification and hit-to-lead development. In this environment, traditional resource intensive natural product programmes that are based on extract library screening, bioassay-guided isolation, structure elucidation and subsequent production scale up face a distinct competitive disadvantage when compared with approaches that utilize defined synthetic chemical libraries. However, emerging trends coupled with unrealized expectations from current R&D strategies are prompting a renewed interest in natural products as a source of chemical diversity and lead generation. No doubt natural products and their derivatives have historically been invaluable as a source of therapeutic agents. However, in the past decade, research into natural products in the pharmaceutical industry has declined, owing to issues such as the lack of compatibility of traditional natural product extract libraries with high throughput screening. Koehn and Carter (2005) have recently reviewed, recent technological advances that help to address these issues, coupled with unrealized expectations from current lead-generation strategies, have led to a renewed interest in natural products in drug discovery (Fig. 1).

The remarkable chemical diversity encompassed by natural products continues to be of relevance to drug discovery. Although today's drug discovery engine operates at an accelerated pace compared with the era in which natural products were pre-eminent sources of drug leads, numerous approaches have been developed to capture their intrinsic value. Crucial breakthroughs in separation and structure determination technologies have lowered the hurdles inherent in screening mixtures of structurally complex molecules. A greater understanding of the exquisite specificity ingrained in secondary metabolites through the evolutionary process has focused attention on their roles as mediators of protein–protein interactions in vital cellular processes, and advances in synthetic chemistry have revolutionized the processes of material supply and the modulation of biological activity through structural modifications. It seems that no compound is too complex to be recreated in the laboratory. Furthermore, our ability to model the binding of these evolved, privileged structures with their targets enables the design of simpler mimetics that have superior properties. Efforts to expand the impact of natural chemical diversity on the drug discovery process follow two main chemistry-driven paths. One seeks to simplify crude mixtures, as well as enhance the impact of minor components in assays, through the creation of fractionated natural product libraries. The other approach uses the power of combinatorial synthesis to amplify the structural context in which the unique features of natural products are expressed. The confluence of these technologies with advances in genomics, metabolic engineering (Khosla and Keasling 2003) and chemical synthesis offer exciting new possibilities to exploit the remarkable chemical diversity of nature's 'small molecules' in the quest for new drugs (Koehn and Carter 2005).

4. MICROBIAL GENOMICS AND DRUG DISCOVERY

With the lack of success of older antibiotic identification strategies, novel ways to identify new targets for the discovery or design of novel inhibitors have been sought. Leonardo da Vinci once remarked that "We know more about the movement of celestial bodies than about the soil underfoot." You could argue that his insight still holds true the best part of 500 years later. But new genomic technologies mean that the microscopic bodies that enliven soil may be about to get the attention they deserve knowing which processes soil microbes are responsible for, and how, is increasingly important to everyone from farmers to climate planners. Microbiologists want to see how the organisms communicate with each other and refine the niches in which they live. Drug developers want to know how the soil microbes poison each other with antibiotics, and other commercially minded researchers think they could discover useful industrial enzymes or additives for biofuels. But beyond a gross understanding of inputs and outputs, the specific ecological roles of microbial species or communities in the dirt remain elusive. The main stumbling block has been that up to 99% of soil microbes cannot be grown in laboratory cultures—the traditional way to study microorganisms. But as genetic technology has improved, it has provided ways around this. First, it managed to reveal the general level of biodiversity through the sheer quantity of different sequences, and then it allowed scientists to trawl for individual genes. This effort was given a major impetus in 1995 by the early arrival of the era of microbial genomics. It was widely believed up to that point that the first full sequence of a bacterial chromosome to be delivered would be that of *E. coli* K-12, which was largely performed in an ordered sequence from overlapping lambda phage clones (Blattner et al. 1997). The focus on *E. coli* was understandable, as a vast amount of gene mapping and microbial physiology had been performed in this organism, and many genes had highly reliable functional annotations. The success of the "shotgun" random sequencing strategy, followed by computer assembly of the short sequences into larger "contigs" and finally the finishing strategy to yield a complete closed genome of *Haemophilus influenzae* was an unexpected and largely unanticipated event (Fleischmann 1995). The overwhelming speed of this approach led to the rapid adoption of this strategy which was subsequently applied to a large number of prokaryotic genomes (Fraser and Fleischmann 1997). The accumulation of this data has occurred at a truly breathtaking rate, and has been further accelerated by the introduction of a new generation of capillary electrophoresis-based DNA sequencers with increased capacity (Heller 2001).

Now technology offers the promise of extracting not just the genomes of the creatures in the soil, but - in a sense - the genome of the soil itself. Many antibiotics have been derived from soil microbes that do grow in cultures, such drugs were an obvious thing to look for in those that do not. So far, Handelsman and her collaborators have turned up antibiotics called turbomycin A and B (Gillespie et al. 2002). Companies were swiftly set up to capitalize on the promise of 'functional screens' for possible antibiotics or industrial compounds. But none of the dozen or so compounds found through metagenomic studies of soil seems set to be useful. One of the problems is that the screens ignore a lot of the soil's natural diversity; genes from many of the microbes may simply not be expressed when cloned into *E. coli*. In fact, terragine is one of them; it was isolated from genes inserted into recombinant streptomycetes. The ability to compare, annotate, and design experiments across multiple microbial genomes via computational methods, has had a profound influence on the conduct of microbiological experimentation. Pharmaceutical firms were quick to understand that this sudden influx of microbial data could be mined for sets of novel targets for both antimicrobial and vaccine developments (Loferer 2000). In addition to performing a vital function in identifying significant numbers of novel targets, the microbial genome efforts would serve as a test bed for development of computational tools and experimental genomic strategies such as micro arrays and proteomic analysis, along with a host of innovative but unproven drug discovery tools generated mainly by small biotechnology firms. The microbial genomics experience, dealing with smaller genomes and organisms that have accessible genetic systems, would prepare for the larger assault on identifying targets from the human genome and developing drugs for these targets (Fig. 1).

4.1 Novel Target Identification through Genomics

Characterized gene targets can be sought using strategies to identify taxon-specific genes employing substantive techniques, most directly between a specific pathogen and human genome. However, until the complete human genome is available, this is likely to be a complex and incomplete strategy. Thus, there are several different ways to approach the experimental issues of identifying novel antimicrobial targets. There are differences in opinion as to what constitute valid targets for novel antibiotics and how to select targets less likely to engender resistance development. The next section will attempt an overview of several experimental strategies employed to discover novel, validated targets. An approach to find new targets is to identify genes whose products (protein or RNA) are essential for bacterial survival under a standard set of conditions. A number of methods, some highlighted below, have been developed to identify this subset of genes. Virtually all existing antibacterial compounds interact with targets that inhibit or kill bacteria that are growing on artificial laboratory media. Indeed, the infrastructure of clinical microbiology susceptibility testing is currently predicated on the performance of tests on artificial growth media (Jorgenson and Ferraro 1998). A valid extension of this line of reasoning would be to identify additional compounds that inhibit growth by interfering with the functions of an essential bacterial gene product. Alternatively, some groups have chosen to seek out targets among the genes essential for bacterial invasion and virulence. These pathogenicity related targets are often expressed only within the context of the host (Isaacson 1997) and several methods to identify these targets are also discussed in the following sections. It is believed that pathogenicity targets expressed only in the host may be less likely to engender resistance. The following sections will discuss strategies for identification of genes essential under *in vitro* test conditions, and genes identified as essential within an infected animal or cell.

4.2 Advanced Bioinformatics Tools

Use of either *in vitro* or *in vivo* systems to identify essential genes will usually turn up many potential candidates for further exploration. Among the genes identified will be both known function (annotated) genes and a significant number of genes of unknown function. In order to better rationalize the next experimental steps, it is often advantageous to return to computational analysis to extract further information about the subset of genes identified as potential antibiotic targets from the essentiality tests. Two examples of sophisticated programs that place genomic information in a rational framework will be cited.

Because of the extraordinary increase in DNA sequencing, the ability to manually annotate the generated sequence data has fallen far behind. A flexible system designed to address this problem and automatically annotate sequence data is MAGPIE (Multipurpose Automated Genome Investigation Environment). The system, running in a UNIX environment, is designed to automatically assign as many features as possible by gathering evidence about genes from a number of program sources (e.g., BLAST, FASTA, ProSearch, Medline). MAGPIE analyzes the information from multiple tools using rules, and properties are assigned based on level of confidence. The user is presented with a graphical view of regions of the genome, with putative reading frames and sets of annotated evidence for each (Gaasterland and Sensen 1996; Gaasterland and Ragan 1998). MAGPIE is designed to be highly portable, and the data storage format allows easy incorporation of new analysis tools or revisions of existing versions.

Another advanced bioinformatics program is the PathoLogic software from DoubleTwist. This program automatically constructs, based on the *E. coli* (EcoCyc database) genome and well-characterized metabolic pathway data, metabolic pathways from the genomic data of an input organism. These computationally derived pathways are based on the EcoCyc framework (Ouzounis and Karp 2000) and are constructed in phases. Initially, information is extracted from the annotations of the input genome, and a database populated with gene, protein names and functions. In the next linking stage, the polypeptides with enzymatic activity are linked to reaction objects, and finally protein complexes that groups the reactions linked to one or more polypeptides are identified. A provision is made for a graphical metabolic pathway editor that permits entry of organism unique pathways into the database. Another feature allows an assessment of the evidence for the existence of a particular pathway in an organism (Fig. 1).

The initial output is a metabolic overview, which somewhat resembles the old biochemical pathway charts. This overall representation of metabolism can be examined in greater detail via a graphical "point and click" feature, which gives increasingly detailed information (down to reaction component structures) at several levels about a selected point. This feature is used to navigate and query the database in an intuitive fashion. The tool identifies "holes", that is enzyme steps missing from the annotation, which may reflect unidentified enzymes in the database or variation in the pathway. Singleton steps where the majority of steps in a pathway are missing are most probably errors in functional annotation. Expression data from chip experiments may also be loaded into the database and visualized (Fig. 1).

There are many additional advanced tools (e.g., KEGG (Ogata 1998), EcoCyc and WIT (Overbeek et al. 2000)) for bioinformatic analysis, and new and more sophisticated versions are in development. Other examples of novel bioinformatics tools are those that have been devised to investigate microbial regulatory motifs (McGuire et al. 2000) and predict regulons (Tan et al. 2000). Increasingly sophisticated tools will be essential for information integration and visualization as the quantity and diversity of data (e.g., genome annotation, expression levels, proteomic results and structural genomic data) continue to grow.

5. DRUG DISCOVERY TECHNOLOGIES FOR GENOMICS-BASED NOVEL ANTIBIOTIC DISCOVERY

The success of antibiotics has been both a blessing and a curse. On the one hand, they have all but eradicated bacterial infectious diseases that once ravaged humankind, and on the other hand, their indiscriminate use has created selection pressure for multidrug resistant pathogens. A particular concern is the pervasive use of human therapeutics as feed additives for promoting growth and preventing infections. This has promoted both the exchange of antibiotic-resistance genes on plasmids or transposons and the spontaneous mutation of drug targets. Disturbingly, recent evidence suggests the predilection of resistant bacteria to develop additional mutations to compensate for lack of fitness can also perpetuate resistance. Till date, bacteria have evolved a plethora of resistance mechanisms to foil antibiotics, including reduced drug uptake into the cell, active efflux of the drug from the cell, modification of antibiotic target to reduce binding, inactivation of antibiotic by enzymatic modification, sequestration of antibiotic by protein binding, metabolic bypass of the inhibited reaction, binding of specific immunity protein to the antibiotic, and overproduction of the antibiotic target (Harvey and Mason 1998). A sea change in antibiotics research has resulted from the introduction of genomics technologies to discovery programs. Microbial genomics provides insights into the way microbes interact with their hosts and with the drugs used to treat them. Most importantly it is revealing new microbial targets for developing entirely new classes of antimicrobial agents. Once genomes have been sequenced, homology searches can be carried out to identify putative targets; microarrays can be used to analyze the expression of genes in response to infection or drug treatment. Even before sequence is available, genome-tagging approaches allow the identification of genes associated with virulence or host invasion (Fig. 1).

The discovery roadmap described above is designed to identify a number of novel antibiotic targets necessary for cell survival, either *in vitro* or in the infected host. Principles, both experimental and computational to identify and characterize these new targets have been outlined above. The goal, however, is to discover effective inhibitors of these targets as lead compounds for novel class antibiotics. As mentioned previously, searches for valid antibacterial targets will identify candidates not only among those genes with characterized functions, but also among the significant number of predicted ORFs that have as yet no functional assignments. In the case of those gene products with known functions, it is often straightforward to set up a biochemical assay for the function which can readily be adapted to HTS. This is usually followed by screening vast libraries of compounds to detect significant inhibition of enzyme function with these targets. To increase the chances of success, the chemical compound collections synthesized by traditional medicinal chemistry are being supplemented by new methods that allow parallel synthesis of large, diverse libraries of novel compounds. An excellent case study of a genomics-driven approach to antibacterial drugs is the work performed on the bacterial fatty acid pathway (Payne et al. 2001). In this approach, informatics analysis of multiple microbial genomes identified the complement of common *fab* genes, which were subjected to gene knockout tests of essentiality. Enzymatic assays were established and both single component and coupled enzymatic reactions were adapted to HTS. Several novel, selective inhibitors of the bacterial fatty acid pathway were identified and optimized from these screens.

In the case of targets where function is not known, it is necessary to set up HTS systems that identify small molecules which interact with the gene products of these novel targets based on other indirect detection methods. These include novel physical and biological technologies to detect protein-inhibitor interactions. Model strategies to accomplish the goal of identifying small molecule inhibitors are discussed in the following sections. The following is by no means a comprehensive list of technologies available; rather it serves to point out many potential approaches that can be employed.

Genome sequencing, bioinformatics, combinatorial chemistry and high throughput screening have led to exaggerated claims that we can soon expect new classes of clinically useful antimicrobials. At present, however, there are no products, even in early development, derived from these technologies. The promise, as detailed by various companies five or more years ago, was that these new technologies would result in a plethora of new targets and new compounds. Furthermore, it was argued that resistance to these compounds would be unlikely, because mutations in the targeted genes that are responsible for function are not compatible with viability. Concern has been expressed that too much focus has been concentrated on the throughput, with not enough focus on understanding the output. Indeed, because the DNA sequencing field has progressed so rapidly, large data sets are subject to errors, ambiguities and incompleteness. However, the maturity and status of microbial genomics is improving as people are trained, databases are constructed and improved analytical tools are invented to handle the flood of information (Potera 2002). This has resulted in several new bacterial targets being identified, many of which are considered essential to the life of bacteria. At present, the field could be described by the statement 'target rich and compound poor' (Fig. 2).

6. TIME TO THINK OUTSIDE THE BOX

New drugs are desperately needed to treat serious as well as common infections (e.g., blood, heart, and urinary tract infections; pneumonia; childhood middle-ear infections; boils; food poisoning; gonorrhea; sore

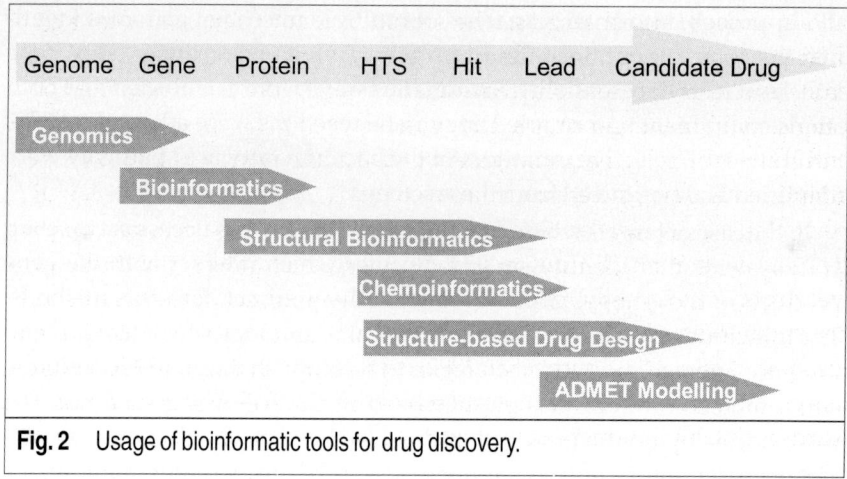

Fig. 2 Usage of bioinformatic tools for drug discovery.

throat, etc.). The bacteria that cause these infections are becoming increasingly resistant to the antibiotics that for years have been considered standard of care, and the list of resistant pathogens keeps growing. It is not possible to predict when an epidemic of drug-resistant bacteria will occur— but we do know it will happen.

Following the initial discoveries of antibiotics in the last century, a handful of classes were deployed and it was widely believed for a while that bacterial infections had been brought under permanent control. It is now clear that use of these compounds has had a profound ecological impact on the pathogens that they were designed to treat. Multidrug resistance is recognized as a real and growing problem, and any single control strategy is unlikely to be universally successful. The case has been made that as part of a strategy, it is vital to supplement the existing antibiotic armamentarium with fundamentally novel compounds to ease the problem of resistance (Cassell and Mekalanos 2001). This paper has attempted to outline some examples of a path forward using the knowledge recently made available by the large-scale sequencing of microbial genomes. Genomics offers us the unique opportunity to identify fundamentally new, validated targets that should, in turn, react with classes of molecules not as yet discovered to be potential antibacterials. Initial molecules so identified may be relatively weak inhibitors and have escaped detection in earlier screens, and it may be necessary to attempt to modify them to improve potency and "drug-like" characteristics.

Alternatively, the newly identified targets may enable our chemistry colleagues to build novel classes of antibiotics based on detailed structural analysis of these novel targets. The simultaneous use of microbial and human genomic information offers the potential to reduce adverse events by identifying potential cross-reactivity against microbial targets and human proteins. Microbial genomics may assist the discovery process by identifying all of the bacterial gene products that may be involved in antibiotic transport and efflux from bacterial cells. Genomic studies will further clarify how resistance to novel classes arise and the fitness costs to the organism that result from resistance. These studies may also suggest novel strategies to greatly reduce emergence of resistance in the first place. One such example is the mutant prevention concept (MPC) in fluoroquinolones (Dougherty et al. 2002; Blondeau et al. 2001). This concept offers guidance, rooted in the molecular biology of drug-target interactions, towards reducing the emergence of quinolone-resistant populations.

The underlying hypothesis is that it might be possible to avoid encountering previously generated resistance by seeking new targets for antimicrobial agents, such as proteins that are crucial for the survival of the bacterium (Rosenberg and McDevitt 2001). Some pharmaceutical companies are screening potential targets for enzyme inhibitors that block vital metabolic pathways that are not hit by existing antimicrobials (Connelly 2001). Many new approaches are being investigated, often by start-up companies. These include efflux-pump inhibitors and quorum-sensing signaling systems (Lazazzera 2000), which are used to sense and respond to changes in cell population density. In this era of genomic science, new technology has caused a rapid growth in knowledge about biological function integrated with DNA sequence data. Genome sequencing, bioinformatics, combinatorial chemistry and high-throughput screening have led to exaggerated claims that we can soon expect new classes of clinically useful antimicrobials (Bax and Mullan 1999). At present, however, there are no products, even in early development, derived from these technologies. The promise, as detailed by various companies five or more years ago, was that these new technologies would result in a plethora of new targets and new compounds. Furthermore, it was argued that resistance to these compounds would be unlikely, because mutations in the targeted genes that are responsible for function are not compatible with viability. Concern has been expressed that too much focus has been concentrated on the throughput, with not enough focus on understanding the output (Mardis et al. 2002). Indeed, because the DNA sequencing field has progressed so rapidly, large data sets are subject to errors, ambiguities and incompleteness.

However, the maturity and status of microbial genomics is improving as people are trained, databases are constructed and improved analytical tools are invented to handle the flood of information (Potera 2002). This has resulted in several new bacterial targets being identified, many of which are considered essential to the life of bacteria. At present, the field could be described by the statement 'target rich and compound poor'. Other new approaches include antisense peptide nucleic acids that can specifically inhibit *E. coli* gene expression and growth (Good 2001). Chemical modifications improve the potency by at least two orders of magnitude, while retaining target specificity. This new functional genomic method opens up exciting possibilities for drug discovery. A recent article (Potera 2002; Good et al. 2001) suggests that rapid advances are being made in part owing to bioterrorism. Using bacterial genetics and genomics, structural biology and assay development, it is hoped that new products will arise that will increase the useful life of approaches that search for inhibitors of cell wall targets (Green 2002). This is an exciting new area that has yet to bear fruit in the antibacterial field. Inhibition of an enzyme pathway does not equate to a new antimicrobial agent and many more steps must be successfully completed before a new antimicrobial drug is produced (Bax 2000).

Is the production of more antibacterial drugs, particularly those that are structural derivatives of previous ones, the best way forward? Each new antimicrobial agent is a precious resource, which will, in a relatively short period of time, become ineffective due to the emergence of bacterial resistance. One day, this resource will run out. Should we not devote more resources to trying to find ways to prolong the life of each of these uniquely beneficial compounds? This will require more research, with much of it in the area of basic bacterial physiology, and some in usage. A particularly important target in this respect is clinically latent bacteria. The molecular basis of tolerance to antimicrobial agents in latent bacteria should be examined in detail with, for example, gene knockouts, genomics and proteomics. Antimicrobial agents should be produced that specifically target clinically latent bacteria. More research into the use of combinations of antibacterial agents could be another way to prolong the life of these drugs. As much antimicrobial resistance originates from commensal bacterial flora, more research should also be devoted to ways of reducing the emergence of resistance in normal flora. It is important to test current antimicrobial agents for their ability to kill clinically latent bacteria. This could lead to the development of some existing antibacterial drugs that could be used in combination with other new agents that do not have a high-level of resistance plasmids in the environment to develop shorter courses of therapy. Potential benefits of this approach would be to increase patient compliance, decrease the potential for resistance development, reduce drug costs and increase efficacy, ultimately extending the life of these antimicrobial agents.

The remarkable chemical diversity encompassed by natural products continues to be of relevance to drug discovery. Although today's drug discovery engine operates at an accelerated pace compared with the era in which natural products were pre-eminent sources of drug leads, numerous approaches have been developed to capture their intrinsic value. Crucial breakthroughs in separation and structure determination technologies have lowered the hurdles inherent in screening mixtures of structurally complex molecules. A greater understanding of the exquisite specificity ingrained in secondary metabolites through the evolutionary process has focused attention on their roles as mediators of protein–protein interactions in vital cellular processes, and advances in synthetic chemistry have revolutionized the processes of material supply and the modulation of biological activity through structural modifications. It seems that no compound is too complex to be recreated in the laboratory. Furthermore, our ability to model the binding of these evolved, privileged structures with their targets enables the design of simpler mimetics that have superior properties. Efforts to expand the impact of natural chemical diversity on the drug discovery process follow two main chemistry-driven paths. One seeks to simplify crude mixtures, as well as enhance the impact of minor components in assays, through the creation of fractionated natural product libraries. The other approach uses the power of combinatorial synthesis to amplify the structural context in which the unique features of natural products are expressed. The confluence of these technologies with advances in genomics, metabolic engineering and chemical synthesis offer exciting new possibilities to exploit the remarkable chemical diversity of nature's 'small molecules' in the quest for new drugs.

The increasing volume of genomics research coupled with information from metagenomics may uncover new proteins, enzymes, pathways, and metabolic products that could become useful tools for microbiology research and, if we are fortunate, novel chemotherapeutic agents. Advances in genomics, transcriptomics and proteomics will also help to elucidate the nature and functions of gene products that have unknown functions at present. From a utilitarian perspective, the biotechnological spin-off could be even more exciting. It will allow us to address many current problems in agriculture, nutrition and medicine, which are all domains relying on complex microbial flora that have been, so far, used largely empirically. These new approaches will clearly reshape many aspects of human activities, when we finally have the tools to explore the fantastic reservoir of biochemical know-how in the microbial world. Bacteria were on this planet long before we arrived and, no doubt, will be here long after we have disappeared. Learning more about the prokaryotic world by smart, modern microbiology methods may help to extend our tenure here; a tenure that is inextricably linked to a fascinating, dynamic, but ill-understood-prokaryotic life-support system.

ACKNOWLEDGEMENTS

Thanks are due to Director, National Botanical Research Institute, Lucknow for providing necessary support. This study was supported by New Millennium Indian Technology Leadership Initiative (NMITLI) project, "Biofungicides from bacteria for controlling plant diseases", Council of Scientific & Industrial Research (CSIR), New Delhi awarded to CSN.

REFERENCES

Ajay WPW, Murcko M (1998). Can we learn to distinguish between 'Drug-like' and 'Non-drug-like' molecules? J Med Chem 41: 3314–3324.

Allan JD, Eliopoulos GM, Moellering RC Jr. (1987). In Contempory Issues in Infectious Diseases Vol. 6 (eds Root, R. K., Trunkey, D. D. & Sande, M. A.) 263–284 (Churchill Livingstone, New York).

Anantharaman V, Aravind L, Koonin EV (2003). Emergence of diverse biochemical activities in evolutionarily conserved structural scaffolds of proteins. Cur Opin Chem Biol 7: 12–20.

Anthony C, Hu H, Bax R, Page C (2002). The future challenges facing the development of new antimicrobial drugs. Nature Reviews 1:895-910.

Barber M, Rozwadaska-Downzenko M (1948). Infection by penicillin-resistant *Staphylococci*. Lancet ii: 641.

Bax R, Mullan N, Verhoef J (2000). The millennium bugs — the need for and development of new antibacterials. Int. J Antimicrob Agents 16: 51–59.

Bax RP, Mullan N (1999). Response of the pharmaceutical industry to antimicrobial resistance. Balliere Clin Infect Dis. 5: 289–304.

Bentley R, Bennett JW (1999). Constructing polyketides: from collie to combinatorial biosynthesis. Annu Rev Microbiol 53, 411–446.

Blattner FR, Plunkett G, Bloch CA, Perna NT, Burland V, Riley M, Collado-Vides J, Glasner JD, Rode CK, Mayhew GF, Gregor J, Davis NW, Kirkpatrick HA, Goeden MA, Rose DJ, Mau B, Shao Y (1997). The Complete Genome Sequence of *Escherichia coli* K-12. Science 277:1453.

Blondeau JM, Zhao X, Hansen G, Drlica K (2001). Mutant prevention concentrations of fluoroquinolones for clinical isolates of *Streptococcus pneumoniae*. Antimicrob Agents Chemother 45: 433-438.

Breinbauer R, Vetter IR, Waldmann H. From protein domains to drug candidates- Natural products as guiding principles in the design and synthesis of compound libraries. *Angew. Chem. Int. Ed.* 41, 2878–2890 (2002).

Bull AT, Ward AC, Goodfellow M (2000). Search and Discovery strategies for biotechnology: the paradigm shift. Microbiol Mol Biol Rev 64(3): 573-606.

Burke JP (1998). Analysis and description of the importance of natural products as scaffolds for drug design. JAMA 280:1270.

Cassell GH, Mekalanos JJ (2001). Concurrent ... infectious tracheobronchitis in dogs. J Amer Med Assoc 285:601.

Chopra I, Hodgson J, Metcalf B, Poste G (1996). New approaches to the control of infections caused by antibiotic resistant bacteria. An industry perspective. JAMA 275: 401–403.

Chothia C (1992). One thousand families for the molecular biologist. Nature 357:543–544

Coates A, Hu Y, Bax R and Page C (2002) the future challenges facing the development of new antimicrobial drugs. Nature Reviews 1:895-910.

Connelly GP (2001). Substrate profiling of new enzymes for drug discovery. Fed Res Progress   NDN-049-0330-1033-1.

Davies J (1994). Inactivation of antibiotics and the dissemination of resistance genes. Science 264:375.

Davies J (1999). In praise of Antibiotics. ASM News 65(4): 1-9.

Dougherty TL, Barrett JF, Pucci MJ (2002). Microbial Genomics and Novel Antibiotic Discovery: New Technology to Search for New Drugs Current Pharmaceutical Design 8:1119-1135.

Feher M, Schmidt JM (2003). Property distributions: Differences between drugs, natural products, and molecules from combinatorial chemistry. J Chem Inf Comput Sci 43:218–227.

Fleischmann RD, Adams MD, White O, Clayton RA, Kirkness EF, Kerlavage AR, Bult CJ, Tomb JF, Dougherty BA, Merrick JM, McKenney K, Sutton G, FitzHugh W, Fields C, Gocayne JD, Scott J, Shirley R, Liu LI, Glodek A, Kelley JM, Weidman JF, Phillips CA, Spriggs T, Hedblom E, Cotton MD, Utterback TR, Hanna MC, Nguyen DT, Saudek DM, Brandon RC, Fine LD, Fritchman JL, Fuhrmann JL, Geoghagen NSM, Gnehm CL, McDonald LA, Small KV, Fraser CM, Mith HO, Venter JC (1995). Whole Genome Random Sequencing and Assembly of *Haemophilus influenzae* Rd. Science 269:496.

Fleming A (1929). On the antibacterial action of cultures of a penicillium, with special reference to their use in the isolation of *B. influenzae*. Br J Exp Pathol 10:226–236.

Frank EK and Carter GT (2005). The evolving role of natural products in drug discovery. Nature Reviews 4:206-220.

Fraser CM, Fleischmann RD (1997). Strategies for whole microbial genome sequencing and analysis. Electrophoresis 18:1207.

Gaasterland T, Ragan MA (1998). Constructing multigenome views of whole microbial genomes. Microb Comp Genomics 3:177.

Gaasterland T, Sensen CW (1996). MAGPIE: Using multiple tools for automated genome interpretation. Trend Genetics 12:76.

Gillespie DE, Brady SF, Bettermann AD, Cianciotto NP, Liles MR, Rondon MR (2002). Isolation of antibiotics turbomycin A and B from a metagenomic library of soil microbial DNA. Appl Environ Microbiol 68:4301–4306.

Good L, Awasthi SK, Dryselius R, Larsson O, Nielsen PE (2001). Bactericidal antisense effects of peptide–PNA conjugates. Nature Biotechnol 19:360–364.

Gootz TD (1990). Discovery and development of new antimicrobial agents. *Clin. Microbiol Rev* 3:13–31.

Gould IM (1999). A review of the role of antibiotic policies in the control of antibiotic resistance. J Antimicrob Chemother 43:459.

Green DW (2002). The bacterial cell wall as a source of antibacterial targets. Expert Opin Ther Targets **6**:1–19.

Harvey J, Mason L (1998). The use and misuse of antibiotics in agriculture. Part 1. Current usage. Soil Association, Bristol, UK.

Heller C (2001). Principles of DNA separation with capillary electrophoresis Electrophoresis 22, 629.

Henkel T, Brunne R, Muller H, Reichel F (1999). Statistical investigation of structural complementarity of natural products and synthetic compounds. Angew Chem Int Ed Engl 38:643–647.

Holm L, Sander C (1996). Mapping the protein universe Science 273:595–603.

Hopwood DA (1985). Production of 'hybrid' antibiotics by genetic engineering. Nature 314:642–644.

Isaacson RE (1997). Novel targets for antibiotics - an update. Exp Opin Invest Drugs 6:1009.

Jorgenson JH, Ferraro MJ (1998). Antimicrobial susceptibility testing: general. principles and contemporary practices. Clin Infect Dis 26:973.

Khosla C, Keasling JD (2003). Metabolic engineering for drug discovery and development. Nature Rev Drug Discov 2:1019–1025.

Kirsop BE (1996). The convention on biological diversity: some implications for microbiology and microbial collections. J Indust Microbiol Biotech 17:505–511.

Knowles DJ (1997). New strategies for antibacterial drug design. Trends Microbiol 5:379–383.

Koehn FE, Carter GT (2005). The evolving role of natural products in drug discovery. Nature Rev 4:206-220.

Kohli RM, Walsh CT, Burkart MD (2002). Biomimetic synthesis and optimization of cyclic peptide antibiotics. Nature 418:658–661.

Lazazzera BA (2000). Quorum sensing and starvation: signals for entry into stationary phase. Curr Opin Microbiol 3:177–182.

Lederberg J (2000). Infectious History. Science 288:287.

Lee ML, Schneider G (2001). Scaffold architecture and pharmacophoric properties of natural products and trade drugs: Application in the design of natural product-based combinatorial libraries. J Comb Chem 3:284–289.

Levy SB (1998). Multidrug Resistance – A Sign of the Times. N. Engl. J. Med. 338:1376.

Lipinski CA, Lombardo F, Dominy BW, Feeney PJ (1997). Experimental and computational approaches to estimate solubility and permeability in drug discovery and development settings. Adv Drug Del Rev 23:3–25.

Loferer H (2000). Mining bacterial genomes for antimicrobial targets. Mol Med Today 6:470.

MacNeil IA (2001). Expression and isolation of antimicrobial small molecules from soil DNA libraries. J Mol Microbiol Biotechnol 3:301–308.

Mardis E, McPherson J, Martienssen R, Wilson RK, McCombie WR(2002). What is finished, and why does it matter? Genome Res 12:669–671.

McGuire AM, Hughes JD, Church GM (2000). Conservation of DNA regulatory motifs and discovery of new motifs in microbial genomes. Genome Res 10:744.

Moir DT, Shaw KJ, Hare RS, Vovis GF (1999). Genomics and Antimicrobial Drug Discovery. Antimicrob Agents Chemother 43: 439.

Monroe S, Polk R (2000). Antimicrobial use and bacterial resistance. Curr Opin Microbiol 3:496.

Nash JM (2001). The Antibiotic Crisis. Times Pacific Magazine 2:1-5.

Neu HC (1983). Structure–activity relations of new β-lactam compounds and *in vitro* activity against common bacteria. Rev Infect Dis 5:S319–S337.

Neu HC (1992). The crisis in antibiotic resistance. Science 257:1064.

Newman D, Cragg G, Kingston D (2003). In: The Practice of Medicinal Chemistry (Ed. Wermuth CG) Academic, London 91–109.

Newman DJ, Cragg GM, Snader, KM (2003). Natural products as a source of new drugs over the period 1981–2002. J Nat Prod 66:1002–1037.

Nizet V (2001). Innate antimicrobial peptide protects the skin from invasive bacterial infection. Nature 414:454–457.

Ogata H, Goto S, Fujibuchi W, Kanehisa M (1998). Computation with the KEGG pathway database. Biosystems 47:119.

Omura S, Ikeda H, Malpartida F, Kieser HM, Hopwood DA (1986). Production of new hybrid antibiotics, mederrhodins A and B, by a genetically engineered strain. Antimicrob Agents Chemother 29:13–19.

Ouzounis CA, Karp PD (2000). Global Properties of the Metabolic Map of *Escherichia coli*. Genome Res 10: 568.

Overbeek R, Larsen N, Pusch GD, D'Souza M, Selkov E Jr, Kyrpides N, Fonstein M, Maltsev N, Selkov E (2000). Nucleic Acids Res 28:123.

Payne DJ, Warren PV, Homes DJ, Ji Y, Lonsdale JT (2001). Drug Disc Today 6: 537.

Pfeifer BA, Admiraal SJ, Gramajo H, Cane DE, Khosla C (2001). Biosynthesis of complex polyketides in a metabolically engineered strain of *E. coli*. Science 291: 1790–1792.

Potera C (2002). Microbial genomics grows in maturity and status. ASM News 68: 271–276.

Projan SJ (2003). Infectious diseases in the 21st century: increasing threats, fewer new treatments and a premium on prevention. Cur Opin Pharmacol 3:457–458.

Projan SJ (2003). Why is big pharma getting out of antibacterial drug discovery? Cur Opin Microbiol 6:427–30.

Rosenberg M, McDevitt D (2001). Exploiting genomics to discover new antibiotics. Trends Microbiol 9:611–617.

Sadowski J, Kubinyi H (1998). Scoring scheme for discriminating between drugs and nondrugs. J Med Chem 41:3325–3329.

Salem GM, Hutchinson EG, Orengo CA, Thornton JM (1999). Correlation of observed fold frequency with the occurrence of local structural motifs. J Mol Biol 287:969–981.

Silverman L, Campbell R, Broach JR (1998). New assay technologies for high throughput screening. Curr Opin Chem Biol 2:397–403.

Stahura F, Godden JW, Ling X, Bajorath J (2000). Distinguishing between natural products and synthetic molecules by descriptor Shannon entropy analysis and binary QSAR calculations. J Chem Inf Comput Sci 40:1245–1252.

Zhang C, DeLisi C (1998). Estimating the number of protein folds. J Mol Biol 284:1301–1305.

Multidrug-resistant Tuberculosis: An Emerging Problem

M. V. BASIL and M. BOSE

1. INTRODUCTION

The emergence of resistance to antimicrobials is a natural biological occurrence. Resistance may be either a characteristic associated with an entire species or acquired through mutation or gene transfer. Resistance genes encode information on a variety of mechanisms that microorganisms use to withstand the inhibitory effects of specific antimicrobials. These mechanisms can confer resistance to other antimicrobials of the same class and sometimes to several different antimicrobial classes (Mitchison 1950). With increasing antimicrobial use and misuse over the years, resistance to antimicrobial agents has emerged in viruses, bacteria, fungi and protozoa, posing new challenges for both clinical management and control programmes (Canetti 1965; Dooley et al. 1992; Small et al. 1993; Alland et al. 1994).

2. MECHANISMS OF ANTI-TUBERCULOSIS DRUG RESISTANCE AND FACTORS ASSOCIATED WITH ITS EMERGENCE

Resistance of *M. tuberculosis* to anti-tuberculosis drugs is man-made. Wild isolates of *M. tuberculosis* that have never been exposed to anti-tuberculosis drugs are virtually never clinically resistant. There are a few exceptions, but these exceptions are not thought to contribute greatly to the overall burden of

Deptt. of Microbiology, Vallabhbhai Patel Chest Institute, University of Delhi, Delhi.

resistance. For instance, isolates of M. *tuberculosis* from Chennai, India, have been found to have a higher average level of resistance to para-aminosalicylic acid (PAS) than isolates from patients in the United Kingdom. The former also have higher minimum inhibitory concentrations (MIC) of thioacetazone than British isolates and isolates from other parts of India (CDC 1991). This resistance is called **natural resistance**. Most bovine isolates are naturally resistant to PAS and pyrazinamide (PZA), and most mycobacterium other than M. *tuberculosis* (MOTT) are resistant to the standard anti-tuberculosis drugs (Frieden 1993). Exposure to a single drug, whether as a result of poor adherence to treatment, inappropriate prescription, irregular drug supply, or poor drug quality, suppresses the growth of bacilli susceptible to that drug but permits the multiplication of pre-existing drug-resistant mutants (Fischl et al. 1992).

The patient then develops **acquired resistance**. Subsequent transmission of such bacilli to other persons may lead to disease that is drug-resistant from the outset, an occurrence known as **primary resistance**. Recently, the terms "resistance among previously treated cases" and "resistance among new cases" have been used for acquired and primary resistance, respectively (Monno et al. 1991). The emergence of drug-resistant M. *tuberculosis* has been associated with a variety of factors such as the lack of a standardized therapeutic regimen, poor programs implementation, frequent or prolonged shortages of drugs, inadequate resources, use of poor quality drugs and non-compliance of patients (Cohn et al. 1997).

3. MULTIDRUG-RESISTANT TUBERCULOSIS

Tuberculosis (TB) was declared a global emergency by the World Health Organisation (WHO) on World TB Day in 1993. This unprecedented declaration was prompted predominantly by two developments; the resurgence of TB in the West from the mid 80s, where the disease had been showing a steady decline from the beginning of the century (Kochi 1991) and a number of outbreaks of multidrug-resistant tuberculosis (MDR-TB) in USA and in many other parts of the world, in the late 1980s and early 1990s. Multidrug-resistant tuberculosis has been a topic of growing importance in the last decade. It has been described as the third epidemic, complicating the epidemics of human immunodeficiency virus (HIV) and the re-emergence of TB in the West (Neville et al. 1994). A series of outbreaks of MDR-TB in hospitals, prisons and shelters for homeless persons in and around New York in the early 90s were a cause of great alarm. A majority of the affected patients were HIV infected and the mortality was extremely high. Some patients were resistant to six or seven drugs (Beck-Sague et al. 1992; Fischl et al. 1992; Frieden et al. 1993; Valway et al. 1994). In a review by WHO of a series of 63 surveys of drug-resistant TB carried out worldwide between 1985 and 1994 the rates of primary resistance to isoniazid ranged from 0-17%; to streptomycin, 0-24%; to rifampicin, 0-3%; and to ethambutol, 0-4%. The rates of acquired resistance were isoniazid, 4-54%; streptomycin, 0-19%; rifampicin 0-15%; and ethambutol, 0-14%. The highest rates of MDR-TB were from Nepal (48%), Gujarat state (India) (34%), New York city (USA) (30%), Bolivia (15%) and South Korea (15%). The WHO and the International Union Against Tuberculosis and Lung Disease (IUATLD) subsequently established a global project of drug resistance surveillance in 35 countries, including India, during 1994 to 1997. The prevalence of primary multidrug resistance was 1.4%. Among patients with histories of treatment for one month or more, the prevalence of resistance to any of the four first line drugs was 36%, and the prevalence of multidrug resistance was 13%. Particularly high prevalence of multidrug resistance was found in the former Soviet Union, Asia, the Dominican Republic, and Argentina. The report concludes that resistance to anti-TB drugs was found in all 35 countries and regions surveyed, suggesting again that this is a global problem (Pablos-Mendez et al. 1998).

Drug resistance in TB is not a new phenomenon. It was recognized very soon after the introduction of effective anti-TB drugs that *M. tuberculosis* can rapidly become resistant to the drugs used against it. Resistance could develop to a single drug or to multiple drugs. However, the definition of multidrug resistance has undergone a change in perception over the years. In the pre-rifampicin era the term multidrug resistant tuberculosis meant disease caused by organisms that were resistant to at least two of the first line anti-TB drugs then available, usually to isoniazid and streptomycin. Today, the term is used to signify disease due to *M. tuberculosis* that is resistant to the two most effective current anti-TB drugs, isoniazid and rifampicin, with or without resistance to other drugs (Veen 1995).

4. PREVALENCE OF DRUG-RESISTANT TUBERCULOSIS IN INDIA

The Indian Council of Medical Research (ICMR) conducted studies on the incidence of drug resistance in *M. tuberculosis* isolates in 1965-1967, in 9 cities in different geographical parts of the country. Culture of sputum samples from patients of tuberculosis and drug susceptibility tests were carried out at the Tuberculosis Research Center (TRC, Chennai). The first study was on newly diagnosed pulmonary tuberculosis patients who had not received any previous treatment, and the second study was on all patients, new and old. The first study showed that the level of primary drug resistance ranged from 11-20% for isoniazid, 8-20% for streptomycin and 4-11% for both drugs (ICMR 1968). Overall, 20% of patients had resistance to one or more drugs ranging from 14-29% in the nine centers. The second study showed that 15-69% of patients exhibited isoniazid resistance, 12-63% streptomycin resistance and 5-58% double drug resistance (ICMR 1969). The TRC, Chennai has been conducting controlled clinical trials in chemotherapy of tuberculosis since 1956. Patients are admitted to these trials on the basis of sputum cultures positive for *M. tuberculosis.* Drug susceptibility tests are also done on all positive cultures. Hence, drug resistance data are available over a 40 year period (1956 to 1997) in patients admitted to the clinical trials. In newly diagnosed patients without prior treatment, resistance to isoniazid rose from 5 to 13%, to streptomycin from 3 to 12% and to both drugs from 1-6% in this period (Paramasivan 1998). Rifampicin resistance in newly diagnosed patients was less than 1% in the period 1985-90 and about 2% in 1990-95. In another comprehensive review of the drug resistance scenario in India, it was reported that the overall primary resistance to isoniazid ranged from 6-13%, except among the rural population of Kolar, Karnataka, with a figure of 27%, to streptomycin from 1-6% and to rifampicin from 0-2%. Acquired resistance ranged from 35-67% for isoniazid, 26-27% for streptomycin and 3-37% for rifampicin. The combined resistance to rifampicin and isoniazid ranged from 0-6% (Paramasivan 1998).

In a study conducted in Pune, 30 isolates of *M. tuberculosis* were collected from HIV seropositive tuberculosis patients and 40 isolates from HIV seronegative tuberculosis patients. Of the 30 isolates from HIV infected patients, 10% were multidrug-resistant. Of the 40 *M. tuberculosis* isolates from HIV seronegative individuals, 2.5 % isolates were multidrug resistant (Pereira et al. 2005).

5. MECHANISM OF DRUG RESISTANCE

Drug resistance in *M. tuberculosis* occurs by random, single step, spontaneous mutations at a low but predictable frequency, in large bacterial populations. There is no convincing evidence for the role of plasmids or transposons as in other bacterial pathogens. It has been estimated that the probability of incidence of drug-resistant mutants in a population of *M. tuberculosis* is depicted by the formula $P=1-(1-r)^n$, where P is the probability of incidence of drug-resistant cases, r is the probability of incidence of drug-resistant mutants and n is the number of bacilli in the lesion (Shimao 1987). The value of r for rifampicin is 10^{-8}, that for isoniazid, streptomycin, kanamycin, ethambutol and para-aminosalicylic (PAS) acid is 10^{-6}. Thus, for isoniazid and streptomycin, there is likely to be one resistant bacillus in a population of one million sensitive bacilli. The corresponding figure for rifampicin is one in one hundred million. The probability of resistance occurring to two drugs at the same time, is the product of the probabilities of drug resistance to a single drug.

6. GENETIC BASIS OF DRUG RESISTANCE

While it was generally known that resistance in *M. tuberculosis* occurs due to spontaneous mutation, the exact genetic basis of this mutation has been discovered only recently. In the last seven years, there have been many dramatic discoveries in molecular biology which have greatly increased the understanding of the mechanism of drug resistance. The most important of these was by Ying Zhang of the Royal Postgraduate Medical School, London. In 1992, he and his colleagues were able to transform an isoniazid resistant strain of *M. smegmatis* by repeated splicing of DNA segments from *M. tuberculosis*, till the strain was rendered completely isoniazid sensitive (Zhang et al. 1992). The segment of DNA which was responsible for restoring drug sensitivity was also able to restore catalase-peroxidase activity. This segment has been identified as the *katG* gene which encodes for catalase-peroxidase activity. It is well-known that many strains of *M. tuberculosis* resistant to isoniazid show reduced catalase-peroxidase activity. However, not all resistant strains of *M. tuberculosis* are catalase-peroxidase deficient. Some strains are catalase-peroxidase positive. Thus, other mechanisms in drug resistance are also possible. Subsequently, a mutation was discovered in the *inhA* gene, which is required for the synthesis of mycolic acid, an essential component of the mycobacterial cell wall, and which may be a target of activity for isoniazid in some resistant strains of *M. tuberculosis*.

In the last few years, there has been considerable progress in our understanding of the mechanisms of action and resistance to anti-tuberculosis agents. To date, there is information about 11 genes involved in resistance to all major anti-TB drugs in *M.tuberculosis*. Mutations in *katG*, *inhA* and *ahpC* genes are found in upto 90% of isoniazid resistant strains, rifampicin resistance is associated with *rpoB* mutations (> 96%), pyrazinamide resistance with *pncA* mutations (72-97%), ethambutol resistance with mutations in *embB* (47-65%), streptomycin resistance with *rrs* or *rpsL* mutations (70%) and fluoroquinolone resistance with *gyrA* substitutions (75-94%). Additional genes and mechanisms may play a role, particularly in association with lower levels of resistance (Alcaide and Telenti 1997). The presence or absence of mutations in drug resistance associated genes does not necessarily indicate susceptibility or resistance to the corresponding antibiotic. Reports of resistant isolates without mutations in drug resistance associated genes and reports of susceptible isolates with mutations in drug resistance associated genes indicate a partial understanding of resistance mechanisms (Ramaswamy and Musser 1998; Ramaswamy et al. 2003).

7. THE THREAT OF DRUG RESISTANCE FOR TB CONTROL

Although tremendous efforts to control TB have been undertaken at the global and national level over the past decade, almost 2 million patients still die every year. According to the recent estimates, one-third of the human population (about 1.86 billion people) was infected with *M. tuberculosis* worldwide in 1997. TB is principally a disease of poverty, with 95% of cases and 98% of deaths occurring in developing countries. Of these, more than half the cases occur in five South- East Asian countries. In 1997, nearly 1.87 million people died of TB and the global case fatality rate was 23%. This figure exceeded 50% in some of the African countries where human immunodeficiency virus (HIV) is highly prevalent. It is estimated that between 2002 and 2020, approximately 1000 million people will be newly infected, over 150 million people will get sick, and 36 million will die of TB if proper control measures are not instituted (Sharma and Mohan 2004). The World Health Organization estimates that upto 50 million persons worldwide may be infected with drug-resistant strains of TB. Also, 300,000 new cases of MDR-TB are diagnosed around the world each year and 79% of the MDR-TB cases now show resistance to three or more drugs (WHO 2003).

Studies have also indicated that drug-resistant TB is more difficult to cure. The cure rates among patients harbouring multidrug-resistant isolates range from 6% to 59%. Due to all these factors, the forty-fourth World Health Assembly (WHA) in 1991 recognized the growing importance of TB and declared TB a public health emergency. In 1993, WHO formulated the DOTS (Directly observed treatment – short course) strategy, which, if implemented correctly, is one of the most cost-effective public health interventions available (WHO 2003).

8. DETECTION OF MDR-TB

8.1 Detection of Drug Resistance

Traditionally, Lowenstein-Jensen (LJ) culture has been used for drug sensitivity testing using (*i*) absolute concentration method; (*ii*) the resistance ratio method; and (*iii*) the proportions method (Vareldzis et al. 1994, Sharma and Mohan 2004). In absolute concentration method, the minimal inhibitory concentration (MIC) of the drug is determined by inoculating the control media and drug containing media with a carefully controlled inoculum of *M. tuberculosis*. Media containing several sequential two-fold dilutions of each drug are used. Resistance is indicated by the lowest concentration of the drug which will inhibit growth (defined as 20 colonies or more at the end of four weeks) (Vareldzis et al. 1994). In resistance ratio method, MIC of the isolate is expressed as a multiple of the MIC of a standard susceptible strain, determined concurrently, in order to avoid intra and inter-laboratory variations.

These two methods require stringent control of the inoculum size and hence are not optimal for direct sensitivity testing from concentrated clinical specimens. In the proportion method, the ratio of the number of colonies growing on drug containing medium to the number of colonies growing on drug free medium indicates the proportion of drug-resistant bacilli present in the bacterial population. Below a certain proportion called critical proportion, a strain is classified as susceptible, and above that as resistant (Vareldzis et al. 1994; Sharma and Mohan 2004).

Traditional methods of drug susceptibility tests require 4 to 6 weeks, leading to undue delay in the initiation of appropriate treatment and during this time the patient continues to transmit drug-resistant infection in the community. There is thus a need for more rapid methods of detection of drug resistance. In this respect, some recent developments are of very great relevance.

8.1.1 BACTEC

The BACTEC method is an automated radiometric technique, which depends on the detection of radiolabelled CO_2 as a measure of growth index for growing microorganisms. Using this system it is possible to provide drug susceptibility results for *M.tuberculosis* in three weeks (Snider et al. 1981; Good and Mastro 1989; Huebner et al. 1993). A comparison with conventional procedures for its ability to detect drug-resistant strains shows excellent correlation.

8.1.2 MGIT

The mycobacteria growth indicator tube (MGIT) system (Becton-Dickinson) is a rapid, nonradioactive method for detection and susceptibility testing of *M. tuberculosis* (Adjers-Koskela and Katila 2003). The MGIT system relies on an oxygen-sensitive fluorescent compound contained in a silicone plug at the bottom of the tube which contains the medium to detect mycobacterial growth. The medium is inoculated with a sample containing mycobacteria and with subsequent growth mycobacteria utilise the oxygen and the compound fluoresces. The fluorescence thus produced is detected by using an ultraviolet transilluminator. Studies carried out both with cultures and direct clinical samples showed comparable results with the BACTEC and the proportions method (Bemer et al. 2002).

8.1.3 PhaB

The phaB determines the protection and amplification of a phage by mycobacteria in clinical specimens (Eltringham et al. 1999; Eltringham et al. 1999). The clinical sample suspected of containing viable mycobacterial cells is infected with the phages. Once the infection has occurred, the cells are washed with a viricidal solution destroying all extracellular phages. The intracellular phages will replicate, lysing *M. tuberculosis* cells, and new phage particles will be released into the media. The number of particles will be proportional to the number of cells infected, which will be quantified by counting plaques using a layer of *M. smegmatis*. If sample aliquots are treated with antibiotics prior to infection with the phage, plaques will appear only if the mycobacterial cells are resistant (Eltringham et al. 1999; Eltringham et al. 1999; Wilson et al. 1997).

For the detection of *M. tuberculosis* a commercial system has been developed, the FASTPlaqueTB assay (Biotec Laboratories, South Africa) with promising results using sputum and urine samples, offering a viable diagnostic alternative for low income countries, with a turnaround time of 1 day from sample collection (Marei et al. 2003). For the diagnosis of drug resistance, phaB results have been satisfactory using clinical isolates and sputum specimens (Eltringham et al. 1999; Eltringham et al. 1999; McNerney et al. 2000; McNerney 2002).

8.1.4 Luciferase Reporter Phages (LRPs)

LRPs are phages harboring the *fflux* (firefly luciferase) gene, which produces visible light when expressed in the presence of luciferin (enzyme substrate) and cellular ATP. LRPs are able to infect, replicate and express their genome

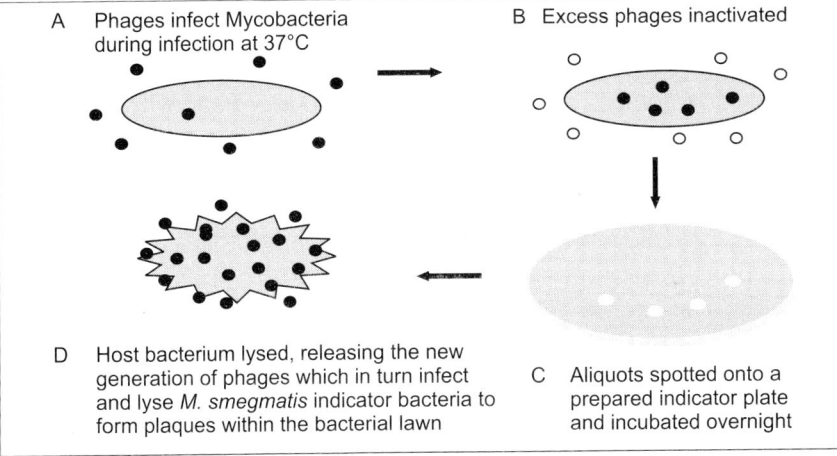

A   Phages infect Mycobacteria during infection at 37°C

B   Excess phages inactivated

D   Host bacterium lysed, releasing the new generation of phages which in turn infect and lyse *M. smegmatis* indicator bacteria to form plaques within the bacterial lawn

C   Aliquots spotted onto a prepared indicator plate and incubated overnight

Fig. 1   Detection of *M. tuberculosis* by PhaB assay: A. Phages are added to the decontaminated clinical sample. Phages infect viable mycobacterial cells present in the solution. B. Ferrous ammonium sulphate is added (viricidal solution) to destroy all non-infecting phages. C &D. Phages begin to replicate leading to cell lysis and release of the new phage particles into the medium. This solution is added onto a lawn of sensor cells (*M. smegmatis*) to produce plaques. In case the sample is negative, no plaques are formed.

(including the *fflux* gene) within viable mycobacterial cells. If an LRP-infected clinical sample releases light after addition of luciferin, the presence of viable *M. tuberculosis* is detected (Riska et al. 1997). If the sample is incubated with antibiotics followed by LRP infection, light emission is proportional to the resistance level (Carriere et al. 1997; Riska et al. 1998; Riska et al. 1999). LRP assays can be used as well for the identification of species from the *M. tuberculosis* complex (Riska et al. 1997; Bardarov et al. 2003). Luciferase activity can be monitored with a luminometer or with photographic film. The luminometer offers higher sensitivity, quantitative results within 54 hours, simple procedure and analysis, and has shown promising results with clinical isolates and sputum samples (Banaiee et al. 2001; Bardarov et al. 2003).

Photographic detection is achieved by using the 'Bronx box' which is an inexpensive light-tight box with a Polaroid cassette that carries a photographic film able to detect light emitted by the infection carried out in microtiter plates. The Bronx box yields qualitative results in 94 hours (Hazbon et al. 2003). Both techniques have shown promising results showing high agreement with BACTEC, MGIT and agar proportion method as reference (Banaiee et al. 2001; Bardarov et al. 2003; Hazbon et al. 2004).

8.2 Molecular Methods

M. tuberculosis acquires drug resistance by antibiotic selection of mutations that occur randomly at chromosomal loci. No plasmids or transposable elements are involved in this process. Nucleotide changes (point mutations, small deletions or insertions) confer resistance to single drugs, and the stepwise accumulation of these mutations leads to MDR-TB. Drug-resistant strains emerge when chemotherapy is intermittent or otherwise inadequate, underscoring the importance of an early detection of drug resistance. The increase in drug-resistant TB and the limited number of therapeutic agents is renewing the effort in the study of the molecular mechanisms of resistance. Rifampicin resistance is an excellent marker for multidrug-resistant *M. tuberculosis*, as 90% of rifampicin-resistant *M. tuberculosis* strains are also isoniazid resistant and, hence, are classified as multidrug-resistant (Telenti et al. 1993). Rifampicin resistance is also amenable to detection by rapid genotypic assays, because approximately 95% of all rifampicin-resistant strains contain mutations localized in an 81-bp core region of the bacterial RNA polymerase gene, *rpoB* (Musser 1995, Riska et al. 2000).

Moreover, virtually all mutations that occur in this region result in rifampicin resistance. By contrast, nearly all rifampicin-susceptible *M. tuberculosis* isolates have the same wild-type nucleotide sequence in this region (Musser 1995; Riska et al. 2000). Various molecular methods have been developed to rapidly detect mutations in the *M. tuberculosis rpoB* core region, including the line probe assay (De Beenhouwer 1995), single-strand conformational polymorphism (SSCP) PCR (Telenti et al. 1993; Bobadilla 2001), and real-time PCR (Torres et al. 2000; García de Viedma et al. 2002; Torres et al. 2003). Some of the most important techniques for detection of drug resistance associated mutations are described below.

8.2.1 DNA Sequencing

Sequencing is the most accurate and reliable method for mutation detection and it is used as the gold standard technique. It allows detecting both previously recognized and unrecognized mutations. Except for *rpoB*, DNA sequencing is unlikely to be used in routine detection of drug resistance mutations because it requires several sequencing reactions per isolate becoming labor-intensive and costly (Hazbón 2004).

8.2.2 The Line Probe Assay (LiPA)

LiPA (Inno-Genetics N.V., Zwijndrecht, Belgium) is used to identify *M. tuberculosis* species and the presence of mutations in the *rpoB* core region.

This region is amplified and biotin-labeled by PCR. This amplicor is detected after hybridization with a strip in which five probes for wild-type *rpoB* sequences, four probes for specific *rpoB* mutations, a conjugate control, and *M. tuberculosis* control probes were immobilized. Bound amplicons are then detected with a color reaction. Banding pattern interpretation allows *M. tuberculosis* complex identification and detection of *rpoB* mutations. The assay is designed to specifically detect four of the most common *rpoB* mutations as well as different mutations in the core region. The test can be performed on cultures or directly from clinical specimens in less than 48 hours. Overall concordance of the LiPA test with phenotypic susceptibility testing and direct sequencing, when performed from cultures, varies from 92.2% to 99.0% (Riska et al. 1997, Hirano et al. 1999, Traore et al. 2000), making the LiPA test a useful method for rapid detection of rifampicin resistance.

8.2.3 DNA Microarrays

DNA microarrays are based on the principle of hybridization. They allow analysis of large amounts of DNA sequences with a single hybridization step. PCR amplicons labeled with fluorophore moieties are generated from the sample to be hybridized to a large collection of probes bound to a solid surface. The bound amplicons emit a fluorescent signal that is scanned with an epifluorescent microscope. Probes are designed to hybridize to fully complementary amplicons. Wild-type and mutant probes are included in the array to determine the presence of specific mutations. Microarrays have been used mainly for species identification and for detection of mutations associated with rifampicin resistance and has had excellent concordance with sequencing results (Gingeras et al. 1998; Troesch et al. 1999). New arrays are being developed covering several mutations in *katG*, *inhA*, *rpoB*, *rpsL* and *gyrA* proving that this approach has a strong potential for the screening of drug resistance associated mutations in *M. tuberculosis* (Gingeras et al. 1998). The widespread application of microarrays is limited because they are still under a research and development stage, require expertise and sophisticated equipment and are costly.

8.2.4 Molecular Beacons

Molecular beacons are hairpin-shaped probes able to detect the presence of specific nucleic acids. They are designed in such a way that the loop portion of the molecule is a probe sequence complementary to a target nucleic acid molecule. The stem is formed by the annealing of complementary arm sequences on the ends of the probe sequence. A quenching moiety is attached

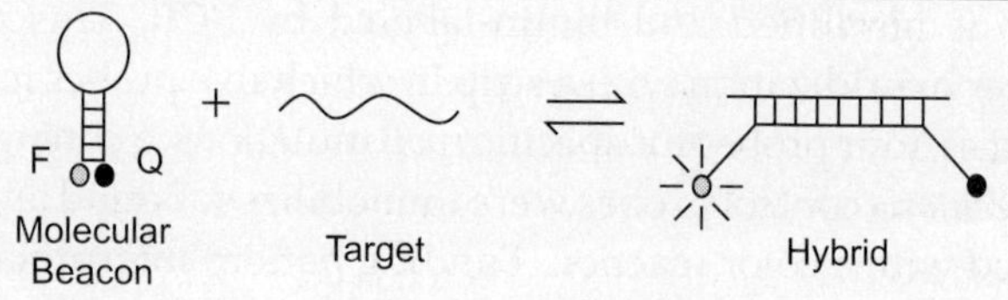

Fig. 2 A molecular beacon has a loop portion complementary to a target nucleic acid molecule and a stem composed of 2 complementary tails, one with a quenching moiety (Q) and the other with a fluorescent moiety (F). When closed, fluorescence is quenched by the quencher. When the probe encounters a target molecule, the beacon opens and hybridizes to it allowing fluorescence to be detected

to the end of one arm and a fluorescent moiety is attached to the end of the other arm. The stem keeps these two moieties in close proximity to each other, causing the fluorescence of the fluorophore to be quenched by energy transfer. When the probe encounters a target molecule, it forms a hybrid that is longer and more stable than the stem hybrid and its rigidity and length preclude the simultaneous existence of the stem hybrid (Figure 2). Thus, the molecular beacon undergoes a spontaneous conformational reorganization that forces the stem apart, and causes the fluorophore and the quencher to move away from each other, leading to the restoration of fluorescence that can be detected.

This is monitored in real-time, where the fluorescence increases every cycle in proportion to the amplification of the hybridizing target, which is not detected in cases when the target is not complementary to the beacon (Marras et al. 2003). Beacons are highly sensitive and specific; a single mismatch in the target sequence diminishes the beacon-target hybrid stability, allowing the detection of point mutations. Beacon assays are performed in sealed wells preventing amplicon contamination, they are easily implemented, automated and can be used in high throughput analysis. In the case of rifampicin, a set of 5 beacons has been designed to cover the *rpoB* core region in a single reaction, with excellent results (Piatek et al. 2000; El-Hajj et al. 2001; Varma-Basil et al. 2004). The assay is sensitive enough to detect 2 bacilli, offers results in 3 hours from sputum collection and identifies the *M. tuberculosis* species proving its strong potential in clinical settings. A different set of molecular beacons has been designed to screen for mutations in the regions with higher frequency of mutations associated with drug resistance for INH [*katG* (position 315), the promoter region of *inhA*, the *oxyR-ahpC* intergenic region and positions 66, 269, 312 and 413 of kasA] (Piatek et al. 2000).

8.2.5 FRET Probes

Fluorescence resonance energy transfer (FRET) probes, also known as light cycler probes, are highly specific probes able to detect the presence of mutations in real-time PCR reactions (Torres et al. 2003; Garcia de Viedma et al. 2002). The detection system of light cycler probes is based on FRET with two different specific oligonucleotides. Hybridization probe 1 is labeled with fluorescein, and hybridization probe 2 is labeled with the fluorophore Light Cycler Red 640. Both probes can hybridize in a head-to-tail arrangement, bringing the two fluorescent dyes into close proximity. A transfer of energy between the two probes results in emission of red fluorescent light. The level of fluorescence is proportional to the amount of DNA generated during the PCR process. In case the template is mutant, the probe-target stability will decrease and the probe will remain in solution hindering FRET, resulting in a flat signal. FRET probe assays have been developed for the rapid detection of resistance to isoniazid and rifampicin. The technique yields concordant results with sequencing data in less than 2 hours (Torres et al. 2003; Garcia de Viedma et al. 2002). This technique has the advantage that the region covered is larger than the region screened by molecular beacons, but because it requires the use of two fluorophore-labeled probes, the cost is doubled.

8.2.6 TaqMan Probes

TaqMan probes are dual labeled hydrolysis probes (Roche Molecular Systems, Inc.). These probes utilize the 5' exonuclease activity of the enzyme Taq polymerase for measuring the amount of target sequences in the samples. TaqMan probes consist of a 18-22 bp oligonucleotide probe which is labeled with a reporter fluorophore at the 5' end and a quencher at the 3' end.

Till the time probe is not hydrolyzed to its target sequence in a reaction, the quencher and the fluorophore remain in proximity to each other, separated only by the length of the probe. This proximity, however, does not completely quench the flourescence of the reporter dye and a background flourescence is observed. During PCR, the probe anneals specifically between the forward and reverse primer to an internal region of the PCR product. The polymerase then carries out the extension of the primer and replicates the template to which the TaqMan is bound. The 5' exonuclease activity of the polymerase cleaves the probe, releasing the reporter molecule away from the close vicinity of the quencher. The fluorescence intensity of the reporter dye, as a result increases. This process repeats in every cycle and

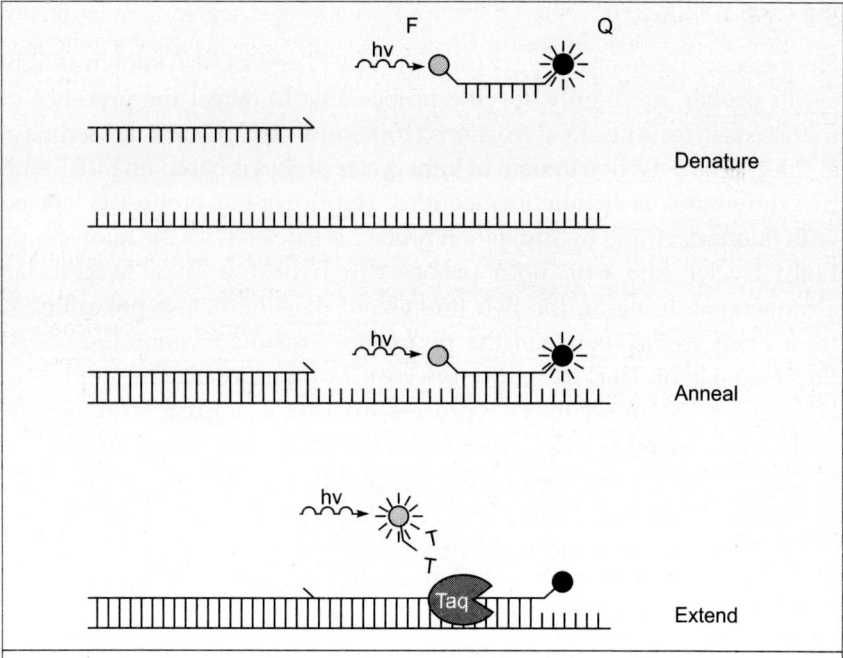

Fig. 3 TaqMan probes: The probes have a fluorophore (F) at the 5' end and a quencher (Q) at the 3' end. During the extension phase of a PCR reaction, the 5' exonuclease activity of the polymerase cleaves the probe, releasing the flourophore. The fluorescence intensity of this reporter molecule then increases.

does not interfere with the accumulation of PCR product. TaqMan probe assays have been developed for the rapid detection of mutations in *rpoB*, *katG* and *embB* genes leading to resistance to rifampicin, isoniazid, and ethambutol, respectively (Wada et al. 2004).

8.2.7 Single-strand Conformation Polymorphism (SSCP)

This technique determines the presence of mutations in specific DNA regions by their migration patterns in polyacrylamide gels. The target region of a gene is amplified by PCR and the product is denatured into two single-stranded molecules and subjected to non-denaturing polyacrylamide gel electrophoresis. Under non-denaturing conditions, the single-stranded DNA (ssDNA) molecule has a secondary structure determined by the nucleotide sequence, buffer conditions, and temperature. Mutant ssDNAs migrate to different positions than the wild type control. SSCP has been used to detect point mutations in *rpoB*, with promising results (Scarpellini et al. 1997). However, there are reports of the presence of silent mutations

(mutations that do not change the amino acid sequence) in this region not resulting in rifampicin resistance, leading to false positives by SSCP (Kim et al. 1997). In addition, this technique has the risk of amplicon contamination due to the extensive post-PCR manipulation required.

8.2.8 Branched Migration Inhibition (BMI)

The method is based on the detection of inhibition of branch migration in four-stranded cruciform structures. The cruciform structures are formed by annealing of partial duplexes prepared by amplification of test and reference nucleic acid sequences using specially designed PCR primers. If test and reference sequences are identical, the cruciform structures resolve into undetectable duplexes; if a mutation is present, the cruciform structures are stable and can be detected in an ELISA reaction. This novel mutation detection method, termed branch migration inhibition (BMI), has been shown to be equally effective in the detection of any mutation, including single-base substitutions, insertions, and deletions. The method is particularly suitable for the genotypic determination of drug susceptibility where the resistance phenotype is associated with multiple mutations in a known genetic locus (Liu et al. 2000).

9. MANAGEMENT

Multidrug-resistant tuberculosis is an iatrogenic problem. Therefore, the prevention of the emergence of drug resistance is of paramount importance. The best strategy to prevent the emergence of drug resistance is to prescribe standardized and well tested regimens in the proper dosage for the proper duration, and to ensure that the patient completes the full course of treatment till he or she is cured. The tendency is often to blame the patient for the development of drug resistance due to poor adherence to treatment. But it is a fact that physicians often contribute to this problem by prescribing inappropriate combination of drugs with inadequate dosage. A study on the prescribing practices of private practitioners by the Foundation for Research in Community Health (FRCH), showed that, in Mumbai, 100 practitioners prescribed a total of 80 different regimens and in Pune, 113 practitioners prescribed 90 different regimens for the treatment of tuberculosis (Uplekar and Shepard 1991). This problem of poor prescription is not confined to the developing countries. A study in the USA that evaluated the prescribing practices of physicians in New Jersey showed that 36% of 1,230 patients were not initially treated with four or more drugs and a substantial proportion of physicians did not initially treat their TB patients according to the recommendations of the Centres for Disease Control/American Thoracic Society (Liu et al. 1998).

It has been well established that the currently advocated short course regimen of six months for the treatment of pulmonary tuberculosis, with isoniazid, rifampicin, pyrazinamide and ethambutol or streptomycin for the first two months followed by isoniazid and rifampicin for the next four months cures 94-97% of patients with resistance to isoniazid or streptomycin or to both drugs (Mitchison and Nunn 1986). Studies have shown that in patients with resistance to streptomycin, isoniazid, or both drugs, only 2%, 8%, and 17%, respectively failed to respond when treated with appropriate regimens. When resistance is demonstrated to both isoniazid and rifampicin (MDR-TB), however, the outlook is very different. Of 38 patients, 35 failed to respond (Mathew et al. 1993).

CONCLUSION

In conclusion, the management of MDRTB is a challenging problem. The treatment is less effective, more toxic and much more expensive compared to the treatment of drug susceptible tuberculosis. Treatment needs to be supervised as far as possible, at least in the initial stages, or till the sputum becomes negative for acid fast bacilli.

REFERENCES

Adjers-Koskela K, Katila ML (2003). Susceptibility testing with the manual mycobacteria growth indicator tube (MGIT) and the MGIT 960 system provides rapid and reliable verification of multidrug-resistant tuberculosis. J Clin Microbiol 41: 1235-1239.

Alcaide F, Telenti A (1997). Molecular techniques in the diagnosis of drug-resistant tuberculosis. Ann Acad Med Singapore 26: 647-650.

Alland D, Kalkut GE, Moss AR, McAdam RA, Hahn JA, Bosworth W, Drucker E, Bloom BR (1994). Transmission of tuberculosis in New York City— An analysis by DNA fingerprinting and conventional epidemiological methods. N Eng J Med 330: 1710–1716.

Bardarov S Jr, Dou H, Eisenach K, Banaiee N, Ya S, Chan J, Jacobs WR Jr, Riska PF (2003). Detection and drug-susceptibility testing of M. tuberculosis from sputum samples using luciferase reporter phage: comparison with the Mycobacteria Growth Indicator Tube (MGIT) system. Diagn Microbiol Infect Dis 45:53-61.

Banaiee N, Bobadilla-Del-Valle M, Bardarov S, Jr., Riska PF, Small PM, Ponce-De-Leon A, Jacobs WR Jr, Hatfull GF, Sifuentes-Osornio J (2001). Luciferase reporter mycobacteriophages for detection, identification, and antibiotic susceptibility. Testing of Mycobacterium tuberculosis in Mexico. J Clin Microbiol 39: 3883-3888.

Beck-Sague C, Dooley SW, Hutton MD, Otten J, Breeden A, Crawford JT, Pitchenik AE, Woodley C, Cauthen G, Jarvis WR (1992). Hospital outbreak of multidrug-resistant

M. tuberculosis infection. Factors in transmission to staff and HIV infected patients. J Am Med Assoc 268: 1280-1286.

Bemer P, Palicova F, Rusch-Gerdes S, Drugeon HB, Pfyffer GE (2002). Multicenter evaluation of fully automated BACTEC Mycobacteria Growth Indicator Tube 960 system for susceptibility testing of *Mycobacterium tuberculosis*. J Clin Microbiol 40 : 150-154.

Bobadilla M, Ponce-de-Leon A, Arenas-Huertero C (2001). *rpoB* gene mutations in rifampin-resistant *Mycobacterium tuberculosis* identified by polymerase chain reaction single-strand conformational polymorphism. Emerg. Infect Dis 7:1010-1013.

Canetti G (1965). Present aspects of bacterial resistance in tuberculosis. Am Rev Respir Dis 92: 687-703.

Carriere C, Riska PF, Zimhony O, Kriakov J, Bardarov S, Burns J, Chan J, Jacobs WR Jr (1997). Conditionally replicating luciferase reporter phages: improved sensitivity for rapid detection and assessment of drug susceptibility of *Mycobacterium tuberculosis*. J Clin Microbiol 35: 3232-3239.

Centers for Disease Control (CDC) (1991). Nosocomial transmission of multidrug-resistant tuberculosis among HIV-infected persons — Florida and New York, 1988–1991. Morb Mortal Wkly Rep 40:585-591.

Cohn D, Bustreo F, Raviglione MC (1997). Drug resistance in tuberculosis: review of the worldwide situation and WHO/IUATLD global surveillance project. Clin Infect Dis 24 : S121-S130.

Dooley SW, Jarvis WR, Marlone WJ, Snider DE (1992). Multidrug-resistant tuberculosis. Ann Intern Med 117: 257–259.

De Beenhouwer H, Lhiang Z, Jannes G, Mijs W, Machtelinckx L, Rossau R, Traore

H, Portaels F (1995). Rapid detection of rifampicin resistance in sputum and biopsy specimens from tuberculosis patients by PCR and line probe assay. Tuber Lung Dis 76: 425-430.

El-Hajj HH, Marras SA, Tyagi S, Kramer FR, Alland D (2001). Detection of rifampin resistance in *Mycobacterium tuberculosis* in a single tube with molecular beacons. J Clin Microbiol 39: 4131-4137.

Eltringham IJ, Drobniewski FA, Mangan JA, Butcher PD, Wilson SM (1999). Evaluation of reverse transcription-PCR and a bacteriophage-based assay for rapid phenotypic detection of rifampin resistance in clinical isolates of *Mycobacterium tuberculosis*. J Clin Microbiol 37: 3524-3527.

Eltringham IJ, Wilson SM, Drobniewski FA (1999). Evaluation of a bacteriophage- based assay (phage amplified biologically assay) as a rapid screen for resistance to isoniazid, ethambutol, streptomycin, pyrazinamide, and ciprofloxacin among clinical isolates of *Mycobacterium tuberculosis*. J Clin Microbiol 37: 3528-3532.

Fischl MA, Uttamchandani RB, Daikos GL, Poblete RB, Moreno JN, Reyes RR, Boota AM, Thompson LM, Cleary TJ, Lai S (1992). An outbreak of tuberculosis caused by multiple-drug-resistant tubercle bacilli among patients with HIV infection. Ann Intern Med 117: 177-183.

Frieden TR, Sterling T, Pablos-Mendez A Kilburn JO, Cauthen GM, Dooley SW (1993). The emergence of drug-resistant tuberculosis in New York City. N Engl J Med 328: 521-526.

Garcia de Viedma D, del Sol Diaz Infantes M, Lasala F, Chaves F, Alcala L, Bouza E (2002). New real-time PCR able to detect in a single tube multiple rifampin resistance mutations and high-level isoniazid resistance mutations in *Mycobacterium tuberculosis*. J Clin Microbiol 40: 988-995.

Gingeras TR, Ghandour G, Wang E, Berno A, Small PM, Drobniewski F, Alland D, Desmond E, Holodniy M, Drenkow J (1998). Simultaneous genotyping and species identification using hybridization pattern recognition analysis of generic Mycobacterium DNA arrays. Genome Res 8: 435-448.

Good RC, Mastro TD (1989). The modern mycobacteriology laboratory: How it can help the clinician. Clin Chest Med 10: 315-322.

Hazbon MH, Guarin N, Ferro BE, Rodriguez AL, Labrada LA, Tovar R, Riska PF, Jacobs WR Jr (2003). Photographic and luminometric detection of luciferase reporter phages for drug susceptibility testing of clinical *Mycobacterium tuberculosis* isolates. J Clin Microbiol 41: 4865-4869.

Hazbon MH (2004). Recent advances in molecular methods for early diagnosis of tuberculosis and drug-resistant tuberculosis. *Biomédica* 24: 149-162.

Hirano K, Abe C, Takahashi M (1999). Mutations in the *rpoB* gene of rifampin-resistant *Mycobacterium tuberculosis* strains isolated mostly in Asian countries and their rapid detection by line probe assay. J Clin Microbiol 37: 2663-2666.

Huebner RE, Good RC, Tokars JI (1993) Current practices in mycobacteriology: Results of a survey of state public health laboratories. J Clin Mycobacteriol 31: 771-775.

ICMR (1968). Prevalence of drug resistance in patients with pulmonary tuberculosis presenting for the first time with symptoms at chest clinics in India I. Findings in urban clinics among patients giving no history of previous chemotherapy. Indian J Med Res 56: 1617.

ICMR (1969). Prevalence of drug resistance in patients with pulmonary tuberculosis presenting for the first time with symptoms at chest clinics in India.II. Findings in urban clinics among all patients, with or without history of previous chemotherapy. Indian J Med Res 57: 823.

Kim BJ, Kim SY, Park BH, Lyu MA, Park IK, Bai GH, Kim SJ, Cha CY, Kook YH (1997). Mutations in the *rpoB* gene of *Mycobacterium tuberculosis* that interfere with PCR-single-strand conformation polymorphism analysis for rifampin susceptibility testing. J Clin Microbiol 35: 492-494.

Kochi A (1991). The global tuberculosis situation and the new control strategy of the World Health Organization. Tubercle 72: 1-6.

Liu YP, Behr MA, Small PM, Kurn N (2000). Genotypic determination of *M. tuberculosis* antibiotic resistance using a novel mutation detection method, the branch migration inhibition *M. tuberculosis* antibiotic resistance test. J Clin Microbiol 38: 3656-3662.

Liu Z, Shilkret KL, Finelli L (1998). Initial drug regimens for the treatment of tuberculosis: Evaluation of physician prescribing practices in New Jersey, 1994 to 1995. Chest 113: 1446-1451.

Marei AM, El-Behedy EM, Mohtady HA, Afify AF (2003). Evaluation of a rapid bacteriophage-based method for the detection of *Mycobacterium tuberculosis* in clinical samples. J Med Microbiol 52: 331-335.

Marras SA, Kramer FR, Tyagi S (2003). Genotyping SNPs with molecular beacons. Methods Mol Biol 212: 111-128.

Mathew R, Santha T, Parthasarathy R et al. (1993). Response of patients with initially drug-resistant organisms to treatment with short-course chemotherapy. Indian J Tuberc 40: 119.

McNerney R (2002). Phage tests for diagnosis and drug susceptibility testing. Int J Tuberc Lung Dis 6:1129-1130.

McNerney R, Kiepiela P, Bishop KS, Nye PM, Stoker NG (2000). Rapid screening of *Mycobacterium tuberculosis* for susceptibility to rifampicin and streptomycin. Int J Tuberc Lung Dis 4: 69-75.

Mitchison DA (1950). Development of streptomycin resistant strains of tubercle bacilli in pulmonary tuberculosis. Thorax 4: 144-161.

Mitchison D, Nunn AJ (1986). Influence of initial drug resistance on the response to short course chemotherapy of pulmonary tuberculosis. Am Rev Respir Dis 133: 423-430.

Monno L, Angarano G, Carbonara S, Monno L, Angarano G, Carbonara S, Coppola S, Costa D, Quarto M, Pastore G (1991). Emergence of drug-resistant *Mycobacterium tuberculosis* in HIV-infected patients. Lancet 337: 852.

Musser JM (1995). Antimicrobial agent resistance in mycobacteria: molecular genetic insights. Clin Microbiol Rev 8: 496-514.

Neville K, Bromberg A, Bromberg R, Bonk S, Hanna BA, Rom WN (1994). The third epidemic - multidrug-resistant tuberculosis. Chest 105: 45-48.

Pablos-Mendez A, Raviglione MC, Laszlo A, Binkin N, Rieder HL, Bustreo F, Cohn DL, Lambregts-van Weezenbeek CS, Kim SJ, Chaulet P, Nunn P (1998). Global surveillance for antituberculosis drug resistance, 1994-1997. World Health Organization-International Union Against Tuberculosis and Lung Disease Working Group on Anti-Tuberculosis Drug Resistance Surveillance. N Engl J Med 338: 1641-1649.

Paramasivan CN (1998). An overview of drug resistant tuberculosis in India. Lung India 16: 21.

Pereira M, Tripathy S, Inamdar V, Ramesh K, Bhavsar M, Date A, Iyyer R, Acchammachary A, Mehendale S, Risbud A (2005). Drug resistance pattern of *Mycobacterium tuberculosis* in seropositive and seronegative HIV-TB patients in Pune, India. Indian J Med Res 121: 235-239.

Piatek AS, Telenti A, Murray MR, El-Hajj H, Jacobs WR, Jr, Kramer FR, Alland D (2000). Genotypic analysis of *Mycobacterium tuberculosis* in two distinct populations using molecular beacons: implications for rapid susceptibility testing. Antimicrob Agents Chemother 44: 103-110.

Ramaswamy S, Musser JM (1998). Molecular genetic basis of antimicrobial agent resistance in *Mycobacterium tuberculosis*: 1998 update. Tuber Lung Dis 79: 3-29.

Ramaswamy SV, Reich R, Dou SJ, Jasperse L, Pan X, Wanger A, Quitugua T, Graviss EA (2003). Single nucleotide polymorphisms in genes associated with isoniazid resistance in *Mycobacterium tuberculosis*. Antimicrob Agents Chemother 47: 1241-1250.

Riska PF, Jacobs WR Jr., Bloom BR, McKitrick J, Chan J (1997). Specific identification of *Mycobacterium tuberculosis* with the luciferase reporter mycobacteriophage: use of p-nitro-alpha-acetylaminobeta- hydroxyl propiophenone. J Clin Microbiol 35: 3225-3231.

Riska PF, Jacobs WR, Jr (1998). The use of luciferase-reporter phage for antibiotic-susceptibility testing of mycobacteria. Methods Mol Biol 101: 431-455.

Riska PF, Su Y, Bardarov S, Freundlich L, Sarkis G, Hatfull G, Carriere C, Kumar V, Chan J, Jacobs WR, Jr (1999). Rapid film-based determination of antibiotic susceptibilities of *Mycobacterium tuberculosis* strains by using a luciferase reporter phage and the Bronx Box. J Clin Microbiol 37: 1144-1149.

Riska PF, Jacobs WR, Jr., Alland D (2000). Molecular determinants of drug resistance in tuberculosis. Int J Tuber Lung Dis 4: S4-S10.

Scarpellini P, Braglia S, Brambilla AM, Dalessandro M, Cichero P, Gori A, Lazzarin A (1997). Detection of rifampin resistance by single-strand conformation polymorphism analysis of cerebrospinal fluid of patients with tuberculosis of the central nervous system. J Clin Microbiol 35: 2802-2806.

Sharma SK, Mohan A (2004). Multidrug-resistant tuberculosis. Indian J Med Res 120: 354-376.

Shimao T (1987). Drug resistance in tuberculosis control. Tubercle 68: 5-18.

Small PM, Shafer RW, Hopewell PC, Singh SP, Murphy MJ, Desmond E, Sierra MF, Schoolnik GK (1993). Exogenous reinfection with multidrug-resistant *M. tuberculosis* in patients with advanced HIV infection. N Eng J Med 328: 1137–1144.

Snider DE Jr, Good RC, Kilburn JO, Laskowski LF Jr, Lusk RH, Marr JJ, Reggiardo Z, Middlebrook G (1981). Rapid drug susceptibility testing of *M. tuberculosis*. Am Rev Respir Dis 123: 402-406.

Telenti A, Imboden P, Marchesi F, Lowrie D, Cole S, Colston MJ, Matter L, Schopfer K, Bodmer T (1993). Detection of rifampicin resistance mutations in *M. tuberculosis*. Lancet 341: 647-650.

Torres MJ, Criado A, Palomares JC, Aznar J (2000). Use of real-time PCR and fluorimetry for rapid detection of rifampin and isoniazid resistance-associated mutations in *M. tuberculosis*. J Clin Microbiol 38: 3194-3199.

Torres MJ, Criado A, Ruiz M, Llanos AC, Palomares J C, Aznar J (2003). Improved real-time PCR for rapid detection of rifampin and isoniazid resistance in *M. tuberculosis* clinical isolates. Diagn Microbiol Infect Dis 45: 207-212.

Traore H, Fissette K, Bastian I, Devleeschouwer M, Portaels F (2000). Detection of rifampicin resistance in *Mycobacterium tuberculosis* isolates from diverse countries by a commercial line probe assay as an initial indicator of multidrug resistance. Int J Tuberc Lung Dis 4: 481-484.

Troesch A, Nguyen H, Miyada CG, Desvarenne S, Gingeras TR, Kaplan PM, Cros P, Mabilat C (1999). Mycobacterium species identification and rifampin resistance testing with high-density DNA probe arrays. J Clin Microbiol 37: 49-55.

Uplekar MW, Shepard DS (1991). Treatment of tuberculosis by private general practitioners in India. Tubercle 72: 284-290.

Valway SE, Greifinger R, Papania M, Kilburn JO, Woodley C, DiFerdinando GT, Dooley SW (1994). Multidrug-resistant tuberculosis in the New York State prison system, 1990-1991. J Infect Dis 170: 151-156.

Varma-Basil M, El-Hajj H, Colangeli R, Hazbon MH, Kumar S, Bose M, Bobadilla-del-Valle M, Garcia LG, Hernandez A, Kramer FR, Osornio JS, Ponce-de-Leon A, Alland D (2004). Rapid detection of rifampicin resistance in *M. tuberculosis* isolates from Mexico and India with molecular beacons. J Clin Microbiol 42: 5512-5516.

Veen J (1995). Drug-resistant tuberculosis: Back to sanatoria, surgery and cod liver oil? (Editorial). Eur Respir J8: 1073-1075.

Vareldzis BP, Grosset J, de Kantor I, Crofton J, Laszlo A, Felten M, Raviglione MC, Kochi Asss (1994). Drug-resistant tuberculosis : Laboratory issues. World Health Organization recommendations. Tuber Lung Dis 75 : 1-7.

Wada T, Maeda S, Tamaru A, Imai S, Hase A, Kobayashi K (2004). Dual-probe assay for rapid detection of drug-resistant *Mycobacterium tuberculosis* by real-time PCR. J Clin Microbiol 42: 5277-5285.

World Health Organization. Global TB Control Report, 2003.

Wilson SM, al-Suwaidi Z, McNerney R, Porter J, Drobniewski F (1997). Evaluation of a new rapid bacteriophage-based method for the drug susceptibility testing of *Mycobacterium tuberculosis*. Nat Med 3: 465-468.

Zhang Y, Heym B, Allen B, Young D, Cole S (1992). The catalase-peroxidase gene of *M. tuberculosis*. Nature 358: 591-593.

Mycobacterium tuberculosis: Pathogenesis, Diagnosis, Chemotherapy and Strategy for Vaccination

P. S. BISEN[1*] and R. P. TIWARI [2]

1. INTRODUCTION

Tuberculosis is a chronic granulomatous disease caused by tubercle bacteria known as *Mycobacterium tuberculosis* and is transmitted from infectious persons through aerosol. It is a facultative intra-cellular organism that is readily phagocytosed, but is resistant to intra-cellular killing by macrophase. It grows intracellularly and can remain dormant for years if the host has good immunity. Hence, many infected individuals do not develop clinical manifestations of the disease. It is slow growing aerobic, non-motile, non-capsulated, non-sporing, obligate parasite and opportunistic intracellular pathogenic microorganism. It is known as acid-fast bacilli, which contains waxy coat of complex lipids and carbohydrates, due to which it makes their membrane impermeable to many drugs (multilayered cell wall) composed of mycolic acid and lipids. The bacteria have marked ability for surviving in dormant state and develop resistance. It creates

---

1. Institute of Biotechnology and Allied Sciences, Seedling Academy of Design, Technology and Management, Korebariyan, Jagatpura, Jaipur-302004, India.
2. Diagnostic Division, Nicholas Piramal India Limited, TTC Industrial Area Pawane-400705, Navi Mumbai, India.
* **Corresponding Author:** Prof. Prakash S. Bisen, Principal Director, Institute of Biotechnology and Allied Sciences, Seedling Academy of Design, Technology and Management, Korebariyan, Jagatpura, Jaipur-302004, India. E-mail: psbisen@gmail.com

global epidemic disease, which occurs in many parts of the body, but most commonly found (80%) in the lungs.

Mycobacterium was the first bacterial human pathogen described and continues to produce devastating illness till today. The discovery of the causative organism for tuberculosis dates back to 1882, when Robert Koch described the isolation of tubercle bacillus. Since then, a large number of Mycobacterial species responsible for causing pulmonary and extrapulmonary infections have been identified in humans as well as in animals (Wolinsky and Schaefer 1973). Tuberculosis displays all the principal characteristic features of the global epidemic diseases and is rampant throughout the world (Center for disease control and prevention, Morbidity Mortality weekly report, 1993 a, b).

Every third person on this earth is believed to be infected with *Mycobacterium tuberculosis*, leading to eight million cases of active tuberculosis per year and approximately three million deaths annually. Tuberculosis thus, has become a major concern worldwide. Significant mortality and morbidity has been reported from various parts of the world including both developed as well as developing nations (Pao et al. 1990; Kolk et al. 1992; Bisen et al. 2003; Garg et al. 2003; Tiwari et al. 2005; Tiwari et al. 2006). Without increased commitment and action at National and International levels, tuberculosis will claim about 30 million more lives in the next decade and there will be about 90 million new cases of active tuberculosis. Such a vast epidemic creates challenges as it raises the demand for public health solutions.

The currently available remedies for fighting tuberculosis are inadequate. The ultimate goal of biomedical TB research around the world should be to reduce the public health burden of this disease by developing improved diagnostic and therapeutic intervention strategies (Ginsberg et al. 1998).

The emergences of epidemic multidrug resistant strains of *Mycobacterium tuberculosis* in conjunction with HIV infection represent a global major health problem. Globally 8 million new cases of tuberculosis currently diagnosed annually with a mortality rate of 3 million (Kochi 1991). In recent years, a large number of Mycobacterial species associated with disease have been identified (Wolinsky et al. 1973) which includes two major pathogenic organisms *Mycobacterium tuberculosis* and *Mycobacterium leprae*, which are known to form tuberculosis and leprosy in human, respectively.

1.1 Evolution and History

Mankinds have known tuberculosis for ages. TB of spine or Potts disease was found in Egyptian mummies. It existed even in Vedic age, i.e., nearly 5000 years ago. Tuberculosis was mentioned in Rig-Veda as **"Rajayakshma"**. Two famous Indian physicians **Charak** and **Susruta** have also given detailed description of the disease. **Hippocrates**, the father of modern medicine who lived in Greece about 2500 years ago, has also described tuberculosis in his book. Historical evidences, thus suggest that people have suffered from tuberculosis since ages.

The actual cause of tuberculosis was discovered on 24th March 1882 by a German scientist **Robert Koch** at Berlin Physiological Society who isolated "Tubercle bacillus" and proved it to be the cause of the disease by his famous **"Koch's postulates"**. Since the bacilli was discovered on 24th March, the day is commemorated as World TB day and the bacillus is also known as Koch's bacillus in honor of the great scientist.

The variation of properties within the genus Mycobacterium is enormous and is reflected in the range of virulence, habitat, growth rate, nutritional requirements and antigenicity. Many of the unique characteristics of Mycobacteria are found to be in their very complex lipid rich cell walls.

The tubercle bacilli are slender, straight or slightly curved rods with rounded ends of the size of 2-4 micron x 0.3-0.5 micron. They have tendency to be arranged in clumps for growth with the optimum pH and temperature 6.8 and 37°C, respectively. The generation time is about 18 hours. Heat and sunlight kill the bacillus quickly but it can remain alive for several days in the dark.

1.2 The Disease

Millions of people have died from tuberculosis, a chronic leading infectious killer of youth and adults and the second most common infectious disease worldwide (WHO report 1996, 2002). WHO also estimates that *Mycobacterium tuberculosis* infects 1.7 billion people every year, which is equal to 33% of the entire world population (Todar's Textbook of Bacteriology Online). Of the 20 million of those infected with active cases, the annual mortality is approximately 3 million. Besides this, globally about 4.6 million cases are dually infected with HIV and tuberculosis (Cole et al. 1998).

It is estimated that 50 million people may already be infected with MDR strains of tuberculosis. Each year 8 million cases of disease arise and 30 million people suffer from active tuberculosis in the world with an annual mortality of 3 million. In India, TB is known as the 'king of diseases' having long survival in dormant state and marked ability to develop resistance. Annually, 500,000 people die of this disease in India. According to RNTCP status report of India 2001, everyday more than 20,000 people become infected with TB and approximately 20 lakh people develop TB yearly leading to a high mortality than even HIV, STD, malaria, leprosy and other tropical diseases. However, a delayed or missed diagnosis of tuberculosis contributes to *Mycobacterium tuberculosis* transmission and mortality (Lienhardt et al. 2001).

Among more than 1 billion people in India today, every second adult is infected with the tuberculosis bacterium. Each year, more than 2 million people develop active tuberculosis and upto 500,000 people die. The problem facing effective tuberculosis control in our country appears at first glance to be much the same as those that too often exist throughout much of eastern Asia.

India accounts for nearly one-third of the global burden of tuberculosis, and the disease is one of India's most important public health problems. In India tuberculosis kills 14 times more people than all tropical disease, combined, 21 times more than malaria, and 400 times more than leprosy. Every sputum positive patients can infect 10-15 individuals in a year. Globally, TB is a leading killer of women (RNTCP, status report, 2001). However, unless urgent action is taken, more than 40 lakh people in India will die from tuberculosis in the next decade.

There are real and potential conflicts between the interests of private physicians and public sector. The quality and supply of drug is erratic. Errors in diagnosis, based predominantly on X-ray and sometimes symptoms alone, are common. While infectious cases are frequently missed, some people are mistakenly diagnosed with TB and inappropriately treated (WHO 1997). The burgeoning pool of infection and the resulting risk of becoming infected with this pool are tremendous in the country (WHO 1997). There is a dire need to come out with strategies to ward off such risk, which can potentially identify and treat the affected ones with best possible precision. It is unfeasible to achieve containment in heavily afflicted populations like India without knowing the etiology of an organism and proper diagnosis of the disease caused by it.

2. CHARACTERISTIC FEATURES

M. tuberculosis is a multifaceted pathogen capable of causing acute disease and an asymptomatic latent infection. Unlike mucosal colonizers, latent

M. tuberculosis resides in the deep tissues and persists for years, before resulting in reactivation of tuberculosis. Currently, 1.7 billion people are estimated to harbor latent *M.tuberculosis* infection. These individuals carry 2-23% lifetime risk of developing (reactivation) TB. The ability of tubercle bacillus to survive even in harsh environmental conditions has been documented by Corper et al. (1933). They inoculated a culture bottle with *M.tuberculosis*, incubated it at 37°C for 12 years, and then re-examined it. The old culture yielded viable microorganisms in the sediment of the bottle. Indeed the population of latently infected individuals constitutes a major impediment to TB control efforts (Nikki et al. 1998). The incidence of HIV has made tuberculosis problem more acute. The incidence of HIV associated tuberculosis is increasing worldwide since the beginning of AIDS epidemic, and is expected to increase even further during the foreseeable future, especially in developing countries like India. Furthermore, the accelerating and amplifying influence of HIV infection is contributing to the increasing incidence of disease caused by multi-drug resistant strains of *M. tuberculosis* (John et al. 1998).

The characteristic features of the tubercle bacillus include its slow growth, dormancy, complex cell envelope, intracellular pathogenesis, and genetic homogeneity. The generation time of *M.tuberculosis* in synthetic medium or infected animals is approximately 24 hours. This contributes to the chronic nature of the disease that imposes lengthy treatment regimens and represents a formidable obstacle for researchers. The state of dormancy in which the bacillus remains quiescent within infected tissue may reflect metabolic shutdown resulting from the action of a cell-mediated immune response that contain unpredicted infection. As immunity wanes through aging or immune suppression, the dormant bacteria reactivate, causing an outbreak of disease often many decades after initial infection (Chan et al. 2000). The slow growth rate and the rigorous containment facilities required are two major hindrances to study this dangerous organism.

2.1 Cell Structure and Metabolism

Mycobacteria are rod-shaped, Gram-positive aerobes or facultative anaerobes non-sporing and non-motile (Fig. 1). They do not form capsules, endospore or conidia but have the lipids termed mycosides. As deduced from its genome, *M. tuberculosis* has the potential to manufacture all the machinery necessary to synthesize all of its essential vitamins, amino acids, and enzyme co-factors.

On the other hand, the inability to culture *M. leprae,* suggests that it has lost many of its metabolic capabilities, and is now an obligate parasite,

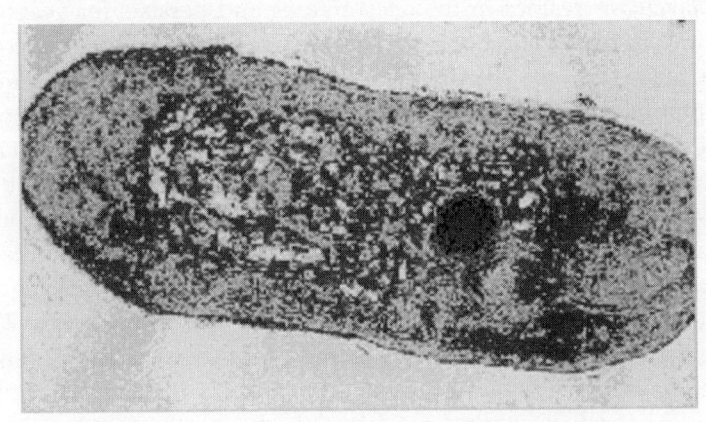

Fig. 1 Electron micrograph of *M. tuberculosis*

dependent on its host for most of its nutritional needs (Cole et al. 1998) This goes in accordance with its severely degenerated genome. Additionally, *M.tuberculosis* has an unusual cell wall, with an additional layer beyond the peptidoglycan layer, which is rich in unusual lipids, glycolipids, and polysaccharides (Fig. 1).

2.2 Taxonomic Status

The generic name *Mycobacterium* (Fungus-Bacterium) was first coined by Lehmann and Newmann in the 1st edition of their book, Atlas of Bacteriology, in 1896 because of the mould-like pellicular growth of the tubercle bacillus on liquid media. The variations of properties within the genus Mycobacterium is enormous and is reflected in the range of virulence, habitat, growth rate, nutritional requirements and antigenicity.

Many of the unique characteristics of Mycobacteria are found to be in their very complex lipid rich cell walls and many of them are pathogenic to humans as well as animals. Mycobacteria are usually acid and alcohol fast at some stage of growth, rarely exhibit grossly visible aerial hyphae, produce acid from sugar by oxidation and with exception of those that do not grow *in vitro* can be divided into rapid and slow growing strain. The original description of the genus was based mainly on morphological and staining properties—features that are considered sufficiently diagnostic to distinguish Mycobacteria from Actinomycetes (Skerman et al. 1980).

Further publications of newly recognized species, primarily of rapid growers have now brought the number to 54. Of these 54, 17 species are pathogenic and others are non-pathogenic. Members of genus Mycobacterium exhibit a great deal of variability with regard to growth rate, disease-causing capability and physiology. Different Mycobacterial species have been described and classified on the basis of rate of growth rate and chromogenecity, etc. With the exception of *M.leprae*, which is still to be cultivated *in vitro*, Mycobacteria can be assigned into two groups; primarily on the basis of growth rates. The so-called rapid growers include strains of those species which under optimal conditions of nutrition and temperature produce colonies within seven days, whereas, their slow growing counterparts require an additional view or more to form such colonies under such conditions. The optimum temperature of both species ranges from 28-45 °C.

2.3 Mycobacterial Genome

M. tuberculosis H37Rv genome comprises 4,411,529 base pairs and around 4000 genes (3924 predicted protein coding genes). The genome has a very high guanine + cytosine content (65.6 %) that is reflected in the biased amino acid content of the proteins. There is a significant preference for the amino acids Ala, Gly, Pro, Arg and Trp, which are all encoded by G+C rich codons, and a comparative reduction in the use of amino acids encoded by A+T rich codons such as Asn, Ile, Lys, Phe and Tyr. *M. tuberculosis* radically differs from other bacteria in that a very large portion of its coding capacity is devoted to the enzymes involved in lipogenesis and lipolysis. In total there are approximately, 250 distinct enzymes involved in fatty acid metabolism in *M.tuberculosis* compared with only 50 in *E. coli* (Cole et al. 1998).

2.4 Growth

M. tuberculosis grows well on liquid and solid media such as:

Liquid media:

A. Middle brook 7H9 and 7H12B broth supplemented with 10% ADC (albumin dextrose and catalase) at 37°C.
B. Sautons media for 6-7 weeks at 37°C (Figure 2).

Solid media:

A. Middle brook 7H10-7H11 agar supplemented with 10% OADC (Oleic acid, albumin dextrose and catalase) at 37°C

2.4.1 Pigmentation

Depending on the pigment produced as well as growth rate (Runyon et al. 1959) sub-classifies the non-tuberculosis bacteria into four subgroups:

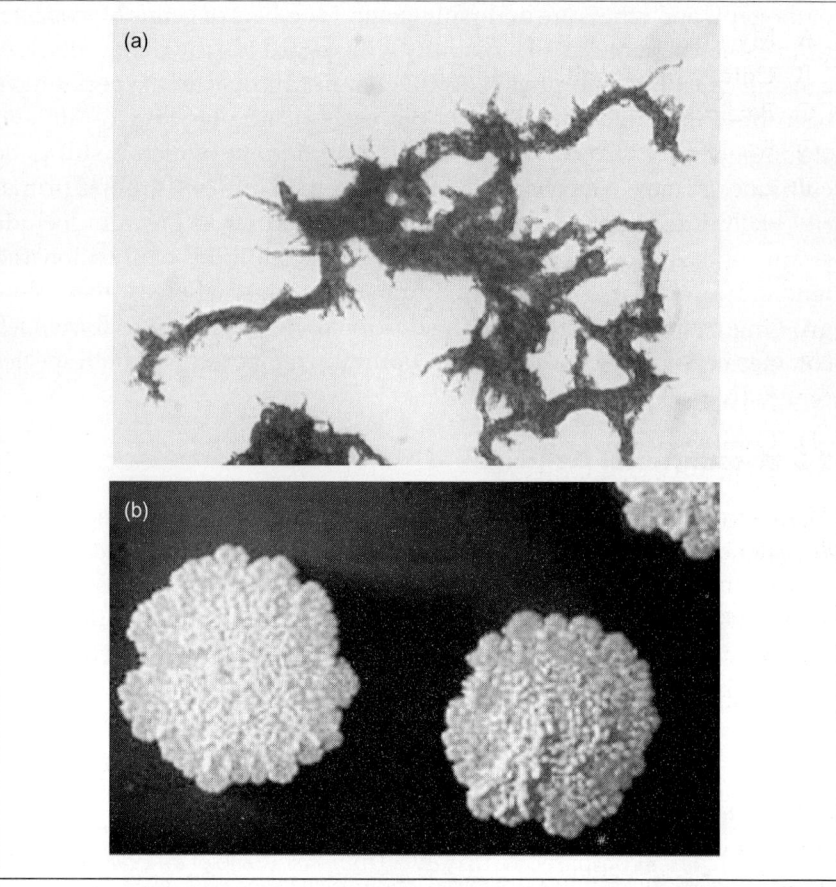

Fig. 2 Growth of *M. tuberculosis* culture on an (a) on Liquid medium and (b) Solid medium

A. Sub Group1: Photomorphogens: These organisms grow slowly and when exposed to light turns yellow, e.g., *M. kansasii, M. mannum, M. simiae.*

B. Sub Group 2: Scotochromogen: These organisms grow slowly and form yellow pigment even in dark. For the culture grown in light the pigment is orange, e.g., *M. scrofulaceum, M. szulgai, M. gordonae.*

C. Sub Group 3: Nonchromogens: Culture of these organisms is colorless and slightly pigmented; growth is slow and inhibited at 22° C but is common at 44°C, e.g., *M. avium, M. intracellularae, M. ulcerans.*

D. Sub group 4: Rapid growers; on primary culture growth may be slow, but in subcultures, growth is evident in 2-3 days. Main organisms in this sub group or category are *M. fortuitum* and *M. chelonae.*

2.4.2 Resistance

A. Mycobacteria can be killed at 60°C for 15-20 min.
B. Culture may be killed on direct sunlight upto 2 hrs.
C. Bacilli may remain viable in droplet nuclei for 8-10 days.
D. Bacilli in sputum may remain active for 20-30 min.
E. Culture may remain viable at room temperature for 6-8 months and at -20°C for 2 years.

2.4.3 Resistance to Chemical Disinfectants

A. Phenol 5%
B. Sulfuric acid 15%
C. Nitric acid 3%
D. Oxalic acid 5%
E. Sodium hypo chloride 4%

For 5 minutes the organism can be destroyed by tincture of iodine and 80% ethanol in 2-10 min. However, 80% ethanol is recommended.

2.4.4 Primary Risk Factor

A. Poverty
B. Malnutrition
C. Over crowding condition (Over population)

TB spreads more easily among family members living in the same house and breathing the same air and spread by aerosol droplets and causes irreversible lung destruction. If it escapes from the lung cause systemic disease affecting many organs including bones, joints, spleen, gastrointestinal tract and brain.

2.5 Epidemiology and its Association with HIV

The reservoir of tuberculosis is primarily humans and infection is acquired by inhalation of the bacilli. These bacilli are suspended as droplet nuclei that result from aerosolization of respiratory secretion of the patients with infectious form of tuberculosis (Fig. 3A, 3B). The incubation period of Mycobacterial infection varies considerably from 4-6 weeks, the latent infection may persist for years. Until recently, tuberculosis all over the world had shown consistent decline. This declining trend of tuberculosis has attributed to the improved socio-economic condition, isolation of infected individuals, improved chemotherapy and prophylaxis. But with the onset of

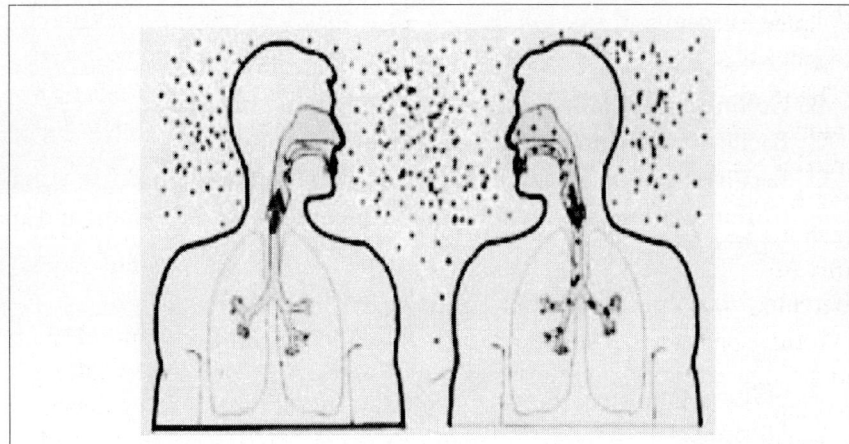

Fig. 3a Transmission of tuberculosis infection from person to person-through aerosol liberate from TB +ve individual inhaled by the normal individual and acquires infection.

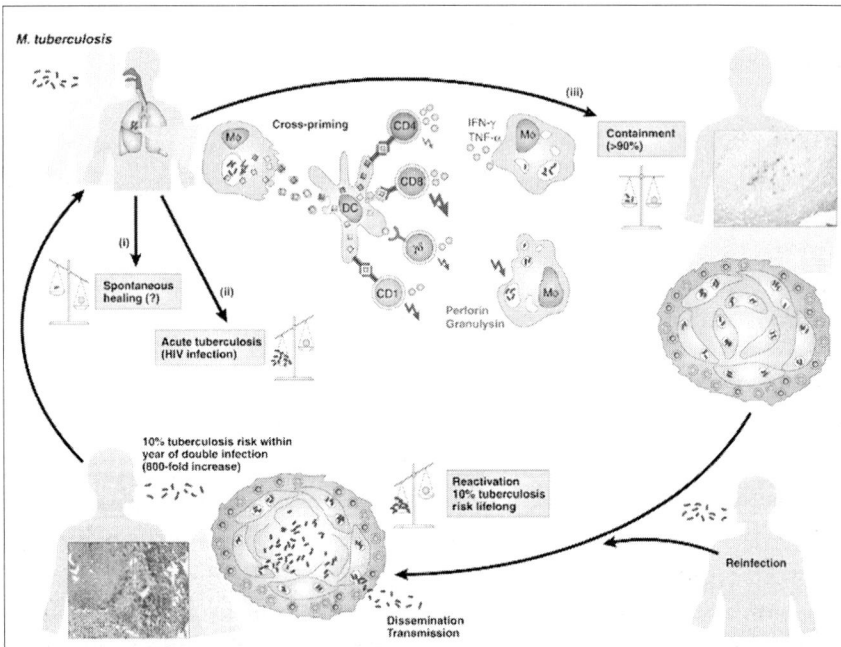

Fig. 3b Different outcome of *M. tuberculosis* infection and underlying immune mechanisms. *M. tuberculosis* enters the host within inhaled droplets.

HIV infection, the incidence of tuberculosis due to reactivation of latent infections has shown an upward trend in HIV positive populations.

The intimate connection between HIV infection and tuberculosis was demonstrated early in Africa and among the intravenous drug user in the United States (Stonburner et al. 1987; De Cock et al. 1992). However, it is now clear worldwide that, tuberculosis is the most common opportunistic infection in AIDS patients. It has been estimated that 20% of HIV patients in Latin America, one-third in Africa and other developing nations have tuberculosis (De Cock et al. 1992).

Worldwide, tuberculosis assumed to be a clinical complication of HIV, and thought to be due to endogenous reactivation of *M. tuberculosis* resulting from the loss of immune control to a latent infection. However, evidences suggest that the risk of primary transmission of tuberculosis by *M. avium, M. intracellulare* mycobacteria had increased over the past few years (Bloom and Murray 1992). Besides this, the increasing Multidrug resistance strains (MDR) of *M.tuberculosis* also pose a threat in primary transmission of disease. Thus, although active tuberculosis and transmission of MDR strain are currently thought to be AIDS associated phenomena, it is possible that both eventually emerge as potential threats in broader population. (Kartikeyan et al. 2007)

2.6 Ecology

It is thought that the well-known infectious agents such as *M. tuberculosis* and *M. leprae* are evolved from a soil bacterium. More specifically, *M. tuberculosis* arose from a soil bacterium that evolved to infect cows, and then transferred to humans about the time of animal domestication. *M. tuberculosis* and *M. leprae* both grow remarkably slow in which *M. tuberculosis* doubles its population every 18-24 hours, while *M. leprae* doubles its population about every 14 days. This extremely long generation time probably contributes to the chronic nature of both diseases.

2.7 Pathology

M. tuberculosis and M. leprae are considered as chronic pathogens, causing diseases that takes months, sometimes years, to develop and without treatment, eventually result in a slow and excruciatingly painful death. Most *Mycobacteria* except these two pathogens do not cause diseases, and hardly typical of the genus, and are consequently called 'wayward sons of honorable parents.' In recent years, with the continuous use of penicillin these Mycobacteria developed resistance against it. These resistant forms of pathogens allowed diseases to become serious threats to humanity once again (Lurie 1964). In 1993 with rates of reported cases of tuberculosis on the rise, the World Health Organization declared it a global emergency and began to make efforts to heighten public awareness (WHO report 1996, 1997, 2002, 2003). In addition to an increase in tuberculosis, it has formed a deadly partnership with the AIDS virus. The two diseases feed off of each other. The depleted immune system caused by AIDS causes increased susceptibility to tuberculosis, which in turn accelerates the progress of AIDS.

Tuberculosis sets up camp in the lungs of its host, where it, for the most part, coexists with its host while lazily following its daily division routine, yet causing no outward symptoms, but rather just accumulating numbers in order to launch its assault on its unsuspecting host. Macrophages that normally ingest pathogens in order to destroy them are made into a cozy home by the tubercle bacillus. Tuberculosis is transmitted from person to person through air, and requires a six to twelve month regimen of at least two drugs to rid its host of the infection. Drug-resistant strains of the pathogen are mainly the result of patients not following directions in taking their medication, and thus do not kill off the disease entirely.

2.8 Pathogenesis

In humans, infection with M. tuberculosis is associated with a wide spectrum of outcomes ranging from containment of infection and development of protective immunity and development of tuberculosis as a consequence of rapid initial mycobacterial replication or reactivation of latent infection after a period of mycobacterial dormancy. The unique ability of M. tuberculosis associated with a pathogenesis allowing such diverse modes of outcome to be dependent on M. tuberculosis virulence, which in turn is comprised of and combines the following three different components. (i) The ability of M. tuberculosis to survive and replicate within host mononuclear phagocytes; (ii) the specific property of M. tuberculosis dormancy, which allows the organism to survive for prolonged periods of time within the tissues of the host; and (iii) the intense interaction of M. tuberculosis with the host which contributes both to pathogenesis and protective immunity (Bloch 1964).

2.9 Infection with *M. tuberculosis*

In an M.tuberculosis-naive individual, subsequent to an aerosol infection with a few (upto 5) bacilli, a primary focus of infection is established which is initially mainly featured by the intracellular multiplication of the organism within the host's most proficient phagocytes, the alveolar macrophages. In fact, it appears that *M. tuberculosis* has adapted to employ any of several cellular molecules abundantly present on the surface of macrophages, such as the complement receptors (CR) (CR3 and CR4), fibronectin receptors, and mannose receptors, to gain access to the intracellular space of these professional "killer" cells (Fig. 4A and B).

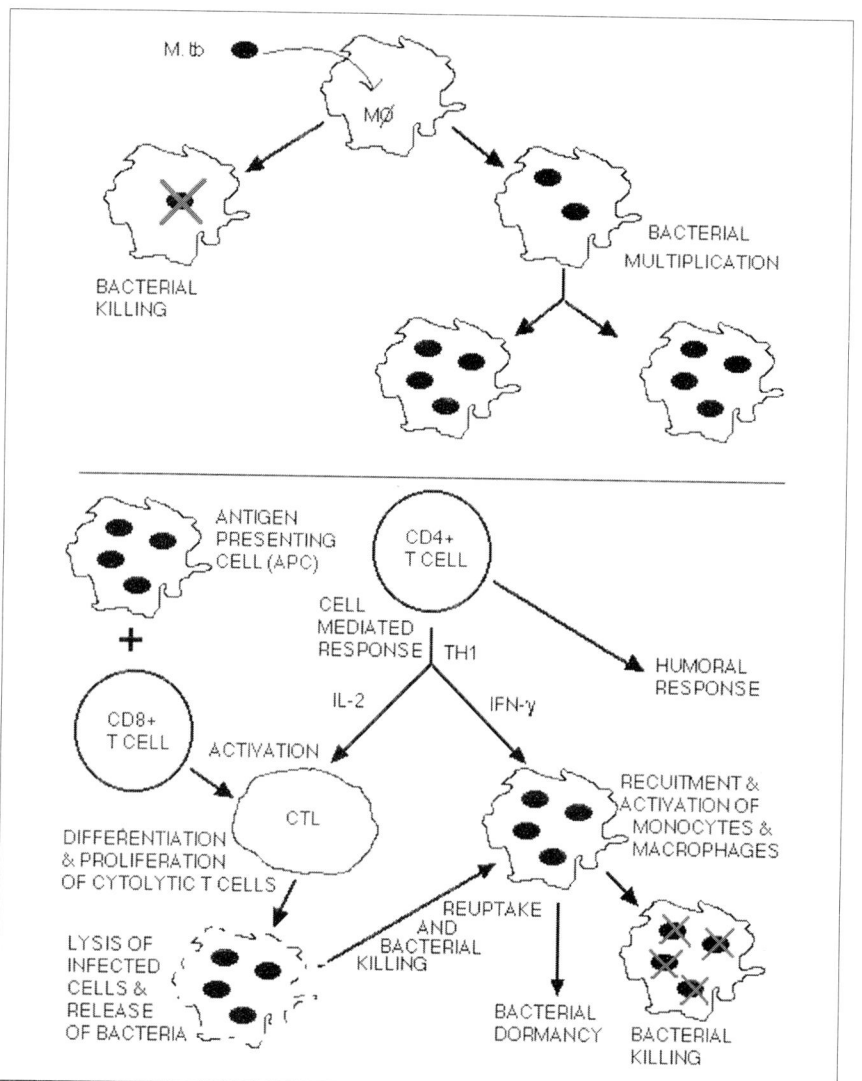

Fig. 4A   Flow chart showing the progression of *M. tuberculosis bacillus* after Inhalation, entry into macrophages and activation of immune.

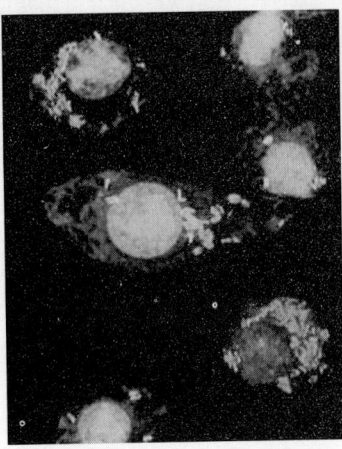

Fig. 4B Infected Mouse macrophages with *M. tuberculosis.*

Through particular modes of intracellular trafficking and a combination of events, which may include alkalinization of the phagosome, prevention of phagosome-lysosome fusion, and escape from the phagosome, *M. tuberculosis* evades powerful intracellular killing mechanisms. Despite induction of several potent macrophages activating molecules such as tumor necrosis factor- α (TNFα) and interleukin-1 (IL-1) by *M. tuberculosis*, intracellular replication ensues. The biological basis for the rapid intracellular replication of *M. tuberculosis* is not fully understood, however, it is most likely multifactorial. Potent macrophage deactivating molecules such as transforming growth factor beta (TGFb) and IL-10 are induced by *M. tuberculosis*, which counteract all known macrophage activating cytokines (such as TNF-a and IFN-g), and microbicidal molecules (reactive oxygen and nitrogen intermediaries). In addition, *M. tuberculosis* is well-equipped to scavenge these potent microbicidal molecules by moieties such as mycobacterial sulfatides and cell wall Lipoarabinomannan. Further, the organism's ability to alter intracellular iron concentrations, as by mycobactins, may allow modulation of the production of macrophage activating cytokines.

Soon after establishment of a focus of *M. tuberculosis* infection, with the recruitment of blood mononuclear cells, the process of *M. tuberculosis* "sensitization" of CD4 and CD8 lymphocytes is initiated. However, due to yet to be defined reasons, protective immunity requires upto 3 weeks to become adequately vigorous to contain *M. tuberculosis* growth. Meanwhile, uncontrolled intracellular replication continues which eventually may culminate in the rupture of infected phagocytes and spread of infection to other cells. Importantly, phagocyte rupture allows the initiation of both extracellular growth, and tissue damage culminating in caseation necrosis. As *M. tuberculosis* growth expands, lymphohematogenous spread allows the seeding of both pulmonary (upper lobes of the lung) and extra pulmonary sites, Mycobacterial replication is controlled, and most infected individuals develop a robust life-long immunity to *M. tuberculosis*.

Protective immunity (Fig. 5) involves the host's capacity to produce T-cell cytokines, that expand *M. tuberculosis* antigen reactive T-cells (IL-2) and induce macrophage activation (IFNγ), and ultimately to develop microbicidal granulomas. The production and activity of cytokines that are crucial to the development of Th1 responses, such as by macrophage IL-12, are critical to the final containment of infection. However, regardless of the development of systemic protective immunity, about 5-10% of *M. tuberculosis*-infected subjects retain the capacity to reactivate mycobacterial growth and development of tuberculosis after a short (1-2 years) or a prolonged (lifetime) period of latency. The ability of *M. tuberculosis* to persist in a dormant state within host tissues is not well-understood, but most probably relates to the capacity of the pathogen to switch its metabolism from a rapid aerobic growth to a slow anaerobic growth. However, the structure of the initial glaucomatous response and/or the nature of cytokines induced by *M. tuberculosis* are important in the initiation or termination of dormancy is not known. However, conditions that weaken cell-mediated immunity increase the chance of terminating the latent state of infection and development of tuberculosis.

The factors that are important in the maintenance of protective immunity against *M. tuberculosis* infection after primary infection (i.e., immunologic surveillance) are not known. In experimental animals, live but not dead organisms induce protective immunity, and two separate T-cell subsets confer protective immunity and delayed type hypersensitivity. It is possible that the sustenance of successful immunological surveillance after *M. tuberculosis* infection is contingent on a dynamic interaction *in situ* (granuloma) which is permissive to the continuous sensitization of *M. tuberculosis* reactive T-cells, which may, in turn, be dependent on the low grade replication of a few remaining bacilli, periodically. Under this scenario, the co-incidence of the breakdown of the immune system by co-conditions (such as HIV infection) with the low grade periodic Mycobacterial replication within "healed" *M. tuberculosis*-infected foci is likely to be conducive to initiation of the process of reactivation and

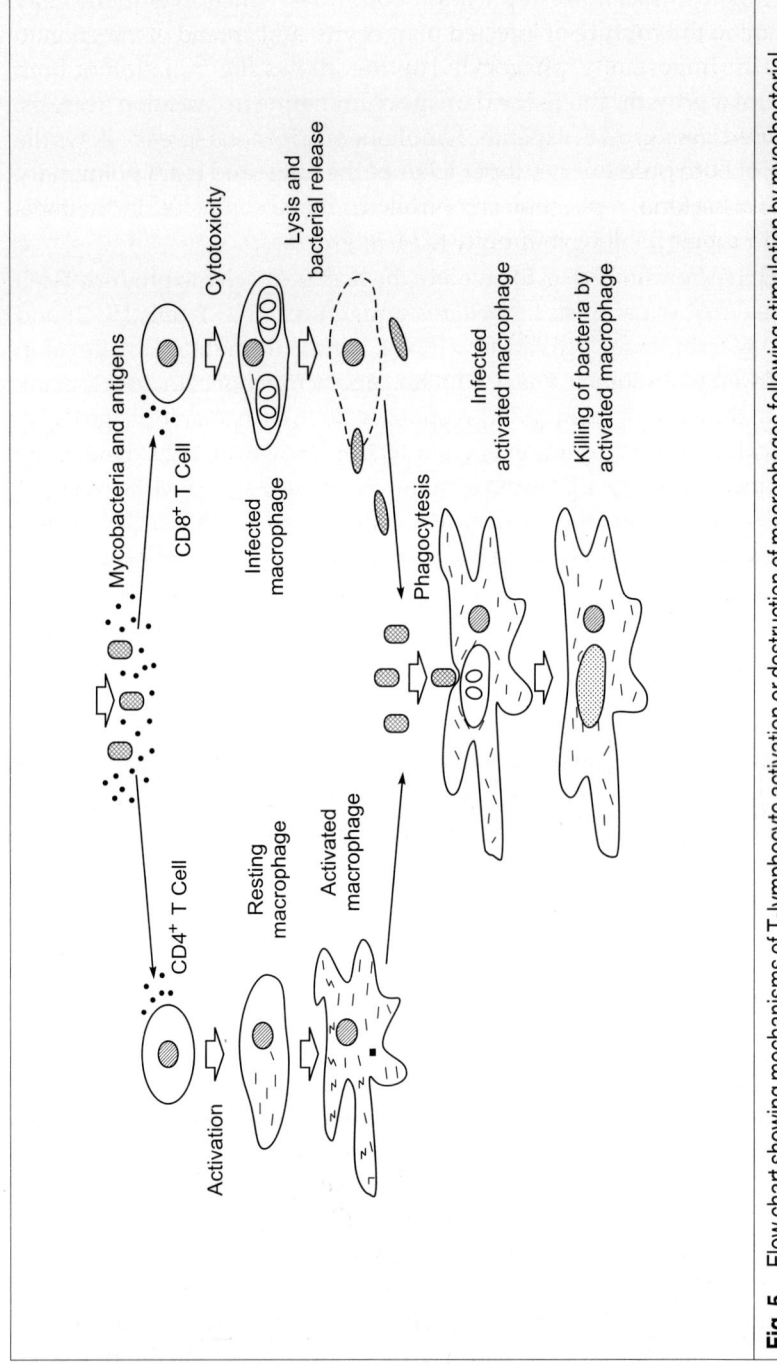

Fig. 5 Flow chart showing mechanisms of T-lymphocyte activation or destruction of macrophages following stimulation by mycobacterial antigens.

exponential growth of *M. tuberculosis*. Whereas the actual events around reactivation are poorly understood, cytokines induced by the organism or its products may be instrumental in this process.

With this regard again the capacity of *M. tuberculosis* to induce cytokines that are suppressive to T-cell function and deactivate macrophages and damage the tissues is noteworthy. Activation of macrophages can result in bacterial killing while cytotoxicity may release bacteria from phagocytes and allow their engulfment and destruction by activated macrophages.

2.9.1 Stage of Tuberculosis Infection

2.9.1.1 Stage-I (Onset: Initial uptake of bacteria)

The first stage begins with inhalation of *M. tuberculosis* in the form of aerosol droplets into the alveolus (Dannenberg et al. 1989) where it is taken up by alveolar macrophages in the lung. This is called early infection and often heals. The macrophages activated by the inhaled particulate material possess potent microbicidal activities, which often arrested the tuberculosis infection at this stage. However, virulent Mycobacteria survive and multiply within weak alveolar cell and then spread to adjacent non-activated macrophages to form a focus of infection.

2.9.1.2 Stage-II (Symbiosis)

This is known as dormant TB and germs remain in the body and may be widespread, but seem to have no effect on the health of their host. Mycobacteria survive in the macrophages, which fail to destroy them and finally burst. Its bacillary load is then ingested by other alveolar macrophages and non-activated macrophages entering from blood stream. A symbiotic relation between macrophage and bacilli develops which lasts between 7-21 days (Alison et al. 1962; Luire 1964). Inside the macrophage, *M. tuberculosis* survive either by avoiding the microbicidal mechanism or due to their inherent resistance to toxic effects by inhibiting the phagosome-lysosome fusion in which mycobacteria survives within the non-acidic vacuoles and lysosome vacuoles that lack protein ATPase pump (Armstrong and Hart 1971; Crowle et al. 1991; Sturgill-Koszycki et al. 1994). The diverse way by which *M. tuberculosis* survives in macrophage and the factors that influence the course of this interaction have been identified which include, bacterial products (Superoxide dismutase) released into the cytoplasm or into the phagosome membrane.

Several surface receptors for mycobacterial binding, including complement receptor CR1 (CD35), CR3 (CD11b/CD18), and CR4 (CD11c/

CD18), the mannose receptor and the antibody Fc receptor (Chan et al. 1989) have been identified. The particular receptor engaged by individual bacterium influences the subsequent intracellular events. The transcriptional regulation associated with different stages of infection has also been identified as one of the key features in the pathogenesis of Mycobacteria following phagocytosis.

2.9.1.3 Stage-III (Initial Caseous Necrosis and Immune response)

In this stage TB may develop infection in lungs and other parts of the body refer to active tuberculosis. The growth of *M. tuberculosis* in macrophages results in the formation of small necrotic lesions with a solid caseous center (2-3 weeks after inhalation of bacilli). The tubercle bacillus survives in the solid caseous lesions, but it apparently cannot multiply there because of anoxic conditions, reduced pH, and the presence of inhibitory fatty acids. At this stage, the cell mediated immune response gets activated which initiates the proliferation of T-lymphocytes and macrophages that accumulate around caseous center to prevent the extension of lesion. Depending upon the host, infection may result in development of weak cell mediated immune response in susceptible host (Dannenberg 1991), which leads to tissue damage resulting in enlargement of caseous lesion. Additionally the tubercle bacilli enter in the draining tracheobronchial lymph nodes and from the small hematogenous lesion, which spreads and extensive caseation develops in many organs resulting in military tuberculosis.

On the other hand, development of (resistant host) strong cell mediated immunity (CMI) results in the inhibition of bacillary multiplication (Dannenberg 1991) by the specifically activated T-lymphocytes, which release IFN-γ that activate macrophages. The cytotoxic cells then destroy the bacilli and the caseous lesion, thereby controlling the progress of the disease.

2.10 Immune Response against Mycobacteria

A typical immunological behavior seen in progression of Mycobacterial disease consists of 3 phases viz., (a) At in first phase, invading Mycobacteria are recognized as being foreign (b) In second phase, the necessary defense mechanisms are staked and employed (c) In the third phase, the actual intracellular struggle between Mycobacteria and Macrophage takes place.

Mycobacteria in common with other intracellular parasites, owe their virulence to their ability to survive within hosts macrophages. The mechanism of uptake of Mycobacteria by mononuclear phagocytic cells influences the efficacy with which the macrophage can kill the bacteria. Uptake of Mycobacteria uses several specific macrophage receptors and is independent of Fc receptors, thereby avoiding triggering an antimicrobial oxidative burst (Warwick et al. 1994). The other way of entry of Mycobacteria to macrophages is mediated by the binding of fibronectin and vitronecting receptors with the fibronectin proteins secreted by Mycobacteria.The exact mechanism by which Mycobacteria survive within macrophages is still a riddle, but Hart & colleagues reported that virulent *M.tuberculosis* can inhibit phagosomal fusion, possibly through the action of surface sulfatides (Warwick et al. 1994).

Although opsonization of Mycobacteria with serum antibody partially reverses the inhibition of fusion virulent *M.tuberculosis* continues to replicate within the fused phagosomes, emphasizing that Mycobacteria have multiple survival mechanisms (Warwick et al. 1994). Mononuclear phagocytes ingest *M. tuberculosis* via compliment receptors (CR'S) such as CR3 (CD11b/CD18) and CR4 (CD11c/CD18), which are the members of the beta-2 (leukocyte) integrin family of cellular adhesions, primarily recognize C3bi-coated particles and a variety of microbial ligands and other glycoproteins. This family of receptors is expressed on all mononuclear phagocytes, assuring access of the pathogens to their intracellular niche. To the potential advantage of the pathogen, ligation of CR3 does not uniformity trigger toxic host cell responses and has recently been shown to selectively suppress interleukin 12 (IL-12) production, an important mediator of the cellular immune response to *M. tuberculosis*.

It has been found that Lipoarabinomannan (LAM) present in Mycobacterial cell walls, a potent inducer of TNF-α in human and murine macrophages (Moreno et al. 1989) plays an important role in the early host response to Mycobacteria and synergies with T-cell derived IFN-γ to stimulate bactericidal mechanisms. There are, however, important differences in the activity of LAM derived from virulent and avirulent *M. tuberculosis*. Purified LAM from the virulent strain H37Rv is a 100 times less potent at inducing secretion of IFN-α and transcription of TNF-α, mRNA *in vitro* than is LAM derived from the avirulent H37Ra strain (Warwick et al. 1994). It is observed that the failure of LAM from virulent *M. tuberculosis* to stimulate these early events in Macrophage activation enables the organisms to multiply within the infected macrophage and so is an important determinant of Mycobacterial virulence.

In summary LAM and Phenolic glycolipid-1 (PGL-1) are well-characterized cell wall glycolipids that contribute to the persistence of *M. tuberculosis* and *M. leprae* both by modulating cytokine effects and by blocking microbial oxygen radicals and studied as serological marker (Sada et al. 1990, 1983).

2.10.1 Role of Humoral Immune Response

M. tuberculosis infection induces humoral response (antibodies) in infected host that are capable of binding to various Mycobacterial antigens; the majority of antibodies have been found to be directed towards the cell wall antigens. Grange 1984 reported the induction of various classes of immunoglobulins in tuberculosis patients to Mycobacterial antigens. The IgM antibodies appear in the initial stages followed by rise and persistence of IgG antibodies. But their role in providing immunity against Mycobacterial infections suggests that antibodies do not protect the host from tuberculosis (Reggiardo and Middlebrook 1974 a,b). However, there have been few reports (Kaplan and Chase 1980) in which it has been shown that antibodies to Mycobacterial antigens lead to the clinical improvement of disease. The high level of antibody, (IgA) concentration in patients has been demonstrated to correlate with the increase in severity of the disease. It may be concluded that contribution of antibodies in inducing immunity against tuberculosis if any, is not clear.

2.10.2 Reactivation of Tuberculosis

Tuberculosis remains a major cause of morbidity and mortality worldwide, infecting approximately one-third of the world's population. In humans, the majority of *M. tuberculosis* infections are initially controlled and a clinically latent infection, characterized by the persistence of low numbers of possibly dormant bacilli is established. Latently infected individuals have an approximately 10% lifetime risk of progressing to a highly contagious disease state called reactivation tuberculosis. In recent years, HIV infection in AIDS with severe CD4 T cell deficiency has emerged as the biggest risk factor for reactivation of latent tuberculosis.

2.11 Antigenic Components of Mycobacteria

The fact that dead bacilli can increase the resistance of animals has led to the speculation that chemical constituents of these bacilli possess the immunogenic properties, depending on their chemical nature. The Mycobacterial antigens have been divided into different categories:

2.11.1 Polysaccharides

Predominant source of polysaccharides is probably the cell wall (Daniel and Janicki 1978). *Arabinoglactan, arabinomannan, mannan, and glucans* are the important polysaccharides recovered from cell walls of Mycobacteria. These polysaccharides do not elicit delayed type skin reaction (Misaki et al. 1977), but antibodies against them have been detected in sera obtained from individuals infected with Mycobacteria. Arabinoglactans from *M. tuberculosis* are known to induce immunosuppression and have been shown to inhibit macrophage function *in vitro* (Ellner and Danial 1979). However, there are no reports for the induction of protective immunity against tuberculosis with polysaccharides.

2.11.2 Lipids

Mycobacterium is rich in lipids comprising of 4% of its dry weight. The chemistry of Mycobacterial lipid constituents and their biological functions have been extensively investigated by (Goren 1982, 1979). Of the various lipids, phospholipids constitute the major polar lipid and comprise 2-4 % of dry weight of cell wall. Major phospholipids include cardiolipin, Phosphatidylmannosides (PIMS) and phosphatidylethanolamine (Subrahmanyam 1964; Mehta and Khuller 1988; Singh and Khuller 1993). PIMS have been reported to be the major phospholipids antigen of Mycobacteria, which elicits both humoral, and cell mediated immune response (Singh and Khuller 1993) when tagged with methylated BSA. This complex has also been observed to induce protective immunity in immunized animals against experimental tuberculosis (Singh and Khuller 1993).

2.11.3 Proteins

The search for improved diagnostic reagents, such as more species-specific skin test antigens and serological markers, has been a driving force in many of the efforts devoted to the fractionation and identification of Mycobacterial protein antigens.

M. *tuberculosis* H37Rv genome comprises 4,411,529 base pairs, around 4000 genes (3924 predicted protein coding genes) and has a very high guanine + cytosine content (65.5%) that is reflected in the blast amino acid content of the proteins. *M. tuberculosis* radically differs from other bacteria in that a very large portion of its coding capacity is developed to the enzymes involved in lipogenesis and lipolysis (Cole et al. 1998).

Koch 1891 made the first attempt to purify the components that were released in culture medium of *M. tuberculosis* which is referred as "Kochs old tuberculin" prepared from the 8-week old medium and used for diagnosis of tuberculosis. It elicits a delayed type of skin hypersensitivity (DTH) reaction called Koch's phenomenon. Later, Siebert and Munday (1932) isolated the culture filtrate protein from *M. tuberculosis* by ammonium sulphate precipitation and termed as Tuberculin purified protein derivatives (PPD) which is still used for skin testing of tuberculosis infection.

Numerous scientists looked for diagnostic reagents that would be able to differentiate between infections with live actively replicating mycobacteria and those sensitized with killed or dormant mycobacteria. In a pioneer study, the culture filtrate proteins of young BCG cultures were fractionated by their chromatographic principle (size, charge and hydrophobicity) and the screening of the fractions was based on comparative analysis of either serological responses or DTH response of two groups of guinea pigs immunized with either live or killed BCG. They identified them as having the ability to differentiate infection from, sensitization in DTH reaction (Roman et al. 1993 a, b).

As gene expression varies according to culture conditions, a shift in temperature or the addition of a physiological stress stimulus like ethanol will increase the synthesis of a certain set of stress proteins (Young and Garbe 1991a). The cultivation of *M. tuberculosis* in medium deficient of Zn or phosphate will induce the synthesis of yet another set of proteins (Anderson et al. 1990; Espitia et al. 1992).

By use of 2D gel electrophoresis, a 58-kDa (novel) protein inducing significant release of TNF-α from human macrophages was identified (Wallis et al. 1993). TNF-α is a potent macrophage activating cytokine and may, therefore, be important in mycobacterial immunity and pathogenicity. The influence of surface exposed antigens is of major importance in the immune responses to mycobacteria.

A 40 KDa protein antigen of *M. tuberculosis* was identified by sequence homology and functional analysis to be an L-alanine dehydrogenase. The protein was initially identified by monoclonal antibody HBT-10 that binds to the *M. tuberculosis* strains and *M. marinum* but not to BCG (Ljungquist et al. 1988). HBT-10 reactive protein found to be secreted by *M. tuberculosis* despite the fact that no signal peptide was identified (Anderson et al. 1992a). The heat shock protein encoded by *hsp* genes have all been shown to interact with the immune system of animals as well as humans. 65 KDa *hsp* proteins have been mapped at the epitope level with respect to both B and T lymphocyte recognition (Kale et al. 1990; Munk et al. 1990).

A direct approach to the study of membrane-integrated proteins was made by the isolation of a 19 KDa protein from membrane fractions of *M. tuberculosis* (Lee et al. 1992). The amino acid sequence appeared unique and immunoblotting results showed the proteins to be immunogenic and present in substantial amounts in virulent *M. tuberculosis* strains.

Lipoproteins are generally associated with the cell surfaces and are known in *M. tuberculosis* as well. Young and Garbe (1991a) demonstrated (3H) palmitate microporation in four *M. tuberculosis* proteins of molecular masses 19, 26, 27 and 38KDa indicating that these proteins are post translationally modified by acylation. The function of the 38-kDa proteins of *M. tuberculosis* is to bind and make phosphate available to the bacteria. The protein is anchored in the cell wall by lipid moiety. Another protein identified as a lipoprotein is the 19-kDa protein. This protein is structurally much more conserved within the Mycobacterial species as judged by the cross reactivity of the monoclonal antibody. 38 KDa proteins from *M. tuberculosis* predominately possess species-specific epitopes, because hyperimmune sera from rabbit immunized with mycobacteria other than *M. tuberculosis* do not react with the affinity purified 38 KDa proteins from *M. tuberculosis*. Peptides derived from this molecule in preliminary studies have shown some potential as an alternative for tuberculin purified protein derivatives (PPD) (Vordermeier et al. 1991, 1992).

The 30/31 KDa proteins are exported to the exterior due to true signal sequences that are cleaved off before release of the protein to the exterior but a fraction of the protein remains attached at least temporarily to the bacterial cell surface (Rambukkana et al. 1991). Several researchers have assessed diagnostic identity of this protein complex. Some of the studies have been conducted with affinity-purified material consisting of all three components of 30/31 KDa proteins. 30/31 KDa protein complex helps mycobacteria to bind with fibronectin coated surfaces. Antigen 85 complex proteins are major secretory products of *M. tuberculosis* and induce strong cellular and humoral immune responses in infected experimental animals and human beings. Also, it has been reported that antigen 85 (30/31 Kda) circulates as complexes with fibronectin and Immunoglobulin G (Bentley et al. 1999) in the serum of an infected subject. Another major component of the secretory protein product of *M.tuberculosis* is a protein with molecular weight of 24 KDa. This protein elicits strong DTH responses.

2.12 Virulence Factors of *M. tuberculosis*

M. tuberculosis adopts different modes to survive inside professional phagocytes and avoid the different defence mechanisms of the host. After interacting with receptors at the macrophage surface, *M. tuberculosis* is phagocytosed and remains in a phagosome that does not fuse with lysosomes and does not acidify. The exclusion of the proton ATPase and the retention of the TACO protein on the phagosomal membrane might explain this phenomenon. Also the route of entry of the bacterium into the cell might be important. It involves the opsonic interaction with C3b following the association of C2a with *M. tuberculosis* and its subsequent cleavage in C3 convertase in the absence of other complement factors. Infection of macrophages by *M. tuberculosis* leads to down regulation of the antigen-presenting molecule CD1 from the cell surface of antigen presenting cells, and to an inhibition of antigen presentation by the MHC-II molecules.

An inhibition of the synthesis of IL12, which is required for the induction of Th1 response, is also observed, as well as an inhibition of the activation of bactericidal responses by IFN-α. The mycobacterial factors that are responsible for above phenomena are unknown. A search for these factors was undertaken through the isolation of mutants impaired in *in vitro* growth. Genetic tools were developed enabling the transfer of DNA into mycobacteria and the inactivation of genes, either by allelic replacement or transposon mutagenesis. The construction of mutant libraries after Signature Transposon Mutagenesis (STM) allowed the screening of avirulent mutants in a mouse model. Using these technique, thirteen different virulence loci was identified. Three of them are located in a cluster of genes that are responsible for the synthesis of phthiocerols and mycocerosic acid or their transport to the bacterial cell surface. In addition, putative genes coding for a transcription regulator, a lipase and transporters were also found to play a role in virulence. In another set of experiments, the inactivation (by allelic replacement) of gene erp, which encodes a surface antigen, was found to lead also to an avirulent phenotype. The roles of these different gene products in virulence mechanisms still remain to be understood. However, several applications can already be considered from the identification of these virulence loci. Several of the mutants are also avirulent in guinea pigs and provide a level of protection similar to that conferred by BCG against *M. tuberculosis* infection.

2.12.1 Cell Envelope

The Gram-positive bacterium with a G + C rich (65.6%) genome, contain an additional layer beyond the peptidoglycan that is exceptionally rich is usual lipids, glycolipids and polysaccharides. Novel biosynthetic pathways generate cell wall components such as Mycolic acids, Phenothiocerol, Lipoarabinomannans and arabinogalactan and several these may contribute to Mycobacterial longevity, trigger inflammatory host reactions, and act in pathogenesis.

The Mycobacterial cell envelope is considered to contain a unique outer membrane permeability barrier on a covalently bound monolayer of mycolic acids, waxes and biologically- active glycolipids (Reggiardo et al. 1974a,b). Little is known about the mechanisms involved in life within the macrophase or the extent and nature of the virulence factors produced by the bacillus and their contribution to disease.

2.12.1.1 Structure and function of the mycobacterial cell envelope

The mycobacterial cell envelope consists of an outer layer, a cell wall and a plasma membrane. Pathogenic mycobacterial species have the capacity to survive in macrophages of the infected hosts. In connection with this observation, is the presence around pathogenic mycobacteria of an outer layer, also called capsule that appears in electron microscopy as an electron transparent zone. This capsule may play an important role in the protection of bacilli against the toxic effects of some enzymes and other detrimental substances produced by phagocytes. This capsule was assumed to be composed of lipids until recently. In order to gain a new insight into the relationships between pathogenic mycobacteria and the infected hosts we reinvestigated the chemical nature of the outermost layer of the mycobacterial envelope which comes into contact with the cells and tissues of the host and controls access to the interior of the bacteria by nutrients and drugs. The outermost components of mycobacterial envelope consist most exclusively of proteins and polysaccharides; small amounts of lipids, including the ubiquitous Phosphatidylinositol mannosides (PIM), are also surface-located.

The majority of the cell envelope lipids are located in deeper compartments of the capsule. A high-molecular weight α-glucan, Arabinomannan and Mannan were structurally characterized. Analysis of enzyme activities present in the capsule of pathogenic and non-pathogenic mycobacterial species demonstrated the presence of selective key enzymes on the surface of *M. tuberculosis* and *M. bovis*; these include superoxide dismutase, catalase, peroxidase and glutamine synthetase, such enzymes may play a role in the pathogenicity of the tubercle bacillus. These enzymes were absent from the surface of *M. smegmatis* but were detected in inner compartments of the capsule, suggesting that the architecture of the cell envelope may be an important factor in the pathogenicity. Mycolic acids, high molecular weight (C60-C90) α-alkyl, β-hydroxy fatty acids, are specific and major substances produced by all mycobacteria; they are principally found covalently-linked to the bacterial cell wall whose lipids are arranged to form a bilayer and which forms a permeability barrier of extremely low fluidity distinct from the plasma membrane.

To determine the role of mycolic acids in the cell wall diffusion barrier, mutants of *M. tuberculosis*, which make less amounts of cell wall-linked mycolic acids or produce only α-mycolic acids have been constructed and analyzed for the fluidity of their cell walls. A mutant that elaborates 40% less covalently linked mycolic acids than the parent strain takes up more rapidly radiolabelled glycerol and chenodeoxycholate. A mutant that produces no oxygenated mycolic acids at all shows profound alterations in the fluidity of its envelope and is attenuated in the mouse model of infection. Altogether, these data clearly demonstrate that the cell wall mycolates of mycobacteria represent a permeability barrier for nutrients and may play an important role in adverse conditions.

2.12.1.2 Cell wall

The cell wall of *M. tuberculosis* is similar to that of other Mycobacteria because it has a thickness of about 20 nm as seen through an electron microscope, different than other Gram-negative and Gram-positive bacteria (Figs. 6a, b, c), it also appears to consist of an inner, electron dense layer surrounded by an outer electron transparent layer (Payne et al. 1982).

M. tuberculosis is capable of synthesizing a variety of other compounds associated with outer layer of cell wall, these include sulfatides and sulfated Trehalose, which might be involved in the virulence of *M. tuberculosis* because they cause the prevalence of phagosome-lysosome fusion in macrophages infected with *M. tuberculosis*. Because of the cell wall's thickness and high lipid content, the wall is mostly impermeable to hydrophilic molecules. However, the presence of proteins in the walls, act as pores through which some hydrophilic molecules are allowed to pass.

Because *M. tuberculosis* is not a toxin producer, its pathogenicity in humans is linked to its capacity to survive and replicate inside the human host, as well as its ability to avoid the host's immune system. Mycobacteria, Nocardia and corynobacteria are unique in that they are able to synthesize Mycolic acid, which are found only in certain components of the cells, specially the cell wall, TM, TD and glycerol mycolates. TD from Mycobacteria is a toxic glycolipid, which affects the function of the mitochondrial membrane by inducing a decrease in respiration and a loss of oxidative phosphorylation. This action might be partly responsible for the pathogenicity of *M. tuberculosis* (James et al. 1982).

2.12.1.3 Acid fastness of mycobacterial cell wall

A bacterial cell wall is composed of a thin, inner layer of peptidoglycan and large amount of glycolipids such as mycolic acid, arabinogalactan-lipid

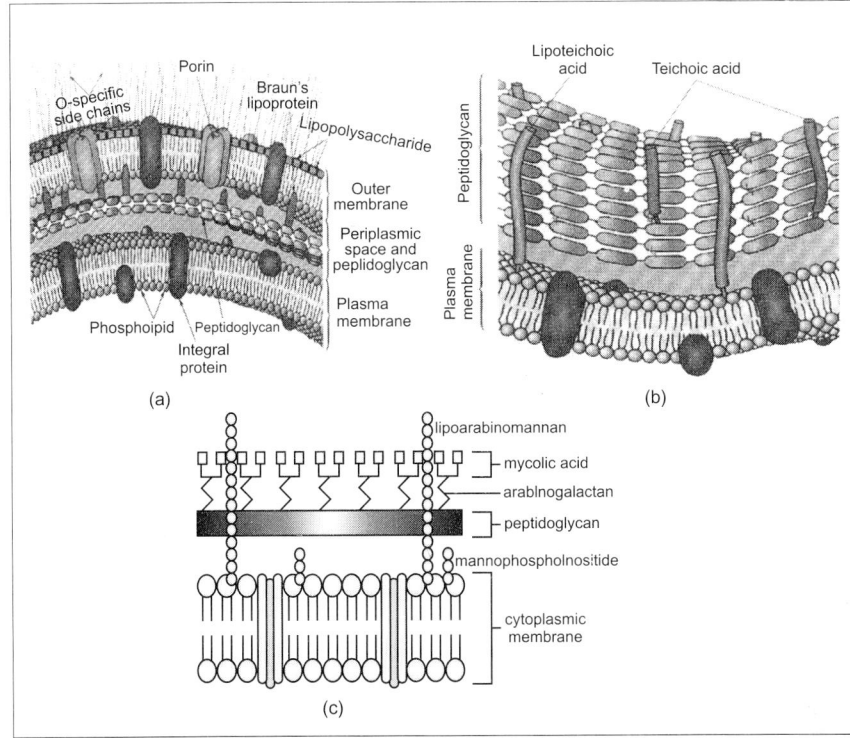

Fig. 6 Structure of: (a) Gram-negative, (b) Gram-positive bacterial cell wall and (c) *M. tuberculosis* (Acid Fast) cell wall.

complex, and Lipoarabinomannan. During the acid-fast staining procedure, the acid-fast cell wall enables the bacterium to resist decolorization (Fig. 7) with acid alcohol and retain the original stain carbol fuchsin. Bacteria with an acid-fast cell wall resist decolorization with an acid-alcohol mixture during the acid-fast stain procedure, retain the initial dye carbol fuchsin and appear. Common acid-fast bacteria of medical importance include *M. tuberculosis*, *M. leprae*, and *M. avium-intracellulare complex*.

2.12.2 Structure

The acid-fast cell wall contains only small amounts of peptidoglycan but large amounts of glycolipids such as mycolic acid. A waxy lipid called mycolic acid makes up approximately 60% of the wall and makes the cell wall relatively impermeable (Peptidoglycan - prevents osmotic lysis).

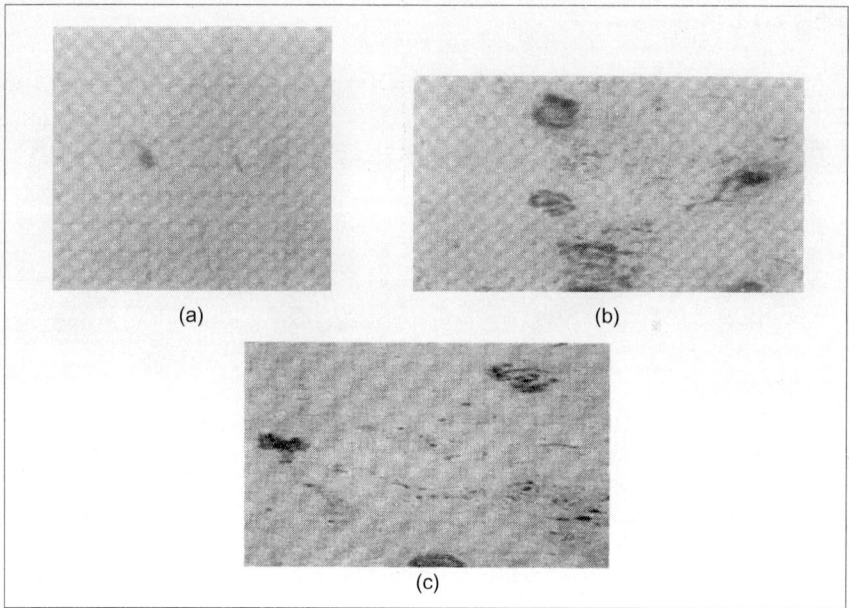

(a)

(b)

(c)

Fig. 7   Diagram of acid-fast stained smear showing pink colored bacilli of *M. tuberculosis*

The cell wall structure of *M. tuberculosis* deserves special attention because it is unique among prokaryotes and it is a major determinant of virulence for the bacteria. The complex cell wall contains peptidoglycan, but otherwise it is composed of complex lipids. Over 60% of the Mycobacterial cell wall is lipid. The cell wall consists of three major components:

2.12.2.1 Mycolic acid and other glycolipids

The mycolic acid and other glycolipids make the acid-fast cell wall rather impermeable and impede the entry of chemicals. This causes the bacteria to grow slowly and be more resistant to chemical agents and lysosomal components of phagocytes than most bacteria. They are unique alpha-branched lipids found in cell walls of Mycobacterium. They make up 50% of the dry weight of the Mycobacterial cell envelope. Mycolic acids are strong hydrophobic molecules that form a lipid shell around the organism and affect permeability properties at the cell surface. Mycolic acids are thought to be a significant determinant of virulence in *M. tuberculosis*; probably, they prevent attack of the Mycobacteria by cationic proteins, lysozyme and oxygen radicals in the phagocytic granule. They also protect extracellular Mycobacteria from complement deposition in serum.

2.12.2.2 Cord factor

It is a factor, which is most abundantly produced in the virulent strains of *M. tuberculosis* responsible for the serpentine cording and is toxic to mammalian cells. Cord factor is most abundantly produced in virulent strains of *M. tuberculosis*.

2.12.2.3 Wax-D

Wax-D in the cell envelope of *M. tuberculosis* is the major component of Freunds Complete Adjuvant (CFA). The high concentration of lipids in the cell wall of *M. tuberculosis* has been associated with these properties of the bacterium:

- Impermeability to stains and dyes
- Resistance to many antibiotics
- Resistance to killing by acidic and alkaline compounds
- Resistance to osmotic lysis via complement deposition
- Resistance to lethal oxidation and survival, inside the macrophages.

2.12.3 Lipids Composition and Virulence

Attempts to investigate the biochemical basis of pathogenicity of tubercle bacilli have suggested that lipids are possibly responsible for tuberculosis infection (Bloch et al. 1964) and morbid effects of Mycobacteria. Similarly, the lipid components of several other pathogens have been found to be correlated to virulence (Cox and Best 1972), and specific lipid components, viz., cord factor and sulpholipid, have been considered to be toxic to animal tissue (Goren and Brennan 1979).

The first step of infection in intracellular survival of the phagocytic cell has been reported to be regulated by cell surface glycolipid virulence factor which may lead to a novel approach for development of TB vaccine in future (Steven et al. 2000).

3. DIAGNOSIS

Current techniques for the diagnosis of *Mycobacterium tuberculosis* are beset by serious limitations and the rapid and specific diagnosis of this disease is one of the most pressing needs in efforts to control and eradicate it. The diagnosis of this devastating disease often relies on clinical acumen, and there is a serious scarcity of reliable tools to make an authentic diagnosis. The current methods used in clinical laboratories are growth dependent and may even take upto 6-8 weeks to report negative / positive results.

With such worst epidemic looming large on the earth, and an enormous amount of search since the time of Koch, we still have no simple, sensitive, and specific test that will differentiate most or all patients with active tuberculosis, from those with quiescent lesions, previous vaccination or other diseases or even from those who are completely healthy (Grange and Loszlo 1990). The timely and accurate identification of persons infected with *Mycobacterium tuberculosis* (screening) and the rapid laboratory confirmation of tuberculosis are two key ingredients of effective public health measures to combat the tuberculosis epidemic. Attempts to control tuberculosis disease are severely hampered by current diagnostic procedures.

Tuberculosis (TB) of the central nervous system (CNS) is the most serious complication of extra-pulmonary tuberculosis. The granulomatous infection of CNS caused by *Mycobacterium tuberculosis* (MTB) predominantly involves the brain and meningitis, but occasionally it affects the spinal cord. CNS tuberculosis may take several forms, including tubercular meningitis (TBM), tuberculomas, tubercular abscesses and rarely myeloradiculopathy. TBM is the most common clinical manifestation of involvement of the CNS in TB (Health India 2001). Before the era of human immunodeficiency virus (HIV), TBM affected children, in particular, those living in developing countries. In those countries HIV infection has made young adults the target population for TBM (Thwaites et al. 2000).

According to the WHO, globally more than 1.3 million new pediatric TB cases are noticed each year leading to 170,000 deaths (aged <1-5 years) annually (WHO Report 2002).

The timely diagnosis is key to the management of TBM: the case fatality rate for untreated TMB is almost 100%, and delays in treatment often lead to permanent neurological damage. The major cause of morbidity and mortality of TBM include communication hydrocephalus, vasculitis with resulting infarction, and ventriculitis (Katrak et al. 2000).

Although TB is a curable disease with the use of appropriate antibiotics, the major hurdle in the treatment lies in the late diagnosis of the disease due to unavailability of simple and cost-effective diagnostic products. Thus, there is an urgent need to develop strategies to effectively tackle its detection, and thereafter its treatment which can be overcome through the simple rapid and cost-effective diagnostic test so that the treatment of TB and prevention of *M. tuberculosis* transmission can be reduced (Tiwari et al. 2005, 2006).

Due to unavailability of reliable diagnostic measures especially for tuberculosis, clinicians still rely on conventional methods like ZN (Ziehl-Neelsen) staining/flurochrome staining, sputum culture, and other phenetic criteria, which are shown quite helpful, but they have their own limitations. The reasons for this are manifold. For microscopy, a minimum number of 10^4/ml bacilli are required and cannot discriminate *M. tuberculosis* from other *Mycobacteria*. Of the estimated 8 million new cases of tuberculosis/year, some are either smear negative pulmonary or extra-pulmonary infection, and hence in many cases diagnosis cannot be confirmed at the time of presentation (Nunn et al. 1994). As mentioned above, sputum culture is most reliable method, but it is lengthy, cumbersome and requires mycobacterial culture facility.

Radiometric liquid (BACTEC) and biphasic (MB chek) culture system have improved both the recovery rates and speed of isolation, but these systems still cannot influence beside decision-making. The most commonly used procedure is 'Mantoux test' which doesn't make a conclusive evidence of active disease, while negative tests do not exclude it. Positive tuberculin test may be due to active tuberculosis, past infection, past BCG vaccination or sensitization by environmental Mycobacteria. So this test will be more helpful where BCG vaccination is no longer used routinely. In Indian context, these type of tests hardly hold the ground good.

Other tests based on hematological features such as a high erythrocyte sedimentation rate (ESR), a mild anemia, elevated lymphocyte count, assay of acute phase reactants are no more than suggestive. The introduction of fibreoptic bronchoscope has rendered lung biopsy as well as bronchial lavage and brushing a simple and safe procedure, although unfortunately the equipment are not universally available.

Recently, with the advent of recombinant DNA technology people have been trying to develop the kits based on polymerase chain reaction (PCR) in diagnosis of tuberculosis. A number of investigators have reported the detection of specific sequence for *Mycobacterium tuberculosis* directly in clinical specimen by PCR method (Miller et al. 1994). It has been found that a significant proportion of respiratory samples and patients with pulmonary tuberculosis, who are subsequently culture positive are initially smeared negative. Pulmonary tuberculosis in two-thirds of such patients can be diagnosed by PCR, but despite its sensitivity, it is highly expensive (e.g., Amplicor PCR is done at the cost of $15/test Bennedsen et al. 1996). Hence, the application of PCR based diagnosis is beset with the need for expertise and is also very costly. Moreover, the amplification of dead bacterial DNA and absence of amplifyable *Mycobacterium tuberculosis* DNA from blood is a problem, except for HIV cases with bacterimia. Therefore, the development of a cost-effective sero diagnostic test for TB is a need of the hour, which can be easily incorporated into a clinical laboratory routine.

Added to this is the problem of cross-reaction to atypical Mycobacteria and individual host immune response. The aspiration of gastric contents for examination by smear culture can't be employed on a large scale. Gastric lavage method can't be used because of the difficulties in collecting proper swab specimens.

A murine monoclonal antibody against LAM was used for detection of antigen in sputum samples by capture ELISA. Similarly, an Immunodiagnostic test for the detection of soluble non-protein Mycobacterial antigen by reverse passive haemagglutination with IgM murine monoclonal antibody has been exploited (Chandramuki et al. 1985). The detection of IgG antibody to *M. tuberculosis* antigen-5 by indirect ELISA was done (Radhakrishnan et al. 1990). A Rabbit IgG to *M.tuberculosis* as the primary antibody were used for the development of a simple immunocytochemical method for the demonstration of Mycobacterial antigen in cerebrospinal fluid (CSF) specimens of patients with tuberculosis meningitis. Rapid dot immunobinding assay (Dot-Iba) was developed while using a culture proven specimen (CSF) from the patients, coupled with activated cynogen bromide sepharose 4B and affinity purified Rabbit antibody to 14 KD antigens was used for the development of Dot-Iba in nitrocellulose membrane (Sumi et al. 1999). The diagnostic potential of three immunological tests, namely, detection of H37 Rv antigens of *M. tuberculosis* in CSF, detection of antibodies (IgG) against H37Rv in CSF and detection of antibodies (IgG) against H37Rv in serum for diagnosis of tuberculosis meningitis in children were developed by Dot ELISA (Shrivastava et al.1998). Although various attempts have been made for diagnosis of pulmonary and extra-pulmonary tuberculosis but none hold the ground satisfied (Garg et al. 2003). Since all of the them use antigens that are not specific to *M. tuberculosis* complex but share antigenic similarity with other members of the genus.

The emergences of epidemic multidrug resistant strains of *Mycobacterium tuberculosis* in conjunction with HIV infection represent a global major health problem. Globally, 8 million new cases of tuberculosis currently diagnosed annually with a mortality rate of 3 million (Kochi 1991). In recent years, a large number of disease associated Mycobacterial species have been identified (Wolinsky et al. 1973), which include two major pathogenic organisms, namely, *Mycobacterium tuberculosis* and *Mycobacterium leprae*, which are known to form tuberculosis and leprosy in human, respectively.

Mycobacteria are complex unicellular organisms having resilient cell wall structure and can suppress host immune response by the action of their immunomodulatory cell wall constituents. It is reported that Mycobacteria initially infect the macrophage and start their metabolic process within the cells. The lipid rich outer surrounding capsule protects bacteria from the toxic radicals and hydrolytic enzyme produced by macrophage. The cell envelop of *Mycobacterium tuberculosis*, a Gram-positive (acid fast) bacteria with G+C rich genome comprises 4,411529 base pairs, around 4000 genes with an additional layer beyond the peptidoglycane that is exceptionally rich in unusual lipids, glycolipid and polysaccharides (Cole et al. 1998). This layer protects the *M. tuberculosis* from the hydrolytic enzymes and toxic radicals produced by macrophages.

Since, *Mycobacterium tuberculosis* is one of the small groups of species and able to survive inside the phagocytic cell of the host, it is likely that its envelope has special properties, which defend the bacteria against host Microbicidal process. The reason of epidemic problem created by *Mycobacterium tuberculosis* is probably associated with the massive amount of complex lipid present in its cell wall. It is reported that the antigen of Mycobacteria are largely in the cell wall, which has complex structure with 30% of lipid contents and 20 μM thickness.

The first step of infection in intracellular survival is the phagocytic cell and this process has been regulated by cell surface glycolipid virulence factor based on genetically heterogeneity of *M. tuberculosis* among the strains as per DNA fragmentation pattern. These glycolipids have been found to have high serodiagnosis discriminating activity (sensitivity and specificity) and these may play a role in pathogenicity and virulence of intracellular pathogens (Chan et al. 1989; Suit et al. 1990).

Tuberculosis meningitis (TBM), which is one of the common clinical manifestations of extrapulmonary tuberculosis, has shown upward trend in developing countries (Sumi et al. 1999). The granulomatous infection of CNS caused by *M. tuberculosis* (MTB) predominantly involves the brain and meningitis, but occasionally it affects the spinal cord. TBM is the most common clinical manifestation of tuberculosis (Health India 2001). Before the era of human immunodeficiency virus (HIV), TBM affected children, in particular, those living in developing countries. In those countries HIV infection has made young adults the target population for TBM (Thwaites et al. 2000). The rate of TB and TBM have increased globally especially in developing countries of Africa and Asia that are affected by the HIV pandemic. TBM is more common in patients who are immunosuppressed, such as older persons, young children, patients with HIV, patients with diabetes and patients taking steroids or cytotoxic drugs (Garg et al. 1999).

The case fatality rate for untreated TMB is almost 100%, and delays in treatment often lead to permanent *neurological* damage (Katrak et al.2000).

Therefore, the timely diagnosis of *M. tuberculosis* from the clinical samples is the key to effective management of tuberculosis and more so extra-pulmonary tuberculosis, especially of TBM.

Several serodiagnostic tests have been developed for the diagnosis of disease.

3.1 Serological Test based on Protein Antigens

Mycobacterium consists of multiple antigenic components that have been difficult to separate and characterize. Interests in developing a reliably specific and sensitive serodiagnostic test encouraged the initial searches for tuberculosis specific monoclonal antibody. Preliminary approaches firstly identified the 38 kDa, 19 kDa and 16 kDa proteins as prominent immunogens (Ivanyi et al. 1985). Antibody to 38 kDa antigens are elevated in multibacillary pulmonary tuberculosis patient, while antibodies to HSP 71 have been found equally increased in sputum smear negative pulmonary disease patients (Elsaghier et al. 1991,1992). Antibodies to the 16 kDa antigens were selectively increased in chronically exposed household contacts of patients or hospital workers (Jacktt et al. 1988). Antibodies to Lipoarabinomannan LAM and the 16 kDa antigens were elevated in the cerebrospinal fluid of the patients with tuberculosis meningitis (Chandramukhi et al. 1989). Recurrent and extensive radiographic pulmonary tuberculosis with poor prognosis was associated with high anti 38 KDa and low anti 16 kDa antibody levels. Patients with less pulmonary cavitations had high anti 19 kDa titer, while bacteriological relapse, during treatment was indicated by a rise in antibodies to 16 kDa antigens.

Though antigen of first choice for diagnostic is undoubtedly the 38 kDa protein so far, but patients with *Lepromatous leprosy* show considerable amount of anti 38 kDa antibody (Bothamley et al. 1991). Of other antigens, the 19 kDa protein is compromised by its cross reactivity with *M. avium* (Nair et al. 1992), while antibody to the 16 kDa protein can be identified also in a significant fraction of infected healthy subjects. Antibodies to 30-31 kDa fibronectin binding proteins, which are cross reactive with other species of Mycobacteria, have been demonstrated in two-thirds of the tuberculosis patients (Turner et al. 1988; Espitia et al. 1992; Van voren et al. 1992).

The pronounced immunogenicity of HSP 71 during paucibacillary infection could be attributed to enhance secretion during intracellular replication, to surface expression enabling recognition by B cells and/or to the adjuvant independence of this antigen (Barris et al. 1992). Antibodies to HSP 71 are directed to both linear and conformational epitopes localized at the polymorphic and largely species-specific carboxyl terminal part of molecules (Elsaghier et al. 1992). The proportion between antibodies to the respective epitopes is yet to be determined, but the bulk of the antibodies to most antigens in the sera of tuberculosis patients thought to be directed to conformational epitopes (Verbon et al. 1992).

There have been numerous unsuccessful attempts to develop clinically useful serodiagnostic kit for tuberculosis. Although the large number of published reports clearly show that antibody levels are significantly higher in patients as a group than in a population, little consideration is given to the value of tests in various operational situations. The crucial factor from the diagnostic point of view is the degree of overlap between those with and without active disease. In the case of tuberculosis, there is apparently no dominant specific antigen, indeed most of the antibody response is directed towards shared mycobacterial antigens (Granze et al. 1990). A combination of tests using several antigens may be needed in view of the large individual variations demonstrable by western blotting analysis (Verbon et al. 1990).

An Enzyme Linked Immunosorbent Assay for detection of serum IgG antibodies, based on Antigen 60 was devised. The false positive rate in non-tuberculous group was 29.4%. The positive rate in active tuberculous group was 83.8% (Wang et al. 1994). A 38 kDa protein antigen from *M. tuberculosis* was purified by monoclonal antibody Moab TB71-based affinity column. This molecule carried two overlapping epitopes recognized by monoclonal antibodies TB-71 and TB-72, which are expressed substantially more strongly by *M. tuberculosis* than by *M. bovis*. However, cross-reactivity between these two species was revealed on the 38 KDa proteins by a rabbit anti-BCG serum.

Li et al. 1998, evaluated anti-A60 and anti-38 ELISA and found although these ELISA are simple, rapid and inexpensive methods, it is helpful but limited in the diagnosis of tuberculosis. Unlike antibodies, T cell proliferation responses to the 38 kDa proteins were expressed equally by 60% of tuberculosis patients and healthy BCG subjects. Results suggest the immunodominance of the species-specific B cell and cross-reactive T cell stimulatory epitopes (Young et al. 1986).

Human T cells reactive to *M. tuberculosis* were cloned from peripheral blood mononuclear cells of four tuberculosis patients. These T cell clones were tested for their reactivity to 65, 19, 14 kDa proteins expressed from clones that were isolated from genomic library. It was exhibited that some patients with tuberculosis exhibited a T cell response to the 65 and 19 kDa proteins (Oftung et al. 1987).

An ELISA was established for the measurement of IgG antibodies in human serum to the 30 kDa native antigens of *M. tuberculosis*. The test had a sensitivity of 70% in patients with sputum positive active pulmonary tuberculosis and specificity in controlled subjects of 100%. Less favorable test characteristics were obtained for patients with miliary and pleural tuberculosis showing 22% and 40% sensitivity, respectively.

Serological surveys were performed in groups of patients with active sputum smear-positive or smear-negative pulmonary tuberculosis, healthy household contacts and controls. Sera were tested for titers of antibodies, (which bound to each of the five purified mycobacterial antigens) by enzyme linked immunoassay, and for competition of binding to single epitopes, using six radiolabelled monoclonal antibodies directed towards the corresponding molecules. For smear positive samples, the best sensitivity (83%) was achieved by exclusive use of the 38 kDa antigen or its corresponding monoclonal antibodies. For smear-negative samples, levels of antibodies binding to the 19 kDa antigens showed a lower sensitivity of 62% compared with the control group or 38% compared with the contact group. Titers of antibody binding to the 14 kDa antigens were raised in *M. bovis* BCG-vaccinated contacts, indicating that the greatest potential of this antigen may be in the detection of infection in a population for which tuberculin testing is unreliable (Jackett et al. 1988).

Antibody reactivity generally is inversely correlated with blastogenic responses: tuberculous sera had high titer antibody to *M.tuberculosis* culture filtrate in a range from 35 kDa to 180 kDa (Havlir et al. 1991). Based on amino acid sequence of BCG, a synthetic peptide was synthesized and reacted with BCG and H37Rv and analyzed by western blot analysis (Minden et al. 1986).

Sera from patients with lepromatous and tuberculoid leprosy when examined in immunoblotting analysis for antibodies to *M.tuberculosis* culture filtrate antigens; it was found that 30 kDa and 31 kDa proteins were present in 88% and 81%, respectively. Antibody to a 32 kDa was found in 12% and 38% of lepromatous and tuberculoid patients, respectively (Hela et al. 1988). The gene was identified as a λgt 11 clone encoding 1008 base pair coding region. The comparison of 32 kDa antigens of *M. tuberculosis* with the *M. bovis* BCG alpha antigen revealed 73.8% homology between DNA sequences and 72.8% homology between amino acid sequences (Borremans et al. 1989).

Proteins secreted by strains of *M. tuberculosis* during short-term zinc deficient batch culture were identified in order to define antigens likely to be relevant to the pathogenesis of human disease. Secreted protein patterns of *M. tuberculosis* were quite similar to those of BCG, but differed by the absence of 46 kDa dimeric protein specific to BCG and by the presence in large amounts of 23 kDa protein which when denatured gave 13 kDa subunits. There have been reports regarding the secreted antigens to which antibody do not yet appear to be produced experimentally.

Five actively secreted proteins MPT 32, MPT 45, MPT 53, MPT 63 and MPT 46 were purified from *M. tuberculosis* culture fluid. Polyclonal antisera identified all proteins to be immunogenic. The MPT 45 protein was shown to correspond to the C-component of the antigen 85 complex. The 27 kDa MPT 51 protein was found to cross-react with three components of the antigen 85 complex and the N terminal amino acid sequences of MPT 51 and MPT 59 showed 60% homology indicating, thereby a family of closely related secretory proteins in mycobacteria (Nagai et al. 1991).

Recently, Gamma interferon assay has been assessed as a potential candidate to replace Mantoux skin test. The assay was evaluated among immigrants, health care workers, *M. tuberculosis* and *M. avium* complex (MAC) patients. The efficacy of assay was not significantly different from that of Mantoux test in cases of active tuberculosis, and it detected three of the seven cases of MAC colonization. In patients with active tuberculosis, the assay had a sensitivity of 77%, not significantly higher for extra-pulmonary than pulmonary cases (83% versus 74%). Quantifier sensitivity was not significantly different for smear-negative or smear-positive cases (80% versus 71%). The assay, however, needs laboratory facilities to stimulate viable lymphocytes and Enzyme Immuno Assay to quantify IFN-α. More experience is needed from long-term studies to determine whether the management of contacts of individuals with infectious tuberculosis, based in part on Mantoux results, can be extrapolated to the QIFN results.

3.2 Recombinant DNA Approaches

Recombinant DNA strategy has been extensively used to survey the *M. tuberculosis* genome, for sequences that encode specific antigens. A recombinant shotgun λgt11 expression library containing these fragments was made (Young et al. 1987; Shinnick et al. 1987a; Bisen et al. 2003). Majority of the clones selected by using polyclonal serum expressed three antigens that were previously identified by using mouse monoclonal antibodies, thus indicating the immunodominance of these proteins. Fusion proteins formed can be detected by Coomassie Blue staining following electrophoresis on SDS-PAGE. Agar blocks containing single plaques can be loaded directly onto stacking gel, thus avoiding the need for extensive sample preparation (Saul et al.1986). Also the use of recombinant approach had been immensely successful in revealing biology of mycobacteria.

The recombinant DNA expression library from an *M. leprae* was constructed to isolate genes encoding the five most immunogenic protein antigens of leprosy bacilli. Killed *M. leprae* can induce delayed type hypersensitivity reaction in healthy individuals. Screening of crude phage lysate of *E. coli* was done by using *M. leprae* T cell specific clones isolated from *M. leprae* vaccinated volunteers (Mustafa et al. 1986). By this method it was found that nearly half of the *M. leprae* specific T cell clones are stimulated to proliferate by lysates containing an epitope of a *M. leprae* protein of relative molecular mass of 18 kDa. The gene encoding a major 28 kDa antigen of *M. leprae* was cloned and sequenced. The gene was expressed under its own promoter in *M. smegmatis*. The amino acid sequence of epitopes recognized by monoclonal antibodies against 28 kDa antigens has been determined. By using a single probe method, the gene for an immunogenic MPB 57 protein of *M. bovis* BCG was cloned. The cloned gene was expressed in an *E. coli* expression vector to give a mature protein, which reacted with a polyclonal antibody raised against MPB 57.

The monoclonal antibody A, protein of *M. bovis* BCG was over expressed in *E. coli* K-12 cells. The presence of antibodies to MabA in human sera was determined by an enzyme-linked immunoassay. In about 80% of serum samples from tuberculosis patients and in about 60% of samples from BCG vaccinated individuals, significant levels of anti-MabA antibodies were found (Thole et al. 1987). This protein was found to be immunologically important component of a variety of mycobacteria including *M. tuberculosis*, *M. leprae* and *M. bovis* BCG. By using recombinant DNA approach it was found that one epitope seen by an antibody was specific to *M. tuberculosis*.

From a λgt11 expression library of *M. paratuberculosis*, three portions of the gene encoding 34 kDa proteins were isolated. Out of three, one (clone) expressed a polypeptide, which cross-reacted with all tested *M. paratuberculosis* strains, but not with many strains of the *M. avium*, *M. intracellulare* and *M. scrofulaceum* group. The recombinant polypeptide was used as a reagent for an enzyme linked immunoassay for diagnosis of John's disease in cattle, which is caused by *M. paratuberculosis*. This assay correctly diagnosed all the tested blood samples from infected cattle at all stages of disease (Kesel et al. 1993).

Analysis of 34 kDa protein of *M. leprae* was shown to be isologous to these immunodominant 34 kDa antigens of *M. paratuberculosis* by performing sequence analysis and alignments (Silbaq et al. 1998).

Recently a gene encoding 40 kDa protein antigens of *M. tuberculosis* has been sequenced and demonstrated that this protein is an L-alanin dehydrogenase, which may play role in cell wall synthesis. Being L-alanine, an important constituent of the peptidoglycan layer, it is thought that this portion may play a role in virulent interaction with the *M. tuberculosis* infected individuals (Anderson et al. 1992b).

A 65 kDa protein has been identified as one of the medically important antigen of *M.tuberculosis*. The gene encoding this antigen was isolated from a gene expression library using monoclonal antibodies directed against the 65 kDa protein antigens as the specific probe. The nucleotide sequence of this gene was determined, and a 540 amino acid sequence was shown to correspond to that of 65 kDa antigens by constructing a plasmid in which this open reading frame was fused to the LacZ gene. The resulting fusion protein reacted specifically with the anti 65 kDa protein antibodies. A second open reading frame was found downstream of the 65 kDa genes, which could encode a protein of 517 amino acids. This putative protein contained 29 tandemly arranged partial or complete matches to a pentapeptide sequence. The amino acid sequences of this antigen display greater than 95% homology for *M. leprae, M. tuberculosis, M. bovis* BCG also displays almost 95% homology (Young et al. 1987)

The screening of expression libraries of mycobacteria by monoclonal antibodies allows the isolation of antigen expressing clones and subsequent analysis of their amino acid sequences. Other probes such as polyclonal antibodies, T cells and oligonucleotides deduced from partial amino acid sequences have been used to isolate and characterize mycobacterial antigens. Purification of such antigens from over expressing bacteria has been initiated through immunological evaluation of T-cell as well as B-cell responses induced by these antigens. The combined use of sub gene libraries and synthetic peptides allow the exact characterization of B-cell and T-cell determinants involved (Ivanyi and Thole 1994).

Very recently *M.tuberculosis* NTI-6479 expression library in ZAP vector was prepared. The library was probed with pooled sera from chronically infected tuberculosis patient's yielded 176 recombinants with a range of signal intensities. Sera obtained from patients who were clinically diagnosed to be in the early phase of *M. tuberculosis* infection were used to probe the 176 recombinants obtained so far. Interestingly, some recombinants that gave very strong signals in the original screen did not react with early phase sera, while others whose signals were extremely weak in the original screen gave very intense signals with serum from recently infected patients, indicating thereby the differential nature of either the expression of these antigens or the immune response elicited by them as a function of disease progression (Rama Rao et al. 1996).

A phage expression library of M. *tuberculosis* H37Rv DNA was immunoscreened by using an anti-culture filtrate rabbit antiserum. MTC 28, a novel 28 kDa proline-rich secreted antigen specific for the M. *tuberculosis* complex was cloned and expressed. Of 20 positive clones isolated, 6 were analyzed and found to express the genes for two known components of the early culture filtrate, the secreted 45/47 kDa antigen complexes and the KatG protein, and two novel genes. Molecular cloning and nucleotide sequence of one of the new genes encoding a culture filtrate protein of 310 amino acid residues. The gene was designated MTC 28. The deduced polypeptide sequence contained at amino terminal, a highly hydrophobic 32 amino acid with secretion signal. Purified recombinant MTC 28 antigen evoked strong delayed-type hypersensitivity and antibody responses in experimental guinea pigs. The strong immunological activity of MTC 28 and the absence of B-cell and T-cell epitopes cross-reactive with a common environmental Mycobacterial species, such as M. *avium*, make this novel antigen an attractive reagent for Immunodiagnostic of tuberculosis (Manca et al. 1997b).

Plum et al. 1997, identified a gene of M. *avium*, which is expressed when this bacilli is grown in human macrophages. A genomic library was made to clone the gene. The library was screened with cDNA derived from mRNA obtained from growing M. *avium* growing in human macrophages. The gene was designated as mig (macrophage induced gene). *Mig* gene has a signal peptide of 19 amino acid residues and is secreted as 30 kDa proteins in acidic environment. The gene has an open reading frame of 295 amino acids with a strong bias for mycobacterial codon usage. *Mig* protein was expressed in M. *smegmatis* and was found to be of limited advantage for survival of bacillus inside the macrophages.

Screening of M. tuberculosis H37Rv genomic library cloned a novel antigen MTB 12 secreted by M. *tuberculosis* with probe from M. *leprae* homologues of the MTB 12 gene. The probe was generated by amplifying this fragment from M. *leprae* genomic DNA by PCR. For serological screening, a polyclonal rabbit antiserum (prehybridized with E.*coli* lysate) raised against culture filtrates of M. *tuberculosis* was employed. The molecular analysis revealed that this protein is initially synthesised as a 16.6 kDa precursor protein containing a 48 amino acid hydrophobic leader sequence. The mature processed form of protein has a molecular mass of 12.5 KDa. Recombinant MTB 12 containing an N-terminal six-histidine tag was expressed in E. *coli* and purified by affinity chromatography. Recombinant MTB protein elicited *in vitro* proliferative responses from the peripheral blood mononuclear cells of a number of purified protein derivative positive

(PPD+) human donors but not from purified protein derivative negative (PPD-) donors (John et al. 1998).

M. tuberculosis complex specific antigen was found to be a suitable candidate for a diagnostic reagent. Resorting to shotgun library approach cloned MPT 63 a novel antigen secreted by *M. tuberculosis*. A genomic DNA library was constructed in bacteriophage λ ZAP II (Stratagene) with DNA extracted from *M. tuberculosis* H37Rv grown on 7H11 Middlebrook agar. Plaques were screened with monospecific anti-MPT 63 antibodies raised in rabbit. The DNA sequence and resulting protein sequence was deciphered. MPT 63 was expressed in *E. coli* as a recombinant protein. The gene was found to be present only in *M. tuberculosis* and *M. bovis* strains and absent in Mycobacterial species that do not belong to the TB complex, most notably, a common environmental species such as *M. avium*. MPT 63 was found to lack epitopes that cross-react serologically with *M. avium* antigens, as shown by indirect and competitive ELISA (Manca et al. 1997a).

The complete genome of *M. tuberculosis* was deciphered by sequencing cosmids and bacterial artificial chromosomes harboring Mycobacterial DNA cassettes. A systematic sequence analysis of selected cosmids, BACS and random small insert clones from a whole-genome shotgun library were utilized to accomplish this task. 3.2 Mb of sequence was generated from cosmids and the remainder was obtained from selected BAC clones and 45,000 whole genome shotgun clones. This culminated in a composite sequence of 4,411,529 base pairs, with G+C content 65.6 %. This is reflected in the biased amino acid content of the proteins. This represents the second largest bacterial genome sequence currently available (after that of *E. coli*). The genome is rich in repetitive DNA, particularly insertion sequences and in new multigene families and duplicated house-keeping genes. The G+C content is relatively constant through the genome. The complete sequence, a list of annotated cosmids and linking regions can be found on websites (http://www.sanger.ac.uk) (http://www.pasteur.fr/mycdb/) (Cole et al. 1998).

Recently, two serine protease antigens MTB 32A and MTB 32B of *M. tuberculosis* were characterized, cloned, and expressed. Genomic DNA from *M. tuberculosis* H37Ra and Erdman was fragmented for library generation by sonication, yielding DNA fragments in a size range of 300 to 4000 bp. The ends were blunted with Klenow polymerase, ligated to EcoRI adaptors, and sub-cloned into EcoRI predigested λZAP bacteriophage arms. Phages were packaged with Gigapack II packaging extracts. Rabbit anti-culture filtrate protein sera raised in rabbit was preabsorbed with E.coli antigens and used for screening shotgun library. The genes were cloned into pET17b vector, and transferred to BL21 (DE3) pLysE (novagen) for over expression of proteins. The recombinant proteins were over expressed and explored for diagnosis and protection against tuberculosis. The gene sequence as well as protein sequence from avirulent strain H37Ra, virulent Erdmann strain, and virulent H37Rv strain, clinical isolate CSU93 were compared and found identical. MTB 32A stimulated peripheral blood mononuclear cells (PBMC) from healthy purified protein derivative positive (PPDT) individuals proliferate and secrete IFN-γ MTB 32A was considered suitable as a candidate for inclusion in subunit vaccine.

Molecular characterization and Human T-cell responses to a member of a novel *M. tuberculosis* mtb 39 gene families was studied in an extensive work performed by Dillon et.al. *M. tuberculosis* H37Ra genomic DNA was isolated and sheared by sonication to a size range of 1-4 Kb. The *M. tuberculosis* H37Rv genomic DNA was partially digested with Sau3AI and genomic libraries were constructed in lambda ZAP II at EcoRI cloning site. Expression screening was performed using a pool of patient sera pre absorbed with *E. coli*. This resulted in the identification and purification of seven immunoreactive clones. The expression screening of genomic *M. tuberculosis* library with tuberculosis patient sera was used to identify novel genes that may be used diagnostically or in the development of tuberculosis vaccine. This strategy helped cloning a novel gene termed Mtb 39a that encodes a 39 kDa protein. Purified recombinant Mtb 39a elicited strong T-cell proliferates and gamma-interferon responses in peripheral blood mononuclear cells from 9 of 12 PPD-positive individuals tested. Further approach led to mapping of 10 distinct T-cell epitopes. The protein demonstrated increased protection from *M. tuberculosis* challenge in experimental mice immunized with mtb 39a DNA (Dillon et al. 1999).

3.3 Methods used for Diagnosis

The conventional methods for detecting tuberculosis are time and labour consuming AFB staining is considered to be insensitive (requiring 10,000 organism/ml of sputum for smear positive result with 100X microscope. ELISA-KP 90 is also known to be of low sensitivity and specificity (cut-off value > 1.0 +ve, and < 0.8 -ve test result) and requires sophisticated infrastructure as also the hypersensitivity based Tuberculin Skin Test (Montaux test), which lacks sensitivity and specificity in BCG vaccinated patient. In the same way MYCODOT is inconvenient for HIV correlated individuals and BACTEC-460 and Roche molecular system (PCR based product) are though sensitive requires very costly infrastructure and technical expertise. Though the various technologies have been developed so for detection of either antigen or antibodies to diagnose the TB affected population, but they have certain limitations (Garg et al. 2003) and categorized as traditional and non-traditional methods. Some of the common methods are:

3.3.1 Traditional Methods and their Limitations

3.3.1.1 Radiological examination (Chest X-ray)

X-ray is used in the diagnosis of pulmonary tuberculosis. It is often inadequate due to the non-specific nature of the finding (Sao et al. 1992). Also the new problem of HIV infected TB patents has created a complications in the use of this method. Since HIV infection can change the classical appearance of the chest, there is no current role for the CT scan in the evaluation of the asymptomatic TB infected child with a normal chest radiograph, even active multiplication of Mycobacteria occurs with or without the presence of radiographic abnormalities or clinical symptoms. For example, gastric aspirate cultures yield M. tuberculosis from a small proportion of recently infected children with normal chest radiographs even active multiplication of Mycobacteria occurs with or without the presence of radiographic abnormalities or clinical symptoms. For example, gastric aspirate cultures yield *M. tuberculosis* from a small proportion of recently infected children with normal chest radiographs.

A posterior-anterior view of the chest is the standard radiograph needed for the detection and description of chest abnormalities. In some instances, other views (e.g., lateral, lordotic) or additional studies (e.g., CT scans) may be necessary.

In pulmonary TB, chest radiograph abnormalities often occur in the apical and posterior segments of the upper lobe or in the superior segments of the lower lobe. However, lesions may appear anywhere in the lungs and may differ in size, shape, density and cavitation, especially in HIV-infected and other immunosuppressed persons.

In HIV-infected persons with pulmonary TB, the chest radiograph may have an unusual appearance. For example, TB may cause infiltrates without cavities in any lung zone, or it may cause mediastinal- or hilar-lymphadenopathy. In HIV-infected persons, almost any abnormality on a chest radiograph may indicate TB. Conversely, the radiograph of an HIV-infected person with TB disease may even appear entirely normal. An isolated granuloma on a chest radiograph is not considered a fibrotic lesion compatible with previous TB. A radiologist or clinician expert in the interpretation of radiographs should interpret all radiographs. Abnormalities on chest radiographs may be suggestive, but are never diagnostic of TB. However, chest radiographs may be used to rule out the possibility of pulmonary TB in a person who has a positive reaction to the tuberculin skin test but no symptoms of disease.

3.3.1.2 Tuberculin skin test (Mantoux test)

The tuberculin skin test is used to determine whether a person has TB infection; also, known as the Mantoux tuberculin skin test is the preferred type of skin test because it is the most accurate method. The positive result of Mantoux tuberculin skin test depends on the size of the induration.

In the Mantoux test an intradermal injection of 0.1 ml of purified protein derivative (PPD) tuberculin containing 5 tuberculin units is given into the volar surface of the forearm of suspected tuberculosis person. Then after 48 to 72 hours, the area of induration is measured for swelling and delay hypersensitivity reaction, reported as mm of induration.

3.3.1.3 Morphological Examination: ZN (Ziehl-Neelsen)

Clinicians still rely on conventional methods like ZN (Zeihl-Neelsen) staining/flurochrome staining, sputum culture, and other phonetic criteria, which are shown to be quite helpful, but they have their own limitations.

3.3.1.4 Culture

The current methods used in clinical laboratories are growth dependent and may even take upto 6-8 weeks to report a negative/positive result.

3.3.1.5 Radiometric culture

The BACTEC 460 Radiometric (Becton Dickinson Instrument Systems, Sparks, Md, USA) system is an automated method for detecting $^{14}CO_2$ liberated by bacteria during metabolism and decarboxylation of ^{14}C-labeled substrates. For detection of mycobacteria, the system uses ^{14}C-labeled palmatic acid as the substrate in modified Middlebrook 7H12 broth. With the BACTEC system, the number of positive cultures may or may not be higher than conventional culture on solid media. The BACTEC system requires capital expenditure and the disposal of radioactive waste precludes its use in many laboratories. The use of BACTEC 12B bottles (Becton Dickinson Diagnostic Instrument System, Sparks, MD) in conjunction with Accuprobe for TB (Gene probe Inc., San Diego CA.) can shorten the time for the identification of this organism, but these procedures still require 1-3 weeks to provide results from patient specimens that contain acid-fast organisms. Also the procedure cannot be easily applied in heavily afflicted populations due to burgeoning costs.

The reasons for this are manifold. For microscopy a minimum number of 104/ml bacilli are required and cannot discriminate *M. tuberculosis* from other Mycobacteria. Of the estimated 8 million new cases of tuberculosis per year some are either smear-negative pulmonary or extra-pulmonary infection, and hence in many cases, diagnosis cannot be confirmed at the time of presentation (Nunn and fellten 1994). As mentioned above, sputum culture is the most reliable method in pulmonary tuberculosis in clinical setting, but process is lengthy, cumbersome and requires Mycobacterial culture facility.

3.4 Non-traditional ELISA (Enzyme Linked Immunosorbent Assays)

3.4.1 Based on kp-90

Among serology based assays enzyme linked immunoassays and similar assays (e.g., so called "spot tests" which are a simplified form of ELISA test) are quite popular for detecting infectious agents. All such assays are based on binding effects of antibodies and antigens. In one form of ELISA assay, for example, an antigen produced by a specific organism is used to test for the presence of antibodies for the antigen in the sera of patients, thus providing an indication that patients have been exposed to these organisms. The antigen is immobilized on a solid support and is incubated with serum to be tested. If a target antibody is present in the serum, indicating exposure of the patient to the disease-causing organism, it binds to the layer of antigen. The number of antigen/antibody bound molecular pairs produced in this way depends on the concentration of the antibody in the serum. Tests detecting antigen in the patient specimen can also be employed if specific antibodies are developed utilizing monoclonal antibody techniques. Such testing envisages the presence of a detectable amount of antigen in the patient specimen. One needs to have specific and sensitive antigens or antibody for the development of such assays. A number of proteinacious and non-protein antigens have been explored from time to time for development of such assays, but none holds the ground good.

3.4.2 Molecular Approach: (Polymerase Chain Reaction)

The development of PCR has undoubtedly ushered in a revolution in the probe technology. Recently, a PCR study was carried out in 370 individuals and compared with the standard bacteriological data. No false-positive results were obtained, and overall *M. tuberculosis* was detected in 20 out of

370 individuals screened. Amplicor *M. tuberculosis* test is highly specific and rapid for routine use in a clinical laboratory, but is an expensive affair.

The Gen-Probe Amplified *M. tuberculosis* Direct (MTD) test has been approved for use in United States for the rapid diagnosis of pulmonary tuberculosis in patients with acid-fast smear positive sputum samples since 1996. These findings indicate that low-level false positive MTD results can occur due to the presence of *M. kansasii, M. avium,* and possibly other Mycobacterium species other than *M. tuberculosis* in sputum. The specificity of MTD-2 recently released by Gen-probe will be revealed in matter of time (Jorgensen et al. 1999).

3.4.3 Rapid Test using "Cocktail" of Antigens

A vast array of potential diagnostic antigens is available. But no single reagent gives 100% sensitivity (Bothamley GH 1995). Hence, future research needs to focus on identifying the best possible combination ("cocktail") of antigens of *M. tuberculosis* complex (Bothamley GH 1995; Lyashchenko et al. 1998; Tiwari et al. 2005; Bisen et al. 2003; Bisen and Tiwari 2005).

ASSURE TB rapid test (Genelabs Diagnostics Pte Ltd., Singapore) is an indirect solid-phase immunochromatographic assay for detecting antibodies in clinical samples (plasma, serum, or whole blood). The test employs an antibody-binding protein conjugated to colloidal gold particle and a "cocktail" of novel antigens – Mtb 11 (CF-10), Mtb 8, Mtb 48, Mtb 81 and 38 kDa protein – immobilized on the membrane, in lateral flow devices. Mtb 81 has been found to be a promising antigen in the serodiagnosis of HIV-TB co-infection (Houghton et al. 2002).

In the recently patented Liposomal Agglutination Card test or **TB screen test** (Bundelkhand University, Jhansi, India & National Research Development Corporation, Government of India), a "cocktail" of purified cell wall-associated antigens (H37Rv) of *M. tuberculosis* has been incorporated onto the surface of liposome particles. This "cocktail" of antigens reacts with specific antibodies present in clinical samples to produce a blue agglutination (Figure 8). The test differentiates healthy controls and BCG-vaccinated individuals from those with active tuberculosis (Tiwari et al. 2005;Bisen and Tiwari 2005). The TB screen test has the potential for use on a mass scale in developing countries because of its low cost (about 12 US cents per test), rapidity (four minutes), high sensitivity (94%) and high specificity (98.3%) (Tiwari et al. 2005) and can be utilized for mass screening of heavily afflicted population followed by further transmission of infection in the community can be checked by routine screening as alternate diagnostic products (e.g., AFB) and effective chemotherapy could be initiated.

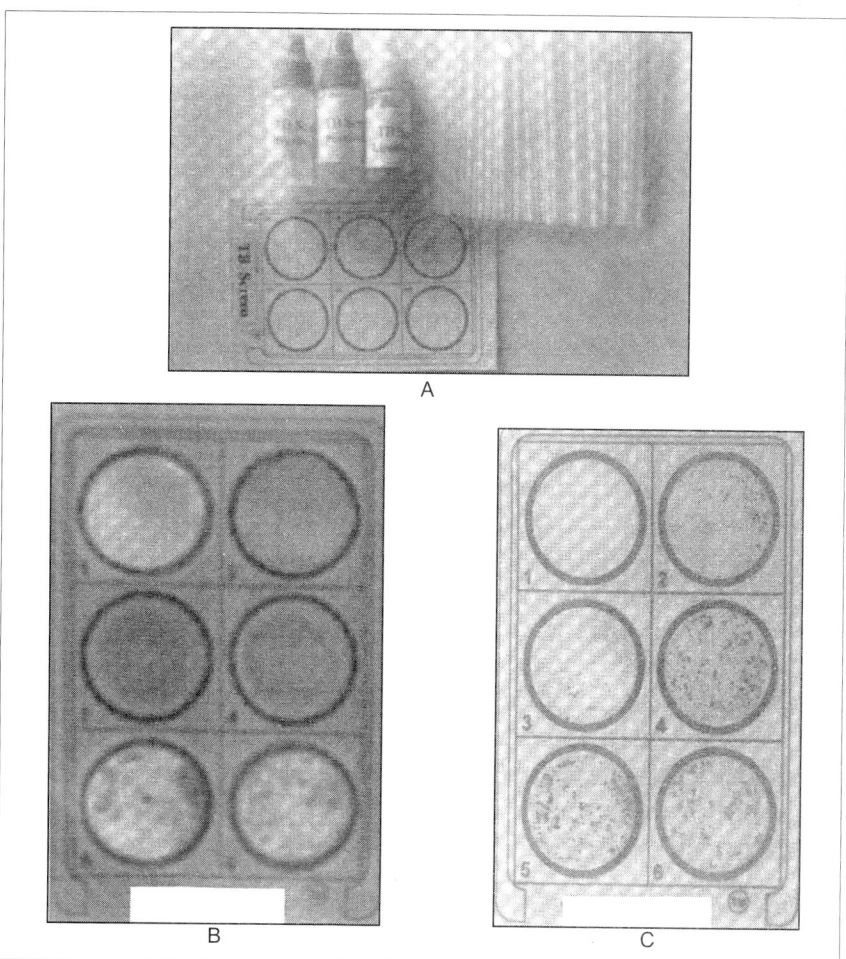

Fig. 8   Accessories for antibody detection kit (A) (TB Screen test); Negative control, Positive control and liposomal antigen suspension (Bottles from left to right, respectively); test card and mixing sticks (B, C) Results of specimens.

4. THERAPEUTIC ASPECTS

4.1 Chemotherapy and Vaccination Strategy

Although TB is a curable disease with the use of appropriate antibiotics, the major hurdle in the treatment lies in the late diagnosis of the disease due to unavailability of simple and cost-effective diagnostic products. Thus, there is an urgent need to develop strategies to effectively tackle its detection and thereafter its treatment.

Currently, more than 15 antimycobacterial drugs are available for tuberculosis patients including Ist, IInd, IIIrd and IVth generation drugs such as Isoniazide (INH), Refampcin, Pyrazinamide, streptomycin, Ethambutanol, AKT-4, etc. to take care the eradication of tuberculosis infection, but are indispensable to prevent the progression of the disease specially those caused by MDR strains of *M. tuberculosis*. The following drugs (essential antituberculosis) are commercially available in global market for prevention of mycobacterium tuberculosis infection:

(A) Isoniazid (INH): Isoniazid is a first line of drug for chemotherapy of TB, symbolized as H and available in tablet form 100-300mg, injection 25 mg/ml in 2 ml ampule and can be administered oral or intramuscular (IM), which has been discovered in 1952-1954 and reported as great efficacy acceptability of low toxicity. It is a hydrazide of isonicotinic acid (synthetic) with potential tuberculostatic and tuberculocidal activities and is rapidly absorbed and diffuses readily in to all fluids and tissues. The mode of action (MOA) is typically found to attack on cat G gene and inhibit the protein and enzyme of catalase involved in the synthesis of mycolic acid. It is largely excreted in the urine within 24 hrs, mostly as inactive metabolite.

(B) Refampicin (Refadine): It is a semi-synthetic derivative of rifamycin, a complex macrocyclic antibiotic available in early 1963s symbolized as R and available in capsule or tablet form 150, 300 mg and can be administered oral or intramuscular (IM), that inhibits (DNA dependent RNA polymerase, thus inhibit RNA synthesis) ribonucleic acid synthesis in a broad range of microbial pathogen. It has bactericidal action and a potent sterilizing effect against tubercle bacilli in both cellular and extracellular location. This has been reported as bactericidal antimycobacterial agent.

Refampcin is lipid soluble following oral administration and rapidly absorbed and distributed throughout the cellular tissues and body fluids; if the meanings are inflamed, significant amounts enter the cerebrospinal fluid. It is extensively recycled in the entrohepatic circulation, and metabolites formed by deacetylation in the liver are eventually excreted in the faeces. Since resistance readily develops, Refampicin must always be administered in combination with other effective antimycobacterial agents.

The mode of action is to inhibit the DNA dependent RNA polymerase leading to inhibition of RNA synthesis and useful drug for meningial tuberculosis, since it is widely distributed into body fluids such as cerebrospinal fluid (CSF).

(C) Pyrazinamide: It is a synthetic analog of nicotinamide that is only weakly bactericidal against *Mycobacterium tuberculosis*, but has potent sterilizing activity, particularly in the relatively acidic intracellular environment of macrophages and in areas of acute inflammation. It is highly effective drug during the first 2 months of treatment while acute inflammatory changes persist and its use has enabled treatment regimens to be shortened and the risk of relapse to be reduced. It is symbolized as Z available in tablet form with 500 mg and administered orally. It is redially absorbed from the gastrointestinal tract and is rapidly distributed throughout all tissues and fluids. It is metabolized mainly in the liver and is excreted largely in the urine. The mode of action of the drug is to act on intracellular pathogen even low pH 5.2-5.4 and inhibit fatty acid synthesis I.

(D) Streptomycin: It is a derivative of *streptomyces griseus*, became available in late 1944 administered IM or IV and is being used as antimycobacterial agent and sensitive Gram-negative infection. It acts (inhibits protein synthesis machinery) on protein synthesis machinery by binding with 30s subunit, decreases the fidelity of mRNA and not absorbed from the gastrointestinal tract, but after intramuscularly administration, it diffuses readily in to the extracellular component of most body tissues and it attains bactericidal concentration, particularly in tuberculous cavities. It is excreted unchanged in the urine.

(E) Ethambutal: The drug has been discovered in 1956-1959 and symbolized as E; available in tablet form in 100-400 mg and administered orally. It is a synthetic congener of 1,2-ethanediamine that is generally considered to have a bacteriostatic effect when administered at the recommended dose and is used in combination with other antituberculosis drugs to prevent or delay the emergency of resistant strains. It is readily absorbed from the gastrointestinal tract and excreted in the urine both unchanged and as inactive hepatic metabolites. The mode of action of the drug is to inhibit (arabinan synthesis both in AG and LAM) incorporation of mycolic acid in to the cell wall of Mycobacteria, which ultimately disrupt the outer cell membrane and increases activity of other drugs.

4.2 Vaccine

Vaccine is a preparation that contains an antigen, consisting of whole disease-causing organisms (killed or weakened) or parts of such organisms, which is used to confer immunity against the disease that the organisms cause. Vaccine preparation can be natural synthetic or derived by recombinant DNA technology.

Before vaccines were invented the only way to create immunity into body was to suffer through about of the disease in question. Once endured, immune system could fight off any future infections before they took hold B-cells in the bloodstream, responsible for fighting of the disease, retain memory of the disease. If the disease returned, the immune system launched a quick attack before the disease could take hold. Vaccine produces the same effect without making the patient suffer through the disease. By introducing a disease into the bloodstream, B-cells are stimulated into action, creating antibodies and a memory record of the pathogen, resulting in strong immunity against respective disease.

4.2.1 BCG Vaccine

The only available vaccine against tuberculosis is the Bacillus Calmette Gurein (BCG) vaccine isolated from *Mycobacterium bovis*. This vaccine generally induces high level of protection in animal models of tuberculosis. However, in humans its efficacy is highly variable, ranging from no protection to almost complete protection, depending on the population tested. The BCG vaccine that is used as prophylactic agent to prevent the occurrence of infection in the immunized individual is administered at the time of birth of the child. The vaccine though effective in imparting but its efficacy has been questioned in several clinical trails conducted in past.

In 1908, Camille Guerin and Albert Calmette initiated their attempts to produce an anti-TB vaccine from a virulent bovine strain. In 1921, vaccination with BCG, an attenuated vaccine, was introduced. The efficiency of the BCG vaccine has been questioned, since its early use and therefore, a large number of trials have been carried out to determine its efficacy. In these studies it was found that the BCG vaccine protected efficiently against leprosy (Fine and Rodrigues 1990) as well as childhood manifestations of TB (disseminated TB) (Rodrigues et al.1993). However, the protective efficacy against pulmonary TB was limited.

Many hypotheses have been suggested to explain the low protective efficacy of BCG against pulmonary TB. These hypotheses include inappropriate treatment and storage of the vaccine, the use of different strains of BCG and lack of an effective stimulation of the optimal blend of T cell populations and, in particular, that of the $CD8^+$ T cells (Hess and Kaufmann 1999). In addition to these hypotheses, the currently used intradermal route of immunization has been suggested as another factor influencing the capacity of BCG to induce optimal immunity in the lungs. In this regard, (Intranasal, i.n.) route of immunization has recently been evaluated as a possible route for BCG delivery, in mouse experimental models. Results from this study showed a high degree of protection against challenge with *M. tuberculosis* in BALB/C mice, following (i.n) BCG vaccination (Falero-Diaz et al. 2000). In a similar model, (i.n) vaccination with BCG conferred as good, if not better protection than subcutaneous (Subcutaneous; s.c.) route, against challenge with virulent *M. bovis* (Lyadova et al. 2001).

4.2.2 Prospects for New Vaccines

Given the limitations of BCG in protection against adult pulmonary TB, there is a considerable scope for improved vaccination strategies. Immunological research has a key position in understanding the pathogenesis of TB, and thereby in developing novel designs for effective prophylactic vaccination, immunodiagnostic tools and immunotherapeutic agents. Two approaches have been considered for vaccine development. One involves the replacement of BCG by a more potent vaccination inducing immune responses capable of either complete elimination of the bacilli, or of reliable containment of persistent infection.

The second approach involves the post-exposure vaccination to boost immunity in individuals whose natural immunity has already been primed by infection or BCG vaccination (Young and Stewart 2002). Indeed, over the past decade research efforts have been directed to evaluate potential vaccine candidates as well as alternative routes of vaccine delivery, such as the i.n. route, in order to improve protection.

4.2.3 New Vaccine Candidates

A wide range of potential vaccine candidates have been generated and subjected to tests for protective efficacy in experimental model of infection. New vaccine candidates include live attenuated vaccines, subunit vaccines and DNA vaccines.

4.2.4 Live Attenuated Vaccines

Advances in the techniques required to genetically modify *Mycobacteria*, as well as increase in the knowledge of the pathogenesis of the microorganism, have made possible to delete genes encoding for potential virulence factors in *M. tuberculosis*, thereby enabling the generation of attenuated mutants. In addition to attenuated strains of *M. tuberculosis*, natural attenuated *Mycobacteria*, such as *M. vaccae* and *M. microti*, are being studied as possible vaccine candidates (Nor and Musa 2004). Another approach has been the improvement of the BCG immunogenicity by the addition of genes encoding cytokines, such as IFN-γ (Murray *et al.* 1996) or *Mycobacterial* proteins, such as the antigen 85 complex (Ag85) (Horwitz et al. 2000).

4.2.5 Subunit Vaccines

Subunit vaccines are currently the most widely studied. This type of vaccine has been focused in particular on proteins present in filtrates prepared from *in vitro* cultures of *M. tuberculosis*, although non-secreted antigens have also been shown to induced protective responses in experimental studies (Coler et al. 2001; Skeiky et al. 2000). The most extensively studied antigens are members of the Ag85 complex, a family of mycolyl transferases enzymes involved in cell wall biosynthesis and present in culture filtrates (Belisle et al. 1997). The Ag85 has been reported to induce strong activation of T cells in several studies (Andersen et al. 1995; Mustafa et al. 1998). Other antigens being studied are:

(i) Early secreted antigenic target (ESAT-6), which has been reported to be absent from all BCG vaccine strains and to induce very strong T cell and antibody responses (Brodin et al. 2004).

(ii) Heat-shock proteins (HSP), such as HSP-65 and HSP-70, found to induce a prominent immune response at both, the antibody and the T cell levels (Silva 1999).

(iii) PstS-1 (38 kDa protein), glycoprotein exposed on the surface of the bacillus and reported to be a powerful B and T cell antigen (Bothamley et al. 1992; Lefevre et al. 1997).

(iv) 19 kDa protein, a lipoprotein found to induce the expression of IL-12 and iNOS in monocytes and dendritic cells through its binding to TLR2 (Brightbill et al. 1999; Thoma- Uszynski et al. 2000) and to promote neutrophil activation (Neufert et al. 2001).

4.2.6 DNA Vaccines

Administration of naked DNA has the potential of eliciting both, cellular and humoral immunity against encoded antigens. Several *Mycobacterial* antigens, including the *PstS-1*, *HSP-65* and the Ag85 have been studied and found to induce protection in animal models (Fonseca et al. 2001). Although the results are promising, concerns about the safety of DNA vaccination have been raised, mainly regarding the possibility of DNA integration into the host genome affecting oncogenes or tumor suppressor genes and, thereby inducing the development of cancer. However, the risk of integration has been reported to be low under a variety of experimental conditions (Manam et al. 2000).

The WHO funded Global fund for AIDS, Tuberculosis and Malaria (GFATM) has allocated USD 2 billion, while Bill and Melinda's foundation pledged USD 89 million for research on tuberculosis vaccine. CDRI Lucknow, India reported the development of a new vaccine against tuberculosis, based on a related non-pathogenic Mycobacterium and is under clinical trial. A genetically modified BCG vaccine (containing new TB specific gene) has been developed at Delhi University shown promising results during pre-clinical studies. The DNA vaccines (based on two genes found in *Mycobacterium tuberculosis*) are under development at IISC Banglore.

The Invitrogen, an American company have used DNA that codes for a *Mycobacterium leprae* antigen, to prepare a vaccine against tuberculosis. This new DNA vaccine produces an immunogenic peptide, which stimulates the T-cell of the hosts immune system to produce gamma-interferon. This vaccine was found to be effective in killing the tubercle bacilli in mice that were heavily infected with *Mycobacterium tuberculosis*. Scientists believe that if this vaccine is used along with anti-tubercular drug, it could produce faster cure in tuberculosis-afflicted patients.

It is now well evident that tuberculosis poses a significant health rate threat to mankind. Multidrug-resistant strains are on the rise, and *Mycobacterium tuberculosis* infection is often associated with human immunodeficiency virus infection. Satisfactory control of tuberculosis can only be achieved using a highly efficacious vaccine.

Currently, two main vaccination strategies are being pursued. The first strategy uses subunit vaccines in the form of protein-adjuvant formulation, naked DNA, or recombinant bacterial or viral carriers that express defined antigens. Promising results have been obtained, but so far no vaccine candidate tested in animal models has proven to be better than BCG. The second strategy, comprising viable *Mycobacterial* vaccines, either attenuated viable *M. tuberculosis* or BCG, or recombinant BCG over expressing certain antigens or immunomodulators, is also being pursued and shows promise.

SUMMARY

Tuberculosis is caused by acid fast bacteria (microorganism) known as *Mycobacterium tuberculosis* transmitted from infectious person to normal healthy individuals due to factor (poverty, overcrowding population, illiteracy, poor working condition) that facilitate the spread of tuberculosis, affecting children and adults in various stage of their life resulting as creation of havoc situation for human kind and became a leading infectious killer worldwide.

The problems are more critical due to correlation of tuberculosis with HIV (immunocompromised) patients and MDR tuberculosis increases more burdens due to patients who take the prescribed drugs irregularly have an increased chance of developing acquired drug-resistant tuberculosis and becoming chronic cases, compared with those who adhere to the prescribed treatments. If an adequate, retreatment regimen is not given at the stage, multidrug-resistant disease often develops. Though the second line antituberculosis drugs are available for the treatment of the patients with MDR tuberculosis. Drug resistance has serious consequences, not only for individual patients, but also for tuberculosis control in general. Patients with untreatable, multidrug-resistant tuberculosis become reservoirs for the disease, spreading resistant infection primilarly to their children.

Over the last 30-40 years have seen substantial improvements in tuberculosis treatment, with the development of highly efficacious treatment regimens. As a result of large-scale programmes of carefully controlled clinical trials, social and institutional awareness through the DOTS, RNTCP, NGOs and various organizations governed by WHO, UNICEF, etc. nationally and internationally.

The duration of treatment has been progressively reduced from 18-24 months to 6 – 9 months and now trying for 1 month. However, in order for these intensive short course regimens to be effective, antituberculosis drugs must be available to patients followed by efficient and effective distribution of the drugs and their monitoring honestly. This needs to be ensured that the patients continue to receive these drugs until cure. To eliminate the burden of tuberculosis infection in developing countries the ground label work need to be done by respective government with the help of primary health care centers, social workers and hospitals for health checkup of entire population, must be mandatory once in a year and necessary intervention strategy need to be implemented to avoid further transmission of infection in community and advise to take the BCG vaccine (vaccinated at the time of infancy) to protect with infection in childhood.

CONCLUSION

Current knowledge of evolutionary biology and genetics makes it clear that the conflict against TB is the survival of both host and pathogenesis. The article outlines the state of art, diagnosis, existing chemotherapy, vaccines and their associated problems. The implementation of WHO recommended DOTS. DOTS plus scheme appear to be the only satisfactory intervention to reduce the present burden in global perspective. However, present regimens are not convenient and convincing to facilitate effective treatment in resource developing nation. Therefore, the attention need to be focused on the development of more effective drugs with an excellent activities against MDR tuberculosis. The strategy for administration of antituberculous drugs and protease inhibitor in co-infected patients need to be explored by using effective drugs in shortening the current therapy period (6-12 months), minimize the adverse effect of the combination therapy in HIV positive subjects followed by effective vaccine. TB is a completely (100%) curable disease, provided the right diagnosis followed by effective chemotherapy and vaccination strategies employed at the right time.

ACKNOWLEDGEMENTS

We thank Dr. Narendra S. Hattikudur, General Manager, Diagnostic division, Nicholas Piramal India Ltd., Mumbai for their cooperation for writing of the manuscript, invaluable scientific advice from Dr. G.V. Kadival, Managing director V.K. Immunotech Ltd., G-6 Kothari compound Tikujiniwadi Road, Manpada Thane, Mumbai, India. Dr. N.K. Shah, Head, Department of biotechnology, MITS Gwalior M.P., India for critical counsel for reviewing of the article, and suggestions and Dr. Niyam Charan Sharma, Former Vice President, Department of Biotechnology, Cadila Pharmaceuticals Ltd. Ahmedabad, India for encouraging us to write an article are gratefully acknowledged.

REFERENCES

Alison MJ, Zappasodi P, Lurie MB (1962). Host parasite relationship in natively resistant and susceptible rabbits on quantitative inhalation of tubercle bacilli. Am Rev Respir Dis 85: 553-569.

Andersen P, Andersen AB, Sorensen AL, Nageli S (1995). Recall of long-lived immunity to *Mycobacterium tuberculosis* infection in mice. J Immunol 154 : 3359-3372.

Anderson AB, Anderson P, Ljungquist L (1992a). Structure and function of a 40,000 molecular weight protein antigen of *M. tuberculosis*. Infect Immun 60: 2317-2323.

Anderson AB, Ljungqvist L, Olsen M (1990). Evidence that protein antigen-b of *Mycobacterium tuberculosis* involved in phosphate metabolism. J Gen Microbiol 136: 477-480.

Anderson P, Askgaard D, Gottshan A, Bennedsen J, Nagai S, Heron I (1992b). Identification of immunodominant antigens during infection with *M. tuberculosis*. Scand J Immmunol 36:823-831.

Armstrong JA, Hart PDA (1971). Response of cultured macrophage to *Mycobacterium tuberculosis* with observations on fusion of lysosome with phagosomes. J Exp Med 134: 713-740.

Barrios C, Lussow AR, Van Embden J, Vander Zee R, Rappuoli R, Costantino P, Louis JA, Lambert PH, Del Giudice G (1992). Mycobacterial heat shock proteins as carrier molecules. II. The use of the 70-kDa mycobacterial heat shock protein as carrier for conjugated vaccines can circumvent the need of adjuvant and Bacillus Calmette Gurein priming. Eur.J.Immunol. 22:1365-1372.

Belisle JT, Vissa VD, Sievert T, Takayama K, Brennan PJ, Besra GS (1997). Role of the major antigen of *Mycobacterium tuberculosis* in cell wall biogenesis. Science 276: 1420-1422.

Bennedsen J, Thompson V, Pfyffer GE, Funke G, Fledman K, Beneke A, Jenkins PA, Hegginbothom M, Fahr A, Hengstler M, Cleator G, P. Klapper P, Wilkins GL (1996). Utility of PCR in diagnosing pulmonary tuberculosis. J. Clin. Microbiol. 29:1711-1719.

Bentleyhibbert SI, Quan X, Newman T, Huygen K, Godfrey HP (1999). Pathophysiology of antigen 85 in patients with active tuberculosis: antigen 85 circulates as complexes with fibronectin and immunoglobulin G. Infect Immun 67(2): 581-588.

Bisen PS, Garg SK, Tiwari RP, Tagore PR, Chandra R, Karnik R, Thaker N, Desai N, Ghosh PK, Fraziano M, Colizzi V (2003). Analysis of shotgun expression library of *Mycobacterium tuberculosis* genome for immunodominant polypeptide. Potential use in serodiagnosis. J Clin Diagn Lab Immunol 6: 1051-1058.

Bisen PS, Tiwari RP (2005). A diagnostic kit for detecting pulmonary and extra-pulmonary tuberculosis. World Intellectual Property Organization. PCT-WO2005/080987 A1. An International Patent published on 1st September 2005 (esp@cenet).

Bloch H (1964). Bacterial Pathogenicity and Chemotherapy: *In* Chemotherapy of tuberculosis" VC Barry (ed.), Butterworths Londan p38.

Borremans M, De Wit L, Volkaert G, Ooms J, De Bruyn J, Huygen K, Van Vooren JP, Stelandre M, Verhofstadt R, Content J (1989). Cloning, sequence determination, and expression of a 32 kilodalton-protein antigen gene of *M. tuberculosis*. Infect Immun 57: 3123-3130.

Bothamley GH (1995). Serological diagnosis of tuberculosis. Eur Respir J 8 (20): 676-688.

Bothamley GH, Rudd R, Festenstein F, Ivanyi J (1992). Clinical value of the measurement of *Mycobacterium tuberculosis* specific antibody in pulmonary tuberculosis. Thorax 47: 270-275.

Bothemley G, Swansonberk J, Britoon W, Ivanyi J (1991). Antibodies to *M. tuberculosis* specific antigen in lepromatous leprosy. Clin Exper Immunol 86:426-432.

Brightbill HD, Libraty DH, Krutzik SR, Yang RB, Belisle JT, Bleharski JR, Maitland M, Norgard MV, Plevy SE, Smale ST, Brennan PJ, Bloom BR, Godowski PJ, Modlin R L (1999). Host defense mechanisms triggered by microbial lipoproteins through toll-like receptors. Science 285: 732-736.

Brodin P, Rosenkrands I, Andersen P, Cole ST, Brosch R (2004). ESAT-6 proteins: protective antigens and virulence factors? Trends Microbiol 12: 500-508.

Center for Disease Control and Prevention. (1993a). Estimates of future global tuberculosis mortality and morbidity. Morbidity Mortality weekly Rep. 42:961-964.

Centre for Disease Control and Prevention (1993b). Diagnosis of tuberculosis by nucleic acid amplification methods applied to clinical specimens. Morbid Mortal Weekly Rep 82: 686-690.

Chan ED, Heifets L, Iseman MD (2000). Immunologic diagnosis of tuberculosis: a review. Tuberc. Lung Dis 80: 131-140.

Chan J, Fujiwara T, Brannan P, Mc Neil M, Turco SJ, Sibille JC, Snapper M, P Aisen P, Bloom BR (1989). Microbial glycolipids: possible virulence factors that scavenge oxygen radicals. Proc Natl Acad Sci USA 86: 2453-2457.

Chandramukhi A, Allen PR, Keen M, Ivanyi J (1985). Detection of Mycobacterial antigen and antibodies in the cerebrospinal fluid of patients with tuberculosis meningitis. J Med Microbiol 20(2): 239-247.

Chandramukhi A, Bothemley GH, Brennam PJ, Ivanyi J (1989). Levels of antibodies to define antigens of *M. tuberculosis* in tuberculosis meningitis. Journal of Clinical Microbiology 27:821-825.

Cole ST, Brosch R, Parkhill J, Ganrnier T, Churcher C, Harris D, Gordon SV, Eiglmeier K, Gas S, Barry CEI, Takaia F, Badcock K, Basham D, Brown D, Chillingworth T, Connor R,Davies R, Devlin K, Feltwell T, Gentles S, Hamlin N, Holroyd S, Hornsby T, Jagels K, Krogh A, Mc Lean J, Moulev S, Murphy L, Oliver K, Osborne J, Quail MA, Rajandream MA, Rogers J, Rutter S, Seeger K, Skelton J, Squarers R, Sulston JE, Taylor K, Whitehead S, Barell BG (1998). Deciphering the biology of *M.tuberculosis* from the complete genome sequence. Nature 393:537-544.

Coler RN, Campos-Neto A, Ovendale P, Day FH, Fling SP, Zhu L, Serbina N, Flynn JL, Reed SG, Alderson MR (2001). Vaccination with the T cell antigen Mtb 8.4 protects against challenge with *Mycobacterium tuberculosis*. J Immunol 166: 6227-6235.

Cox RA, Best GK (1972). Cell wall composition of two strains of *Blastomyces dermatitidis* exhibiting differences in virulence mice. Infect Immun 5: 447-453.

Crowle AJ, Dopl R, Ross E, May MH (1991). Evidence that vesicles containing living, virulent *Mycobacterium tuberculosis* or *Mycobacterium avium* in culture of human macrophages are not acidic. Infect Immun 59: 1823-1831.

Daniel TM, Janicki BW (1978). Mycobacterial antigens: A review of their isolation, chemistry and immunological properties. Microbial Rev 42: 85-113

Dannenberg AM Jr (1989). Immune mechanisms in the pathogenesis of pulmonary tuberculosis Review of Infectious Disease 11:369-377.

Dannenberg AM Jr (1991). Delayed type hypersensitivity and cell mediated immunity in tuberculosis. Immunol Today 12: 228-233.

De Cock KM, Soro B, Sanhoaty M, Lucas SB (1992). Tuberculosis and HIV infection in sub-Saharan Africa JAMA. 268: 1581-1587.

Dillon CD, Alderson MR, Day CH, Lewinsohn DM, Coler R, Bement T, Campos-Neto A, Skeiky YAW, Orme IM, Roberts A, Steen S, Dalemans W, Badaro R, Steven GR (1999). Molecular Characterization and human T-cell responses to a member of a novel *M.tuberculosis* mtb39 gene family. Infect Immun 67(6): 2941-2950.

Ellner JJ, Danial TM (1979). Immunosuppression by mycobacterial arabinomannan. Clin Exper Immunol 35: 250-251.

Elsaghier A, Lathigra R, Ivanyi J (1992). Epitope-scan of *M. tuberculosis* specific carboxy-terminal residues of the hsp-71 stress protein. Mol Immunol 29: 1153-1156.

Elsaghier A, Winkins AFEGL, Malhotra PK, Jindal S, Ivanyi J (1991). Elevated antibody levels to stress protein HSP 70 in smear-negative tuberculosis. Immunology Infectious Disease 1:323-328.

Espitia C, Scirrlto E, Bottasso D, Gonzalez amaro R, Hernandez pando R, Nacilla R (1992). High antibody levels to the *Mycobacterial* fibronectin-binding antigen of 30-31kDa in tuberculosis lepromatous leprosy. Clin Exper Immunol 87:362-367.

Falero-Diaz G, Challacombe S, Banerjee D, Douce G, Boyd A, Ivanyi J (2000). Intranasal vaccination of mice against infection with *Mycobacterium tuberculosis*. Vaccine 18: 3223-3229.

Fine PE, Rodrigues LC (1990). Modern vaccines. Mycobacterial diseases. Lancet 335: 1016-1020.

Fonseca DP, Benaissa-Trouw B, van Engelen M, Kraaijeveld CA, Snippe H, Verheul AF (2001). Induction of cell-mediated immunity against *Mycobacterium tuberculosis* using DNA vaccines encoding cytotoxic and helper T-cell epitopes of the 38-kilodalton protein. Infect Immun 69: 4839-4845.

Garg RK (1999). Tuberculosis of the central nervous system. Post-grad Med J Mar 75 (881): 133- 140.

Garg SK, Tiwari RP,Tiwari D, Singh R, Malhotra D, Ramnani VK, Prasad GBKS, Chandra R, Fraziano M, Colizzi V, Bisen PS (2003). Diagnosis of Tuberculosis: Available Technologies, Limitations and Possibilities. J Clin Lab Anal 16:1-7.

Ginsberg AM (1998). The tuberculosis epidemic Scientific Challenges and opportunities. Public Health Report March-April 113(2): 128-36.

Goren MB (1982). Immunoreactive substances of Mycobacteria. Am Rev Respir Dis 125: 50-69.

Goren MB, Brennan PJ (1979). Mycobacterial lipids: chemistry and biological activities, *In'* Tuberculosis'(G.P.Youman ed.), W.B. Saunders, Philadelphia, p.136.

Grange JM (1988). Mycobacteria and Human Disease. Published by Edward-Arnold Pvt. Ltd.

Grange JM, Laszlo A (1990). Serodiagnostic tests for tuberculosis a need for assessment of their operational predictive accuracy and acceptability. WHO Bulletin OMS. 68:571-576.

Havlir DV, Ellner JJ, Chervenak K, Boom WH (1991). Selective expansion of human gamma delta T cells by monocytes infected with live *M. tuberculosis.* J Clin Invest 87: 729-733.

Health India (2001). All you wanted to know about tuberculosis. Available at http://www. Health India.org/tb.html.Accessed 16 July.

Hella SR, Shinnick TM, Cohen ML (1988). Serological responses of patients with lepromatous and tuberculoid leprosy to 30,31 and 32 KDa antigens of *M. tuberculosis.* J Clin Microbiol 26: 2200-2202.

Hess J, Kaufmann SH (1999). Live antigen carriers as tools for improved antituberculosis vaccines. FEMS Immunol Med Microbiol 23: 165-173.

Horwitz MA, Harth G, Dillon BJ, Maslesa-Galic S (2000). Recombinant bacillus calmette-guerin (BCG) vaccines expressing the *Mycobacterium tuberculosis* 30-kDa major secretory protein induce greater protective immunity against tuberculosis than conventional BCG vaccines in a highly susceptible animal model. Proc Natl Acad Sci USA 97: 13853-13858.

Houghton RL, Lodes MJ, Dillon DC, Reynolds LD, Day CH, McNeill PD, Hendrickson RC, Skeiky YAW, Sampaio DP, Badaro R, Lyashchenko KP, Reed SG (2002). Use of multiepitope polyproteins in serodiagnosis of active tuberculosis. Clin Diag Lab Immunol 9(4): 883-891.

Ivanyi J, Morris JA, Keen M (1985). Studies with Mab to *Mycobacteria*, (Eds. Macario AJL and Macario EL) Page, 59-90. Mab Against Bacteria, Academic Press, London.

Ivanyi J, Thole J (1994). *In* Tuberculosis: Pathogenesis, Protection, and Control (ASM press, Washington DC, (Edited by BR Bloom).

Jackett PS, Bothamley GH, Batra HV, Mistry A, Young DB, Ivanyi J (1988). Specificity of antibodies to immunodominant mycobacterial antigens in pulmonary tuberculosis. Jour Clin Microbiol 26:2313-2318.

James OK, Takayama K, Armstrong Emma L (1982). Synthesis of Trehalose dimycolate (Cord factor) by a cell free system of *Mycobacterium smegmatis*. Biochem and Biophys Res Commu108 (1): 132-139.

Jorgensen JH, Salinas JR, Paxson R, Magnon K, Patterson JE, Patterson TF (1999). False-positive Gen-Probe direct *Mycobacterium tuberculosis* amplification test results for patients with pulmonary *M. kansasii* and *M.avium* infections. J Clin Microbiol 37(1): 175-178.

Kale AB, Kiessling R, Wanembeden JDA, Thole JER, Kumaratne DS, Pisa P, Wondimu A, Ottenhoff THM (1990). Induction of antigenic specific CD4 + HLA DR restricted cytotoxic T lymphocyte (CTL) as well as non-specific non-restricted killer cell by the recombinant mycobacterial 65 KDa heat shock protein. European J Immunol 20:369-377.

Kaplan MH, Chase MW (1980). Antibodies to mycobacteria in human tuberculosis I. Development of antibodies before and after antimicrobial therapy. J Infect Dis 142: 825-834.

Kartikeyan S, Bharmal RN, Tiwari RP, Bisen PS (2007) HIV and AIDS. Basic Elements and Priorities. Springer Biomedical Unit, Dordrecht, The Netherlands. ISBN 10:1-4020-5788-1, ISBN-13:978-1-4020-5788-5. (In Press).

Katrak SM, Shembalar PK, Bijwe SR, Bhandarkar LD (2000). The clinical, radiological and pathological profile of tuberculous meningitis in patients with and without human immunodeficiency virus infection. J Neurol Sci 181: 118-126.

Kessel MD, Gilot P, Missone M, Coene M, Cocito C (1993). Cloning and expression of a portion of the 34-Kilodalton proteins Gene of *Mycobacterium paratuberculosis*: Its application to serological analysis of Johne's disease. J Clin Microbiol 31: 947-954.

Kochi A (1991). The global tuberculosis situation and the new control strategy of the world health organization. Tubercle 72: 1-6.

Kolk AHJ, Schuitema ARJ, Kuijper S, van Leeuwen J, Hermans PWM, van Embden JDA, Hartskeerl RA (1992). Detection of *M. tuberculosis* in clinical samples by using polymerase chain reaction and the non-radioactive detection system. J of Clin Microbiol 30:2567-2575.

Lee BY, Hepta SA, Brennan PJ (1992). Characterization of the major membrane protein of the virulent *M. tuberculosis*. Infect Immunol 60:2066-2072.

Lienhardt C, Rowley J, Manney K, Lahai G, Needham D, Milligan P, Mc.Adam KP (2001). Factors affecting time delay to treatment in a tuberculosis control programme in a sub-saharan African country: the Experience of The Gambia. Int J Tubercle and Lung Disease 5:233-239.

Ljungquist L, Worsaae A, Heron I (1998). Antibody response against *M. tuberculosis* in 11 strains of inbred mice: novel Mab specification generated by fusion, using spleens from BALB10 and CBA Mice. J Infect Immune. 56:1994-1998.

Lurie MB (1964). Resistance to tuberculosis experimental studies in native and acquired defensive mechanism. Harvard University Press, Cambridge, Mass.

Lyadova IV, Vordermeier HM, Eruslanov EB, Khaidukov SV, Apt AS, Hewinson R G (2001). Intranasal BCG vaccination protects BALB/c mice against virulent *Mycobacterium bovis* and accelerates production of IFN-gamma in their lungs. Clin Exp Immunol 126: 274-279.

Lyashchenko K, Manca M, Colangeli R, Heijbel A, Williams A, Gennaro ML (1998). Use of *Mycobacterium tuberculosis* complex-specific antigen cocktails for a skin test specific for tuberculosis. Infect Immunol 66: 3606-3610.

Manam S, Ledwith BJ, Barnum AB, Troilo PJ, Pauley CJ, Harper LB, Griffiths TG, Niu Z, Denisova L, Follmer TT, Pacchione SJ, Wang Z, Beare CM, Bagdon WJ, Nichols W W (2000). Plasmid DNA vaccines: tissue distribution and effects of DNA sequence, adjuvants and delivery method on integration into host DNA. Intervirology 43: 273-281.

Manca C, Lyashchenko K, Colangeli R, Gennaro ML (1997b). MTC28, novel 28-kDa proline-rich secreted antigen specific for the *M.tuberculosis* complex. Infect Immun 65(12): 4951-4957.

Manca C, Lyashchenko K, Wiker HG, Usai D, Colangeli R, Gennaro ML (1997a). Molecular cloning, purification, and serological characterization of MPT 63, a novel antigen secreted by *M. tuberculosis*. Infect Immun 65(1): 16-23.

Mehta PK, Khuller GK (1988). Protective immunity to experimental tuberculosis by manophosphoinositides of mycobacteria. Med Microbial Immunol 177: 265-284.

Minden P, Houghten RA, Spear JR, Shinnick. MT (1986). A chemically synthesized peptide, which elicits humoral and cellular immune response to Mycobacterial antigens. Infect Immun 53:560-564.

Misaki A, Azuma I, Yammamura Y (1977). Structural and immunochemical studies on D-arabino-D-mannans and D-Mannan of *Mycobacterium tuberculosis* and other *Mycobacterium* species. J Biochem 82: 1759-1770.

Moreno C, Taverne J, Mehlert A, Bate CAW, Brealey RJ, Meager A, Rook GAW, Playfair JHL (1989). Lipoarabinomannan from *M. tuberculosis* induces the production of tumor necrosis factor from human and murine macrophages. Clin Exp Immunol 76: 240-245.

Munk ME, Shinnick TM, Kaufmann SHE (1990). Epitopes of mycobacterial heat shock protein 65 for human T cells comprise different structure. Infect Immun 180: 272-277.

Murray JF (1998). Tuberculosis and HIV infection: A global perspective. Respiration 65(5): 335-342.

Mustafa AS, Amoudy HA, Wiker HG, Abal AT, Ravn P, Oftung F, Andersen P (1998). Comparison of antigen-specific T-cell responses of tuberculosis patients using complex or single antigens of *Mycobacterium tuberculosis*. Scand J Immunol 48: 535-543.

Mustafa AS, Gill HK, Nerland A, Britton WJ, Mehra V, Bloom BR, Young RA, T. Godal T (1986). Human T-cell clone recognizes a major *M.leprae* protein antigen expressed in *E. coli*. Nature 309: 63-66.

Nagai S, Wiker HG, Harboe M, Kinomoto M (1991). Isolation and partial characterization of major protein antigens in the culture fluid of *M. tuberculosis*. Infect Immun 59:372-382.

Neufert C, Pai RK, Noss EH, Berger M, Boom WH, Harding CV (2001). *Mycobacterium tuberculosis* 19-kDa lipoprotein promotes neutrophil activation. J Immunol 167: 1542-1549.

Nor NM, Musa M (2004). Approaches towards the development of a vaccine against tuberculosis: recombinant BCG and DNA vaccine. Tuberculosis (Edinb) 84: 102-109.

Nunn P, Fellten M (1994). Surveillance of resistance to tuberculosis drugs in developing countries. Tubercle Lung Disease 75:163-167.

Oftung F, Mustafa AS, Husson R, Young RA, Godal T (1987). Human T cell clones recognize two abundant *M. tuberculosis* protein antigens expressed in *Escherichia coli*. J Immunol 138:927-931.

Pao CC, Yen TSB, You JB, Mao JS, Fiss EH, Chang CH (1990). Detection and identification of *M.tuberculosis* by DNA amplification. J Clin Microbiol 28:1877-1880.

Parrish NM, Dick JD, Bishai WR (1998). Mechanisms of Latency in *M. tuberculosis*. Trends in Microbiology 6: 107-112.

Payne SN, Draper P, Rees RJW (1982). Serological activity of purified glycolipid from *Mycobacterium leprae*. Int J Lepr 50: 220-221.

Porcelli SA (2000). Albert Einstein College of Medicine Nature.404: 884-888.

Radhakrishnan VV, Annamma M, Sobha S (1990). Corelation between culture of *M. tuberculosis* and IgG antibody to *M.tuberculosis* antigen-5 in the cerebrospinal fluid of patients with tuberculosis meningitis. J Infect Dis 21 (3): 271-277.

Rama Rao A, Satchidanandam V (1996). Analysis of a genomic DNA expression library of *M. tuberculosis* using tuberculosis patient sera: evidence for modulation of host immune response. Infect Immun 64: 3765-3771.

Rambukkana A, Das PK, Chand A, Bass JG, Groothvis DG, Kolk AHJ (1991). Sub cellular distribution of M.ab defined epitopes on immunodominant *M.tuberculosis* protein in the 30KDa region. Identification and localization of 29/33KDa doublet protein on mycobacterial cell wall. Scand J Immunol 33:763-775.

Reggiardo Z, Middlebrook G (1974b). Serologically active glycolipid families from *Mycobacterium bovis* BCG. II. Serologic studies on human sera. Am J Epidemiol 100:477-486.

Reggiardo Z, Middlebrookn G (1974a). Serologically active glycolipid families from *Mycobacterium bovis* BCG. I. Extraction, purification and immunologic studies. Am J Epidemiol 100: 469-476.

Revised National Tuberculosis Control Program (RNTCP). Status Report "TB India 2001" edited by Central TB Division, Directrate general of Health Services, Ministry of Health and Family Welfare, Nirman Bhavan New Delhi, India.

Rodrigues LC, Diwan VK, Wheeler JG (1993). Protective effect of BCG against tuberculous meningitis and miliary tuberculosis: a meta-analysis. Int J Epidemiol 22: 1154-1158.

Roman F, Angier J, Pescher P, Marchal G (1993a). Isolation of a proline rich mycobacterial protein eliciting delayed type hypersensitivity reaction only in guinea pigs immunized with living mycobacteria. Proc Natl Acad Sci USA 90:5322-5326.

Roman G, Laqueyrerie F, Militzer P, Pescher P, Chavarol P, Lagrandieie M, Auregan G, Gheorghin M, Marchal G (1993b). Identification of an *M. bovis* BCG 45/47RDG antigen complex, an immunodominant target for antibody response after immunization with living bacteria. Infect Immunol 61:742-750.

Sada E, Brennan PJ, Herrera T, Torres M (1990). Evaluation of Lipoarabinomannan for the serological diagnosis of tuberculosis. J Clin Microbiol 28:2587-2590.

Sada E, Ruiz-Palacios G.M, Lopez-Vidal Y, Ponce de Leon S (1983). Detection of mycobacterial antigens in cerebrospinal fluid of patients with tuberculous meningitis by enzyme-linked immunosorbent assay. Lancet. 2:651-652.

Sao T, Juang TCY, Sai T, Lan YC, Lee CH (1992). Whole lung tuberculosis: A disease with high mortality which is frequently misdiagnosed.Chest101: 1309-1311.

Saul Allan, Yeganeh F (1986). Electrophoretic identification of fusion proteins expressed in single recombinant -bacteriophage plaques. Analytical Biochem 156: 354-356.

Shinnick TM, Krat C, Schadow S (1987a). Isolation and restriction site maps of the Genes encoding five *M. tuberculosis* proteins. Infect Immun 55(7): 1718-1721.

Silbaq Fauzi S, Sangnae C, Cole ST, Brennan PJ (1998). Characterization of a 34-kDa protein of *Mycobacterium leprae* that is isologous to the immunodominant 34 kDa antigen of *Mycobacterium paratuberculosis*. Infec Immun 66 (11): 5576-5579.

Silva CL (1999). The potential use of heat-shock proteins to vaccinate against Mycobacterial infections. Microbes Infect 1: 429-435.

Singh AP, Khuller GK (1993). Enhancement of immunogenicity of immnuoprotective glycerophospholipid antigen of mycobacteria using liposomes containing Lipid A. J Lip Res 3(2): 303-316.

Sirugue SA, Sixou S, Lakhdar-Ghazal S, .Tocanne FJF, Aneelle GL (1990). Mycobacteria glycolipids as potential pathogenicity effectors: alteration of model and nutral membranes. Biochemistry 29: 8498-8502.

Skeiky YA, Ovendale PJ, Jen S, Alderson MR, Dillon DC, Smith S, Wilson CB, Orme IM, Reed SG, Campos-Neto A (2000). T cell expression cloning of a *Mycobacterium tuberculosis* gene encoding a protective antigen associated with the early control of infection. J Immunol 165: 7140-7149.

Skerman VDB, Mcgown V, Sneath PHA (1980). Approved lists of bacterial names. Int J Syst Bacteriol 30: 225-420.

Srivastava KL, Bansal M, Gupta S, Srivastava R, Kapoor RK, Wakhlu I, Srivastava BS (1998). Diagnosis of tuberculous meningitis by detection of antigen and antibodies in CSF and sera. Indian Pediatr 9: 841-850.

Stoneburner RL, Ruiz MM, Milburg JA, Schultz S, Vennema A, Morse DL (1987). Tuberculosis and acquired immunodeficiency syndrome. New York City, Morbid Mortal Weekly Rep. 36: 786-796.

Sturgill-Koszycki S, Schlesinger PH, Chakoraborty P, Haddin PL, Collin HL (1994). Lack of acidification in *Mycobacterium tuberculosis* phagosome produced by exclusion of proton ATPase. Science. 263: 678-681.

Subrahmanyam D (1964). Studies on the phosphoglycerides of *Mycobacterium tuberculosis*. Can J Biochem 42:1195-1201.

Sumi MG, Mathai A, Sarada C, Radhakrishnan VV (1999). Rapid diagnosis of tuberculous meningitis by a dot immunobinding assay to detect mycobacterial antigen in cerebrospinal fluid specimens. J Clin Microbiol 37:3925-3927.

Thole JE, Wilco R, Keulen J, Arend H, Kolk J, Dick G, Groothuis G, Berwald LG, Tiesjema RH, Van Embden A (1987). Characterization, sequence determination, and Immunogenecity of a 64-Kilodalton protein of Mycobacterium bovis BCG expressed in Escherichia coli K. Infect Immun 55 (6): 1466-1475.

Thoma-Uszynski S, Kiertscher SM, Ochoa MT, Bouis DA, Norgard MV, Miyake K, Godowski PJ, Roth MD, Modlin RL (2000). Activation of tolllike receptor 2 on human dendritic cells triggers induction of IL-12, but not IL-10. J Immunol 165: 3804-3810.

Thwaites G, Chau TT, Mai NT, Drobniewski F, McAdam K, Farrar J (2000). Tuberculous meningitis. J Neurol Neurosurg Psychiatry 68:289-299.

Thwaites G, Chau TT, Mai NT, Drobniewski F, McAdam K, Farrar J (2000). Tuberculous meningitis. J Neurol Neurosurg Psychiatry 68:289-299.

Tiwari RP, Hattikudur NS, Bharmal RN, Kartikeyan S, Deshmukh Neeta M, Bisen PS (2006). Modern approaches to a rapid diagnosis of Tuberculosis: Promises and challenges ahead. Tuberculosis available online 6 Oct 2006 doi: 10.1016/j.tube. 2006.07.005.

Tiwari RP, Tiwari D, Garg SK, Chandra R, Bisen PS (2005). Glycolipids of *Mycobacterium tuberculosis* strain H37Rv are potential serological markers for diagnosis of active tuberculosis. J Clin Diagn Lab Immunol.12 (3): 465-473.

Turner M, Vanvoren JP, Bruyan JD, E. Pdierekx SE, Yernault JC (1998). Humoral immune response in human TB: immunoglobulins G1A and M directed against the P32 protein antigen of *Mycobacterium bovis*, Bacillus, Calmette, Guerin. J Clin Microbiol 26:1714-1719.

Van Vooren JP, Drowart A, De Bruyn J, Launois P, Millan J, .Dellaporte E, Develoux M, Yernault JC, Huygen K (1992). Humoral immune responses against the 85A and 85B antigens of *Mycobacterium bovis* BCG in patients with leprosy and tuberculosis J Clin Microbiol 30: 1608-1610.

Verbon A, Hartskeerl RA, Schuitema A, Kolk AHJ (1992). Characterization of B cell epitopes on the 16kDa antigen of *M. tuberculosis*. Clin Exp Immunol 89: 395-401.

Verbon. A, Kuijper S, Jansen HM, Speelman P, Kolk AHJ (1990). Antigens in culture supernatant of *M. tuberculosis*: epitopes defined by monoclonal and human antibodies. J Clin Microbiol 136:955-964.

Vordermeier H, Harris DP, Mehrotra PK, Romen E, Elsaghier A, Morino C, Ivanyi J (1992). *M. tuberculosis* complex specific T cell stimulation and DTH reaction induced with a peptide from the 38kDa protein Scand J Immunol 35:711-718.

Vordermeier HM, Harris DP, Romen E, Lathigra R, Morino C, Ivanyi J (1991). Identification of T cell stimulatory peptides from the 38kDa proteins of M.tuberculosis. J. Immunol 147:1023-1029.

Wallis RS, Paranjpe R, Phillips M (1993). Identification by two-dimensional gel electrophoresis of a 58-kDa, tumour necrosis factor-inducing protein of *M. tuberculosis*. Infect Immun 61: 627-632.

Wang FD, Su WJ, Jang TN, Liu CY (1994). Serological evaluation for tuberculosis by antigen 60 ELISA tests. Chung Hua I Hsueh Tsa Chih 54: 300-305.

Warwick –Davies J, Dhillon J, O'Brien L, Andrew PW, Lowrie DB (1994). Apparent killing of *M. tuberculosis* by cytokine-activated human monocytes can be an artefact of a cytotoxic effect on the monocytes. Clin Exp Immunol 96: 214-217.

Webb JR, Vedvick TS, Alderson MR, Ovendale PJ, Johnson SM, Reed SG, Skeiky YAW (1998). Molecular Cloning, Expression, and Immunogenicity of MTB12, a novel low - molecular-weight antigen secreted by *M.tuberculosis*. Infect Immun 66 (9): 4208-4214.

Wolinsky E, Schaefer WB (1973). Proposed numbering scheme for Mycobacterial serotypes by agglutination. Int J Syst Bacteriol 23:182-183.

World Health Organization (1996). The World Health Report. World Health Organization, Geneva, WHO.

World Health Organization (1997). Anti-tuberculosis drug resistance in the world. The WHO /IUATLD global project on anti-tuberculosis drug-resistance surveillance. 1994-1997. World Health Organization, Geneva, Switzerland.

World Health Organization (2002). Strategic framework to decrease the burden of TB/ HIV, Geneva, WHO.

World Health Organization (2003). Global Tuberculosis Control: Surveillance, Planning, Financing, WHO report 2003. (WHO/CDS/TB/2003.316), WHO, Geneva, Switzerland.

Young D, Kent L, Rees A, Lamb J, Ivanyi J (1986). Immunological activity of a 38-Kilodalton protein purified from *M. tuberculosis*. Infect Immun 54:177-183.

Young DB, Garbe TR (1991). Heat shock proteins and antigens of *M. tuberculosis*. Infect Immun 59: 3086-3093.

Young DB, Kent L, Young RA (1987). Screening of a recombinant mycobacterial DNA library with polyclonal antiserum and molecular weight analysis of expressed antigens. Infect. Immun. 55:1421-1425.

Young DB, Stewart GR (2002). Tuberculosis vaccines. Br Med Bull 62: 73-86.

Helicobacter pylori Infection: Diseases and Cure: A Molecular Approach

A. SHARMA and S. K. DHAR*

1. INTRODUCTION

The human species live in a dynamic state of co-existence with several forms of microbes. Nearly all of these host microbe relationships are mutually beneficial. On occasion, however, mutualism gives way to parasitism, and this leads to damage to the human host. The interaction of such microbes turned pathogen could further be headed towards complex cellular mechanisms resulting severe alternation in the functioning of host's cells. This kind of interaction between microbes and human provides a multidimensional scope for research and, hence, to find new therapeutic options to treat the severe diseases caused due to such interaction. One of the most relevant widespread examples is the interaction of bacteria *Helicobacter pylori* with the human host.

Helicobacter pylori are Gram-negative spiral-shaped bacteria that infects almost half of the world's population. They are found in the gastric mucous layer or adherent to the epithelial lining of the stomach. This infection can contribute to the development of diseases in the stomach and duodenum. However, the host response play major role in the pathogenesis of *H. pylori* and as predictor of disease. The bacteria cause more than 90% of duodenal ulcers and upto 80% of gastric ulcers (Zevering et al. 1999). The pathogen is

Special Centre for Molecular Medicine, Jawaharlal Nehru University, New Delhi-110067, India.
*Corresponding author: skdhar2002@yahoo.co.in

also responsible for causing gastric cancer that is the second most common cancer in the world. *H. pylori* have attracted researchers to understand molecular biology aspects of infection and pathogenesis. Extensive research work has been carried out since *H. pylori* became recognized as a significant human pathogen. In particular, advances in molecular bacteriology, including the complete determination of the *H. pylori* genome (Tomb et al. 1997), have provided tools to understand the pathogenesis of disease. Research will continue to get the most effective treatment of *H. pylori*, especially in the light of growing resistance against certain antibiotics. Efforts in this direction will help to develop effective screening tests, therapies, or means of preventing infection, such as vaccine.

2. HISTORY

The presence of gastric spirochaeta organism was first documented over a century ago. It was felt that these spiral bacteria were merely contaminated and the medical community generally ignored the reports. In 1983, Warren and Marshall described *Helicobacter pylori*. Marshall and Warren were able to demonstrate a strong association between the presence of *H. pylori* and the finding of inflammation on gastric biopsy (Marshall and Warren 1984). Recently, these two Australian Scientists were honored with Nobel Prize for the year 2005 in Physiology and Medicine for their discovery that inflammation in the stomach (gastritis) as well as ulceration of the stomach or duoclenum (peptic ulcer disease) is the result of an infection of the stomach caused by the bacterium *Helicobacter pylori*. The importance of their discovery lies in the fact that after their findings of the real cause of the gastric diseases as bacteria, a new initiative for the effective therapy of these diseases were started. The bacterium was initially called *Campylobacter pyloridis*, and later on placed in a separate genus *Helicobacter*. The name *pylori* come from the Latin word *pylorus*, and refer to the circular opening leading from the stomach into the duodenum. Since then many discoveries have been made which have revolutionized our understanding of *H. pylori* infection.

2.1 The Bacteria

H. pylori are spiral-shaped, Gram-negative bacteria, about 3 micrometers long with a diameter of about 0.5 micrometer. It has 4-6 flagella (Fig. 1). The bacterium is microaerophilic, i.e., it requires oxygen but at lower levels than those contained in the atmosphere. Under conditions of environmental

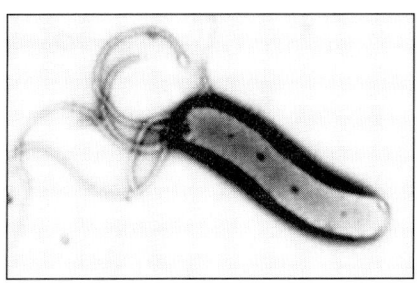

Fig. 1 *Helicobacter pylori*: Gram-negative, spiral-shaped bacteria with 4-6 flagella.

stress, *Helicobacter* gets converted from a spiral to a coccoid form (Chan et al. 1994). These forms of *H. pylori* are more resistant and dormant and may not pose infection risk.

2.2 The Infection

Humans appear to be the principle host of *H. pylori*. *H. pylori* can thus be regarded as indigenous or "normal" flora, which most humans acquire within the first few years of childhood and then carry for life; and colonization increases progressively with age. The mode of transmission is believed to be person to person by fecal-oral, oral-oral or gastro-oral route. Waterborne transmission of *H. pylori*, via contaminated well water and river water, also appears likely. *H. pylori* infection usually persists for many years, leading to various gastric diseases in due course of time. It was thought previously that the bacterial population cannot persist the acidic environment of the stomach but later on *H. pylori* was found to survive in the stomach mucosal layer. It employs multiple mechanisms to survive and get established in hostile stomach environment. Motility provided by flagella allows it to swim to less acidic pH area beneath the gastric mucous. Urease enzyme plays important role to protect the bacteria from acidic gastric environment by degrading urea and buffering the acidic surroundings thereby permitting *H. pylori* to survive in the acidic environment of the stomach (Suerbaum et al. 2002).

$$[C=O\,(NH_2)_2] + H^+ + 2H_2O \xrightarrow{\text{Urease}} HCO_3^- + 2\,(NH_4^+)$$

Thus, Urease hydrolyzes urea into carbon dioxide and ammonia, which helps in creating a protective anti-acid barrier around *H. pylori* as ammonia neutralizes the acidic pH of gastric acid. The Urease enzyme activity is regulated by a unique pH gated urea channel-Ure I that opens at low pH and shuts down the influx of urea under neutral conditions in order to avoid over alkalizations. This is an excellent way of maintaining intracellular and extracellular pH. Urease production and motility are essential for the initial establishment in host system followed by adherence to the gastric epithelial cells and colonization of the stomach. Bacteria with better adherence properties could colonize with high densities and will be more resistant to be cleared from the stomach. The binding of *H. pylori* with the epithelial cells involves glycolipids and several proteins. The best characterized adhesin, Bab is a protein that binds to a Lewis antigen of human cells. Although three *bab* alleles have been identified (*babA1*, *babA2*, and *babB*), only the *babA2* gene product is necessary for Lewis b binding activity (Ilver et al. 1998, Kim et al. 2001). Bab A2 is a membrane-bound adhesin encoded by the *H. pylori* strain-selective gene *babA2* that binds the blood group antigen Lewis b present on gastric epithelial cell membranes, and *H. pylori* strains possessing *babA2* are associated with increased risk for gastric adenocarcinoma (Gerhard et al. 1999). Some of these steps of the establishment of *H. pylori* in the host are shown in Figure 2.

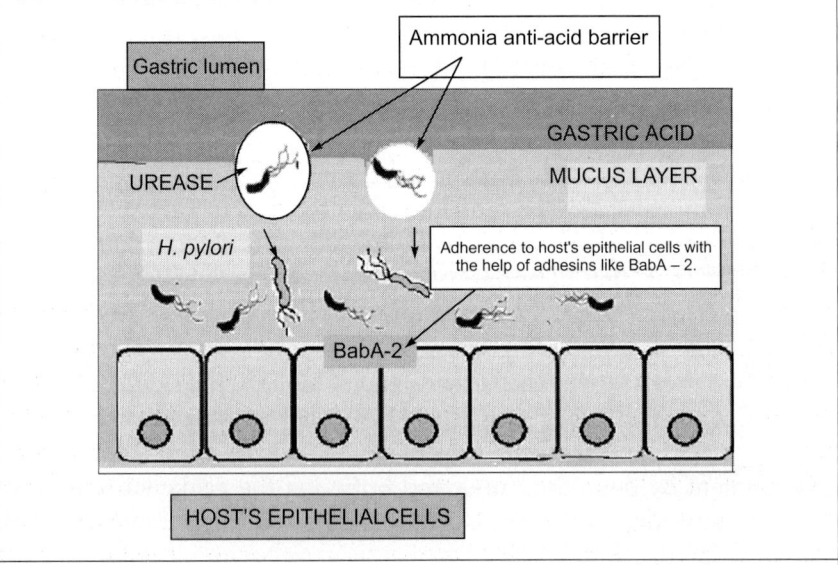

Fig. 2   Establishment of *H. pylori* in the host.
Helicobacter pylori, by means of their flagella, are able to swim through the mucus layer of the stomach and adhere to the epithelial cells beneath the mucous layer where the pH is near neutral. *H. pylori* produce an enzyme urease that hydrolyses urea into carbon dioxide and ammonia which neutralizes the gastric acid and thus creates an anti acid barrier around *H. pylori*.

3. EPIDEMIOLOGY

In Western countries, *H. pylori* associated gastritis ultimately causes peptic ulceration in about 30% of infected persons. There is increased prevalence with advancing age; over 50% of asymptomatic persons over 60 years of age show evidence of *H. pylori* infection. In underdeveloped parts of the world, the prevalence of *H. pylori* colonization is nearly 80% by age of 30. Densely populated region and poor socioeconomic condition is associated with high infection rate. The infection is more common in crowded living conditions with poor sanitation. The fact that the majority of infections develop during childhood and that *de novo* infection after the age of 20 is rare suggests that sanitary conditions during childhood play a crucial role in the acquisition of infection. In countries with poor sanitation, 90% of the adult population can be infected. Interfamilial spread appears to play a central role in transmission of infection in both developing and developed countries. Males in many populations also appear to have 20% to 30% higher rate of infection than females (Replogle et al. 1995). The situation in India is not fully clear, a majority of population is infected at a young age, probably by feco-oral or oral-oral route, and the association with duodenal ulcer and gastric ulcer is high although the prevalence of gastric cancer is low (Guha et al. 1997). Gastric cancer caused by *H. pylori* has been found to be common in Japan with the highest stomach cancer risk closely followed by Korea. (Lee et al. 2002). Japanese investigators have noted that more than 65% of Japanese above the age of 50 years are infected with *H. pylori*. However, in Japanese population association between VacA, CagA or babA2 and clinical outcome is not very clear. These results contradict recent study (Blaser et al. 1995) showing high prevalence due to these factors in western population. Therefore, it is possible that the prevalence of gastric carcinoma in Japanese population is due to virulence factors other than found in pathogenic strains in western population (Mizushima et al. 2001).

4. PATHOGENESIS

There is a strong association between gastric *H. pylori* infection and gastroduodenal diseases. The presence of the bacterium invariably causes the surrounding mucosa to become inflamed, and the degree of gastritis present is positively correlated with the extent of *H. pylori* colonization. Although most people infected with *H. pylori* remain asymptomatic, in some the gastritis progresses to more severe forms of gastric diseases. Atrophy, characterized by distortion and destruction of the glands, can develop.

Furthermore, *H. pylori* induced epithelial cell degeneration results in ulceration. In addition, the infection for a long time could lead to development of cancer. The organism is present in ~92% of patients with active chronic gastritis, ~88-100% with duodenal ulcer, ~58-100% with gastric ulcer and ~46-94% with gastric cancer (Micev et al. 2001). Disease outcome depends on various factors including bacterial genotype and on host physiology, genotype, and dietary habits. A schematic model of disease progression in the host is demonstrated in Figure 3.

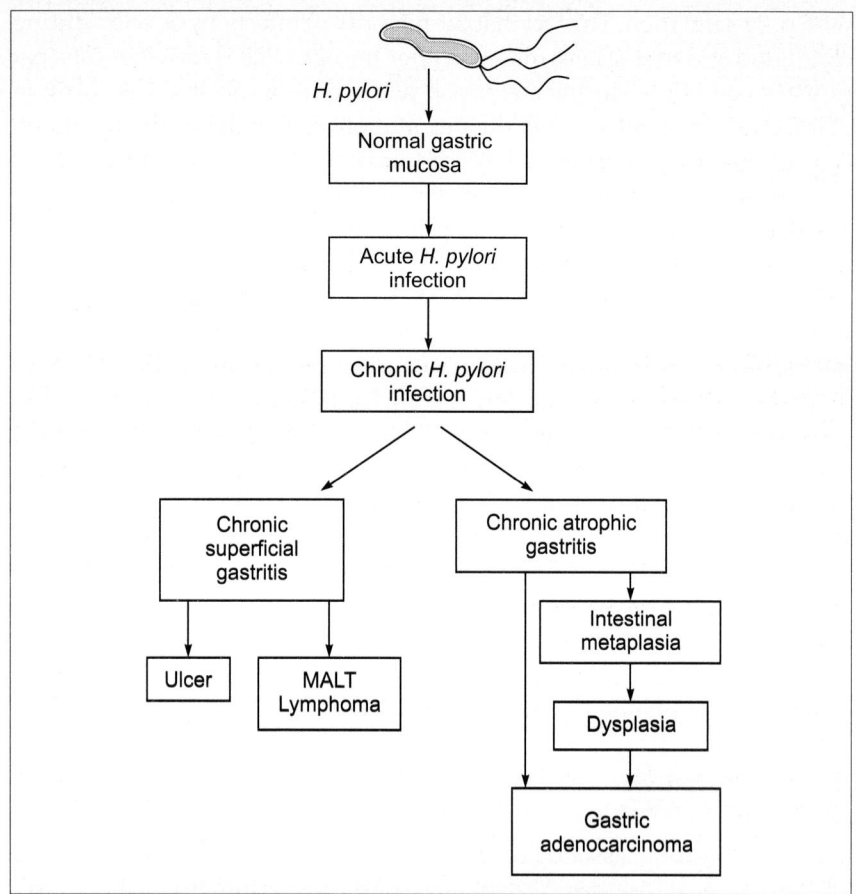

Fig. 3   Disease Progression.
After few months of infection the person develops chronic gastritis which remains for the whole life. In some cases it turns to peptic ulcers and lymphoma of mucosa associated lymphoid reticular tissue (MALT) while in another cases it turns to atrophic gastritis leading towards gastric adenocarcinoma.

4.1 Gastritis

Gastritis commonly refers to inflammation of the lining of the stomach, but the term is often used to cover a variety of symptoms resulting from stomach lining inflammation and symptoms of burning or discomfort. The causal relationship between *H. pylori* and chronic superficial gastritis is well established. The evidence for the association is based on: (1) Virtually all *H. pylori*-positive patients demonstrate antral gastritis, (2) Eradication of *H. pylori* infection results in resolution of gastritis and (3) The lesion of chronic superficial gastritis has been reproduced following intragastric administration of the organism in animal models and in human volunteer (Angie Man-Chi Eng 1994). Atrophic gastritis is believed to be a precancerous lesion that leads to carcinogenesis.

4.2 Peptic Ulcer

An ulcer is an erosion of tissue, somewhat circular in shape in the wall of the stomach, occurs due to the damage to mucous layer mainly caused by *H. pylori* invasion and it increases with continuous exposure to acidity. With enough acidity and continued absence of mucous, the ulcer may grow deep enough to reach the underlying blood flow. When this happens, the ulcer is said to have perforated the stomach wall, allowing stomach acid and pepsin directly to act on the tissue underneath. The ulcer formation could become very severe in some cases. The most common ulcer symptom is gnawing or burning pain in the stomach. It may last from minutes to hours and may be relieved by eating or by taking antacids. Less common ulcer symptoms include nausea, vomiting, and loss of appetite. In severe cases, bleeding can also occur; prolonged bleeding may cause anemia leading to weakness and fatigue.

It is important to note at this point that the second most common cause of peptic ulcer is use of non-steroidal anti-inflammatory drugs NSAIDs that act by inhibiting inflammatory cyclooxygenase enzyme COX1 that catalyze inflammatory prostaglandin formation. The normal function of prostaglandins is to increase mucous and bicarbonate production, inhibits stomach acid secretion, and increase blood flow within the stomach wall. By acting on COX-1, NSAIDs restrict these self-protection mechanisms, allowing stomach ulcers to develop. It is primarily through this mechanism, not a local acid effect, that NSAIDs cause stomach ulcers.

4.3 Gastric Cancer

H. pylori infection in its severe form can turn in gastric carcinogenesis. In 1994, the International Agency for Research on Cancer Working Group

(IARC) stated, "*H. pylori* plays a causal role in the chain of events leading to cancer" referring to adenocarcinoma and lymphoma of the stomach as well as to the more benign mucosal-associated lymphoid tissues (MALT) (Nomura et al. 1991).

4.3.1 Gastric Adenocarcinoma

H. pylori are considered to be definite carcinogen that leads to adenocarcinoma of the stomach. Chronic infection with *H. pylori* leads to a persistent, often lifelong inflammatory reaction in the stomach. After many years of chronic inflammation, atrophic gastritis develops. Atrophic gastritis is associated with cystic degeneration, impaired mucus production and eventually intestinal metaplasia. The risk of progression of cancer becomes high by this stage. The cascade of events from chronic inflammation to gastric atrophy and increased bacterial production of toxic nitrogenous by-products is the leading mechanism for damage of the gastric epithelial cell DNA that leads to abnormal proliferation of host cells and gastric cancer.

4.3.2 Gastric Lymphoma

The normal stomach does not have any lymphoid follicles. In the presence of infection, however, reactive T-cells, plasma cells and other T-cells organize in the area of the infection and subsequently develop lymph tissue closely resembling follicles. Mucosa associated lymphoid tissue (MALT) may undergo malignant change causing a lymphoma of the stomach. *H. pylori* have been found to be associated with development of MALT and subsequent transformation to malignant lymphoma. Although the link between inflammation and cancer has been well-established, the mechanism involved is still to be understood fully.

5. DIAGNOSIS

There are several invasive as well as non-invasive methods available to diagnose *H. pylori* infection. Histological evaluation, culture, polymerase chain reaction (PCR) and rapid urease tests are typically performed on tissue obtained from endoscopy. Alternatively, simple breath tests, serology, and stool assays are sometimes used and trials investigating PCR amplification of saliva, feces, and dental plaque to detect the presence of *H. pylori* are ongoing. These several methods have some advantages and also some disadvantages depending on the factors like reliability, costs and diagnosis time.

5.1 Invasive Tests

5.1.1 Rapid Urease Test

H. pylori is a urea–producing organism and this rapid urease tests provide a simple, rapid and cost-effective method for the detection of *H. pylori*. Samples obtained on endoscopy are placed in urea-containing medium, if urease is present, the urea will be broken down to carbon dioxide and ammonia, with a resultant increase in the pH of the medium and a subsequent color change in the pH-dependent indicator. This is a good tool for endoscopist in making decision as to whether or not therapy should be prescribed. Two rapid urease tests are commercially available, the HUT-test and the CLO-test, both tests show comparable sensitivity and specificities. In recent years, new commercial strip test, the Pyloritek is designed to give a result in 1 hour, without requiring special incubation temperature and also allowing several biopsies to be tested at the same time. This test is limited by the possibility of false positive results. Decreased urease activity, caused either by recent ingestion of antibiotic agents, bismuth compounds, proton pump inhibitors, or by bile reflux, can contribute to these false-positive results.

5.1.2 Culture

Culture is not routine diagnosis for *H. pylori* infection because of the availability of other invasive tests to detect the organism in patients. Culturing *H. pylori* in a culture medium involves some difficulties and also the required facilities of culture may not be available in all the hospitals and clinics. One of the major advantages of culture is that it allows sensitivity testing of *H. pylori* to the drug agents used in the treatment.

5.1.3 Polymerase Chain Reaction (PCR)

With the advent of PCR, many exciting possibilities emerged for diagnosing and classifying *H. pylori* infection. The detection of *H. pylori* in gastric biopsy and gastric juice by PCR is more sensitive method compared to other invasive methods. The test can be used particularly for post treatment diagnosis of *H. pylori*. The limitation of this test is risk of false positive results that may be either from residual *H. pylori* DNA on the fiber optic endoscopes because of inadequate cleaning and disinfection or from cross-contamination during processing of specimens in this laboratory. At present, PCR is still technically demanding and not generally available as a routine diagnosis tool.

5.2 Non-invasive Tests

5.2.1 Serology

During *H. pylori* infection, the immune system provokes local and systemic antibody responses and transient rise in IgM, followed by a rise in specific IgA and IgG. These antibodies can be detected in serum or whole-blood samples easily by Enzyme-linked immunosorbent assay (ELISA) or latex agglutination that have been developed in most clinical laboratories. The serological tests have been appreciated well and several ELISA kits have been tested with sensitivity and specificity in the range of 90-95% and 80-90%, respectively. However, this method is not a useful way of confirming eradication of *H. pylori* since several different samples and changes in titers of specified amounts over time would be needed.

5.2.2 Urea Breath Test

The (^{13}C)-urea breath test is highly sensitive and specific for the detection of *H. pylori* infection. In contrast to serology that cannot distinguish between past and present infection, the urea breath test is a measure of current *H. pylori* infection. For the urea breath test, a capsule or water that contains urea is given to patient. The urea is tagged with radioactive carbon (carbon-14) or a form of carbon that is not radioactive (carbon-13). If *H. pylori* bacteria are present in stomach, they will break down the urea, eventually causing exhalation of carbon dioxide that contains the tagged carbon (Fig. 4). If *H. pylori* is not present in the stomach, the urea does not break down and instead passes out of your body in urine. This test is accurate and safe.

5.2.3 Stool Antigen Test

Stool antigen test detects substances that trigger the immune system to fight an *H. pylori* infection (*H. pylori* antigens). In an *H. pylori* infection, these antigens are shed in a person's feces (stool). Stool antigen testing is less expensive than the other tests and the results can be obtained quickly (in about 3 hours). Stool antigen testing may be done to support diagnosis of *H. pylori* infection or to determine whether treatment for an *H. pylori* infection has been successful.

6. THERAPY

Therapy for *H. pylori* poses several unique challenges. Determining the optimum treatment of *H. pylori* infection is difficult, because the organism lives in an environment not easily accessible to many medications and emerging bacterial resistance present added challenge. Treatment of

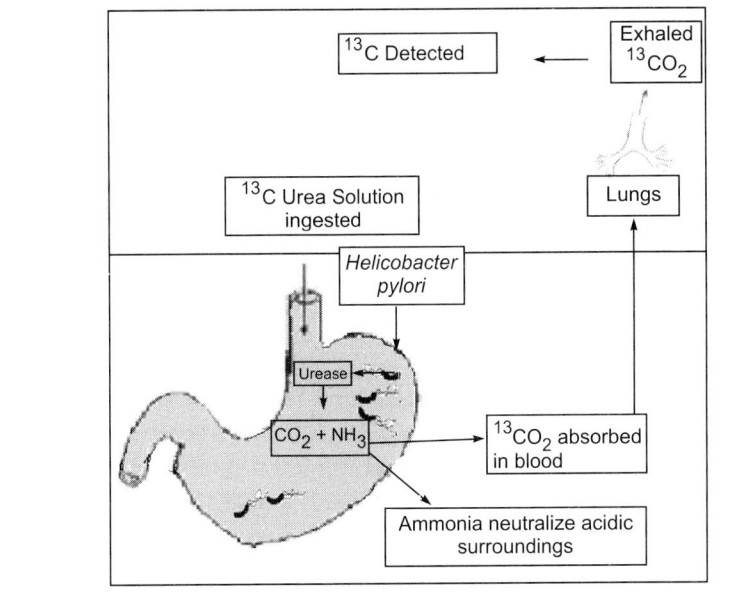

Fig. 4 DIAGNOSIS – **Urea breath test.**
Urea breath test is based on the principle that a solution of urea labeled with carbon-13 will be rapidly hydrolyzed by the urease enzyme of *H. pylori*. The resulting CO_2 is absorbed across the gastric mucosa and hence, via the systemic circulation, excreted as $^{13}CO_2$ in the expired breath. The 13C-urea breath test is non-radioactive and can be used as a screening test for *H. pylori*.

patients infected with *H. pylori* requires a multidrug regimen. Peptic ulcers are treated with drugs that kill the bacteria, reduce stomach acid secretion, and protect the stomach lining. Antibiotics are used to kill the bacteria. H_2 blockers work by blocking histamine, which stimulates acid secretion. Proton pump inhibitors suppress acid production by halting the mechanism that pumps the acid into the stomach. Bismuth subsalicylate is used to protect the stomach lining from acid. It also kills *H. pylori*. So the treatment involves a combination of antibiotics, acid suppressors and stomach protectors. Antibiotic regimens recommended for patients may differ across regions of the world because different areas have begun to show resistance to particular antibiotics.

6.1 Triple Therapy

A triple antimicrobial regimen containing bismuth subsalicylate, tetracycline, and metronidazole has been used extensively and is referred as triple therapy. Two-week triple therapy reduces ulcer symptoms, kills the bacteria, and prevents ulcer recurrence in more than 90 percent of patients. Resistance to antimicrobials, in particular to metronidazole, is an important problem and a cause for treatment failure in some cases. The antibiotics used in triple therapy may cause mild side effects such as nausea, vomiting, diarrhea, dark stools, and dizziness. Nevertheless, recent studies show that 2 weeks of triple therapy is ideal. Another option is 2 weeks of dual therapy. Dual therapy involves two drugs: an antibiotic and a proton pump inhibitor to suppress stomach acid production. It is not as effective as triple therapy.

6.2 Quadruple Therapy

Scientists have experimented with quadruple therapy, which adds an anti-secretory drug, to suppress the acid secretion, to the standard triple therapy treatment. The drugs given in this combination are omeprazole, CBS, tetracycline and metronidazole. One study showed this therapy to be effective with only a week's course of treatment in more than 96% of patients. The goal of ongoing research in this field is to develop the most effective therapy combination that can work in one week of treatment or less.

7. MOLECULAR MECHANISM OF *H. PYLORI* INFECTION

Research on *H. pylori* has advanced from establishing its connection with the gastric diseases to the exact determination of the bacterial factors responsible for pathogenesis and understanding their mechanism of action. Though disease outcome is a result of interplay between several bacterial factors and their interaction with the host system, but it is the bacterial factor that initiates pathogenic reaction. Some of the virulence factors and their interaction with the host are shown in Figure 5.

7.1 Virulence Factors

Many factors contribute to the virulence of *H. pylori* including urease, flagella, adhesins, the vacuolating cytotoxin VacA and the protein CagA (Table 1). Knowledge about their mode of action will be certainly helpful in designing new therapeutic options against *H. pylori* infection.

7.1.1 Urease

H. pylori usually prefer to colonize in extreme low pH of stomach lumen and urease enzyme plays important role for the survival of the bacteria from acidic gastric environment. These genes encode for UreA (26.5 kDa) and

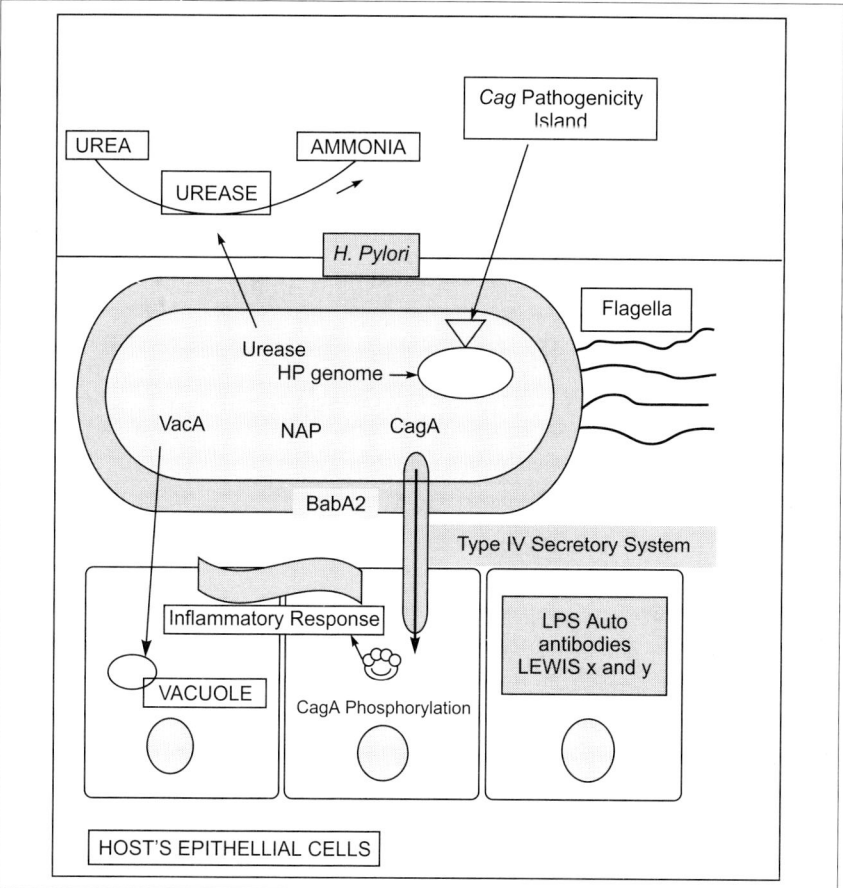

Fig. 5   *H. pylori* Virulence Factors and their Interaction with Host cells.
H. pylori adhere with the host cell surface through adhesins like BabA. The CagA
protein, translocated by type IV secretory system (thick arrow) encoded by Cag
pathogenesis island (PAI). CagA phosphorylation and the inflammatory response
are also indicated.

UreB (60.3 kDa) proteins. Ni^{2+} uptake is required for the urease activity.
Urease acts by hydrolyzing urea and generating ammonia in the cytosol
and periplasmic buffer. This helps forming a neutral layer around the
bacterial surface and thus provides protection from the effect of highly
acidic conditions. Urease activity may also contribute to toxicity by the
production of ammonia that is toxic to the cells. It is one of the major
antigens recognized by the human sera (reviewed in Dhar SK et al. 2003).

Table 1 Virulence factors of *H. pylori*

Factors	Function	Distribution
Urease	Buffer acidic pH	All strains
Flagella	Motility	All strains
NAP	Neutrophil activation	All strains
BabA	Adhesion for Lewis B antigen	Prevalent in type I strains
LPS	Low toxicity	All strains
Lewis xy antigen	Molecular mimicry	Some strains
IceA	Homolog of Nla III restriction endonuclease	Some strains
VacA	Cytotoxicity (two alleles)	All strains
Cag PAI	31 genes coding for type IV secretion system	Type I strains
CagA	Immunodominant antigen (Part of *Cag* PAI)	Type I strains
PicB	Equivalent to *CagE*, a key component of type IV secretion system	Type I strains

Note: Type I strains: *H. pylori* strains containing *Cag*-PAI

7.1.2 Flagella and Adhesions

Flagella and adhesions are required for the bacterial motility and attachment to the host cells. To avoid prolonged exposure to acid and getting discharged in the intestine, *H. pylori* must reach the stomach epithelial cell line under the thick mucous layer. Non-motile mutant bacteria cannot colonize the stomach suggesting that flagellar movement is essential for the bacteria to cross the mucosa barrier. *H. pylori* produce adhesion molecules which bind to membrane associated lipids and carbohydrates. The binding involves several proteins and glycolipids and these molecules have been shown to be responsible for binding. Out of them BabA, an outer membrane protein of *H. pylori* binds to Lewis B type antigen of human cells (Ilver et al. 1998).

Recently, Mahdavi (Mahdavi et al. 2002) identified a sialic acid–binding adhesion (SabA) which binds to sialylated Lewis x antigen expressed during chronic inflammation, might also contribute to virulence.

H. pylori protein (Hpa) also binds to sialylated glycoconjugates. Recently, two identified closely related outer membrane proteins AlpA and AlpB required for adherence to gastric epithelial cells were identified. The pattern of AlpAB-dependent adherence of *H. pylori* to the gastric epithelial surface showed a clear difference to the BabA2-mediated adherence to Lewis, suggests the involvement of a different type of receptor (Odenbreit et al. 1999). Additional bacterial proteins also help in establishing contact with the host cells. It is unlikely that mutation in a single gene can create adhesion defective mutant bacteria incapable to adhere with host cells.

7.1.3 Cag Pathogenicity Island (PAI)

Pathogenicity islands are particular regions of the prokaryotic genome responsible for coding virulence factors of pathogenic bacteria (namely, adhesins, toxins, invasins, protein secretion systems, iron uptake systems, and others). The *Helicobacter pylori* genome also reported to contain such pathogenicity island region referred as *cag* pathogenicity Island. Several forms of the gastroduodenal diseases are associated with the presence of a 40kb pathogenecity island containing more than thirty genes (Censini et al. 1996). Genes in the pathogenicity island are involved in different pathogenic processes including the induction of host cells to release pro-inflammatory chemokines and modification of intracellular signaling (Crabtree et al. 1995).

Majority of these genes encode components of a type IV secretion system that is capable of translocating CagA into epithelial cells where it interferes with cellular signal transduction processes. Depending on the presence or absence of this region, *H. pylori* strains are termed as CagA+ or CagA-. Infection with CagA+ strains of *H. pylori* increases the virulence of the bacterium (Gold et al. 1999). Data obtained from the epidemiological studies strongly supports that severe gastric diseases are always associated with CagA+ strains. CagA positive *H. pylori* strains are known to induce interleukin-8 (IL-8) secretion from gastric epithelial cells. However, recent studies have indicated that it is not CagA itself but the products of other genes in the *cag* pathogenicity island (PAI) that are responsible for IL-8 induction. The exact function of CagA protein is not clearly understood so far. CagA protein is injected from *H. pylori* into host cells using the type IV secretion system (Censini et al. 1996). Upon transfer into the host cells, CagA is tyrosine phosphorylated (Odenbreit et al. 2000). Introduction of CagA+ in the host cells also triggers morphological changes similar to those induced by growth factors (Segal et al. 1999). Recently, SHP-2 tyrosine phosphatase has been identified as an intracellular target of *H. pylori* CagA protein. CagA formed a physical complex with the SH2-containing tyrosine phosphatase SHP-2 in a phosphorylation dependent manner and stimulated the phosphatase activity leading to the dephosphorylation of some cellular proteins. Wild type but not phosphorylation resistant, CagA induce abnormal proliferation and movement of gastric epithelial cells, promoting the acquisition of a humming bird phenotype (Higashi et al. 2002).

7.1.4 Vacuolating Cytotoxin A (VacA)

H. pylori vacuolating cytotoxin (VacA) forms large cytoplasmic vacuoles, which eventually fill up the entire cyctosol (Cover et al. 1992). The molecular mass of VacA is 95kDa and it is present in half of the *H. pylori* isolates. It is one of the major virulence factors of *H. pylori,* which is a strong immunogen in human. VacA gene encodes a protein approximately 140 kDa in mass. An amino terminal signal sequence and a carboxy-terminal fragment are proteolytically cleaved to produce a ~88kDa mature toxin (Cover et al. 1992). Mature VacA is released and placed in between the mucus layer and the apical domain of the stomach epithelial cells. Upon insertion into the plasma membrane, VacA forms anion specific channels of low conductance. These channels release bicarbonate and organic anions from the cell cytosol to support bacterial growth. The toxin channels are slowly endocytosed and finally reach late endosomal compartments.

The presence of vacuolar ATPase proton pump on the membrane of the endosome increases hydrogen ion concentration inside the lumen of the endosome. These proton pumps are essential for the formation of the vacuoles. In the presence of weak bases like ammonia generated by the *H. pylori* urease, active acidotropic NH_4^+ ions are accumulated inside the endosome.

This leads to water influx and vesicle swelling leading to the vacuole formation. By an unknown mechanism, VacA alter the tight junction and increase the permeability of iron and nickel ions through the paracellular route. These nutrients are essential for *H. pylori* growth. VacA induced vacuolization results in decreased proteolytic activity in the endocytic pathway and proteolysis of antigen in the antigen–processing compartment of antigen presenting cells. As a result, it inhibits the stimulation of T-cell clones specific for epitope generation in the antigen–processing compartment.

7.1.5 Neutrophil Activating Protein (NAP)

H. pylori induced inflammation is associated with the infiltration of phagocytes (mainly neutrophils) to the gastric mucosa. The tissue damage may be attributed to the combined effects of bacterial factors and host inflammatory mediators. Recently, a protein capable of promoting neutrophil infiltration and adhesion to endothelial cells was identified. The purified protein was termed as HP-NAP. It is a 150kDa dodecameric (composed of 15 kDa subunits) iron–binding protein that promotes adhesion of PMNs (polymorphonuclear lymphocytes) to endothelial cells (Yoshida et al. 1993). HPNAP plays an important role in immunity. Vaccination of mice with HP-NAP induced protection against *H. pylori* challenge. This is consistent with the finding of HP-NAP specific antibodies present in the majority of *H. pylori* infected patients (Satin et al. 2000).

Therefore, HP-NAP is a virulence factor important for the *H. pylori* pathogenic effects at the site of infection and a candidate antigen for vaccine development.

7.1.6 Type IV Secretion System

H. pylori type I strain is characterized by the presence of Cag Pathogenicity Island (PAI), which contains 31 open reading frames. Some of these ORFs share significant homology with the virulence (vir) genes *virB4, virB7, virB8, virB9, virB10 , virB11* and *virD4* of the so-called *VirB/D* complex of type IV secretion systems known from *Agrobacterium tumefaciens* and *Borodetella pertussis* (Odenbreit et al. 2000; Christie et al. 2000). The functions and the localization of some of these homologous genes are known. *VirB2* is the structural subunit of the conjugative pilus, whereas *virB4, virB9 and virD4* have a functional ATPase activity.

The known functions of type IV secretion systems vary in different organisms. In *E. coli*, the system codes for the conjugative pilus and the necessary components to transfer DNA from one bacterium to the other. In *A. tumufaciens*, it codes for the system that transfers the Ti plasmid DNA from the bacterium to the nucleus of the plant cell. In *B. pertussis*, it codes for the apparatus that allows secretion of pertussis toxin into the medium. In *H. pylori*, the *Cag* PAI induces epithelial cells to secrete interleukin 8 (IL-8), a mediator of inflammation, by activating nuclear factors kappa B complexes.

It also induces: Remodeling of the cell surface and pedestal formation, Tyrosine phosphorylation of a 145 kDa host protein (Segal et al. 1996), and activation of the transcription factor AP-1 and expression of the proto-oncogenes c-*fos* and c-*jun* by activation of the ERK/MAP kinase cascade, result in ELK-1 phosphorylation and increased c-*fos* transcription (Covacci et al. 2003*).

8. H. PYLORI AND GASTRIC CANCER

There are two possible theories on how *H. pylori* cause cancer. Cancer is either a result of inflammation causing damage and regeneration of the gastric epithelial cells or a result of an increased pH in the stomach. These factors allow the bacteria to colonize in the stomach and produce the cancer-causing compounds that induce mutations in the gastric epithelium. *H. pylori* cause the infiltration of the gastric mucosa by neutrophils, macrophages, and T and B-lymphocytes. This infiltrate has no effect on removing the *H. pylori* from the gastric epithelial tissue. The inflammation response starts with the colonization of the epithelial cells and these cells produce chemotaxis molecules to activate the neutrophils to come at the infection site. Along with the neutrophils, other inflammatory cells come to the mucosa during the infection. The migration of this infiltrate causes some cell damage.

The combined effect of the inflammatory infiltrate and presence of toxic molecules cause damage to gastric epithelium cells resulting in genetic mutations. Several inflammatory mediators, such as hydrogen peroxide, can cause damage to the gastric epithelium. Other molecules such as reactive oxygen and reactive nitrogen molecules, made by neutrophils, are released when the neutrophils enter the stomach and undergo lysis due to the low pH levels. When this occurs, the reactive oxygen and nitrogen molecules accumulate in the gastric epithelial cells and damage the DNA.

DNA damage affects on tumor suppressor genes and proto-oncogenes that leads to the loss of function to suppress uncontrolled cell growth. In addition, another way that mutations can arise is due to the continual cellular regeneration of the gastric epithelium, where increased cell division can decrease DNA stability and lead to gastric cancer. These mutations can contribute to the progression of malignant transformation leading to gastric cancer.

Current studies indicate that activation of COX-2 is involved in the carcinogenic cascade triggered by *H. pylori* in the stomach. Proinflammatory cyclooxygenase (COX) enzymes (COX-1 and 2) catalyze inflammatory prostaglandin formation. COX-1 is expressed constitutively, whereas COX-2 is induced by cytokines and its expression in the gastric epithelial cells is increased when co-cultured with *H. pylori*. COX-2 products are involved in the regulation of cellular proliferation, apoptosis, angiogenesis and adhesion processes. A link between *H. pylori* and gastric COX-2 expression has been obtained from studies which demonstrated that *H. pylori* infection leads to increased expression of this enzyme in the gastric mucosa, whereas eradication of the bacterium reverts this effect.

Thus, we see that some classes of irritants, together with the tissue injury and ensuing inflammation, enhance cell proliferation. Thus, chronic injury or inflammation over decade leads to a sustained expansion of tissue proliferative zones and predisposes to neoplastic progression.

9. OVERALL BIOLOGICAL EFFECTS

We have seen that how multiple virulence factors modulate host in different ways giving rise to *H. pylori* pathogenesis. Host factors also play important role in determining the extent and severity of the damage of the host tissue. Biological effects of *H. pylori* on the host cells include the cellular proliferation, inflammation and apoptosis. *H. pylori* infection is associated with enhanced cellular proliferation of host cells. However, the mechanism behind the proliferation is not clearly understood. Expression of various cell cycle markers associated with the proliferation was checked in the host cells following co-culture of gastric epithelial cells with bacteria. In a recent report, cyclin D1 transcription in gastric cancer (AGS) cells was enhanced by co-culture with *H. pylori* (Hirata et al. 2001). This activation of cyclin D1 was partly dependent on the Cag pathogenicity island but not on VacA. Among various cyclins, cyclin D1 regulates passage through the restriction point and entry into the S-phase. Furthermore, cyclin D1 over expression shortens the G1 phase and increases the rate of cellular proliferation.

Co-culture of *H. pylori* with epithelial cells was reported to reduce expression of the cell cycle regulatory protein p27, which leads to epithelial cell G1 arrest. Cell proliferation may also be induced by *H. pylori* as a result of host response to bacteria. Production of gastrin by mucosa G cells increases in the presence of *H. pylori*. *In vitro*, gastrin enhances the proliferation of gastric epithelial cells. Transgenic mice over expressing gastrin yielded gastric adenocarcinoma more frequently and in less time in the presence of *H. pylori* than in the absence of bacteria suggesting that high gastrin level and *H. pylori* colonization may cooperatively enhance the incidence of gastric *cancer* (Wang et al. 2000).

10. IMMUNE RESPONSE TO INFECTION: IMMUNOPATHOGENESIS

Infection of *H. pylori* induces a weak and inappropriate host's immune response that fails to eradicate the bacterial population; in fact, it helps in the pathogenesis, making the term "immunopathogenesis" very relevant in case of *H. pylori* infection. The basic and initial damage occurs to host due to the *H. pylori* presence is the gastric inflammation. Actually it is the host's immune response against *H. pylori* infection that causes gastric inflammation in virtually all infected persons. This inflammatory response initially consists of the recruitment of neutrophils, followed by T and B lymphocytes, plasma cells, and macrophages, along with epithelial-cell damage.

The host response is triggered primarily by the attachment of bacteria to epithelial cells. The pathogen can bind to class II major-histocompatibility-complex (MHC) molecules on the surface of gastric epithelial cells, inducing their apoptosis. The gastric epithelium of *H. pylori* infected persons has enhanced levels of interleukin-1*b*, interleukin-2, interleukin-6, interleukin-8, and tumor necrosis factor TNF alpha. Among these, interleukin-8, a potent neutrophil-activating chemokine expressed by gastric epithelial cells, apparently has a central role in the inflammatory immune reaction. *H. pylori* strains carrying the *cag*-PAI induce a far stronger interleukin-8 response than *cag*-negative strains.

The inflammatory response is also due to the presence of other factors such as outer membrane proteins. Yamaoka et al. has described a new inflammation related gene *oipA* that promotes IL-8 production (Yamaoka et al. 2002). This group has demonstrated that the vast majority of *cag*+ strains isolated from patients in eastern Asia have an in-frame copy of *oipA*, and when co-cultured with gastric epithelial cells *in vitro*, these strains induce high levels of IL-8.

10.1 Specific Immune Response

IgG, IgA and IgM are the specific antibodies against *H. pylori* detected in gastric mucosa and sera of the infected patients. Oral administration of these antibodies in animals has resulted in eradication of bacteria (Kleanthous et al. 1998).

The specific immune response is complex and varies among infected humans. The systemic response is characterized by a marked increase in plasma IgG, which remains present for months after the infection has been cured. The local response includes the production of IgA, which binds to the surface antigens of *H. pylori in vitro*.

During specific immune response, T helper cells (Th) expressing CD4 can be broadly divided into two functional subsets, type 1 (Th1) and type 2 (Th2) T-helper cells, each having distinct pattern of cytokine secretion. Th1 cells produce IL-2 and IFN γ and promote cell-mediated immune responses, whereas Th2 cells secrete IL-4, IL-5, IL-6, and IL-10 and induce B-cell activation and differentiation.

Regulatory T cells (Treg) also seems to be involved in the induction of immune response directed against *H. pylori*, as CD4+ CD25+ Treg cells limit the host inflammatory response by decreasing activation of interferon gamma (IFNγ) producing CD4+ T cells and thus deactivating CD4+ T cells within inflamed mucosa (Raghavan et al. 2003).

Surface receptors on dendritic cells, such as C-type lectins, are also involved in regulating the host response to *H. pylori*. C-type lectins recognize carbohydrate structures and, upon binding, internalize pathogens for antigen processing and presentation to T cells.

Several observations suggest pathogenic role of specific antibodies in infected persons. It has been indicated that *H. pylori* Lewis antigen may cause induction of auto reactive antibodies. This can be understood with the findings that *H. pylori* Lewis antigen expression leads to induction of immune response that enhances gastric inflammation and injury (Appelmelk et al. 2000). *H. pylori* LPS may have an important role in autoimmune responses in the gastric mucosa. The structure of LPS O-specific antigen in different *H. pylori* strains is similar to that of host Lewis X or Lewis Y blood group antigens (Aspinall et al. 1996) that are expressed in normal human gastric mucosa. This molecular mimicry may account for some of the gastric auto-antibodies induced by *H. pylori*.

Appelmelk and colleagues (Appelmelk et al. 1996) showed that the H+, K+ proton pump has LewisY epitopes and anti-LewisY antibodies induced by *H. pylori* might contribute to the development of atrophic gastritis. Recent studies suggest that peptide epitopes on gastric H+, K+ ATPase may also be the target of autoimmune responses in chronic gastritis (Appelmelk and Negrini 1997). Lewis antigens are expressed more frequently on *cagA* positive than on *cagA* negative strains. This suggests that *cagA* positive strains could potentially be more likely to stimulate autoimmune responses. The observations in animal models suggest that it is the inappropriate host T-cell response towards *H. pylori* that facilitates the development and chronicity of gastric inflammation and leads to severe pathological consequences.

There is evidence that *H. pylori* infection is associated to a T helper (h) 1-type response with overproduction of IFNγ (Fan et al. 1994) which gives rise to hyperactivation of macrophages, neutrophils and eosinophils, which, in turn, damage epithelia by secretion of free radicals, enzymes and induction of programmed cell death or apoptosis.

Ammonia generated from the hydrolysis of urea by Urease has been suggested as an inducer of apoptosis of gastric epithelial cells in the course of *H. pylori* infection (Igarashi et al. 2001). However, recent findings demonstrated that ammonia accelerated TNF gamma induced apoptosis in gastric epithelial cells, while ammonia or urease molecules by themselves are not so efficient in the induction of apoptosis. In continuation, *H. pylori* LPS has been shown to induce not only apoptosis of epithelial cells but also of lymphocytes and plasma cells (Fan et al. 2000).

11. MOLECULAR CELL BIOLOGY OF *H. PYLORI*

Ongoing research developments to understand the interactive mechanism of *H. pylori* with the human hosts indicates about the uniqueness of this pathogen in several ways by which the bacteria establishes itself in the hosts system and successfully causes the severe forms of gastric pathology. This uniqueness that makes it such a successful parasite lies in some of the important features of the cellular functioning of the bacteria.

However, not much extensive research work has been reported so far to understand the basic molecular and cell biology of *H. pylori*, the efforts in this direction could lead towards elimination of *H. pylori*. Research in basic biology of *H. pylori* could further advance into applied clinical research, by inhibiting important molecular and metabolic pathways to alter the functioning of the pathogenic cell. A study of basic biological features could reveal out some new virulence factors too.

11.1 Basic Biology

A few studies have been done to reveal out the basic biology aspects such as DNA repair, replication, recombination and transcription. *H. pylori* recombinational repair and transcription studies contribute to better understanding of its pathogenesis. Recombination could be the main factor responsible for the genetic diversity of *H. pylori*. Some progresses have been done in this regard. Recently, HpMutS2 was purified and shown that HpMutS2 inhibits DNA strand exchange reactions *in vitro*. Thus, MutS2 protein is candidate for controlling recombination and therefore genetic diversity in bacteria (Pinto et al. 2005). It was reported that *H. pylori* elicit an oxidative stress during host colonization and ultimately pathogen DNA is subjected to lethal or mutational damage. (O'Rourke et al. 2003). It was also found that *H. pylori* strains lacking a functional endonuclease III (HpNth) showed elevated spontaneous and induced mutation rates and were more sensitive than the parental strain to killing by exposure to oxidative agents or activated macrophages.

Few works have been done in *H. pylori* transcription studies mediated by iron that contribute to better understanding of its pathogenesis. The global response of iron starvation in *H. pylori* has been characterized by spotted DNA microarrays. Comparatively, the expression levels before and after growth-limiting iron starvation in *H. pylori* affected the three classes of genes: those affected specifically in the exponential phase or the stationary phase and those affected in both growth phases. Many of these genes are iron-regulated led to the detection of several putative Fur boxes, thus suggesting a direct role for Fur in iron-dependent regulation of these genes. Microscopic analysis revealed that iron limitation marked affects on the ability to swim and survive for longer period in stationary phase than do exponential-phase bacteria. Proteins involved in iron metabolism are suggested to represent major virulence determinants of *H. pylori* (McGee et al. 1999), and it was shown that ferrous iron uptake mediated by the transport protein Feo B is a prerequisite for the establishment of *H. pylori* infection *in vivo* (Velayudhan et al. 2000). It is concluded that *H. pylori* are heavily dependent on ferrous iron, which is stabilized by the low pH and by the low oxygen concentration of the human stomach.

In contrast to eukaryotic ferritins, which are well characterized concerning their structure, biocatalytic functions, and regulation, our knowledge of prokaryotic ferritins is still relatively limited (Andrews et al. 1998). The ferritin Pfr, the major iron storage protein of *H. pylori* was shown to be essential for iron storage in subcellular granules and for iron resistance of *H. pylori* (Bereswill et al. 1998).

Recently, the genome sequences of two unrelated isolates of *H. pylori*, 26695 and J99 have been determined. The 26695 and J99 circular chromosomes are 1, 667, 867 bp (Tomb et al. 1997) and 1, 643, 831 bp (Alm et al. 1999) in size, respectively. The gene content of two sequenced *H. pylori* genomes suggests that the basic mechanism of chromosomal replication is similar to that of other eubacteria.

However, detailed genomic analysis revealed few surprising features, in particular in the initiation of DNA replication. The typical eubacterial block of replication genes, *dnaA-dnaN-recF-gyrB* does not exist; the *dnaA* gene is located ~600kb away from the *dnaN-gyrB* genes, while the *recF* gene is missing. The *dnaC* gene encoding DnaC protein, which delivers the DnaB helicase to the prepriming complex, is absent and initiator protein DnaA does not interact with DnaB helicase.

Present drug regimen for *H. pylori* is not successful to eradicate the pathogen from host because the organism lives in an environment not easily accessible to many medications and emerging bacterial resistance to antibiotics present added challenges. It is known that some of the replication proteins like bacterial gyrases are good candidates for drug targets. Therefore, a thorough understanding of DNA replication initiation in *H. pylori* might be useful to find new targets for therapy.

Recently, Anna Zawilak (Zawilak et al. 2001) and co-authors characterized the key elements for chromosomal replication initiation in *H.*

pylori. The initiator protein DnaA was cloned and characterized and the putative *oriC* region were reported. They also reported functional domain mapping of DnaA. Based on DNaseI and gel retardation analysis they have demonstrated that the C-terminal domain of *H. pylori* DnaA protein is responsible for DNA binding. By using *in silico* and *in vitro* studies, they have further shown that the putative *oriC* region containing five DnaA boxes are located upstream of the *dnaA* gene.

In another study, the same group has shown the amount of the DnaA protein are approximately 3000 molecules per single cell, and a large quantity of the protein is present in the inner cell membrane. Through gel retardation assay and surface plasmon resonance (SPR) studies they have demonstrated that *H. pylori* DnaA protein exhibits lower DNA-binding specificity than *E. coli* DnaA (Zawilak et al. 2003).

In vivo and *in vitro* characterization of replicative DnaB helicase has also been reported (Soni et al. 2003). The characterization of DnaB helicase could be an important step in understanding the basic mechanism of helicase loading in *H. pylori* as the bacteria lacks a helicase loader DnaC homolog from the genome sequence unlike that of *E. coli*. In fact, *H. pylori* DnaB have been shown to bypass *E. coli* DnaC activity *in vivo* (Soni et al. 2005) suggesting that HpDnaB is a unique protein.

11.2 Genomics and Proteomics

The study in genomics and proteomics aims to analysis of genome, to assign functions to unknown proteins, to define new metabolic pathways, to explore interaction between proteins and to analyze the regulatory mechanisms that respond to various stimuli from the host or the environment.

Genomic procedures would be helpful to understand some of the complex interactions that take place between *H. pylori* and its natural or experimental host. One particularly useful research tool for such study is whole genome profiling with micro arrays. Micro arrays can be used to study a variety of aspects of pathogens including gene expression profiling, comparative genomics, the bacterial response to interaction with host cells and finally the host response to infection.

In a study by Israel et al., the genetic make up of two *H. pylori* clinical isolates, one from a gastric ulcer (GU) patient, and the other from a duodenal ulcer (DU) patient, were analyzed using micro array techniques (Israel et al. 2001). It was found that the DU strain lacked a portion of *Cag* PAI genes while the GU strain had an intact island. This is an important step to detect genetic differences that can be attributed to differences in pathogenesis.

Israel et al. carried out a separate study to check whether changes could be detected in the genome of *H. pylori* in the same host with the progression of the infection. They were able to identify changes within the entire genome of sequential isolates of the sequenced strain J99, obtained from the same patient after 6 years. There were multiple deletions and acquisitions of genes detected particularly in the J99 plasticity zone detected by micro array analysis. This study supports the notion that there is microevolution of *H. pylori* strains within the host (Israel et al. 2001). This leads to the interesting conclusion that inside its host *H. pylori* is in a continuous state of genetic flux that may allow it to rapidly adapt to changing conditions.

Another study examined differential expression of genes in *H. pylori* between 0 and 2 h after exposure of cells to acid (pH 5.0) in mid log phase. A surprisingly large number of genes were found to be expressed differentially before and after acid exposure (Israel et al. 2001). Among the induced genes were the urease operons, a number of enzymes in metabolic pathways involved in ammonia production and genes involved in flagellar biosynthesis. Two virulence associated proteins, VacA and CagA were repressed at both the RNA and protein level. There were changes in a large number of genes that regulate the synthesis of the cell envelope and outer membrane proteins. These are important findings that could lead to the understanding of the regulation of differential gene expression under various conditions.

About one-third of the total proteins of *H. pylori* are of unknown function. Assigning functions to unknown proteins, identifying proteins with novel functions and establishing functional connections between proteins or metabolic pathways are now major focus of genomics and proteomics studies.

A global protein-protein interaction map has been established based on the yeast two-hybrid technique that included most of the *H. pylori* open reading frames. This study validated many known protein-protein interactions and also gave many clues regarding new interactions that would be important to understand basic biology of *H. pylori* and pathogenesis (Rain et al. 2001).

The proteome of an organism represents the total proteins expressed by a genome. The basis of the proteome technology is the separation of proteins by two-dimensional electrophoresis (2-DE) followed by their individual identification by mass spectrometry. Proteomics offers another application towards the rapid and reliable identification of intact bacterial cells by mass spectrometry on the basis of the detection of specific biomarker molecules.

To specifically identify cell surface proteins, Sabarth et al. used selective biotinylation and affinity purification followed by 2D-gels (Sabarth et al. 2002a). Their methods revealed 82 cell surface protein species, of which 18 were identified. Ten were among the most highly expressed proteins in *H. pylori*. Three proteins were predicted to be secreted proteins, while only one predicted outer membrane protein was identified. Future studies using these methods to identify the subset of proteins expressed in different strains and under different growth conditions will be quite important. In order to identify the molecules secreted by *H. pylori,* a more sensitive method of precipitating proteins from the culture medium was employed (Bumann et al. 2002). Of the 20 proteins identified, 10 contain signal sequences while three were flagellar proteins.

Another study identified potential vaccine candidates that are highly immuno-reactive, secreted or surface exposed and contains potential T-cell epitopes (Sabbarth et al. 2002a and 2002b). They identified 15 ideal targets, out of which six had previously been shown to afford protection.

These current advances in genomics and proteomics will help to have a more advanced viewpoint of *H. pylori* in particular its interaction with the host, genetic diversity and the mechanisms underlying its evolution.

12. MOLECULAR APPROACH TOWARDS PREVENTION AND CURE

The current therapy for *H. pylori* infection is efficacious, but faces problems such as antibiotic resistance, side effects and possible recurrence of infections. The developments in the field of molecular biology of *H. pylori* infection offer new alternatives for the eradication of *H. pylori*. The present level of understanding about complex interaction between the host and the pathogen suggests that many factors play role in this mechanism of pathogenesis. The identification of these factors and the peculiar knowledge about their way of functioning will help in designing novel drugs to inhibit their action and thus pathogenesis. Prevention of mere existence of the bacteria in the human system even as a commensal flora will be a better strategy. For this purpose vaccination is the most promising option.

Studies have shown that therapeutic immunization is possible in a natural setting of *Helicobacter* infection, confirming that the inability of the natural immune response to clear *Helicobacter* infection can be overcome. Thus, development of vaccines to enable the immune system to generate an effective immune response will be the ultimate way for the global eradication of the pathogen. Another new dimension in this relevance is the development of DNA vaccines against *H. pylori* antigen. These developments suggest that the molecular approach towards the infection and cure will lead to new findings which will be applied in future for prevention of *H. pylori* infection (Fig. 6).

12.1 Vaccines

The eradication of *H. pylori* infection has its own limitations and difficulties because of growing resistance against antibiotics. The special requirements

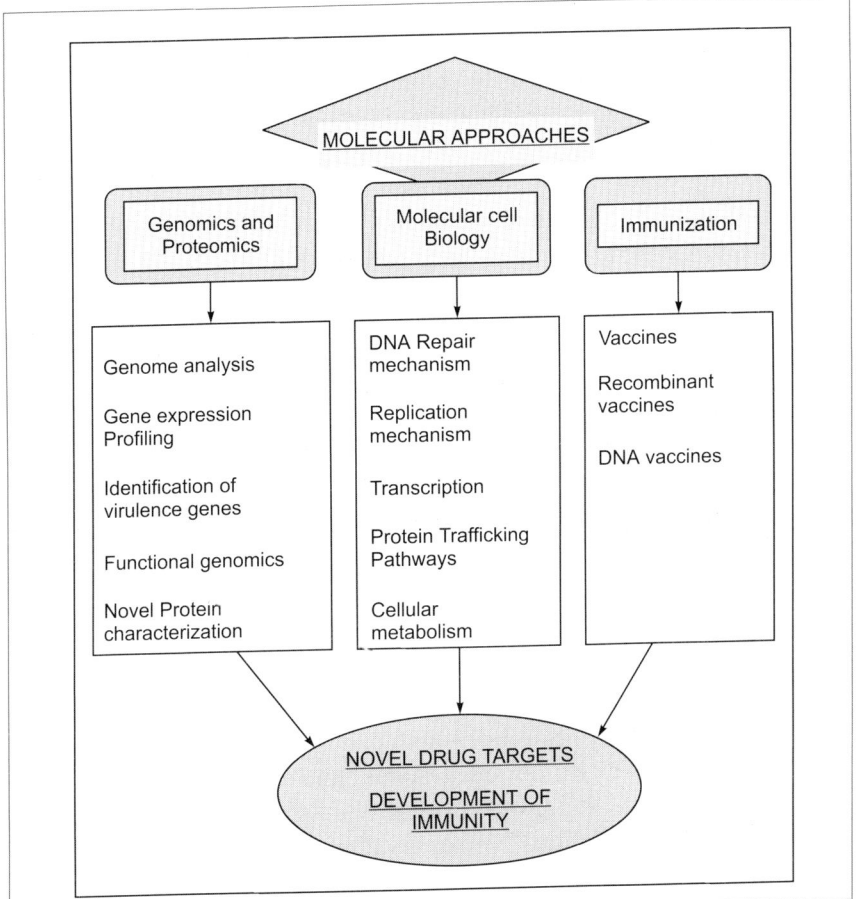

Fig. 6 Molecular Approaches for Effective Prevention and Treatment of *H. pylori*. While immunization aims at prevention of the infection by development of immunity induced by advanced vaccine, research in genomics and proteomics and the findings about basic biology of *H. pylori* could lead towards identification of new drug targets against *H. pylori* infection.

for diagnosis and the costly therapy specially in relevance with the developing and underdeveloped countries where the prevalence of the disease is about 90% creates a need for alternative strategies to control the *H. pylori* infection.

Experience in the control of other infections suggests that vaccination may be an alternative. The development of an efficacious vaccine against *H. pylori* offers several advantages. It appears to be the only means of global control of *H. pylori* infection. An effective vaccine should be able to induce a specific non-pathogenic humoral and T-cell response.

Various approaches have been followed towards the development of vaccines against *H. pylori*, most of which are based on the use of selected antigens known to be involved in the pathogenesis of the infection, and also to confer protection prophylactically and/or therapeutically in animal models of infection.

A Th1-type CD4 + T-cell response generated due to natural infection of *H. pylori* has not been able to eradicate the bacterial infection. A Th2-type response is required for protection. Effective immunization against *H. pylori* infection can induce a switch towards a Th2-type response. In order to develop an effective vaccine, efforts have been focused primarily on the identification of antigens of *H. pylori* able to confer protection. Several protective antigens have been identified, including urease, the cytotoxin VacA, two heat shock proteins (HspA and HspB), the cytotoxin-associated antigen (CagA), the neutrophil activating protein (NAP), and catalase. The list of protective antigens is still growing. Urease is currently the most promising candidate, and its value as a vaccine antigen has been confirmed by numerous studies in mice, ferrets and non-human primates (Michetti et al. 1997). Urease is reported to be a safe antigen, as it does not induce adverse reactions when administered orally with an adjuvant to mice. Urease is also safe when administered to non-human primates. The urease was given with a mucosal adjuvant, the heat labile enterotoxin (LT) of *Escherichia coli*. Urease given with LT induced an immune response and a reduction in the density of gastric *H. pylori* infection, an indication that vaccination may lead to cure of human *H. pylori* infection (Lembo et al. 2001).

With regard to the antigen preparation used for immunization, an association of two or three antigens may elicit a stronger immune response than the use of single antigens. Development of effective vaccines with many antigens including whole inactivated cells, bacterial lysate and several purified antigen have been proved successful (Telford et al. 1996).

The evidences that T-cells play an important pathogenic role in the course of *H. pylori* infection lead to the design of specific vaccine. Adoptive transfer of T-cells from donor mice immunized with *H. pylori* leads to a reduction of bacteria in the stomach of recipients (Mohammadi et al. 1997).

In the mouse model, other vaccines have been employed to obtain protection against *H. pylori*. *H. pylori* (Sydney strain) whole-cell sonicate combined with cholera toxin (CT) when orally administered to mice, induced a local secretory IgA response followed by protection (Goto et al. 1999).

An 18Kd outer membrane antigen from *H. pylori*, identified as lipoprotein 20, has been used to generate specific monoclonal antibodies (MAb) that are able to reduce bacterial number in *H. pylori*-infected mice (Keenan et al. 2000).

Advancement in vaccine immunization depends on new modes of delivery of the antigen. Future perspectives are the use of live attenuated vectors for delivering *H. pylori* antigens into the host. Recent results in mice indicate that immunization with genetically engineered bacteria expressing *H. pylori* antigens elicits a protective immune response against gastric *Helicobacter* infections. Immunization with live bacterial vaccines usually requires only one or two doses, does not depend on the addition of extra mucosal adjuvant and the vaccine can be produced at very low cost. (Michetti et al. 1997). The vaccine will contain two parts, a subunit vaccine that contains denatured cytotoxin proteins, and a live attenuated bacterial vaccine using mutated *H. pylori* bacteria. The subunit vaccine will be used to stimulate professional antigen presenting cells (APC), dendritic cells, B cells, and macrophages to induce the production of antibodies specific for proteins like VacA protein. The live attenuated bacterial vaccine will be a mutated *H. pylori* strain that expresses multiple adhesion molecules, urease, and a modified VacA protein. Therefore, the live attenuated mutated bacterial vaccine will provide several antigens for an adaptive response along with antibodies against the VacA cytotoxin.

The risk of reversion of mutated form to a wild-type form is there but the chance of such occurrences appears to be very less. A much extensive and detailed experimental study is required in this context. Animal models have their limits; however, well-designed clinical trials will help to clear that whether the promising results obtained in animals apply to humans. If successful, vaccination may indeed be able to end the coexistence of humans and *Helicobacter pylori*.

12.2 DNA Vaccines

Recently, DNA vaccine has been demonstrated to induce both humoral and cellular immunity and it is becoming a promising treatment for viral, bacterial and parasitic pathogens. It has been widely used in laboratory animals and non-human primates to induce humoral and cellular immune responses. Clinical trials have shown that DNA vaccine is safe and well-tolerated. Moreover, some reports have indicated that it could produce long-lasting immunity. The vaccine is a recombinant plasmid with heat stability. It can be used not only for protection but also for treatment.

DNA vaccines might offer great advantages for prophylactic *H. pylori* eradication especially in countries where the pathogen is acquired since childhood. Since, DNA vaccination has the tendency to shift the T-cell response towards the Th1 type, the anti *H. pylori* vaccine constructs should be designed to obtain a Th2 response.

However, at present, research on *H. pylori* vaccine generation mostly focuses on the recombinant proteins of *H. pylori* as the antigen of the vaccine in combination with mucosal adjuvants. It has been shown that live attenuated bacteria carrier including attenuated strains of *Salmonella* and *Shigella in vitro* could deliver DNA vaccines to human cells. Bacterial DNA vaccine delivery has also been demonstrated *in vivo* in several experimental animal models of infectious diseases and tumors. Dr. Can Xu constructed a live recombinant attenuated *Salmonella typhimurium* DNA vaccine strain expressing HpaA protein. It was found that recombinant attenuated *Salmonella typhimurium* DNA vaccine carrying *H. pylori hpaA* gene could express HpaA protein with immunogenicity (Can X$_4$ et al. 2005).

DNA vaccines have been developed against various pathogens, including viruses, bacteria and parasite pathogens. Thus, DNA vaccines may be an alternative to classical protein vaccine. Artur Dzwonek compared the efficacy of conventional protein vaccination with that of genetic vaccination against experimental infection with *H. pylori* in mice (Dzwonek et al. 2004). These studies indicate that genetic immunization may be as effective as conventional oral immunization in protection against experimental *H. pylori* infection. The expression plasmid vectors that are used in genetic immunization not only encode antigenic protein(s) but also serve as an adjuvant, whereas oral vaccination against *H. pylori* requires the use of antigenic proteins together with specific adjuvants. In animal models, the most potent mucosal adjuvants are bacterial toxins, such as cholera toxin, which usually produce severe side effects in humans. Thus, the immunization with DNA vaccines offers practical advances in vaccination over immunization by oral route. However, the risk factors as well as several other aspects have to be considered before the human clinical trials. Further research is going on in the promising field of DNA vaccines to explore its protective and therapeutic effect.

12.3 Identifying New Drug Targets

As the currently used therapy is not very successful in complete elimination of *H. pylori*, we need to identify new bacterial factors that can be targeted by drugs either to kill the bacterial population or to inhibit the pathogenesis process inside the human hosts.

Similarly, in the context of emerging drug resistance, new therapeutic regimens incorporating existing antibiotic agents and newly developed compounds will be an important step. In this regard, Nitazoxanide could be an effective agent that may play a role in future therapies in combination with omeprazole and clarithromycin (Megraud et al. 2001).

One of the important newly recognized drug targets for treatment of *H. pylori* infection is fumarate reductase (FRD). *H. pylori* FRD appears to play an important role in the energy metabolism of *H. pylori*. In addition, FRD is also found to be essential for the colonization of *H. pylori* in the mouse model. FRD has been reported as a promising chemotherapeutic target against *H. pylori* (Mendz et al. 1995; Ge Z 2002).

Rabeprazole sodium is a new substituted benzimidazole proton pump inhibitor that has been shown to inhibit H^+, K^+-ATPase in animal models. This drug is more effective in blocking acid secretion than omeprazole, lansoprazole, or pantoprazole (Williams et al. 1999). Clinical trials are needed to assess the clinical importance of these findings.

Another interesting area would be the screening of traditional herbs for anti *H. pylori* activity. With the growing incidences of drug resistance and the side effects of the antibiotics, these traditional medicines could be of great help. Unfortunately, the researchers have not exploited these areas yet. There are few reports in the literature from Taiwan (Wang YC et al. 2005) and Greece (Stamatis G et al. 2003) where the potential of such medicinal herbs normally used for gastric ailments have been used successfully to screen for anti *H. pylori* activity. Once such activity is established, the next step would be to find out the active ingredient followed by its mode of action and the target enzymes or molecules in *H. pylori*.

Mastic gum, a resin produced by the *Pistacia lentiscus* tree has been reported to be an effective agent for anti *H. pylori* activities. This resin is also effective in the treatment of benign gastric ulcers (Huwez et al. 1998).

There are reports in the literature that curcumin, the active ingredient of turmeric, prevents growth of *H. pylori* CagA+ strains *in vitro* (Foryst et al. 2004). The mapping of the complete genome of *H. pylori* has opened the door for new era in chemotherapeutic drugs. Combined approaches of genomics, proteomics and transcriptomics will be helpful to determine essential gene products as targets for novel therapeutic agents. Comparative genomics can allow researchers to predict the essential genes, such as motility or acid resistance that would associate with virulence factors. In addition we can find out open reading frames that have little homology with host proteins and discover putative targets for either vaccine development or antimicrobial action specific to *H. pylori*.

A relatively new method called Signature Tagged Mutagenesis (STM) can be used for identifying genes that are essential for the process of infection (Shea et al. 2000). The bacterial genes are mutated by the insertion of a transposon carrying a uniquely tagged DNA sequence, or by other insertional mutagenesis method. It is a negative selection method for virulence gene identification based on the mutation. STM has been used to identify the genes essential for *H. pylori* colonization (Kavermann et al. 2003). Genes identified in this way may be used as potential therapeutic targets if inhibitors are designed against them.

Micro array technology can also be used to identify genes up regulated during conditions associated with *in vivo* growth. Modern proteomics methods such as 2-DE techniques, MALDI-TOF, Mass Spectrometry (MS) and protein chip array may be proved useful in identifying newly expressed proteins of *H. pylori* in different stress conditions. These techniques have the potential to identify secreted proteins that can be recognized by human and murine sera leading to the establishment of new targets for therapy.

CONCLUSION

The bacterium *Helicobacter pylori* forms a unique relationship with the human species. It constitutes the most complex interactive mechanism with the host cells. The immune reaction against the bacteria actually helps in its pathogenesis rather than elimination. It causes all kinds of gastroduodenal pathology in humans from gastric inflammation to peptic ulcer and also the most severe gastric cancer. The increased resistance towards antibiotics and the genome diversity has added to the challenge of effective therapy. These bacteria are subject of interest to molecular biologists as well as physicians because of its complex mechanisms to survive in the host and to alter the host cellular machinery. Understanding the molecular mechanisms of its interaction with the hosts' cells would be helpful to suggest an effective way for its eradication. The ongoing efforts in this direction and the experimental developments have generated hopes for prevention of *H. pylori* infection.

More emphasis should be laid on the advances in genomics and proteomics and thus to gain a deep understanding of biological properties of the bacteria at a fast pace, leading to the designing of novel drug targets for complete eradication of *Helicobacter pylori* infection.

ACKNOWLEDGEMENTS

The authors acknowledge Rajesh K Soni for his help to prepare the manuscript. SKD acknowledges University Grant Commission (Excellence programme) and the Council of Scientific and Industrial Research (CSIR) for funding.

REFERENCES

Alm RA, Ling LS, Moir DT, King BL, Brown ED, Doig PC, Smith DR, Noonan B, Guild BC, deJonge BL, Carmel G, Tummino PJ, Caruso A, Uria-Nickelsen M, Mills DM, Ives C, Gibson R, Merberg D, Mills SD., Jiang Q, Taylor DE, Vovis GF, Trust TJ (1999). Genomic sequence comparison of two unrelated isolates of the human gastric pathogen *Helicobacter pylori*. Nature 397:176-180.

Andrews SC (1998). Iron storage in bacteria. Adv. Microb. Physiol 40:281-351. Angie Man-Chi Eng (1994). A New Bacterium *H. pylori*, in Gastric Diseases and Cancer.. 7th International Conference on Health Problems Related to the Chinese. Beijing. China July 1-3, 1994.

Appelmelk BJ, Monteiro MA, Martin SL, Moran AP, Vandenbroucke-Grauls CM (2000). Why *Helicobacter pylori* has Lewis antigens. Trends Microbiol 8:565–570.

Appelmelk BJ, Negrini R (1997). *Helicobacter pylori* and gastricautoimmunity. Curr Opin Gastroenterol 13: 35–39.

Appelmelk BJ, Simoons-Smit I, Negrini R, Moran AP, Aspinall GO, Forte JG, De Vries T, Quan H, Verboom T, Maaskant JJ, Ghiara P, Kuipers EJ, Bloemena E, Tadema TM, Townsend RR, Tyagarajan K, Crothers JM Jr, Monteiro MA, Savio A, De Graaff J (1996). Potential role of molecular mimicry between *Helicobacter pylori* lipopolysaccharide and host Lewis blood group antigens in autoimmunity. Infect Immun 64: 2031–2040.

Aspinall GO, Monteiro MO (1996). Lipopolysaccharide of *Helicobacter pylori* strains P466 and MO19: Structures of the Oantigen and core oligosaccharide regions. Biochem 35: 2498–504.

Bereswill S, Waidner U, Odenbreit S, Lichte F, Fassbinder F, Bode G Kist, G. (1998). Structural, functional and mutational analysis of the *pfr* gene encoding a ferritin from *Helicobacter pylori*. Microbiol 144: 2505-2516.

Blaser MJ, Perez-Perez GI, Kleanthous H, Cover TL, Peek RM, Chyou PH, Stemmermann GN, Nomura A (1995). Infection with *Helicobacter pylori* strains possessing *cagA* is associated with an increased risk of developing adenocarcinoma of the stomach. Cancer Res 55:2111-2115.

Bumann D, Holland P, Siejak F, Koesling J, Sabarth N, Lamer S, Zimny-Arndt U, Jungblut PR, Meyer TF (2002). A comparison of murine and human immuno proteomes of *Helicobacter pylori* validates the preclinical murine infection model for antigen screening. Infect Immun 70: 6494–6498.

Can X, Zhao-Shen L, Yi-Qi D, Zhen-Xing T, Yan-Fang G, Jing J, Hong-Yu W, Guo-Ming X (2005). Construction of a recombinant attenuated *Salmonella typhimurium* DNA vaccine carrying *Helicobacter pylori* hpaA World J Gastroenterol. 11:114-117.

Censini S, Lange C, Xiang Z, Crabtree JE, Ghiara P, Borodovsky M., Rappuoli R Covacci A (1996). cag, a pathogenicity island of *Helicobacter pylori*, encodes type I-specific and disease-associated virulence factors. *Proc Natl Acad Sci* USA 93:14648-14653.

Chan WY, Hui PK, Leung KM, Chow J, Kwok F, Ng CS (1994). Coccoid forms of *Helicobacter pylori* in the human stomach. *Am J Clin Pathol* 102:503-507.

Christie PJ, Vogel JP (2000). Bacterial type IV secretion: conjugation systems adapted to deliver effector molecules to host cells. *Trends Microbiol* 8: 354-360.

Covacci A, Rappuoli R (2003). *Helicobacter pylori*: After the Genomes, Back to Biology. *J Exp Med* 197: 807-811

Cover TL, Blaser MJ (1992). Purification and characterization of the vacuolating toxin from *Helicobacter pylori*. *J Biol Chem* 267:10570-10575.

Crabtree JE, Farmery SM (1995). *Helicobacter pylori* and gastric mucosal cytokines: evidence that CagA-positive strains are more virulent. *Lab Invest* 73:742-745.

Dhar SK, Soni RK, Das BK, Mukhopadhyay G (2003). Molecular mechanism of action of major *Helicobacter pylori* virulence factors. Mol Cell Biochem 253: 207–215.

Dzwonek A, Mikula M, Woszczyński M, Hennig E, Ostrowski J (2004). Protective effect of vaccination with DNA of the *H. pylori* genomic library in experimentally infected mice. *Cell Mol Biol Lett* 9: 483-495.

Fan XJ, Chua A, Shahi CN, McDevitt J, Keeling PWN, Kelleher D (1994). Gastric T lymphocyte responses to *Helicobacter pylori* in patients with *H pylori* colonisation. Gut 35: 1379-1384.

Fan X, Gunasena H, Cheng Z, Espejo R, Crowe SE, Ernst PB, Reyes VE (2000). *Helicobacter pylori* urease binds to class II MHC on gastric epithelial cells and induces their apoptosis. *J Immunol* 165: 1918-1924.

Foryst-Ludwig A, Neumann M, Schneider-Brachert W, Naumann M (2004). Curcumin blocks NF-kappaB and the motogenic response in *Helicobacter pylori*-infected epithelial cells. *Biochem Biophys Res Commun* 16:1065-1072.

Gerhard M, Lehn N, Neumayer N, Boren T, Rad R, Schepp W, Miehlke S, Classen M, Prinz C (1999). Clinical relevance of the *Helicobacter pylori* gene for blood-group antigen-binding adhesin. *Proc Natl Acad Sci* 96: 12778–12783.

Ge Z (2002). Potential of fumarate reductase as a novel therapeutic target in *Helicobacter pylori* infection. *Expert Opin Ther Targets* 6:135-46.

Gold BD (1999). Pediatric *Helicobacter pylori* infection: clinical manifestations, diagnosis, and therapy. Current Topics in Microbiology & Immunology 241: 71-102.

Goto T, Nishizono A, Fujioka T, Kewaki J, Mifune K, Nasu M (1999). Local secretory immunoglobulin A and postimmunization gastritis correlate with protection against *Helicobacter pylori* infection after oral vaccination of mice. *Infect Immun* 67: 2531-2539.

Guha Mazumder DN, Ghosal UC (1997). Epidemiology of *Helicobactor pylori* in India. *J Gastroenterol* 3: S3-S5.

Higashi H, Tsutsumi R, Fujita A, Yamazaki S, Asaka M, Azuma T, Hatakeyama M (2002). Biological activity of the *Helicobacter pylori* virulence factor CagA is determined by variation in the tyrosine phosphorylation sites. *Proc Natl Acad Sci* 99: 14428-14433.

Hirata Y, Maeda S, Mitsuno Y, Akanuma M, Yamaji Y, Ogura K, Yoshida H, Shiratori Y, Omata M (2001). *Helicobacter pylori* activates the cyclin D1 gene through mitogen-activated protein kinase pathway in gastric cancer cells. *Infect Immun* 69: 3965-3971.

Huwez FU, Thirlwell D, Cockayne A, Ala'Aldeen DA (1998). Mastic gum kills *Helicobacter pylori*. *New Engl J Med* 339: 1946.

Igarashi M, Kitada Y, Yoshiyama H, Takagi A, Miwa T, Koga Y (2001). Ammonia as an accelerator of tumor necrosis factor alpha-induced apoptosis of gastric epithelial cells in *Helicobacter pylori* infection. *Infect Immun* 69: 816-821.

Ilver D, Arnqvist A, Ogren J, Frick IM, Kersulyte D, Incecik ET, Berg DE, Covacci A, Engstrand L, Boren T (1998).*Helicobacter pylori* adhesin binding fucosylated histo-blood group antigens revealed by retagging. *Science* 279: 373-377.

Israel DA, Salama N, Krishna U et al. (2001). *Helicobacter pylori* genetic diversity within the gastric niche of a single human host. *Proc Natl Acad Sci USA* 98: 4625–4630.

Kavermann H, Burns BP, Angermuller K et al. (2003). Identification and characterization of *Helicobacter pylori* genes essential for gastric colonization. J Exp Med. 197:813–822.

Keenan J, Oliaro J, Domigan N, Potter H, Aitken G, Allardyce R, Roake J (2000). Immune response to an 18-kilodalton outer membrane antigen identifies lipoprotein 20 as a *Helicobacter pylori* vaccine candidate. *Infect Immun* 68:3337-3343.

Kim SY, Woo CW, Lee YM, Son BR, Kim JW, Chae HB, Youn SJ, Park SMJ (2001). Korean genotyping CagA, VacA subtype, IceA1, and BabA of *Helicobacter pylori* isolates from Korean patients, and their association with gastroduodenal diseases. *Med Sci* 16:579-584.

Kleanthous H, Lee CK, Monath TP (1998). Vaccine development against infection with *Helicobacter pylori*. *Br Med Bull* 54:229-241.

Lee Hyuk-Joon, Han-Kwang Y, Yoon-Ok A (2002). Gastric cancer in Korea. Gastric Cancer 5: 177–182.

Lembo A, Caradonna L, Magrone T, Mastronardi ML, Caccavo D, Jirillo E, Amati L (2001). *Helicobacter pylori* infection, immune response and vaccination. Curr Drug Targets Immune. *Endocr Metabol Disord.* 1:199-208.

Mahdavi J, Sonden B, Hurtig M, Olfat FO, Forsberg L, Roche N, Angstrom J, Larsson T, Teneberg S, Karlsson KA, Altraja S, Wadstrom T, Kersulyte D, Berg DE, Dubois A, Petersson C, Magnusson KE, Norberg T, Lindh F, Lundskog BB, Arnqvist A, Hammarstrom L, Boren T. *Helicobacter pyl*ori SabA adhesin in persistent infection and chronic inflammation. *Science.* 297:573-578.

Marshall BJ, Warren JR (1984). Unidentified curved bacilli in the stomach of patients with gastritis and peptic ulceration. *Lancet* 1: 1311–1315.

McGee DJ, Mobley HL (1999). Mechanisms of *Helicobacter pylori* infection: bacterial factors. *Curr Top Microbiol Immunol* 241:155-180.

Mégraud F, Hazell S, Glupczynski Y (2001). Antibiotic susceptibility and resistance. In: Mobley HLT, Mendz GL, Hazell SL. In *Helicobacter pylori*: physiology and genetics. Washington, D.C. ASM Press. 511-530.

Mendz GL, Hazell SL, Srinivasan S (1995). Fumarate reductase: a target for therapeutic intervention against *Helicobacter pylori*. *Arch Biochem Biophys* 321:153.

Michetti P, Kreiss C, Kotloff KL, Porta N, Blanco JL, Bachmann D *et al.*(1997). Oral immunization of *H. pylori* infected adults with recombinant Urease and LT adjuvant *Gastroenterology* 112:A1042.

Micev M, Milena Cosic-Micev (2001). *Helicobacter pylori* infection, duodenal gastric metaplasia and duodenal ulcer. *Arch Gastroenterohepatol* 20: 3- 4.

Mizushima T, Sugiyama T, Komatsu Y, Ishizuka J, Kato M, Asaka M (2001). Clinical relevance of the babA2 genotype of *Helicobacter pylori*. *J Clin Microbiol* 39:2463-2465.

Mohammadi M, Nedrud J, Redline R, Lycke N, Czinn SJ (1997). Murine CD4 T-cell response to *Helicobacter* infection: TH1 cells enhance gastritis and TH2 cells reduce bacterial load. *Gastroenterology* 113: 1848-1857.

Negrini R, Lisato L, Zanella I, Cavazzini L, Gullini S, Villanacci V, Poiesi C, Albertini A, Ghielmi S (1991). *Helicobacter pylori* infection induces antibodies cross-reacting with human gastric mucosa. Gastroenterology 101: 437–445.

Nomura A, Stemmerman GN, Chyou PH, Kato I (1991). *Helicobacter pylori* infection and the risk of gastric carcinoma. *N Engl J Med* 325: 1127-31.

Odenbreit S, Puls J, Sedlmaier B, Gerland E, Fischer W, Haas R (2000). Translocation of *Helicobacter pylori* CagA into gastric epithelial cells by type IV secretion. *Science* 287: 1497-500.

Odenbreit S, Till M, Hofreuter D, Faller G, Haas R (1999). Genetic and functional characterization of the alpAB gene locus essential for the adhesion of *Helicobacter pylori* to human gastric tissue. *Mol Microbiol* 31: 1537-1548.

O'Rourke EJ, Chevalier C, Pinto AV, Thiberge JM, Ielpi L, Labigne A, Radicella JP (2003). Pathogen DNA as target for host-generated oxidative stress: role for repair of bacterial DNA damage in *Helicobacter pylori* colonization. *Proc Natl Acad Sci* 100: 2789-2794.

Pinto AV, Mathieu A, Marsin S, Veaute X, Ielpi L, Labigne A, Radicella JP (2005). Suppression of homologous and homeologous recombination by the bacterial MutS2 protein. *Mol Cell* 17: 113-120.

Raghavan S, Fredriksson M, Svennerholm AM et al. (2003). Absence of CD4+ CD25+ regulatory T cells is associated with a loss of regulation leading to increased pathology in *Helicobacter*-infected mice. Clin Exp Immunol. 132:393-400.

Rain JC, Selig L, de Reuse, H et al. (2001). The protein–protein interaction map of *Helicobacter pylori*. Nature 2001; 409:211–215.

Replogle ML, Glaser SL, Hiatt RA, Parsonnet J (1995). Biologic sex as a risk factor for *Helicobacter pylori* infection in healthy young adults. *Am J Epidemiol* 142: 856-863.
Sabarth N, Lamer S, Zimny-Arndt U, Jungblut PR, Meyer TF, Bumann D (2002a). Identification of surface proteins of *Helicobacter pylori* by selective biotinylation, affinity purification, and two-dimensional gel electrophoresis. *J Biol Chem* 277:27896–27902.

Sabarth N, Hurwitz R, Meyer TF, Bumann D (2002b). Multi-parameter selection of *Helicobacter pylori* antigens identifies two novel antigens with high protective efficacy. *Infect Immun* 2002;70:6499–503.

Satin B, Del Giudice G, Della Bianca V, Dusi S, Laudanna C, Tonello F, Kelleher D, Rappuoli R, Montecucco C, Rossi F (2000). The neutrophil-activating protein (HP-NAP) of *Helicobacter pylori* is a protective antigen and a major virulence factor. *J Exp Med* 191: 1467-76.

Segal ED, Cha J, Lo J, Falkow S, Tompkins LS (1999). Altered states: involvement of phosphorylated CagA in the induction of host cellular growth changes by *Helicobacter pylori*. *Proc Natl Acad Sci* 96:14559-64.

Segal ED, Falkow S, Tompkins LS (1996). *Helicobacter pylori* attachment to gastric cells induces cytoskeletal rearrangements and tyrosine phosphorylation of host cell proteins. *Proc Natl Acad Sci* U S A 93: 1259-64.

Shea JE, Santangelo JD, Feldman RG (2000). Signature tagged mutagenesis in identification of virulence genes in pathogens. *Curr Opin Microbiol* 3: 451–458.

Soni RK, Mehra P, Choudhury NR, Mukhopadhyay G, Dhar SK (2003). Functional characterization of *Helicobacter pylori* DnaB helicase. *Nucleic Acids Res* 31: 6828-6840.

Soni RK, Mehra P, Mukhopadhyay G, Dhar SK (2005). *Helicobacter pylori* DnaB helicase can bypass *Escherichia coli* DnaC function *in vivo*. *Biochem J* 389: 541-548.

Stamatis G, Kyriazopoulos P, Golegou S, Basayiannis A, Skaltsas S, Skaltsa H (2003). *In vitro* anti-*Helicobacter pylori* activity of Greek herbal medicines. *J Ethnopharmacol* 88: 175-179.

Suerbaum S, Michetti P (2002). *Helicobacter pylori* infection. *N Engl J Med* 347: 1175-1186.

Telford JL, Ghiara P (1996). Prospects for the development of a vaccine against *Helicobacter pylori*. *Drugs* 52: 799-804.

Tomb JF, White O, Kerlavage AR, Clayton RA, Sutton GG, Fleischmann RD, Ketchum KA, Klenk HP, Gill S, Dougherty BA et al. (1997). The complete genome sequence of the gastric pathogen *Helicobacter pylori*. Nature 388: 539–547.

Velayudhan J, Hughes NJ, McColm AA, Bagshaw J, Clayton CL, Andrews SC, Kelly DJ (2000). Iron acquisition and virulence in *Helicobacter pylori*: a major role for FeoB, a high-affinity ferrous iron transporter. *Mol Microbiol* 37: 274-286.

Vollmers HP, Dammrich J, Ribbert H, Grassel S, Debus S, Heesemannn J, Muller-Hermelink HK (1994). Human monoclonal antibodies from stomach carcinoma patients react with *Helicobacter pylori* and stimulate stomach cancer cells in vitro. *Cancer* 74: 1525–1532.

Wang TC, Dangler CA, Chen D, Goldenring JR, Koh T, Raychowdhury R, Coffey RJ, Ito S, Varro A, Dockray GJ, Fox JG (2000). Synergistic interaction between hypergastrinemia and *Helicobacter* infection in a mouse model of gastric cancer. *Gastroenterology* 118: 36-47.

Wang YC, Huang TL (2005). Screening of anti-*Helicobacter pylori* herbs deriving from Taiwanese folk medicinal plants. *FEMS Immunol Med Microbiol* 43: 295-300.

Williams MP, Pounder RE (1999). The pharmacology of rabeprazole. *Aliment Pharmacol Ther* 13: 3-10.

Yamaoka Y, Kita M, Kodama, T et al. (2002). *Helicobacter pylori* infection in mice: Role of outer membrane proteins in colonization and inflammation. Gastroenterology 123: 1992-2004.

Yoshida N, Granger DN, Evans DJ Jr, Evans DG, Graham DY, Anderson DC, Wolf RE, Kvietys PR (1993). Mechanisms involved in *Helicobacter pylori* induced inflammation. *Gastroenterology*.105:1431-1440.

Zawilak A, Cebrat S, Mackiewicz P, Krol-Hulewicz A, Jakimowicz D, Messer W, Gosciniak G, Zakrzewska- Cz erwinska J (2001). Identification of a putative chromosomal replication origin from *Helicobacter pylori* and its interaction with the initiator protein DnaA. *Nucleic Acids Res* 29: 2251-2259.

Zawilak A, Durrant MC, Jakimowicz P, Backert S, Zakrzewska-Czerwinska J (2003). DNA binding specificity of the replication initiator protein, DnaA from *Helicobacter pylori*. *J Mol Biol* 334: 933-947.

Zevering Y, Jacob, Meyer (1999). Naturally acquired human immune responses against *Helicobacter pylori* and implications for vaccine development. *Gut* 45: 465-474.

Pathogenesis of *Salmonella typhi* and Typhoid Fever: Multiple Drug Resistance, Vaccines and Therapeutics

S. K. JAIN*, A. MADANI and N. HAMID

1. INTRODUCTION

Typhoid fever is a major health problem, especially in developing world where there is substandard water supply and lack of sanitation. Due to its changing modes of presentation, and the development of multidrug-resistance, typhoid fever is becoming increasingly difficult to diagnose and treat. Human typhoid is caused by *Salmonella typhi*, while *S. typhimurium* is a causative agent for typhoid like disease in rodents. A *Salmonellae* species are Gram-negative bacteria that are highly pathogenic, infecting a wide range of hosts and causing a variety of diseases. The annual global burden of typhoid is estimated to be about 16 million cases, leading to 600,000 deaths (Rupali 2004). Other *Salmonellae* serovars such as *S. typhimurium, S. enteritis* cause infections in domestic animals which can be transmitted to humans and represent a serious concern for the food industry (Mastroeni 2002). It is endemic in many developing countries including areas of Africa, Asia and South America (Levine et al. 1999). Children in endemic areas, travelers and personals working in microbiological laboratories are particularly at higher risk of contracting the disease.

Address of the corresponding author: Department of Biotechnology, Faculty of Science, Hamdard University, Hamdard Nagar. New Delhi 110062 India. Tel: 91-11-2605 9688, Fax 2605 9663, E-mail: skjain@jamiahamdard.ac.in

1.1 Pathogenesis of *Salmonella*

Salmonellae are highly adapted to humans and cause several diseases. The organism gains entry into the host by oral route following ingestion of contaminated food and water. Part of the inoculums that survives the acidity of stomach gets accumulated in intestine and invades the enterocytes (Wray and Sojka 1978). The pathogen proliferates within the phagocytes or in extra cellular spaces (Finlay and Falkow 1989). In intracellular location, the bacterium is protected from the complement system and antibodies (Collins and Campbell 1982). The onset of typhoid fever is characterized by a long incubation period ranging from one week to one month (Ivanoff et al. 1994). After entry it invades and multiplies in liver and spleen, then migrates to gall bladder and ultimately to intestine (Richter-Dahlfors et al. 1997). The infection often results in intestinal ulceration, which may be fatal if left untreated (Takeuchi 1967; Finlay et al. 1988). The pathogenesis of virulent *Salmonella* is highly dependent on rapid multiplication in the extra-cellular location of host tissue. Early studies on interaction of *S. typhi* with cultured epithelial cells have found it to be highly invasive; the invasion being an active process requiring live bacteria (Yabuchi et al. 1986). The *S. typhi* was found to target ileal M-cells as the site of entry into deeper tissues in murine ileal loop (Jones et al. 1995). It was found that M-cells are actively invaded by *S. typhimurium*, whereas the non-invasive mutants of *S. typhimurium* are not internalized by these cells suggesting that invasion contributes to this process (Clark et al. 1994). However, some *Salmonella* species are able to survive and propagate within phagocytes and therefore, are resistant to humoral response. Phagocytes can transport these bacteria to the general circulation that ultimately results in development of enteric fever (Chong et al. 1996).

Most non-typhoidal *Salmonellae* enter the body when contaminated food is ingested (Fig. 1). Person-to-person spread of *Salmonellae* also occurs. To be fully pathogenic, *Salmonellae* must possess a variety of attributes called virulence factors. These include (1) the ability to invade cells, (2) a complete lipopolysaccharide coat, (3) the ability to replicate intracellularly, and (4) the elaboration of toxin (s). After ingestion, the organisms colonize the ileum and colon, invade the intestinal epithelium and proliferate within epithelium and lymphoid follicles. The mechanism by which *Salmonellae* invade the epithelium is only partially understood and involves an initial binding to specific receptors on the epithelial cell surface followed by invasion. Invasion occurs by the organism inducing the enterocytes membrane to undergo "ruffling" and thereby to stimulate pinocytosis of the organisms (Figs. 2 and 3). Invasion is dependent on rearrangement of the cell cytoskeleton and

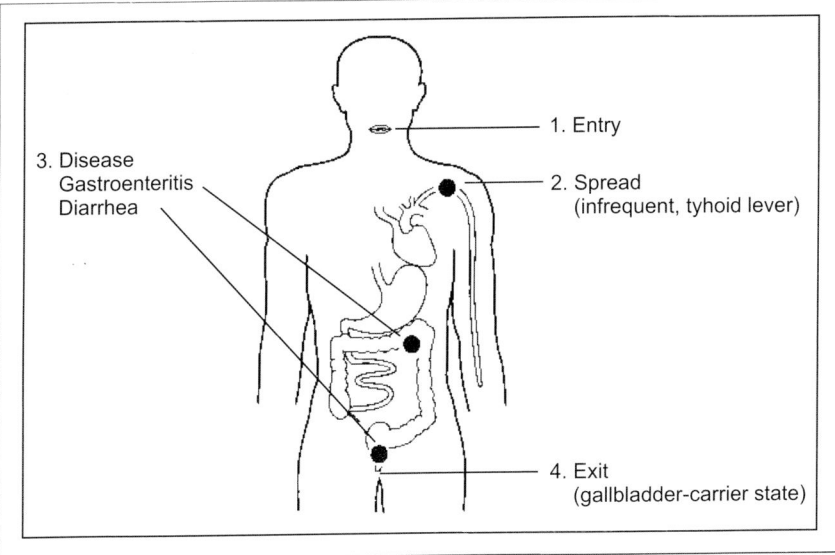

Fig. 1   Route for pathogenesis of *Salmonella typhi.*

probably involves increase in cellular concentration of inositol phosphate and calcium. Attachment of bacteria and their invasion are under distinct genetic control and involve multiple genes in both chromosomes and plasmids.

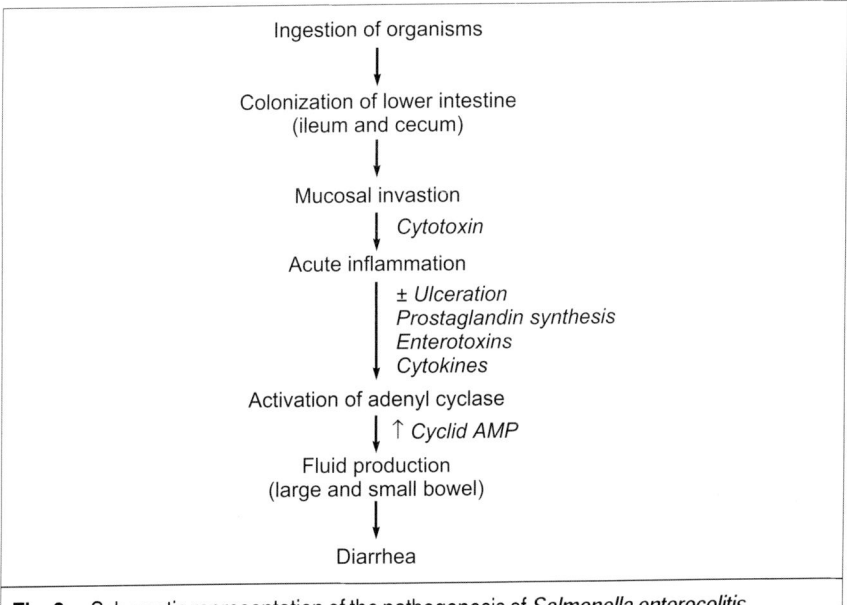

Fig. 2   Schematic representation of the pathogenesis of *Salmonella enterocolitis.*

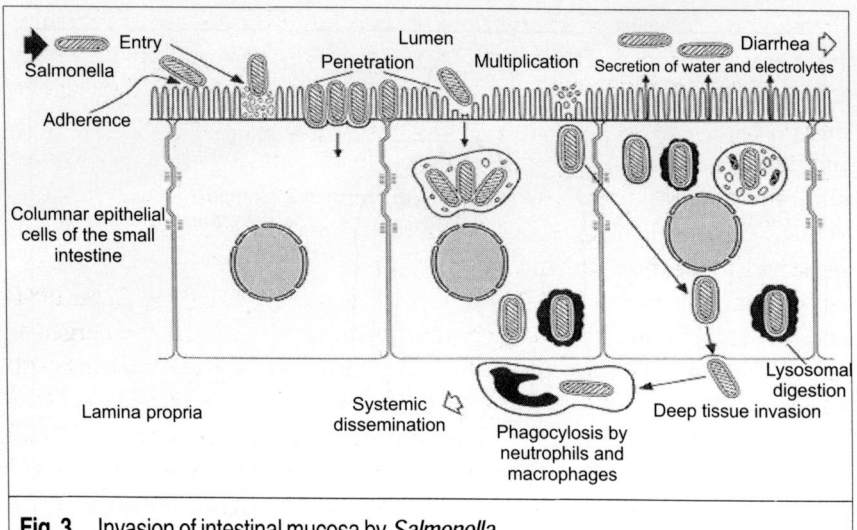

Fig. 3   Invasion of intestinal mucosa by *Salmonella*.

1.2 Interaction of *Salmonella* with Host Epithelial Cells

Much effort has been dedicated to understand the interaction of *Salmonellae* with the host epithelial cells. The mechanism by which enteric pathogens pass through intestinal epithelium is the subject of considerable interest. After oral ingestion of *S. typhi* in humans, the penetration of the gut epithelium by the bacterium is an important initial step for successful progression of a systemic infection. It was observed that *Salmonellae* invades enterocytes and actively infects M-cells of Peyer's patches in animal models (Garcia-del Portillo et al. 1994; Jones et al. 1994; Wallis and Galyov 2000; Garcia-del Portillo 2001). When studied in cell culture models, it was suggested that massive cytoskeleton rearrangements are induced after contact of the bacterium with host cell surface (Zierler and Galan 1995). This leads to the formation of pseudopods or "so-called" membrane ruffles on host cell membranes enclosing bacterium and finally leading to its uptake into phagosomal compartment (Ginocchio et al. 1994).

2. THE EMERGENCE OF ANTIBIOTIC RESISTANCE IN *SALMONELLA*

Antimicrobial chemotherapy is a critical aspect of the management of typhoid fever for both the individual patient and the community. Further difficulties in treatment are likely to emerge in the wake of drug resistance for developed countries where typhoid fever is largely a disease of returning travelers as well as for those developing countries where typhoid fever is still endemic (Fig. 4). Resistance to antimicrobial chemotherapeutic agents and possible reversal of the resistance has become a significant issue for the medical professionals (Morell 1997; Parry et al. 1998). However, chloramphenicol resistance spreading as epidemics (Colquhoun and Weetch 1950; Butler et al. 1973) and multidrug resistance (MDR), defined as resistance to the three first-line agents commonly used to treat typhoid fever, i.e., chloramphenicol, ampicillin and co-trimoxazole (Ling and Chau 1984) and low-level fluoroquinolone resistance (Wain et al. 1997) have emerged as new challenges for antimicrobial chemotherapy (White et al. 1999). In South Africa in 1991 death of patients was reported in three out of six cases of MDR typhoid died because of a delay in administration of effective antimicrobial therapy (Coovadia et al. 1992). During an outbreak of typhoid fever caused by quinolone (nalidixic acid) resistant *S. typhi* in Tajikistan, there were 10,677 cases with 108 deaths between January 1996 and June 1997 (Mermin et al. 1999). These outbreaks serve as reminder of the incredible potential of this organism to cause widespread morbidity and mortality. More information regarding the epidemiology and mechanisms of resistance in

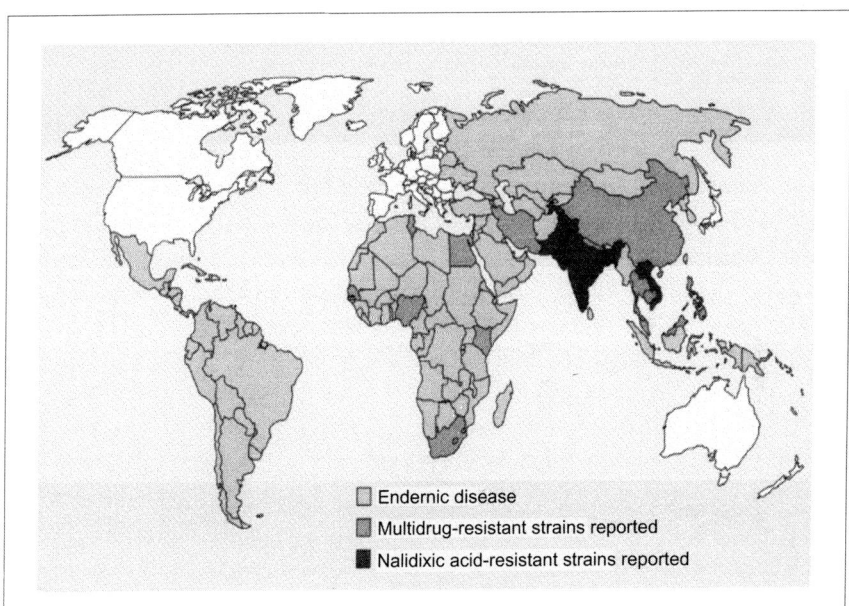

Fig. 4   Global distribution of resistance to *Salmonella enterica* serotype typhi, 1990–2002. All shaded areas are areas of endemic disease. (Parry et al., 2002)

typhoid fever is still needed. Here, we review the current data available on resistance in typhoid fever caused by S. typhi. (Fig. 4) (Parry et al. 2002)

3. HISTORICAL ASPECTS OF THE DEVELOPMENT OF DRUG RESISTANCE

A report of appearance of chloramphenicol resistance in S. typhi was published as early as in 1950 only after 2 years of first report of its use in typhoid therapy (Colquhoun et al. 1950). However, most isolates remained sensitive to chloramphenicol throughout the 1950s and 1960s. In 1972 a major typhoid outbreak occurred in Mexico city that was caused by S. typhi strains which were resistant to chloramphenicol, tetracycline, streptomycin and the sulphonamides. This resistance pattern was shared with Shigella dysenteriae and Shigella flexneri isolated from a concurrent outbreak of bacterial dysentery (Gangarosa et al. 1972).

4. MOLECULAR MECHANISM OF DRUG RESISTANCE

By transferring drug resistance from Salmonellae to E. coli it was shown that a plasmid was associates with the outbreak of typhoid fever and this was believed to be shared by the resistant Shigella isolates. However, when these two plasmids were typed by analyzing the incompatibility (Inc) group in comparison with a standard set of plasmids, it was shown that they were different. The S. typhi plasmids were of the IncH incompatibility group whereas the Shigella plasmids were of IncO group (Datta and Olarte 1974).

By 1984, resistance to each of the three first line agents (chloramphenicol, ampicillin and co-trimoxazole) was present in the S. typhi population. The resistance genes were located in a variety of different plasmids, and no single plasmid was shown to encode resistance to all three drugs simultaneously. However, in 1988 there was an outbreak of 230 cases of typhoid fever in the Kashmir valley, caused by MDR S. typhi that was resistant to all three first-line drugs (Kamili et al. 1993). MDR S. typhi isolated in Shanghai between 1988 and 1989 was shown to carry the genes for resistance on a single plasmid which could be transferred to E. coli (Zhang 1991). Other early outbreaks of MDR S. typhi were reported from Pakistan (1990) (Mirza and Hart 1993) and The Mekong Delta (1993) (Tran et al. 1995).

The majority of MDR plasmids isolated from S. typhi worldwide are now of the incompatibility group HI1 (Hampton et al. 1998; Harnett et al. 1998) and are highly transmissible. There are, however, some regions where *S. typhi* remains susceptible. For example, in Indonesia chloramphenicol was still used to treat typhoid fever in 1990s, (Van den Bergh et al. 1999) and 96% of *S. typhi* isolates were found to be sensitive to chloramphenicol. This is despite the fact that in 1978–1979, 14% of isolates in a Jakarta hospital were resistant to chloramphenicol (Komalarini et al. 1980).

4.1 Resistance to Fluoroquinolones and Cephalosporins

Although high-level resistance to these agents has not been widely reported in *S. typhi*, resistance does occur in similar bacteria. Testing the fluoroquinolones resistance is particularly problematic, in that reduced susceptibility to these agents is not always reported. This is because testing an isolate with reduced sensitivity to the fluoroquinolones by an antibiotic disc method usually produces a zone of clearance that is interpreted as fully susceptible. To overcome this difficulty, nalidixic acid discs should be used to screen for reduced susceptibility, and then the minimum inhibitory concentration (MIC) determined, for example, by the use of E-strips. There are ongoing discussions about introducing a new "intermediate" result or lowering the fluoroquinolone breakpoints (Crump et al. 2003).

Fluoroquinolone resistance is usually associated with point mutations that affect the structure of DNA gyrase. In *S. typhi*, mutations occur within a domain of the *gyrA* gene (which encodes the A subunit of DNA gyrase) termed the quinolone-resistance determining region. Point mutations conferring an amino acid substitution at positions serine 83 or aspartate 87 have been found in quinolone (nalidixic acid)-resistant *S. typhi* (Wain et al. 1997) but these isolates show susceptibility, at least in the laboratory, to the fluoroquinolones (e.g., ciprofloxacin). This resistance is not to plasmids. However, transferable resistance to nalidixic acid and the fluoroquinolones has been reported on a broad host range plasmid, from a clinical isolate of *Klebsiella pneumoniae* (Martínez 1998). The protein responsible, *qnr* (quinolone resistance) is thought to protect DNA gyrase, the target site for the fluoroquinolones, by yet unidentified mechanism. Although the prevalence of this resistance mechanism in clinical isolates is unknown, these findings imply that such a plasmid may enhance the development and spread of resistance to these valuable antimicrobial agents. The plasmid-borne *qnr* gene was found to be located within a mobile element which is similar to the mobile elements found on the resistance plasmids of *S. typhi* (Tran and Jacoby 2002). Should the *qnr* gene become widespread, it is likely to be acquired by the plasmids of *S. typhi* (Wain et al. 2003). This observation could have major implications in typhoid management.

Resistance to the third generation cephalosporins is usually mediated by extended spectrum β-lactamases which are commonly plasmid-borne (Jacoby and Sutton 1991) and may also be associated with mobile elements. Such elements, called integrons, have been identified in *S. typhi* (Partridge et al. 2001). Resistance to third generation cephalosporins has been reported in other *Salmonellae* (Shannon French 1998) but very few reports have been published describing such resistance in *S. typhi* (Saha et al. 1999).

5. THE CURRENT SITUATION OF MDR

Asia: MDR typhoid fever is now endemic in India, Pakistan, Bangladesh, Vietnam, Malaysia, Indonesia, China and Tajikistan and is present in migrant workers in the Middle East, Egypt and Thailand (Ling and Chau 1984; Threlfall et al. 1992).

Africa: Typhoid has been common in South Africa for some time and MDR *S. typhi* strains have been widely reported (Hampton et al. 1998). Until recently there was little published data about the presence of MDR typhoid elsewhere in sub-Saharan Africa. However, in last few years the studies in Kenya (Kariuki et al. 2000) and Ghana (Mills-Robertson et al. 2002) have shown that MDR typhoid fever is indeed present there. The prevalence of reduced susceptibility to fluoroquinolones is even more difficult to assess, but data from returning travellers suggests it is less common in Africa than in Asia.

Americas: Some of the early outbreaks of chloramphenicol-resistant typhoid were documented in Mexico and Peru (Olarte and Galindo 1973). However, although there are reports of MDR *S. typhi* in Central and South America, the levels of typhoid fever in general seem to be lower than in Asia and Africa. This probably reflects improved public health measures, many of which were implemented to prevent cholera.

Europe: Typhoid fever is no longer endemic in most of Europe although outbreaks still occur from carriers who are food handlers (Gruner et al. 1997), from imported contaminated food or as a result of a breakdown in hygiene allowing contamination of drinking water with sewage (Usera et al. 1995). Mediterranean countries have higher levels than most Northern European countries, probably because of infections associated with contaminated seafood (Stroffolini et al. 1992).

5.1 MDR – Future Challenges

The two main challenges are to limit the spread of existing resistance and to detect novel resistance early enough to prevent the spread of new resistant genes. In order to succeed, appropriate and timely diagnosis and therapy is needed in regions where typhoid fever is more common. The promise of new technology such as DNA microarrays, which can simultaneously screen hundreds of genes, needs to be exploited within the microbiological setting before we can fully evaluate drug resistance within the bacterial population.

5.2 Vaccines and Therapeutics

For successful development of a defined vaccine, a clear understanding of both the cellular and humoral components of the immune responses elicited during infection and identification of antigens that trigger protective response is essential. The candidate molecules for vaccine can be grouped into the following categories:

1. Crude fractions
2. Live pathogens rendered avirulent by genetic manipulations
3. Subunits (including extracts or metabolic products)
4. Recombinant and synthetic antigens
5. Recombinant bacteria/virus expressing specific antigens
6. Anti-idiotype vaccines
7. Naked DNA vaccines

Though a great deal of progress has been made in identification and production of potentially protective antigens, our understanding of the mechanisms of protective immunity against pathogenic diseases in human is still relatively poor. The integration of research on pathogenesis of bacterial infections with studies on the effect of vaccines and antibiotics will provide an unprecedented insight to counteract bacterial infections (Rupali 2004). Any molecule used in vaccination must stimulate protective immunity without simultaneously causing untoward immuno pathological or immunosuppressive reactions. To circumvent some of the adversities, various approaches of immunological manipulations, either alone or in combination with chemotherapy and vaccination has been explored. Use of well-defined recombinant protein in combination with appropriate adjuvant is more likely to overcome many limitations and provide species or strain specificity. The nucleic acid vaccines have strong potential and may become the vaccines of future. The nucleic acids used for vaccination are expressed by the host cells and elicit specific cytotoxic T-cells (CTL) that are capable of killing intracellular pathogens. Besides, T-helper cells and antibodies are also formed. Although vaccines have been developed against many major diseases that can be kept under control, there is a growing need for improvement of existing vaccines in terms of increased efficacy and improved safety as well as development of complementary new vaccines. Better technological possibilities combined with increased knowledge in related fields such as immunology and molecular biology allow for new vaccination strategies.

Many studies have been undertaken to understand the host response against *Salmonellae* infection. However, several aspects of immune response against this microorganism are yet to be understood. In particular, the detailed mechanism of protective immunity is uncertain and the identity of the protective antigen(s) has not been fully determined. The development of effective vaccines for protection against diseases caused by *Salmonellae* sp. has been an extensively studied (Nasser 2002). In immunization programmes, killed bacteria or inactivated whole cell vaccines are used routinely. Live vaccines are more effective in induction of host immunity against salmonellosis than the killed vaccines due to higher ability of live vaccines to induce both antibodies and cell-mediated immunity (Gherardi et al. 2000). Killed vaccines, on the other hand, cannot replicate and are therefore, non-infectious but are less effective than live vaccines in inducing protective immunity. The ability to elicit antibodies in addition to cell mediated immunity is important for optimal protection induced by *Salmonellae* vaccines (Mastroeni et al. 1993). Some of the outer membrane proteins seems to have strong immunogenic potential (Foulaki et al. 1989; Muthukkumar and Muthukkaruppan 1993). However, only limited success has been achieved using porins as the immunizing agent (Kussi et al. 1981). Recent studies have shown that some non-porin OMPs may evoke strong immune response and confer 100% protection against *Salmonellae* challenge in animals. Some of these proteins have very strong potential for the development of a subunit vaccine against typhoid (Chatfield et al. 1991; Hamid 2001 Thesis).

5.3 Current Status of Typhoid Management

Presently a number of vaccines against typhoid are available; however, there is a need for better and improved new generation vaccines against *Salmonellae*. The mechanism of protection conferred by these vaccines has not been fully elucidated. The selection of vaccine strains for typhoid is empirical, the reasons for the success and failure of a vaccine are poorly understood and the protective capacity of a vaccine can only be ascertained by challenge of natural infection (Mastroeni et al. 1994). There are two fundamental aspects for vaccine development against *Salmonellae* infections including typhoid fever; one is the identification and characterization of molecular components of bacteria that confer a long-term immunological protection and the other being the mutations that lead to attenuation of the bacterial strains which are capable of eliciting a strong immune response. The vaccine development requires a thorough understanding of genetics as well as physiological responses of *Salmonella* under different conditions during infection (Calva and Puente 1995). Antigens that are currently being exploited include the outer membrane proteins (OMPs), the heat shock proteins and the Vi capsular antigen. Some of the currently available vaccines against typhoid have been described below:

(i) Whole cell parenteral vaccine: The inactivated whole cell parenteral vaccines including heat killed, phenol preserved (used since 1896) and acetone-inactivated vaccines have been shown to confer moderate level of protection (upto 60-70%) and have a vaccine efficacy for upto seven years (controlled field trials in Yugoslavia and Guyana sponsored by WHO in between 1960 to 1970) (Yugoslav 1964; Hefjee et al. 1966; Levine et al. 1989; Chilean Typhoid Committee 1990). However, the use of these vaccines is associated with adverse reactions at high frequencies and therefore, these vaccines could never get accepted as public health tools (Bhutta 1996) such as Polish Typhoid Committee (1966). Approximately, 25% of the recipients of these vaccines developed fever and systemic reactions, and 15% could not go to work (Hefjee et al. 1966; Levine et al. 1994) Yugoslav Typhoid Commission 1964.

(ii) Vi polysaccharide vaccine: *Salmonellae* are covered by Vi-polysaccharide, a homopolymer of N-acetyl-galactosamine and uronic acid, which correlates with virulence (Heyns et al. 1959; Ashcroft et al. 1964). The major protective antigen(s) of *S. typhimurium* are likely to be located in the polysaccharide of the bacterial envelope (Heyns and Keissling 1967). A poor serological response to Vi antigen has been shown in acute typhoid fever, while high profile response occurs in chronic carriers. The Vi antigen can protect the O-antigen of *S. typhi* from agglutination by antibodies and it has been proposed that Vi antibodies play an important role in protection against typhoid fever (Heyns et al. 1959; Ashcroft et al. 1964; Polish Typhoid Committee 1966; Heyns et al. 1967; Klugman et al. 1987; Levine et al. 1989; Chilean Typhoid Committee 1990; Ding et al. 1990; Levine et al. 1995; Klugman et al. 1996). However, it has been found that bacteria isolated from the blood of patients with typhoid fever invariably bear a capsular polysaccharide, the Vi-antigen (Felix and Pitt 1934). Vi antigen has been recognized as an important virulence factor in *S. typhi* because Vi^+ strains have been found to be more virulent than Vi^- mutants in human volunteers

(Landy 1954; Hornick et al. 1970). The protective role of Vi antigen was first proposed by Landy (1954) who found that a highly purified form of Vi antigen obtained by denaturing technique was unable to protect volunteers. However, when the Vi polysaccharide purified using non-denaturing techniques with CTAB, it was protective and could be used as parenteral vaccine (Polish Typhoid Committee 1966). A Vi based anti-typhoid vaccine was prepared under trade name "Typhim Vi" Robbins (1990). In another study, it was found that two separate non-denatured Vi polysaccharide (Vi-PS) preparations isolated independently at NIH (USA) and at the Merieux Institute (France) contained 5% and 0.2% of contaminating LPS, respectively. Both these preparations elicited high titers of Vi antibodies in about 90% of recipients, while less pure lots also stimulated O-antibodies in more than 80% of recipients (Wong and McClelland 1992). The Vi antibodies generated by these preparations persisted for at least three years (Tacket et al. 1986).

In two independent randomized field trials held in South Africa and Nepal, it was observed that a single intramuscular injection of purified *S. typhi* Vi polysaccharide could confer good degree of protection (64% to 72%). These trials were conducted to assess the safety and efficacy of Typhim Vi vaccine produced by Merieux Institute of Medicine for use in Chile, Congo, Cote d'Ivoire (Ivory Coast), France, Republic of Korea, Netherlands, Togo, Peru, United Kingdom and Philippines. The results of these trials clearly established the efficacy of typhoid vaccine based on Vi-antigen (Klugmen 1987; Szu et al. 1987, Hessel et al. 1999). Further, it was found that conjugated Vi-PS vaccines produced higher immune responses, both humoral and cell mediated than the non-conjugated Vi-PS vaccine (Ivanof et al. 1994) (Table 1).

Table 1 Efficacy of heat inactivated, phenol preserved killed whole cell parenteral vaccine and purified Vi capsular polysaccharide parenteral vaccine.

	Heat phenolised vaccine (n=23, 431)	Placebo (n=24, 241)	P value	Vi-vaccine (n=5692)	Placebo (n=5692)	P value
Year	(1983-90)			Year (1983-86)		
No. of cases	49	146	<0.001	30	66	<0.001
incidence/105	210	602			527	
Efficacy	65% (52-75%)			55% (30-70%)	527	

Reference: Ivanoff, B., Levine, M.M. & Lambart,P.H.(1994).

(iii) Killed vaccines: Oral vaccination against typhoid fever was first attempted with killed bacterial cells. But these vaccines conferred low or no efficacy in field trials (Bhutta 1996). Therefore, efforts were made towards the isolation of live attenuated vaccine strains suitable for immunization in animals and humans via oral route (Levine et al. 1995). Among various candidate vaccine strains developed, the most evaluated is the *S. typhi galE* mutant strain Ty21a that were produced by chemical mutagenesis of a pathogenic *S. typhi* strain (Germaniar and Furer 1975). As a result of the mutation in *galE* gene, the activity of enzyme UDP-galactose- 4 epimerase, responsible for the interconversion of UDP-galactose to UDP-glucose is hindered (Fig. 5) (Ivanoff et al. 1994). The non-specific action of nitroso-guanidine resulted in mutants with inability to produce hydrogen sulphide, several nutritional auxotrophies having growth rate approximately half that of the parent strain Ty2 and the lack of Vi-antigen in addition to galE mutation (Ivanof et al. 1994). Galactose, which is important component of smooth LPS O-antigen in wild type *S. typhi,* is produced by the action of UDP galactose-4-epimerase. No synthesis of smooth O-antigen was, therefore, observed when Ty21 strain was grown in absence of galactose. As a result, they become rough and more prone to lysis (Germaniar and Furer 1971).

Ty21a provides significant protection without causing any adverse effect as evident from the results of three placebo-controlled studies utilizing active surveillance methods (Levine et al. 1989). Field trials to assess the efficacy of Ty21a vaccine, involving approximately school children in Chile, and Egypt and 3 years old children in Indonesia (Wahdan et al. 1980; Levine et al. 1990; Simanjuntak et al. 1991) showed no adverse reaction (Bhutta et al. 1996; Levine et al. 1990). Further, it was found that the vaccine formulation,

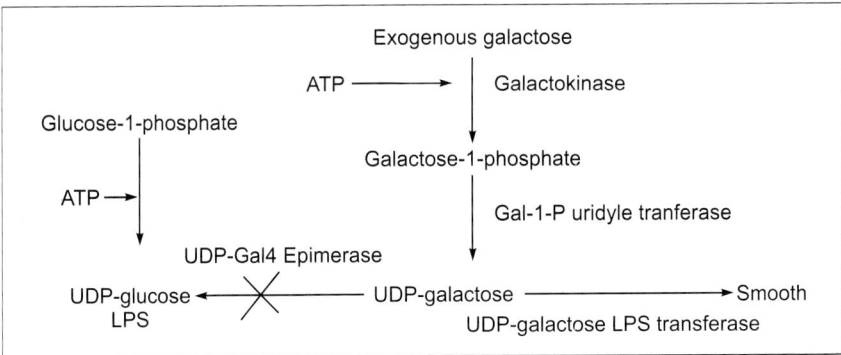

Fig. 5 The incorporation of exogenous galactose by the gale mutant. In absence of UDP galactose-4-epimerase there is an accumulation of galactose-1-phosphate and UDP-galactose (Ivanoff et al., 1994).

number of doses and timing of doses markedly influence the level of protection (Black et al. 1987; Ferreccio et al. 1989; Levine et al. 1989; Levine et al. 1990; Simanjuntak et al. 1991; Levine et al. 1999). The results of a trial in Egypt are summarized in Table 2 (Ivanoff et al. 1994).

Table 2   Efficacy of Ty21a three doses in liquid formulation or in enteric coated capsules given within a week in Egypt (Alexandria) and Chile (Santiago).

	Egypt: liquid formulation Vaccine (n=16, 486)	Chile: Enteric coated capsules Vaccines (n= 22,170)
No. of cases	1	35
Incidence	6	158
Efficacy (%)	95.6 (77-99)[a]	66 (50-77)[a]
P value	<0.001	< 0.00001

(a) 95% confidence interval
Reference: Ivanoff, B., Levine M. M., and Lambert, P. H. (1994).

Another trial was conducted in Santiago (Chile) that utilized lyophilized vaccine in enteric coated and acid resistant capsules (Ivanoff et al. 1994) (Table 3). It was later reported that four doses of Ty21a vaccine in enteric-coated capsules given within eight days were more protective than two or three doses (Ferreccio et al. 1989). Swiss Serum and Vaccine Institute prepared a liquid formulation consisted of two components, a lyophilized vaccine and the buffer (Ivanoff et al. 1994; Levine et al. 1999) that were tried in Santiago (Chile) (Levine et al. 1990) and in Indonesia (Simanjuntak et al. 1991). It was found that the liquid suspension was superior to enteric-coated capsules and the liquid suspension was highly protective in young children as well as in older subjects (Table 4). However, Ty21a vaccine is not always well-tolerated by infants (aged 6 to 24 months) and elicits a much less antibody response than in older children (Murphy et al. 1991). Although Ty21a vaccine is very effective and is well tolerated, it is not precisely clear as to what mutations are responsible for the stable and impressive attenuation of this vaccine strain.

It was thought that a virulence Ty21a is because of the mutation in *galE* gene. The spontaneous *galE* revertants of Ty21a strain were found to be as virulent as the original strain (Silva-Salinas et al. 1985). Furthermore, it was recognized that Ty21a strain has a requirement for valine and leucine and also lacks Vi-antigen (Silva-Salinas et al. 1987). These properties are indicative of multiple mutations in Ty21a strain and it is difficult to highlight the precise basis for attenuation (Hackett 1990).

Table 3 Comparison of the efficacy of Ty21a vaccine in two different formulations administered at different immunization schedule in Santiago.

	Enteric coated capsules n=22170[a]	Gelatine capsules (NaHCO$_3$) n=22379[a]	Placebo n=21906
No. of cases	23	56	68
Incidence [b]	103.7	250.5	310.4
Efficacy (%)	66.6	19.4	-

(a) Shot interval=3 doses, 2 doses apart.
(b) Incidence per 100,000 population
Reference: Ivanoff, B., Levine M. M., and Lambert, P. H. (1994)

Table 4 Comparison of protective efficacy of Ty21a vaccine in different formulations [Liquid Enteric coated capsule (Ecc) and Gelatin capsules], compared with Placebo in Egypt, Chile and Indonesia.

Countries	Vaccine formulation	Efficacy	Relative efficacy	Placebo*
Egypt	Liquid[+] NaHCO$_3$	96%	-	50
Chile (Occidente)	Ecc/Gelatine cap.	66% Ecc, 19% Gelatine cap.	Ecc> Gelantin cap.	110
Chile (Oriente)	Liquid/Ecc	35% Ecc, 76% Liquid	Liquid> Ecc	100
Indonesia	Liquid/Ecc	42% Ecc, 53% Liquid	Liquid> Ecc	810

* Figures of the incidence of typhoid fever, among those using placebo per 100,000 persons per year.
Reference: . Ivanoff, B., Levine M. M., and Lambert, P. H. (1994).

It has also been reported that *galE* mutation in *S. typhi* strain Ty2 had no effect on its virulence in humans (Hone et al. 1988). However, the double mutants defective in *galE* and Vi-antigen are highly attenuated in mice but have no observable attenuation in humans. It was suggested that the mutants may be capable of scavenging enough galactose in the host to synthesize smooth LPS, thus reverting the mutation (Hone et al. 1987; Hackett 1990). This clearly indicates that *galE* mutation is not responsible for attenuation in human (Hone et al. 1988; Hone et al. 1987). The drawbacks of Ty21a vaccine include: (a) its requirement for multiple doses to stimulate protection, (b) the fragility of vaccine strain in the fermentation and lyophilization process for large-scale manufacture and, (c) the fact that a non-specific method of mutagenesis results in multiple genetic changes in

Ty21a strain. The difference between two popular typhoid vaccines has been shown in Table 5 (Levine et al. 1995).

Table 5   Comparison of the characteristics of the live oral vaccine Ty21a and Vi polysaccharide vaccine.

	Ty 21 a	*Vi polysaccharide*
Type of vaccine	Live	Subunit
Route of administration	Oral	Parenteral
Immunization schedule	3 or 4 doses (given every other day)	1 dose
Cold chain required	yes	no
Well tolerated	yes	yes
Efficacy	60-96%	64-72%
Duration of efficacy	7 years	3 years
Evidence of a herd immunity	yes	?
Interference with use of serum Vi antibody as a screening test to detect chronic typhoid carriers	no	yes

5.4 New Generation of Attenuated Vaccines

To overcome problems associated with Ty21a, genetic engineering is being applied to create new attenuated strains of S. *typhi*, which might serve as oral vaccines. Non-revertable defined mutations such as inactivation of the genes encoding enzymes involved in various biochemical pathways (Salva-Salinas et al. 1987) and regulatory systems (Curtiss and Kely 1987), other regulatory genes (Pickard et al. 1994) and various important virulence properties (Miller et al. 1993) can be introduced to produce live vaccine strains (Chatfield et al. 1989). The attenuation potential of these mutations is assessed by feeding mutated S. *typhimurium* strains to mice and comparing the results obtained with isogenic wild type strains. However, there are several examples where specific mutations that successfully attenuated S. *typhimurium* but failed to attenuate S. *typhi* (Hone et al. 1987; Hone et al. 1992; Tacket et al. 1992a). Live attenuated *Salmonella* that carry lesions in genes encoding aromatic amino acid biosynthetic pathway become dependent on p-amino benzoic acid (PABA) that is not available in mammalian tissue (Stocker 1990). Anti-Salmonella vaccines formulated with such strains are effective in mice (Hoiseth and Stocker 1981), cattle (Smith et al. 1984; Jones et al. 1991) and chickens (Barrow et al. 1991) and also have shown promise in human trials (Tacket et al. 1992b). Another important deletion mutation is in gene *purA* which leads to specific requirement for adenine or adenosine. These nutritional requirements render the strain unable to sustain growth in mammalian tissues (Mac Fralane et al. 1987; O'Collaghan et al. 1988).

In addition, recombinant *Salmonella* vaccines have been shown to protect against a broad range of pathogens in animal models and preliminary results from clinical trials demonstrate that protective immunity against heterologous antigens is achievable (Dunstan et al. 1996). A significant development involves the use of *Salmonellae* for delivery of DNA vaccines. Use of bivalent *Salmonellae* strains to deliver DNA vaccines allows the induction of immunity against the *Salmonellae* carrier, the heterologous antigen(s) expressed by *Salmonellae*, and the antigen(s) encoded by the DNA vaccines. Further, it should be possible to simultaneously express the heterologous antigens integrated into *Salmonellae* chromosome as well as cloned into the plasmid(s) carried by the *Salmonella*. It may even be possible to 'prime' and 'boost' within a single orally delivered immunization by delivering the DNA vaccine to appropriate antigen presenting cells (APCs) for expression of antigen, and co-delivering an expressed antigen. This DNA vaccine delivery system is very promising and offers a new and exciting approach to mucosal vaccine delivery (Mollenkopf et al. 2001; Niethammer et al. 2001). To date, however, there have been no reports of human trials of recombinant *Salmonella* for the delivery of DNA vaccines.

5.4.1 Outer Membrane Proteins of Salmonella as Typhoid Vaccine

Salmonellae have inner cell membrane, cell envelope and the outer membrane in between a peptidoglycan cell wall structure. The inner membrane is the typical phospholipid bilayer structure, enclosing cytoplasm and other cell inclusions, while the outer membrane consists of two leaflets of different characters. The outer membrane is 7.5 nm thick, constituted by lipopolysaccharides (LPS), proteins, phospholipids (PLs) and enterobacterial common antigen (ECA). The inner leaflet is composed of PLs and outer leaflet is mainly LPS (Samuelson et al. 2002). As the exposed cell wall components of Gram-negative bacteria directly interact with host cells and antibodies, an immune response evoked against them would be able to recognize the invading pathogens and confer protection. The LPS and protein components of outer membrane elicit immune response. The proteins (OMPs) that include integral membrane proteins as well as the proteins that are anchored to the outer membrane account for approximately 50% of outer membrane (Lin et al. 2001). Outer membrane proteins span the membrane several times and differ from other membrane spanning proteins in the sense that the spanning structure is mainly anti-parallel β barrel structure (Samuelson et al. 2002). Characterized by β-barrel structures, integral OMPs are essential for maintaining the integrity and selective permeability of bacterial membrane. In addition, OMP production is often regulated by environmental cues also play important role in bacterial pathogenesis by enhancing adaptability of bacterial pathogens to various environments. Thus, OMPs represent important virulence factors and are important for bacterial adaptation to host niches, which are usually hostile to invade pathogens. Some of these roles have been enumerated in Table 6. As discussed, OMPs are potential candidate for the development of subunit vaccine against typhoid.

Table 6 Mechanism of OMP-mediated bacterial adaptive response to host environment.

Adaptive response	Mechanisms	OMPs involved
Iron uptake	Expression of OMPs directly binding to host proteins such as transferrin, lactoferrin or hemoprotein.	Tbp, Lbp, HemR
	Synthesis of high-affinity iron-siderophores and expression of OMPs and their binding to siderophore complexes.	FhuA, FepA
Antimicrobial peptide resistance	Direct degradation of antimicrobial peptides through production of outer membrane associated proteases.	OmpT, PgtE
	Modification of bacterial surface through production of OMPs with enzymatic activities.	PagP
Serum resistance	Prevention of activation of complement cascades by binding to factor H or C4bp, down-regulators of complement activation.	Por1A, Por1B, OspE
Multidrug resistance	Unknown	OmpX
	Key roles in multidrug efflux systems.	TolC, OprM
Bile resistance	Modulate membrane permeability by regulating the production of specific porins.	OmpC, OmpU
	Key roles in multidrug efflux systems.	TolC

6. MANAGEMENT OF TYPHOID WITH HERBAL DRUGS

The increase in bacterial resistance to antibiotics has made it ever more difficult to prevent and treat the infections. Therefore, the study of pathogenic mechanisms of bacterial infection at molecular level is not only of scientific interest but also has implications in typhoid therapy. Also the herbs and plant products offer a possible alternate to antibiotics and other chemical drugs. The ethylacetate-insoluble fraction of the methanol extract of "Ogwu Odenigbo" a popular Nigerian traditional herbal medicine for typhoid fever that is prepared from the stem bark of *Cleistropholis patens* Benth (Annonaceae) was reported to be highly effective (Ebi and Kamalu, 2001). Methanol extracts of this plant parts are commonly used in Cameroon for the treatment of typhoid fever. The used formulations are: (1) Formulation A, comprising *Cymbogogon citratus* leaves, *Carica papaya* leaves and *Zea mays* silk. (2) Formulation B, comprising *C. papaya* roots, *Mangifera indica* leaves, *Citrus lemon* fruit and *C. citratus* leaves. (3) *C. papaya* leaves. (4) *Embelia coccinea* whole plant. (5) *Comelina bengalensis* leaves. (6) *Telfaria occidentalis* leaves. (7) *Gossypium arboreum* whole plant. Generally, Formulation A elicited therapeutic effect at 0.02 to 0.06 mg/ml, formulation B elicited bactericidal activity at 0.06 to 0.25 mg/ml. *C. bengalensis* leaves on the other hand, showed the lowest activity with a concentration range of 0.132 to 2.0 mg/ml and 1 to 4 mg/ml in MIC and MBC assays, respectively. *S. paratyphi* was most sensitive to the formulations while *S. typhimurium* was the least sensitive and higher drug concentrations (upto 4 mg/ml) were required.

It is concluded that plant extracts with low MIC and MBC values (1mg/ml and lower) may contain compounds with therapeutic potential activity (Nkuo-Akenji et al. 2001). It has been reported that black tea (*Camelia sinensis*) extracts show antibacterial activity against *Salmonellae* serotypes causing enteric fever (Ciraj et al. 2001). Various plant extracts have been reported to show antibacterial activity against *Salmonella typhimurium* ATCC 23564 viz., *Calotropis procera, Terminalia arjuna, Terminalia belerica, Eclipta alba, Sapium indicum, Vetivera zizanioids, Mesua ferrea, Cinnamomum tamala, Acacia arabica, Acacia catechu, Tamarindus indica, Allium sativum, Aloe barbadensis, Myristica fragrance, Syzgium aromaticum, Punica granatum, Nigella sativa, Solanum nigrum, Carum copitacum, Tribulus terrestris* (Ahmad et al. 1998). Many other medicinal plants, especially *Euphorbia hirta, Citrus aurantifolia, Cassia oxidentalis, Cassia occidentalis* and *Cassia eucalyptus* have also been claimed to be effective in the treatment of typhoid fever (Evans et al. 2002). *Cassia occidentalis* roots, *Heimia salicifolia* aerial parts, *Punica granatum* fruit pericarp and *Rosa borboniana* flowers produced active extracts with high potential. Taking the MDR of *Salmonella typhi* into account, these findings could be useful in the search for new clinically antimicrobials (Perez and Anesini 1994).

A crude drug formulation comprised extracts of several medicinal herbs displayed inhibitory action against *Salmonella typhi in vitro*. Subsequently, mice were challenged with a virulent strain of *Salmonella typhi* (Ty2) and the protective effect of the formulation was evaluated with post-infective, pre-infective, single and multiple dose schedules, administered either orally or subcutaneously. A schedule that included multiple divided doses prophylactically administered had a significant therapeutic effect (Sohni et al. 1995).

We have evaluated a large number of plants and have found four plants viz. *Embelica officinalis* (Amla), *Terminalia chebula* (Hara), *Terminalia belerica* (Bahera) and *Punica granatum* (Anar) to have strong anti-Salmonella activity. These can be used successfully for conferring prevention against experimentally induced Salmonellosis in management of typhoid fever (Table 7).

Table 7   Anti-Salmonella activity of certain plants with known antimicrobial properties.

S. No.	Plant: Botanical Name	Common Name	Anti-Salmonella activity
1.	*Plumbago zylanica*	Chita	+
2.	*Terminalia belerica*	Bahera	+++++
3.	*Punica granatum*	Anar	+++++
4.	*Asparagus resemosus*	Satawar	-
5.	*Holarrhena antidysentrica*	Kurchi	-
6.	*Picrorhiza kurroa*	Kutki	-
7.	*Terminalia chebula*	Harda	++++
8.	*Withania somnifera*	Ashwagandha	-
9.	*Embelica officinalis*	Amla	++++
10.	*Tinospora cordifolia*	Gilo	-
11.	*Vitis vinifera*	Munnaka	-
Std.	*Ciproflaxacin*	Antibiotic	+++++

6.1 Nitric Oxide and Host Defense against Typhoid

Nitric oxide (NO) is a gaseous, inorganic, free radical, molecular species that is produced in most of the biological systems. It regulates a diverse array of physiological functions and acts as an inter and extracellular messenger in most mammalian organs (Moncada and Higs 1993; Ignarro 2000). A wide variety of cells, such as leukocytes, hepatocytes, vascular smooth muscle cells, and endothelial cells, can produce NO during enzymatic conversion of L-arginine to L-citrulline by nitric oxide synthase (NOS). A large amount of NO generated by the inducible isoform of NOS (iNOS) has been demonstrated to have a beneficial effect in host defense mechanisms against various pathogenic bacteria (Granger et al. 1988; James 1995; Nathan 1997;

Nathan and Shiloh 2000). It has been presumed that iNOS expression can provide antimicrobial activity through formation of reactive nitrogen oxides derived from NO (Yoshida 1993; Brunelli et al. 1995; De Groote 1995; Kuwahara et al. 2000; Miyamoto et al. 2000). For example, peroxynitrite (ONOO$^-$), a potent oxidant formed from NO and superoxide radical (O$_2^-$), is microbicidal for various bacteria, including *Salmonella enterica* serovar typhimurium (Brunelli et al. 1995; DeGroote 1995; De Groote 2000; Kuwahara 2000), and nitrosothiols, one-electron oxidized derivatives of NO, have potent bacteriostatic activity against serovar typhimurium (Akaike 2000; De Groote et al. 2000; Miyamoto et al. 2000).

The pharmacological inhibition of either NO or superoxide production resulted in a remarkable enhancement of *Salmonella* growth and increased mortality in murine salmonellosis, suggesting that both NO and superoxide contribute critically to host defense against *Salmonella serovar typhimurium* (Umezawa et al. 1997). It was found that the NO biosynthesis blocks the pathogenesis of *Salmonellae in vivo* (MacFarlane et al. 1997).

We have found that pre-treatment of animals with NO donors such as arginine and citrulline confers significant protection against *S. typhimurium* challenge. These compounds help in accelerated bacterial clearance and potentiate the immune response. The studies have shown that the NO-donors not only kill bacteria but also protect the tissue from the ill effects of bacterial infection. These compounds seem to have great potential for reducing the dose and duration of chemotherapy for the management of typhoid fever.

CONCLUSION

Typhoid fever remains a major health problem in India and other developing countries. It results in considerable mortality and morbidity. Further the cost of typhoid fever including the cost due to loss of many days, is very high to the society. The occurrence of MDR strains of *Salmonella* have made the management of typhoid more difficult. A number of antibiotics that had high therapeutic efficacy few years back, are not effective today. Currently, no typhoid specific herbal drugs are available in either Ayurvedic or Unani system of medicine. However, few of the medicinal plants have shown strong anti-Salmonella activity. These can be exploited for development of new herbal medicines for typhoid and may provide an answer to this problem. The efficacy of currently available vaccines against typhoid is less than desired and these are also associated with certain side effects. There is a need for the development of new vaccines against typhoid. The outer membrane proteins of S. typhi are highly immunogenic. Some of the OMPs have strong potential for the development of a new subunit vaccine.

ACKNOWLEDGEMENTS

These studies have been supported by research projects from CCRUM and DST to SKJ. NH is a UGC-CSIR NET JRF and AM is an ICMR SRF. The financial assistance from these agencies is gratefully acknowledged.

REFERENCES

Ahmad I, Mehmood Z, Mohammad F (1998). Screening of some Indian medicinal plants for their antimicrobial properties. J. Ethnopharmacol. 62: 183-9.

Akaike T (2000). Mechanisms of biological S-nitrosation and its measurement. Free Radic Res. Nov 33 (5):461-9. Review.

Ashcroft MT, Ritchie JM, Nicholson CC, Stuart CA (1964). Antibody responses to vaccination of British guiana school children with heat-killed-phenolozed and acetone-killed lyophilized typhoid vaccines. Am J Hyg. Sep 80:221-8.

Barrow PA, Hassan JO, Lovell MA, Berchieri A (1990). Vaccination of chickens with aroA and other mutants of Salmonella typhimurium and S. enteritidis. Res Microbiol. 141 (7-8): 851-3.

Bhutta ZA (1996). Impact of age and drug resistance on mortality in typhoid fever. Arch Dis Child. 75(3):214-7.

Black RE, Levine MM, Clements ML, Losonsky G, Herrington D, Berman S, Formal SB (1987). Prevention of shigellosis by a Salmonella typhi-Shigella sonnei bivalent vaccine. J Infect Dis. 155 (6):1260-5.

Brunelli L, Crow JP, Beckman JS (1995). The comparative toxicity of nitric oxide and peroxynitrite to Escherichia coli. Arch Biochem Biophys. 10; 316 (1):327-34.

Butler T, Linh NN, Arnold K, Pollack M (1973). Chloramphenicol resistant typhoid fever in Vietnam associated with R factor. Lancet. 3; 2(7836): 983-5.

Calva E, Puente JL (1995). Southeast. J. Trop. Med. Public. Health. 26 (Suppl.2); 1-6.

Chatfield SN, Dorman CJ, Hayward C, Dougan G (1991). Role of ompR-dependent genes in Salmonella typhimurium virulence: mutants deficient in both OmpC and OmpF are attenuated in vivo. Infect. Immun 59: 449-452.

Chatfield SN, Strugnell RA, Dougan G (1989). Live Salmonella as vaccines and carriers of foreign antigenic determinants. Vaccine. 7(6):495-8. Review.

Chilean Typhoid Committee, Black RE, Levine MM, Ferreccio C, Clements ML, Lanata C, Rooney J, Germanier R (1990). Infect Immun. 8; 81.

Chong C, Bost K, Clements JD (1996). Differential production of interleukin-12 mRNA by murine macrophages in response to viable or killed Salmonella spp. Infect. Immun. 64; 1154-1160.

Ciraj AM, Seema DS, Bhat GK, Shivananda PG (2001). Nalidixic acid screening test for the detection of decreased susceptibility to ciprofloxacin in Salmonella typhi. Indian J Pathol Microbiol. 44 (4):407-8.

Clark MA, Jepson MA, Simmons NL, Hurst BH (1994). Preferential interaction of *Salmonella typhimurium* with mouse Peyer's patch M-cells. Res. Microbiol.145: 543-552.

Collins FM, Campbell SG (1082). Immunity to intracellular bacteria. Vet Immunol Immunopathol. 3(1-2):5-66. Review.

Colquohon J, Weetch RS (1950). Resistance to chloramphenicol developing during treatment of typhoid fever. Lancet. 25;2(22): 621-3.

Coovadia YM, Gathiram V, Bhamjee A, Garratt RM, Mlisana K, Pillay N, Madlalose T, Short M. (1992). An outbreak of multi-resistant *Salmonella typhi* in South Africa. Q J Med. 82(298):91-100.

Crump JA, Youssef FG, Luby SP, Wasfy MO, Rangel JM, Taalat M, Oun SA, Mahoney FJ (2003). Estimating the incidence of typhoid fever and other febrile illnesses in developing countries. Emerg Infect Dis. 9 (5):539-44.

Curtiss R, Kelly SM (1987). *Salmonella typhimurium* deletion mutants lacking adenylate cyclase and cyclic AMP receptor protein are avirulent and immunogenic. Infect Immun. 55 (12):3035-43.

Datta N, Olarte J (1974). R factors in strains of *Salmonella typhi* and *Shigella dysenteriae* 1 isolated during epidemics in Mexico: classification by compatibility. Antimicrob Agents Chemother. 5 (3):310-7.

De Groote MA, Fang FC (1995). NO inhibitions: antimicrobial properties of nitric oxide. Clin Infect Dis. Suppl 2:S162-5. Review.

De Groote MA, Granger D, Xu Y, Campbell G, Prince R, Fang FC (1995). Genetic and redox determinants of nitric oxide cytotoxicity in a *Salmonella typhimurium* model. Proc Natl Acad Sci U S A. 3;92(14):6399-403.

Ding HF, Nakoneczna I, Hsu HS (1990). Protective immunity induced in mice by detoxified salmonella lipopolysaccharide. J Med Microbiol. 31 (2); 95 -102.

Dunstan CA, Salafranca MN, Adhikari S, Xia Y, Feng L, Harrison JK (1996). Identification of two rat genes orthologous to the human interleukin-8 receptors. J Biol Chem. 20; 271(51):32770-6.

Ebi GC, Kamalu TN (2001). Phytochemical and antimicrobial properties of constituents of *Ogwu Odenigbo*, a popular Nigerian herbal medicine for typhoid fever. Phyt other Res. 15(1):73-5.

Evans CE, Banso A, Samuel OA (2002). Efficacy of some nupe medicinal plants against *Salmonella typhi*: an *in vitro* study. J. Ethnopharmacol 80 (1):21-4.

Felix A (1950). Note on a case of Brill's disease in London.Lancet. 1950 Jul 15;2(3):99-100.

Ferreccio C, Levine MM, Rodriguez H, Contreras R (1989). Comparative efficacy of two, three, or four doses of TY21a live oral typhoid vaccine in enteric-coated capsules: a field trial in an endemic area. J Infect Dis. 159 (4):766-9.

Finlay BB, Starnbach MN, Francis CL, Stocker BA, Chatfield S, Dougan G, Falkow S (1988). Identification and characterization of TnphoA mutants of *Salmonella* that are unable to pass through a polarized MDCK epithelial cell monolayer. Mol Microbiol. 2(6):757-66.

Finlay BB, Falkow S (1989). *Salmonella* as an intracellular parasite. Mol Microbiol. 3(12):1833-41. Review.

Foulaki K, Gruber W, Schlecht S (1989). Isolation and immunological characterization of a 55-kilodalton surface protein from *Salmonella typhimurium*. Infect Immun. 57 (5):1399-404.

Gangarosa EJ, Bennett JV, Wyatt C, Pierce PE, Olarte J, Hernandes PM, Vazquez V, Bessudo D (1972). An epidemic-associated episome? J Infect Dis. 126 (2):215-218.

Garcia-del Portillo F (2001). *Salmonella* intracellular proliferation: where, when and how? Microbes Infect. 3(14-15):1305-11. Review.

Garcia-del Portillo F, Pucciarelli MG, Jefferies WA, Finlay BB (1994). *Salmonella typhimurium* induces selective aggregation and internalization of host cell surface proteins during invasion of epithelial cells. J Cell Sci. 107 (Pt 7):2005-20.

Germanier R, Furer E (1971). Immunity in experimental salmonellosis. II. Basis for the avirulence and protective capacity of gal E mutants of *Salmonella typhimurium*. Infect Immun. 4(6):663-73.

Germanier R, Fuer E (1975). Isolation and characterization of Gal E mutant Ty 21a of *Salmonella typhi*: a candidate strain for a live, oral typhoid vaccine. J Infect Dis. 131 (5):553-8.

Gherardi MM, Ramirez JC, Esteban M (2000). Interleukin-12 (IL-12) enhancement of the cellular immune response against human immunodeficiency virus type 1 env antigen in a DNA prime/vaccinia virus boost vaccine regimen is time and dose dependent: suppressive effects of IL-12 boost are mediated by nitric oxide. J Virol. 74(14):6278-86.

Ginocchio CC, Olmsted SB, Wells CL and Galan JE (1994). Contact with epithelial cells induces the formation of surface appendages on *Salmonella typhimurium*. Cell 76, 717-24.

Granger DL, Hibbs JB Jr, Perfect JR, Durack DT (1988). Specific amino acid (L-arginine) requirement for the microbiostatic activity of murine macrophages. J Clin Invest. 81(4):1129-36.

Gruner EM, Flepp U, Gabathuler KL, Thong M, Altwegg (1997). J Microbiol Meth. 28; 3, 179–185.

Hackett J (1990). *Salmonella*-based vaccines. Vaccine. 8(1):5-11.

Hamid T (2001). Biological characterization of the outer membrane proteins of S. *typhi* and S. *typhimurium* and studies on their role in protection against typhoid. Thesis submitted to Centre for Biotechnology, Hamdard University, New Delhi.

Hampton MD, Ward LR, Rowe B, Threlfall EJ (1998). Molecular fingerprinting of multidrug-resistant *Salmonella enterica* serotype Typhi. Emerg Infect Dis. 4(2):317-20.

Harnett N, McLeod S, AuYong Y, Wan J, Alexander S, Khakhria R, Krishnan C (1998). Molecular characterization of multi-resistant strains of *Salmonella typhi* from South Asia isolated in Ontario, Canada. Can J Microbiol. 44(4):356-63.

Hejfec LB, Salmin LV, Lejtman MZ, Kuz'minova ML, Vasil'eva AV, Levina LA, Bencianova TG, Pavlova EA, Antonova AA (1966). A controlled field trial and laboratory study of five typhoid vaccines in the USSR. Bull World Health Organ. 34(3):321-39.

Hessel L, Debois H, Fletcher M, Dumas R (1999). Experience with *Salmonella typhi* Vi capsular polysaccharide vaccine. Eur J Clin Microbiol Infect Dis. 18(9):609-20.

Heyns, K. and Kiessling, G. (1967). *Carbohydrate Res.* 3; 340.

Heyns, K., Kiessling, G., Lindenberg, W., Paulsen, H. and Webstar. M.E. (1959). *Chem. Ber.* 92; 2435.

Hoiseth SK, Stocker BA (1981). Aromatic-dependent *Salmonella typhimurium* are non-virulent and effective as live vaccines. Nature. 21;291(5812):238-9.

Hone D, Morona R, Attridge S, Hackett J (1987). Construction of defined galE mutants of *Salmonella* for use as vaccines. J Infect Dis. 156(1):167-74.

Hone DM, Attridge SR, Forrest B, Morona R, Daniels D, LaBrooy JT, Bartholomeusz RC, Shearman DJ, Hackett J. A (1988). galE via (Vi antigen-negative) mutant of *Salmonella typhi* Ty2 retains virulence in humans. Infect Immun. 56 (5):1326-33.

Hone DM, Harris AM, Chatfield S, Dougan G, Levine MM (1991). Construction of genetically defined double aro mutants of *Salmonella typhi*. Vaccine. 9(11):810-6.

Hornick RB, Greisman SE, Woodward TE, Du Pont HL, Dawkins AT and Snyder MJ (1970). Typhoid fever: pathogenesis and immunologic control. 2. N. Engl. J. Med. 283, 739-46.

Ignarro LJ (2000). Introduction and overview, p. 3-19. *In* L. J. Ignarro (ed.), Nitric oxide: biology and pathobiology. Academic Press, San Diego, Calif.

Ivanoff B, Levine MM Lambert PH (1994). Vaccination against typhoid fever: present status. Bull. World Health Organization. 72; 957-971.

Jacoby GA, Sutton L (1991). Properties of plasmids responsible for production of extended-spectrum beta-lactamases. Antimicrob Agents Chemother. 35(1):164-9.

James SL (1995). Role of nitric oxide in parasitic infections. Microbiol Rev. 59(4):533-47. Review.

Jones BD, Ghori N Falkow S (1994). *Salmonella typhimurium* initiates murine infection by penetrating and destroying the specialized epithelial M cells of the Peyer's patches. J. Exp. M. 401, 15-23.

Jones B, Pascopella L, Falkow S (1995). Entry of microbes into the host: using M cells to break the mucosal barrier. Curr Opin Immunol. 7(4):474-8.

Jones PW, Dougan G, Hayward C, Mackensie N, Collins P, Chatfield SN (1991). Oral vaccination of calves against experimental salmonellosis using a double aro mutant of *Salmonella typhimurium*. Vaccine. 9(1):29-34.

Kamili MA, Ali G, Shah MY, Rashid S, Khan S, Allaqaband GQ (1993). Multiple drug-resistant typhoid fever outbreak in Kashmir Valley. Indian J Med Sci. 47(6):147-51.

Kariuki S, Gilks C, Revathi G, Hart CA (2000). Genotypic analysis of multidrug-resistant *Salmonella enterica* Serovar typhi, Kenya. Emerg Infect Dis. 6 (6):649-51.

Klugman KP, Gilbertson IT, Koornhof HJ, Robbins JB, Schneerson R, Schulz D, Cadoz M, Armand J (1987). Protective activity of Vi capsular polysaccharide vaccine against typhoid fever. Lancet. 21;2 (8569):1165-9.

Klugman KP, Koornhof HJ, Robbins JB, Le Cam NN (1996). Immunogenicity, efficacy and serological correlate of protection of *Salmonella typhi* Vi capsular polysaccharide vaccine three years after immunization. Vaccine. 14 (5):435-8.

Komalarini S, Njotosiswojo S, Rockhill RC, Lesmana M (1980). Chloramphenicol resistant strains in salmonellosis in Jakarta. Southeast Asian J Trop Med Public Health. 11(4): 539-42.

Kuusi N, Nurminen M, Sarvas M (1981). Immunochemical characterization of major outer membrane components from *Salmonella typhimurium*. Infect Immun. 33(3):750-7.

Kuwahara H, Miyamoto Y, Akaike T, Kubota T, Sawa T, Okamoto S, Maeda H (2000). *Helicobacter pylori* urease suppresses bactericidal activity of peroxynitrite via carbon dioxide production. Infect Immun. 68(8):4378-83.

Landy M (1954). Studies on Vi antigen. VI. Immunization of human beings with purified Vi antigen. Am J Hyg. 60(1):52-62.

Levine MM, Ferreccio C, Abrego P, Martin OS, Ortiz E, Cryz S (1999). Duration of efficacy of Ty21a, attenuated *Salmonella typhi* live oral vaccine. Vaccine. 1;17 Suppl 2:S22-7.

Levine MM, Ferreccio C, Black RE, Germanier R (1987). Large-scale field trial of Ty21a live oral typhoid vaccine in enteric-coated capsule formulation. Lancet. 9;1 (8541):1049-52.

Levine MM, Tacket C, Galan J, Barry E, Noriegan F, Chatfield SN, Szetien M and Dougan G (1995). *Southeast. Asian. J. Trop. Med. Public Health*. 26 (Suppl.2): 264.

Levine MM, Taylor DN, Ferreccio C (1989). Typhoid vaccines come of age. Pediatr Infect Dis J. 8(6):374-81. Review.

Levine MM, Hone D, Tacket C, Ferreccio C, Cryz S (1990). Clinical and field trials with attenuated *Salmonella typhi* as live oral vaccines and as "carrier" vaccines. Res Microbiol. 141(7-8):807-16. Review.

Lin FY, Ho VA, Khiem HB, Trach DD, Bay PV, Thanh TC, Kossaczka Z, Bryla DA, Shiloach J, Robbins JB, Schneerson R, Szu SC (2001). The efficacy of a *Salmonella typhi* Vi conjugate vaccine in two to five year old children. N Engl J Med. 2001;344:1263-1269.

Ling J, Chau PY (1984). Plasmids mediating resistance to chloramphenicol, trimethoprim, and ampicillin in *Salmonella typhi* strains isolated in the Southeast Asian region. J Infect Dis. 149(4):652.

MacFarlane AS, Schwacha MG, Eisenstein TK (1998). *In vivo* blockage of nitric oxide with aminoguanidine inhibits immuno suppression induced by an attenuated strain of *Salmonella typhimurium* potentiates Salmonella infection and inhibits macropahge and polymorphonuclear leukocyte influx into the spleen. Infect. Immun 67: 891-898.

Martinez DE (1998). Mortality patterns suggest lack of senescence in hydra. Exp Gerontol 33(3):217-25.

Mastroeni P (2002). Immunity to systemic Salmonella infections. Curr Mol Med. 2(4):393-406. Review.

Mastroeni P, Villarreal-Ramos B, Harrison JA, Demarco de Hormaeche R, Hormaeche CE (1994). Toxicity of lipopolysaccharide and of soluble extracts of *Salmonella typhimurium* in mice immunized with a live attenuated aroA salmonella vaccine. Infect Immun 62(6):2285-8.

Mc farland, W.C. and Stocker, B.A.D. (1987). *Microbiol. Path.* 3; 129.

Mermin JH, Villar R, Carpenter J, Roberts L, Samaridden A, Gasanova L, Lomakina S, Bopp C, Hutwagner L, Mead P, Ross B, Mintz ED. (1999). A massive epidemic of multidrug-resistant typhoid fever in Tajikistan associated with consumption of municipal water. J Infect Dis. 179(6):1416-22.

Miller SI, Loomis WP, Alpuche-Aranca C, Behalau I Haiyman, F. (1993). Vaccine. 11; 122.

Mills-Robertson F, Addy ME, Mensah P, Crupper SS (2002). Molecular characterization of antibiotic resistance in clinical *Salmonella typhi* isolated in Ghana. FEMS Microbiol Lett. 8;215(2):249-53.

Mirza SH, CA Hart (1993). Ann Trop Med Parasitol. 87; 4, 373–377.

Miyamoto Y, Akaike T, Alam MS, Inoue K, Hamamoto T, Ikebe N, Yoshitake J, Okamoto T, Maeda H (2000). Novel functions of human alpha(1)-protease inhibitor after S-nitrosylation: inhibition of cysteine protease and antibacterial activity. Biochem Biophys Res Commun. 27;267(3):918-23.

Mollenkopf HJ, Groine-Triebkorn D, Andersen P, Hess J, Kaufmann SH (2001). Protective efficacy against tuberculosis of ESAT-6 secreted by a live *Salmonella typhimurium* vaccine carrier strain and expressed by naked DNA. Vaccine. 16;19(28-29):4028-35.

Moncada S, Higgs A (1993). The L-arginine-nitric oxide pathway. N Engl J Med. 30; 329(27):2002-12. Review.

Morell (1997). Science. 278: 5338; 575b–5576.

Murphy L, Grez L, Schlesinger et al (1991). Immunogenicity of *Salmonella typhi* Ty21a vaccine for young children. Infect Immun. 59; 4291–4293.

Muthukkumar S, Muthukkaruppan VR (1993). Mechanism of protective immunity induced by Porin-lipopolysaccahride against murine salmonellosis. Infect. Immun 1993; 61: 3017-3925.

Nasser MW (2002). Mechanism of pathogenesis of salmonellae: role of NO donors in protection against typhoid. Thesis submitted to Centre for Biotechnology, Hamdard University, New Delhi.

Nathan C (1997). J. Clin. Investig. 100; 2417-2423.

Nathan C, MU Shiloh (2000). Proc. Natl. Acad. Sci. USA. 97; 8841-8848.

Neithammer AG, Primus FJ, Xiang R, Dolman CS, Ruehlmann JM, Ba Y, Gillies SD, Reisfeld RA (2001). Vaccine.12: 20; (3-4), 421-9.

Nkuo-Akenji T, Ndip R, McThomas A, Fru EC (2001) Anti-Salmonella activity of medicinal plants from Cameroon. Cent Afr J Med. 47(6):155-8.

O'Callaghan D, Maskell D, Liew FY, Easmon CS, Dougan G (1988). Characterization of aromatic- and purine-dependent *Salmonella typhimurium*: attention, persistence, and ability to induce protective immunity in BALB/c mice. Infect Immun. 56(2):419-23.

Olarte J, Galindo E (1973). *Salmonella typhi* resistant to chloramphenicol, ampicillin, and other antimicrobial agents: strains isolated during an extensive typhoid fever epidemic in Mexico. Antimicrob Agents Chemother. 4 (6):597-601.

Parry C, Wain J, Chinh NT, Vinh H, Farrar JJ (1998). Quinolone-resistant *Salmonella typhi* in Vietnam. Lancet. 25;351 (9111):1289.

Parry CM, Hien TT, Dougan G, White NJ, Farrar JJ (2002). Typhoid fever. N Engl J Med. 28; 347(22):1770-82. Review.

Partridge SR, Recchia GD, Stokes HW, Hall RM (2001). Family of class 1 integrons related to In4 from Tn1696. Antimicrob Agents Chemother. 45(11):3014-20.

Perez C, Anesini C (1994). *In vitro* antibacterial activity of Argentine folk medicinal plants against *Salmonella typhi*. J Ethnopharmacol. 44(1):41-6.

Pickard D, Li J, Roberts M, Maskell D, Hone D, Levine M, Dougan G, Chatfield S (1994). Characterization of defined ompR mutants of *Salmonella typhi*: ompR is involved in the regulation of Vi polysaccharide expression. Infect Immun. 62(9):3984-93.

Polish Typhoid Committee (1966). Bull. W.H.O. 34; 211.

Richter-Dahlfors A, Buchan AM, Finlay BB (1997). Murine salmonellosis studied by confocal microscopy: *Salmonella typhimurium* resides intracellularly inside macrophages and exerts a cytotoxic effect on phagocytes in vivo. J Exp Med. 18;186(4):569-80.

Robbins A (1990). Modern vaccines. Progress towards vaccines we need and do not have. Lancet. 16;335 (8703):1436-8. Review.

Rupali P, Abraham OC, Jesudason MV, John TJ, Zachariah A, Sivaram S, Mathai D (2004). Treatment failure in typhoid fever with ciprofloxacin susceptible *Salmonella enterica* serotype Typhi. Diagn Microbiol Infect Dis. 49(1):1-3.

Saha SK, Talukder SY, Islam M, Saha S (1999). A highly ceftriaxone-resistant *Salmonella typhi* in Bangladesh. Pediatr Infect Dis J. 18(4):387.

Samuelson P, Gunneriusson E, Nygren PA, Stahl S (2002). Display of proteins on bacteria. J Biotechnol. 26;96(2):129-54. Review.

Shannon K, French G (1998). Multiple-antibiotic-resistant salmonella. Lancet. 8;352(9126):490.

Silva-Salinas BA, Rodriguez-Aguayo L, Maldonado-Ballesteros A, Valenzuela-Montero ME, Seoane-Montecinos M (1985). Properties of 2 gal+ derivatives of the vaccine mutant gal E Ty21a strain of *Salmonella typhi*. Bol Med Hosp Infant Mex. 42(4):234-9.

Silva-Salinas, BA, Gunenlee C, Mena G C, Cabell OF (1987). J. Infect. Dis. 155; 1077.

Simanjuntak CH, Paleologo FP, Punjabi NH, Darmowigoto R, Soeprawoto, Totosudirjo H, Haryanto P, Suprijanto E, Witham ND, Hoffman SL (1991). Oral immunisation against typhoid fever in Indonesia with Ty21a vaccine. Lancet. 26;338(8774):1055-9.

Smith BP, Reina-Guerra M, Stocker BA, Hoiseth SK, Johnson E (1984). Aromatic-dependent Salmonella dublin as a parenteral modified live vaccine for calves. Am J Vet Res. 45(11):2231-5.

Sohni YR, Kaimal P, Bhatt RM (1995). Prophylactic therapy of Salmonella typhi septicemia in mice with a traditionally prescribed crude drug formulation. J Ethnopharmacol. 45(2):141-7.

Stocker BA (1990). Aromatic-dependent Salmonella as live vaccine presenters of foreign epitopes as inserts in flagellin. Res Microbiol. 141(7-8):787-96.

Stroffolini T, Manzillo G, De Sena R, Manzillo E, Pagliano P, Zaccarelli M, Russo M, Soscia M, Giusti G (1992). Typhoid fever in the Neapolitan area: a case-control study. Eur J Epidemiol. 8(4):539-42.

Szu SC, Stone AL, Robbins JD, Schneerson R, Robbins JB (1987). Vi capsular polysaccharide-protein conjugates for prevention of typhoid fever. Preparation, characterization, and immunogenicity in laboratory animals. J Exp Med. 1;166(5):1510-24.

Tacket CO, Ferreccio C, Robbins JB, Tsai CM, Schulz D, Cadoz M, Goudeau A, Levine MM (1986). Safety and immunogenicity of two Salmonella typhi Vi capsular polysaccharide vaccines. J Infect Dis. 154(2):342-5.

Tacket CO, Hone DM, Curtiss R 3rd, Kelly SM, Losonsky G, Guers L, Harris AM, Edelman R, Levine MM (1992a). Comparison of the safety and immunogenicity of delta aroC delta aroD and delta cya delta crp Salmonella typhi strains in adult volunteers. Infect Immun. 60(2):536-41.

Tacket CO, Hone DM, Losonsky GA, Guers L, Edelman R, Levine MM (1992b). Clinical acceptability and immunogenicity of CVD 908 Salmonella typhi vaccine strain. Vaccine. 10(7):443-6.

Takeuchi A (1967). Electron microscope studies of experimental Salmonella infection. I. Penetration into the intestinal epithelium by Salmonella typhimurium. Am J Pathol. 50(1):109-36.

Threlfall EJ, Ward LR, Rowe B, Raghupathi S, Chandrasekaran V, Vandepitte J, Lemmens P (1992). Widespread occurrence of multiple drug-resistant Salmonella typhi in India. Eur J Clin Microbiol Infect Dis. 11(11):990-3.

Tran JH, Jacoby GA (2002). Mechanism of plasmid-mediated quinolone resistance. Proc Natl Acad Sci USA. 16;99(8):5638-42. Tran TH, Bethell DB, Nguyen TT, Wain J, To SD, Le TP, Bui MC, Nguyen MD, Pham TT, Walsh AL, et al (1995). Short course of ofloxacin for treatment of multidrug-resistant typhoid. Clin Infect Dis. 20(4):917-23.

Umezawa K, Akaike T, Fujii S, Suga M, Setoguchi K, Ozawa A, Maeda H (1997). Induction of nitric oxide synthesis and xanthine oxidase and their roles in the antimicrobial mechanism against Salmonella typhimurium infection in mice. Infect Immun. 65(7):2932-40.

Usera MA, Echeita A, Aladuena A, Alvarez J, Carreno C, Orcau A, Planas C (1995). Investigation of an outbreak of water-borne typhoid fever in Catalonia in 1994. Enferm Infecc Microbiol Clin. 13(8):450-4. Spanish.

Van den Bergh ET, Gasem MH, Keuter M, Dolmans MV (1999). Outcome in three groups of patients with typhoid fever in Indonesia between 1948 and 1990. Trop Med Int Health. 4(3):211-5.

Wahdan MH, Serie C, Germanier R, Lackany A, Cerisier Y, Guerin N, Sallam S, Geoffroy P, el Tantawi AS, Guesry P (1980). A controlled field trial of liver oral typhoid vaccine Ty21a. Bull World Health Organ. 58(3):469-74.

Wain J, Diem Nga LT, Kidgell C, James K, Fortune S, Song Diep T, Ali T, O Gaora P, Parry C, Parkhill J, Farrar J, White NJ, Dougan G (2003). Molecular analysis of incHI1 antimicrobial resistance plasmids from Salmonella serovar Typhi strains associated with typhoid fever. Antimicrob Agents Chemother. 47(9):2732-9.

Wain J, Hoa NT, Chinh NT, Vinh H, Everett MJ, Diep TS, Day NP, Solomon T, White NJ, Piddock LJ, Parry CM (1997). Quinolone-resistant *Salmonella typhi* in Viet Nam: molecular basis of resistance and clinical response to treatment. Clin Infect Dis. 25(6):1404-10.

Wallis TS, Galyov EE (2000). Molecular basis of Salmonella-induced enteritis. Mol Microbiol. 36(5):997-1005. Review.

White DG, S Zhao, R Sudler, S Ayers, S Friedman, S Chen *et al* (2001). N. Engl J. Med. 345; 16, 1147–1154.

Wong KK, McClelland M (1992). Dissection of the *Salmonella typhimurium* genome by use of a Tn5 derivative carrying rare restriction sites. J Bacteriol. 174(11):3807-11.

Wray C, Sojka WJ (1978). Experimental *Salmonella typhimurium* infection in calves. Res Vet Sci. 25(2):139-43.

Yabuuchi E, Ikedo M, Ezaki T (1986). Invasiveness of *Salmonella typhi* strains in HeLa S3 monolayer cells. Microbiol Immunol. 30(12):1213-24.

Yoshida K, Akaike T, Doi T, Sato K, Ijiri S, Suga M, Ando M, Maeda H (1993) Pronounced enhancement of NO-dependent antimicrobial action by an NO-oxidizing agent, imidazolineoxyl N-oxide. Infect Immun. 61(8):3552-5.

Yugoslav Typhoid Commission (1964) Bul. W.H.O. 30; 623.

Zhang L (1991). Mechanism of multiresistant *Salmonella typhi* Zhonghua Yi Xue Za Zhi. 71(6):314-7, 24. Chinese.

Zierler MK, Galan JE (1995). Contact with cultured epithelial cells stimulates secretion of *Salmonella typhimurium* invasion protein InvJ. Infect Immun. 63(10):4024-8.

History and Taxonomy of *Candida* and Candidiasis

R. GAUTAM, A. P. GARG and A. GARG

1. INTRODUCTION

The first well-documented scientific reference to "thrush" comes from papers of Pepys in 1665, however, the original description of a case of thrush
should be credited to Hippocrates, who predates Pepys. It was Berg (1846)
who used scientific approach to study the thrush but still confusion existed
about the identification of the organism causing the disease. At that time, the
major problem associated with defining the yeast, was because of the lack of
scientific literature for the identification of the pathogens. Langenbeck
(1839) was the first who observed a fungus associated with thrush, but its
identity was incorrect, hence, its importance in the disease could not be
realized. Gruby (1842) studied the organism isolated by Langenbeck but
came to the conclusion that it was a species of *Sporotrichum*. The same organism was also studied in detail by Robin, who reclassified it as *Oidium
albicans* (Robin 1853). This was the first use of the name *albicans* (which
means "to whiten") with the fungus of thrush.

The incorrect identification of the causative organism of thrush
perpetuated by the insistence of several key taxonomists that the thrush
fungus should be a species of *Monilia,* a fungus that is mostly found
associated with plant materials (Garg and Sharma 1984; Bhatnagar and
Garg 1990), but the organism described by Berg and others was

Department of Microbiology, C.C.S. University, Meerut-250005 India.

morphologically different from that of *Monilia*, (Winner and Harley 1954). The controversy still persisted until Berkhout and others in 1923 proposed that the organism of thrush lesions was clearly not a species of *Monilia* and, as a consequence, she proposed that the favored generic name should be *Candida*, which is derived from the Latin phrase *toga candida* (a special white robe worn by candidates for the Roman Senate). The derived name refers to the whitish colonies on agar or the oral lesions of aphthae or thrush. Other investigators attempted to classify the thrush agent by using biochemical and serological assays (Castellani 1920), however, Benham (1931) focused her research on the use of more rigid criteria in the identification of the fungus, concluding that, of all the yeasts like fungi, only *C. albicans* was pathogenic for laboratory animals, although, she was not entirely correct in her concept. Eventually, early investigators arrived at correct conclusions, and given the lack of sophisticated methodologies, their observations have stood the test of time, especially in regard to the observations on the morphology of *C. albicans* as well as generalizations about the status of immune competence in patients and the severity of candidiasis.

2. TAXONOMY OF *CANDIDA*

Wickerham, 1952 reviewed its taxonomy, and in 1958 an excellent text, "The Chemistry and Biology of Yeasts," edited by Cook, appeared. There are some pathogenic species of *Candida* that are still classified in Deuteromycetes, a class not known to reproduce sexually. On the other hand, the teleomorph stages of several pathogenic (Table 1) and non-pathogenic *Candida* species have been described, but these organisms (especially the pathogens) still retain their designation as *Candida* sp. *C. krusei*, for example, is not routinely referred as *Issatchenkia orientalis, though* its teleomorph stage is known. The anamorph stage (asexual) is still popular even for those species of *Candida,* whose sexual stage (teleomorph) are known, however this policy, whether taxonomically correct or not, creates a unity for this group of pathogens and reduces stress for researchers who must learn new names of pathogens. The older classification of many species of *Candida* in Deuteromycetes was mainly based on the fact that they lacked a sexual stage. The known sexual stages of species are ascomycetous (Kurtzman and Fell 1998).

Further, comparative sequence data from a number of studies support the idea that both the sexual and asexual *Candida* species are related phylogenetically to Ascomycete rather than to Basidiomycete. For example, parsimony analyses of nucleotides from SSU rDNA sequences of 75

Table 1   A partial list of pathogenic species of *Candida*

Reported by 1988[a]	Current list (Ananomorph stage)	Sexual (teleomorph form)
C. albicans	C. albicans	Not described
C. tropicalis	C .tropicalis	Not described
C. glabrata	C. glabrata	Not described
C. parapsilosis	C. parapsolosis	Not described
C. krusei	C. krusei	Issatchenkia orientalis
C. guilliermondii	C. gulliermondii	Pichia guilliermondii
C. kefyr[b]	C. kefyr	Kluyveromyces marxianus
C. viswanathii	C. viswanathii	Not described
	C. lusitaniae	Clavisopora lusitaniae
	C. dubliniensis	Not described
	C. famata	Debaryomyces hansenii
	C. inconspicua	Not described
	C. utilis	Pichia jadinii

[a] Odds, 1998
[b] Formerly *C. pseudotropicalis*

representative Ascomycetes and Basidiomycetes clearly validate the taxonomic alignment of species of *Candida* among the Ascomycetes (Figure 1) (Garg et al. 1995). Additionally, there are other characteristics that distinguish *Candida* species as an ascomycetous and not basidiomycetous yeast. For example, they are urease-negative, non-encapsulated, fermentative, and non-inositol assimilative; make β-glucans in their cell wall; and do not produce starch or carotenoid pigments. Budding in *Candida* species results in one bud per locus, typical of an Ascomycete, whereas in the Basidiomycetes, cell buds are repeatedly formed from a single locus. Also, the molecular structure of the cross walls (septa) differ, being transparent and homogenous in the Ascomycetes (such as *Candida)* while tripartite in Basidiomycetes (Kurtzman and Fell 1998). *Cryptococcus neoformans,* another human pathogen, is a basidiomycetous yeast that is urease-positive, encapsulated, and non-fermentative; assimilates inositol; and produces unicellular, budding yeasts only (Rippon 1974).

A scheme that illustrates the current views on the taxonomy of the an-amorphic *Candida,* their teleomorphic stages, as well as a close relative of *Candida (Saccharomyces)* is presented in Table 2 (Kurtzman and Fell 1998). In some instances, assignments of members of the order Saccharomycetales are uncertain, and those genera are indicated by a question mark. The teleomorph forms of *Candida* species are diverse structurally but tentatively

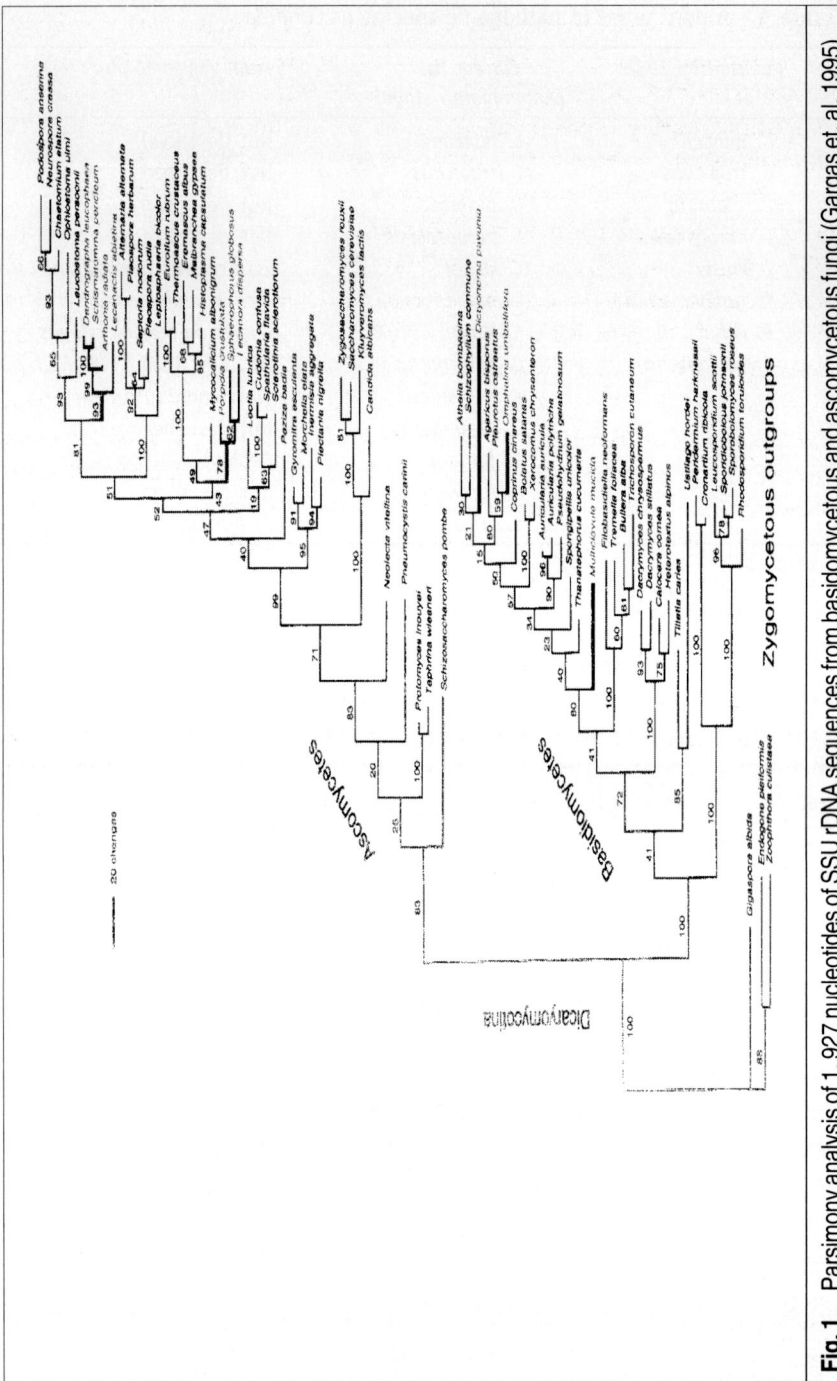

Fig. 1   Parsimony analysis of 1, 927 nucleotides of SSU rDNA sequences from basidiomycetous and ascomycetous fungi (Gargas et. al. 1995)

are assigned to two families of Ascomycetes *i.e.,* (Table 2) (Gargas et al. 1995). Sexual reproduction (conjugation) is induced on media such as V8 agar, yeast-malt extract agar, or on McClary's acetate medium and occurs between compatible mating types (heterothallic species), an example being *Pichia guilliermondii (C. guilliermondii),* but other species are self-sterile (homothallic), as in the case of *Kluyveromyces marxianus (C. kefyr).* Interestingly, conjugation in the homothallic species, *Debaryomyces hansenii (C. famata* and others), occurs between a cell and its bud. This type of conjugation is referred to as heterogamous conjugation that occurs between cells of different size and shape *versus* isogamous conjugation that occurs between cells of similar size and shape. Often sexual reproduction is described as unconjugated, indicating that diploids persist before they eventually undergo meiosis to form sexual spores (ascospores). Both asci and ascospores vary in shape from species to species; asci can be persistent (*Debaryomyces, Issatchenkia*), deliquescent (*Clavispora),* or both *(Pichia).* The number of ascospores that form from a diploid cell following meiosis is also species dependent but usually it is from 1 to 4 (Kurtzman and Fell 1998). One of the more interesting of the teleomorphs, is *Metschinkowia* species, in which asci can arise from spheroidal chlamydospores (Kurtzman and Fell 1998).

Table 2 The classification of the anamorphic *Candida,* teleomorph stages of *Candida* and *Saccharomyces* (Kurtzman and Fell 1998).

Phylum: Ascomycota
Class: Hemiascomycetes
Order: Saccharomycetalesb
Family: Metschnikowiaceae
Genus: *Clavispora*
Family: Saccharomycetaceae
Genus:? *Debaryomyces*
? *Issatchenkiac*
? *Pichia*
Kluyveromyces
Saccharomyces
Family: Candidaceae (Anamorphic)
Genus: *Candida*

? uncertain affilation

In our department, we have characterized Candida albicans with other species on the basis of information collected from GenBank (www.ncbi.nlm.nih.gov) and Saccharomyces Genome Database (www.yeastgenome.org). We have searched, retrieved and compared 18s rRNA sequence of *S. cerevisiae* with other Candida species. There were 94% and 84% identities

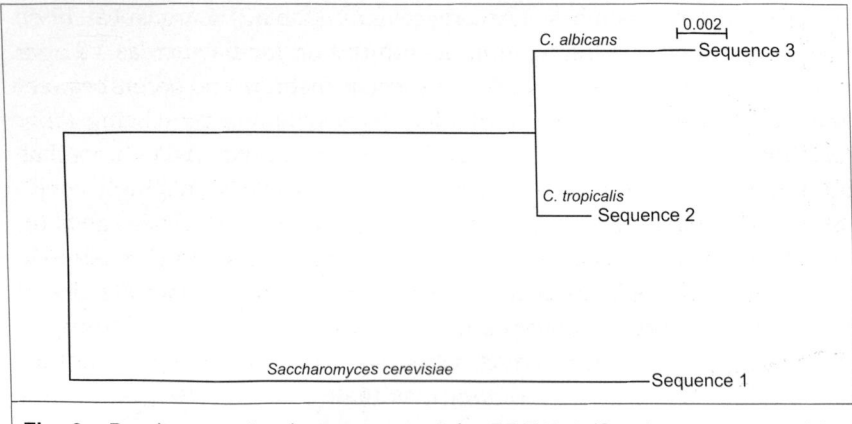

Fig. 2 Dendrogram showing sequence 1 for RDN18-1 (Saccharomyces cervisiae S288C), sequence 2 for Ca.rrn 001 (Candidaaldicans SC5314) and sequence 3 for Candida tropicalis MYA-340Y

with *Candia albicans,* found in Ca.rrn001 and Cao19.4552, respectively. With help of ClustalX 1.83, a dendogram was developed as shown in the Fig. 2.

2.1 Pathogenic *Candida* Species and their Prevalence in Patients

Odds (1998) has listed eight pathogenic species of *Candida* that are found in nature (Table 1) and after surveillance, studies have included *C. famata, C. inconspicua, C. utilis, C. lusitaniae,* and other species from blood and other sites of patients with candidiasis as pathogenic, however, in lower frequencies. Also, a new species, *C. dubliniensis,* which is very similar phenotypically to *C. albicans* (produces germ tubes and chlamydospores among other similar traits), has now been identified (Sullivan et al. 1995; Sullivan et al.1997; Salkin et al. 1998; Sullivan and Coleman 1998) and is included among the list of pathogens.

The phylogenetic relationships among the most common of the *Candida* spp. that are pathogenic, as determined by DNA sequences of the V3 variable region of the large ribosomal subunit gene, are shown in Figure 2. A bootstrap analysis of the 10 sequences indicates that *C. dubliniensis* is grouped separately from the cluster of *C. albicans* and *C. stellatoidea* and that the other species tested are more divergent.

Several medical microbiologists have suggested (Hazen 1995; Abi-Said et al. 1997; Abu-Elteen et al. 1997; Pfaller 1999; Costa et al. 2000; D'Agata et al. 2000; Fernandez et al. 2000; Hung et al. 2000) that the predominant human pathogen among species of *Candida* is *C. albicans,* regardless of the clinical setting and locations of the study. Thus, *C. albicans* is the most often observed pathogen in the AIDS patient (mucosal disease) as well as from septicemias (fungemias) associated with cancer chemotherapy, transplant patients, premature infants, surgical infections, or other predisposing conditions. Next in frequency among the *Candida* pathogens is *C. tropicalis* (Borg-von et al.1993; Knoke et al. 1997; Nucci et al. 1998; Pfaller et al. 1999; Costa et al. 2000), *C. parapsilosis* (Hazen 1991; Nucci et al. 1998; Kao et al. 1999; Pfaller et al. 1999; Costa et al. 2000), or *C. glabrata* (Borg-von et al. 1993; Abu-Elteen et al. 1997; Pfaller *et al.* 1999) at least in regard to blood isolates. These three species alternate as the second most common pathogen, depending on the source of the data of interest, a higher mortality is often reported for septicemias caused by *C. glabrata,* but the reasons for this observation are not well understood (Rennert et al. 2000). Likewise, *C. dubliniensis* has emerged as a new and important pathogenic species (Schorling 2000). Finally, a group of *Candida* species usually reported as infrequent pathogens (*C. inconspicua, C. lusitaniae, C. guilliermondii, C. kerry, C. famata,* and *C.utilis*) are now added to the list of human pathogens. These species have recently received attention because these are increasing fast in frequency among all clinical isolates (Borg-von et al. 1993; Collazos et al. 1999; Hazen 1995; Wingard et al. 1993).

2.2 The Uncommon Human *Candida* Pathogens

The species of *C. inconspicua, C. krusei, C. famata* (formerly *Torulopsis candida*), *C. guilliermondii, C. viswanathii, C. kefyr* (formerly *C. pseudotropicalis*), *C. utilis,* and *C. lusitaniae* are considered less common causes of disease in humans. All of these species grow as yeasts, but the type of pseudohyphae varies from species to species. On Dalmau plates of cornmeal agar, e.g., *C. krusei, C. kefyr,* and *C. lusitaniae* produce "typical" pseudohyphae, whereas strains of *C. guilliermondii, C. famata,* and *C. lusitaniae* produce "primitive" pseudohyphae that are composed of ovoid cells in chains (Kurtzman and Fell 1998). However, strain variability in this property has been reported. For example, in other accounts, *C. guilliermondii* has been reported to produce pseudomycelium that is fine in texture and short (Rippon 1974). *C. krusei* produces flat, dry colonies with a prominent surface film that extends several millimeters up on the inner surface of the culture. This phenotype and its elongated yeast (bud) cells provide a fairly easy way to identify *C. krusei.* Culture phenotypes of *C. krusei* and *C. inconspicua* are identical, and separation of these two species is obtained only by molecular methodologies, including randomly amplified polymorphic DNA, Southern hybridization using specific probes, or karyotype analyses (Baily et al. 1997).

As stated above, these species are uncommon pathogens, and probably for that reason, their study has lagged behind than those which are most common. Nevertheless, there is indication that they cause disease and some of these studies are discussed. For example, the total number of *C. krusei* isolates from bronchial secretions has statistically increased from initial surveillance studies over time from 1987 to 1992. The frequency of isolation of *C. krusei* compared to all *Candida* species from blood cultures of patients has ranged from 1.0 to 5.8% of all *Candida* isolates (Pfaller 1999; Wingard et al. 1991), whereas the frequencies of *C. famata* (2%) and *C. inconspicua* (1 to 4.1 %) indicate their low level of recovery from human disease. *C. famata* was first reported in 1986 in a case of catheter-induced septicemia. In addition to these surveillance studies, *C. guilliermondii* has been reported as a cause of a fungemia in cancer patients; *C. kefyr* was the cause of esophagitis and vaginitis, *C. utilis* in chronic urinary tract infections, *C. viswanathii* in meningitis, and *C. famata* from infections in a surgical patient, peritonitis, and fungemias . *C. inconspicua* has also been reported from patients with human immunodeficiency virus (HIV) infections (Baily et al. 1997) and in hematological malignancies .

Certainly, the list of uncommon pathogens of humans described above is incomplete. Hazen, for example, also has included *C. ciferri* (nails, ear), *C. zeylanoides* (groin, nails), *C. rugosa* (blood, burn wounds, catheter), *C. haemulonii* (toe, nails, blood), *C. norvegensis* (blood, peritoneal fluid), and *C. pulcherrima* (nails) as emerging disease-causing species (Hazen et al. 1991). In addition, *C. chiropterum, C. humicola,* and *C. pintolopesii* are also referenced as isolates from human disease (Hazen et al. 1991). Of this latter group of fungi, the teleomorph stages of *C. ciferrii, C. lipolytica, C. norvegensis, C. pulcherrima* and *C. pintolopesii* have been described (Hazen et al. 1991).

3. ECOLOGY OF *CANDIDA*

Natural Habitats: Candida is yeast and the most common cause of opportunistic mycoses worldwide. It is also a frequent colonizer of human skin and mucous membranes. *Candida* is a member of normal flora of skin, mouth, vagina and stool (Bhatnagar 2004). As pathogen and also as a colonizer, it is found in the environment, particularly on leaves, flowers, water and soil. (Garg and Sharma 1984)

Usually *Candida albicans* live in a vaginal atmosphere, because pH is acid here (5.0 - 4.0). In truth the *"true home"* of *Candida* is the little intestine mucous; localization can be considered a *"beach house"*, coetaneous localization (presence of fungus spots when you are exposed to sun) can represent the *"mountain house"*, or *"Thrush"* in mouth cavity, the *"lake house"*.

Since last 50 years of research on yeast, a number of new habitats have been established for several species of *Candida*. (Table 3) Excellent original papers on the ecology of yeasts are *"A taxonomic study of soil yeasts"* by Bouthilet 1951 and *"Ecology of yeast"* by Lund 1958 and several reviews given important literature. It should be emphasized that the common assumption of so many medical writers that the genus *Candida* is found primarily associated with the animal body is justified by scant evidence. Even less justified is the assumption that all species of *Candida* are potentially pathogenic. Di Menna (1954) has made an excellent comparative study of the non-pathogenic yeasts of human skin and the alimentary tract; her results parallel those of Connell and Skinner, 1953. Both investigations indicate that human skin surfaces are true habitats for certain yeasts, including *Candida* spp., and there seems to be a characteristic skin flora. Yeasts seem to live and multiply on the skin and to persist there for some time, and they appear to be as normal, for instance, *Micrococcus* or *Staphylococcus* spp. In the carefully performed study by Di Menna (1954), 120 strains of yeasts isolated from human skin and 388 from air and from room surfaces were compared. Like Connell and Skinner, 1953, she found a quantitative and qualitative difference in the flora, but to a lesser extent. She did not find, however, the high incidence of fermenting species or the low incidence of *Rhodotorula* spp. as reported by Connell and Skinner.

Table 3 Habitats for Species of *Candida* was Recorded (Skinner and Fletcher 2000)

Candida Spp.	Isolated from (habitat)	Place and Reference
C. brumptii	Soil	New Zealand (Menna and Parle 1954)
C. guilliermondii	Fruit processing plant	U.S.A (Recca and Mrak 1952)
C. guilliermondii	Soil	Denmark (Lund 1954)
C. guilliermondii	Insect	U.S.A.(Phaff and Knapp 1956) and Denmark (Lund 1954)
C. krusei	Flowers, mushrooms, dung,	Denmark (Lund 1954)
C. krusei	Citrus fruit processing	U.S.A (Recca and Mrak 1952)
C. krusei	Plant insects	U.S.A (Dobzhansky et al. 1956)
C. krusei	Turbid beer	England (Wiles 1953)
C. lipolytica	Soil	New Zealand (Menna 1955)
C. lipolytica	Beer	Germany and England (Windsch 1954)

C. lipolytica	Air	U.S.A. (Connel and Skinner 1953)
C. mycoderma	Tomato, fermenting coffee	Mexico (Ruiz 1949)
C. mycoderma	Dried egg yolk,	Denmark (Lund 1954)
C. mesenterica	Mushrooms, dung, soil	Denmark (Lund 1954);
C. mesenterica	Insects	U.S.A (Phaff and Knapp 1956)
C. monosa	Insects	Hawaii (Hedrick and Burke 1950); olive infusion liquor, Italy
C. parapsilosis	Soil,	Denmark (Lund 1954) New Zealand
C. parapsilosis	Citrus fruit processing plant	U.S.A (Recca and Mrak 1952)
C. parapsilosis	Flowers fruit root crops, dung mushrooms	Denmark (Lund 1954)
C. macedoninesis	Soil	New Zealand (Menna 1955)
C. intermedia	Insects	Hawaii (Hedrick and Burke 1950) and U.S.A. (Miller and Mrak 1953)
C. zeylanoides	Insects, fruit	Denmark (Lund1954)
C. zeylanoides	Air	U.S.A (Connel and Skinner 1953)
C. reukaufii	Insects	Demark (Lund 1954)
C. catenulate	Insects	U.S.A (Dobzhansky et al. 1956)
C. humicola	Breeding place of fruit flies	U.S.A (Carson et al. 1956)
C. albicans	Various plant surface	India (Garg and Sharma 1984; 1985)
C. albicans	Sapropel from fresh water lakes	U.S.S.R. (Messineva and Skadooskii 1947)

C. albicans is the only species of the genus which is definitely pathogenic although there is evidence that certain ubiquitous species may very rarely be involved in pathological lesions. Although, *C. albicans* is very common on the mucous membranes and in the feces (Marples and Menna 1952) of a considerable percentage of healthy individuals, until recently it had been found only in habitats connected with the animal body. *C. albicans* is perhaps the easiest species of the genus to identify, and is the one that interests most workers. If it were present in non-animal habitats in any abundance, it would have been found, where it has been specifically sought in any habitat other than the animal body (where it is abundant) it has not been found, or found only very rarely indeed. Persons cited above who found *C. albicans* in non-animal habitats seemed inclined to regard this organism as adventitious, which appears most likely.

Akiba and Iwata, 1954, have reported an interesting, almost unbelievable parasitism involving *"Bacterium candidoestruens"* a Gram-negative rod similar to *Serratia marcescens*. This serratia like organism has been reported to invade the living cells of *C. albicans* and *C. tropicalis* as well as certain species of *Saccharomyces*, causing rupture of the yeast cells after 48 h, with release of the "Bacterium." Since the latter organism produces a brilliant red pigment, individual yeast cells and colonies, thus, parasitized take on a red color. This interesting observation, to the knowledge of the reviewers, is the first reported hyperparasitism of this type and must necessarily await confirmation from other laboratories. Virtanen (1951) reports a symbiosis between *Candida* spp. (*C. albicans* and *C. tropicalis*) and *Staphylococcus aureus* and *S. albus*, where associated growth of the latter. When grown in association with some strains of *Lactobacillus bulgaricus*, *C. krusei* will stimulate the production of lactic acid by the former, presumably as a result of the concomitant production of growth factors (Swaby 1945).

Several reports of a similar relationship involving *C. albicans* and oral *lactobacilli* have recently appeared. Young et al. (1951) have observed a rise in the incidence of *C. albicans* in the mouth of healthy young adults. Further, it was shown that 86% of these Candida-positive mouths were also carriers of oral *lactobacilli* (Young et al. 1956). Studying a possible interaction of the two organisms, Young et al. (1956) demonstrated an inhibitory effect of an oral *lactobacillus*, similar to *Lactobacillus acidophilus*, on media capable of supporting good growth of both organisms. These investigators concluded that it was a pH effect due to the production of lactic acid, however, in vitamin-deficient media in which single omissions of thiamine, folic acid, riboflavin, and niacin were made, growth of *L. acidophilus* occurred only in mixed cultures with *C. albicans*. Wilson and Goaz have also shown that young active cultures of *C. albicans* liberate sufficient vitamins B to meet the growth requirements of oral lactobacilli, proportional to the amount of thiamine, riboflavin, calcium pantothenate, and pyridoxamine liberated in such appropriately deficient media when cultured with *C. albicans*. With the exception of lysine there is no liberation of amino acids by young cultures of *C. albicans*. However, chemically defined, amino acid deficient media, preconditioned by 6-to 9-day growth of *C. albicans,* provide all amino acids required by an oral strain of *Lactobacillus casei*. Bioassay by single omission techniques indicates liberation of the five essential amino acids; cysteine, glutamic acid, lysine, phenylalanine and tryptophan. In addition, significant amount of arginine, isoleucine, leucine, tyrosine and valine were also liberated. Further, a reduction in the length of *L. casei* was induced in a mutually adequate medium when preconditioned by 3-day growth of *C. albicans*. A similar reduction could be affected by addition of an enzymatic digest of a growth stimulating peptide. These investigations suggest that in the normal mouth a counterbalance exists between these organisms, with *C. albicans* providing growth factors for the oral lactobacilli and the latter controlling the yeasts population by lactic acid production (Young et al. 1951).

Browne 1960 has found that *C. albicans* may play a role in the ecology of several human pathogens, including *Corynebacterium diphtheriae* and the β-hemolytic *Streptococci*. Testing the survival of these organisms on cotton swabs dried at room temperature, it was found that neither a strain of the β-hemolytic *Streptococcus* nor of *C. diphtheriae* survived more than a few days. When broth cultures of these organisms each mixed with similar cultures of *C. albicans* were tested under identical conditions, *C. diphtheriae* and the β-hemolytic streptococci were recovered from swabs left at room temperature for periods of upto 19 weeks. No *C. diphtheriae* of β-hemolytic streptococci were recovered from swabs prepared in the same manner when *C. albicans* was omitted from the mixture. Further neither 2% yeasts extract nor heat-killed suspensions of *C. albicans* provided conditions for survival of *C. diphtheriae* over an extended period.

4. METHODS TO STUDY *CANDIDA*

4.1 Isolation

The fact that *Candida* spp. are so easily isolated on acidified (pH 4.0 to 4.5) glucose agar, even from materials heavily contaminated with bacteria, has rendered new methods or new media all but useless and practically none has been reported. If however, the worker is attempting to isolate pathogenic fungi including *Candida* spp. from animal sources a strongly acid medium is not recommended, since many pathogenic fungi are unable to grow in such medium. In such cases non-acid ingredients to prevent growth of bacteria are necessary. Thompson, in 1945, proposed the use of penicillin and streptomycin to keep down the growth of most bacteria. Fleury's 1952 and Littman 1947 modifications and the original Thompson medium were used by one of the present authors (C.E.S) and found to be about equally good. The use of cycloheximide (Clark and Wallace 1955) may be helpful in isolating *Candida* spp. from sources heavily contaminated with molds.

Generally clinical samples are collected in Sabouraud's broth and YEPD broth and incubated at 37°C. For pure culture YEPD Agar and Sabouraud dextrose agar is commonly used for the isolation of pathogenic species of *Candida*. We have found that use of chloroamphenicol @50mg/L in Sabouraud's agar normally prevents the growth of Bacteria. The advantage of chloroamphenicol is that it can used prior to autoclaving (Garg and Gautam 2005). In our laboratory we are also using extract of soya nutries (Savour India Pvt. Ltd.) in place of peptone for routine isolation of *Candida* spp. and other fungi.

4.2 Identification of *Candida* Species

Yeasts are normally isolated from sterile sites such as blood, cerebrospinal fluid, and synovial fluid and should be identified in the laboratory or be sent to a reference laboratory. Approximately, 70 to 80% of yeasts isolated in a clinical microbiology laboratory are *C. albicans.* If the germ tube test is positive, further testing is not normally required. Traditionally, methods for the identification of *Candida* specie rely on a combination of morphological and biochemical characteristics (Barnett et al. 2000; Buckely 1989).

The principal morphological and biochemical tests include:

- Appearance and color of colonies on primary isolation media (Sabourauds, CHROM agar).
- Size and shape of cells.
- Production of hyphae and/or pseudohyphae.
- Ability to produce germ tubes.
- Ability to produce chlamydoconidia (chlamydospores).
- Assimilation of carbohydrates.
- Assimilation of nitrate.
- Fermentation of sugars.

Some of the traditional identification methods, particularly the biochemical tests, are cumbersome and time-consuming to perform. Consequently, a number of rapid tests and commercial identification systems have been developed and introduced in microbiology laboratories.

Candida species generally grow well on common mycological and bacteriological media, including Sabouraud's glucose agar, sheep blood agar and horse blood agar. Some species of *Candida,* such as *C. albicans* and *C. dubliniensis,* grow even in the presence of cycloheximide. Colonies of *Candida* species grown at temperatures between 25 and 37°C have a smooth to wrinkled textured colony of white to beige color when grown on glucose-peptone agar (Sabouraud). Different *Candida* species in a mixed specimen may be distinguished on media containing chromogenic compounds such as CHROMagar Candida medium (Odds and Bernaert 1994). However, CHROMagar is only useful for the presumptive identification of certain medically important species, including *C. albicans, C. tropicalis, C. glabrata, C. krusei, C. dubliniensis,* and to an extent, *C. parapsilosis* (Sullivan and Coleman 1998; Sullivan and Coleman 1998). Following 48 h of growth at 37°C on this medium, *C. albicans* appears light blue-green. Other pathogenic species of *Candida* appear as purple or pink. The recently described chlamydospore and germ-positive yeast *C. dubliniensis* that has been recovered primarily from the oral cavities of human immunodeficiency virus (HIV)-infected individuals usually appears as dark green colonies on this medium and can be easily distinguished from other *Candida* species (Sullivan and Coleman 1998). However, it must be emphasized that CHROMagar Candida agar is useful only for identifying colonies of *C. dubliniensis* following primary plating from clinical specimens as isolates can lose their dark green coloration on this medium following subculture or storage (Sullivan and Coleman 1998). On Czapek Dox agar using the Dalmau technique *Candida* species (with the exception of *C. glabrata)* usually produce abundant hyphae growing from the edge of the original inoculum. Clusters of oval to spherical blastospores (blastoconidia) may be seen along the length of hyphae. The arrangement of hyphae and blastospores is often characteristic of a particular species.

Chlamydospore Production: With the Dalmau techniques, large, highly refractile, thick-walled chlamydoconidia may be seen terminally or on short lateral branches in *C. albicans* isolates (Figure 3). Approximately, 60% of clinical isolates of *C. albicans* produce chlamydoconidia (Buckely 1989). The test provides a rapid presumptive identification step for *C. albicans* and *C. dubliniensis.* Alternative media to Czapek Dox for inducing chlamydospore production include cornmeal agar and rice-agar-Tween (Buckely 1989).

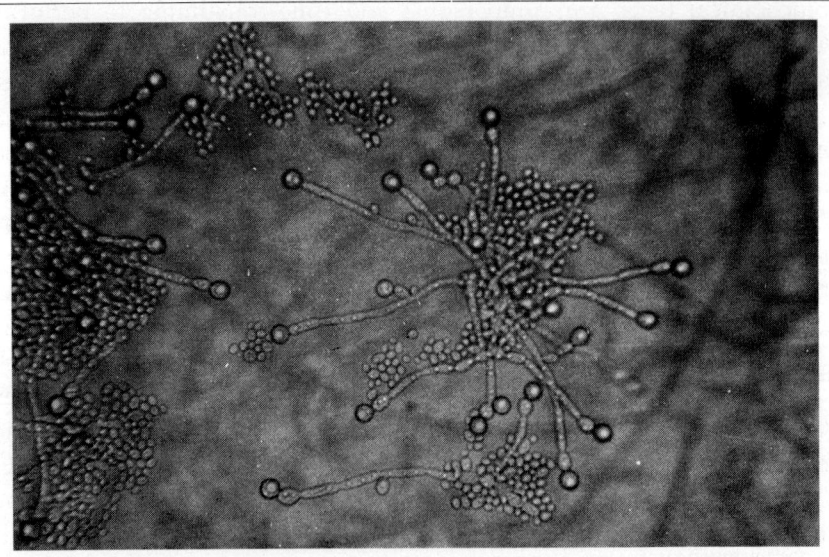

Fig. 3   Thick-walled chlamydoconidia seen terminally and on short lateral branches in *Candida albicans* isolates, on corn meal agar.

Germ Tube Production: The germ tube test provides a rapid identification test for *C. albicans* that can be carried out on primary or purified cultures. The formation of germ tubes in serum or similar media is characteristic of 95 to 97% of clinical isolates of *C. albicans* and *C. dubliniensis*.

4.3 Commercial Kits for Identifying *Candida* Species

For laboratories where the identification of yeasts constitutes a small part of the routine workload, the use of commercially produced auxanographic kits is particularly attractive. All of these kits will identify *Candida* species within 24 to 36 h, and many of them will provide an identification profile that can be used for epidemiological strain typing purposes. The following products have found a useful place in the routine laboratory, although the list is not exhaustive: API ID 32C, API 20C AUX (commonly used in the United States), Fungicrom, Auxacolor, Candifast and Fungitest.

The development of commercial kits has provided the routine laboratory with the opportunity to standardize the identification of *Candida* species, however, for many laboratories, the costs involved may exclude the use of kits, but this decision must be weighed against the efforts in preparing media and reagents.

4.4 Newly Developed Rapid Diagnostic Culture Medium

In our laboratory we have also developed a rapid diagnostic culture medium for pathogenic *Candida* spp. We have conducted an experiment on 276 clinical isolates of candidiasis from L.L.R.M. Medical College, Meerut and AIIMS, New Delhi, which gives positive results within 2 to 4 h for *C. albicans*, *C. famata*, *C. glabrata*, *C. haemulonii*, *C. kefyr*, *C. krusei*, *C. parapsilosis*, *C. tropicalis* (Garg et al. 2006) (Figure 4).

4.5 Preservation of Cultures

Yeast cultures, including *Candida* spp., can be preserved by many methods including stab and liquid cultures contained in tightly sealed tubes or bottles and stored at between 1 and 4°C. Liquid cultures are recommended (Moriss 1958) since fewer spores are formed, and therefore, the likelihood of change in the culture due to segregation of characters is reduced.

Cultures of *Candida* spp., including *C. albicans*, *C. guilliermondii*, *C. stellatoidea*, and *C. tropicalis*, can be preserved for many months when covered with mineral oil and left at room temperature. Viability of some cultures, which do not survive cold storage, can be maintained by this method (Ajello et al. 1951). A few investigators, including Fennell et al.

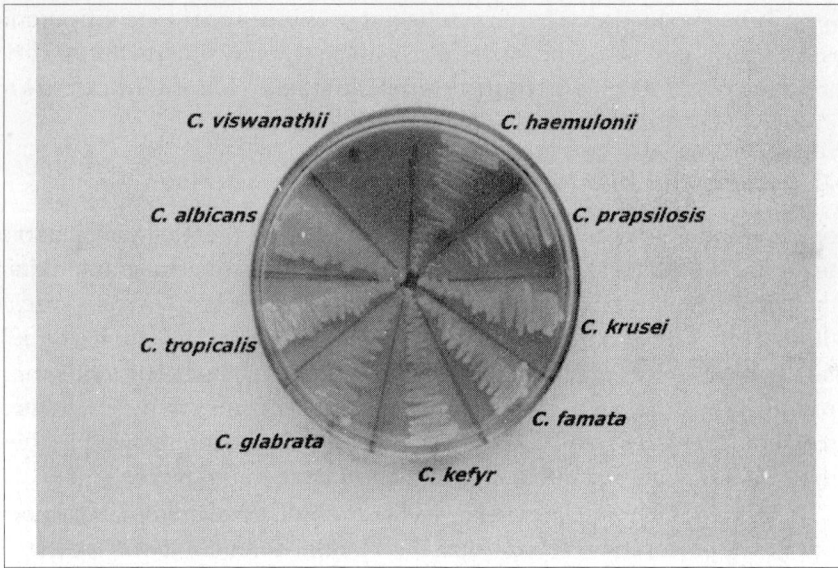

Fig. 4   Rapid diagnosis Pathogenic *Candida* species culture medium.

(1950), and Wickerham and Andreason (1942) recommend lyophilization as the method of choice for preservation of certain fungi and yeast, especially *C. albicans*. There have been few reports of physiological changes in yeast preserved by lyophilization (Atkin 1949) and of low viability after storage. There is an unfortunate dearth of information on the use of this technique (Morris 1958).

For the preservation of *Candida* spores, 15% glycerol at -20°C is one of the method of choice in most laboratories whose spores can survive for 6-12 months. Liquid paraffin is also used for the preservation of *Candida* spores.

5. MORPHOLOGY AND PHYSIOLOGY OF *CANDIDA*

Since 1947, there are many papers and reviews written by well-known scientists about morphology and physiology of *Candida*. In this text, we are going to discuss only some important findings that are based on morphology and physiology of *Candida*, which are helpful in the identification and characterization of *Candida* spp.

5.1 Macroscopic and Microscopic Features

The colonies of *Candida* spp. are cream to yellowish in color, grow rapidly and mature in 48 h (Figure 5). The texture of the colony may be pasty, smooth,

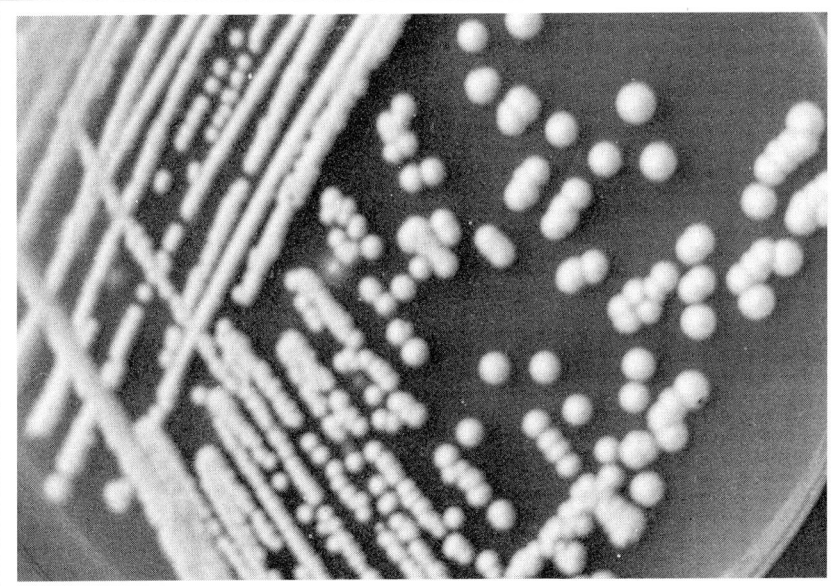

Fig. 5 Cream colored to yellowish colonies of *Candida albicans* on YEPD agar plates after 48 hours.

glistening or dry, wrinkled and dull, depending on the species (Larone 1995).

The microscopic features of *Candida* spp. also show species-related variations. All species produce blastoconidia singly or in small clusters. Blastoconidia may be round or elongate. Most species produce pseudohyphae, which may be long, branched or curved. True hyphae and chlamydospores are produced by strains of some *Candida* spp. Although, they are the members of the same genus, the various species do have some degree of unique behavior with respect to their colony texture, microscopic morphology on cornmeal Tween 80 agar at 25°C (Dalmau method) and fermentation or assimilation profiles in biochemical tests.

6. DISTINCT MORPHOGENIC STATES OF *CANDIDA ALBICANS*

A striking feature of *C. albicans* biology is its ability to grow in a variety of morphological forms. These range from unicellular budding yeast to true hyphae with parallel-sided walls (Figure 6) (Merson et al. 1989; Odds 1988). In between these two extremes, the fungus can exhibit a variety of growth forms that are collectively referred to as pseudohyphae. In these forms, the

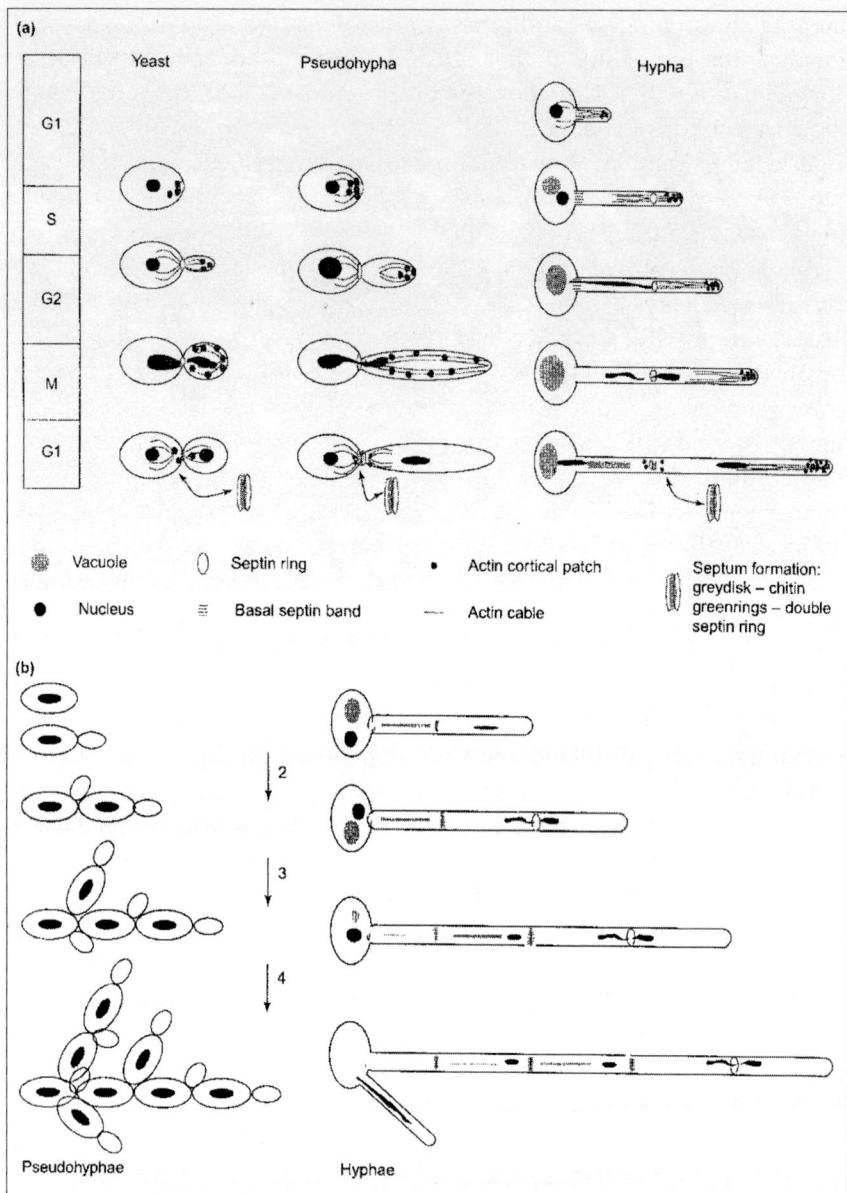

Fig. 6   Representation of cell cycles of yeast, hyphae and pseudohyphae.
a. Cell cycle of yeast and the first cell cycle of hyphae and
b. Pseudohyphae induced from unbudded yeast cells.
c. Representation of cell cycles 2-4 in hyphae and pseudohyphae. (Sudbery et. al. 2000)

daughter bud elongates and, after septum formation, the daughter cell remains attached to the mother cell. As a result, filaments composed of elongated cells with constrictions, at the septa are formed. The elongation of buds in pseudohyphae can be so extreme that these filaments can superficially resemble hyphae. Because of this, it is often useful to be able to refer to pseudohyphae and hyphae collectively and we will use the term 'filamentous' for this purpose.

A change in cellular morphology also occurs during reversible colony switching (Sultsky et al. 1985; Soll 2002). This is an epigenetic phenomenon; yeast cells normally form smooth, white dome-shaped colonies. However, at low frequency, *C. albicans* strains can spontaneously and reversibly convert to a variant colony shape (i.e., star, ring, irregular wrinkle, hat, stipple and fuzzy) in which the colonies are composed of a mixture of yeast and filamentous cells. A simpler, biphasic white-opaque switching system, found in some strains of *C. albicans* that are homozygous for the mating-type-like locus (Miller and Johnson 2002) involves a switch between white domed colonies and opaque flat colonies (Sultsky et al. 1987). White colonies contain oval cells with the normal yeast morphology. Opaque colonies contain oblong-shaped cells that are twice the size of white-phase cells and have pimples on the surface of the cell wall. The importance of the opaque form has increased through the recognition that this form alone is competent to undergo mating (Miller and Johnson 2002).

The ability to switch between yeast, hyphal, and pseudohyphal morphologies is often considered to be necessary for virulence, although formal proof remains lacking. Both hyphae and pseudohyphae are invasive; they invade the agar substratum when they grow in the laboratory. One view is that this property could promote tissue penetration during the early stages of infection, whereas the yeast form might be more suited for dissemination in the bloodstream. The filamentous forms might also be important for the colonization of organs, such as the kidney.

6.1 Cell Shape Distinguishes Different Forms

The traditional criterion for distinguishing between different forms is cell shape (Merson and Odds 1989). Hyphae that developed from an unbudded yeast cell (also termed a blastospore) have no constriction at the neck of the mother cell and have parallel sides along their entire length. The formation of unconstricted filaments in response to serum is the basis of the 'germ tube test', which is used in clinical diagnoses to distinguish *C. albicans* from other *Candida* species (Heddon and Buck 1980); although *C. dubliniensis*, the nearest relative to *C.albicans*, also forms unconstricted hyphae (Gilfillan 1998). Pseudohyphal cells have a constriction at the neck of the mother cell and the bud and at every subsequent septal junction. By contrast, both the width and length of a pseudohyphal cell can vary enormously, so that at one extreme they resemble hyphae and at the other extreme they can resemble yeast cells with elongated buds. However, a characteristic feature of pseudohyphae is that the width of the compartments that make up the filaments is not constant, being wider at the center than at the two ends (Merson and Odds 1989). The morphological index (MI) is a metric that distinguishes pseudohyphae from hyphae on the basis of the dimensions of the cellular compartments, which takes into account the cell length and the ratio of the maximum to the minimum cell width (Merson and Odds 1989). The difference between hyphae and pseudohyphae is defined by a numerical threshold chosen on the basis of whether cells are separable by sonication. Determination of the MI is quite laborious because each cell requires three individual measurements (length, maximum width and minimum width) and enough cells (usually 200-400) need to be characterized for statistical significance.

6.2 Experimental Guidelines for Distinguishing Hyphae and Pseudohyphae

There are several features listed below that distinguish hyphae and pseudohyphae. It would be unrealistic to address all of these characteristics. However, there are three simple tests for analyzing the filamentous forms of *C. albicans*.

(i) Cell shape measurements.

- Are there constrictions at septal junctions?
- Are the sides of the elongated compartment parallel?
- Is the width characteristic of hyphae (~ 2 μm) or pseudohyphae (≥ 3 μm)?

(ii) Location of the first septin ring.

- Is the first septum/septin ring located at the bud neck (pseudohyphae) or within the germ tube (hyphae)?

(iii) Location of the first mitosis.

- Does the first mitosis occur across the bud neck (pseudohyphae) or within the germ tube (hyphae)?

6.3 Experimental Conditions for Germ Tube Formation

There are several features essential for *in vitro* morphological transition of *C. albicans* (Odds 1988). The earliest workers with *C. albicans* found that changes in the growth environment of *C. albicans* led to changes in cell shape. Some environmental factors that favour the filamentation in *C. albicans* are: temperature of 37-40°C, pH around 7.0 and an initial blastospore concentration not exceeding 10^6/ml. Also, certain chemicals, such as N-acetyl-D-glucosamine, amino acids, biotin, sulfhydryl compounds, heme group, zinc, serum, etc. are required. The production of germ tubes in serum has remained the method of choice for identifying *C. albicans* in clinical specimens. Although serum is excellent at promoting the yeast to mycelial conversion, unfortunately, its chemical complexity makes it unsuitable for studying the molecular basis for inducing the hyphal form.

C. albicans seems to be able to regulate its own morphology under at least two circumstances: as a slow response to adverse nutrient environments, in general, and as a rapid response to certain physicochemical conditions. In the latter case, filamentous forms develop under the conditions of temperature and pH that are found in host tissues *in vivo*. Several specific media have been developed for the culture of *C. albcans*. Certain defined media will support both growth forms, the determining factor being either temperature or pH. In general, the budding phenotype is prevalent at low temperatures or low pH, while the hyphal phenotype is prevalent at high temperatures and high pH. A synthetic medium (composed of a mixture of amino acids) has been described by Lee in which a switch in pH suffices to guarantee the production of either the yeast form (pH 4.5) or the hyphal form (pH 6.5), both being cultured at 37°C (Soll 2002).

7. PHYSIOLOGY OF *CANDIDA*

7.1 Metabolic Products and Nutrition

Diddens and Lodder 1942, described the ability of various species of *Candida* to utilize a few selected carbohydrates. Wickerham and Burton, 1948 have greatly expanded these tests and studied a larger number of additional carbohydrates, organic acids and alcohols as to their ability to serve as sole carbon sources. Connel and Skinner (1951) utilizing same method could find any advantage to the use of these compounds as a tool for taxonomy. Indeed, Thjoota and Torheim (1955) indicated that acid rather than gas production might be used as a taxonomic factor. Van Niel (1940) analyzed whether the organism produces acid from a certain carbohydrate than to determine whether it produces sufficient gas to appear in a gas trap.

The ability of *Candida* spp. to utilize carbon compounds has been in relation to the possible use of species, especially *C. arborea* as food of fodder yeasts. Particular interest has been shown in the possible utilization of natural occurring sugars, especially pentoses, by this species. Although several species of *Candida* will utilize pentoses. Windich, 1948 suggest the use of *C. tropocalis* for the production of fodder yeast from wood product fermentations. Furthermore, he shows that pentose-utilizing species assimilate xylose and only xylose-assimilating species will assimilate arabinose or rhamnose. The bioconversion of xylose to xylitol under experimental conditions with pH values of 4.0, 5.5 and 7.0 and tetracycline concentrations of 20 and 40 mg/L was carried out to verify the influence of these parameters on *Candida guilliermondii* metabolism for xylitol production (Martinez 1999).

Candida utilis CBS 621 exhibits the Kluyver effect for maltose, i.e., this yeast can respire maltose and is able to ferment glucose, but is unable to ferment maltose. When glucose was pulsed to a maltose-grown, oxygen-limited chemostat culture of *C. utilis*, ethanol formation from glucose started almost instantaneously, indicating that the enzymes needed for alcoholic fermentation are expressed in maltose-grown cells. However, the addition of glucose inhibited maltose metabolism (Weusthuis et al. 1994).

Candida utilis NRRLY-900 was grown in aerobic continuous culture in a minimal salts medium with sucrose (1% w/v) as the carbon source. Increasing the concentration of zinc in the medium from 2.3 microM results in an increase in the apparent critical dilution rate from 0.3 to 0.47h-1, and in the maximum biomass productivity form 1.5g dry weight per litre per hour (at D=0.33-1) to 2.56g per litre per hour (at D = 0.45-1).

C. boidinii S2 was shown to have the ability to utilize each of two pectic compounds, pectin and polygalacturonate, as a carbon source. This strain produced pectin-depolymerizing enzymes (Tomoyuki et al. 2000).

The salt tolerance of the respiratory yeast *Candida tropicalis* and the fermentative yeast *Saccharomyces cerevisiae* have been compared in glucose media. *C. tropicalis* showed a better adaptation to Na^+ and Li^+ and maintained higher intracellular $K^+:Na^+$ and $K^+:Li^+$ ratios than *S. cerevisiae*.

Novel biotechnology tools like immobilization were also applied for the isolation and incorporation of such food components in ordinary foods. The synthesis of nutraceuticals was reported to be successful by empoloying immobilized lipases, such as those from *Candida antartica* and *Lactobacillus ruteri*.

There have been many studies on vitamin production by species of *Candida*, most of them by fodder yeast, especially *C. arborea* and *C. guilliermondii*. *Candida guilliermondii* Wickerham was found to be superior to the other yeasts tested for growth yields and riboflavin production when cultivated on different carbohydrate and hydrocarbon containing media. Among the refined petroleum fractions and cheap hydrocarbons tested, solar was selected as a carbon source best suited for the fermentation process. The highest growth yield and riboflavin output (10.64 mg/100 ml) by *C. guilliermondii* Wickerham were achieved by placing aliquots of 20 ml medium (pH 6.0) in 100 ml Erlenmeyer flasks and incubating the inoculated medium at 30°C for 14 days in the dark (Sabry et al. 1989).

Pyke (1958) has reviewed much of the information regarding the industrial use of *Candida* species in his general discussion of the technology of yeast. Excellent and more complete reviews of food yeast, including the use of *C. arborea* and *C. tropocalis*, published by Irvin (1954). Most recent edition of Prescott and Dunn's Industrial Microbiology reviews yeast technology, but it is more valuable as a source of patent information.

It is important to note *Candida* is now considered among the most important technological yeasts. In fact, Windisch (1954) lists *Sacharomyces cerevisise*, *Torulopsis* (*Candida*) *utilis*, and *Candida tropicalis* as the most important industrial fermenters.

The manufacture of fodder yeast from waste sulfite liquor using *C. tropicalis* has been described by Yamaguchi and Mukaiyama (1981), Windisch (1954) suggests the use of *C. pseudotropicalis* in the fermentation of whey. Preliminary yeast worker attempts to increase the food value of whey by yeasts fermentation results that *C. krusei* was not satisfactory as other yeasts, such as *Torulopsis cermoris* . This is understandable since that *C. krusei* does not utilize, lactose whereas *T. cremoris*, which, according to Lodder and Kreger-van Rij (Lodder and Kreger 1952), is identical with *C. pseudotropicalis*, is the only lactos-fermenting species of *Candida*.

Parmalee and Nelson (1949) and Peters and Nelson (1948) reported the use of *C. lipolytica* in the manufacture of blue cheese from pasteurized milk. Their result shows that cultures of *C. lipolytica* improve flavor and increase volatile acid concentration. The 11 "strains" used, varied both in flavour improvement and volatile acid production. There was no flavor improvement, however, all lipolytic bacteria tested. Peters and Nelson (1948) have extensively studied those nutritional and environment growth factors, which influence lipase production by *Mycotorula lipolytica* (*C. lipolytica*). In general, those environmental factors, which favour the development of short cells and, therefore result in rapid growth, reduce the lipase activity. Lipase production in a defined synthetic medium is dependent upon the presence of thiamine or pyridoxamine, either of which satisfies the requirement. The yeast *Candida rugosa* produces multiple extracellular lipases. The production of extra and intracellular lipases was investigated in continuous cultures using a sole or different mixtures of carbon sources. Also, the effect of different C : N ratios were tested. Lipase productivity in continuous cultures increased by 50% compared with data obtained from batch fermentations and depended on the dilution rate applied. Maximum yields relative to consumed substrate were obtained with oleic acid at low dilution rate. It was found that during nitrogen limitation, lipase activity was suppressed. All carbon source mixtures tested allowed both cell growth and lipase production, but extra and intracellular lipase activities were affected by the combination of substrates used. Maximum extracellular lipolytic productivity was attained with lactic and oleic acid mixtures, probably due to the non-repressor effect of these carbon sources (Montesinos et al. 2003). The chemical composition of the biomass also depended on the type of substrate used and was related to the accumulation of lipidic compounds as intracellular inclusions, which were observed when oleic acid was used as the carbon source. The results obtained were compared with previous data from batch and fed-batch cultures in order to select the best process strategies for the lipase production with *C. rugosa*. The best lipase yields were obtained in fed-batch fermentations using oleic acid (Montesinos et al. 2003).

Several species of *Candida* are capable of producing relatively large amounts of fat. The most productive of these are *C. pulcherrima*, *C. utilis* and *C. reukaufii*. Good yields of fat have been obtained from sucrose or xylose by fermentation with *C. reukaufii*. Oil similar to olive oil is synthesized in abundance under cultural conditions of abundant vitamins, especially biotin, low nitrogen, intensive aeration, and at a pH above 4.5 to 5.0. Fat synthesis follows protein formation and if the amount of protein formed is low, fat synthesis increased.

Several species of *Candida* tested have been shown to produce sterols. The yield of sterol varies with the species and the medium in which it is grown. Typical yields reported by Appleton et al. (1955), *C krusei*, 0.3 %, and *C. parapsolosis*, 0.06%, respectively. Sterols were not produced by *C. albicans* or *C. tropicalis*. *Saccharomyces* spp. appears to be the largest group of sterol-producing yeasts with yields as high as 9% in some species. Salt-tolerant species of *Candida* have been reported in Japan in Unishiokara a Japanese fishery fermentation product (Zenitani 1954).

An excellent series of reports on the occurrence of *Candida* and brewing yeasts have been published by Wiles and associates (Wiles 1950; Wiles 1951; Wiles 1953). Species encountered, included *C. lipolytica, C. krusei, C. guilliermondii, C. pelliculosa*, and *C. mycoderma*. Various *Candida* and *Brettanomyces* species cause turbidity in English top fermented beer (Wiles 1950) *C. pelliculosa* is responsible for fruity flavors and *C. krusei* as well as *Brettanomyces* species give strong yeasts odors (Wiles 1951). *C. mycoderma* a film-forming yeasts isolated as contaminants of brewery yeasts (Wiles 1950; Wiles 1951). The ability of these cultures to use 35 different carbon sources was studied and of these succinic acid, mannitol and sorbitol were found taxonomically useful in the identification of film-forming yeasts. Certain metabolic aspects of Mycoderma (*Candida*), specifically the relationship of thiamine concentration to the formation of alcohol, had been previously studied by Walker and Ramathandran (1949). *C. rugosa* has been reported to be among those yeasts which cause clouding of dry white wines in California. *C. krusei* and *C. tropicalis* are among those yeasts important in spoilage of rice stored in a sealed bin (Teunisson 1954).

C. albicans is described for fermentation of fresh water lake sediment with other yeasts. Compounds isolated from salt-water plants are also utilized by *Candida* that *C. krusei* and *C. solani* adaptively utilize the seaweed constituents fucoidin, fucose, laminarin and sodium alginate (Messineva and Skadooskii 1947).

Candida and other yeasts to hydrolyze pectin, of all yeasts tested only *Saccharomyces fragilis* and its imperfect form *Candida pseudotropicalis*, produce an extracellular, non-adaptive polygalacturonase, free from pectin esterase. The polygalacturonase has optimal temperature and pH of 55°C to 60°C and 3.5 to 4.0, respectively. *C. tropicalis* is the best source of enzyme, trehalase although it is produced as an adaptive enzyme. Methods of purification and the kinetics of the enzyme are optimal temperature and pH, which are 48°C and 4.1 to 5.3, respectively .

There has been little biochemical study of *Candida* spp., expect for *C. albicans*. These investigations concerned directly with applications to technology, taxonomy, and identification. Many reports have recently been made but no attempt will be made to give complete coverage to those purely biochemical aspects of the genus. In this regard Bruchmann (1956) has demonstrated the presence of aldolase, carboxylase and glucose dehydrase in *C. reukaufii*. *C. albicans* contains an active cystinase (Romano and Nikerson 1954). Bruyn (1954) reportedly shows the first example of the biological oxidation of an unsaturated compound ($R-CH=CH_2$) give rise to the corresponding glycol ($R-CHOHCH_2OH$), using *C. lipolytica* and with L (-) hexadecane-1 as the substrate.

Several investigators have studied pH as physical factors. The range of pH over which *C. albicans* will grow has always important because of this organism's pathogenic ability. In respect to pH it acts more like the saprophytic yeasts and molds. Although it was reported in the review (Sckinner 1947) that the optimal pH was 7, recent reports indicate that best growth is obtained between 5.1 and 6.4 although varying degrees of growth do occur between pH 2.2 and 9.6. The effect of Dilute solutions varies among the *Candida* species, *C. robusts* being more resistant than *C. albicans, C. tropicalis, C. krusei,* and *C. rugosa.* Chirstophersen and Precht (1954) have studied the effect of incubation temperature on cell size of *Torulopsis kefyr* (*Candida pseudotropicalis*). Oxygen tolerance, and effect of temperature and relative humidity on growth of *C. pseudotropicalis* have been studied by Bottomely et al. (1950) in relation to mold growth in stored corn. These results show that growth of *C. pseudotropicalis* ceases in corn stored at a relative humidity of 80 per cent and all oxygen concentrations tested (0.1-21%), suggesting a marked lack of tolerance to low oxygen tensions.

Thiaminase activity has been reported in *C. albicans* and certain other fungi, but not in other including *Cryptococcus.* There appears to be equilibrium between synthesis and inactivation of thiamine in those thiamine producing species. The possible relationship between certain nutritional conditions and the presence of *C. albicans* in the intestinal tract of man and animals is noteworthy. This would seem to be fruitful area of investigation in view of the numerous reports of *C. albicans* in the intestine of man and many other animals.

REFERENCES

Abi-Said DE, Uzon O, Raad I, Pinzcowski H, Vartivarian O (1997). The epidemiology of hematogenous candidiasis caused by different *Candida* species. Clin Infect Dis 24: 1122-1128.

Abu-Eiteen KH, Abdul Malek, Abdul Wahid (1997). Prevalence and susceptibility of vaginal yeast isolates in Jordan. Mycoses 40:179-185.

Ajello L (1956). Soil as natural reservoir for human pathogenic fungi. Science 123: 876-879.

Akiba T and Iwata, K (1954). On the destructive invasion of a new species, *"Bacterium candidoestruens,"* into *Candida* cells. *Japan. J. Exptl. Med.,* 24,159-166.

Appleton, G S, Kieber. R J, and Payne, W J (1955). The sterol content of fungi. II. Appl. Microbiol., 3, 249-251.

Atkin L, Moses W and Gray PP (1949). The preservation of yeast cultures by lyophilization. *J. Bacteriol,* 57, 575-578.

Baily G G, Moore C B, Essayag S M, Wit S, Burnie J P and Denning D W (1997). *Candida inconspicua,* a fluconazole-resistant pathogen in patients infected with human immunodeficiency virus. *Clin. Infect. Dis.* 25:161-163.

Benham R W (1931). Certain monilias parasitic on man : their identification by morphology and by agglutination. *J. Infect. Dis.* 49: 183-215.

Berg FT 1846. *Om Torsk Hos Barn.* L. Hjerta, Stockholm, Sweden.

Bhatnagar B and A P Garg (1990). Qualitative and Quantitative analysis of microfungi colonizing leaf-litter of *Oryza sativa* L. *J. Indian Bot. Soc.* Vol, 69: 81-87.

Bhatnagar G (2004). Reproductive tract infections in women. Ph.D Thesis, Department of Microbiology, C.C.S. University, Meerut. (INDIA).

Borg-von Zeppelin M, Eiffert H, Kann M and Ruchel R (1993). Changes in the spectrum of fungal isolates: results from clinical specimens gathered in 1987/88 compared with those in 1991/92 in the university hospital Gottingen, Germany. Mycoses 36:247-253.

Bottomley R A, Christensen C M and Geddes W F (1950). The influence of various temperatures, humidities, and oxygen concentrations on mold growth and biochemical changes in stored yellow corn. Grain storage studies. IV. Cereal Chem., 27, 271-296.

Bouthilet R J (1951). A taxonomic study of soil yeasts. Mycopathol. *et Mycol. Appl.*, 6, 79-85.

Buckely HR (1989). Identification of yeasts, p.97-109. *In* E. G. V. Evans and M.D. Richardson (ed.), Medical Mycology- A Practical Approach. IRL Press at Oxford University Press, Oxford, United Kingdom.

Carson H L, Knapp E P, and Phaff H J (1956). Studies on the ecology of Drosophila in the Yosemite region of California. III. *Ecology*, 37, 538-544.

Castellani A (1920). The etiology of thrush. J Trop Med Hyg 23:17-22.

Christophersen J and Precht H (1954). Uber den Einfluss der Wachstumstemperatur auf die Grosse von Hefezellen. Zentr. Bakteriol. Parasitenk., Abt. II, 108, 1-6.

Clark D S and Wallace R H (1955). Candida malicola nov. sp. isolated from apples. *Can. J. Microbiol.*, 1, 275-276.

Collazos J, Mayo J and Martinez E, (1999). Changing spectrum of HIV infection and its associated conditions in Spain: the end of the beginning? *AIDS Patient Care Sex. Transm. Dis.* 13:347-353.

Connell G H and Skinner C E (1951). The origin of some species of the genus Candida in non-pseudomycelium-producing anascosporogenous yeasts. *Mycopathol. et Mycol. Appl.*, 6, 65-71.

Connell G H and Skinner C E (1953). The external surface of the human body as a habitat for non-fermenting non-pigmented yeasts. *J. Bacteriol.*, 66, 627-633.

Costa S F, Marinho I, Araujo E A, Manrique A E, Medeiros E A and Levin A S (2000). Nosocomial fungaemia: a 2-year prospective study. *Hosp. Infect.* 45:69-72.

D'Agata E M, Mount D B, Thayer V and Schaffner W (2000). Hospital-acquired infections among chronic hemodialysis patients. *Am. Kidney Dis.* 35:1083-1088.

Di Menna M E (1955). A search for pathogenic species of yeasts in New Zealand soils. *J. Gen. Microbiol.*, 12, 54-62.

Di Menna M E and Parle J M (1954). Some extra-human occurrences of pathogenic yeast. *Proc. Univ. Otago Med. School*, 32, 2.

Diddens H A and Lodder J (1942). Die anaskosporogenen Hefen, II. Halfte.North Holland Publ., Amsterdam.

Dobzhansky T H, Cooper D M, Phaff H J, Knapp E P, and Carson H J (1956). Studies on the ecology of Drosophila in the Yosemite region of California. II. *Ecology*, 37, 544-550. edited by A. H. Cook. Academic Press, Inc., New York.

Fennell DI, Raper KB, and Flickinger H (1950). Further investigations on the preservation of mold cultures.*Mycologia*, 42, 135-147.

Fernandez M, Moylett E H, Noyola D E and Baker C J (2000). Candidal meningitis in neonates: a year review. *Clin. Infect. Dis.* 31:458-463.

Garg A P and Sharma PD (1984). Ecology of phylloplane and litter fungi of triticale. Nord. *J. Bot.* 4: 707-715.

Garg AP, Gautam R and Almamary AG 2006. A selective cultural medium for rapid isolation of *Candida* spp. from clinical samples. In the process of publication, ISHAM 2006.

Gargas A, DePriest P T, Grube M and Tehler A (1995). Multiple origins of lichen symbioses in fungi suggested by SSU rDNA phylogeny. Science 268:1492-1495.

Gilfillan GDet al (1998). *Candida dubliniensis*: phylogeny and putative virulence factors. *Microbiology* 144, 829-838.

Gruby M (1842). Recherches anatomiques sur ue plant cryptogame qui contitute le vari muguet des enfants. *C.R. Acad. Sci.* (Paris) 14: 634-636.

Hazen K C (1995). New and emerging yeast infections. *Clin. Microbiol. Rev.* 8:462-478.

Hazen KC, Theisz G W and Howell SA (1991). Chronic urinary tract infection due to *Candida utilis*. *J. Clin. Microbiol.* 37:824-827.

Heddon DM and Buck JD (1980). A re-emphasis–germ tubes diagnostic for *Candida albicans* have no constrictions. *Mycopathologia* 70,95-101

Hedrick L R and Burke G C (1950). Yeasts from Hawaiian fruit flies. Their identification and ability to produce riboflavin. *J. Bacteriol.*, 59, 481-484.

Hung C C, Chen M Y, Hsieh S M, Sheng W H and Chang S C (2000). Clinical spectrum, morbidity, and mortality of acquired immunodeficiency syndrome in Taiwan: a 5-year prospective study. *J. Acquir. Immune Defic. Syndr.* 24: 378-385.

Irvin R (1954). Commercial yeast manufacture, ch. 9. In Industrial fermentations, Vol. I. Edited by L. A. Underkofler and R. J. Hickey. Chemical Publishing Co., Inc., New York.

Kao A S, Brandt M E, Pruitt W R, Conn I A, Perkins B A, Stephens D S, Baughman W S, Reingold A I, Rothrock G A, Pfaller M A, Pinner R W and Hajjeh R A (1999). The epidemiology of candidemia in two United States cities: results of population-based active surveillance. *Clin. Infect. Dis.* 29:1164-1170.

Knoke M, Schulz K and Bernhardt H, (1997). Dynamics of *Candida* isolations from humans from 1992-1995 in Greifswald, Germany. *Mycoses* 40:105-110.

Kurtzman C P and Fell J W (1998). *The Yeasts, a Taxonomic Study,* 4th ed. Elsevier Science, BV, Amsterdam, The Netherlands.

Langenbeck B (1839). Auffinggung von Pilzen aus der Schleimhaut der Speiserohne einer Typhus-Leiche. *Neue Not.Geb. Nature-u-Helik (Froriep)* 12:145-147.

Lodder J and Kreger-Van Rij N J W (1952). The yeasts. A taxonomic study. North Holland Pub. Co., Amsterdam.

Lund A (1954). Studies on the ecology of yeasts. Munksgaard, Copenhagen.

Lund A (1958). Ecology of yeasts, ch. 2. In The chemistry and biology of yeasts. Edited by A. H. Cook. Academic Press, Inc., New York.

Marples M J and Di Menna M E (1952). The incidence of Candida albicans in Dunedin, New Zealand. *J. Pathol. Bacteriol.*, 64, 497-502.

Merson-Davies LA and Odds FC (1989). A morphology index for cell shape in *Candida albicans*. *J.Gen.Mocrobiol.*135, 3143-3152

Messineva M A and Skadooskii S N (1947). Growth of microorganisms on sapropel.Yeast fermentation of sapropel in the laboratory and in situ. Mikrobiologiya, 16, 43-49.

Miller MG and Johnson AD (2002). White-opaque switching in *Candida albicans* is controlled by mating-type locus homeodomain proteins and allows efficient mating. *Cell* 100,293-302.

Miller W H and Mrak E M (1953). Yeasts associated with dried-fruit beetles in figs. *Appl. Microbiol.*, 1, 174-178.

Montesinos JL, Dalmau E, and Casas C (2003). Lipase production in continuous culture of *Candida rugosa J. of Chem. Tech. & Bio.* Vol. 78, 7, 753 – 761.

Morris E O (1958). Yeast growth, ch. 6. In The chemistry and biology of yeasts, Edited by A. H. Cook. Academic Press, Inc., New York.

Morris E O (1958). Yeast growth, ch. 6. In The chemistry and biology of yeasts.

Nucci M, Silveira M, Spector N, Silveira F, Velasco E, Martins C, Derossi A, Columbo A, and Pulcheri W. (1998). Fungemia in cancer patients in Brazil: predominance of nonalbicans species. *Mycopathologia* 141:65-68.

Odds F C (1988). Biological aspects of pathogenic *Candida* spp. , 7-15. In *Candida and Candidosis.* Bailliere Tindall, London, United Kingdom.

Odds F C, and Bernaert R (1994). CHROMagar *Candida*, a new differential isolation medium for presumptive identification of clinically important *Candida* species . *J.Clin. Microbiolo.* 32:1923-1929

Parmalee C E and Nelson F E (1949). The use of Candida lipolytica cultures in the manufacture of blue cheese from pasteurized homogenized milk. *J. Dairy Sci.,* 32, 993-1000.

Peters II and Nelson F E (1948). Factors influencing the production of lipase by Mycotorula lipolytica. *J. Bacteriol.,* 55, 581-591.

Pfaller M A, Jones R N, Doern G V, Fluit A C, Verhoef J, Sader H S, Messer S A, Houston A, Coffman S, and Hollis R J (1999). International surveillance of blood stream infections due to *Candida* species in the European SENTRY program: species distribution and antifungal susceptibility including the investigational triazole and echinocandin agents. SENTRY Participant Group (Europe). *Diagn. Microbiol. Infect. Dis.* 35:19-25.

Pyke M (1958). The technology of yeast, Ch. 10. In The chemistry and biology of yeasts. Edited by A. H. Cook. Academic Press, Inc., New York.

Recca J and Mrak E M (1952). Yeasts occurring in citrus products. *Food Technol.*, 6, 450-454.

Rennert G, Rennert HS, Pitlik S, Finkelestein R and Kitzes-Cohen R (2000). Epidemiology of candidiemia-a nationwide survey in Israel. Infection 28:26-29.

Rippon J W (1974). Candidosis, p. 175-204. *In* J. W. Rippon (ed.), *Medical Mycology: The Pathogenic Fungi and the Pathogenic Actinomycetes.* W. B. Saunders Co., Philadelphia, Pa.

Robin CP (1853). *Odium albicians* as the cause of oral thrush. Saccordo's Syll Fung VIII:918, XV:230.

Ruiz Oronoz M (1949). Estudio de una variedad de Mycoderma cerevisiae Desmasi6res, aislada del jitomate (*Lycopersiuim esculentum*). *An. Inst. Biol. Univ. Mdx.*, 20, 43-56.

Sabry SA *et al.* (1989). Production of riboflavin (vitamin B2) by hydrocarbon-utilizing yeasts, *Microbiologia*, 5(1), 45 – 52.

Skinner CE (1947). The yeast-like fungi: Candida and Brettanomyces. *Bacteriol. Revs.*, 11, 227-274.

Skinner CE and Fletcher DW (2000). A review of genus *Candida*. Department of Bacteriology and Public Health 24:397-416.

Slutsky B et al (1985). High-frequency switching of colony morphology in *candida albicans. Science* 230, 666-669.

Soll DR (2002). *Candida and Candiadiasis.* (Calderone, R., ed.), pp.123-142, ASM Press, Washington.

Sullivan D, and Coleman D (1998). *Candida dubliniensis:* characteristics and identification. *J. Clin. Microbiol.* 36: 329-334.

Sullivan DJ, Westerneng T J, Haynes K A, Bennett D E, and Coleman D C (1995). *Candida dubliniensis* sp. nov: phenotypic and molecular characterization of a novel species associated with oral candidosis in HIV-infected individuals. *Microbiology* 141(Pt. 7): 1507-1521.

Swaby R J (1945). Production of lactic acid from wheat starch. J. Australian *Inst. Agr. Sci.*, 2, 179-187.

Teunisson D J (1954). Yeasts from freshly combined rough rice stored in a sealed bin. *Appl. Microbiol.*, 2, 215-220.

Thjotta T and Torheim B J (1955). Studies on the fermentation of sugars in the Candida group. *Acta Pathol. Microbiol.Scand.*, 36, 237-249.

Thompson L (1945). Note on a selective medium for fungi. Proc. Staff Meetings Mayo Clinic, 20, 248-249.

Tomoyuki Nakagawa, Tatsuro Miyaji, Hiroya Yurimoto, Yasuyoshi Sakai, Nobuo Kato, and Noboru Tomizuka (2000). A Methylotrophic Pathway Participates in Pectin Utilization by Candida boidinii. *Appl Environ Microbiol.* 66(10): 4253–4257.

Virtanen I (1951). Observations on the symbiosis of some fungi and bacteria. *Ann. Med. Exptl. Biol. Fenniae*, 29, 352- 358.

Weusthuis RA, Visser W, Pronk JT, Scheffers WA, Van Dijken JP (1994). Effects of oxygen limitation on sugar metabolism in yeasts: a continuous-culture study of the Kluyver effect. Microbio.140 (Pt 4): 703-15.

Wickerham L J (1952). Recent advances in the taxonomy of yeasts. *Ann. Rev. Microbiol.* 6, 317-332.

Wickerham L J and Andreason A A (1942). The lyophil process: its use in the preservation of yeasts. *Wallerstein Labs. Communs.*, 5, 165-169.

Wiles A E (1950). Studies of some yeasts causing spoilage of draught beers. *J. Inst. Brewing*, 56, 183-193.

Wiles A E (1950). Wild yeasts. *Brewers' Guild J.*, 36, 141-159.

Wiles A E (1951). Frets and hazes. *Brewers' Guild J.*, 37, 114-131.

Wiles A E (1953). The identification and significance of yeasts encountered in the brewery. *J. Inst. Brewing*, 59, 265-284.

Windisch S (1954). Uber einige Kahmhefen als Bierschidlunge. Brauerei, 8, 139-142.

Wingard J R, Merz W G, Rinaldi M G, Miller C B, Karp J E, and Saral R (1993). Association of *Torulopsis glabrata* infections with fluconazole prophylaxis in neutropenic bone marrow transplant patients. *Antimicrob. Agents Chemother.* 37:1847-1849.

Wingard JR, Merz WG, Rinaldi M G, Johnson TR, Karp JE, and Saral R (1991). Increase in *Candida krusei* infection among patients with bone marrow transplantation and neutropenia treated prophylactically with fluconazole. N. *Engl.* J. *Med.* 18:1274-1277.

Winner HI, and Hurley R (1954). *Candida albicans.* Little, Brown and Company, Boston, Mass.

Yamaguchi M and Mukaiyama T (1981). The stereoselective synthesis of D- and L-ribose. Chemistry.

Young G, Krasner R I, and Yudkofsky P L (1956). Interactions of oral strains of Candida albicans and lactobacilli. J. Bacteriol., 72, 525-529.

Young G, Resca H G, and Sullivan M T (1951). The yeasts of the normal mouth and their relation to salivary activity. *J. Dental Research*, 30, 426-430.

Zenitani B (1954). Yeasts occurring in fishery fermentation products. V. On the morphological characteristics of asporogenous yeasts. Sci. Bull. Fac. Agr. Kyushu Univ., 14, 589-594.

Types of Candidiasis and their Management

A. P. GARG and R. GAUTAM

1. INTRODUCTION

Various species of *Candida* cause a variety of infections in several parts of human body, which are commonly reffered as candidiasis.

2. OROPHARYNGEAL CANDIDIASIS

Oropharyngeal candidiasis (OPC) is the general term given to the oral infection caused by *Candida*. This condition is also often referred to informally as thrush. OPC is considered as an opportunistic infection, whereas asymptomatic oral colonization may also occur. OPC implies the presence of signs and symptoms of infection.

2.1 Epidemiology

Candida spp. are part of the normal mouth flora in 25-50% of healthy individuals (Nguyen and Yu 1995). Such carriage is referred to as asymptomatic colonization. *Candida albicans* is the most frequent colonizer (70-80%) but any of the non-*albicans Candida* species may also be seen. Salivary flow, salivary pH and glucose concentrations are the major factors that influence the frequency of oral candidal colonization. Specifically, carriage rates are higher in:

Department of Microbiology, C.C.S. University, Meerut-250005, India.

- HIV-infected patients and patients with AIDS. In these patients, the rate of carriage is a function of the level of immunosuppression (Ryder et al. 1998).
- Patients heavily treated with fluconazole more oftenly carry non-*albicans Candida* spp. that are resistant to the azole antifungal agents (Baddour, 1995; Lozano et al. 1998; Marco et al. 1998).
- Hospitalized patients regardless of underlying disease (Bow et al. 1995).
- Denture users (Brummer et al. 1993).
- Diabetic patients (Sugar et al. 1990).
- Cancer patients (Blizzard et al. 1968).

2.2 Clinical Manifestations

The most frequent and familiar form of OPC is pseudomembranous candidiasis (or thrush). The general characteristics of the three main forms of OPC are following:

A. Pseudomembranous Oropharyngeal Candidiasis is commonly called *thrush* and presents with raised confluent white to creamy elevated patches on a hyperemic base. This can be better appreciated when plaques are removed and the base bleeds. The lesions can involve any part of the mouth. Patients affected with pseudomembranous OPC complain of mouth soreness or pain, burning tongue, taste changes, or dryness. Chronic oral discomfort associated with this form of candidiasis, especially in patients with AIDS may impair the intake of adequate oral nutrition and contribute to weight loss and inanition. The presence of odynophagia (pain on swallowing) suggests extension of the process to the form known as Esophageal Candidiasis.

B. Erythematous or Atrophic Oropharyngeal Candidiasis presents with diffuse redness of the palate and the dorsum of the tongue. When the infection is principally on the tongue, the term candidal glossitis is used. Chronic atrophic candidiasis is associated with an indolent course. Some patients have no symptoms, whereas others complain of a metallic taste or a local burning sensation. Denture wearers seem especially prone to this condition, in which case it is referred to as denture-induced stomatitis or denture sore mouth.

C. Angular Cheilitis form of OPC has also been called "perleche." It is characterized by inflammation of the angles of the mouth, with or without fissuring. Other causes of this picture include iron deficiency and

Staphylococcus aureus infection. Patients are usually very symptomatic with local soreness, tenderness, pain, or burning.

2.3 Therapies

Topical therapy usually suffices for milder forms of the disease, however, in diseased patients with immunosuppression (most notably, disease in HIV/ AIDS patients), and disease in which there are symptoms that suggest esophageal involvement e.g., pain on swallowing, are best treated with systemic therapy. Different therapeutic options available for treatment of oropharyngeal candidiasis are summarized in Table 1.

Table 1   Therapeutic options for the treatment of Oropharyngeal candidiasis

Drug	Dosage	Comments
Nystatin	Suspension: 500,000 Units by swish & swallow qid x 7-14 days.	It has an unpleasant taste and may cause nausea and GI disturbance. Vaginal tablets in combination with unsweetened mints or chewing gum are better tolerated.
Fluconazole	100 mg/d x 7-14 days (a loading dose of 200 mg has been recommended for immuno-suppressed patients and/or severe OPC).	Fluconazole is superior to nystatin, clotrimazole, and ketoconazole. High doses (upto 800 mg/day) can be used in difficult cases. Success has been obtained even in cases of *in vitro* resistance.
Itraconazole	Suspension: 200 mg (20 ml) qd by swish & swallow without food x 7-14 days. Capsules: 200 mg/day (taken with food) x 2-4 weeks.	The capsules have limited bioavailability and their absorption is improved if they are taken with a fatty meal. The efficacy of the capsules is thought about equal to that of ketoconazole.
Ketoconazole	200-400 mg/day x 7-14 days.	This agent has limited bioavailability, requires an acidic environment for best absorption, and causes liver toxicity. As it is less efficacious than fluconazole and itraconazole suspension, it is less frequently used.

2.4 Oropharyngeal Candidiasis (OPC) in Patients with HIV-AIDS

In HIV epidemic, OPC was identified as not only an immunologic marker of HIV infection but also a rapid progression to AIDS. Oropharyngeal candidiasis occurs in about one-third to one-half of HIV-seropositive persons and in upto 90% of the patients with AIDS, making it the most frequent opportunistic infection in AIDS (Feigal et al. 1991; Sangeorzan 1994). Symptomatic OPC is rare until the CD4+ T-cells fall to < 500 cells/mm^3. As AIDS progresses, either because of lack of adequate therapy or resistance to antiretroviral therapy, the severity and permanent nature of AIDS-associated immunosuppression make it difficult to permanently eradicate. The natural history in patients with advanced HIV infection is one of the remission and then relapse.

Prophylaxis with azole antifungals to prevent OPC among HIV-infected individuals has been extensively assessed (Goldman et al. 2000.) Such therapy is effective, but this form of long-term treatment has created the perfect setting for emergence of resistance (Maenza et al. 1995; Martins et al. 1997). *Candida* spp. resistant to antifungal agents become common after several years of therapy (Pelletier et al. 2000).

3. ESOPHAGEAL CANDIDIASIS

Esophageal candidiasis is an infection of the esophagus tube that connects the mouth to the stomach. It is caused by an overgrowth of *Candida*, a fungus that is normally found in the mouth, gastrointestinal tract (G.I.) and vagina, as well as on the skin. Esophageal candidiasis is an AIDS-defining illness.

3.1 Epidemiology

The risk factors for candidal esophagitis are the same as those for OPC. The incidence of overall rate of esophagitis among patients infected with HIV has been reported to be as high as 15 to 20% (Bammert and Fostel 2000). These rates are very much affected by use of effective anti-retroviral therapy.

3.2 Clinical Manifestations

Patients with candidal esophagitis usually have symptoms like dysphagia (a perception of difficulty with swallowing), odynophagia (pain during swallowing), substernal chest pain that is not clearly related to swallowing, and/or a feeling of obstruction in the chest. Candidal esophagitis has been associated with fever (Meletiadis et al. 2002), but this is uncommon. Upon endoscopy, white exudative lesions similar to those of OPC are seen. They may be small or widely spread and contiguous.

3.3 Candidal Esophagitis in Patients with HIV/AIDS

It is sometimes said to be the "most frequent invasive candidal disease in patients with AIDS", however, biopsy studies have shown that *Candida* in the esophagus is never invasive in the true sense (Gonzalez et al. 1992), rather the process is limited to the surface of the esophagus. Candidal esophagitis is, however, a significant disease and was the most common AIDS-defining disease in two cohorts of HIV-infected women (Meletiadis et al. 2002).

4. VULVOVAGINAL CANDIDIASIS

Genital candidiasis may affect the vulva and vaginal area, as well as the perianal and crural folds, causing candidal intertrigo. Vulvovaginal candidiasis (VVC) is a mucosal infection caused by several species of *Candida*, however, *C. albicans* is the most common yeast obtained from vaginal fungal cultures (Sobel 1999; Bhatnagar 2004). The emergence of *C. glabrata* vaginal colonization has been described as a significant problem in HIV-infected women receiving fluconazole prophylaxis.

Causes of vaginal discharge include infection with:

- *Candida* (a yeast)
- *Trichomonas* (a small parasite)
- Bacterial vaginosis (an imbalance of the amounts of bacteria which live in the vagina).

4.1 Growth of *Candida*

About 20% of non-pregnant women aged 15 to 55 harbour the yeast *Candida albicans* in the vagina but most have no symptoms and it is harmless to them. Overgrowth of vaginal candidiasis causes a heavy white curd-like vaginal discharge, a burning sensation in the vagina and vulva and/or an itchy rash (Bhatnagar 2005). This is also known as thrush or monilia.

Overgrowth of *candida* occurs most commonly with:

- Pregnancy
- A course of broad spectrum antibiotics such as tetracycline or coamoxiclav
- Diabetes
- Iron deficiency anaemia
- Immunological deficiency
- On top of another skin condition, often psoriasis, lichen planus or lichen sclerosus.
- Other illness

It may sometimes be aggravated by sexual intercourse.

The doctor diagnoses the condition by inspecting the affected area and taking a swab. In recurrent cases the swab should be repeated after treatment to see whether *Candida albicans* is still present, or if a non-albicans *Candida* is present. These are more resistant to treatment. Swab results can be misleading, however, because the yeast can be present without causing symptoms and it can only be cultured if a certain amount is present.

Some women appear to be hypersensitive to the organism, so although its presence cannot be proven by the laboratory, the yeast is still responsible for the subject's symptoms, especially if they recur each month.

4.2 Treatment

There are a variety of effective treatments for candidiasis. Topical antifungal vaginal tablets are usually recommended in mild cases, sometimes a single treatment is all that is necessary. In addition anti-yeast cream may advised to be applied to the vulva or oral antifungal medicines are also recommended. The creams can be used safely in pregnancy, but the tablets are best avoided. Not all itchy genital rashes are due to *Candida*, so if the treatment is unsuccessful it may be because you have another reason for pruritus vulvae or vulvodynia.

Table 2 Conventional multi-day antifungal treatments for *Candida vulvovaginitis*

Drug	Drug format	Dosage
Butoconazole	2% cream	5 grams qd hs x 1-7 days
Clotrimazole	1% cream, 100 mg tablets	5 grams qd hs or 100 mg qd hs x 1-7 days
Miconazole	2% cream, 100 mg & 200 mg suppositories	5 grams qd hs, 100 mg supp qd hs, 200 mg supp qd hs x 1-7 days
Nystatin	100,000 units vaginal tablets	1 tab/day x 7-14 days
Terconazole	0.4% cream, 0.8% cream, 80 mg suppositories	0.4% cream qd, 0.8% cream hs or 80 mg supp qd hs x 1-7 days

Table 3 One-day regimen for uncomplicated *Candida vulvovaginitis*

Drug	Drug format	Dosage
Topical		
Clotrimazole	500 mg tablets	500 mg qd x 1 day
Tioconazole	6.5% ointment	5 grams x 1 application
Butoconazole	2% cream (Sustained-released formulation)	5 grams x 1 application

Oral

Fluconazole	150 mg tablet	150 mg x 1 tablet
Itraconazole	100 mg capsules	200 mg BID x 1 day

4.3 Recurrent Candidiasis

Occasionally, the fungal infection persists despite adequate conventional therapy. In some women, this may be a sign of diabetes or immune deficiency, and appropriate tests should be done. It is now thought that women who experience recurrent vulvo-vaginal symptoms do so because of persistent yeast infection, rather than re-infection. The aim of treatment in this situation, is therefore, to avoid the overgrowth of *Candida* that leads to symptoms, rather than necessarily being able to achieve complete eradication or cure. Changes to the diet or treatment to remove yeasts in the bowel and elsewhere or treating your sexual partner do not appear to be effective in reducing recurrent episodes.

There are evidences that the following measures can be helpful:

- Cotton underwear and loose fitting clothing avoid pantihose.
- Non-soap cleanser or bath oil.
- Wipe from in front backwards and wash after going to the toilet.
- Treat before each menstrual period and before antibiotic therapy.
- Apply hydrocortisone cream to reduce itching.
- Cut out sugary foods.
- Eat acidophilus yogurt (but this sometimes aggravates thrush).
- Prolonged courses of topical antifungal agents.
- Oral antifungal treatment to remove yeasts in the bowel and elsewhere.
- Treatment of sexual partner.

4.4 Cyclic Vulvovaginitis

Cyclic vulvovaginitis is a recurrent burning and itching sensation that occurs at the same stage of every menstrual cycle. The pain specifically worsens just before or during the menstrual bleeding, or may settle during bleeding. Cyclic vulvovaginitis can lead to localised or generalised vulvodynia. Cyclic vulvovaginitis occurs in women of all ages that are still menstruating.

4.4.1 Symptoms of Cyclic Vulvovaginitis

- Between cyclical flare-ups the patient may have no symptoms.

- Intense burning irritation and itching just before or during menstrual bleeding.
- Pain can be aggravated by sexual activity and is usually worse the day after intercourse.

4.4.2 Causes of Cyclic Vulvovaginitis

Cyclic vulvovaginitis is possibly caused by a hypersensitivity to *Candida* antigen. Many women have recurrent yeast infections. Vaginal smears and cultures should be performed to determine the cause. If a culture taken during a symptomatic phase comes back negative, a swab and a scraping during the asymptomatic phase should be taken and cultured. Negative culture does not rule out cyclic vulvovaginitis as a cause of symptoms. One should report the specific strain of *Candida* and its drug sensitivities so that the most appropriate treatment may be selected.

4.4.3 Management of Cyclic Vulvovaginitis

The main goal of treatment is to remove the cause. Topical and oral antifungal agents are generally used to treat the flare-ups. The oral agents ketoconazole, itraconazole and fluconazole may also be prescribed to prevent further candidal infection, taken intermittently as recommended by the doctor. The irritation may settle with hydrocortisone cream; it is not usually helpful or desirable to use more potent topical steroids. Some patients find the tricyclic antidepressant drug amitriptyline helpful.

5. CHRONIC MUCOCUTANEOUS CANDIDIASIS

CMC is a collection of syndromes in which patients have chronic and/or recurrent infections of the skin, nails, and mucous membranes with *Candida* species (Odds 1988). Four subtypes of CMC have been described: CMC with familial susceptibility, CMC with endocrinopathy, CMC with both familial susceptibility and endocrinopathy, and CMC with onset later than 10 years of age (Higgs and Wells 1994). Several disorders may coexist or accompany patients with CMC, such as infectious diseases other than HIV infection, autoimmune diseases, endocrinopathies, thymoma, dental enamel dysplasia, vitiligo, and alopecia totalis (Krikpatrick 1994). The central defect in patients with CMC is their inability to develop effective cell-mediated immune responses against *Candida* (Lehner et al. 1972). Characteristically, patients with CMC have negative delayed hypersensitive skin reactions against *C. albicans* and decreased lymphokine production after stimulation by *Candida* antigens. Most CMC patients have normal serum and saliva immunoglobulin concentrations and normal or elevated concentrations of *Candida* serum antibodies (Budtz 1974). The importance of Th1 cytokines in resistance and Th2 in susceptibility to *Candida* infections has been demonstrated in murine models. Investigations by (Kobrynski et al. 1996) showed that a decrease in CD4+/ CD29+, T -helper inducer cells along with T -helper cell dysregulation could lead to defective memory responses to antigens in CMC patients and a decrease in cell-mediated immunity due to inhibition of Th1 cells by increased levels of IL-4. These findings were supported by Lilic et al. (1996) who found that *Candida* antigens trigger a predominantly Th2 instead of a Th1 cytokine response in patients with CMC. Treatment strategies should combine treatment with antifungal agents and immunologic reconstitution. Prolonged antifungal treatment is required and is temporarily effective with azole antifungal agents fluconazole, ketoconazole, or itraconazole, but relapses are frequently seen when therapy is stopped (Rex et al. 2000). As in OPC in HIV-infected patients, development of resistance is of concern with long-term suppressive therapy as shown for ketoconazole (Hay and Clayton 1998; Horsburgh et al. 1983).

6. CANDIDIASIS OF SKIN

Candida species, mostly *C. albicans,* but others such as *C. tropicalis, C. guilliermondii* and *C. parapsilosis* as well, can cause superficial infections in both the skin and the nails (Odds 1998). The organism has a predilection for warm and moist areas (intertriginous areas), which are created by maceration and occlusion. In HIV-infected patients candidal skin infections have been frequently observed together with other superficial mycotic infections (Aly and Berger 1996). Candidal skin infection is a common cause of diaper rash in infants. The macular erythematous rash is most marked in the intertriginous areas of the gluteal crease, perineum, and inguinal fold. A similar distribution of intertrigo occurs in the elderly and obese and diabetic, adult patients. The infection may begin in adults with vesicopustules that enlarge and progress to maceration and finally erythema. In women, inframammary areas may be affected, and frequent exposure of the hands to water may result in candidiasis of the digital folds of the hands. Itching and burning are the common symptoms. Diagnosis of cutaneous candidal skin infections is made by scraping the infected areas and examining the material with 10 to 15% potassium hydroxide (Drake et al. 1996). The most important aspect of treatment is keeping the infected areas dry and exposed to the air. Second, topical antifungal agents are the therapy of choice for uncomplicated cutaneous candidiasis. Topical azoles and clotrimazole, miconazole, and nystatin are effective when applied as creams or powder combined with or without a corticosteroid several times a day.

7. CANDIDEMIA

Candida infections other than those on the skin and mucous membranes are referred to as invasive candidiasis. Invasive candidiasis includes candidemia, acute or chronic hematogenously disseminated candidiasis, and infection of a single deep organ, either by hematogenous seeding or direct inoculation. The vast majority of cases of invasive candidiasis are the result of bloodstream invasion and hematogenous spread of the organism. Candidemia is defined as the isolation of *Candida* spp. from at least one blood culture specimen. Candidemia may or may not be accompanied by clinical signs of infection, e.g., fever, chills, or the sepsis syndrome. Candidemia may give rise to hematogenous spread of *Candida* to one or multiple organs.

7.1 Epidemiology

Candidemia has become one of the most important nosocomial infections over the last two decades. An increment in its frequency of more than 400 times as agent causing bloodstream infections, was noted during the 1980s (Baley 1981), During the 1990s *Candida* spp. were the fourth most common agent causing nosocomial bloodstream infection (Duran 1989). *C. albicans* has historically been the most frequent cause of candidemia. However, the frequency of candidemia due to non-*albicans* species of *Candida* rose during the 1990s (Poirot 1985; Welsh et al. 1995; Welsh and Ely 1999) and recent surveys find that *C. albicans* causes only about half of all cases of candidemia (Oldfield et al. 1990; Pelz et al. 2000). The species that have grown in frequency have been *C. tropicalis*, *C. glabrata*, and *C. parapsilosis* (Hamilton et al. 1991; Johnson et al. 1998). While *C. tropicalis* and *C. glabrata* have increased in frequency in the adult populations, *C. parapsilosis* seems to have a special affinity for pediatric and especially neonatal settings (Welsh et al. 1995; Law et al. 1997). Among bone marrow transplant patients, *C. krusei* has also been noted to increase in frequency (Welsh et al. 1995).

7.2 Portal of Entry

The majority of disseminated *Candida* infections originate from the gastrointestinal tract. Krause et al. 1969 reported that a healthy person with a normal gastrointestinal tract who drank a suspension containing a large amount of *Candida albicans* developed candidemia and candiduria. Gastrointestinal surgery, especially when there is intra-abdominal sepsis, may lead to neutrophil dysfunction that also contributes to susceptibility to candidemia (Simms and Amico 1969). Likewise, patients with leukemia or lymphoma receive chemotherapeutic regimens that cause serious damage to the intestinal mucosa, leading to *Candida* colonization and subsequent bloodstream invasion (Bow et al. 1995).

7.3 Clinical Manifestations

The clinical manifestations of candidemia range from nothing more than fever to overt and life-threatening sepsis. When candidemia is detected, it is also helpful to consider it in the context of one of the four overlapping forms of invasive candidiasis.

7.4 Candidemia in Patients with HIV/AIDS

Despite oropharyngeal candidiasis being the most frequent fungal infection-affecting patients with HIV and AIDS, the invasive forms of candidiasis are rare in this immunosuppressed population. Isolated case reports and a single retrospective study have been published on candidemia in HIV-infected adults (Lackner et al. 1992; Kuehnert et al. 1998). These cases have occurred in patients with end-stage-AIDS (CD4+ counts < 50 cells/ mm^3) and extensive prior therapy with fluconazole. Other than that, the risk factors for HIV infected individuals who develop candidemia are similar to those in non-HIV infected individuals: central venous catheters, parenteral nutrition and previous use of broad-spectrum antibiotics. Hence, this complication appears to represent a typical nosocomial infection rather than a condition related to the presence or absence of HIV. From a therapeutic perspective, the frequent use of azoles in HIV-infected patients means that amphotericin B should be used as the first-line antifungal agent when invasive candidiasis is seen (Lackner et al. 1992).

7.5 Candidemia in Cancer Patients

Disseminated candidiasis is one of the main causes of death in neutropenic cancer patients (Bodey et al. 1992; Bodey 1984). In autopsy surveys, the prevalence of disseminated candidiasis among cancer patients found post-mortem ranged from 2% in Western Europe to 26% in Japan (Bodey et al. 1992). A prospective, multicenter surveillance study by the Invasive Fungal

Infection Group of the European Organization for Research and Treatment of Cancer (EORTC) found a crude 3D-day mortality of 39% among candidemic cancer patients (Viscoli et al. 1999). From this study, it appears that candidemia is not limited to patients with advanced, terminal disease. Half of the patients with candidemia had non-advanced disease, and candidemia developed relatively shortly after the diagnosis of cancer (Viscoli et al. 1999). In most studies, *C. albicans* is still the most frequent cause of candidemia in patients with solid tumors, there is an increase in the prevalence of non-*albicans* species of *Candida,* predominantly among patients with hematologic malignancies (Viscoli et al. 1999).

7.6 Candidemia in Solid Organ Transplant Recipients

The combination of prolonged immunosuppressive therapy with specific surgical procedures renders solid-organ transplant recipients at high risk for developing disseminated candidiasis. Renal transplantation is associated with the lowest incidence of fungal infection of all solid-organ transplants. Candidiasis is a major problem in liver transplant patients, accounting for 15 to 30% of the life-threatening infections in this population (Collins et al. 1994). Intra-abdominal candidiasis results from the sequence of opening the colonized upper small bowel and extensive intraoperative bleeding. Specific risk factors for development of candidemia in liver transplant recipients are renal failure, length of the transplant operation, retransplantation or reoperation, and cytomegalovirus (CMV) infection (Collins et al. 1994). Candidemia occurs relatively early in the post transplant period and appears to be unrelated to the level of immunosuppression or antirejection therapy (Collins et al. 1994). A substantial percentage of candidal infections in transplant recipients may originate from the donor organ. For example, heart-lung transplant recipients have a high incidence of airway colonization with *Candida,* probably originating from the donor trachea (Kusne et al. 1988). Likewise, the donor duodenal segment used in liver or pancreas transplant procedures is frequently colonized with *Candida* spp. and may be a source of infection (Potter et al. 1989).

7.7 *Candida* Spp. Causing Candidemia

Although *C. albicans* remains the most important cause of invasive and disseminated candidiasis, the incidence of other species, such as *C. tropicalis, C. glabrata,* and *C. krusei,* has been increasing (Pfaller et al. 1999; Viscoli et al. 1999). *C. albicans* accounts for 50 to 55% of bloodstream isolates, *C. tropicalis* and *C. glabrata* for 10 to 20% each, *C. parapsilosis* for 10 to 15%, and C. *krusei* for 2 to 4%. C. *tropicalis* appears to be more virulent than C. *albicans,* as there is a much higher frequency of invasive infections among immunocompromised hosts who are colonized with C. *tropicalis* (Wingard 1995). The increased incidence of C. *tropicalis,* particularly in patients with cancer, may be due to the increased invasiveness of the organism when the gastrointestinal mucosa is damaged and the intestinal flora is suppressed by antibiotics (Viscoli et al. 1999). Candidemia due to C. *parapsilosis* is generally associated with catheters, hyperalimentation, or prosthetic devices. The propensity of C. *parapsilosis* for intravenous catheters is due to its specific properties: *in vitro,* C. *parapsilosis* proliferates better on and is more adherent to acrylic than other *Candida* spp. in a glucose-rich solution (Weems et al. 1987). Also, C. *parapsilosis* is the most common species of *Candida* to be isolated from the hands of health care workers in the ICU, especially those who wear gloves (Rangel et al. 1999).

7.8 Therapy

Treatment of candidemia has two fundamental elements. First, the most likely source of the fungemia should be identified and treated, if possible. In the non-neutropenic patient, an intravascular catheter is the most likely source and removal of such catheters is often valuable. Other common sources are the urinary tract and localized infections (Ramos et al. 2004). If possible, the local conditions at these sites should be treated e.g., eliminate urinary obstruction, drain an abscess. Second, an antifungal agent should be employed. Of the available antifungal agents, only amphotericin B and fluconazole are readily available in the parenteral formulations required for treating critically ill patients. To date, two large randomized trials (Perfect et al. 1986) and two large observational series (Nakamura et al. 2000) have examined the relative utility of these agents and found that they are generally comparable, particularly as treatment of C. *albicans* fungemia in non-neutropenic patients. Outside of this setting, the choice of therapy is less certain and follows the general rules for treatment of invasive candidiasis. The recently licensed intravenous preparation of itraconazole is currently being studied as therapy for invasive candidiasis, but no data on its relative utility are yet available.

8. URINARY CANDIDIASIS

Candiduria refers to the presence of *Candida* spp. in the urine. This finding is for the most part a benign process associated with the use of urinary catheters and antimicrobial therapy (Kao et al. 1999). Nevertheless, candiduria may be one of the most challenging of the candidal infections. The challenge comes from the fact that finding *Candida* spp. in the urine can be either completely insignificant e.g., due to contamination or asymptomatic colonization or be a marker of a very serious entity such as invasive renal parenchymal disease related to disseminated candidiasis or postlaparotomy peritonitis (Wenzel 1994; Benedetti 1996; Kao et al. 1999). In between these two extremes are other important clinical possibilities e.g., candidal cystitis and fungus balls that also require specific treatment.

8.1 Causes of Candiduria

- Contamination from external genitalia, more frequent in girls/women due to vaginal colonization, rising in colony counts tend to be $< 10^3$ cells/mL.
- Seen with indwelling urethral catheter, antibiotic treatment, old age, and pregnancy; colony counts as high as 10^5 cells/mL in catheterized patients.
- Urethritis, occurs in both men and women usually related to balanitis and vulvovaginitis, respectively.
- Cystitis related to indwelling urethral catheters and diabetes, many cases are asymptomatic; visualization via cytoscopy is the only way to confirm.
- Fungus ball formation (bladder, renal pelvis) is commonly associated with urinary tract obstruction.
- Upper urinary tract candidiasis due to ascending infection variety of pathologic findings can occur, including: papillary necrosis, perinephric abscess, emphysematous pyelonephritis and calyceal invasion, usually associated with urinary tract obstruction.
- Hematogenous candidiasis with renal involvement, upper urinary tract candidiasis as part of the syndrome of acute disseminated candidiasis. The kidney is one of the most frequently involved organs in this form of disease; colony counts in the urine may range from zero to 10^5 cells/mL.

8.2 Candiduria and *Candida* Species

Candida albicans causes ~50% of cases of Candiduria (Shaw et al. 2000). *Candida glabrata* has been consistently responsible for ~25% of cases of candiduria (Shaw et al. 2000). *Candida tropicalis* is the third most common agent. Around 5% of patients with candiduria will have 2 or more species simultaneously. (Kao et al. 1999; Bhatnagar 2004)

8.3 Clinical Manifestations

Clinical manifestations are of little help for physicians dealing with funguria. Symptoms like dysuria, frequency and urgency were presented only in 2-4% of patients with candiduria studied by Kauffman et al. 1993. Patients with urethritis and cystitis, if symptomatic, will present with the same symptomatology as is seen with a bacterial urinary tract infection. However, catheterized patients rarely have these complaints. And, when they do, it is not possible to decide if the symptoms are caused by the catheter itself or are related to lower urinary tract infection. Even patients with potentially significant candiduria are usually asymptomatic. Further confusing the situation, many patients with candiduria and fever have indeed quite convincing conditions that explain their hyperthermia e.g., pneumonia, bacteremia, bacterial urinary tract infection (Faergemann 1993). The one notable exception is that candiduria in the presence of a fungus ball of the urinary collecting system should be taken quite seriously due to the possibility of urinary obstruction.

8.4 Therapies

Fluconazole, oral or intravenous, 400 mg loading dose, then 200 mg/day for 14 days. Therapy can be possible unless the patient is unable to take oral therapies. Amphotericin B, bladder irrigation or intravenously, bladder irrigation is classically performed by irrigating with a solution of 50 mg amphotericin B in 1 litre of sterile water over a 24-48 h period. Intravenous therapy is appropriate for cases of invasive candidiasis. A group of researchers, randomized 316 patients with asymptomatic candiduria to receive fluconazole for 14 days. (Shaw et al. 2000)

Bladder irrigation with amphotericin B is a time-honored therapy of unknown value. Relapse or recurrence occurs in between 20 and 40% of cases treated with amphotericin B bladder irrigation treatments (Faergemann 1993; Hoerauf 1998). In conclusion, resolution of candiduria is accomplished in about 50% of cases using any of these treatment modalities. Failure rates seem to be higher among patients with diabetes (Kao et al. 1999). Relapses are particularly frequent among patients who remain catheterized. (Shaw et al. 2000)

9. INTRA-ABDOMINAL CANDIDIASIS

Infection of the pancreas with *Candida* spp. is another form of intra-abdominal candidiasis that was considered extremely unusual for many years but now it is being recognized as an important, albeit infrequent, problem. All cases of pancreatic candidiasis are related to some form of pancreatic injury. Indeed, its retroperitoneal location seems to protect the pancreas from hematogenous or contiguous infections, but once some type of injury has occurred a predisposition to this type of complication becomes manifest. Bacterial infections are much more frequent, with enteric microorganisms being the ones usually recovered. Cultures show a polymicrobial flora in upto 50% of cases. Candidal infection may occur either as part of a mixed infection or by itself.

The presence of *Candida* in pancreatic cultures, especially when part of a mixed flora, has often been disregarded (Karyotakis 1993). However, evidence of the importance of this organism in pancreatic infection processes continues to grow (Grauer et al. 1993). *Candida* spp. have been implicated in some cases of the following varieties of pancreatic diseases.

9.1 Acute Pancreatitis

This may occur in pancreatitis of any etiology, including gallstone pancreatitis, alcoholic, traumatic, post-surgical, post-traumatic or idiopathic. Candidal infection may be superimposed as well on any of the possible complications of pancreatitis, including:

- Necrotizing Pancreatitis (Grauer et al. 1993).
- Pancreatic Pseudocysts (Chelkowski 1987; Puccia et al. 1998).
- Pancreatic Abscesses (Reinel and Clarke 1992).

9.1.1 Intraductal Pancreatic Obstruction (Fungus ball) (Chesney et al. 1978)

The route of infection is not always clear, but the most likely alternatives include exposure to a colonized gastrointestinal tract, including contaminated bile (via reflux), exposure to exterior sources via abdominal drains or opened packing/marsupialization and hematogenous dissemination during an episode of candidemia.

9.1.2 Risk Factors

Patients with pancreatitis are frequently exposed to several of the usual risk factors for invasive candidiasis, especially:

- Use of multiple and broad spectrum antibiotics.
- Central venous lines.
- Multiple deep abdominal surgeries. Recent pancreatic surgery has been specifically associated with pancreatic candidiasis. (Wheat 1992; Buzzini and Martini 2001)
- Total parenteral nutrition.

Both endoscopic retrograde cholangiopancreatography (ERCP) (Grauer et al. 1993) and pancreatic transplant (Henderson 1981) have been associated with candidal infection of the pancreas.

9.2 Therapies

Surgical procedures: Extensive surgical debridement of retroperitoneal tissue with scheduled re-explorations is recommended for the treatment of necrotizing pancreatitis and pancreatic abscesses (Boutati and Anaissie 1997; Albisetti et al. 2004). A few authors have reported success when performing CT-guided drainage of small abscesses or pseudocysts infected with *Candida* (Chelkowski 1987).

Antifungal therapy: A review of the literature on candidal pancreatic abscesses described 100% mortality in four patients who did not receive antifungal therapy for this condition. In contradistinction, six other patients in the same report were treated with surgery plus Amphotericin B, and all six survived (Karyotakis et al. 1987). The limited published experience on antifungal treatment for pancreatic candidiasis mostly focuses on use of amphotericin B. Precise doses and length of therapy have not been studied, but we would recommend doses of at least 0.5 mg/kg/day for 10-14 days. Because of its favorable pharmacokinetics and toxicity profile, fluconazole is an attractive alternative. However, very limited experience exists on the use of this agent for pancreatic candidiasis.

10. HEPATOSPLENIC CANDIDIASIS

Hepatosplenic candidiasis also called chronic disseminated candidiasis is predominantly an infection of neutropenic patients, especially those with leukemia. Although the infection is initiated while neutropenia is present, it is frequently not detected until after neutropenia has resolved. Presenting signs and symptoms include fever, abdominal pain, hepatomegaly, and splenomegaly. Liver function tests and creative protein levels are frequently elevated. Imaging of the abdomen usually reveals multiple focal lesions in the liver, spleen and kidney. Ultrasonography is less sensitive than either computed tomography scanning or magnetic resonance imaging for detecting these lesions (Anttila et al. 1997; Sallah et al. 1998). Although finding multiple lesions in the liver and spleen in a patient with leukemia who has recently recovered from neutropenia is suggestive of hepatosplenic candidiasis, it is important to make a microbiologic diagnosis to guide the choice of therapy. For example, fungi such as *Aspergillus* spp. and *Blastoschizomyces* spp. can also cause hepatosplenic infections (Martino et al. 1990). The infecting microorganism can sometimes be isolated by blood culture. In the remaining patients, a lesion should be aspirated either percutaneously or by laparoscopy.

Treatment of hepatosplenic candidiasis requires long-term antifungal therapy; however, the optimal treatment regimen remains controversial and should be individualized to the patient (Edwards et al. 1977). Fluconazole is a reasonable choice for a patient who is clinically stable and likely infected with a susceptible organism. Some mycologists recommend treating all patients with amphotericin B for 1 to 2 weeks before switching to fluconazole. Amphotericin B is also appropriate for patients who are likely infected with an organism that is likely resistant to fluconazole. Lipid formulations of amphotericin B are attractive for the treatment of hepatosplenic candidiasis because they have a reduced chance of causing nephrotoxicity. Amphotericin B lipid complex has been used to treat hepatosplenic candidiasis in both adults and children (Walsh et al. 1997).

Once antifungal therapy is initiated, it should be continued until the lesions either resolve or calcify.

11. ENDOCARDITIS

Candida species are the most common cause of fungal endocarditis (Rubinstein and Lang 1995; Ellis et al. 2001). Risk factors for candidal endocarditis includes prosthetic valve surgery, pre-existing valvular heart disease, intravenous catheters, broad-spectrum antibiotics, intravenous drug abuse, and immunosuppression (Rubinstein and Lang 1995; Ellis et al. 2001). Prosthetic valve surgery is the greatest risk factor for candidal endocarditis, and over 50% of patients with this infection have had such surgery (Ellis et al. 2001). Endocarditis usually occurs within the first year of surgery but may occur years later.

Patients with candidal endocarditis have a clinical presentation similar to bacterial endocarditis. However, there appears to be a higher incidence of large thrombi that may embolize to the major blood vessels. In addition, it has been reported that endophthalmitis occurs more frequently (Rubinstein and Lang, 1995). Blood cultures grow *Candida* species in approximately 85% of the patients; in the remaining cases the mycologic diagnosis is made by either culture or histology of the infected valve (Ellis et al. 2001). The majority of patients are symptomatic for approximately 1 month before the diagnosis of endocarditis is made.

The standard treatment of candidal endocarditis is amphotericin B for at least 6 weeks combined with valve replacement and debridement of all infected tissue. In patients who are not surgical candidates, chronic suppressive therapy with fluconazole has been successful (Nguyen et al. 1996), and a few patients with native valve endocarditis have even been cured with fluconazole alone. Even when antifungal therapy is combined with valve replacement, there is approximately a 30% chance of relapse (Ellis et al. 2001).

12. PERICARDITIS AND MYOCARDITIS

Purulent pericarditis due to *Candida* species is a rare, life-threatening infection. Risk factors include immunosuppression, cardiac surgery, and broadspectrum antibiotics. The signs and symptoms of candidal pericarditis are non-specific, and the diagnosis is made by either culture of the pericardial fluid or histologic examination of the pericardial tissue. Treatment usually requires pericardiectomy combined with antifungal therapy, usually amphotericin B (Kraus et al. 1988).

Candidal myocarditis is also rare. It is seen most commonly in patients who are immunocompromised due to neutropenia or corticosteroids (Atkinson et al. 1984). Intravenous drug-abuse and broad-spectrum antibiotics are also predisposing factors for this infection (Cavaliere 1980; Parker 1980). In some cases, myocarditis occurs concomitantly with endocarditis. Myocarditis may also be a manifestation of hematogenously candidiasis. Arrhythmias due to abnormalities of the conduction system are common (Parker 1980). The majority of cases of candidal myocarditis are diagnosed at autopsy. However, this infection should be suspected in patients with candidemia and cardiac conduction abnormalities. These patients should receive aggressive antifungal therapy.

13. ENDOPHTHALMITIS

Candida can be implicated as causes of both exogenous endophthalmitis and endogenous endophthalmitis due to hematogenous seeding of the eye. Candidal *exogenous* endophthalmitis is rare. Reported cases have been linked to eye surgery keratoplasty, cataract extraction, traumatic ulcers, and contaminated ophthalmic irrigation solutions (Behrens et al. 1991; Kauffman et al. 1993; Langenhaeck et al. 1993). Most cases of exogenous endophthalmitis have a bacterial etiology.

Candida is recognized as the most common cause of *endogenous* endophthalmitis. Embolic seeding of *Candida* spp. into the retina may cause both retinal and vitreal lesions that can leave important visual sequela, an appropriate immune response seems to be necessary for this condition (Henderson et al. 1980). Neutropenic patients rarely develop clinically apparent candidal endogenous endophthalmitis.

13.1 Epidemiology

Upto 78% of cases of autopsy-proven invasive candidiasis show signs of retinal involvement (Edwards et al. 1974). Studies performed during the 1980s reported variable incidence of candidal endophthalmitis (range 5-78%, median 29%) among non-neutropenic populations with, or at high risk for, candidemia (Henderson et al. 1981; Brooks 1989). It seems that the early use of antifungal agents in patients with candidemia has reduced both the incidence and severity of this complication (Barza 1998). Data from a randomized trial comparing fluconazole with amphotericin B for the treatment of candidemia in non-neutropenic patients, revealed an incidence of classic progressive candidal endophthalmitis of only 1% (Rex et al. 1994).

13.2 Clinical Manifestations

Typical lesions of candidal endophthalmitis are whitish chorioretinal spots with filamentous borders protruding into the vitreous and causing vitreal haze. These lesions can be single or multiple, isolated or confluent, and preceded or not by other less specific findings, including cotton wool spots, retinal hemorrhages, and white centered hemorrhages also known as Roth spots.

In an outstanding review of 76 cases of *Candida* endophthalmitis, (Edwards et al. 1974) noticed that one-quarter of cases also had exudates with associated hemorrhages. However, because of the unspecificity of these retinal lesions and the overlap of underlying medical conditions (diabetes, hypertension, bacterial sepsis, retinopathy of AIDS, etc.) that produce the same type of lesions among patients at high risk to develop candidemia, they are of limited clinical applicability (Rodriguez et al. 1997).

Occasionally embolic lesions can affect other eye structures producing conjunctivitis (Edwards et al. 1974), episcleritis (Sorrell et al. 1984), iritis

(Meyers et al. 1973), or iris abscesses (Meyers et al. 1973). Severe vitritis and anterior chamber involvement are classically seen in advanced stages of the disease, usually related to a delay in initiation of appropriate therapy. (Aguilar 1979)

13.3 Specific Diagnostic Strategies

A funduscopic examination with pupillary dilatation should be routinely performed in patients with candidemia. While some authors have suggested use of eye examinations in cases of suspected invasive candidiasis (Wenzel 1994). It is preferable this examination be done by an ophthalmologist, as small lesions are easily overlooked.

If an eye examination is performed early during the course of invasive candidiasis, most identified lesions will be limited to the retina. If disease is found that extends beyond the retina, a diagnostic and therapeutic vitrectomy is useful. Centrifugation of vitreous specimen or passage through a millipore filter for subsequent stain and culture is recommended. Calcofluor white staining can help to identify small numbers of fungi.

Blood cultures should be performed as well. However, if the patient is presenting with symptomatic eye disease, these cultures are usually negative. Concomitant endocarditis is possible, especially among IV drug abusers. (Bisbe et al. 1992)

13.4 Therapies

All data for the treatment of candidal endophthalmitis come from case reports and animal models. No data is available to recommend with certainty which agent, procedure, or length of therapy is the best. Systemic fungal therapy as for invasive candidiasis with close ophthalmologic should follow-up. If lesions extend into the vitreous or progress into the vitreous despite adequate antifungal therapy, therapeutic pars plana vitrectomy should be considered. (Snip and Michels 1976)

Amphotericin B has poor vitreal penetration. Intravitreal administration of this drug is recommended at the time of performing vitrectomy. A dose of 5 micrograms should be used. (Stern et al. 1977)

Fluconazole penetrates freely into the eyes, specially if tissue is inflamed. Although, animal models have not been very encouraging about the efficacy of this drug (Jones et al. 1981; Akler et al. 1995). Recently reported on 16 infected eyes, 15 of which were cured after receiving fluconazole as sole therapy at ~200 mg/d for approximately two months.

14. MENINGITIS

Candidal meningitis most frequently occurs in premature infants and in patients who have had a recent neurosurgical procedure (Fernandez et al. 2000). It is seen less commonly in patients who are immunocompromised due to neutropenia, AIDS, or hereditary neutrophil function defects (Agus et al. 2000). In premature infants, meningitis arises via hematogenous seeding. Thus, the risk factors for meningitis in neonates are similar to those for hematogenously disseminated candidiasis: low birth weight, broad-spectrum antibiotics, total parenteral nutrition, and an indwelling central venous catheter (Fernandez et al. 2000). Meningitis in neurosurgical patients arises from direct inoculation of the organism into the central nervous system around the time of surgery. The majority of these patients have had either a craniotomy or cerebrospinal fluid (CSF) shunt (Grees and Gordan 1999). Additional risk factors include broad-spectrum antibiotics and antecedent bacterial meningitis (Nguyen and Yu 1995).

The clinical presentation of candidal meningitis depends on the host. In neonates, there is frequently metabolic acidosis and respiratory decompensation, which is manifested by an increase in oxygen requirement and apnea. On CSF examination, pleocytosis is inconsistent and the predominant leukocyte type can range from mononuclear cells to neutrophils. Although hypoglycorrhachia is common, it is sometimes absent. CSF Gram stain is usually negative, however, the culture grows *Candida* species in about 75% of cases. *Candida* species can also be cultured from the blood in the majority of neonates (Fernandez et al. 2000). Adults with candidal meningitis usually have a subacute onset of symptoms, which include fever, headache, and altered mental status. Neck stiffness occurs infrequently.

Candidal meningitis should be treated initially with amphotericin B (Nguyen and Yu 1995). Flucytosine should be added when feasible. However, many low-birth weight infants are unable to take oral medications safely. In these patients, the use of amphotericin B alone is curative approximately 90% of the time (Fernandez et al. 2000). Because relapses can occur, therapy should be continued for 4 weeks after clinical resolution of the infection. There is much less experience with fluconazole; it should be reserved for follow-up or suppressive therapy. Liposomal amphotericin B has been used successfully to treat a small number of neonates with candidal meningitis (Scarcella et al. 1998). In the rabbit model of candidal meningitis, liposomal amphotericin B is equivalent in efficacy to amphotericin B deoxycholate and superior to both amphotericin B lipid complex and amphotericin B colloidal dispersion (Groll et al. 2000). If prosthetic material is present in the central nervous system, it should be replaced or removed whenever possible.

15. PNEUMONIA

Candidal pneumonia is a rare infection that is frequently over diagnosed. When the infection occurs, it develops either as a result of hematogenous seeding or by aspiration of infected secretions (Haron 1993). Risk factors include neutropenia, candidal esophagitis, use of broad-spectrum antibiotics and hematologic malignancy. When the pneumonia arises from hematogenous seeding, a miliary pattern may be seen on chest radiograph. These patients may also have candidemia or evidence of infection in other organs. Patients who develop *candidal* pneumonia from aspiration of infected secretions usually have nonspecific patchy or lobar infiltrates that may be either unilateral or bilateral (Haron 1993). In these patients, the infection usually remains localized to the lung.

To diagnose candidal pneumonia, it is essential to obtain histological evidence of fungal invasion of the lung tissue (Haron 1993; Masur et al. 1977). Because *Candida* species commonly colonize the oropharynx and tracheobronchial tree of hospitalized patients, the diagnosis of candidal pneumonia cannot be made by sputum Gram stain or culture. Furthermore, even the isolation of a large number of organisms from bronchial brushings or bronchoalveolar lavage specimens in non-neutropenic patients is almost always indicative of colonization rather than true infection (Rello et al. 1998).

REFERENCES

Aguilar GL, Blumenkrantz MS, Egbert PR and McCulley JP (1979). *Candida* endophthalmitis after intravenous drug abuse. Arch Ophthalmol 97:96-100.

Akler ME, Vellend H, McNeely DM, Walmsley SL, and GoldWL (1995). Use of fluconazole in the treatment of candidal endophthalmits. Clin Infect Dis 20: 657-664.

Albisetti M, Lauener RP, Gungor T, Schar G, Niggli FK and Nadal D (2004). Disseminated Fusarium oxysporum infection in hemophagocytic lymphohistiocytosis. Infection 32:364-366.

Baddour LM (1995). Long-term suppressive therapy for Candida parapsilosis-induced prosthetic valve endocarditis. Mayo Clin Proc 70:773-775.

Baley JE, Annable WL and Kllegman RM (1981). *Candida* endophthalmitis in the premature infant. J Pediatr 98:458-461.

Bammert GF and Fostel JM (2000). Genome-wide expression patterns in *Saccharomyces cerevisiae:* Comparison of drug treatments and genetic alterations affecting biosynthesis of ergosterol. Antimicrob Agents Chemother 44: 1255-1265.

Barza M (1998). Editorial response: Treatment options for *Candida* endophthalmitis. Clin Infect Dis 27:1134-1136.

Behrens-Baumann W, Ruechel R, Zimmermann O, and Vogel M (1991). *Candida tropicalis* endophthalmitis following penetrating keratoplasty. Br. J Opthalmol 75:565.

Benedetti EAC, Gruessner C Troppmann, Papalois BE, Sutherland DE, Dunn DL, and Gruessner RW (1996). Intra-abdominal fungal infections after pancreatic transplantation: incidence, treatment, and outcome. J Amer Coll Surgeons 183:307-316.

Bhatnagar G (2004). Reproductive tract infections in women. Ph.D Thesis, Department of Microbiology, C.C.S. University, Meerut. (INDIA).

Bisbe J, Miro JM, Latorre X, Moreno A, Mallolas J, Gatell JM, Delabellacasa JP and Soriano E (1992). Disseminated candidiasis in addicts who use brown heroin - report of 83 cases and review. *Clin. Infect. Dis.* 15:910-923.

Blizzard R M, and Gibbs J H (1968). Candidiasis: studies pertaining to its association with endocrinopathies and pernicious anemia. *Pediatrics: 231-237.*

Bodey G P (1984). Candidiasis in cancer patients. *Am. J. Med.* 77(4D): 13-19.

Bodey G, Buehmann B, Duguid W, Gibbs D, Hanak H, Hotchi M, Mall G, Martino P, Meunier F, Milliken S, Naoe S, Okudaira M, Scevola D and Van't Wout J (1992). Fungal infections in cancer patients: an international autopsy survey. *Eur. J. Clin. Microbiol. Infect. Dis.* 11:99-109.

Boutati E I and Anaissie E J (1997). *Fusarium,* a significant emerging pathogen in patients with hematologic malignancy: Ten years' experience at a cancer center and implications for management. *Blood.* 90:999-1008.

Bow E J, Loewen R, Cheang M S, and Schacter B (1995). Invasive fungal disease in adults undergoing remission-induction therapy for acute myeloid leukemia: the pathogenetic role of the antileukemic regimen. *Clin. Infect. Dis.* 21:361-369.

Brooks R G (1989). Prospective study of Candida endophthalmitis in hospitalized patients with candidemia. *Arch. Intern. Med.* 149:2226-2228.

Brummer E, Castaneda E, and Restrepo A (1993). Paracoccidioidomycosis: An update. *Clin. Microbiol. Rev.* 6:89-117.

Budtz-Jorgensen E (1990). Histopathology, immunology, and serology of oral yeast infections. Diagnosis of oral candidosis. *Acta Odontol. Scand.* 48:37-43.

Buzzini P and Martini A (2001). Large-scale screening of selected *Candida* maltosa, Debaryomyces hansenii and Pichia anomala killer toxin activity against pathogenic yeasts. *Med. Mycol.* 39:479-482.

Cavaliere A (1980). *Candida myocarditis* in young heroin addict. Pathol. Res. Pract. 168; 224-228.

Chelkowski J, Samson R A, Wiewiorowska M and Golinski P 1987. Ochratoxin A formation by isolated strains of the conidial stage of Aspergillus glaucus Link ex Grey (= Eurotium herbariorum Wiggers Link ex Gray) from cereal grains. *Nahrung.* 31:267-9.

Chesney P J, Justman R A and Bogdanowica W M (1978). Candida meningitis in newborn infants: A review and report of combined amphotericin B-flucytosine therapy. *The Johns Hopkins Medical Journal.* 142:155-160.

Collins L A, Samore M H, Roberts M S, Luzzati R, Jenkins R L, Lewis W D and Karchmer A W (1994). Risk factors for invasive fungal infections complicating orthotopic liver transplantation. *J. Infect. Dis.* 170:644-652.

Drake L A, Dinehart S M, Farmer E R, Goltz R W, Graham G F, Hordinsky M K, Lewis C W, Pariser D M, Skouge J W, Webster S B, Whitaker D C, Butler B, Lowery B J, Elewski B E, Elgart M L, Jacobs P H, Lesher J L, Jr. and Scher R K. (1996). Guidelines of care for superficial mycotic infections of the skin: onychomycosis. *J. Am. Acad. Dermatol.* 34(2 Pt. 1): 116-121.

Duran J A, Malvar A and Pereiro M (1989). Fusarium moniliforme keratitis. *Acta. Ophthalmol. (Copenh).* 67:710-3.

Edwards J E, Foos RY Jr, Montgomerie J Z and Guze L B (1974). Ocular manifestations of Candida septicemia: Review of seventy six cases of hematogenous *Candida* endophthalmitis. *Medicine.* 53:47-75.

Faergemann J (1993). Pityriasis versicolor. *Semin Dermatol.* 12:276-9.

Feigal D W, Katz M H, Greenspan D, Westenhouse J, Winkelstein W, Lang W Jr, Samuel M, Buchbinder S P, Hessol N A, Lifson AR, Rutherford G W, Moss A, Osmond D, Shiboski S and Greenspan JS (1991). The prevalence of oral lesions in HIV-infected homosexual and bisexual men: three San Francisco epidemiological cohorts. *AIDS.* 5:519-525.

Goldman M, Cloud G A, Smedema M, LeMonte A, Connolly P, McKinsey DS, Kauffman C A, Moskovitz B and Wheat L J (2000). Does long-term itraconazole prophylaxis result in in vitro azole resistance in mucosal *Candida albicans* isolates from persons with advanced human immunodeficiency virus infection? *Antimicrob. Agents Chemother.* 44:1585-1587.

Gonzalez Dieguez C C, Barrios Alonso V, Torrealday Taboada H, Tamariz Martel A, Hellin Sanz T, Brito Perez J M and Quero Jimenez M (1992). Intracardiac mycetoma induced by central catheterization. *An Esp Pediatr.* 37:63-5.

Grauer M E, Bokemeyer C, Welte T, Freund M and Link H (1993). Successful treatment of Mucor pneumonia in a patient with relapsed lymphoblastic leukemia after bone marrow transplantation. *Bone Marrow Transplant.* 12:421-424.

Hamilton J R, Noble A, Denning D W and Stevens D A (1991). Performance of cryptococcus antigen latex agglutination kits on serum and cerebrospinal fluid specimens of AIDS patients before and after pronase treatment. *J. Clin. Microbiol.* 29:333-9.

Haron E, Vartivarian S, Anniassie E, Dekmezian R, Bodey GP (1993). Primary candida Pneumonia. Experience at a large cancer center and review of the literature. Medicine Baltimore.

Hay R, and Clayton Y M (1988). Fluconazole in the management of patients with chronic mucocutaneous candidosis [letter]. *Br. J. Dermatol.* 119:683-684.

Henderson D K, Edwards J E J, and Montgomerie J Z (1981). Hematogenous Candida endophthalmitis in patients receiving parenteral hyperalimentation fluids. *The Journal of Infectious Diseases.* 143:655-661.

Henderson D K, Hockey L J, Vukalcic L J and Edwards J E J (1980). Effect of immunosuppression on the development of experiemental hematogenous Candida endophthalmitis. *Infect. Immun.* 27:628-631.

Higgs J M, and Wells R S (1974). Classification of chronic mucocutaneous candidiasis with observations on the clinical picture and its therapy. *Hautarzt.* 25:159-165.

Hoerauf A, Hammer S, Muller-Myhsok B and Rupprecht H (1998). Intra-abdominal Candida infection during acute necrotizing pancreatitis has a high prevalence and is associated with increased mortality. *Crit Care Med.* 26:2010-2015.

Horsburgh C R Jr and Kirkpatrick C H (1983). Long term therapy of chronic mucocutaneous candidiasis with ketoconazole: experience with twenty-one patients. *Am. J. Med.*74:23-29.

Johnson E M, Szekely A and Warnock D W (1998). *In vitro* activity of voriconazole, itraconazole and amphotericin B against filamentous fungi. *J Antimicrob Chemother.* 42:741-745.

Kao A S, Brandt M E, Pruitt W R, Conn L A, Perkins B A, Stephens D S, Baughman W S, Reingold A L, Rothrock G A, Pfaller M A, Pinner R W and Hajjeh R A (1999). The epidemiology of candidemia in two United States cities: Results of a population-based active surveillance. *Clin Infect Dis.* 29:1164-1170.

Karus WE, Valenstein PN and Corey GR (1988). Purulent Pericarditis caused by *Candida*; report; for three cases and identification of high risk population as an aid to early diagnosis. Rev. Infect. Dis. 10:34-41.

Karyotakis N C, Anaissie E J, Hachem R, Dignani M C and Samonis G (1993). Comparison of the efficacy of polyenes and triazoles against hematogenous *Candida krusei* infection in neutropenic mice. *J. Infect. Dis.* 168:1311-1313.

Kauffman C A, Bradley S F and Vine A K (1993). Candida endophthalmitis associated with intraocular lens implantation: Efficacy of fluconazole therapy. Mycoses. 36:13-17.

Kirkpatrick C H (1994). Chronic mucocutaneous candidiasis. J. *Am. Acad. Dermatol.* 31(3 Pt. 2):S14-S17.

Kobrynski LJ, Tanimune L, Kilpatrick L, Campbell D E and Douglas S D (1996). Production of T-helper cell subsets and cytokines by lymphocytes from patients with chronic mucocutaneous candidiasis. *Clin. Diagn. Lab. Immunol.* 3: 740-745.

Kuehnert M J, Clark E, Lockhart S R, Soll D R, Chia J and Jarvis W R (1998). *Candida albicans* endocarditis associated with a contaminated aortic valve allograft: implications for regulation of allograft processing. *Clin Infect Dis.* 27:688-91.

Kusne S, Dummer J S, Singh N, Iwatsuki S, Makowka L,, Esquivel C, Tzakis A G, Starzl T E and Ho M (1988). Infections after liver transplantation. An analysis *of* 101 con secutive cases. *Medicine* (Baltimore) 67:132-143.

Lackner H, Schwinger W, Urban C, Muller W, Ritschel E, Reiterer F, Kuttnig-Haim M, Urlesberger B and Hauer C (1992). Liposomal amphotericin-B (AmBisome) for treatment of disseminated fungal infections in two infants of very low birth weight [see comments]. *Pediatrics.* 89:1259-61.

Langenhaeck M, and T. Zeyen (1993). Fungal endophthalmitis: A report of two cases. Bull. Soc. Belge. *Ophthalmol.* 250:63-66.

Law D, Moore C B and Denning D W (1997). Activity of SCH 56592.

Lehner T, Wilton J M and Ivanyi L (1972). Immunodeficiencies in chronic muco-cutaneous candidosis. *Immunology* 22:775-787.

Lilic D, Cant A J, Abinun M, Calvert J E and Spickett G P (1996). Chronic mucocutaneous candidiasis. I. Altered antigen-stimulated IL-2, IL-4, IL-6 and interferongamma (IFN-gamma) production. *Clin. Exp. Immunol.* 105: 205-212.

Maenza J R, Merz W G, Romagnoli M J, Keruly J C, Moore R D and Gallant J E (1995). Prevelance of fluconazole-resistant candidiasis in HIV-infected patients. *33rd Annual Meeting of the Infectious Diseases Society of America,* Abstract No. 323.

Marco F, Pfaller M A, Messer S A and Jones R N (1998). Activity of MK-0991 (L-743,872), a new echinocandin, compared with those of LY303366 and four other antifungal agents tested against blood stream isolates of Candida spp. *Diagn Microbiol Infect Dis.* 32:33-37.

Marriott D J, Jones P D, Hoy J F, Speed B R and Harkness J L (1993). Fluconazole once a week as secondary prophylaxis against oropharyngeal candidiasis in HIV-infected patients. *Med. J. Aust.* 158:312-316.

Martins M D, Lozano-Chiu M and Rex J H (1997). Point prevalence of carriage of fluconazole-resistant *Candida* in the oropharynx of HIV-Infected patients. *Clin. Infect. Dis.* 25:843-846.

Masur H, Rosen PP, ana Armstrong D (1977). Pulmonary disease caused by *Candida* species. Am. J. Med. 63:914-925.

Meyers B R, Lieberman T W, and Ferry A P (1973). Candida endophthalmitis complicating candidemia. *Ann. Intern. Med.* 79:647-653.

Nguyen M H and Yu V L (1995). Meningitis caused by Candida species: an emerging problem in neurosurgical patients. *Clin. Infect. Dis.* 21:323-327.

Nguyen MT, Wiess PJ, La Barre RC, Miller LK, Oldfield EC and Wallace MR (1996). Orally transmitted amphotericin B in the treatment of oral Candidiasis in HIV infected patients by azole resistant *Candida albicans* [letter]. AIDS 10:1745-1747.

Odds F C (1998). Candidosis of the skin, nails and other external sites, p. 136-142. In *Candida* and Candidosis. *A Review and Bibliography,* 2nd ed. Balliere Tindall, London, United Kingdom.

Odds F C (1988). Chronic mucocutaneous candidosis, p. 143-152. In *Candida* and *Candidosis. A Review and Bibliography,* 2nd ed. Balliere Tindall, London, United Kingdom.

Oldfield E C 3rd, Garst P D, Hostettler C, White M and Samuelson D (1990). Randomized, double-blind trial of 1- versus 4-hour amphotericin B infusion durations. Antimicrob. *Agents Chemother.* 34:1402-1406.

Pelletier R, Peter J, Antin C, Gonzalez C, Wood L and Walsh T J (2000). Emergence of resistance of *Candida albicans* to clotrimazole in human immunodeficiency virus-infected children: *In vitro* and clinical correlations. *J Clin Microbiol.* 38:1563-8.

Pelz R K, Hendrix C W, Swoboda S M, Merz W G and Lipsett P A (2000). A double-blind placebo controlled trial of prophylactic fluconazole to prevent *Candida* infections in critically ill surgical patients. *Crit. Care Med.* 27 (Suppl.):A33 (abstract 43).

Perfect J R, Savani D V and Durack D T (1986). Comparison of itraconazole and fluconazole in treatment of cryptococcal meningitis and *Candida pyelonephritis* in rabbits. Antimicrob. Agents Chemother. 29:579-583.

Pfaller M A, Messer S A, Hollis R J, Jones R N, Doern G V, Brandt M E and Hajjeh R A (1999). Trends in species distribution and susceptibility to fluconazole among blood stream isolates of *Candida* species in the United States. *Diagn. Microbiol. Infect. Dis.* 33:217-222.

Poirot J L, Laporte J P, Gueho E, Verny A, Gorin N C, Najman A, Marteau M and Roux P (1985). A propos d'un cas de mycose profonde a Fusarium verticilloides [A case of deep mycosis due to Fusarium verticilloides]. Medecine et Maladies Infectieuses. 10:529-532.

Potter van Loon B J, Lichtendahl A T, Mulder S S, Van Houten H, Gooszen H G and Van den Broek PJ (1989). Microbial contamination of donor duodenum in whole-pancreas transplantation. *Diabetes* 38(Suppl. 1):242-243.

Puccia R, Carmona A K, Gesztesi J L, Juliano L and Travassos L R (1998). Exocellular proteolytic activity of Paracoccidioides brasiliensis: cleavage of components associated with the basement membrane. Med Mycol. 36:345-348.

Ramos J M, Cuenca-Estrella M, Gutierrez F, Elia M and Rodriguez-Tudela J L (2004). Clinical case of endocarditis due to Trichosporon inkin and antifungal susceptibility profile of the organism. *J Clin Microbiol.* 42:2341-2344.

Rangel-Frausto M S, Wiblin T, Blumberg H M, Saiman L, Patterson J, Rinaldi M, pfaller M, Edwards J E, Jr, arvis WJ, Dawson J, and R. P. Wenzel. 1999. National epidemiology of mycoses survey (NEMIS): variations in rates of bloodstream infections due to *Candida* species in seven surgical intensive care units and six neonatal intensive care units. *Clin. Infect. Dis.* 29:253-258.

Reinel D and Clarke C (1992). Comparative efficacy and safety of amorolfine nail lacquer 5% in onychomycosis, once-weekly versus twice-weekly. *Clin Exp Dermatol.* 17 (Suppl 1):44-49.

Rello J, Esandi ME, Diaz E, Mariscal D, Gallego M, Valles J (1998). The role of the *Candida* sp. isolated from bronchoscopic sample in non-neutropenic patients. Ches + 114:146-149.

Rex J H and Walsh T J (1999). Editorial response: Estimating the true cost of amphotericin B. *Clin Infect Dis.* 29:1408-1410.

Rex J H, Bennett J E, Sugar A M, Pappas P G, Van der Horst C M, Edwards J E, Washburn R G, Scheld W M, Karchmer A W, Dine A P, Levenstein M J, Webb C D, the Candidemia Study Group, and the NIAID Mycoses Study Group (1994). A randomized trial comparing fluconazole with amphotericin B for the treatment of candidemia in patients without neutropenia. *N. Engl. J. Med.* 331:1325-1330.

Rex J H, Walsh T J, Sobel J D, Filler S G, Pappas P G, Dismukes W E, and Edwards J E (2000). Practice guidelines for the treatment of candidiasis. *Clin. Infect. Dis.* 30: 662-678.

Rubinstein E and Lang R (1995). Fungal endocarditis Eu. Heart. J. 16:84-89.

Ryder N S, Wagner S and Leitner I (1998). *In vitro* activities of terbinafine against cutaneous isolates of *Candida albicans* and other pathogenic yeasts. *Antimicrob. Agents Chemother.* 42:1057-1061.

Sangeorzan J A, Bradley S F, He X, Zarins L T, Ridenour G L, Tiballi R N and Kauffman C A (1994). Epidemiology of oral candidiasis in HIV-infected patients: Colonization, infection, treatment and emergence of fluconazole resistance. *Am. J. Med.* 97:339-346.

Shaw C K L, McCleave M and Wormald P J (2000). Unusual presentations of isolated sphenoid fungal sinusitis. *J Laryngol Otol.* 114:385-388.

Simms H H, and D'Amico R (1992). Intra-abdominal sepsis alters tumor necrosis factor-alpha and interleukin-1 beta binding to human neutrophils. *Crit. Care Med.* 20:11-16.

Snip R C and Michels R G (1976). Pars plana vitrectomy in the management of endogenous Candida endophthalmitis. *A S O.* 82:699-704.

Sobel J D and Vazquez J A (1999). Fungal infections of the urinary tract. *World J Urol.* 17:410-414.

Sorrell T C, Dunlop C, Collignon P J and Harding J A (1984). Exogenous ocular candidiasis associated with intravenous heroin abuse. *Br J Ophthalmol.* 68:841-5.

Stern G A, Fetkenhour C L and O'Grady R B (1977). Intravitreal amphotericin B treatment of Candida endophthamitis. *Arch. Ophthalmol.* 95:89-93.

Sugar A M, Saunders C and Diamond R D (1990). Successful treatment of Candida osteomyelitis with fluconazole. A non-comparative study of two patients. *Diagn. Microbiol. Infect. Dis.* 13:517-520.

Viscoli C, Girmenia C, Marinus A, Collette L, Martino P, Lebau B, Spence D, Krcmery V, De Pauw B E, Meunier F and The Invasive Fungal Infection Group of the EORTC (1999). Candidemia in cancer patients. A prospective, multicenter surveillance study in Europe by the Invasive Fungal Infection Group (IFIG) of the European Organization for Research and Treatment of Cancer (EORTC). *Clin. Infect. Dis.* 28:1071-1079.

Weems J, Jr, Chamberland M E, Ward J, Willy M, Padhye A A and Solomon S L (1987). *Candida parapsilosis* fungemia associated with parenteral nutrition and contaminated blood pressure transducers. *J. Clin. Microbiol.* 25: 1029-1032.

Welsh O, Salinas M C and Rodriguez M A (1995). Treatment of eumycetoma and actinomycetoma. *Curr Top Med Mycol.* 6:47-71.

Welsh R D and Ely R W (1999). Scopulariopsis chartarum systemic mycosis in a dog. *J Clin Microbiol.* 37:2102-2103.

Wenzel R P (1994). Epidemiology of nosocomial *Candida* infections. *Infect. Dis. Clin. Pract.* 3 (Suppl. 2): S56-S59.

Wheat L J, Connolly-Stringfield P, Williams B, Connolly K, Blair R, Bartlett M and Durkin M (1992). Diagnosis of histoplasmosis in patients with the acquired immunodeficiency syndrome by detection of histoplasma capsulatum polysaccharide antigen in bronchoalveolar lavage fluid. *Am. Rev. Respir. Dis.* 145:1421-1424.

Wingard J R (1995). Importance of *Candida* species other than *C. albicans* as pathogens in oncology patients. *Clin. Infect. Dis.* 20:115-125.

Control of *Candida* Infections and Mechanism of Action of Respective Drugs

R. GAUTAM and A. P. GARG

1. INTRODUCTION

Over the past two decades many *non-albicans Candida* (NCAC) species have emerged as significant pathogens of clinical importance (Fridkin and Jarvis 1996; Hazen 1995; Pfaller and Wezel 1992). The NCAC species are heterogeneous group of organisms that are fundamentally different from each other and from *C. albicans* at the biological level (Barnett et al. 2000; De Hoog and Guarra 1995). This diversity in the range of *Candida* species now associated with human infection has provided new challenges in the diagnosis and treatment of candidiasis and in the study of their virulence and biology. Although close to 200 *Candida* species have been described, relatively few of these are of medical significance (Barnett et al. 2000; De Hoog and Guarra 1995; Odds 1988; Odds 1987).

In the early 20th century, *C. albicans* was considered the only *Candida* species of real medical importance, although *C. stellatoidea* (now considered synonymous with *C. albicans*), *C. parapsilosis*, *C. tropicalis*, and *C. guilliermondii* were considered occasional pathogens (Hazen 1995; Odds 1988). However, during the 1980s and 1990s, the frequency with which these organisms were recovered from sites of infection increased and species that were previously considered non-pathogenic, such as *C. glabrata*, *C. krusei*, and *C. lusitaniae* emerged as pathogens (Pfaller and Wezel 1992). In

Department of Microbiology, C.C.S. University, Meerut-250005, India addition in 1995, a novel *Candida* species, called *C. dubliniensis,* was identified for the first time. The reasons for this shift in the epidemiology of *Candida* infections may be linked to the general increase in the incidence of mycoses that occurred from the 1960s onward (Banerjee et al. 1991; Jarvis 1995). The advent of new medical therapies and procedures to treat cancer, the increase in invasive medical procedures, and the widespread use of broad-spectrum antibiotics have all been linked with this increase in fungal infections. The emergence of human immunodeficiency virus (HIV) and AIDS has also increased the incidence of mucosal candidiasis caused by *Candida* species (Coleman et al. 1993; Fischer-Hoch and Hutwanger 1995).

Many of these organisms are commensals, or at least transient commensals of the human alimentary canal. *C. albicans,* the most common human commensal of the genus *Candida,* is generally regarded as the most virulent of the *Candida* species. The NCAC species are generally less pathogenic or non-pathogenic, and accordingly, infections caused by these species are usually seen only in severely immunocompromised hosts or in individuals with serious underlying disease (Pfaller et al. 1999; Coleman et al. 1993). Hence, the increased longevity of severely immunologically compromised individuals in modern hospitals has allowed many of these species to emerge and cause disease.

2. ANTIFUNGAL DRUGS AND THEIR MODE OF ACTION

Antifungal treatments against *Candida* infections are hampered by several factors including the limited number of active agents, the emergence of refractory fungal species and the development of resistance. This situation has triggered the search for new antifungal agents with novel modes of action. Different cellular processes involved in the biosynthesis of components required for the growth of fungal cells have been targeted by antifungal agents (Fig. 1).

Ergosterol biosynthesis is specific to fungi and is necessary for their growth; scientists responsible for the design and isolation of antifungal agents have largely exploited this feature. As a result, most of the antifungal agents used in medicine target the ergosterol biosynthetic pathway and include polyenes, azoles, allylamines, or morpholine derivatives (Fig. 2). Among currently used antifungal agents, 5-fluorocytosine (5-FC) is one of the few that does not directly interfere with the ergosterol biosynthetic pathway.*

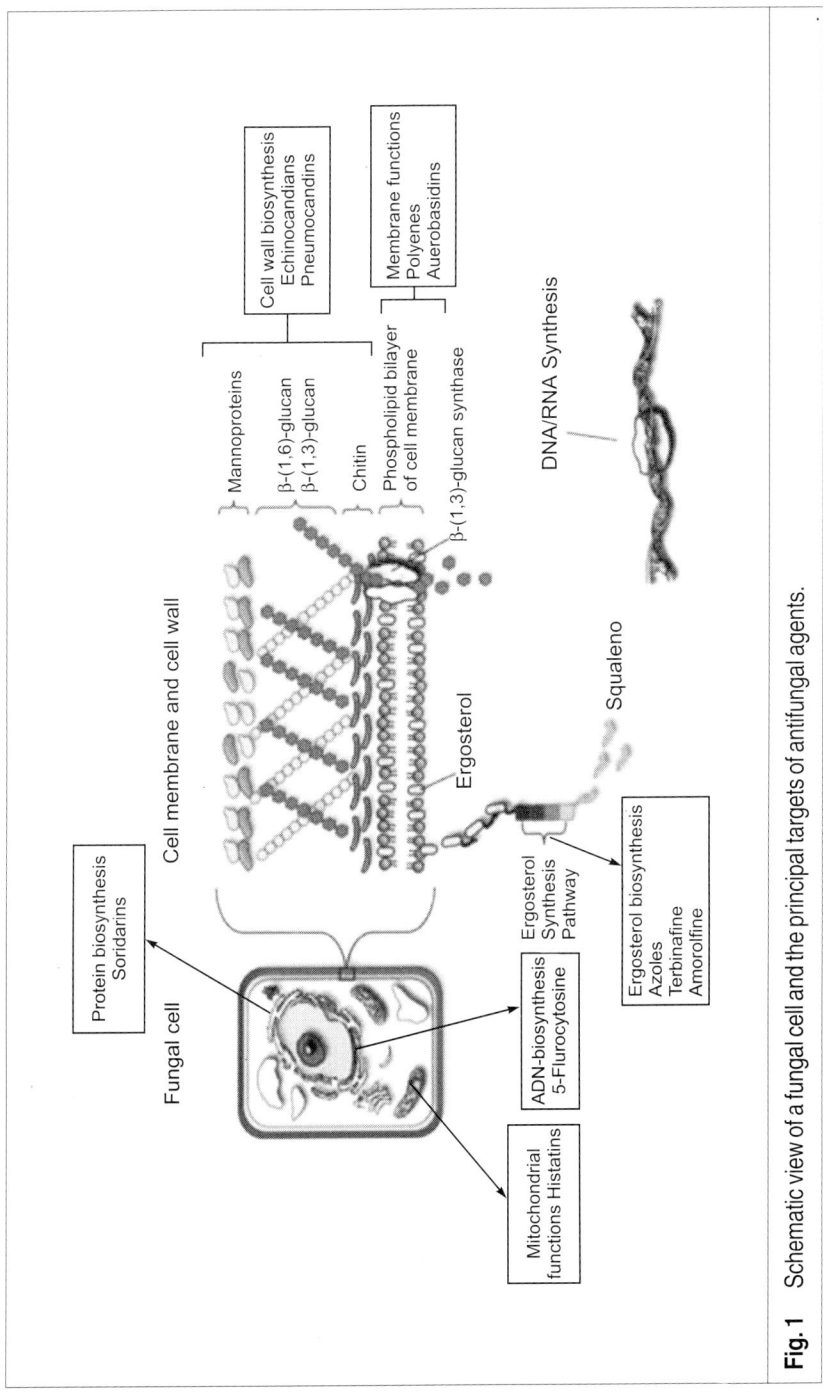

Fig. 1   Schematic view of a fungal cell and the principal targets of antifungal agents.

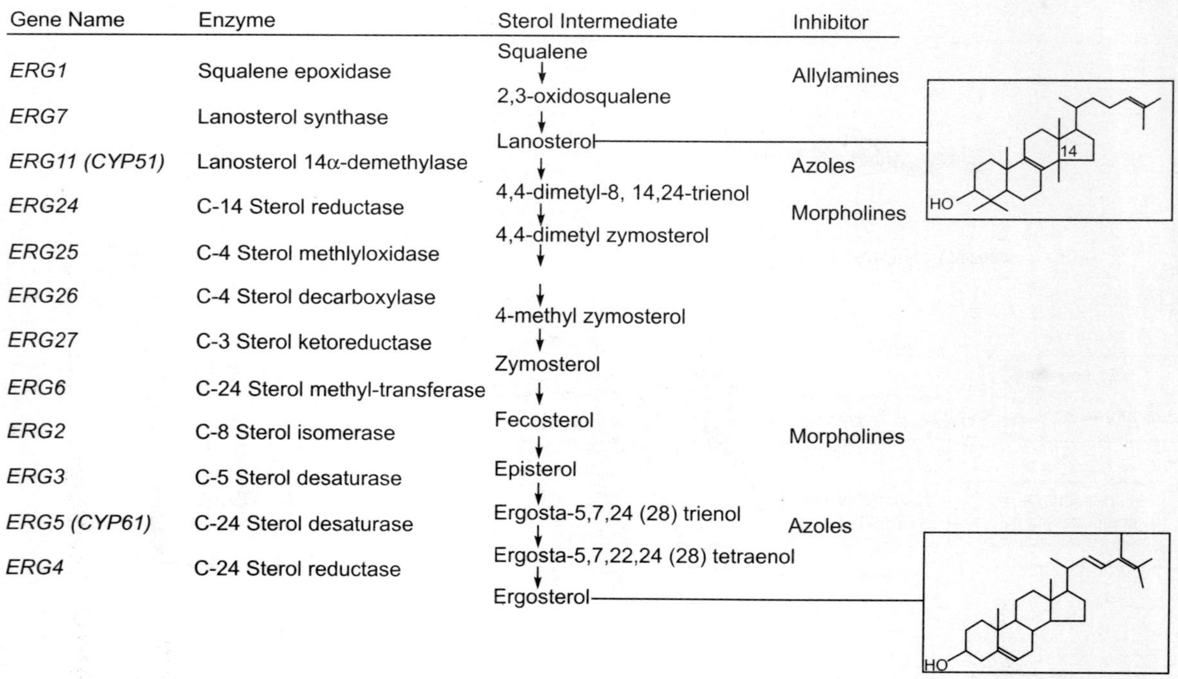

Gene Name	Enzyme	Sterol Intermediate	Inhibitor
ERG1	Squalene epoxidase	Squalene	Allylamines
		↓ 2,3-oxidosqualene	
ERG7	Lanosterol synthase	↓	
ERG11 (CYP51)	Lanosterol 14α-demethylase	↓ Lanosterol	Azoles
ERG24	C-14 Sterol reductase	↓ 4,4-dimetyl-8, 14,24-trienol	
ERG25	C-4 Sterol methyloxidase	↓ 4,4-dimetyl zymosterol	Morpholines
ERG26	C-4 Sterol decarboxylase	↓	
ERG27	C-3 Sterol ketoreductase	↓ 4-methyl zymosterol	
ERG6	C-24 Sterol methyl-transferase	↓	
ERG2	C-8 Sterol isomerase	↓ Zymosterol	Morpholines
ERG3	C-5 Sterol desaturase	↓ Fecosterol	
ERG5 (CYP61)	C-24 Sterol desaturase	↓ Episterol	Azoles
ERG4	C-24 Sterol reductase	↓ Ergosta-5,7,24 (28) trienol	
		↓ Ergosta-5,7,22,24 (28) tetraenol	
		↓ Ergosterol	

Fig. 2 Ergosterol biosynthesis pathway as a target of antifungal agents (Sanglard and Bille 2002).

2.1 Polyenes

Polyenes belong to a class of natural antifungal compounds discovered in the early 1950s. One of the most successful polyene derivatives, amphotericin B (AmB) (Fig. 2) is produced by *Streptomyces nodosus* (Gallis et al. 1990). AmB can form soluble salts in both basic and acidic environments, is neither orally nor intramuscularly absorbed, and is virtually insoluble in water. Although AmB is an effective antifungal drug, it possesses several drawbacks limiting its clinical utility. Systemic and renal problems are often encountered with AmB (Aramwit et al. 2000). To reduce its unwanted side effects, AmB has been formulated in liposomes (Adler 1994), lipid complexes

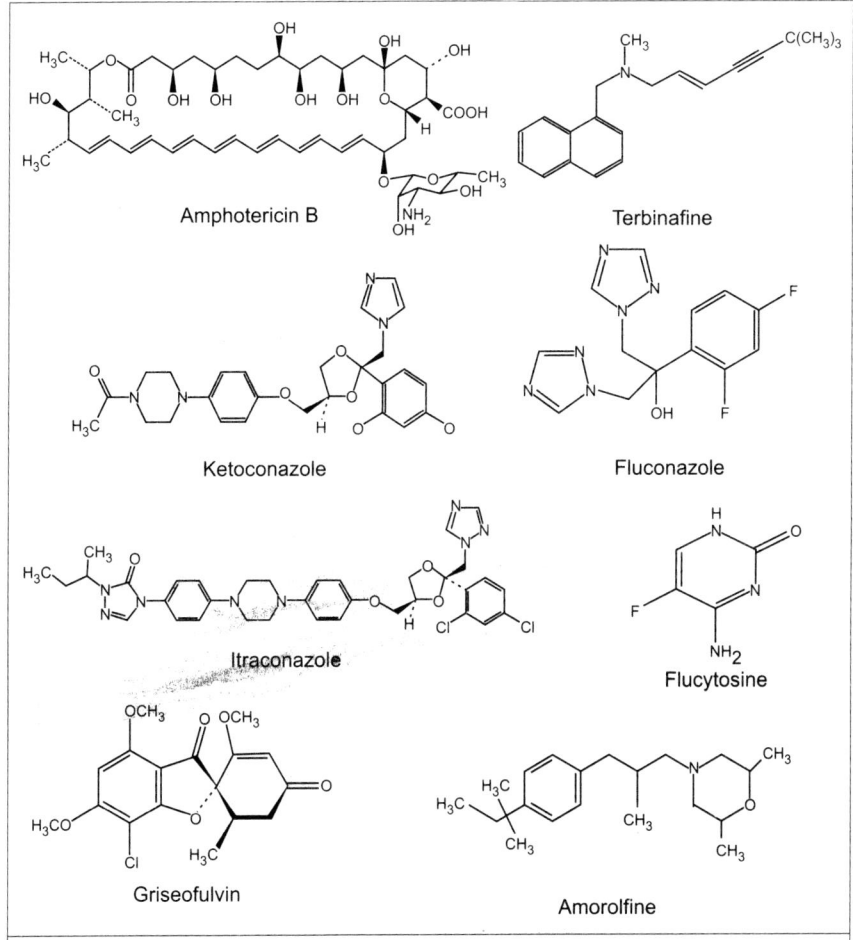

Fig. 3 Chemical structure of antifungal agents in current use (Sanglard and Bille 2002).

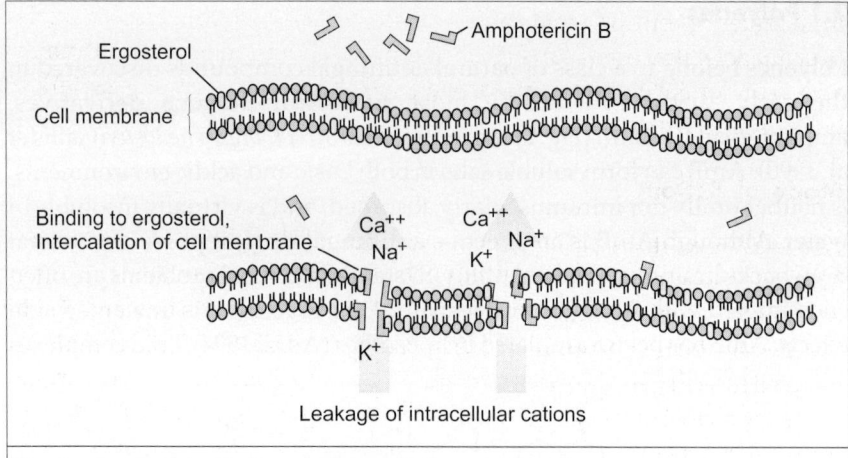

Ergosterol

Cell membrane

Amphotericin B

Binding to ergosterol, Intercalation of cell membrane

Ca^{++} Ca^{++}
Na$^+$ Na$^+$
                    K$^+$

K$^+$

Leakage of intracellular cations

Fig. 4   Amp B act by binding to ergosterol in fungal cell membrane

(Rapp et al. 1997), and colloidal suspensions (Hostetler et al. 1992) to allow the use of higher doses of AmB against fungal pathogens and at the same time reduce its toxic effects to mammalian cells.

Mode of Action

The primary mode of action of polyenes including AmB or nystatin is to bind ersgosterol in the membrane bilayer of susceptible fungi (Fig. 4). This interaction is thought to result in the production of aqueous pores consisting of polyene molecules linked to the membrane sterols. This configuration gives rise to a pore-like structure in which the polyene hydroxyl residues face inward, leading to altered permeability, leakage of vital cytoplasmic components (mono- or divalent cations), and death of the organism (Bolard 1991). Some lines of evidence also suggest that AmB exerts oxidative damage to the cell, contributing to its toxic effect (Sokol et al. 1986). Polyenes can bind also other sterols such as cholesterol, which accounts for much of their human toxicity (Bolard et al. 1980). The advantage of AmB, however, is its much higher affinity for ergosterol-containing membranes than for cholesterol-containing membranes.

2.2 Azoles

Azole antifungal agents discovered in the late 1960s are synthetic compounds belonging to the largest group of antifungal agents. Azole antifungal agents used in medicine are categorized into N-1 substituted imidazoles (ketoconazole, miconazole, clotrimazole) and triazoles (fluconazole, itraconazole) (Fig. 3). The new generation of azole antifungal agents under development (posaconazole, ravuconazole, voriconazole) also belongs to triazoles.

Mode of Action

Azoles have a cytochrome P450 as a common cellular target in yeast or fungi. This cytochrome P450 is named CYP51A1, according to an internationally recognized nomenclature (Nelson et al. 1996), and is involved in the 14a-demethylation of lanosterol (Vanden et al. 1994). CYP51A1 is also now referred to as Ergllp, the product of the *ERGll* gene. The unhindered nitrogen of the imidazole or triazole ring of azole antifungal agents binds to the heme iron of the cytochrome P450 as a sixth ligand, thus inhibiting the enzymatic reaction. The affinity of imidazole and triazole derivatives is not only dependent on this interaction, but is also determined by the N-1 substituent, which is actually responsible for the high affinity of azole antifungal agents to their target (Vanden et al. 1993). Each of these agents has distinct pharmacokinetics, and antifungal efficacy is quite different between yeast and fungal species of medical relevance (Lyman and Walsh 1992). Recently, another target for azole antifungal agents was described as sterol Li22-desaturase, which is a cytochrome P450 enzyme involved in the last step of ergosterol biosynthesis, i.e., the saturation of ergosta 5, 7-dienol into ergosterol (Kelly et al. 1997) (Fig. 2). Another report from Vanden Bossche et al. 1993 suggests an additional target of azole antifungal agents, which is an NADPH-dependent 3-ketosteroid reductase that catalyzes the last step of demethylation at the C-4 methyl group of yeast sterols (Fig. 5) Accumulation of 3-keto steroids (obtusifolione or 14a-methlylfecosterone) has been detected in *Histoplasma capsulatum* and *C. neoformans* upon incubation of cells with itraconazole as a result of NADPH-dependent 3-ketosteroid reductase inhibition and provides until now only indirect evidence that azole antifungal agents inhibit this enzyme (Vanden et al. 1990).

2.3 5-FC

5-FC (Fig. 3) belongs to the class of pyrimidine analogs that was developed in the 1950s as a potential antineoplastic agent. Abandoned as an anticancer drug due to its lack of activity against tumors, it showed, however, good *in vitro* and *in vivo* antifungal activity. Because it is highly water-soluble, it can be administered by oral and intravenous (IV) routes (Polak 1990).

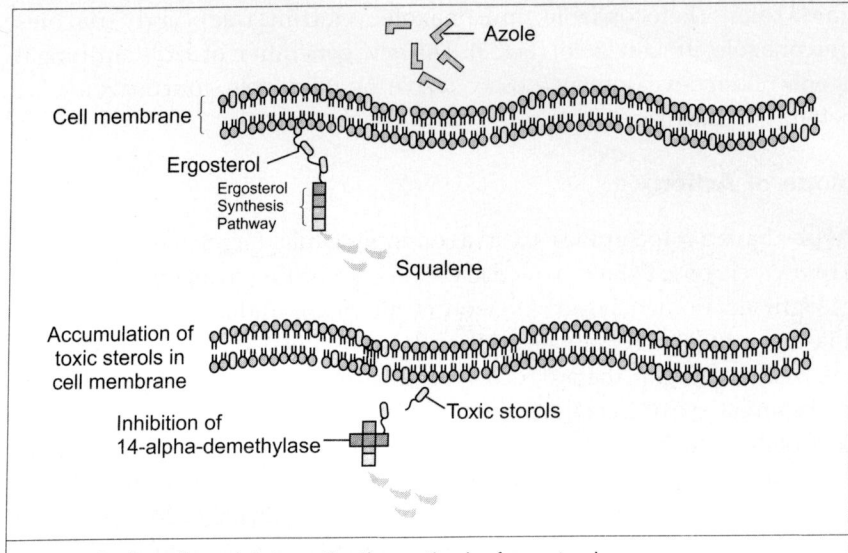

Fig. 5   Azole antifungals interrupting the synthesis of ergosterol.

Mode of Action

5-FC is taken up by fungal cells by a cytosine permease and is deaminated by a cytosine deaminase to 5-fluorouracil (5-FU), a potent antimetabolite. 5-FU can be converted to a nucleoside triphosphate and, when incorporated into RNA, causes miscoding. In addition, 5-FU can be converted to a deoxynucleoside, which inhibits thymidylate synthase and, thereby DNA synthesis. 5-FC shows little toxicity in mammalian cells, since cytosine deaminase is absent or poorly active in these cells. 5-FU is, however, a potent anticancer agent but is impermeable to fungal cells (Diasio et al. 1978). The conversion of 5-FC to 5-FU is possible by intestinal bacteria, and therefore, 5-FC can show toxicity in oral formulations (Diasio et al. 1978). (Figure 6)

2.4 Allylamines

The allylamines are synthetic compounds discovered in the 1970s. Terbinafine and naftifine (Fig. 3) belong to this class of antifungal agents. Other compounds structurally related to allylamines have been described (butenafine, a benzylamine; tolnaftate, a thiocarbamate). Historically, the antifungal activity of allylamines began by accident, when it was discovered that naftifine, the first allylamine synthesized by Sandoz Research, had antifungal activity although it was first identified by a central nervous system discovery program.

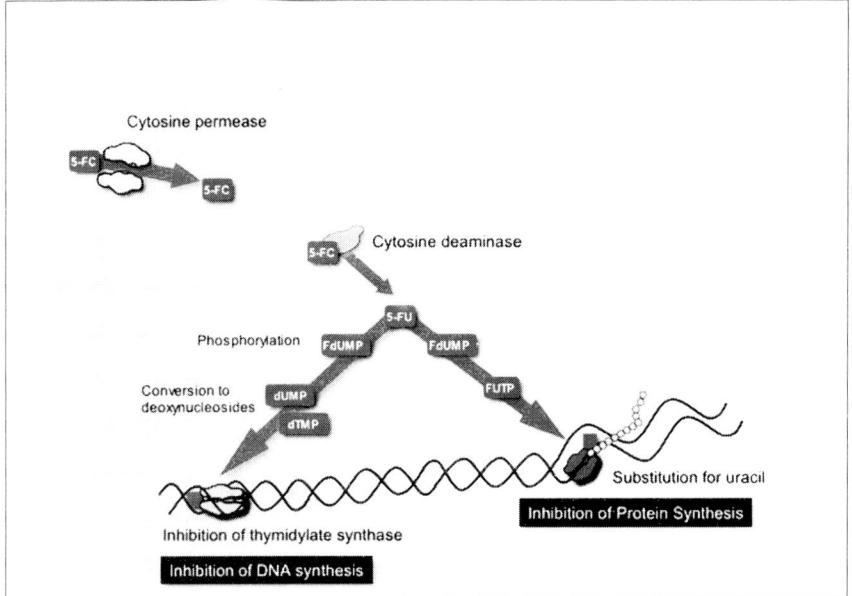

Fig. 6 5-FU is phosphorylated and incorporated into RNA where it causes miscoding and halts protein synthesis. (5-FC, 5—fluorocytosine; 5-FC, 5-flurouracil; FdUMP, 5-fluorodeoxy uridine; FUMP, 5-fluorouridine monophosphate; FUDP, 5-fluorouridine diphosphate; FUTP, 5-fluorouridine triphosphate; dUMP, deoxyuridine monophosphate; dTMP, deoxythymidine monophosphate)

Mode of Action

Allylamines, as well as benzylamines and thiocarbamates, target squalene epoxidase, which is the first postsqualene enzyme of the ergosterol biosynthetic pathway (Fig. 2). This enzyme is encoded by *ERG1*, an essential gene in *Saccharomyces cerevisiae* (Pfaller *et al.* 1999). *ERG1* has been isolated in *C. albicans* (Favre and Ryder 1997). Squalene epoxidase is a membrane-bound microsomal enzyme and requires molecular oxygen, NADPH, and flavin adenine dinucleotide for activity. Allylamines are reversible, non-competitive inhibitors of squalene epoxidase. The inhibitory effect of allylamines in yeast and fungi is thought to be due to ergosterol depletion and squalene accumulation. Among allylamines, benzylamines, and thiocarbamates, only terbinafine is used as a systemic agent.

2.5 Morpholines

The morpholines (amorolfine, fenpropimorph, tridemorph) (Fig. 3) are totally synthetic compounds that (with the exception of amorolfine) are mostly used in agricultural fungicide preparations.

Mode of Action

Amorolfine inhibits two enzymes in a single biosynthetic pathway. The enzymes involved are the sterol Δ^{14} reductase and the Δ^{7-8} isomerase, which reside downstream of the azole target enzyme (Fig. 2). These enzymes are encoded by the *ERG24* and *ERG2* genes, respectively (Hartman and Sanglard 1997). Current data suggest that the primary target of amorolfine is the sterol 814 reductase, because of the earlier position of this enzyme in the ergosterol biosynthetic pathway, even if the sterol Δ^{14} reductase exhibits IC_{50} is much higher than measured for the sterol Δ^{7-8} isomerase (Mercer 1991). Inhibition of yeast growth by morpholine derivatives results in the formation of ignosterol, a sterol by-product occurring when the *ERG24* gene is deleted in *S. cerevisiae* (Crowley et al. 1996).

2.6 Griseofulvin

Griseofulvin (Fig. 3) is the unique representative of a class of drugs produced by the molds *Penicillium griseofulvum* and *Penicillium janczewski*. Griseofulvin is one of the first antifungal agents discovered, as early as 1939 (Oxford et al. 1939). Its use as an orally active antifungal agent against ringworm in humans was first described by Williams et al. 1958.

Mode of Action

Griseofulvin inhibits mitosis by interfering with microtubule function. It does not bind directly to tubulin but instead interacts with microtubule-associated proteins (Kerridge and Vanden, 1990).

3. ANTIFUNGALS WITH ESTABLISHED MODE OF ACTION

Among the antifungal agents in development with an established mode of action, azole antifungal agents have a preponderant position. In an attempt to increase the efficacy of azoles, several new derivatives are currently being developed and they exhibit enhanced spectra of activity and improved pharmacokinetics. The agents at the front of the developmental process are three azole derivatives, i.e., voriconazole, posaconazole and ravuconazole (Fig. 7).

3.1 Voriconazole

Voriconazole (Fig. 7) is the most advanced investigational new azole agent. It is structurally related to fluconazole, however, it is more potent in

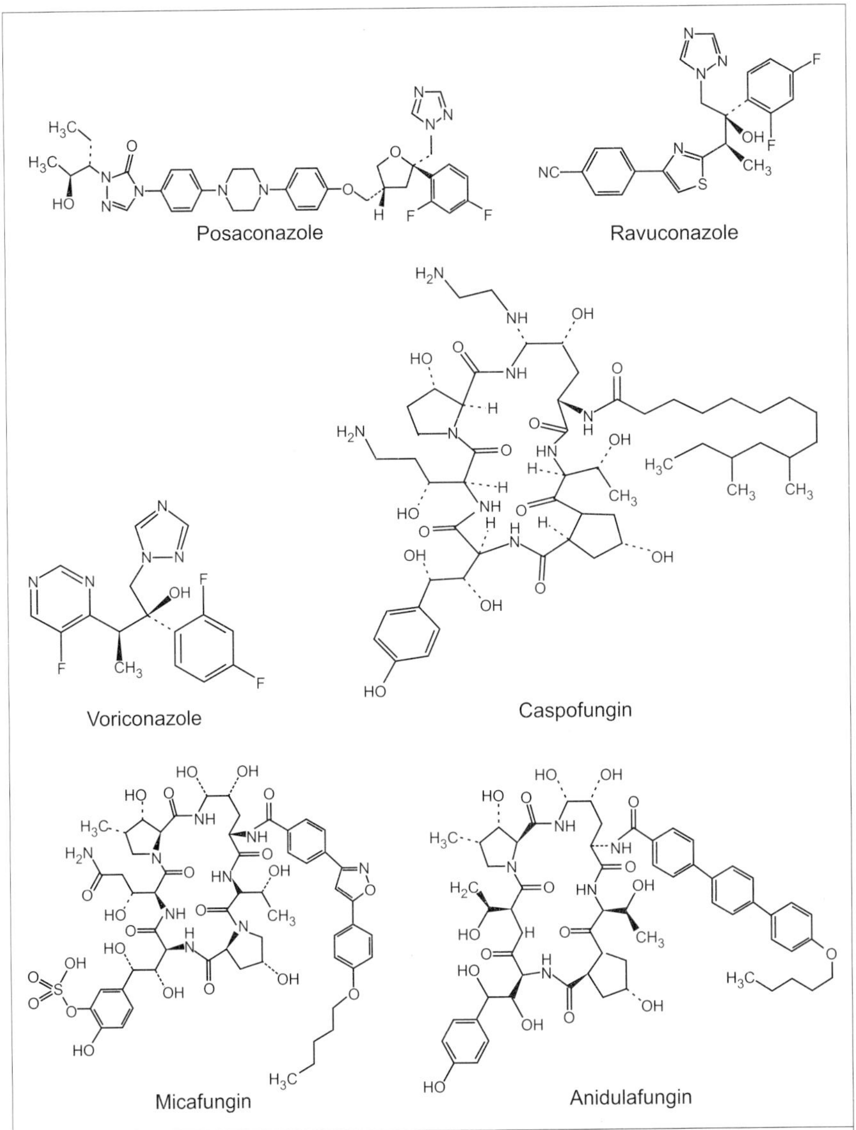

Fig. 7 Chemical structure of investigative antifungal agents (Sanglard and Bille 2002).

inhibiting *in vitro* ergosterol biosynthesis in *C. albicans*, *C. krusei*, or *A. fumigatus* than its companion compound. The *in vitro* activity of voriconazole against *Candida* species is excellent with an MIC_{90} in the range of 0.5 µg/ml. Voriconazole is most active against *C. albicans* (MIC_{90}, 0.06µg/

ml) and is least active against *C. glabrata* (MIC$_{90}$, 2.0 µg/ml). Voriconazole is active against dimorphic fungi (*Blastomyces dermatitidis C. immitis* and *H. capsulatum*) and against a wide range of filamentous fungi and dermatophytes, especially *A. fumigatus* (Espinel 1998). Voriconazole is available as IV and oral administrations. Following oral administration, voriconazole is rapidly and extensively absorbed, with a bioavailability of 90%, a time to maximum concentration of drug in serum (T$_{max}$) of less than 2 h, and a half-life of about 6 h. Voriconazole is generally well-tolerated by patients, although reversible side effects such as visual disturbance or elevated liver transaminases have been reported. Phase II and phase III clinical trials have been recently completed, and they showed comparable or higher efficacy of voriconazole as compared to other standard antifungal drug treatments such as AmB, fluconazole, or itraconazole (Klepser et al. 2000). Voriconazole has shown similar efficacy as fluconazole in the treatment of esophageal candidiasis in immunocompromised adults (Dupont et al. 2000). Voriconazole was at least equal to liposomal AmB as empirical treatment in febrile neutropenic adults (Walsh et al. 2000).

3.2 Posaconazole

Posaconazole (UK-I09496) (Fig. 7) is structurally related to itraconazole. It is developed as an oral agent and for IV use as a water-soluble prodrug derivative (SCH-59884). SCH-59884 is dephosphorylated *in vivo* to an active intermediate ester, which is in turn hydrolyzed to posaconazole (SCH-56592). Posaconazole has comparable activity to voriconazole against several yeast (*Candida* spp. in particular) and fungal pathogens in term of MIC but is more active than itraconazole, especially against *Aspergillus* spp., with geometric mean MICs 3 to 4 times lower than those of itraconazole (Cacciapuoti et al. 2000). The long half-life of posaconazole after oral administration in animal models (18 to 22 h) constitutes a major difference from voriconazole. Posaconazole has been found equally effective as fluconazole in the treatment of esophageal candidiasis in human immunodeficiency virus (HIV) infected patients in two studies (Nieto *et al.* 2000). In a small open study of fungal infections refractory to classic antifungal drugs, the response rate of *Candida* infections with posaconazole was 75% (Hachem et al. 2000).

3.3 Ravuconazole

Ravuconazole (BMS-2077147) (Fig. 7) is the latest investigated azole agent undergoing clinical development. It is less structurally related to other azole derivatives. Like voriconazole and posaconazole, ravuconazole exhibits high *in vitro* activity against most *Candida* and *Aspergillus* spp. with comparable MIC_{90} (MIC_{90}, 0.5 µg/ml) for *Candida* spp. (Pfaller et al. 1999). No clinical trial data have been yet published with ravuconazole.

Among the few *in vitro* studies comparing the activities of the three new investigative agents against a high number of fungal pathogens, the general trend is that their antifungal activity is decreased in yeast species displaying high MICs for conventional azole derivatives such as fluconazole or itraconazole. The resistance that developed following the use of these agents seems, therefore, to have an impact on MICs of the new agents. When MICs of yeast species are elevated for both fluconazole and itraconazole, the MICs for the new investigative agents are usually higher than those observed for most susceptible isolates (Pfaller et al. 1998).

4. ANTIFUNGAL DRUGS ACTING AT THE LEVEL OF PROTEIN SYNTHESIS

Protein synthesis is an attractive target in the development of antimicrobial agents. One major problem, however, is the high similarity of both fungal and mammalian protein synthesis. At the level of ribosome protein biosynthesis, the structural and sequence similarity between fungal and mammalian ribosomal RNAs, subunits, and soluble factors is remarkable, with the exception of elongation factor 3 (EF3). Sordarins were found as successful compounds targeting fungal protein biosynthesis. Sordarins are natural fungal products from *Graphium putredinis* (Dominguez et al. 1998) and were found as highly specific inhibitors of translation elongation factor 2 (EF2) in yeasts. EF2 in *S. cerevisiae* is expressed from the *EFT1* and *EFT2* genes (Capa et al. 1998). Several sordarin derivatives were obtained, and some of them, such as GM 222712 and GM 237354 (Fig. 4B), showed excellent *in vitro* activities against a variety of pathogenic fungi, including *Candida* spp., *C. neoformans, P. carinii,* and some filamentous fungi (*Aspergillus flavus* and *A. fumigatus*). For example, the MIC_{90} of GM 222712 and GM 237354 was 0.004 and 0.015 µg/ml, respectively, for *C. albicans.* However, their activities against *C. glabrata* and *C. parapsilosis* are poor and absent for *C. krusei* (Capa et al. 1998). The *in vivo* activity of GM193663 and GM237354 was studied recently in mouse models of disseminated candidiasis and aspergillosis and in a rat model of pneumocystosis. The sordarin derivatives were successful in the treatment of candidiasis and pneumocystosis but gave less satisfactory results for pulmonary aspergillosis (Martinez et al. 2000). *C. immitis* was susceptible both *in vitro* and *in vivo* to the GM compounds, which were found to be equivalent to or superior than fluconazole in a murine model of systemic coccidioidomycosis (Clemons and Stevens 2000). Although their development in clinical trials has not been undertaken yet, sordarins showed that protein synthesis is a valuable target for the design of future antifungal agents.

5. ANTIFUNGAL DRUG RESISTANCE

Drug resistance is a complex manifestation of factors in both host and pathogen. From a clinical perspective, drug resistance may be defined as the persistence or progression of an infection despite appropriate drug therapy. The clinical outcome of treatment depends not only on the susceptibility of the pathogen to a given drug but also on factors including pharmacokinetics, drug interactions, immune status and patient compliance, as well as several specific conditions such as the occurrence of biofilms on surfaces of catheters and prosthetic valves (White et al. 1998).

5.1 Development of Resistance to Antifungal Drugs

The emergence of drug resistance in all pathogenic microorganisms, including fungi, is an evolutionary process initiated by exposure to antimicrobial agents (Levin et al. 1999). Resistance evolves because antimicrobial agents are rarely deployed in a way that completely eradicates the pathogen population, with survivors subject to natural selection. Whenever the pathogen population remains large enough over the course of drug treatment, the evolution of resistance is all but inevitable.

The evolution of drug resistance depends on genetic variability, the ultimate source of which is mutation. Once a mutation conferring resistance arises in the pathogen, its fate is determined by key population processes, such as selection, genetic drift, recombination, and migration (including transmission between hosts). The emergence and spread of drug resistance depends on more than the variety of different possible mutations that enable the pathogen to avoid, remove, or inactivate a drug. These resistance mutations interact with the rest of the genome to determine the composite phenotype (Hartman et al. 2001; White et al. 2001). In the evolutionary process the most important component of a resistance phenotype is reproductive output, or fitness. The relative fitness of sensitive and resistant genotypes determines how quickly resistance will spread in a pathogen population exposed to a drug and whether resistance will persist in the absence of the drug. Although biochemical and genetic mechanisms of antifungal drug resistance are well-documented, the evolutionary processes by which these mechanisms spread and persist in pathogen populations await investigation.

Fungi are major pathogens of agricultural plants (Pennisi 2001) and important opportunistic pathogens of humans (Georgopapadakou 1998; Odds 2000), recently ranking as the seventh most common cause of infectious disease-related deaths in the United States (McNeil et al. 2001; Pinner et al. 1996). Opportunistic fungal pathogens may cause superficial or invasive infections. The majority of invasive mycoses are caused by *Cryptococcus neoformans* and species of *Candida* and *Aspergillus*. Despite the increasing importance of opportunistic fungal pathogens, the number of effective antifungal drugs remain limited with resistance compromising the effectiveness of all but the newest (Georgopapadakou *et al.* 1998; White *et al.* 1998).

Antifungal drug resistance has been studied most extensively with the diploid pathogenic yeast *Candida albicans*. *C. albicans* is a ubiquitous commensal, residing on the mucosal surfaces of the mouth, digestive tract, or genitourinary system of 1-50/0-60% of healthy humans, depending on the sample group. *C. albicans* is also a good example of an opportunistic pathogen, causing both superficial infections and invasive fungal disease in immunocompromised individuals (Edwards *et al.* 2001; Rex *et al.* 2000). Species of *Candida* are the fourth most common cause of nosocomial bloodstream infections with *C. albicans* the most commonly encountered (Pfaller *et al.* 1998; Rangel-Frausto *et al.* 1999).

C. albicans has been the fungal pathogen of choice for studying drug resistance because it is more easily manipulated and contained than other medically important fungal pathogens, such as *Cryptococcus neoformans, Aspergillus fumigatus,* and *Histoplasma capsulatum* (Georgopapadakou 1998). *C. albicans* is a wet yeast, classified as a relatively low risk to research personnel. Unfortunately, *C. albicans* is, as yet, not amenable to conventional genetic analysis. Although *C. albicans* cells rendered homo- or hemizygous for mating-type-like genes are able to mate, meiosis has not been observed (Hull et al. 2000; Magee and Magee 2000). This disadvantage can be overcome with complementary studies using *S. cerevisiae* as a genetic stand-in. It is notable that some of the key studies of drug resistance (Calabreses et al. 2000; Lamb and Kelly 1999; Lata Panwar *et al.* 2001) dimorphism, and virulence (Lengeler et al. 2000 ; Lorenz and Fink 2001) have exploited this synergy. *S. cerevisiae* stands on its own as a pathosystem based on recent reports of its clinical isolation from a variety of body sites and patient groups, with subsequent successful infection using these isolates in animal models (Goldstein and McCusker 2001).

The evolution of antifungal drug resistance in the pathogen proceeds on temporal, spatial and genomic scales. Temporal scales of evolution range from ancient events associated with speciation to contemporary changes occurring within one patient under treatment. Spatial scales range from large-scale global or regional populations of patients to fine-scale niches within the body of one patient. Genomic scales range from the small-scale interactions of primary resistance mutations with one or a few genes to large-scale interactions with many genes. Large-scale interactions result in substantial changes in genome-wide patterns of gene expression in the pathogen. The effects of small and large-scale interactions are detected as changes in fitness. The introduction and spread in pathogen populations (recruitment) of genetic mechanisms conferring antifungal drug resistance on different temporal, spatial, and genomic scales can be observed through both epidemiological and population genetic studies. Population genetic studies are implemented with samples of the pathogen from patient population with serial isolations of the pathogen from single patient or, alternatively, through experimental evolution of the pathogen in batch cultures, chemostats, or animal models.

5.2 Mechanisms of Resistance to Antifungal Drugs

The problem of antifungal drug resistance observed during the past decade has stimulated studies to understand the molecular mechanisms of resistance to these agents. The understanding of mechanisms of resistance to current antifungal drugs is also important to the design of new compounds. Therefore, it is important to present the current knowledge of this specific subject. In principle, when microbiological resistance to an antifungal drug in a fungal pathogen is detected by the measurement of MICs, it can be further categorized into primary or secondary resistance. Primary resistance, also called intrinsic resistance, is observed usually prior to *in vivo* or *in vitro* drug exposure. Secondary resistance (or acquired resistance) develops, however, upon exposure to an antifungal agent and can be either reversible or stably acquired. Several examples of transient adaptation to azole antifungal agents measured by increasing MICs have been documented, which could be reversed more or less rapidly by several subcultures in azole-free media (Calvet et al. 1997; Marr et al. 1998). *In vitro* acquisition of antifungal resistance is, however, mostly stable, since it is maintained over several generations in drug-free media. The stable resistance phenotype is thought to be the result of genetic alterations and their identification is the focus of different molecular studies.

5.2.1 Mechanisms of Resistance to Polyenes

Microbiological resistance to AmB can be intrinsic or acquired. Intrinsic resistance to AmB is common for some *C. lusitaniae* (Pfaller et al. 1994) and *Trichosporon* species (Walsh *et al.* 1990), while acquired resistance during antifungal treatments with AmB is still rarely reported among fungal and yeast isolates. Some *C. lusitaniae* isolates are also able to develop *in vitro* rapid switches to AmB resistance when exposed to the drug (Yoon et al. 1999). Acquired resistance to AmB is often associated with alteration of membrane lipids and especially sterols. Recently, clinical *C. albicans* isolates resistant to AmB were described lacking ergosterol and accumulating other sterols (313-ergosta- 7 22-dienol and 313-ergosta-8-enol) typical for a defect in the sterol *Ll5, 6* desaturase system (Kelly et al. 1996). Such a defect is known in *S. cerevisiae* harboring a defect of the *Ll5, 6* desaturase gene *ERG3* (Kelly *et al.* 1995). A defect in *Ll8-7* isomerase in a clinical *C. neoformans* isolate from an AIDS patient was linked also with AmB resistance (Kelly *et al.* 1994). A decrease in the content of cell membrane-associated ergosterol can also cause AmB resistance, since AmB requires the presence of ergosterol to cause damage to fungal cells. Different investigators supported this possibility by demonstrating that (i) development of inducible resistance (induced by adaptation mechanism) in a strain of *C. albicans* was accompanied by a decrease in the ergosterol content of the cells (Capek et al. 1974), and (ii) clinical polyeneresistant *C. albicans* isolates obtained from neutropenic patients had a 74 to 85% decrease in their ergosterol content (Dick *et al.* 1980). Another mechanism accounting for the resistance of yeast to AmB is thought to be mediated by increased catalase activity, which can contribute to diminish oxidative damage caused by this agent (Sokol *et al.* 1986).

5.2.2 Mechanisms of Resistance to 5-FC

5-FC is also an antifungal agent against which resistance can be intrinsic or acquired. Resistance may occur due to the deficiency or lack of enzymes implicated in the metabolism of 5-FC or may be due to the deregulation of the pyrimidine biosynthetic pathway, whose products can compete with the fluorinated metabolites of 5-FC (Whelan et al. 1988). Detailed investigations on the molecular mechanisms of resistance to 5-FC have shown that intrinsic resistance to 5-FC in fungi can be due to a defect in the cytosine permease (as observed in *S. cerevisiae* and *C. glabrata* but not in *C. albicans* and *C. neoformans*), whereas acquired resistance results from a failure to metabolize 5-FC to 5-FUTP and 5-FdUMP or from the loss of a feedback control of pyrimidine biosynthesis (Kerridge et al. 1988). Due to the high prevalence of intrinsic resistance to 5-FC, this compound is best utilized in combination with other antifungal agents, for example, with AmB (Polak and Hartman 1991).

5.2.3 Mechanisms of Resistance to Azole Antifungal Agents

Reports on resistance to azole antifungal agents have been rare until the late 1980s. The first cases of resistance were reported in *C. albicans* after prolonged therapy with miconazole and ketoconazole (Holt and Azmi 1978; Horsburgh and Kirkpatrick 1983). Following the use of fluconazole for a wide variety of clinical settings, antifungal resistance to this agent has been more frequently reported (White *et al.* 1998). There are several mechanisms by which yeasts can become resistant to azole antifungal agents. In clinical isolates resistance to azole antifungal agents can be the result of following phenomena: (i) cells can fail to accumulate these agents; (ii) the cellular content of Ergllp can be elevated; (iii) the affinity of Ergllp to these agents can be decreased; (iv) the ergosterol biosynthetic pathway can be altered by inactivation of the sterol *Ll5,6* desaturase.

5.2.3.1 Resistance to azole antifungal agents involving alterations of the cellular target

Alterations of affinity of azole derivatives to Erg11 p is another important mechanism of resistance that has been described in different post-treatment yeast species, namely, *C. albicans* (Vanden et al. 1990) and recently in *C. neoformans* (Lamb et al. 1995). Affinity alterations are thought to be due to mutations in the gene encoding Ergllp *(ERG11)*, which by conformational changes, can affect the binding of azole derivatives. When comparing *ERG11* sequences from matched pairs of azole-susceptible and azole-resistant *C. albicans* isolates, several laboratories have described nucleotide substitutions in *ERG11* alleles from azole-resistant *C. albicans* isolates resulting in amino acid changes. The effect of amino substitutions on azole resistance has not been yet determined systematically. *In vitro* binding assays of azoles with mutant Ergllp protein variants have been carried out for the substitutions R467K, G464S and Y132H, which are found in *ERG11* alleles of azole-resistant isolates (Kelly et al. 1999). Functional expression of PCR–amplified *ERG11* alleles in *S. cerevisiae* followed by azole susceptibility assays have also been performed as a convenient alternative to the first assay to reveal mutations coupled with the development of azole resistance. While some *ERG11* alleles contain a single mutation responsible for azole resistance, other *ERG11* alleles were found to contain several mutations with potential additive effects. The recently reported crystal structure of a cytochrome P450 from *Mycobacterium tuberculosis,* while a soluble ortholog of Erg11 p, will help determine the effect of mutations on the protein conformation and interferences with azole binding (Podust et al. 2001). Upregulation of *ERG11* has been mentioned as a possible cause of azole resistance in a few clinical isolates of *C. albicans* and *C. glabrata.* Upregulation of *ERG11* did not exceed a factor of 3 to 5 in azoleresistant isolates when compared with *ERG11* expression in related azole-susceptible strains (Lopez *et al.* 1998; Marichal et al. 1997). Upregulation of *ERG11* can be achieved in principle by deregulating the gene transcription or by gene amplification. The latter possibility has been demonstrated in a *C. glabrata* isolate resistant to azole derivatives (Marichal *et al.* 1997). Upregulation of *ERG11* can also be obtained by exposure of *C. albicans* to ergosterol biosynthesis inhibitors, especially to azole antifungal agents (Henry et al. 2000). Exposure of *C. albicans* to this type of drug affects the expression of other *ERG* genes, as recently confirmed by genomewide expression studies performed in *C. albicans* (De Backer et al. 2001).

5.2.3.2 Resistance mechanisms involving alterations in the ergosterol biosynthetic pathway

An analysis of the sterol composition of azole-resistant yeasts has provided several hypotheses on specific alterations of enzymes involved in the complex ergosterol biosynthetic pathway. Accumulation of ergosta-7,22-dienol 313-01 was observed in two distinct azole-resistant *C. albicans* clinical isolates, which is a feature consistent with an absence of sterol *Lls,* 6 desaturase activity encoded by *ERG3* (Kelly et al. 1996). Interestingly, azole resistance in these two isolates was coupled with resistance to AmB, which was expected because of the absence of ergosterol in these cells. Some controversy still exists on the role of *ERG3* in development of azole resistance. For example, it has been reported in *C. glabrata* that null mutations in both *ERG3* and *ERG11* were required to establish azole resistance, while a null mutation in the *ERG3* gene alone was not sufficient for the acquisition of azole resistance but was compatible with AmB resistance only (Geber et al. 1995). The *ERG3* defect in *S. cerevisiae* suppresses the lethality resulting from the inactivation of *ERG11* and is linked to the acquisition of resistance to azole derivatives (Kelly et al. 1995). The role of *ERG3* in azole resistance originates also from the observation that treatment of a normal yeast cell with azoles inhibits Erg11 p and thus results in the accumulation of 14a-methylated sterols and 14a-methylergosta-8, 24 (28)-dien-3, 6a-diol (Brajtburg et al. 1990). Formation of this latter sterol metabolite is thought to be catalyzed by the *ERG3* gene product (the sterol *Ll5, 6* desaturase) and thus inactivation of this gene suppresses toxicity and causes azole resistance. This specific mechanism of resistance to azole derivatives seems to mimic azole resistance obtained in laboratory conditions in *S. cerevisiae* by mutations of the *ERG3* gene (Watson et al. 1988). Loss of function mutations in *ERG3* alleles from the *C. albicans* azole-resistant Darlington strain were characterized recently (Miyazaki et al. 1999). Unfortunately, the effects of these mutations on azole resistance were masked by other azole resistance mechanisms in this strain (Miyazaki et al. 1999).

5.2.3.3 Combination of mechanisms

The multiplicity of mechanisms of resistance to azole antifungal agents enables yeast cells to develop a stepwise increase in resistance to these agents. There are still few clinical isolates that have been analyzed at the molecular level for the presence and the combination of these different mechanisms. The combination of resistance mechanisms is a common feature in the stepwise development of azole resistance in *C. albicans* (Franz et al. 1998; Lopez et al. 1998). Azole resistance in *C. glabrata* isolates investigated thus far seems, however, to be caused by a single mechanism i.e., upregulation of multi drug transporters (Sanglard et al. 1999). Development of resistance is based on a unique combination of mechanisms, which enables not only a stepwise increase in resistance to azole derivatives but also an increase in resistance to specific azoles.

5.2.4 Resistance Mechanisms to Allylamines and Morpholines

Resistance to allylamines in yeast and fungi has been rarely reported (Evans 1999). However, the potential of developing resistance by the action of multidrug efflux transporters exists. For example, the overexpression in *S. cerevisiae* of the *C. albicans CDR1* and *CDR2* genes and of the *CaMDR1* gene can confer resistance to terbinafine (Sanglard et al. 1997), showing that this compound is a substrate for these transporters. Consistent with these results is the observation that several *C. albicans* isolates resistant to azole antifungal agents by multidrug transporter gene upregulation are also less susceptible to terbinafine (Ryder et al. 1998).

Acquired resistance to morpholine derivatives has not been reported so far in yeast pathogens, which is probably due to the limited use of this antifungal agent for the treatment of superficial fungal infections. However, resistance to morpholine derivatives can be obtained in *S. cerevisiae* by the overexpression of the *ERG24* gene among those involved in sterol biosynthesis (Lai et al. 1994). Moreover, recent work has pointed out that amorolfine, as in the case of terbinafine, could be a substrate of *C. albicans* multidrug efflux transporters encoded by the *CDR1* and *CDR2* genes (Sanglard et al. 1997). *C. albicans* clinical isolates resistant to azole antifungal agents are also less susceptible to amorolfine (Hiratani and Yamaguchi 1994). A potential exists, therefore, for developing resistance to this agent.

5.3 IMPACT OF RESISTANCE MECHANISMS IN THE DESIGN OF NEW ANTIFUNGAL AGENTS

The discovery of mechanisms of resistance to established or investigative antifungal agents has several potential impacts on the design of new antifungal drugs. A major advance in the understanding of antifungal drug resistance mechanisms was the involvement of multidrug transporters in the development of resistance. It is highly desirable to avoid interference of new antifungal drugs with multidrug transporters, because these transporters, when expressed in cells to high levels, can cause resistance to multiple unrelated substances. Molecular tools are now available to exclude potential new antifungal agents as substrates of multidrug efflux transporters.

The availability of mutated enzymes that confer resistance to their target antifungal agents can also assist in improving existing agents by chemical derivatization. For example, when using the different mutant *ERG* 11 alleles as described in "Mechanisms of resistance to azole antifungal agents" in affinity assays with new azole derivatives, it is possible to estimate their performance with respect to their binding efficiency and design their structures to become insensitive to *ERG* 11 mutations.

6. FUTURE DIRECTIONS AND CONCLUSIONS

Antifungal drug discovery is now exploring the use of genomic information as an additional exploration tool in drug target screening. No doubt future agents will emerge from the use of this important source of information in conjunction with traditional drug screening procedures. Genomic information is now being extended to a significant number of yeast and fungal pathogens including *C. albicans, C. neoformans,* and *A. fumigatus*. Genomic information is likely to assist in the future identification of the mode of actions of antifungal drugs with unknown targets. Several examples of drug target identification have been reported with the help of the *S. cerevisiae* genome and even "off-targets" of several classical antifungal agents have been now characterized (Hughes et al. 2000 ; Marton *et al.* 1998). A genome wide expression study of *C. albicans* genes performed for the first time by De Backer et al. (2001) has confirmed the validity of this approach by investigating the effect of itraconazole on gene expression.

Before the introduction of new antifungal agents will be seen in clinical practice, treatment options will remain with the few existing agents. To extend the range of activities of these agents when needed, it will be important to explore their possible potentation by combination with each other or by combination with companion drugs or with immunomodulators.

REFERENCES

Adler-Moore J (1994). AmBisome targeting to fungal infections. *Bone Marrow Transpl.* 14:S3-S7.

Aramwit P, Yu B G, Lavasanifar A, Samuel J and Kwon G S (2000). The effect of serum albumin on the aggregation state and toxicity of amphotericin B. *J. Pharm. Sci.* 89: 1589-1593.

Banerjee S N, Emori T G, Culver DH, Gaynes RP, Jarvis WR, Horan T, Edwards J R, Tolson J, Henderson T, Martone WJ, and the National Nosocomical infections Surveillance System (1991). Secular trends in nosocomical primary bloodstream infections in the United States, 1980-1989. *Am. J. med.* 91(Suppl. 3B): 86S-89S.

Barnett J A, Payne RW and Yarrow D (2000). Yeasts: Characteristics and the identification, 3rd ed. Cambridge University Press, Cambridge, United Kingdom.

Bolard J, Seigneuret M and Boudet G (1980). Interaction between phospholipid bilayer membranes and the polyene antibiotic amphotericin B: lipid state and cholesterol content dependence. *Biochim. Biophys. Acta* 599:280-293.

Brajtburg J, Powderly W G, Kobayashi G S and Medoff G (1990). Amphotericin B: current understanding of mechanisms of action. *Antimicrob. Agents Chemother.* 34: 183-188.

Cacciapuoti A, Loebenberg D, Corcoran E, Menzel F, Jr., Moss E L, Jr., Norris C, Michalski M, Raynor K, Halpern J, Mendrick C, Arnold B, Antonacci B, Parmegiani R, Yarosh-Tomaine T, Miller G H and Hare R S (2000). *In vitro* and *in vivo* activities of SCH 56592 (posaconazole), a new triazole antifungal agent, against *Aspergillus* and *Candida*. *Antimicrob. Agents Chemother.* 44:2017-2022.

Calabreses D, Bille J, Sanglard D (2000). A novel multidrug efflux transporter gene of the major facilitator super family from *Candida albicans* (FLU 1) conferring resistance to fluconazol. *Microbiology* 146: 2743-54.

Calvet H M, Yeaman M R and Filler S G (1997). Reversible fluconazole resistance in *Candida albicans:* a potential *in vitro* model. *Antimicrob. Agents Chemother.* 41:535-539.

Capa L, Mendoza A, Lavandera J L, Gomez de las Heras F and Garcia-Bustos J F (1998). Translation elongation factor 2 is part of the target for a new family of antifungals. *Antimicrob. Agents Chemother.* 42:2694-2699.

Capek A, Simek A, Bruna L, Svab A and Budesinsky Z (1974). Antimicrobial agents. XX. Ergosterol content of *Candida albicans* cells during adaptation to antimycotics. *Folia Microbiol.* 19:79-80.

Clemons K V and Stevens D A (2000). Efficacies of sord arinderivatives GM193663, GM211676, and GM237354 in a murine model of systemic coccidioidomycosis. *Antimicrob. Agents Chemother.* 44:1874-1877.

Coleman DC, Bennett DE, Sullivan DJ, Gallaghar PJ, Henman MC, Shanley DB and Russell RJ (1993). Oral Candida in HIV infection and AIDS : new perspectives/new approaches. *Crit. Rev.. Microbiol.* 19:61-82

Crowley JH, Smith SJ, Leak F W and Parks L W (1996). Aerobic isolation of *anERG24* null mutant of *Saccharomyces cerevisiae. J. Bacteriol.* 178:2991-2993.

De Backer M D, llyina T, Ma X J, Vandoninck S, Luyten W H and Vanden Bossche H (2001). Genomic profiling of the response of *Candida albicans* to itraconazole treatment using a DNA microarray. *Antimicrob. Agents Chemother.* 45:1660-1670.

De Hoog GS and Guarra J (1995). Atlas of Clinical Fungi. Centraalbureau voor Schimmelcultures, Barn and Delft, the Netherlands, and Universitat Rovira I Virgili, Reus, Spain.

Diasio R B, Lakings D E, and Bennett J E (1978). Evidence for conversion of 5-fluorocytosine to 5-fluorouracil in humans: possible factor in 5-fluorocytosine clinical toxicity. *Antimicrob. Agents Chemother.* 14:903-908.

Dick J D, Merz W G and Saral R (1980). Incidence of polyene-resistant yeasts recovered from clinical specimens. *Antimicrob. Agents Chemother.* 18:158-163.

Dupont B, Ally R, Burke J, Hodges M R and Romero A J (2000). A double-blind, randomized, multicenter trial of voriconazole (VORl) vs. fluconazole (FLU) for the treatment of esophageal candidiasis (EC) in immunocompromised adults, abstr. 706, p. 365. *In Abstr. 40th Intersci. Conf. Antimicrob. Agents Chemother.* American Society for Microbiology, Washington, D.C.

Edwards JE Jr. (2001). Management of severe *Candida* infections: integration and review of current guideline for treatment and prevention. *Curr. Clin. Top. Infect. Dis.* 21:135-47

Espinel-Ingroff A (1998). *In vitro* activity of the new triazole voriconazole (UK-109, 496) against opportunistic filamentous and dimorphic fungi and common and emerging yeast pathogens. *J. Clin. Microbiol.* 36:198-202.

Evans E G (1999). Resistance of *Candida* species to antifungal agents used in the treatment of onychomycosis: a review of current problems. *Br. J. Dermatol.* 141:33-35.

Favre B and Ryder N S (1997). Cloning and expression of squalene epoxidase from the pathogenic yeast *Candida albicans. Gene* 189:119-126.

Franz R, Kelly S L, Lamb D C, Kelly D E, Ruhnke M and Morschhauser J (1998). Multiple molecular mechanisms contribute to a stepwise development of fluconazole resistance in clinical *Candida albicans* strains. *Antimicrob. Agents Chemother.* 42:3065-3072.

Fridkin SK and Jarvis WR (1996). Epidemiology of nosocomical fungal infections. *Clin. Microbiol. Rev.* 9:499-511.

Gallis H A, Drew R H and Pickard W W (1990). Amphotericin B: 30 years of clinical experience. *Rev. Infect. Dis.* 12: 308-329.

Geber A, Hitchcock C A, Swartz J E, Pullen F S, Marsden K E, Kwon-Chung K J and Bennett J E (1995). Deletion of the *Candida glabrata ERG3* and *ERGll* genes: effect on cell viability, cell growth, sterol composition, and antifungal susceptibility. *Antimicrob. Agents Chemother.* 39: 2708-2717.

Georgopapadakou NH (1998). Antifungales: mechanism of action and resistance, estibaleshed and novel drugs. *Curr. Opin. Microbiol.* 1:547-57.

Goldstein AL, McCusker JH (2001). Development of Saccharomyces cerevisiae as a model of pathogen. A system for the genetic identification of gene products required for survival in the mammalian host environment. *Genetics* 159:499-513.

Hachem R Y, Raad II, Afif C M, Negroni R, Graybill J, Hadley S, Kantarjian H, Adams S and Mukwaya G (2000). An open, non-comparative multicenter study to evaluate efficacy and safety of posaconazole (SCH 56592) in the treatment of invasive fungal infections (IFI) refractory (R) to or intolerant (I) to standard therapy (ST), abstr. 232, p. 372. *In Abstr. 40th Intersci. Conf. Antimicrob. Agents Chemother.* American Society for Microbiology, Washington, D.C.

Hartman JL, Garvik B, Hartwell L (2001). Principal for the buffering of genetics variation. *Science* 291: 1001-4

Hartman P G and Sanglard D (1997). Inhibitors of ergostrol biosynthesis as antifungal agents. *Curro Pharmaceutic. Design* 3:177-208.

Hazen KC (1995). New and emerging yeast pathogens. *Clin. Microbiol. Rev.* 8:462-478.

Henry K W, Nickels J T and Edlind T D (2000). Upreguilation of *ERG* genes in *Candida* species by azoles and other: sterol biosynthesis inhibitors. *Antimicrob. Agents Chemo. ther.* 44:2693-2700.

Hiratani T and Yamaguchi H (1994). Cross-resistance of *Candida albicans* to several different families of antifungals with ergosterol biosynthesis-inhibiting activity. *Jpn. J. Anti. biot.* 47:125-128.

Holt R J and Azmi A (1978). Miconazole-resistant *Candida. Lanceti*: 50-51.

Horsburgh C R and Kirkpatrick C H (1983). Long-term therapy of chronic mucocutaneous candidiasis with ketoconazole: experience with twenty one patients. *Am. J. Med.*: 74 (Suppl. 1B): 23-29.

Hostetler J S, Clemons K V, Hanson L H and Stevens D A (1992). Efficacy and safety of amphotericin B colloidal ! dispersion compared with those of amphotericin B deoxychol late suspension for treatment of disseminated murine cryptococcosis. *Antimicrob. Agents Chemother.* 36:2656-2660.

Hughes T R, Marton M J, Jones A R, Roberts C J, Stoughton R, Armour C D, Bennett H A, Coffey E, Dai H, He Y D, Kidd M J, King A M, Meyer M R, Slade D, Lum P Y, Stepaniants S B, Shoemaker D D, Gachotte D, Chakraburtty K, Simon J, Bard M and Friend S H (2000). Functional discovery via a compendium of expression profiles. *Cell* 102:109-126.

Hull CM, Rasiner RM, Johnson AD (2000). Evidence for matting of the "asexual" yeast *Candida albicans* in a mammalian host. *Science* 289:307-10.

Jarvis WR (1995). Epidemiology of nosocomial fungul infections, emphasis on Candida species. *Clin. Infect. Dis.* 20:1526-1530.

Kelly S L, Lamb D C, Corran A J, Baldwin B C and Kelly D E (1995). Mode of action and resistance to azole antifungals associated with the formation of 14a-methylergosta-8, 24 (28)-dien-3 [3,6a-diol. *Biochem. Biophys. Res. Commun.* 207:910-915.

Kelly S L, Lamb D C and Kelly D E (1999). Y132H substitution in *Candida albicans* sterol 14alpha-demethylase confers fluconazole resistance by preventing binding to haem. *FEMS Microbiol. Lett.* 180:171-175.

Kelly S L, Lamb D C, Baldwin B C, Corran A J and Kelly D E (1997). Characterization of *Saccharomyces cerevisiae* CYP61, sterol delta22-desaturase, and inhibition by azole antifungal agents. *J. Biol. Chem.* 272:9986-9988.

Kelly S L, Lamb D C, Kelly D E, Loeffler J and Einsele H (1996). Resistance to fluconazole and amphotericin in *Candida albicans* from AIDS patients. *Lancet* 348: 1523-1524.

Kelly S L, Lamb D C, Taylor M, Corran A J, Baldwin B C and Powderly W G (1994). Resistance to amphotericin B associated with defective sterol delta 8-7 isomerase in a *Cryptococcus neoformans* strain from an AIDS patient. *FEMS Microbiol. Lett.* 122:39-42.

Kerridge D and Vanden Bossche H (1990). Drug discovery: a biochemist's approach. *Handbook Exp. Pharmacol.* 96: 31-76.

Kerridge D, Fasoli M and Wayman J F (1988). Drug resistance in *Candida albicans* and *Candida glabrata*. *Ann. N. Y. Acad. Sci.* 544:245-259.

Klepser M E, Hoffman H L and Ernst E J (2000). Novel triazole antifungal agents. *Exp. Opin. Investig. Drugs* 9: 593-605.

Lai M H, Bard M, Pierson C A, Alexander F, Goebl M, Carter G T and Kirsch D R (1994). The identification of a gene family in the *Saccharomyces cerevisiae* ergosterol biosynthesis pathway. *Gene* 140:41-49.

Lamb D C, Corran A, Baldwin B C, Kwon-Chung J and Kelly S L (1995). Resistant P450Al activity in azole antifungal tolerant *Cryptococcus neoformans* from AIDS patients. *FEBS Left.* 368:326-330.

Lengeler K B, Davidson R C, D'Souza C, Harashima T, Shen W C (2000). Singal transduction eascades regulating fungal development and virulence. *Microbiol. Mol. Biol. Rev.* 64:746-85.

Levin BR, Lipsith M and Bonhoeffer S (1999). Population biology, evolution, and synthesis. *Science* 283:806-9.

Lopez-Ribot J, McAtee R K, Lee I, Kirkpatrick W R, White T C, Sanglard D, and Patterson T F (1998). Distinct patterns of gene expression associated with the development of fluconazole resistance in serial *Candida albicans* isolates from HIV-infected patients with oropharyngeal candidiasis. *Antimicrob. Agents Chemother.* 42:2932-2937.

Lorenz MC, Fink GR (2001). The glyoxylate cycle is required for fungal virulence. *Nature* 412:83-86

Lyman C A and Walsh T J (1992). Systemically administered antifungal agents. A review of their clinical pharmacology and therapeutic applications. *Drugs* 44:9-35.

Magee BB, Magee PT (2000). Induction of mating in *Candida albicans* by construction of MTLa and MTLalpha strains. *Science* 289:310-13.

Marichal P, Vanden Bossche H, Odds F C, Nobels G, Warnock D W, Timmerman V, Van Broeckhoven C, Fay S and Mose-Larsen P (1997). Molecular biological characterization of an azole-resistant *Candida glabrata* isolate. *Antimicrob. Agents Chemother.* 41:2229-2237.

Marr K A, Lyons C N, Rustad T R, Bowden R A, White T C and Rustad T (1998). Rapid, transient fluconazole resistance in *Candida albicans* is associated with increased mRNA levels of *CDR*. *Antimicrob. Agents Chemother.* 42: 2584-2589.

Martinez A, Aviles P, Jimenez E, Caballero J and Gargallo-Viola D (2000). Activities of sordarins in experimental models of candidiasis, aspergillosis, and pneumocystosis. *Antimicrob. Agents Chemother.* 44:3389-3394.

Marton M J, De-Risi J L, Bennett H A, Iyer V R, Meyer M R, Roberts C J, Stoughton R, Burchard J, Slade D, Dai H, Bassett D E Jr, Hartwell L H, P. O. Brown, and S. H. Friend. 1998. Drug target validation and identification of secondary drug target effects using DNA microarrays. *Nat. Med.* 4: 1293-1301.

McNeil MM, Nash SL, Hajjeh RA, Phelan MA and Conn LA et al. (2001). Trends bin mortality due to invasive mycotic diseases in the United States, 1980-1997. *Cli.Infect Dis.* 33:641-47.

Mercer E I (1991). Morpholine antifungals and their mode of action. *Biochem. Soc. Trans.* 19:788-793.

Miyazaki Y, Geber A, Miyazaki H, Falconer D, Parkinson T, Hitchcock C, Grimberg B, Nyswaner K and Bennett J E (1999). Cloning, sequencing, expression and allelic sequence diversity of *ERG3* (C-5 sterol desaturase gene) in *Candida albicans. Gene* 236:43-51.

Nelson D R, Koymans I, Kamataki T, Stegeman J J, Feyereisen R, Waxman D J, Waterman M R, Gotoh O, Coon M J, Estabrook R W, Gunsalus I C, and Nebert D W (1996). P450 superfamily: update on new sequences, gene mapping, accession numbers and nomenclature. *Pharmacogenetics* 6:1-42.

Nieto I, Northland R, Pittisuttithum P, Firnhaber C, Jacobson S, Greaves W, Plott R T and Taglietti M (2000). Posaconazole equivalent to fluconazole in the treatment of oropharyngeal candidiasis, abstr. 1108, p. 372. *In Abstr. 40th Inter Sci. Conf. Antimicrob. Agents Chemother.* American Society for Microbiology, Washington, D.C.

Odds FC (2000). Pathogenic fungi in the 21st century. Trends *Microbiol.*8: 2001.

Odds FC (1987). Candidas infection: an overview. *Crit. Rev. Microbiol.*15: 1-5.

Odds FC (1988). Candida and candidosis: a revive and Bibliography, 2nd ed. Balliere Tindal, London, United Kingdom.

Oxford A E, Raistrick M and Simonart P (1939). Studies in biochemistry of microorganisms LX griseofulvin C17H1606Cl metabolic product of *Penicillium griseoulvin. Biochem. J.* 33:240-248.

Panwar SL, Krishnamurthy S, Gupta, V, Alarco AM, Raymond M, Sanglord D and Prasad R (2001) Ca ALK8, an alkane assimilating cytochrome P450, confers multidrug resistance when expressed in a hypersensitive stain of *Candida albicans. Yeast* 18:1117-29.

Pennisi E (2001). The Push to pit genomics against fungal pathogens. *Science* 292:2273-74.

Pfaller M and Wezel R (1992). Impact of the changing epidemiology of fungal infection in the 1990s. *Eur. J. Clin. Microbiol. Infect. Dis.* 11:287-291.

Pfaller MA, Jones RN, Messer SA, Edmond MB, Wenzel RP (1998). National surveillance of nosocomial bloodstream infection due to *Candida albicans*: frequency of occurrence and anti fungal susceptibility in the SCOPE program. *Digan. Microbiol. Infect. Dis.* 31:327-32.

Pfaller MA, Jones RN, Doern GV, Fluit AC, Verhoef J, Sader HS, Messer SA, Houstons A, Coffman S, Hollis RJ and the SENTRY Participant Group (Europ).1999.

Pinne RW, Teutsch SM., Simonsen L, Klug LA, Graber JM et al. (1996). Trends in infectious diseases moratality in the United States. *JAMA* 275:189-93.

Podust L M, Poulos T L and Waterman M R (2001). Crystal structure of cytochrome P450 140l-sterol demethylase (CYP51) from *Mycobacterium tuberculosis* in complex with azole inhibitors. *Proc. Natl. Acad. Sci. USA* 98:3068-3073.

Polak A (1990). Mode of action studies, p. *153-182.InJ.* F. Ryley (ed.), *Chemotherapy of Fungal Diseases.* SpringerVerlag, Berlin, Germany.

Polak A and Hartman P G (1991). Antifungal chemotherapy. *Progr. Drug Res.* 37:181-269.

Rangel-Frausto MS, Wiblin T, Blumberg HM, Saiman L, Patterson J et al. (1999). National epidemiology of mycoses survey (NEMIS): variation in rate of blood stream infection due to *Candida* species in seven surgical intensive care unit and six neonatal intensive care units. *Clin. Infect. Dis.* 29:253-58.

Rapp R P, Gubbins P O, and Evans M E (1997). Amphotericin B lipid complex. *Ann. Pharmacother.* 31:1174-1186.

Rex JH, Walsh T J, Sobel JD, Filler SG, Pappas PG (2000). Practice gudeline of the treatment of Candidiasis species of. Infectious Diseases Society of America. *Clin. Infect. Dis.* 30:662-78.

Ryder N S, Wagner S and Leitner I (1998). *In vitro* activities of terbinafine against cutaneous isolates of *Candida albicans* and other pathogenic yeasts. *Antimicrob. Agents Chemother.* 42:1057-1061.

Sanglard D, Ischer F, Calabrese D, Majcherczyk P A and Bille J (1999). The ATP binding cassette transporter gene *CgCDRl* from *Candida glabrata* is involved in the resistance of clinical isolates to azole antifungal agents. *Antimicrob. Agents Chemother.* 43:2753-2765.

Sanglard D, Ischer F, Monod M and Bille J (1997). Cloning of *Candida albicans* genes conferring resistance to azole antifungal agents: characterization of *CDR2*, a new multidrug ABC-transporter gene. *Microbiology* 143:405-416.

Sanglard D and Bille J (2002). Current understanding of the modes of action of and resistance mechanisms to conventional and emerging antifungal agents for treatment of Candida infections. Calderone, R. (ed.). *Candida and Candidiasis* American Society for Microbiology Press, pp 349-383.

Sokol-Anderson M L, Brajtburg J and Medoff G (1986). Amphotericin B-induced oxidative damage and killing of *Candida albicans*. *J. Infect. Dis.* 154:76-83.

Vanden Bossche H, Marichal P and Moereels H (1993). Molecular mechanisms of antifungal activity and fungal resistance: focus on inhibitors of ergosterol biosynthesis, p. 179-197. *In* B. Maresca, G. S. Kobayashi, and H. Yamaguchi (ed.), *Molecular Biology and Its Application to Medical Mycology*. Springer-Verlag, Berlin, Germany.

Vanden Bossche H, Marichal P, Gorrens J, Bellens D, Moereels H and Janssen P AJ (1990). Mutation in cytochrome P450-dependent 14a-demethylase results in decreased affinity for azole antifungals. *Biochem. Soc. Trans.* 18:56-59.

Vanden Bossche H, Marichal P and Odds F C (1994). Molecular mechanisms of drug resistance in fungi. *Trends Microbial.* 2:393-400.

Walsh T J, Melcher G P, Rinaldi M G, Lecciones J, McGough D A, Kelly P, Lee J, Callender D, Rubin M and Pizzo P A (1990). *Trichosporon beigelii*, an emerging pathogen resistant to amphotericin B. *J. Clin. Microbiol.* 28: 1616-1622.

Walsh T J, Pappas P, Winston D, Blumer J, Peterson F, Yanovich S, Raffalli S, Stiff P, Greenberg R, Donowitz G, Schuster M, Arndt C, Reinhardt J, Reboli A, Anaissie E, Hadley S, Garber G, Wingard J, Fioritoni G and Lee J (2000). Voriconazole versus liposomal amphotericin B for empirical antifungal therapy of persistently febrile neutropenic patients: a randomized, international, multicenter trial, abstr. L-1. *In Late Breaker 40th Inter. Sci. Conf Antimicrob. Agents Chemother.* American Society for Microbiology, Washington, D.C.

Watson PF, Rose M E and Kelly S 1 (1988). Isolation and analysis of ketoconazole resistant mutants of *Saccharomyces cerevisiae*. *J. Med. Vet. Mycol.* 26:153-162.

Whelan W (1988). The genetic basis of resistance to 5fluorocytosine in *Candida* species and *Cryptococcus neoformans*. *Crit. Rev. Microbial.* 15:45-56.

White TC, Marr KA, Bowden RA (1998). Clinical, cellular, and molecular factors that contribute to antifungal drug resistance *Clin. Microbiol, Rev*11: 382-402.

White KP (2001). Functional genomics and the study of the development, variation and evolution. *Nat. Rev. Genet.*2: 528-37.

Williams D J, Marten R M and Sarkany I (1958). Oral treatment of ringworm by griseofulvin. *Lancet* ii:1212.

Yoon S A, Vazquez J A, Steffan P E, Sobel J D and Akins R A (1999). High-frequency, *in vitro* reversible switching of *Candida lusitaniae* clinical isolates from amphotericin B susceptibility to resistance. *Antimicrob. Agents Chemother.* 43:836-845.

Potential of Enzymes in Therapeutics

R. K. SAXENA*, L. AGARWAL, K. DUTT, S. SAXENA and J. ISAR

1. INTRODUCTION

The science of medicine has now reached a new threshold in this century where diagnosis of disease and its consequent therapy is being developed to be specific, fast and focused. One of the fundamental findings, which galvanized medical science, was that the absence or presence of an enzyme could possibly be the reason for certain diseases or deficiencies because enzymes, which are biological catalysts, are proteinaceous in nature and play a pivotal role in maintaining homeostasis (Lopez 1994). Today, they have gained the floor in medical research by ushering in a whole new era of medical diagnostics and therapeutics.

Even though the potential of using enzymes as drugs was demonstrated as early as in the late 19th century, however, their use apart from as digestive aids was severely limited. In 1906, a Scottish embryologist, John Beard, proposed that the pancreatic enzymes represent the body's main defense against cancer, and would be useful as a treatment for all types of cancer (Beard 1906). This formed the basis of enzyme therapy, which remained largely forgotten for many years before being rediscovered (Shively 1969). Enzyme therapy states an obvious use of enzymes as therapeutics having a broad variety of specific uses such as oncolytics, anticoagulants or thromobolytics and as replacements for metabolic deficiencies. Thus, the manufacture or processing of enzymes for use as drugs has become an important facet of today's pharmaceutical industry (Cassileth 1998).

* Corresponding author: rksmicro@yahoo.co.in

Generally, for an enzyme to be considered as a therapeutic agent it must meet the following norms:

1. The enzyme must reliably reach its site of action in the tissue compartment.
2. The enzyme must function under the environmental conditions found at the presumed site of action, with respect to ionic milieu, substrate and cosubstrate supply, and presence of endogenous inhibitors.
3. Therapeutic effects must be established in controlled trials and related to the particular activity of the enzyme applied.
4. The therapeutic benefit should outweigh the adverse reactions, particularly immunological complications.

These requirements can be met positively by only a few enzymes approved up to now either as the main active principles or as additives in pharmaceutical formulations.

In this chapter, we shall be presenting an account of widely accepted and sufficiently documented therapeutic enzymes complemented with their topical trends in research and usage.

2. ENZYMES AS DIGESTIVE AIDS

On an average a person tends to loose 13% of digestive enzymes every decade. This decrease in enzyme production leads to enzyme deficiency, which brings about digestive disorders, fatigue, impaired mental capabilities and early signs of degenerative aging leading to the necessity of enzyme supplements. This can be prevented if enzyme supplements are provided along with the regular diet. A variety of supplemental enzymes are available in capsule or powder form from different sources and can be ingested orally. Mostly they are hydrolases including lipase, protease, amylase, etc. and are effective in a wide range of conditions including indigestion, pancreatic insufficiency, steatorrhea and lactose intolerance.

Supplementation of diet with small doses of acid-stable lipase from plant source was far more effective than conventional pancreatin (pancreatic lipase) in the treatment of malabsorption, malnutrition and steatorrhea. Proteolytic enzymes such as bromelain and papain assist in digestion. Bromelain has a very wide effective range of activity making it a part of protein supplements. The enzyme is effective as a substrate for pancreatic enzymes in the treatment of pancreatic insufficiency (Balakrishnan et al. 1981, Knill-Jones et al. 1970). Papain has a mild, soothing effect on the stomach and aids in protein digestion. Thus, this enzyme considered "biological scalpel" is necessary for effective digestion and useful for all stomach complaints. The utilization of papain in medicine and pharmacology has been well-documented by nutritional studies. Studies show that the incidence of achlorhydria increases with age, which effectively reduces the ability of pepsin to digest protein. Papain supplements have been used as an aid for protein digestion in these patients. Impaired gastric proteolysis can also be treated with crude hog pepsin (Bunte 1960).

In 1994, amylases were first commercialized for treating digestive disorders. Amylases serving as glucose balancer ®, are effective in acidic environments where other enzymes tend to be less stable. Preparations of lactose are employed as digestive aid, in particular to alleviate symptoms associated with lactose intolerance. Supplementation of sucrase helps to prevent gastrointestinal discomfort.

Today in the market several enzyme supplements are available commercially under many trade names. Most prominent among them are : Phyter-Zyme containing a cocktail of enzymes like bromelain, lipase, α-1, 4 glucan glucohydrolase, catalase, etc. effectively aids in digestion , EnzyMega contains protease (bromelain), amylase, maltase, sucrase, lactase, lipase, cellulase and an ionic blend of 77 naturally reoccurring organic trace minerals. Lipase assisted preparation called Lypozyme results in balancing fatty acids in the body system along with aiding in the proper digestion of fats, carbohydrates and proteins that leads to a feeling of satiety which further controls appetite, lowers cholestrol and triglyceride levels. Another formulation called BalanceZyme Plus containing protease, lipase, amylase, cellulase and plantain leaf is a completely natural support to the endocrine and digestive system.

3. ENZYMES IN CIRCULATORY SYSTEM

A constant dynamic equilibrium between blood coagulation (clotting) and fibrinolysis (the process of dissolving the clotted blood) is extremely important because if the balance is disturbed it either leads to excessive bleeding or thromboses and embolisms. According to clinical studies, the best approach to therapeutic thrombolysis (dissolution of clots) is an intravenous injection of an enzyme capable of converting plasminogen to plasmin (clot dissolving enzyme). This type of therapy is known as thombolytic therapy (thrombus or clot splitting) or fibrinolytic (fibrin splitting) therapy. This form of therapy has the potential to become the initial treatment for all forms of acute deep-vein thrombosis and pulmonary embolism. The most common enzymes used for thrombolytic therapy are proteases, streptokinases and urokinases. Intravenous therapy with plant protease has been established to be more effective than anti-coagulant therapy (e.g., heparin, warfarin) at recanalizing obstructed arteries and in improving blood flow through stenosed arterial segments. Plasmin and thrombin, non-specific proteolytic enzymes are the body's "most important" enzymes of lysis and clotting. Thrombin represents the protease that cleaves fibrinogen, thereby initiating formation of insoluble fibrin and carrying out thrombolysis (Ouriel et al. 1995). It has been used for heart attacks with real success in recent years, which prompted researchers to see if thrombolysis could be of value in ischemic stroke.

Bromelain, helps in the prevention and treatment of cardiovascular disease by reducing platelet aggregation, arterial plaquing and clot formation (Heinicke et al. 1972 and Morita et al. 1979). In the last 20 years "new" thrombolytic enzymes, which have shown some therapeutic use are *arvin*, *Batroxobin*, a protease derived from the venom of *Bothrops atrox* (trade name: Reptilase) and *kallikrein* isolated from hog pancreas (trade names: Padutin, Bioactin). In controlled trials, infusion of such enzymes, yielded substantial defibrinogenation associated with marked and long-lasting clinical improvement in patients suffering from occlusive peripheral artery disease.

Another enzyme, Streptokinase (trade names: e.g., Streptase, Kabikinase, Kinalysin), which is produced by hemolytic bacteria, dissolves fibrin leading to dissolution of pulmonary emboli and venous thromboses. The current drug of choice in thrombolytic therapy is Urokinase (trade names: Actosolv, Ukidan) that is isolated from human urine or tissue culture is at present considered (Flohë et al. 1990). Studies indicate that urokinase can lyse plasma f clots and mediate fibrinolysis in pulmonary microvasculature (Higazi et al. 1998). Since, it specifically cleaves plasminogen to form plasmin with a straight dose-effect relationship it is more easily controlled than streptokinase. But there is still restricted use of urokinase compared to streptokinase because of its comparatively high production cost (Ouriel et al. 1995). Laboratory observations suggest that urokinase may be the most appropriate agent for catheter-directed venous thrombolysis. The introduction of direct catheter installation of thrombolytic enzyme urokinase into the thrombus not only improved the efficacy of thrombolytic therapy but also has reduced the risk of serious bleeding. It is a safe, inexpensive, and convenient enzyme for dissolving fibrin or blood clots after glaucoma surgery i.e., for fibrinolytic therapy. Use of urokinase significantly improved 30-days survival in intraventricular hemorrhage patients where clots remained in the ventricles for months after a hemorrhage as the cerebrospinal fluid has a limited fibrinolytic activity. Clinical studies of fibrinolytic therapy for intraventricular hemorrhage have found a 30-35% reduction in mortality with treatment. It is a lifesaving treatment in cases involving pulmonary embolisms and myocardial infarctions (blockages of vessels in lungs and heart, respectively).

Since 1987, when the first recombinant enzyme drug, Activase® (alteplase; recombinant human tissue plasminogen activator), was approved by the Food and Drug Administration (FDA), several other anti-coagulant or coagulant enzymes have been approved (Vellard 2003).

4. ENZYMES AND CANCER

Proteases have selective anticancer cell activity (selective oncocytolysis) and have been promoted by numerous alternative cancer practitioners for many years where evaluation of the benefit of proteolytic enzymes in patients with advanced pancreatic cancer in a large-scale study is done followed by a smaller study that showed dramatic improvements in patients (Gonzalez and Isaacs 1999). The clinical research that currently exists on proteolytic enzymes suggests significant benefits in the treatment of many forms of cancer. (Leipner and Saller 2000). Specifically, these studies have shown improvements in the general condition of patients, quality of life, and modest to significant improvements in life expectancy in patients with cancers of the breast, lung, stomach, head and neck, ovaries, cervix, and colon; and lymphomas and multiple myeloma. These studies involved the use of proteolytic enzymes in conjunction with conventional therapy (surgery, chemotherapy and/or radiation) indicating that proteolytic enzymes can be used safely and effectively with these treatments.

The earliest observation that turned out to be important for the development of L-Asparaginase as a therapeutic agent against malignant tumours was made in 1922 (Clementi 1922), whereas in 1967 the efficacy of L-Asparaginase in treating leukemias was exhibited. L-Asparaginase is not an anticancer drug, in general, but is effective against certain types of tumours, thus becoming an essential component of chemotherapy strategies. Asparaginase was the first enzyme with antitumor activity to be intensively studied for acute lymphoblastic leukemia in children. This is used in combination therapy (Schemer and Holcenberg 1981) as this enzyme splits asparagine into aspartic acid and ammonium ions, thereby destroying these malignant cells. L-Asparaginase treatment for Acute Lymphoblastic leukemia is a major breakthrough in modern oncology as it induces complete remissions in over 90% children within four weeks (Gallagher et al. 1989). Commercially, it is available under the trade names Oncaspar, Kidrolase and Elspar.

The development of asparaginase from *Escherichia coli* and the serologically different asparaginase from plant pathogen *Erwinia carotovora* has not only added to the choice of antileukaemia drugs and provided a valuable guide to the selection and development of new therapeutic enzymes. Succinylated *Acinetobacter* glutaminase-asparaginase (SAGA) has better broader antitumour activity than *Escherichia coli* L- Asparaginase as proved by pharmacological and toxicological investigations in 20 adult patients. Chondroitinase AC and Chondroitinase B inhibit tumour growth, neovasculization and metastasis by removing chondroitin sulfate proteoglycans (Vellard 2000).

5. ENZYMES AND ARTHRITIS

Sir Archibald Garrod described and named arthritis, which is directly linked to problems with the circulatory system when too many circulatory immune complexes (CICs) collect in joint tissue resulting in arthritic inflammation putting the body in a state of autoimmunity and an often chronic and painful state (Ransberger 1986). It has been found that proteolytic enzymes go to the source of arthritis, reduce CICs and break-up the waste accumulations that cause painful arthritic conditions leading to much more effective treatment as compared to the temporary relief from arthritis pain offered by most prescription and over the counter medication. In a study, 29 arthritic subjects were investigated and were given bromelain therapy. Although this study was uncontrolled, after starting the bromelain therapy soft tissue swelling subsided mostly, or completely, in 72.4% of the subjects proving that oral bromelain therapy is able to produce anti-inflammatory and analgesic actions in humans (Cohen and Goldman 1964). Klein and Kullich (2000) conducted a randomized, double-blind trial comparing therapy with oral enzyme Phlogenzym containing bromelain, trypsin and rutin with diclofenac in treating gonarthritis (arthritis of the knee joint) in 73 patients. It was found that the two treatments were equivalent at a statistically significant level, with the oral enzymes being better tolerated. Other commercial enzyme preparations available are Wobenzym N and Mulsal. The specific combination of systemic oral enzymes present in them is one of the best medicines for both adults and children for the treatment of inflammatory joint conditions and non-inflammatory active osteoarthritis and ankylosing spondilytis.

One of the most important treatments today for arthritis particularly the most difficult to treat joint disease such as rheumatoid arthritis is, systemic oral enzymes. These enzymes are a medication for the basic treatment in arthritis, acting on symptoms and having disease modifying properties. The symptomatic efficacy of systemic oral enzymes is comparable to non–steroidal anti-inflammatory drugs (NSAIDs) as they can treat all joints and have no side effects in contrast to NSAIDs . Virtually all common stages and types of arthritis can be treated using systemic oral enzymes, which also has a preventive and a therapeutic effect on osteoarthritis and also on traumatic events. After surgical intervention, these enzymes support in the regeneration of the joint tissues and also the contralateral joints, which are overused due to the operated joints.

6. ENZYMES IN AIDS

Aids caused by a retrovirus named as human immunodeficiency virus (Montagmier et al. 1984; Sabatier 1987) is one of the most severe infection ever known to have attacked the human population. Although immunological dysfunction is common to all AIDS patients, the clinical spectrum of HIV infection is diverse and multiple organ involvement is frequently evident (Dalgteish et al. 1984). Till today, there is no conclusive treatment to eliminate HIV from the body, however, timely treatment of opportunistic infections can keep one healthy for many years. The commonly available treatment for AIDS is the treatment against opportunistic infections.

Due to their suppressed immune systems, AIDS patients can fall prey to certain illnesses that people with healthy immune responses are rarely susceptible to AIDS wasting syndrome involves major weight loss, chronic diarrhea or weakness, and constant or intermittent fever for at least 30 days. Approved treatments include Marinol (dronabinol), Megace (megestrol), and Serostim (somatropin rDNA for injection).

During the last decade, researchers have developed powerful drugs called Antiretroviral drugs that drastically reduce the viral load in blood. However, they do not permanently cure one from HIV. This line of treatment, called HAART (Highly Active Antiretroviral Therapy) has resulted in a huge reduction of AIDS-related deaths but has serious side effects, is very expensive and is out of reach for a majority of the infected people.

Enzyme therapy has an excellent track record in the treatment of many types of autoimmune diseases as they build up the immune system simultaneously decreasing inflammation with no side effects. Manipulation of enzymes, especially protease results in a class of drugs that have proven to be "highly effective in knocking out the AIDS virus while allowing the immune system to recover". Dr. Stauder and his associates of the Medizinische Enzymforschungsgesellschaft (Germany) also investigated the use of hydrolytic enzymes as adjuvant therapy in AIDS/ARC/LAS patients showing high levels of circulating immune complexes (CIC). Experimental studies proved that certain hydrolytic enzyme mixtures having pancreatin, trypsin, chymotrypsin, bromelain, papain, amylase, and lypase with rutin eliminate IC. Pancreatic enzyme supplements might reduce faecal fat loss in human immunodeficiency virus (HIV)-positive patients experiencing nutrient malabsorption (Greener 2001). This drug called ABT-538 shuts off 99% of the AIDS virus's replication and is "10 times more effective than AZT in halting virus replication". The use of WOBENZYM alone and in combination with anti-retrovirals in selected groups of HIV seropositive individuals with CD4 counts 200 to 500 should be further investigated. Apart from this, lysozyme has been found to exhibit activity against HIV, as has RNAase A and urinary RNAase U, which selectively degrade viral RNA (Vellard 2003).

7. ENZYMES AND ALLERGY

An allergy is a hyperactive response of the immune system to certain substances called "allergens", which are "foreign" to our bodies and can range from food and pollen to dust and drugs. The main symptoms associated with allergy are asthma, itchy watery eyes, itchy running nose allergic salute - pushing up on the nose, causing a white crease to appear across the bridge of the nose, itching, eczema, hives, dark circles under and around the eyes, recurring headache, shortness of breath, wheeze, cough, diarrhea and stomach cramps.

According to many allergies experts, the most common cause of multiple food allergies is having a 'leaky gut' which is an increase in intestinal permeability (Reno and Devrais 1995). Milk allergy happens due to the lack of the immune system to recognize milk proteins as being harmless and can be minimized if not eliminated by digestive enzymes. Findings from animal and human studies confirm that DTP (diphtheria and tetanus toxoids and pertussis) and tetanus vaccinations can also induce allergic responses and can increase the risk of allergies, including allergic asthma. An analysis of data from nearly 14,000 infants and children revealed that having a history of asthma is twice as great among those who were vaccinated with DTP or tetanus vaccines than among those who were not. Work exposure to flour or cotton dust, animal fur, smoke, and a wide variety of chemicals has been linked to increased risk of asthma. An asthma attack usually begins with sudden fits of wheezing, coughing, or shortness of breath. However, it may also begin insidiously with slowly increasing manifestations of respiratory distress. A sensation of tightness in the chest is also common.

Enzyme therapy is effective in fighting allergies because of their ability to break down protein allergens and to block the process that causes an allergic reaction (Mc Cann 1993). Enzymes, which is proven to have an antiallergic response, are pancreatic enzymes, bromelain, lactase, trypsin and chymotrysin. These enzymes are available in the form of enteric-coated tablets, which can be taken with meals. Individuals with low pancreatic enzyme output have an increased chance of suffering from food allergies. In such cases, enzymes supplements augment the body's own pancreatic enzymes. Therefore, the use of pancreatic enzymes has been suggested in the treatment of food allergies. To treat allergic asthma, bromelain has been recommended which reduces the thickness of mucus though clinical actions in asthmatics remain unproven. The use of digestive enzymes, taken with each meal, is also of great benefit. Pure plant enzymes are fully absorbed following oral administration. These have the ability to hydrolyse dietary proteins and polypeptides that have leaked into the bloodstream as food antigens. While the exact mechanism that enables pancreatic enzymes to block allergic reactions from food is unknown, one possibility is that the enzymes improve digestion, thereby breaking down large allergenic proteins into smaller non-allergenic molecules. In addition, there is evidence that a proportion of orally administered enzymes can be absorbed intact into the bloodstream. Once inside the body, they are apparently capable of exerting an anti-inflammatory effect.

8. ENZYMES AND WOUND HEALING

Wound healing is the process of repair that follows injury to the skin and other soft tissues. They may result from trauma or from a surgical incision. In addition, pressure ulcers (also known as decubitus ulcers or bed sores), and burns are also considered wounds. Enzyme therapy greatly accelerates tissue repair and wound healing. Wounds heal more rapidly with reduced scar formation. Proteases in particular are well-known for their ability to speed muscle healing, blood cleansing, enhance circulation, promote joint health, provide yeast control, and other benefits. Use of pancreatic enzymes can be traced back more than 200 years to scientist and physician Jean Senebier (1742-1809) of Switzerland who applied animal gastric juice to poorly healing wounds and open varicose ulcers of several patients, which resulted in healing of inflamed tissues. Bromelain supplementation has also been shown to accelerate the healing of soft-tissue injuries in male boxers (Tassman et al. 1965) and reduce swelling, bruising, healing time, and pain (Blonstein 1960). Most currently available bromelain products are not enteric-coated, and it is not known if such products would be as effective as enteric-coated bromelain. Other enzymes used are fibrinolysin/deoxyribonuclease and collagenase .

These enzymes when combined with broad-spectrum antibiotics in ointment, powder or solution are recommended for removal of fibrin layers from wounds to improve healing. The therapeutic effectiveness and the tolerability of a collagenase-chloramphenicol ointment were studied on a group of 20 patients with lower limb ulcerations of various origins. In a clinical survey conducted in patients with burns, pressure sores, abrasions and trophic ulcers to value the antiseptic and reparative properties of collagenasis (clostridiopeptidiasi A), it was observed that results were highly positive in terms of the healing or cleansing of the wound bearing further witness to the reliability of the enzymes (Pientrantoni et al. 1996). Encouraging results have been observed for collagenase clostridiopeptidase in children with partial thickness burns. Several products of higher quality and purity are now in clinical trials. Repairzyme and Wobenzyme are the commercial enzyme preparations available for treating different wounds while Debrase gel dressing and Vibrilase ™ have been found to be effective in treating burns (Vellard 2003).

CONCLUSION

Since enzymes have a key role to play in the functioning of every organ system, their therapeutic applications range from enzyme-replacement therapies against a number of enzyme deficiency diseases to therapies against acute cardiac infarction. However, major obstacles to be overcome in enzyme therapies include low potency and selectivity, inadequate regulation of activity, instability, immunogenicity and high manufacturing cost. Active research is going on in many parts of the world to find new and more potent enzymes to be used as drugs and supplements leading to the establishment of enzyme therapy.

ACKNOWLEDGEMENTS

The authors acknowledge with thanks Dr. (Mrs.) Neelam Saxena, Dr. Anoop Batra, Gautam Kr. Meghwanshi, Saurabh Saran, Pritesh and Rekha for their help and support in critically evaluating the manuscript.

REFERENCES

Balakrishnan V, Hareendran A, Nair CS (1981). Double-blind cross-over trial of an enzyme preparation in pancreatic steatorrhea. J Asso Phys Ind 29:207-209.

Beard J (1906). The Action of Trypsin upon the Living Cells of Jensen's Mouse Tumor. Br Med J 4:140-141.

Blonstein J(1960). Control of swelling in boxing injuries. Practitioner 203:206.

Bunte H (1960). Langenbecks Arch. Dtsch. Z. Chir. 296: 129-137.

Cassileth B (1998). The alternative medicine handbook. W.W. Norton and Co. New York, USA.

Clementi A (1922). La desemidation enzymatique d L-Asparagine chez les differentes especes anilmales et la signification physiologique de sa presence dans l'organisma. Arch Int physiol. 19: 369-376.

Cohen A, Goldman J (1964). Bromelain therapy in Rheumatoid arthritis. Pennsylvania Med. J. 67: 27-30.

Dalgeish AG, Beverly PCL, Chapman PR (1984). The CD4 (CT4) antigen is an essential component of the receptor for the AIDS retrovirus. Nature. 312: 763-767.

Flohë L, Gunzler WA, GmbH G (1990). Enzymes in therapy. Enzymes in industry: production and applications. Edited by Wolfgang Gerhartz. VCH Verlagsgesellschaft (Weinheim) and VCH Publishers (New York). pp-178-18.

Gallagher MP, Marshall RD, Wilson R (1989). L-Asparaginase a drug for treatment of acute lymphoblastic leukemia. Essays biochem. 24: 1-40.

Gonzalez NJ, Isaacs LL (1999). Evaluation of pancreatic proteolytic enzyme treatment of adenocarcinoma of the pancreas, with nutrition and detoxification support. Nutr Cancer. 33:117-24.

Greener M (2001). AIDS: Pancreatic enzyme supplements might counter nutrient malabsorption. A DGReview of:"Efficacy of oral pancreatic enzyme therapy for the treatment of fat malabsorption in HIV-infected patients". Alimentary Pharmacology and Therapeutics.

Heinicke RM, Van der Wal M, Yokoyama MM (1972). Effect of bromelain on human platelet aggregation. Experientia 28:844-845.

Higazi AA, Bdeir K, Hiss E, Arad S, Kuo A, Barghouti I, Cines DB (1998). Lysis of plasma clots by urokinase-soluble urokinase receptor complexes. Blood 92(6):2075-2083.

Klein G, Kullich W (2000). Short-term treatment of painful osteoarthritis of the knee with oral enzymes; A Randomised, Double-Blind study versus Diclofenac. Clin Drug Invest 19:15-23.

Knill-Jones RP, Pearce H, Batten J (1970). Comparative trial of Nutrizym in chronic pancreatic insufficiency. Brit Med J. 4:21.

Leipner J, Saller R (2000). Systemic enzyme therapy in oncology: effect and mode of action. Drugs 59:769-780.

Lopez DA, Williams RM, Miehlke M (1994). Enzymes: the fountain of life (The Neville Press).

Montagmier L, Chermann J, Cbourne-Sinouse F (1984). A new retrovirus human T-lymphotropic: characterization and possible role in lymphoadenopathy and acquired deficiency syndrome. New York, Coldspring Laboratories.pp- 363-379.

Morita AH, Uchida DA, Taussig SJ (1979). Chromato graphic fractionation and characterization of the active platelet aggregation inhibitory factor from bromelain. Arch Inter Phar Ther 239:340-350.

Ouriel K, Welch EL, Shortell CK, Geary K, Fiore WM, Cimino C (1995) Comparison of streptokinase, urokinase and recombinant tissue plasminogen activator in an *in vitro* model of venous thrombolysis. J. Vasc. Surg. 22(5): 593-597.

Pientrantoni S, Bergantino A, Scorza V (1996). Enzymatic cleaning with collagenase in the local treatment of skin lesions complicated by cab and necrotic debris. Minerva Chir. 51(9): 751-754.

Ransberger K (1986). Enzyme treatment of immune complex diseases. Arthritis Rheuma. 8:16-9.

Reno L, Devrais J (1995). Allergy Free Eating, Celestial Arts, Berkeley, CA pp. 19-20.

Sabatier (1987) Aids and the third world. Published in association with the Norwegian Red Cross. pp- 5-75.

Schemer G, Holcenberg JS (1981). Enzymes as drugs. Edited by J.S.Holcenberg and J.Roberts, Wiley Interscience, New York, pp 455-473.

Schultis K, Wagner E (1968). Dtsch.Med.Wochenschr. 93: 1685-1691.

Shively FL (1969). Multiple Proteolytic Enzyme Therapy of Cancer. Dayton: Johnson-Watson.

Tassman G, Zafran J, Zayon G (1965). A double-blind crossover study of a plant proteolytic enzyme in oral surgery. J Dent Med. 20:51– 54.

Vellard M (2003). The enzyme as drug: application of enzymes as pharmaceuticals. Curr. Opinion in Biotechnol. 14: 444-450.

WarrellRPJr, Chou TC, Gordon C, Tan C, Roberts J, Sternberg SS, Philips FS, Young CW (1980). Phase I evaluation of sucinylated Acinetobacter glutaminase-asparaginase in adults. Cancer Res. 40(12): 4546-51.

Combating Multidrug Resistant Infectious Microbes: A Burgeoning Problem

S. SAXENA

1. INTRODUCTION

1.1 Antimicrobial Resistance : A Global Threat

The discovery of antibiotic "Penicillin" was the triumph over the disease-causing bacteria, is one of the greatest success stories of modern medicine. Over past fifty years, antibiotics have been the corner stone in classical treatment of bacterial infectious disease therapy but today after their widespread use, many antibiotics do not pack the same punch once they did. This is due to the phenomenon of antimicrobial drug resistance. Indeed many dramatic successes achieved during this period led to a common expression among the clinicians that antibiotic based chemotherapy heralded the complete quest over infectious diseases. This overconfidence and conception of re-markable healing power of antibiotics was further driven by a variety of factors ranging from misuse of antibiotics, interaction between patients and the prescribers, economic incentives given by the pharmaceutical company for a particular drug, the regulatory environment and characteristics of the country's growth system.

"Antimicrobial resistance is a global problem that needs urgent action". Infectious diseases are the world's leading killers after cardiovascular diseases as they account for 13.3 billion deaths which is approximately 25%

Department of Biotechnology & Environmental Sciences, Thapar Institute of Engineering & Technology (Deemed University), Patiala, Punjab-147004 India. Email: sanjaibiotech@yahoo.com; ssaxena@tiet.ac.in if the total global deaths (WHO 2002). Antibiotic resistance signifies the "*Par excellence Darwinism*" i.e., the surviving strains have emerged as a result of protection and selection by the antibiotic. It has also revealed the genetic plasticity of the bacteria in terms of acquisition or replacement of genetic traits among the genera and species which evolutionarily are millennia apart.

Patient's perception of a new drug in the market to be more effective than older drugs leads to self-medication. Prescribers' perceptions regarding patient expectations and demands substantially influences prescribing practice. Physicians can be pressured by patient expectations to prescribe antimicrobials even in the absence of appropriate indications. Patient compliance with recommended treatment is another major problem. Patients forget to take medication, interrupt their treatment when they begin to feel better, or may be unable to afford a full course, thereby creating an ideal environment for microbes to adapt rather than be killed. Hospitals, are worldwide major contributors of the problem of antimicrobial resistance. The combination of highly susceptible patients, intensive and prolonged antimicrobial use, and cross-infection have resulted in nosocomial infections with highly resistant bacterial pathogens.

In past sixty years, hundreds of tons of antimicrobial agents have been released on bacterial populations, which have responded to deluge with a progeny, which is no longer susceptible but flourishes i.e., have become drug resistant.

2. NICHE OF MULTIDRUG RESISTANT MICROBES (MDR'S)

2.1 Drugs and Bugs are Ubiquitous

The global society is skeptical about multidrug resistance and therefore not only suffering with the antibiotic fad but is in pursuit to eradicate the infectious bugs from their vicinity by overexploiting antibacterials i.e., the compounds that kill bacteria. These chemicals may be toxic when taken through oral route but commonly find themselves as ingredients of healthcare products like soaps, surgical cloths and antiseptic lotions. They have also been impregnated to items such as textiles, toys, high chairs, mattress pads without having sufficient evidence that addition of such antibacterials would ward off infectious microorganisms.

Thus, nature is the reservoir of antimicrobial drugs as well as bugs; however, the overuse or misuse of these drugs has increased the residual levels in the environmental reservoirs leading to the emergence of super bugs or the multidrug resistant microbes.

2.2 Antibiotics in the Environment

Hospitals and Multidrug resistant Microbes: Hospitals are a breeding ground of antibiotic resistant bacteria. This is driven by failure of hospital hygiene, selective pressure by overuse of particular antibiotic and mobile genetic elements that encode for bacterial resistance mechanisms (Weinstein 2001). At any given point of time approximately 25-30% patients receive systemic antibiotics for treatment of infections or as prophylaxis (Eickoff 1992). This creates enormous selective pressure and the emergence and spread of antibiotic resistant bacteria. Intensive Care Units (ICU) in hospitals worldwide are faced with rapid spread and emergence of antibiotic resistant bacteria. Both Gram-positive and Gram-negative bacteria are reported as important causes of hospital acquired infections (Vincent et al. 1995; Hanberger et al. 1999; Sieradzki et al. 1999). Methicillin resistant *Staphylococcus aureus* (MRSA) is a serious problem in hospitals. This is due to the fact that 50% of healthy populations are carrier of *Staphylococcus aureus* and about 30% of populations are prolonged carriers in their nostrils and skin (Waldvogel 1995). The incidence of Methicillin resistance in *Staphylococcus aureus* on a doubling trend in past 10 years i.e from 20-25% in 1990 to 25- 50% in 1997 (Jones 2001). The first MRSA was reported in 1968 in UK (Barrett et al. 1968) and since then the hospital incidence of MRSA isolates has radically increased. Ciprofloxacin, one of the first Floroquinolones was intensively used for the treatment for MRSA infections (Mulligan et al. 1987). Resistance to ciprofloxacin by MRSA was reported from clinical specimens in early 1990s (Harnett et al. 1991). Within a span of one year between 1999 and 2000 an estimated 125,969 hospitalizations were recorded with diagnosis of MRSA infections including 31440 for septicemia, 29,823 for pneumonia and 64,706 for other infections (Kuehnert et al. 2005). *Streptococcus pneumoniae* is a leading cause of pneumonia, meningitis and middle ear infections in the developed world and by estimates cause more than 1 million childhood deaths per year in developing countries (Mulholland 1999). Penicillin resistant *Streptococcus pneumoniae* (PRSP) once unknown is a common feature in the community today. Overall prevalence of PRSP in 34 medical college centers in United States is to the tune of 12.1%. Worldwide prevalence of PRSP is on a rise as exemplified in England and Wales where it was below 1% in 1990 and was found to be 7.4% in 1997. Resistant strains of PRSP are found to be associated with hospitals and nursing homes (Reacher et al. 2000).

Predominant organisms responsible for multidrug resistant nosocomial infections are *Candida albicans, Staphylococcus aureus, Pseudomonas aeruginosa* in blood streams and *Candida* species, *E. coli* and *Pseudomonas aeruginosa* in respiratory tract and surgical infections (Hseuh et al. 1999). One US based surveillance study indicated that MRSA accounted for 57% infections in ICU's (NNIS 2003). Similarly, a UK based surveillance accounted for 43% *S. aureus* bacteraemias (Livermore 2004).

Vancomycin Resistant Enterococci (VRE) is the second most common organisms recovered from nosocomial urinary tract and wound infections. The reason that these microorganisms thrive is that they acquire resistance to several commonly used antibiotics (Schaberg et al. 1991). Risk factor of appearance of VRE in hospitalized patients is due to heavy use of vancomycin, third generation Cepahlosporins and active β-lactams (Edmond et al. 1995). MRSA and VRE are very hardy surviving on hospital environmental surface as well as on hands and in the gastrointestinal tract of healthcare personnel (Gold and Mollering 1996).

Antibiotics and other antibacterial agents enter natural water systems through residential or commercial discharges including hospital effluents. These first seep into the soil from where they step into surface water, ground water and finally the potential drinking water. Bacteria or the residential microbial community is affected qualitatively as well as quantitatively by the presence of antibiotic residues. Despite the fact that antibiotics are used by human body, cases may vary where 50-90% of an administered drug may be excreted by the body in biologically active form (Raloff 1998). Hospital effluents contribute a large part partially metabolized antibiotics which are discharged into the sewage system. The dilution of hospital effluents by municipal sewage will lower the concentration of antibiotics only moderately, because municipal waste water also contains antibiotic substances and disinfectants from households, veterinary sources and to a minor extent from livestock and account for induction of antibiotic resistance in bacteria surviving in municipal sewage systems. Antibiotics have been detected in the μg/L range in municipal sewage, in the effluent of sewage treatment plants (STPs), in surface water and in ground water (Chitnis et al. 2000). Ciprofloxacin a floroquinolone antibiotic has been detected in a range of 0.7-124.5μg/L and Ampicillin in range of 20-80 μg/L (Hartmann et al. 1999; Kummerer, 2001; 2003). *Enterococcus faecalis* has been accounted for transfer of antibiotic resistance plasmids to other enterococci and *Listeria monocytogenes* in water treatment plants (Marcinek et al. 1998). VRE was first isolated from STP's in England and a small town in Germany and later from pig and poultry farms (McDonald et al. 1997). Hospital effluent has recently shown that sewage systems have a high incidence of coliforms predominated by *Escherichia coli* (Chitnis et al. 2004).

2.3 Agriculture, Food and Multidrug Resistance

To comply with the ever-increasing demand for food and dairy products agricultural practices have evidenced drastic changes. Antibiotics have been used as therapeutic agents to combat bacterial infections and have been misused grossly misused in the field of medicine by the practitioners as well as the patients. Applications of antibiotics in agriculture began in early 1950s wherein they were given in the feeds in the range of 5- 10 ppm mainly as prophylactic and growth promoting agents rather than treating infections (Levy 1992). However, the range of antibiotic use in agriculture is no more sacrosanct as the application rates have increased to 10-20 folds.

Drugs like Bacitracin, Chlortetracycline or Erythromycin are being supplemented in the feed of goats and sheep at rate of 35-100 mg per head which have found to induce an increase of 3% to 5% in weight gain and feed efficiency (Gillespie 1997). Swine feed is being supplemented with 2-500g of Bacitracin, Chlortetracycline,Erythromycin, Penicillin, Streptomycin, Tylosin, Virginiamycin per ton of feed and in range of 1- 400 g for poultry. Of the one million tons of the antibiotics released in last 50 years in the biosphere, 50% are estimated to flow through the veterinary and agricultural channels (Prescott 2000; Mazel and Davies 1999; Levy 1992). Additional problems surface when there are structural similarities between drugs for human and veterinary use, for example, Ormetoprim a veterinary drug is structurally similar to trimethoprim which is used in humans (Fineman 1998). Antibiotics also enter the environment through dusting of fruit trees against diseases as a prophylactic measure, for example, for the control of fire blight in apples (Vidaver 2002).

Antibiotic resistant zoonotic pathogens in particular *Salmonella*, *Escherichia coli* and *Enterococci* in animals used as food is of special concern since these have a high probability of getting transferred to humans (WHO 1997). Food is not only an agricultural or trade related commodity but also a public health issue. This is evident from the serious outbreaks of food borne diseases have occurred in recent years in virtually all continents demonstrating the public health and social aspect of the food borne pathogens. It has also been observed that outbreaks of new pathogens are emerging and long known are expanding their reach in food production and distribution processes which is rapidly evolving to meet the diversified consumer demand and international competition. Food and water borne diseases account for nearly 2.1 million deaths in developing countries and affects 30% population in the industrialized counties (Heymann 2002). Globalization of food trade presents an intercontinental challenge to food safety control agencies in that food can become contaminated in one country and cause outbreaks of food borne illness in another. The escalation of food production and the consolidation of the food industries present opportunities for food borne pathogens to infect large numbers of consumers. The principal invasive intestinal bacterial pathogens of food-animal origin are *Campylobacter, Salmonella, Listeria, Escherichia coli* O157:H7 (and other Shiga toxin–and enterotoxin-producing strains of *E. coli*), *Yersinia* and *Vibrio enteritidis;* ground beef with *E. coli* O157:H7; pork with *Yersinia;* and shellfish with *Vibrio.* Multidrug resistant *Salmonella typhimurium* DT104 emerged in 1988 among cattle in England and Wales before it became common in humans (Akkina et al. 1999). Cattle to person transmission of *E. coli* O157:H7 has been confirmed as the likely source of *E. coli* O157: H7 infections in England (Trevera et al. 1999), Canada (Louie et al. 1999) and the USA. *Campylobacter jejuni* in United States alone is responsible for an estimated 2.4 million cases of illness with an estimated hospitalization rate of 10.2% and fatality rate of 0.1% (Mead et al. 1999; CDC, 2000). Resistance to human isolates of *C. jejuni* increased from 13% to 18% from 1997-98 and Quinolones resistant *C. jejuni* increased from 1.3% to 10.2% between 1992 to 1998 (Smith et al. 1999). VRE from pigs, poultry and humans were isolated from Germany and this emergence appeared to be associated with high volume use of this and other glycopeptides as growth promoters in food animals (Klare et al. 1995). *E. coli* O57:H7 infestation is found in a range of meat and produce products from salami to melons to lettuce. Okinawa, Japan has a local US military base was linked to an *E. coli* O157: H7 infestation in hamburger made from frozen ground beef. Further investigation revealed that the meat was produced some six months earlier in the US. Unwashed apples pressed in a mill were found to contain vancomycin resistant *E. coli* which survived for three weeks in refrigerated unpasturized apple cidar (Besser et al. 1993). Staphylococcal food intoxication is estimated to the tune of 185,000 cases of food borne illness (Mead et al. 1999). *Staphylococcal* food infections have a shorter duration of incubation which could last from 30 minutes to 8 hours after the consumption of the contaminated food. *Staphylococcus aureus* can be transmitted from an infected food handler who may be suffering from a respiratory or superficial skin infection during food production or handling. Possibilities exist that *Staphylococcal* infection may be transmitted via cross-contamination of meat products with skin and hides of animals during slaughter. A variety of proteinaceous foods like meat, meat products, poultry, fish and fish products provide support to *Staphylococcal* growth. *Staphylococci* can tolerate 10-20% salt and 50-60% sucrose concentrations.

Staphylococcus aureus along with *Salmonella infantis* was isolated from a turkey that was responsible for a food borne disease outbreak in Florida prison in 1990 (Meehan et al. 1992). Food contamination by *Staphylococcus* is fatal since it produces a variety of enterotoxins which are not destroyed by cooking being heat stable and therefore do not ensure food safety. Animals are also heavily colonized with *Staphylococcus* leading to contamination during food processing. One report of food poisoning by MRSA is linked to shredded pork barbeque (Jones et al. 2002). Enterotoxin producing MRSA has also been isolated from kitchen workers in hospital kitchen in Brazil (Soares et al. 1997).

Sea foods and integrated fish farming is another major sector in food industry in South East Asia. The major feed additive is manure obtained from the livestock in particular pig and poultry. The manure contains high levels of antibiotic residues which creates a selective pressure in the fish pond leading to the evolution of antibiotic resistant bacteria. The residues also accumulate in the fish and thus, enter the food chain (Peterson et al. 2002).

Plant bacteria like *Salmonella*, *E. coli*, *Listeria species* and *Cyclospora caytenensis* globe trot after insinuating themselves into fruit, vegetables, poultry, beef and dairy products generating trade related infections. The situation becomes more complex when it involves unknown and unpredictable microbes such as *Cyclospora caytanensis*. *C. caytenensis* was first detected in humans in early 70s and was confined to tropical and subtropical regions of the developing world. In 1996, CDC (Centre for Disease Control), USA, received reports of 1500 Cyclosporiasis cases in the United States and Canada (Herwalett 2000). It was concluded by the investigators that the most plausible cause of infection was the use of contaminated water which was used for growing the raspberries. *C. cayetenensis* infections have also been associated with other produce including basil, lettuce and snow peas (CDC 1998; Herwalett 2000).

MRSA poses one of the greatest threats to the mankind as MRSA strains are generally resistant to most commonly used antibiotics. The glycopeptide drug Vancomycin was probably the last hope to combat this organism until recently it was found that low level resistance has been developed (Hiramatsu 2001) and very recently the VRE has recently transferred to *Staphylococcus aureus* (Weigel et al. 2003) The transfer of the genetic element containing the van A vancomycin resistance gene from *Enterococcus faecalis* to *S. aureus* was demonstrated in the laboratory (Noble et al. 1992). The first clinical infection with VRSA was reported in 2002 (CDC 2002).

The heavy input of antibiotics and antimicrobial substances in every sphere of the biosphere has widely contributed to generation of antimicrobial resistance pathogenic microbes or the superbugs in different ecological niche, posing a great threat to the society and public health throughout the globe; and it is beyond expectation that the infections would be cured by first choice of antibiotics.

3. MECHANISMS BY WHICH MICROBES ACQUIRE RESISTANCE

Acquisition of antibiotic resistance operates through four routes viz, (a) does not absorb antibiotic (b) expels it (iii) degrades it (iv) has altered the usual molecular target for the antibiotics and drug does not have any effect. Structure altering or inactivating enzymes like β-lactasmases or aminoglycoside modifying enzymes, altering Penicillin Binding Proteins or cell wall target sites, altered DNA gyrase targets, permeability mutations and ribosomal modifications are some methods by which microbes acquire drug resistance (Levy 2002; Gold and Mollering 1996; Bearden et al. 2001; Aaterson 2001). Resistances can also occur through mutations i.e., point mutations, deletions, inversion, etc. in the bacterial genome. When a mutation for resistance to an antibiotic occurs and if the person is under treatment by the same antibiotic the antibiotic will kill or inhibit all non-resistant (susceptible) bacteria and leave antibiotic resistant bacteria to multiply and flourish. This is referred as selection. These antibiotic resistant bacteria can then transfer resistance to other bacteria through genetic material contributing to spread of the antibiotic resistant bacteria. In case of *Pseudomonas aeruginosa* correlation between high mutation rate and antibiotic resistance has been reported with patients suffering with cystic fibrosis in lungs (Oliver et al. 2000). Antibiotic treatment can also be selected for mutator bacteria present in all wild populations. These have a higher rate of mutations as compared to their normal counterparts. High incidence of mutator types have been reported in *E.coli, Salmonella enterica* and *P. aeruginosa* isolates from patients and natural environments having defective methyl directed mismatch repair (MMR) mechanism (Oliver et al. 2000; LeClerc et al. 1996).

Resistance has a fitness cost for the microorganism, but compensatory mutations occur rapidly and accumulate to overcome this fitness cost and this explains why many types of resistances may never disappear in bacterial population. Resistance gene sequences from drug resistant microbe can also spread via recombination into several class of gene expression cassettes and disseminated through horizontal gene transfer via transduction, conjugation or transformation. Transduction is mediated through bacteriophages and generally has a narrow host range i.e., remains restricted to a particular species only. Conjugation covers a broader range which crosses even the Gram barrier by cell to cell contact via conjugative elements. The capacity of transformation is found in naturally competent cells which have the ability to take naked DNA from the environment and remodel their DNA sequences by formation of mosaic genes (Bachi 2002).

Self-replicating plasmids, prophages, transposons, integrons and resistance islands all represent DNA elements that frequently carry resistance genes into sensitive organisms. These elements add DNA to the microbe and utilize site specific recombinases-integrases for their integration into the genome (Normark and Normark 2002).

Bacteria also develop resistance by either restricting the entry of the drug molecule within them or quickly expelling them out of their systems when entered. Some Gram-negative bacteria are able to change the structure of porins (proteins) channels restricting the entry of the antimicrobial agent. *Pseudomonas* resistance to Imipenam or Clistatin is an example of this mechanism. Resistance in *Pseudomonas aeruginosa* and *Escherichia coli* to Floroquinolones is attributed to their decreased ability to reach the target site in sufficient concentration so as to elicit a response (Walsch 2000; Hooper 2001).

Efflux pumps help in eliminating or expelling the antimicrobial drug which has entered the bacterial cell. Efflux pumps are basically variants of the bacterial membrane pumps responsible for the exchange of nutrients and movement of waste outside the cell. *E. coli, Shigella possess* an active efflux pump as a mechanism of antibacterial resistance. *Staphylococcus aureus* uses an efflux pump to develop resistance towards macrolide class of antibiotics. Last major mechanism to acquire resistance by a bacterium is the alteration or elimination of antibacterial drug target. Alteration in the penicillin binding proteins (PBP) is the major mechanism by which *Streptococcus pneumoniae* and *Staphylococcus aureus* become resistant to penicillin and Methicillin, respectively.

4. STRATEGIES TO TACKLE MULTIDRUG RESISTANT MICROBES

Antimicrobial drug resistance in pathogenic microbes has snowballed as a multifaceted complex public health issue with economic, social and political implications that are global in scope and cross all environmental and ethical boundaries. The multidrug resistant microbes have been categorized according to where and how they are transmitted such as drug resistant infections acquired in the community, in healthcare system, through the food supply or internationally. Thus, antimicrobial drug resistance is a global problem. The eradication of antimicrobial drug resistance is neither a realistic nor a desirable goal as eradication would mean a very significant reduction in use of antimicrobial drugs and benefits obtained from them. Thus, the objective should be effective use of antimicrobials against infection, reduction of morbidity and mortality and further spread of infection.

Dominant strategies which can be taken up to tackle antimicrobial drug resistance should include, surveillance of antimicrobial usage and antimicrobial drug resistance pattern adoption of measures ensuring appropriate and rational use of existing antimicrobials, and finally encouraging development of new antimicrobials and alternative treatments (WHO 2000).

4.1 Surveillance

Monitoring the resistance pattern over a time in clinical isolates at hospital as well as at community level is an important aspect to evaluate the pattern of use of antimicrobials and correlate with the actual consumption. Resistance varies with variation in the geographical location, as per the community structure and the level of health facility available. The local surveillance data, thus, is of prime importance for clinical management and upgradation of clinical guidelines from time to time and for educating the prescribers. The local surveillance data should be able to distinguish between the hospitals acquired infections as well as the community acquired infections.

There is a need of a national reference laboratory to monitor and coordinate epidemiological data of antimicrobial resistance in common pathogens in community, hospitals and other healthcare settings. A strong network of laboratories having external and internal quality assurance should collect and report resistance data and provide quality microbiology diagnostic services. Surveillance on consumption of antimicrobials should emphasize on aggregate data on the most expensive and highly used antimicrobial drugs, compare actual consumption with the expected consumption (morbidity data), monitor diagnostic insecurity, prescribers knowledge and habit, inappropriate promotion of antimicrobials and unrestricted availability of antimicrobials. Defined daily dose (DDD) can be used for the comparison of antimicrobial consumption across institutions, regions and countries.

4.2 National Approaches to Contain Antimicrobial Drug Resistance

A national task force should be consisting of ministry of health, health professionals (doctors, pharmacists, nurses), academia, national drug regulatory authority, pharmaceutical industry, consumer groups and NGO groups involved in healthcare. It should develop incentive-based approach for the optimal use of antimicrobials.

The national task force should promote programs for monitoring population wide infection control programs such as water safety and sanitation, immunization, public education on hygiene and prevention. The national reference laboratory should be able to ensure a reliable, good quality, epidemiologically sound data. This laboratory should establish standardized microbiological and diagnostic methods, provide external quality assurance for all participating laboratories and take part in external quality assurance.

. The public should be educated on prevention of infection and further transmission of infectious microbes using simple, cheap and effective measures. All healthcare professionals such as doctors, pharmacists, nurses and paramedical workers should be made aware of the antimicrobial drug resistance. Postgraduate training and continuing education programs for all in service personnel including intern doctors should be made mandatory. Active surveillance of multidrug resistant microbes to manage outbreaks of nosocomial epidemics is a necessity and therefore a committee responsible for the development of control of infections in hospitals. Interventions like hand washing by staff between patients and before undertaking any procedures like injections; use barrier precautions by wearing gloves and gowns for certain procedures; adequate sterilization or disinfection of supplies and equipments should be followed strictly to prevent the spread of antimicrobial drug resistance.

Misuse of antimicrobials may be curbed by enforcing prescription based dispensing of antimicrobials by pharmacists, which has been written by a prescribed practitioner. Further antimicrobials may be classified as "non-restricted", "restricted" or "very restricted". Antimicrobials designated as "non-restricted" may be prescribed are safe, effective and reasonably priced without the approval of a specialist. Restricted Antimicrobials should be more expensive, have a wider spectrum of activity and used for specific infections. Very restricted antimicrobials should be reserved for life threatening infections only which should be prescribed when culture sensitivity studies indicate the ineffectiveness of non-restricted or restricted class of antimicrobials.

Quality, efficacy and safety as per national standards should be monitored to prevent under dosage which leads to an increase in antimicrobial drug resistance. Manufacture and availability of substandard and counterfeit drugs available globally should be curbed by enhanced vigilance by governments, importers, exporters, retailers, pharmaceutical industry and healthcare professionals.

Non-human applications of antimicrobials are predominantly carried out for mass treatment of the livestock or poultry or for improving their yield and growth in the feed. A surveillance datum needs to be generated on the occurrence of antimicrobial drug resistance in livestock and poultry. Antimicrobials have also been used for curing fish as well as plant diseases. Improved surveillance agricultural settings will allow early detection of resistance trends in pathogens that pose a risk to animal and plant health, as well as in bacteria that enter the food supply. There should be a datum on prophylactic as well as growth promoting use of antimicrobials. Application of human therapeutics which has a growth promoting effect in livestock and poultry should be completely banned. An effective regulatory system should be designed to check the usage of antimicrobials in agriculture.

4.3 New Drugs and New Therapeutic Approaches

There is a need to enhance efforts to develop new therapeutic regimes or antibiotics that could block or circumvent the resistance mechanisms or attack a new target. Previously the Pharma industry was involved in developing newer versions of drugs which were previously isolated from bacteria or fungi (microbial natural products) by using chemical methods. However, this approach was partially and temporarily successful since the resistance mechanisms has co-evolved in the organisms producing it (Nikaido 1998). The pipeline for new drugs is small since the major pharmaceutical industries have abandoned the antibiotic discovery programs. Fortunately small startup companies have taken up this initiative but would require support from pharmaceutical majors (Projan 2003).

Antibiotic rotation with close microbiological monitoring could be a successful strategy to slow the emergence and restrict the spread of multidrug resistant microbes in health care settings (John 2000).

Monotherapy with potential drugs have a possibility of getting resistant by overuse and use in immunocompromised hosts. Therefore, a combination of two or more drugs is currently being used as strategy to restrict the emergence of drug resistance in bacteria.

4.4 Plants as a Source of New Antimicrobials

Plants produce a variety of secondary metabolites which could be directly or indirectly exploited as new antimicrobial drugs. A variety of antimicrobial compounds have been isolated from plants like Thymol (*Thymus vulagris*), Hypericin (*Hypericum perforatum*), Allicin from Garlic (*Allium cepa*) have shown excellent activity against MRSA, *Candida albicans*, *Cryptococcus neoformans*. These could be potential candidates to be developed into next generation of antimicrobials to combat multidrug resistant pathogenic microbes (Esquenazi et al. 2002; Gibbons et al. 2002; Ankri and Mirelman 1999). Clinical microbiologists have become interested in screening plant extracts for new phytochemical entities which could be directly used as antimicrobial drugs by physicians or serve as templates for designing efficient anti-infectives (Cowan 1999). Petalostemumol, from purple prairie clover has been found to have a potential antibacterial activity against *Bacillus subtilis* and *Staphylococcus aureus* (Hufford et al. 1993). Higher plants like *Shorea hemsleyana* have yielded Stilbene tetramers which have a prominent activity against MRSA (Nitta et al. 2002). Mansinone F is a sesquiterpene isolated from *Ulmus davidiana* var. *japonica* which has an excellent activity against 19 MRSA isolates with MIC range of 0.39 - 3.13 µg/ml which can be fairly compared with Vancomycin (0.39-1.56 µg/ml), the most widely used antibiotic to combat MRSA (Shin et al. 2000). There is a vast literature on plant compounds having potential activity against MRSA. However, there are possibilities that these compounds may also develop antimicrobial resistance being a single chemical moiety.

4.5 Resistance Modifying Agents or Bioenhancers

These are compounds which potentate the activity of a conventional drug towards a resistant microbe as a combination. These compounds generally target a resistant mechanism for reversal or act synergistically via an unknown mechanism Augmentin ® is a combination of Amoxicillin (β-lactam) and Clavulanic acid as potassium which is a microbially derived inhibitor of β- lactamases produced by antibiotic resistant bacteria or could act by inhibiting the Nor A efflux mechanism or multidrug resistance in *Staphylococcus aureus* (Gibbons et al. 2003). Plants also serve as a potential source of resistance modifying agents. Diterpenes like totorol potentate the activity of Methicillin via interference of penicillin binding proteins expression (Nicolson et al. 1999). When incorporated at a concentration of 1 mg/ml it increased the activity of Methicillin against MRSA by eight folds. Inhibition of efflux is potentially one way to improve the clinical efficacy of an antibiotic, even in the presence of target based mutations, by increasing intracellular antibiotic concentrations.

Because of emerging resistance to all classes of antibiotics, in particular the fluoroquinolones, there has been a significant focus by the pharmaceutical industry on addressing this problem. Reserpine, a plant alkaloid, is a known inhibitor of both mammalian and Gram-positive bacterial efflux (Mullein et al. 2004). Reserpine was found to enhance the MIC of tetracycline by four folds when studied against a variety of MRSA and Methicillin susceptible *Staphylococcus aureus* (MSSA) strains (Schmitz et al. 1998). 5-Methoxyhydrocarpin (5-MHC) is been synthesized by several medicinal plants producing Berberine which is found to inhibit Nor A multidrug resistant pump of *Staphylococcus aureus* which confers resistance to quinolones and antiseptics (Hsieh et al. 1998). The accumulation of berberine in the bacterial cells was strongly increased in the presence of 5-MHC (Stermitz et al. 2000).

4.6 Herbal Pharmacogenomics and Proteomics

The approaches and strategies currently employed for antibacterial drug discovery have been revolutionized by the use of biotechnological methods in pharmaceutical and medical research. Genomics and Proteomics have been the fastest tools in Biotechnology for the antimicrobial drug discovery. Genomics could probably bring in limelight the prevalence of efflux pumps in all organisms (Saier et al. 1998). These pumps are significant in mediating both innate and acquired resistance to many antimicrobial agents (Lewis 1999). Pharmaceutical companies have now embarked on programs to find inhibitors of these pumps that could enhance the activity of, or prevent the development of resistance to current antimicrobial therapies (Renau et al. 1999).

Genomic approaches could further help in target identification which helps in deciding the utility of a particular drug candidate. This depends upon the nature of the target; if it is highly conserved in host as well as the pathogen then it is very difficult to identify a selective inhibitor. Valid drug targets are, as a first pass criterion, those genes that encode proteins (or structural/catalytic RNA molecules) required for the cell to grow (Black and

Hare 2000). Peptide deformylase (PDF) is a protein target with potential to facilitate the discovery of a broad-spectrum antibacterial drug. PDF is responsible for removing the *N*-formyl group from the N-terminal methionine following translation (Yuan et al. 2001). The ultimate goal of the genomics-driven, target-based approach is to discover novel antimicrobials that operate via a novel mechanism of action from natural resources (McDevitt and Rosenberg 2001).

Proteomics is defined as "the study of the protein products of the genome, and their interactions and functions." Fundamentally, proteomics is composed of two elements: the separation of protein products coupled with a means of identification. Genomics data is helpful in identifying new targets on the protein level. Proteomics has the advantage because of its ability to combine chemistry, biology, bioinformatics and engineering into one discipline. Protein function, tertiary structure, protein-protein and protein-ligand interaction and drug-target affinity can be determined collectively using this approach (Burbaum and Tobal 2002).

Herbal pharmacogenomics and proteomics can widen the scope and screening the natural products and natural product formulations with antimicrobial activity by isolating new targets of antimicrobial action and also help in validating the natural product formulations for minimizing the use of current and conventional antimicrobials.

CONCLUSION

The director general of WHO in his report on antimicrobial resistance has indicated that we are left with only a decade or two to make use of many of the medicines presently available to stop infectious diseases otherwise the world would go back into the preantibiotic era (WHO 2000).

Implementation of regulatory control for rational use of antibiotics in healthcare settings and agriculture would significantly reduce the emergence of antimicrobial drug resistant microbes. Further efforts in discovering novel antimicrobial agents from plants using genomic and proteomic approaches would be helpful in reducing the use of current antimicrobials. Simultaneously, conventional antimicrobials could be reused with a resistance modifying agent or bioenhancers as a formulation to revert drug resistance. Herbal formulations could also prove to be useful in combating antimicrobial drug resistance.

Hence, it can be concluded that antimicrobial drug resistance despite being a global concern can be controlled if there is a concerted global efforts using a combination of one or more strategies discussed above.

REFERENCES

Aaterson DL (2001). Extended-spectrum beta-lactamases: the European experience. Curr Opin Infect Dis 14: 697-701.

Akkina JE, Hogue AT, Angulo FJ et al. (1999). Epidemiologic aspects, control, and importance of multiple-drug resistant *Salmonella typhimurium* DT104 in the United States. J Am Vet Med Assoc 1999; 214: 790–8.

Ankri S, Mirelman D (1999). Antimicrobial properties of allicin from garlic. Microb Infect 1: 125-129.

Bachi BR (2002). Resistance mechanism in gram positive bacteria. Int J Med Microb 292: 27-35.

Bearden DT, Danziger LH (2001). Mechanism of action of and resistance to quinolones. Pharmacotherapy 21: 224S- 232S.

Besser RE, Lett SM, Weber JT, Doyle MP, Barrett TJ, Wells JG, Griffin PM (1993). An outbreak of diarrhea and hemolytic uremic syndrome from *Escherichia coli* O157:H7 in fresh-pressed apple cider. JAMA 269: 2217-2220.

Black T, Hare R (2000). Will genomics revolutionize antimicrobial drug discovery? Curr Opinion Microbiol 3: 522-527.

Burbaum J, Tobal GM (2002). Proteomics and drug discovery. Curr Opin Chem Biol 6: 427-433.

Centres for Disease Control and Prevention (2004). *Escherichia coli* O157:H7 infections associated with ground beef from a US military installation—Okinawa, Japan, Morb Mortal Wkly Rep 2005, 54: 40-42.

Centre for Disease control and Prevention (1998). Outbreak of Cyclosporiasis- Ontario, Canada. Morb. Mortality Weekly Rep. 47: 806-806.

Centre for Disease Control and Prevention (2002). *Staphylococcus aureus* resistant to vancomycin—United States, 2002. Morb. Mort. Wkly Rep, 51: 565–7.

Centre for Disease Control and Prevention (2005). Escherichia coli O157:H7 infections associated with ground beef from a US military installation –Okinawa, Japan Morb. Mortality Weekly Report 54: 40-42.

Chitnis V, Chitnis S, Vaidya K, Ravikant S, Patil S, and Chitnis DS (2004). Bacterial population changes in hospital effluent treatment plant in central India. Water Res 38: 441-447.

Chitnis V, Chitnis D, Patil S, Ravi Kant (2000) Hospital effluent. A source of multiple-drug resistant bacteria. Current Sci 79: 989- 991.

Eickoff TC (1992). Antibiotics and nosocomial infections. In Jr. Bennett and P.S. Branchmann (eds) Hospital Infections, Third Edition Boston MA Little Brown and Company.

Esquenazi D, Wigg MD, Miranda MM, Rodrigues HM, Tostes JB, Rozental S, da Silva AJ, Alviano CS. (2002). Antimicrobial and antiviral activities of polyphenolics from *Cocos nucifera* Linn. (Palmae) husk fiber extract. Res Microbiol 153: 647-52.

Feinmen SE (1998). Antibiotics in animal feed: drug resistance revisited. Am Soc Microbiol News 64: 24-30.

Gibbons S, Oluwatuyi M, Kaatz GW (2003). A Novel Inhibitor of Multidrug E.ux Pumps in *Staphylococcus aureus*. J Antimicrob Chemother 51: 13–17.

Gillespie JR. (1997) Modern livestock and poultry production. Albany (NY). Delmar Publishing; 1997.

Gold HS and Mollering RC (1996). Antimicrobial Drug Resistance. N Engl J Med 335: 1445-1453.

Hanberger H, Garcia-Rodriguez JA, Gobernado M, Goossens H, Nilsson LE, Struelens MJ (1999). Antibiotic Susceptibility Among Aerobic Gram-negative Bacilli in Intensive Care Units in 5 European Countries. JAMA 281: 67-71.

Harnett N, Brown S, Krishan C (1991). Emergence of quinolone resistance of Methicillin resistant *Staphylococcus aureus* in Ontario, Canada. Antimicrob Agents Chemother 35: 1911-1913.

Herwalett BL (2000). *Cyclospora cayetenensis*: a review focusing on outbreaks of Cyclosporiasis in the 1990s. Clin Infect Diseases 31: 1040-1057.

Heymann DL (2002). Food safety, an essential public health priority: Marakkesh, Morocco. In FAO/WHO global forum of food safety regulations 2002, Marrakesh.

Hiramatsu K (2001). Vancomycin resistant *Staphylococcus aureus*: new model of antibiotic resistance. Lancet Inf. Dis 1: 147-155.

Hooper DC (2001). Emerging mechanisms of fluoroquinolone resistance Emerg Infect Dis 7: 337-341.

Hsieh PC, Siegel SA, Rogers B, Davis D, Lewis K (1998). Bacteria lacking a multidrug pump: A sensitive tool for drug discovery PNAS 95: 6602-6606.

Jones TF, Kellum ME, Porter SS, Bell M, Schaffner W (2002). An outbreak of community acquired food borne illness caused by Methicillin resistant *Staphylococcus aureus*. Emerg Inf Diseases 8: 82- 84.

Kimball AM, Taneda K (2004). Emerging Infections and Global Trade: A New Method for Gauging Impact. Rev Sci Tech 23: 753-60.

Klare I, Heier H, Klaus R, Reissbrodt, Witte W (1995). Van A mediated high level glycopeptide resistance in *Enterococcus faecium*. FEMS Microbiol Lett 125: 165-171.

Kuehnert MJ, Hill HA, Kupronis BA, Tokars, JI, Solomon SL, and Jernigan DB (2005). Methicillin resistant *Staphylococcus aureus* hospitalizations. Emerg. Infect. Diseases 11(6): 868-872.

LeClerc J, Li EB, Payne WL, Cebula TA (1996). High mutation frequencies among *Escherichia coli* and *Salmonella pathogens*. Science 274: 1208-1211.

Levy SB (1992). The Antibiotic Paradox: How Miracle Drugs are destroying the Miracle. New York: Plenum Press.

Levy SB (2002). Active efflux, a common mechanism for biocide and antibiotic resistance. J Appl Microbiol. 92 Suppl: 65S-71S.

Lewis K (1999). Multidrug resistance: versatile drug sensors of bacterial cells. Curr Biol 9: R403-R407.

Louie M, Reed S, Louie L, Ziebell K, Rehm K, Borczyk A, Lior H (1999). Molecular typing methods to investigate transmission of *E. coli* O157:H7 from cattle to humans. Epidemiol Infect 123: 17-24.

Lukásová J, Sustackova A (2003). Enterococci and Antibiotic Resistance. Acta Vet Brno 72: 315-323.

Marcinek H, Wirth R, Muscholl-Silberhorn A, Gauer M (1998). *Enterococcus faecalis* gene transfer under natural conditions in municipal sewage water treatment plants. Appl Environ Microbiol 64: 626- 632.

Mazel D, Davies J (1999). Antibiotic resistance in microbes. Cell Mol Life Sci 56: 742-754.

McDevitt D, Rosenberg M (2001). Exploiting genomics to discover new antibiotics. Trends Microbiol 9: 611–617.

McDonald LC, Kuehnert MJ, Tenover FC, Jarvis WR (1997). Vancomycin-resistant enterococci outside the health care setting: prevalence, sources and public health implications. Emerg Infect Dis 3: 311-317.

Mead PS, Slutsker L, Deitz Y, MacCaig LF, Bersee JS, Shapiro G, Griffin PM, Tauxe RV (1999). Food related illness and death in United States. Emerg Inf Dis 5: 607-625.

Mullein S, Mani N, Groassman TH (2004). Inhibition of Antibiotic efflux in bacteria by the Novel multidrug resistance inhibitors Biricodar (VX-710) and Timcodar (VX-853). Antimicrob Agents Chemother 48: 4171- 4176.

Mulligan ME, Ruane PJ, Johnston L, Wong P, Wheelok JP, MacDonald K, Reinhardt JF, Johnson CC, Statner B, Blomquist I, McCarthy J, O'Brein W, Gardner S, Hammer L, Citron Dm (1987). Ciprofloxacin for eradication of Methicillin resistant *Staphylococcus aureus* colonization Am J Medicine 82: 215-219.

Nicolson K, Evans G, O'Toole PW (1999). Potentiation of Methicillin activity against Methicillin-resistant *Staphylococcus aureus* by diterpenes. FEMS Microbiol Lett 179: 233–239.

Nitta T, Arai T, Takamatsu H, Inatomi Y, Murata H, Iinuma M, Tanaka T, Ito T, Asai F, Ibrahim I, Nakanishi T, Watabe K (2002). Antibacterial activity of extracts prepared from Tropical and sub-tropical plants on Methicillin Resistant *Staphylococcus aureus*. J Health Sci 48: 273-276.

Noble WC, Virani Z, Cree RGA (1992). Co-transfer of vancomycin and other resistance genes from *Enterococcus faecalis* NCTC 12201 to *Staphylococcus aureus*. FEMS Microbiol Lett 93: 195–198.

Normark BH, Normark S (2002). Evolution and spread of antibiotic resistance. J Intern Med 252: 91-106.

Oliver A, Canton R, Campo P, Baquero F, Blazquez J (2000). High frequency of hypermutable *Pseudomonas aeruginosa* in cystic fibrosis lung infection. Science 288: 1251-1254.

Peterson A, Andersen JL, Kaewmak T, Somsiri T, and A. Dalsgaard (2002). Impact of Integrated Fish Farming on Antimicrobial Resistance in a Pond Environment. Applied Environ Microbiol 68: 6036-6042.

Prescott JF (2000). Antimicrobial drug resistance and its epidemiology. In Antimicrobial therapy and Veterinary medicine , edn 3 Edited by Prescott JF, Baggot JD, Walker RD Ames Iowa State University Press 2000: 27-49.

Projan SJ (2003). Why is big Pharma getting out of antibacterial drug discovery? Curr Opin Microbiol 6: 427–430.

Raloff (1998). Drugged waters: Does it matter that Pharmaceuticals are tuning up with water supplies. Sci News 153: 187-189.

Reacher MH, Shah A, Livermore DM, Wale MCJ, Graham C, Johnson AP, Heine H, Monnickendam MM, Barker KF, James D, George RC (2000). Bacteraemia and antibiotic resistance of its pathogens reported in England and Wales between 1990 and 1998: trend analysis. BMJ 320: 213-216.

Renau TE, Leger R, Filonova L, Flamme E, Wang M, Yen R, Madsen D, Griffith D, Chamberland S, Dudley MN, Lee VJ, Lomovskaya O., Watkins W, Ohta T, Nakayama K, Ishida Y (2003). Conformationally-restricted analogues of efflux pump inhibitors that potentiate the activity of levofloxacin in *Pseudomonas aeruginosa*. Bioorg Med Chem Lett 13: 2755-2758.

Saier MH Jr, Paulsen IT, Sliwinski MK, Pao SS, Skurray RA, Nikaido H (1998). Evolutionary origins of multidrug and drug-specific efflux pumps in bacteria. FASEB J 12: 265-274.

Schaberg DR, Culver DA, Gaynes RP (1991). Major trends in microbial etiology of nosocomial infection Am J Med 91: 72S-75S.

Schmitz FJ, Fluit AC, Luckefahr M, Engler B, Hofmann B, Verhoef J, Heinz HP, Hadding U, Jones ME (1998). The effect of reserpine, an inhibitor of multidrug efflux pumps, on the *in vitro* activities of ciprofloxacin, sparfloxacin and moxifloxacin against clinical isolates of *Staphylococcus aureus*. J Antimicrob Chemother 42: 807–810.

Shin DY, Kim HS, Min KH , Hyun SS, Kim SA, Huh H, Choi EC, Choi YH, Kim J, Choi SH, Kim WB, and Suh YG (2000). Isolation of a Potent Anti-MRSA Sesquiterpenoid Quinone from *Ulmus davidiana* var. *japonica* 48: 1805-1806.

Sieradzki K, Roberts RB, Haber SW, Tomasz A (1999). The development of vancomycin resistance in a patient with methicillin-resistant *Staphylococcus aureus* infection. N Engl J Med 340:517-23.

Soares MJ, Tokumara Miyazaki NH, Noleto AL, and Figueiredo AM (1997). Enterotoxins production by *Staphylococcus aureus* clones and detection of Brazilian epidemic MRSA clone among the food handlers. J Med Microbiol 46: 214-221.

Stermitz FR, Lorenz P, Tawara JN, Zenewicz LA, Lewis K (2000). Synergy in a medicinal plant: Antimicrobial action of Berberine potentiated by 5'- methoxyhydocarpin a multidrug pump inhibitor. PNAS USA 97: 1433-1437.

Trevera WB, Willshaw GA, Cheasty T, Domingue G, and Wray C (1999).

Transmission of vero cytotoxin producing *E. coli* O157 infection from farm animals in Cornwall and West Devon. Communicable Diseases & Public Health 2: 263-268.

Vidaver A (2002). Use of antimicrobials in plant agriculture. Clin Infect Dis 34: S107-S110.

Vincent JL, Bihari DJ, Suter PM, Bruining HA, White J, Nicolas-Chanoin MH, Wolff M, Spencer RC, Hemmer M (1995). The prevalence of nosocomial infection in intensive care units in Europe. Results of the European prevalence of infections in intensive care (EPIC) study. JAMA 274: 639-44.

Waldvogel, FA (1993). *Staphylococcus aureus*: In GL Mandell, Bennett JE, and Dolin R (eds.) Mendel, Douglas and Bennett's Principle and Practice of Infectious Diseases. Fourth Edition NY Churchill Livingstone Inc., P 1757.

Walsch C (2000). Molecular mechanism that confer antibacterial drug resistance. Nature 406: 775-781.

Weigel LM, Clewell DB, Gill SR, Clark NC, McDougal LK, Flannaagan SE, Kolonay JF, Shetty J, Killgore GE, Tenover FC (2003). Genetic analysis of a high level Vaucomycin resistant isolate of *Staphylococcus aureus*. Science 302: 1569-1571.

WHO (2001). Global strategy for the containment of antimicrobial resistance. World Health Organization; Geneva.

Yuan Z, Trias J, White RJ (2001). Deformylase as a novel antibacterial target. Drug Discov Today 6: 954–61.

Proteomics and Genomics Approach to Study Plant-Microbes Cross Communication

R. MALLA[1], U. POKHAREL[2] R. PRASAD[3] and A. VARMA[4]

1. INTRODUCTION

When phylogeny and ecology are considered, the arbuscular mycorrhizal symbiosis is probably the most important interaction between plants and microbes. Mycorrhizae are obligate biotrophs which colonize a wide range of plant species. These symbionts are mutually beneficial. The fungus delivers phosphate to the plant while receiving carbon and lipid (Bago et al. 2003). Arbuscular mycorrhizae (AM) played a key role in the evolution of vascular land plants and remain a vital component of plant communities (Karandashov et al. 2004). About 80% land plants form AM, which improve their nutrient uptake. AM are an integral part of many ecosystems, and are considered to be particularly advantageous to plants growing in tropical soils, in which nutrients are cycled rapidly and are present in low steady-state concentrations (Parniske 2004). AM fungi belongs to the *Glomeromycota*, an ancient group of fungi that was present about 450 million years ago, and were instrumental for plants to colonize land (Remy et al. 1994; Redecker et al. 2000; Schussler et al. 2001). As few as 150 fungal species are known to form AM in the roots of vast number of plant species. AM fungi cannot complete their life cycle in the absence of a host root. The indis-

1   Tribhuvan University, Trichandra College, King's way, Kathmandu, Nepal.
2   Punjab University, Department of Microbiology, Chandigarh, India.
3 & 4 Amity Institute of Microbial Sciences, Amity University Uttar Pradesh Sector-125, New Super Express Highway, Uttar Pradesh, Noida, India.

pensable compounds that are provided by the host root have not yet been identified. Moreover, the genetics of the *Glomeromycota* are still a mystery; no sexual stages are known and single spores can contain hundreds of nuclei. No evidence for recombination has been found in the fungi suggesting that they reproduce clonally and have been asexual for the entire period of their association with plants. The observation of rDNA polymorphism indicated that nuclei coexisting in a single fungal cell are not genetically identical (Kuhn et al. 2001), a view that has been challenged recently (Pawlowska and Taylor 2004).

Mycorrhizal plants often exhibit an increased nutrient uptake, compared with non-mycorrhizal plants which can be achieved by the development of large extraradical mycelium (Sanders and Tinker 1971). In AM symbiosis exchange of nutrients takes place across the symbiotic interfaces developed during the colonization process of host plant root (Smith and Smith 1990). The growth of AM fungi in host roots is regulated by the supply of photosynthates and the growth of both partners in the symbiosis is well-harmonized (Saito 1997). When fungal hyphae grow within the intercellular spaces of the root cortex, intercellular interfaces are created. When fungal hyphae penetrate the wall of the root cells intracellular interfaces are developed. The intracellular structures include coils of fungal hyphae formed during initial colonization of epidermis and hypodermis in many AM roots and in cortical cells of *paris*- type mycorrhizas. In *arum*- type mycorrhizas, arbuscules are the main intracellular structures (Smith and Smith 1997). Both inter and intracellular interfaces have the same basic structure, they are always composed of the membranes of both symbionts separated by a narrow interfacial space. There are significant differences in composition and structure of both plant and fungal walls and in interfacial apoplast (Bonfante 2001). When AM fungi colonize root cells to form arbuscules the host plasma membrane invaginates and proliferates around the developing fungus. The plasma membrane surrounding the arbuscules is termed periarbuscular membrane. Alexander et al. (1989) have reported a 3.7- fold increase of host plasmalemma in arbuscule containing cells. The formation of periarbuscular membrane must involve massive synthesis of protein and lipids. Some of the proteins being unique to this membrane. The difficulty of isolating the periarbuscular membrane has hampered its biochemical characterization. However, Benabdellah et al. (2000) have shown by 2D-PAGE analysis of plasma membrane proteins isolated from mycorrhizal and non-mycorrhizal roots that development of the symbiosis induces the synthesis of atleast 10 new plasma membrane polypeptides. In addition, the amount of phospholipids is higher in membranes isolated from mycorrhizal roots (Benabdellah 1999).

Cytochemical and immuno-cytochemical analysis have shown that the periarbuscular membrane displays ATPase activity, which is usually absent in peripheral membrane of the colonized cortical root cells, suggesting that this membrane acquire transport functions that are usually absent in the root cortical cells (Gianinazzi-Pearson et al. 1991, 2000). Although immunocytological studies have shown that the periarbuscular membrane shares components with the peribacteroid membrane in the *Rhizobium*-legume symbiosis, the nature of other proteins in this membrane remains unknown.

Use of affinity probes, such as enzymes, lectins or antibodies has shown that the interfacial apoplast between the periarbuscular membrane and the fungal surface presents many molecules typical of the primary cell wall, such as cellulose, β- 1,4- glucans, polygalacturans and hemicellulose (Bonfante and Perotto 1995; Gollotte et al. 1997). However, morphological analysis of the interface material indicates that these components are not assembled into a fully structured cell wall (Bonfante and Perotto 1995). *In situ* hybridization experiments have shown that activation of some genes in arbuscule containing cells such as those encoding proteins rich in hydroxyproline (Balestrini et al. 1997), arabinogalactan proteins (van Buuren et al. 1999) and xyloglucan endotransglycolases (Maldonado-Mendoza and Harrison 1998) might account for the composition and structure of the interfacial matrix.

2. STRUCTURE OF AM FUNGI

On the basis of structure formed by these fungi in colonized roots, the mycorrhizal fungi may be, Ectomycorrhizas (ECM), Arbuscular mycorrhizas (AM), Ericoid mycorrhizas, Arbutoid mycorrhizas, Monotropoid mycorrhizas, Ecto-endomycorrhizas, Orchidaceous mycorrhizas and Jungermannioid mycorrhizas. Today despite the large number of plant species forming AM associations worldwide, only two major morphological types have been defined, the *Arum* and *Paris* type, respectively. Dickson (2004) surveyed 12 plants colonized by six species of arbuscular mycorrhizal fungi to explore the diversity of *Arum* and *Paris* mycorrhizal structures. The survey indicated that there was a continuum of mycorrhizal structures ranging from *Arum* to *Paris* depending upon both the host plant and the fungus. The time course showed that the total colonization increased and the establishment of the various mycorrhizal structures did not appear to change greatly over time. He concluded that the morphological structures in individual plants could be grouped into classes that were more diverse than *Arum* and *Paris* previously recorded plant families by Smith and Smith (1997). Eight classes were recognized, forming a structural sequence between *Arum* and *Paris* type. These are: classic arum with intercellular hyphae (IH), and arbuscules in cortical cells, arum with IH and paired arbuscules in adjacent cortical cells, distinct individual arbuscules on IH and PH (penetrating hyphae), distinct individual arbuscules but on PH, distinct individual arbuscule on PH and IH in outer layers of root, arbusculate coil (AC) and hyphal coils (HC) in inner cortex and IH in outer layers of root, *Arum* and *Paris* (both arbuscules and arbusculate coils in cortical cells), *Paris* (arbusculate coils and hyphal coils in cortical cells).

Arum-type mycorrhizas were formed by all three fungi (*Glomus caledonium, Glomus intraradices* and *Gigaspora rosea*) in Flax, (*Linum usitatissimum*) with paired arbuscule (Smith et al. 2004) as shown previously (Dickson et al. 2003). *Medicago truncatula* as shown previously formed *Arum*-type AM with the two *Glomus* species (Burleigh et al. 2002). The *Paris*-type AM formed in plant/fungus combination was capable of phosphorus transfer, even though there were very few arbuscule coils. This is consistent with mycorrhizal effects on phosphorus uptake in Australian weed *Asphodelus fistulosus* via *Paris*-type AM colonized by *G. coronatumin* (Cavagnaro et al. 2003). At low phosphorus, *A. fistulosus* showed very marked positive responses to colonization both in phosphorus uptake and growth. Both responses decreased with increased P supply. De Grandcourt et al. (2004) studied mycorrhizal colonization, growth, phosphorus content, net photosynthesis and root respiration on seedlings of two-co-occurring species (*Dicorynia guianansis* and *Eperua falcata*) grown at three soil phosphorus concentration with or without inoculation with arbuscular mycorrhiza seedlings of both species were unable to absorb phosphorus in the absence of mycorrhizal association. They exhibited coil *Paris*-type mycorrhizae. Regarding phosphorus acquisition, the two species belong to two different functional groups, *D. guianensis* being an obligate mycotrophic species.

3. FUNCTIONS OF AM FUNGI

The key functions of AM symbiosis can be summarized as follows: (1) Plant establishment and improvement in rooting, (2) Enhancing plant tolerance to biotic and abiotic stresses, (3) Improving nutrient cycling, (4) Enhancing plant community diversity (Zhang et al. 2003). Use of AM fungi in the long-term would thus favor an agricultural system that is both production and protection oriented, thus enhancing and stabilizing agro-ecosystems (Leake 2001; Waceke et al. 2001). Lipid, which is the dominant form of stored carbon in the fungal partner and which fuels spore germination, is made by the fungus within the root and is exported to the extraradical mycelium. The glycoxylate cycle is the central to the flow of carbon in the AM symbiosis (Lammers et al. 2001). As AM is mutualistic symbiont, AM fungi drain up to 20% of photosynthetic carbon (Jakobsen and Rosendahl 1991) and in return, provide plants with large amounts of nutrients (P, N, K, Zn, etc.) and water from the soil. AM fungi also produce glycoprotein extracellularly on the mycelia in the bulk soil, which together with the physical network of hyphae, help to aggregate soil (Wright and Upadhyaya 1996). During arbuscular mycorrhizal development fungal hyphae grow throughout root epidermis, exodermal and cortical cell layers to reach the inner cortex where the symbiosis functional unit the arbuscules develop. Three essential components of a plant signaling network, a receptor-like kinase, a predicted ion-channel and a calmodulin-dependent protein kinase have been identified. A detailed morphological study of symbiotic plant mutants revealed that different subsets of plant genes support the progress of fungal infection in successive root cell layers. Moreover, evidence of a diffusible fungal signaling factor that triggers gene activation in the root has recently been obtained (Parniske 2004).

Wurst et al. (2004) studied the effects of earthworm (*Aporrectodea caliginosa*) and AM, *Glomus intraradices* on the growth and chemistry of *Plantago lanceolata*. They found earthworm did not affect AM root colonization. Earthworms enhanced shoot biomass and VAM reduces root biomass. AM increased plant phosphorus content, but reduced the total amount of nitrogen in leaves. They suggested that earthworm promotes plant growth by enhancing soil nitrogen availability. The colonized plant is better nourished and adapted to its environment. It obtains increased protection against environmental stresses (Sylvia and Williams 1992), including drought (Subramanian et al. 1995), cold (Charest et al. 1993; Paradis et al. 1995), salinity (Giri and Mukerji 2003; Hilderbrandt et al. 2001) and pollution (Tonin et al. 2001; Turnau et al. 2001). In addition, symbiosis tends to reduce the incidence of root diseases and minimizes the harmful effect of certain pathogenic agents (Dehne 1982) by improving the growth and health of colonized plants. At the same time, they obtain increased protection against environmental conditions detrimental to their survival.

4. BIOTECHNOLOGICAL APPLICATIONS

In recent years, mycorrhizal studies have intensified as researchers look to understand and improve plant fitness, to develop safe bioherbicides and to produce compounds for industrial and pharmaceutical applications (Shearer 2001). These fungi can be used as tool for biological hardening of micropropagated plantlets and potent biological control agent against root pathogens. Not much progress has been made in genetic engineering and commercialization of AM fungi in forestry, agriculture and flori-horticulture due to the absence of axenic culture. To express foreign genes of importance to the host plants in a symbiosis specific manner protoplast would serve as excellent biological systems. Viable protoplasts can also be used to exchange metabolites which are not accessible with intact hyphae, root tissues, or fungus under functional symbiosis.

Transformation extends the potential for genetic manipulation not only in established genetic systems but also in fungi less amenable for study by conventional genetic analysis. Fungal transformation typically involves the use of regenerative protoplasts, identification of selectable marker genes and stable integration of the transforming DNA into the chromosome of the organism (Finkelstein 1992). Mycorrhizal fungi are highly recalcitrant to transformation because of the difficulties in culturing and obtaining protoplasts. Very few reports of transformation of ECM have been published (Bills et al. 1999; Peng et al. 1993). Despite the existence of diverse transformation techniques, gene delivery remains a limiting step in innumerable fields of research. Zak (1971) stated that identification of the fungus involved in the natural mycorrhizal structure is sharply limited by existing techniques. The development of detection and identification methods for the AM fungal matter within and outside the host plant root is of crucial importance since these symbiotic fungi cannot be raised axenically. The characterization of enzymes participate in symbiosis is an important aspect in mycorrhizal study. Till date there are no reports of isolation of pure enzymes from AM fungi as it suffers due to the non-cultivability of these fungi *in vitro* condition.

5. NUTRIENT TRANSPORT ACROSS INTERFACES IN AM SYMBIOSIS

Root colonization by the AM fungus is accompanied by the development of a network of fine extraradical hyphae, which increases the absorption rate of slow diffusing nutrients, mainly phosphate (P) from the soil to the plant (Jakoben 1999). Movement of phosphate in the symbiosis begins with its uptake from the soil solution by the phosphate transporters located in the extraradical hyphae, followed by its translocation to the intraradical fungal structures and transfer to the arbuscular interfaces of the colonized root cells. P uptake by extraradical hyphae is followed by the synthesis of large amounts of polyphosphate (polyP), which is partly stored in the fungal vacuoles (Cox et al. 1980; Dexheimer et al. 1996). *In vivo* P^{31} nuclear magnetic resonance has shown that the phosphate ions taken up by the fungus are transformed into polyP with a short chain length (Rasmussen et al. 2000). On the basis of observation, the authors suggested that the process of phosphate translocation along the hyphae of AM fungi may be similar to that of ectomycorrhizal fungi, in which the phosphate appears to be transported mainly as polyP as motile vacuole system. This is supposed to be followed by hydrolysis of polyP by the action of the phosphatases that have been localized in the vacuoles of the intraradical hyphae (Ezawa et al. 1995; Jabari-Hare et al. 1990; Tisserant et al. 1993) and exopolyphosphatases (Ezawa et al. 2001).

The P transfer across the symbiotic interface begins with efflux from the fungus into the interfacial apoplast. According to Clarkson (1985), membrane transport processes for P transport favour absorption over loss. Several hypotheses have been proposed to explain the high P loss from the arbuscule (Harley and Smith 1983; Smith and Read 1997). As individual arbuscule have a lifespan of about 7-11 days, it was accepted in the past that its senescence and partial breakdown would release P to be taken up by the host plant. Nevertheless, measurements of P flux rates across the interface showed that the release of P from senescing arbuscules would be far too slow to account for P transfer to the host (Cox and Tinker 1976). Consequently, it is believed that transporter of phosphate across the arbuscular membrane probably involved specific efflux channels or transporters.

Molecular studies of host plasma membrane H^+- ATPases and phosphate transporters have provided evidence that the subsequent uptake of phosphate by the plant may be an active transport process. Up-regulation of host plasma membrane ATPase gene has been demonstrated in barley, tobacco and tomato plants (Gianinazzi-Pearson et al. 2000; Murphy et al. 1997). In tobacco roots two H^+-ATPase genes were found to be induced in arbuscule-containing cells by using promoter β-glucoronidase (GUS) fusions (Gianinazzi-Pearson et al. 2000). Immuno-cytochemical analysis of these roots detected the H^+-ATPase protein in the periarbuscular membrane.

Rosewarne et al. (1999) have identified a tomato phosphate transporter that may be involved in the uptake of the phosphate that is effluxed across the arbuscular membrane. These observations suggest that the periarbuscular membrane plays an important role in the absorption of the phosphate ions by the host plant that have been transported by the fungus and released into the symbiotic interface through H^+-phosphate symporter activated by the plasma membrane H^+-ATPase.

Ryan et al. (2003) determined the concentration of phosphorus (P), potassium (K) and magnesium (Mg) in AM fungi by cryoanalytical scanning electron microscopy *in situ*. The amount of P in intraradical hyphae was high and formed strong linear relationship with K and Mg. Little Ca was detected. The P was undetectable in uncolonized cortical cell vacuoles but was generally elevated in vacuoles surrounding an arbuscule and in the liquid surrounding hyphae in intercellular space. Their results suggested that both young arbuscules and intercellular hyphae are sites for P-transfer that Mg^{2+} and K are probably balancing cations for P anions in hyphae and that host cells may limit arbuscule lifespan through deposition of material rich in Ca.

The extraradical hyphae of *G. intraradices* hydrolyzed 5-bromo-4-chloro-3-indolyl phosphate and phenolphthalein diphosphate; moreover they transferred significantly more P to roots when they had access to inositol hexaphosphoric acid (phytate). In additions, *G. intraradices* can hydrolyze organic P and the resultant inorganic P can be taken up and transported to host roots (Koide and Kabir 2000).

Casarine et al. (2004) studied a differential effect of ectomycorrhizal symbiosis on soil P mobilization and host P nutrition. They found *Rhizopogon roseolus* had a strong effect on the mobilization of poorly available P, whereas *Hebeloma cylindrosporum* had no effect. However, $CaCO_3$ suppressed the positive effect of *R. roseolus*. With soluble P addition, both ectomycorrhizal species improved shoot P concentration of *Pinus pinaster*. Poulton et al. (2002) studied the mycorrhizal infection and high soil P condition improved several vegetative and reproductive traits including pollen production per flower and per plant. These responses were greater for the male function than the female function.

6. ACQUISITION OF NITROGEN DIRECTLY FROM ORGANIC MATERIAL

AM fungi the *Glomales*, which form mycorrhizal symbiosis are believed to be obligate biotrophs that are wholly dependent on the plant partner for their carbon supply. In contrast other mycorrhizal types have been shown to acquire nitrogen directly from organic sources (Hodge et al. 2001). The arbuscular mycorrhizal symbiosis can both enhance decomposition of and increase nitrogen capture from complex organic material (grass leaves) in soil. Hyphal growth of the fungal partner was increased in the presence of the organic material, independently of the host plant. Recent studies suggest that some plants may circumvent N mineralization carried out by saprotrophs because their ectomycorrhizal fungi have the capacity to hydrolyze protein. Pretreatment of the protein-tannin complex by saprotrophs did make its N available to ectomycorrhizal fungi. However, when the protein-tannin complex was the major N source, *Pisolithus tinctorious* increased shoot P but not N content, even in the presence of saprotrophs. Interaction between saprotrophs and ectomycorrhizal fungi may be different for N and P because of immobilization of N by ectomycorrhizal fungi (Wu et al. 2003).

Hodge (2003) studied the capture of nitrogen by plants from N-rich complex organic materials as well as the subsequent impact on N capture of the addition of a mycorrhizal inoculum *G. hoi*. He found that the N capture from organic material was dependent on root length produced and was always higher when the mycorrhizal inoculum was present. Nitrogen isotope patterns in plants provide insight into plant N dynamics (Hobbie and Colpaert 2003).

Smith et al. (2004) investigated structural and functional diversity in arbuscular mycorrhizal symbiosis involving three plant species and three AM fungi and measured contributions of the fungi to P uptake using comparted pots and P^{33}. The plant/fungus combination varied in growth and P responses. *Linum usitatissimum* responded positively to all fungi, and *Medicago truncatula* to *Glomus intraradices* but not *Gigaspora rosea*. Specific activities of P^{33} in plants and soil indicated that fungal P uptake made substantial contributions to plant/fungus combinations. *G. intraradices* delivered close to 100% of the P in all three plants. This finding was fully in accord with previous conclusions from the data collected at 6 wk (Smith et al. 2003). *G. caledonium* and *Gi. rosea* delivered less P. The amount was not related to colonization or to growth of P responses. They concluded that AM colonization can result in complete inactivation of the direct P uptake pathway via root hairs and epidermis. Gavito et al. (2002) conducted an experiment to test whether P uptake by mycorrhizal hyphae could be enhanced by growing the host plant under CO_2 enrichment. The result indicated that root, not hyphal P uptake was increased by elevated CO_2 in the mycorrhizal treatments and also indicated that enhanced plant C supply does not alter growth or function of arbuscular mycorrhizal fungi. Larsen et al. (1996) found marked reduction of P transport by binomyl in *G. caledonium*-cucumber symbiosis, the activity of fungal alkaline phosphatase inside the roots was unaffected by binomyl, although Thingstrup and Rosendahl (1994) showed the activity of malate dehydrogenase and alkaline phosphatase were inhibited by binomyl.

7. PCR IN ARBUSCULAR MYCORRHIZAL FUNGI

Since the molecular biological techniques have been immersed in the understanding of the arbuscular mycorrhizal symbiosis, cDNA libraries have been established from mycorrhizal RNA using several molecular techniques such as suppressive subtractive hybridization and large numbers of clones are being sequenced to obtain expressed sequenced tags (ESTs). These ESTs can be transferred to arrays and hybridized with cDNA from different sources to obtain RNA accumulation profiles of genes expressed during the mycorrhizal symbiosis (Franken and Requena 2001). The major shortcoming of the analysis of AM fungal gene expression is the obligate symbiotic growth of the fungus. This limits the amount of fungal material available, which can only be obtained either from spores or from extra-radical hyphae. Approximately just 1% of the mRNA extracted from highly colonized roots belongs to the fungal partner, which hinders the discovery of low expression fungal genes. The targets for PCR cloning also involve genes involved in nutrient transport as well as in cell wall metabolism. A fragment of the gene coding for the nitrate reductase apoprotein from the AM fungus *G. intraradices* was PCR amplified and subsequently cloned. *In situ* hybridization and northern blot experiments showed that the corresponding mRNA was accumulated in fungal arbuscules and that its expression inversely correlated with the plant nitrate reductase activity. It was hypothesized that the AM fungi could assist their host in nitrate assimilation in this way during symbiosis (Kaldorf et al. 1998). Several copies of chitin synthase genes from *G. margarita* were found after PCR with degenerate primers. A new interesting target for PCR cloning could be the enzymes involved in lipid biosynthesis. A recent finding suggests that the failure of AM fungi to complete their life cycle in the absence of the plant could be due to a lack of storage lipid biosynthesis during the asymbiotic growth (Bago et al. 1999). It is, therefore, been suggested that the switch of catabolism to anabolism of storage lipid is a key step in AM fungal development. Cloning and expression analysis of genes encoding enzymes involved in lipid biosynthesis could possibly provide support for this hypothesis.

Microsatellite is the new approach in molecular biology developed as marker. Kretzer et al. (2004) for the first time used microsatellite marker to differentiate fungal gene from ectomycorrhizae with great confidence. They developed markers for two sister species of *Rhizopogon, R. vesiculosus* and *R. vinicolor* and used selected markers to investigate genet size and distribution from ectomycorrhizal samples. Both species formed ectomycorrhizas with tuberculate morphology on Douglas-fir. Gene diversity at the newly described microsatellite loci ranged from 0.00 to 0.68 in *R. vesiculosus* and from 0.00 to 0.43 in *R. venicolor*. The largest distance observed between tuberculate ectomycorrhizas of the same genet was 13.4 m for *R. vesiculosus* but only 2 m for *R. vinicor*.

Alkan et al. (2004) applied rapid method to quantify the colonization of the arbuscular mycorrhizal fungus *G. intraradices* in plant using quantitative real-time polymerase chain reaction (qRT-PCR) technique. Specific PCR primers for the fungus (28s rDNA sequence) and host root tissue were developed. High degree of correlation were obtained and is the first report of the application of the qRT-PCR technique for quantification of AMF colonization in plant which should open up the possibility of its wider application in AM research.

Peter et al. (2003) have sequences 1519 and 1681 expressed sequence tags (ESTs) from *Laccavia bicolor* and *Pisolithus microcarpus* cDNA libraries. Based on sequence similarities, 11% *L. bicolor* unique transcripts were also detected in *P. microcarpus*. Similarly, these two shared only a low proportion of common transcripts with other basidiomycetous fungi. Such a low proportion of shared transcripts between basidiomycetes suggests that the variability of expresses transcripts in different fungi and fungal tissues is considerably high.

A cDNA clone showing similarity to a yeast alkaline phosphatase (ALPase) gene, *PHO8* (Keneko et al. 1987) was found in an expressed sequence- tagged (EST) library constructed from the extraradical hyphae of *G. intraradices*. Using this clone Aono et al. 2004, cloned the ALPase gene from the AM fungi *G. intraradices* and *Gi. margarita* for the first time. They identified an EST clone (*GiO8A10*) with similarity to the *Saccharomyces cerevisiae* repressible ALPase encoded by the *PHO8* gene (Keneko et al. 1987). They obtained full length ALPase cDNA from *G. intraradices* and *G. margarita* designated *GiALP* and *GmALP*, respectively. Both *GiALP and GmALP* were constitutively expressed in mycorrhizal roots, irrespective of the growth stage, infection rate and environmental phosphate concentration. The levels of *GiALP and GmALP* transcripts were higher in mycorrhizal roots than in germinating spores and extraradical hyphae. The secondary structures of *GiALP* and *GmALP* and yeast ALPases are very similar. The hydrophobic domain, which is thought to be the transmembrane domain is present in the N- terminal regions of *GiALP, GmALP* and yeast ALPases. The yeast *PHO8* enzyme is localized in vacuoles and attached to tonoplast (Clark et al. 1982). In this regard, the enzymes encoded by *GiALP* and *GmALP* may also be membrane proteins. Yeast ALPase is a N-glycosylated protein composed of two identical subunits with an molecular mass of 66,000 of which the carbohydrate components accounts for about 8% (Onishi et al. 1979). Yeast ALPase is synthesized as an inactive form. The ALPase protein becomes an active form by carbohydrate modification and removal of C-terminal peptide through delivery of ALPase protein from endoplasmic reticulum (ER) to vacuole (Klionsky and Emr 1989). Probably the *GiALP* and *GmALP* proteins in yeast cells may not be processed appropriately into an active form.

8. RANDOM AMPLIFIED POLYMORPHIC DNA (RAPD)

Variation within species can be assayed using the RAPD method (Weish and McClelland 1990; Williams et al. 1990), in which arbitrary short oligonucleotide primers, targeting unknown sequences in the genome are used to generate amplification products that often show size polymorphism within species. RAPD analysis offers the possibility of creating polymorphism without any prior knowledge of the DNA sequences of the organism investigated. The method is fast and economic for screening large number of samples. RAPD band pattern has been used to define some fungal species in which species-specific bands or combinations of bands have been considered. In these techniques, there is the assumption that bands with identical mobility and staining intensity are of the same or very similar sequences. Characterization of species at morphological and protein level is not fully reliable, since environmental conditions influence the nature of the organism to a great extent. The use of molecular markers such as RAPDs along with morphological and protein traits may provide a more clear concept of the species. RAPD markers are randomly distributed throughout the genome and can be efficiently and randomly sampled using established procedure (Williams et al.1990). RAPD procedure developed by Welsh and McClelland (1990) and Williams et al. (1990) that involves simultaneous amplification of several anonymous loci in the genome has been used for genetic, taxonomic and ecological studies of several fungi (Zinno et al. 1998).

PCR based techniques have already been applied to endo and ectomycorrhizal fungi where morphological characters are in conflict, ambiguous and missing (Podila et al. 2004). This approach has allowed the development of molecular tools for their identification and increased the level of understanding in the molecular taxonomy of microorganism (Varma et al. 2004; Solaiman and Abbott 2004). The most commonly used PCR based techniques includes amplification of variable regions in the ribosomal genes, restriction fragment length polymorphism (RFLP), amplification of short repeated sequences (microsatellites) and random amplification of polymorphic DNA (RAPD) (Erlich et al. 1991). These techniques provide a different degree of resolution in the study of genetic polymorphisms. RAPD reveals intraspecific differences by originating DNA fingerprints, which may be unique for a single isolate (Perotto et al. 1996). Identification of individual clones is essential for the better understanding of the diversity, structure and dynamic of populations of ectomycorrhizal fungi. Unfortunately, this approach is time-consuming. RAPD (Welsh and McClelland 1990; Williams et al. 1990) has, therefore, been used for the analysis of population of *Suillus granulatus* (Jacobson et al. 1993) and *Laccavia bicolor* (Buschena et al. 1992). However, this technique has been reported to be very sensitive to experimental variables and RAPD assay conditions described for one species may not be suitable with another.

Junghans et al. (1998) studied twenty *Pisolithus tinctorius* isolates from different geographic locations and different hosts were characterized by the random amplified polymorphic DNA technique. Thirteen arbitrary primers generated 87 DNA fragments, all of them polymorphic. These data were used to calculate genetic distances among the isolates. The pairwise genetic distances ranged from 1 to 100%, with an average of 58.7%. Cluster based on the amplified fragments grouped the isolates according to their host and geographical origins. Group I contained isolates collected in Brazil and group II those collected in the Northern Hemisphere. In addition to the diversity seen at the molecular level, the isolates also showed host specificity. The data also suggest new guidelines for future investigations on the taxonomy and systematic of this important fungus species. Furthermore, these results support future experiments aimed at the selection and development of improved isolates of *P. tinctorius*. Huai et al. (2003) studied the genetic variation and spatial distribution of the ectomycorrhizal fungus *Tricholoma terreum*. The 33 sporophores studied belonged to distinct genotypes, based on analysis of random amplified polymorphic DNA (RAPD) markers. The genets of *T. terreum* were small and not larger than 0.5 m. Two major phenetic groups, i.e., eight individuals in group 1 and 25 in group 2, were identified by principal component analysis and the unweighed pair group method with arithmetic means of simple matching coefficients, respectively. The application of random amplified polymorphic DNA (RAPD) analysis for the identification of ectomycorrhizal symbionts of spruce (*Picea abies*) belonging to the genera *Boletus, Amanita and Lactarius* at and below the species level was investigated. Using both fingerprinting [M13, $(GTG)_5$, $(GACA)_4$] as well as random oligonucleotide primers (V1 and V5), a high degree of variability of amplified DNA fragments (band-sharing index 65-80%) was detected between different strains of the same species, hence enabling the identification of individual strains within the same species. The band-sharing index between different species of the same genus (*Boletus, Russula* and *Amanita*) was in the range of 20-30% and similar values were obtained when strains from different taxa were compared. Thus, RAPD is too sensitive at this level of relationship and cannot be used to align unknown symbionts to a given taxon. They, therefore, conclude that RAPD is a promising tool for the identification of individual strains and could thus, be used to distinguish indigenous and introduced mycorrhizal strains from the same species in natural ecosystem. The genetic variability of *Trichoderma* isolates using the RAPD were analysed by Góes et al. (2002) and found high intra-specific genetic variation among those fungi.

9. RAPD OF *PIRIFORMOSPORA INDICA* AND *SEBACINA VERMIFERA* SENSU MEMBERS OF THE NEW ORDER-*SEBACINALES*

The taxonomy and identification of Arbuscular Mycorrhizal Fungi (AMF) is based on the distinct morphology of their spores, cell wall structures and it is very difficult to distinguish between genera or species when fungi are within the root tissues (Read 1999; Varma 1999). Molecular characterization offers an alternative approach for more reliable and reproductive identification. The prime feature of such approaches are immunological characterization, protein profiles-one- and 2- dimensional –PAGE, isozyme polymorphism and DNA analysis. The analysis of enzymes, isozymes like Laccase, Malate dehydrogenase, Esterase, Peptidase, Peroxidase, Acid and Alkaline phosphatase and non-enzymic proteins and their mobility have displayed clear variations among different species. The introduction of molecular techniques like, DNA-DNA hybridization, restriction fragment length polymorphism (RFLP), random amplified polymorphic DNA (RAPD), DNA amplification fingerprinting (DAF) has been a major force in the areas of systematic and population biology. The PCR based methods have significantly increased the level of understanding in the molecular taxonomy of microorganism (Varma et al. 2004; Solaiman et al. 2004).

PCR based techniques have already been applied to endo and ectomycorrhizal fungi where morphological characters are in conflict, ambiguous and missing (Podila et al. 2004). This approach has allowed the development of molecular tools for their identification (Hartmann et al. 2004). The most commonly used PCR based techniques include amplification of variable regions in the ribosomal genes, restriction fragment length polymorphism (RFLP), amplification of short repeated sequences (microsatellites) and random amplification of polymorphic DNA (Erlich et al. 1991). RAPD reveals intra-specific differences by originating DNA fingerprints, which may be unique for a single isolate (Perotto et al. 1996; Morton and Redecker 2001).

The fungus *Piriformospora indica* (Verma et al. 1998; Varma et al. 2001; Malla et al. 2002; Pham et al. 2004a) has the capability to grow axenically as well as on the roots of a vast majority of plants tested including Liverworts, Pteridophytes and Gymnosperms. The fungus was named *Piriformospora indica* based on its characteristic pear-shaped chlamydospore and found to mimic the capabilities of typical AM fungus. Interestingly, the host spectrum of *P. indica* is very much like arbuscular mycorrhizal fungi (AMF). Pronounced growth promotional effect was seen with terrestrial orchids (Blechert et al. 1999; Rai et al. 2001; Sahay and Varma 1999). Proteomics and genomics data about this fungus has recently been described (Peskan et al. 2004; Kaldorf et al. 2005; Shahollari et al. 2005). *Sebacina vermifera* sensu Warcup and Talbot consists of a broad complex of species possibly including mycobionts of jungermannioid and ericoid mycorrhizas. Extrapolating from the known rDNA sequences in *Sebacinaceae*, it is evident that there is a cosm of mycorrhizal biodiversity yet to be discovered in this group.

Pear-shaped chlamydospores are the characteristic of both the genera, *P. indica* and *S. vermifera* sensu (Fig.1). The cultivability of these fungi provided ample opportunity to study the enzymes like Acid and Alkaline phosphatase (Fig. 2), immunological characterization (Fig. 3), protein profiles-one- and two- dimensional–PAGE (Fig. 4), isozyme polymorphism, molecular marker like random amplified polymorphic DNA (RAPD) to establish variability in between *P. indica* and closely related organism *S. vermifera* sensu (Weib et al. 2004; Malla and Varma 2004; Malla et al. 2004). This communication deals with the potential value of RAPDs tools and techniques enzymes polymorphism for detecting polymorphism among

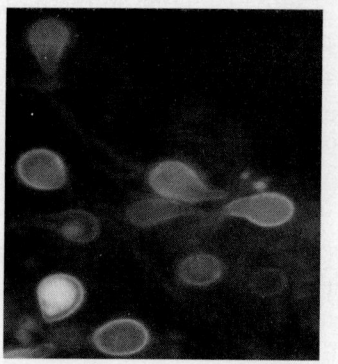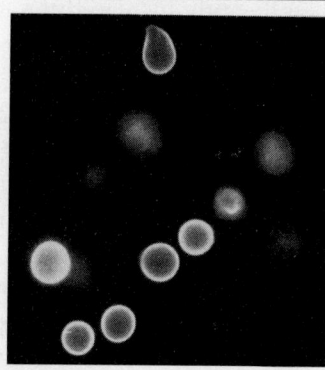

Fig. 1 Morphology of *Piriformospora indica* and *Sebacina vermifera* sensu.
Autofluorescence shown by *P. indica* (Left) and *S. vermifera* (Right) chlamydospores (arrow) seen under Leica microscope (model, 020-518.500) using I 3
filter with excitations range 450-490 λ at 400x. The characteristic pear-shaped
chlamydospores are more frequent in *P. indica* than *S. vermifera* sensu.

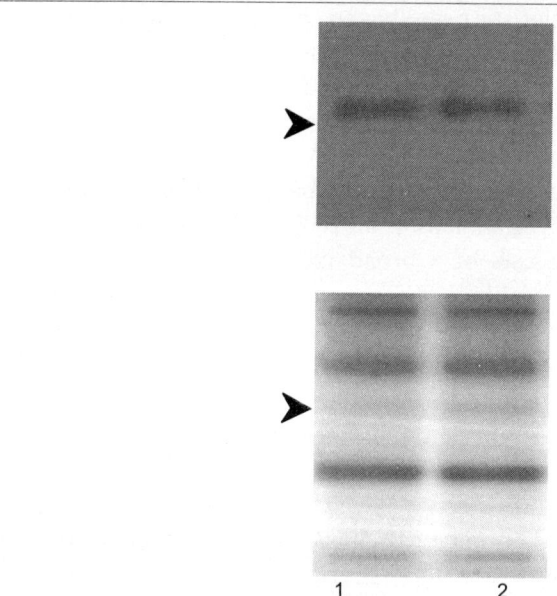

1   2

Fig. 2 Acid phosphatase of *Piriformospora indica* and *Sebacina vermifera* sensu in
native gel.
Native PAGE of ACPase stained with Coomassie blue (bottom) and gel assayed
(top) using *p*-NPP as substrate. Lane 1 *P. indica* and Lane 2 *S. vermifera* sensu.
Separation was done in 10% gel at 4 ºC for 6h.

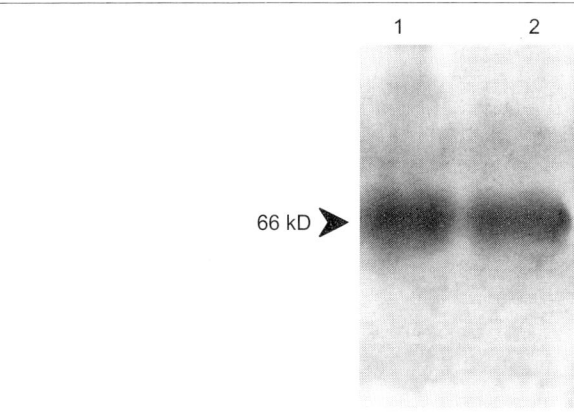

Fig. 3 Western blot analysis of *Piriformospora indica* and *Sebacina vermifera* sensu.
Immunoblot of acid phosphatase developed in *P.indica* and *S. vermifera* sensu
separated by 10% SDS PAGE transferred to nitrocellulose membrane by
electroblotting. Lane 1, *P. indica* reacted with homologous antiserum and Lane 2, *S.
vermifera* sensu cross-reacted with *P. indica* antiserum. The result shows precisely
defined bands representing similarity in their molecular mass supporting
immunologically highly related species.

symbiotic fungal isolates with special reference to *Piriformospora indica*
(Verma, Varma, Kost, Rexer and Franken) and *Sebacina vermifera* sensu
(Warcup and Talbot).

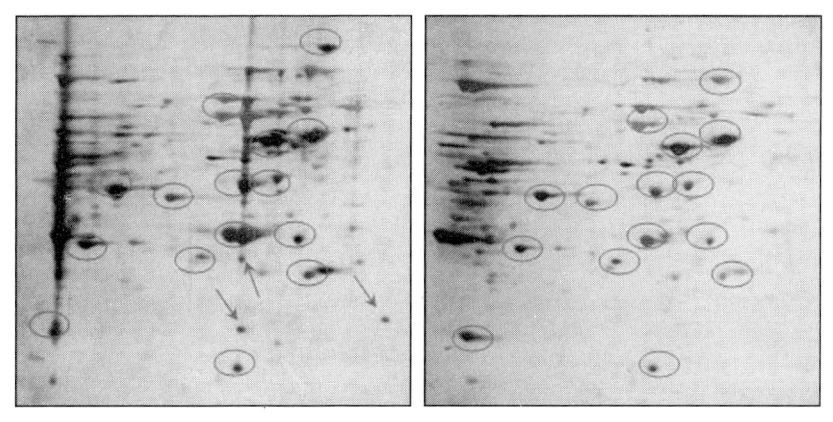

Fig. 4 2D PAGE of *P. indica* (Upper) and *S. vermifera* (Lower).

Actively growing colonies of *P. indica* and *S. vermifera* sensu were transferred in modified aspergillus medium (Hill and Kafer 2001; Pham et al. 2004b), incubated for 10 days at 28 ± 2°C in dark with constant shaking at 120 rpm. The morphological features of the fungi were studied with the aid of Leica microscope (Model, 020-518.500, Germany).

Isolation and purification of fungal DNA (Fig. 5) was done following modified CTAB protocol of Moller et al. (1992). After extraction, DNA was purified by using RNase A. DNA was diluted in TE buffer (1x) and stored at -20^0C for further use. RAPD analysis was done following Zinno et al. (1998). DNA amplification was performed in a total volume of 25 ml containing (ml): 2.5, Buffer (10X without $MgCl_2$), 2.5, $MgCl_2$, 0.8, dNTPs (10 mM), 1.0, primer (30 ng/μl), 0.5 *Taq* polymerase (3U/μl) and DNA concentration ranging from 5-25. Random 10 bp oligonucleotide primers (Operon Technologies Alameda, California) were used to produce amplification. The DNA was amplified in PTC-200 Thermal Cycler (Techne make, U.K.) with the following thermal profiles: 95°C for 5 min, initial denaturation cycle; 94°C for 30 sec, denaturation cycle; 36°C for 2 min, annealing; 72°C for 2 min, Extension; repeated for 36 cycles and final extension was done at 72°C for 5 min. For separation, the amplified DNA samples were mixed with 6x loading dye and were electrophoresed on 1.5% agarose gel in 1% TAE at 3.5 volt/cm for 2 hours, stained with ethidium bromide.

Statistical analysis was performed using the NTSYS-pc program (Rohlf 1992). The degree of genetic relatedness or similarity was estimated using

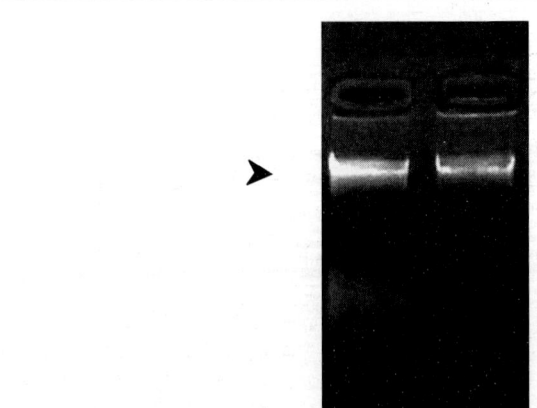

Fig. 5   Genomic DNA of *P. indica* and *S. vermifera* sensu.
P. indica (Left) and *S. vermifera* (Right). Isolation and purification of DNA was done by CTAB protocol.

the Jaccard coefficient. Clustering of similarity matrices was done by UPGMA (Un-weighted Pair Group method with Arithmetic Mean) and projection by TREE program of NTSYS-pc.

Out of seven 10mer oligonucleotide primers, six gave scorable, reproducible DNA products (bands) suitable for establishment of genetic diversity. RAPD patterns of the isolates amplified with primers OPA-10 (Lane 2 and 3), OPD-01 (Lane 4 and 5), OPC-06 (Lane 6 and 7), OPC-10 (Lane 8 and 9), OPC-01 (Lane 10 and 11), OPI-04 (Lane 12 and 13) and OPI-10 (Lane 14 and 15) clearly differentiated between *P. indica* and *S. vermifera* sensu (Fig. 6). Most amplicons obtained with 6 primers were polymorphic

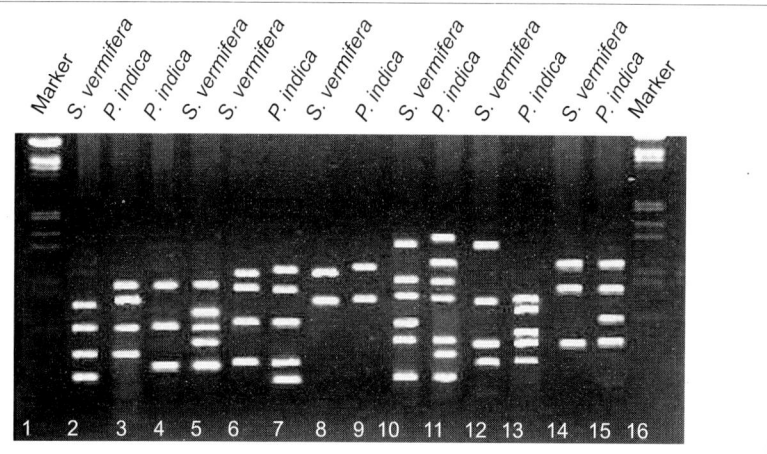

Fig. 6   Random amplified polymorphic DNA (RAPD) analysis of *P. Indica* and *S. vermifera* sensu.
The RAPD analysis of *P. indica* and *S. vermifera* sensu to show genetic variation between these two fungi. Out of seven primers used for amplification six have given a productive polymorphism. Lane 1 and 16 marker. Lane 2 and 3 Primer OPA10. Lane 4 and 5 OPD01. Lane 6 and 7 OPC06, Lane 8 and 9 OPC10, Lane 10 and 11 OPC 01. Lane 12 and 13 OPI04. Lane 14 and 15 OPI10. No polymorphism was observed when the genomic DNA was amplified with OPC10 Lane 8 and 9.

between *P. indica* and *S. vermifera* sensu. The RAPD fragments were used to calculate Jaccard's similarity coefficients. The result showed 58% similarity between these two fungi according to Dendrogram (Fig. 7).

CONCLUSION

Piriformospora indica and *Sebacina vermifera* sensu belonging to same taxonomic group show similar morphology, immunological relationship,

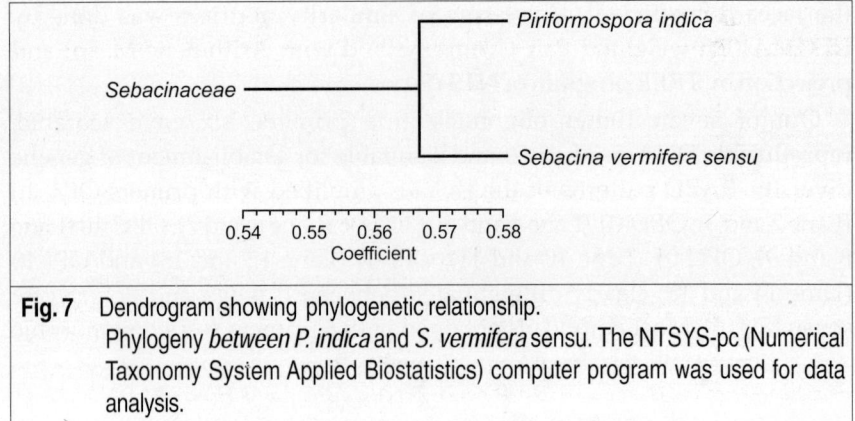

Fig. 7   Dendrogram showing phylogenetic relationship.
Phylogeny *between P. indica* and *S. vermifera* sensu. The NTSYS-pc (Numerical Taxonomy System Applied Biostatistics) computer program was used for data analysis.

functions and isozymes. However, they show distinct genetic variation based on the Random Amplified Polymorphic DNA (RAPD) analysis. Seven random primers of different origin were used. Out of seven, six primers gave scorable, reproducible DNA products (bands) suitable for establishment of genetic diversity. UPGMA cluster analysis clustered the isolates into two distinct groups. An average genetic similarity between both the fungi was 0.58 (i.e., 58%), can be considered to place in species of same roof. Results illustrated the potential value of RAPDs technique for detecting polymorphism among fungal isolates.

The RAPD data confirmed that even among these two species of *Sebacinales* belonging to same morpho-zymographical groups, the level of variation was substantially high according to RAPD. Molecular characterization offers an alternative approach for more reliable and reproductive identification at species level. We, therefore, suggest that along with morphology, immunological characterization, isozyme polymorphism, RAPD analysis are promising tools for the identification of individual strains, and could thus be used to distinguish indigenous and introduced mycorrhizal strains from the same species in natural ecosystem.

ACKNOWLEDGEMENTS

Rajani Malla is thankful to Dr. Ashok K Chauhan, founder president, Amity-RBEF New Delhi, Prof. Shital Raj Basnet and Prof. Bharat Mani Pokharel and Jeevan Sherchan from Tribhuvan University, Kathmandu. Sincere thanks to Dr. Upendra from All India Institute of Medical Sciences, Giang Huang Pham from International Centre for Genetic Engineering and Biotechnology, and Dr. Ravi and Uttam from Banaras Hindu University.

REFERENCES

Alexander T, Toth R, Meier R, Weber HC (1989). Dynamics of arbuscule development and degeneration in onion, bean, and tomato with reference to vesicular-arbuscular mycorrhizas in grasses. *Can J Bot* 67: 2505-2513

Alkan N, Gadkar V, Coburn J, Yarden O, Kapulnik Y (2004). Quantification of the arbuscular mycorrhizal fungus *Glomus intraradices* in host tissue using real-time polymerase chain reaction. *New Phytol* 161: 877-885

Aono T, Maldonado-Mendoza IE, Dewbre GR, Harrison MJ, Saito M (2004). Expression of alkaline phosphatase genes in arbuscular mycorrhizas. *New Phytol* 162: 525-534

Bago B, Pfeffer PE, Douds DD, Brouillette J, Becard G, Shachar-Hill Y (1999). Carbon metabolism in spores of the arbuscular mycorrhizal fungus *Glomus intraradices* as revealed by nuclear magnetic resonance spectroscopy. *Plant Physiol* 121: 263-271

Bago B, Pfeffer PE, Abubaker J, Jun J, Allen JW, Brouillette J, Douds DD, Lammers PJ, Shachar-Hill Y (2003). Carbon export from arbuscular mycorrhizal roots involves the translocation of carbohydrates as well as lipid. *Plant physiol* 131: 1496-1507

Balestrini R, Jose - Estanyol M,Puigdomenech P, Bonfante P (1997). Hydroxyproline-rich glycoprotein mRNA accumulation in maize root cells colonized by an arbuscular mycorrhizal fungus as revealed by *in situ* hybridization. *Protoplasma* 198: 36-42

Benabdellah K (1999). Alteraciones bioquimicas inducidas en membranes de celulas radicales de tomate (*Lycopersicon esculantum* Mill) Por la formacion de micorrizas arbusculares. Ph.D. Thesis. University of Granada, Spain

Benabdellah K, Azcon-Aguilar C, Ferrol N (2000). Alterations in the plasma membrane polypeptide pattern of tomato roots (*Lycopersicon esculantum*) during the development of arbuscular mycorrhiza. *J Exp Bot* 51: 747-754

Bills S, Podila GK, Hiremath S (1999). Genetic engineering of an ectomycorrhizal fungus *Laccaria laccata* for use as a biological control agent. *Mycologia* 91: 237-242

Blechert O, Kost G, Hassel A, Rexer R-H, Varma A (1999). First remarks on the symbiotic interactions between *Piriformospora indica* and terrestrial orchids. In: Varma A and Hock B (eds.) Mycorrhizae 2nd Edition, Springer-Verlag pp 683-688

Bonfante P (2001). At the interface between mycorrhizal fungi and plants: the structural organization of cell wall, plasma membrane and cytoskeleton. In: Hock B (ed) The Mycota: fungal association pp 45-61

Bonfante P, Perotto S (1995). Strategies of arbuscular mycorrhizal fungi when infecting host plants. New Phytol 130: 3-21

Burleigh SH, Cavagnaro TR, Jakobsen I (2002). Functional diversity of arbuscular mycorrhizas extends to the expression of plant genes involved in P nutrition. J Exp Bot 53: 1-9

Buschena CA, Doudrick RL, Anderson NA. (1992). Persistence of *Laccaria spp.* As mycorrhizal symbionts of container–grown black spruce. Can J Res 22: 1883-1887

van Buuren ML, Trieu AT, Blaylock LA, Harrison MJ (1999). Novel genes induced during an arbuscular mycorrhizal symbiosis formed between *Medicago truncatula* and *Glomus versiforme*. Mol Plant-Microb Interact 12: 171-181

Casarine V, Plassard PH, Arvieu (2004). Quantification of ectomycorrhizal fungal effects on the bioavailability and mobilization of soil P in the rhizosphere of *Pinus pinaster*. New Phytol 163: 177-185

Cavagnaro TR, Smith FA, Ayling SM, Smith SE (2003). Growth and phosphorus nutrition of a Paris-type arbuscular mycorrhizal symbiosis. New Phytol 157: 127-134

Charest C, Dalpé Y, Brown A (1993). The effect of vesicular-arbuscular mycorrhizae and chilling on two hybrids of *Zea Mays* L. Mycorrhizae 4: 89-92

Clark DW, Tkacz JS, Lampen JO (1982). Asparigine-linked carbohydrate does not determine the cellular location of yeast vacuolar non-specific alkaline phosphatase. J Biotechnol 152: 865-873

Clarkson DT (1985). Factors affecting mineral nutrient acquisition by plant. Annu Rev Plant Physiol 36: 77-115

Cox G, Tinker PB (1976). Translocation and transfer of nutrients in vesicular- arbuscular mycorrhizas. New Phytol 77: 371-378

Cox G, Moran Kg, Sanders F, Nokolds C, Tinker PB (1980). Translocation and transfer of nutrients in vesicular arbuscular mycorrhizas. Polyphosphate granules and phosphorus translocation. New Phytol 84: 649-659

Dehne HW (1982). Interaction between vesicular-arbuscular mycorrhizal fungi and plant pathogens. Phytopathology 72: 1115-1119

Dexheimer J, Gerard J, Ayatti H, Ghanbaja J (1996). Etude de I. origine et de la re?partition des granules vacuolaires dans les hyphes d'une endomycorhize a? ve?sicules et arbuscules. Acta Bot Gallica 143: 167-180

Dickson S (2004). The Arum-Paris continuum of mycorrhizal symbioses. New Phytol 163: 187-200

Dickson S, Schweiger P, Smith F A, Soderstrom B, Smith SE (2003). Paired arbuscules in the Arum-type mycorrhizal symbiosis with *Linum usitatissimum.* Can J Bot 81: 457-463

Erlich HA, Gelfand DH, Sninsky JJ (1991). Recent advances in the polymerase chain reaction. Science 252: 1643-1651

Ezawa T, Saito M, Yoshida T (1995). Comparison of phosphatase localization in the intraradical hyphae of arbuscular mycorrhizal fungi *Glomus spp* and *Gigaspora spp.* Plant and Soil 176: 57-63

Ezawa T, Smith SE, Smith FA (2001). Differentiation of polyphosphate metabolism between the extra and intraradical hyphae of arbuscular mycorrhizal fungi. New Phytol 149: 555-563

Finkelstein DB (1992). Transformation. In: Finkelstein DB, Ball C (Eds) in Biotechnology of Filamentous Fungi: Technology and Products. Butter-Heinemann. pp 113-156

Franken P, Requena N (2001). Analysis of gene expression in arbuscular mycorrhizas: new approaches and challenges. New Phytol 150: 517-523

Gavito ME, Bruhn D, Jakobsen I (2002). Phosphorus uptake by mycorrhizal hyphae does not increase when the host plant grows under atmospheric CO_2 enrichment. New Phytol 154: 751-760

Gianinazzi-pearson V, Smith SE, Gianinazzi S, Smith FA (1991). Enzymatic studies on the metabolism of vesicular-arbuscular mycorrhizas V is Proton ATPase a component of ATP hydrolyzing enzyme activities in plant fungus interfaces. New Phytol 117: 61-74

Gianinazzi-pearson V, Arnould C, Oufattole M, Arango M, Gianinazzi S (2000). Differential activation of H^+-ATPase genes by an arbuscular mycorrhizal fungus in root cells of transgenic tobacco. Planta 211: 609-613

Giri B, Mukerji KG (2003). Mycorrhizal inoculant alleviates salt stress in *Sesbania aegyptiaca* and *Sesbania grandiflora* under field conditions: evidence for reduced sodium and improved magnesium uptake. Springer-Verlag, Mycorrhiza 14:307-312

Góes LB, Costa ABL, Freire LLC, Oliveira (2002). Randomly amplified polymorphic DNA of *Trichoderma* isolates and antagonism against *Rhizoctonia solani.* Braz Arch Biol technol vol. 45

Gollotte A, Cordier C, Lemoine MC, Gianinazzi- Pearson V (1997). Role of fungal wall components in interactions between endomycorrhizal symbionts. In: HEA Schenk, R Herrman, KW Jeon, NE Muller and Schwemmler (Eds). Eukaryotism and Symbiosis: intertaxonomic Combination versus Symbiotic Adaptation. 417-427

de Grandcourt A, Epron D, Montpied P, Louisanna E, Bereau M, Garbaye J, Guehl JM (2004). Contrasting responses to mycorrhizal inoculation and phosphorus availability in seedlings of two tropical rain forest tree species. New Phytol 161: 865-875

Harley JL, Smith SE (1983). Mycorrhizal Symbiosis. Academic Press, London

Hartmann A, Pukall R., Rothballer M, Gantner S, Metz S, Schloter, Mogge B (2004). Microbial community analysis in the rhizosphere by *in situ* and *ex situ* application of molecular probing, biomarker and cultivation technique. In: Varma A, Abbott L, Werner D and Hampp R (eds) In: Plant Surface Microbiology. Springer-Verlag, Berlin Heidelberg Germany. 449-469

Hilderbrandt U, Janetta K, Ouziad F, Nawrath K, Bothe H (2001). Arbuscular mycorrhizal colonization of halophytes in central European salt marshes. Mycorrhiza 10: 175-183

Hill TW, Kafer E (2001). Improved protocols for aspergillus medium: trace elements and minimum medium salt stock solutions. Fungal Genet News Letter 48: 20-21

Hobbie EA, Colpaert JV (2003). Nitrogen availability and colonization by mycorrhizal fungi correlate with nitrogen isotope patterns in plants. New Phytol 157: 115-126

Hodge A, Campbell CD, Fitter AH (2001). An arbuscular mycorrhizal fungus accelerates decomposition and acquire nitrogen directly from organic material. Nature 413: 297-299

Hodge A (2003). Plant nitrogen capture from organic matter as affected by spatial dispersion, interspecific competition and mycorrhizal colonization. New Phytol 157: 303-314

Huai WX, Guo LD, He W (2003). Genetic diversity of an ectomycorrhizal fungus *Tricholoma terreum* in a *Larix principis-rupprechtii* stand assessed using random amplified polymorphic DNA. Mycorrhiza 13: 265-270

Jabari-Hare SH, Therien J, Charest PM (1990). High resolution cytochemical study of the vesicular-arbuscular mycorrhizal association, *Glomus clarum-Allium porrum*. New Phytol 114: 481-496

Jakoben I (1999). Transport of phosphorus and carbon in arbuscular mycorrhizas. In Mycorrhiza: structure, function, molecular biology and biotechnology A Varma and B Hock (Eds) 305-332

Jakobsen I, Rosendahl I (1991). Carbon flow into soil and external hyphae from roots of mycorrhizal cucumber plants. New Phytol 115: 77-83

Junghans DT, Gomes EA, Guimarães WV, Barros EG, Araújo EF (1998). Genetic diversity of the ectomycorrhizal fungus *Pisolithus tinctorius* based on RAPD-PCR analysis. Mycorrhiza 7: 243-248

Kaldorf M, Sdhmelzer E, Bothe H (1998). Expression of maize and fungal nitrate reductase genes in arbuscular mycorrhiza. Mol Plant-Microb Int. 11: 435-448

Kaldorf M, Koch B, Rexer KH, Kost G, Varma A (2005). Patterns of interaction between *Populus* Esch5 and *Piriformospora indica*: A transition from mutualism to antagonism. Plant Biology 7: 210-218

Karandashov V, Nagy R, Wegmüller S, Amrhein N, Bucher M (2004). Evolutionary conservation of a phosphate transporter in the arbuscular mycorrhizal symbiosis. PNAS 101: 6285-6290

Keneko Y, Hayashi N, Toh-e A, Banno I, Oshima Y (1987). Structural characteristics of the Pho8 gene encoding repressible alkaline phosphatase in *Saccharomyces cerevisiae*. Gene 58: 137-148

Klionsky DJ, EMr SD (1989) Membrane protein sorting: biosynthesis, transport and processing of yeast vacuolar alkaline phosphatase. EMBO Journal 8: 2241-2250

Koide R, Kabir Z (2000). Extraradical hyphae of the mycorrhizal fungus *Glomus intraradices* can hydrolyse organic phosphate. New Phytol 148: 511-517

Kretzer AM, Dunham S, Molina R, Spatafora JW (2004). Microsatellite markers reveal the below ground distribution of genet in two species of *Rhizopogon* forming tuberculate ectomycorrhizas on Douglas fir. New Phytol 161: 313-320

Kuhn G, Hijri M, Sanders IR (2001). Evidence for the evolution of multiple genomes in arbuscular mycorrhizal fungi. Nature 414:745-748

Lammers PJ, Jun J, Abubaker J, Arreola R, Gopalan A, Bargo B, Hernandez-Sebastia C, Allen JW, Douds DD, Pfeffer PE, Shachar-Hill Y (2001). The glyoxylate cycle in an arbuscular mycorrhizal fungus. Carbon flux and gene expression. Plant Physiol 127: 1287-1298

Larsen BJ, Thingstrup I, Jakobsen I (1996). Bonomyl inhibits phosphorus transport but not fungal alkaline phosphatase activity in the Glomus-cucumber symbiosis. New Phytol 132:127-133

Leake JR (2001). Is diversity of ectomycorrhizal fungi important for ecosystem function? Commentary. New Phytol 152: 1-3

Maldonado-Mendoza IE, Harrison MJ (1998). A xyloglucan endotransglycosylase (XET) gene from *Medicago truncatula* induced in arbuscular mycorrhizae. In Abstract 2nd Int. Conf.on Mycorrhiza, pp 5-10

Malla R, Singh A, Md Z, Yadav V, Suniti, Verma A, Rai M, Varma A (2002). *Piriformospora indica* and plant growth promoting Rhizobacteria: An Appraisal. In: Rao GP, Bhat DJ, Lakhanpal TN and Manoharichari C (Eds) Frontiers of fungal diversity in India pp 401-419

Malla R, Varma A (2004a). Phosphatase(s) in Microorganisms. Biotechnological Applications of Microbes. IK International- India, New York and Kluwer academic Press, Holland 125-50

Malla R, Prasad R, Giang PH , Pokharel U, Oelmueller R, Varma A (2004b). Characteristic features of symbiotic fungus *Piriformospora indica*. Endocytobiosis Cell Res 15: 579-600

Moller EM, Bahnweg G, Sandermann H, Geiger HH (1992). A simple and efficient protocol for isolation of high molecular weight DNA from filament fungi, fruit bodies and infected plant tissue. Nucleic Acids Res 20: 6115-6116

Morton JB, Redecker D (2001). Two new families of Glomales, Archaeosporaceae and Paraglomaceae with two new genera *Archaeospora* and *Paraglomus* based on concordant molecular and morphological characters. Mycologia 93:181-195

Murphy J, Langridge P, Smith SE (1997). Cloning plant genes differentially expressed during colonization of roots of *Hordeum vulgare* by the vesicular-arbuscular mycorrhizal fungus *Glomus intraradices*. New Phytol 135:291-301

Onishi HR, Tkacz JS, Lampen JO (1979). Glycoprotein nature of yeast alkaline phosphatase – Formation of active enzyme in the presence of tunicamycin. J Biol Chem 254: 1943-1952

Parniske M (2004). Molecular genetics of the arbuscular mycorrhizal symbiosis. Curr Opin Plant Biol 7: 414-421

Paradis R, Dalpé Y, Charest C (1995). The combined effect of arbuscular mycorrhizas and short-term cold exposure on wheat. New Phytol 129:637-642

Pawlowska TE, Taylor JW (2004). Organization of genetic variation in individuals of arbuscular mycorrhizal fungi. Nature 427:733-737

Peng M, Lemke PA, Shaw JJ (1993). Improved conditions for protoplast formation and transformation of *Pleurotus ostreatus*. Appl Microbiol Biotechnol 40: 101-106

Perotto S, Actis-Perino E, Perugini J, Bonfante P (1996). Molecular diversity of fungi from ericoid mycorrhizal roots. Mol Ecol 5: 123-131

Peškan-Berghöfer T, Shahollari B, Pham H-G, Hehl S, Markent C, Blank V, Kost G, Varma A, Oelmueller R (2004). Association of *Piriformospora indica* with *Arabidopsis thaliana* roots represent a novel system to study beneficial plant-microbe interactions and involve in early plant protein modifications in the endocytoplasmic reticulum and in the plasma membrane. Physiol Plant 122: 465-471

Peter M, Courty PE, Kohler A, Delaruelle C, Martin D (2003). Analysis of expressed sequence tags from the ectomycorrhizal basidiomycetes *Laccaria bicolor* and *Pisolithus microcarpus*. New Phytol 159:117-129

Pham GH, Singh An, Malla R, Kumari R, Saxena AK, Rexer K-H, Kost G, Luis P, Kaldorf M, Buscot F, Herrmann S, Peskan T, Oelmüller R, Mittag M, Declerck S, Hehl S, Varma A (2004a). Interaction of *Piriformospora indica* with diverse microorganisms and plants. In: Varma A, Abbott L, Werner D and Hampp R (Eds) In Plant Surface Microbiology, Springer-Verlag, pp 237-265

Pham GH, Kumari R, Singh An, Malla R, Prasad R, Sachdev M, Kaldorf M, Buscot F, Oelmüller R, Hampp R, Saxena AK, Rexer K-H, Kost G, Varma A (2004b). Axenic Cultures of *Piriformospora indica*. In: Varma A, Abbott L, Werner D and Hampp R (Eds) In: Plant Surface Microbiology. Springer-Verlag, pp 593-613

Podila GK, Lanfranco L (2004). Functional genomic approaches for studies of mycorrhizal symbiosis. Plant surface microbiology. In: Varma A, Abbott L, Werner D, Hampp R. (Eds). Plant Surface Microbiology. Springer-Verlag, Berlin Heidelberg pp 567-592

Poulton JL, Bryla D, Koide RT, Stephenson AG (2002). Mycorrhizal infection and high soil phosphorus improve vegetative growth and the female and male functions in tomato. New Phytol 154: 255-264

Rai MK, Singh A, Arya D, Varma A (2001b). Positive growth responses of *Withania somnifera* and *Spilanthes calva* were cultivated with *Piriformospora indica* in field. Mycorrhiza 11:123-128

Rasmussen N, Lloyd DC, Ratchiffe G, Hansen PE and Jackobsen I (2000). P^{31} NMR for the study of P metabolism and translocation in arbuscular mycorrhizal fungi. Plant Soil 226:245-253

Read DJ (1999). Mycorrhiza: the state of the art. In: Varma A and Hock B (Eds) Mycorrhiza: Structure, Function, Molecular biology and Biotechnology, 2nd Edition, Springer-Verlag, Germany, pp 3-34

Redecker D, Kodner R, Graham LE (2000). Glomalean fungi from the Ordovician. Science 289:1920-1921

Remy W, Taylor TN, Hass H, Kerp H (1994). Four hundred- million-year-old vesicular arbuscular mycorrhizae. Proc Natl Acad Sci 91:11841-11843

Rohlf, FJ (1992). NTSYS-pc. Numerical Taxonomy and Multivariate Analysis System. Vers. 1.70. Exeter Software, Setauket, New York

Rosewarne GM, Barker S, Smith SE, Smith FA, Schachtman DP (1999). A *Lycopersicon esculentum* phosphate transporter (*LePT1*) involved in phosphorus uptake from a vesicular-arbuscular mycorrhizal fungus. New Phytol 144: 507- 516

Ryan MH, McCully ME, Huang CX (2003). Location and quantification of phosphorus and other elements in fully hydrated, soil-grown arbuscular mycorrhizas: a cryo-analytical scanning electron microscopy study. New Phytol 160: 429-441

Sahay NS, Varma A (1999). *Piriformospora indica*: a new biological hardening tool for micropropagated plants. FEMS Microbiol Lett 181:297-302

Saito M (1997). Regulation of arbuscular mycorrhizal symbiosis: hyphal growth in host roots and nutrient exchange. JARQ 31:179-183

Sanders FE, Tinker PB (1971). Mechanism of absorption of phosphate from soil by endogone mycorrhizas. Nature 233: 278-279

Schussler A, Schwarzott D, Walker C (2001). A new fungal phylum, the Glomeromycota: phylogeny and evolution. Mycol Res 105:1413-1421

Shahollari B, Varma A, Oelmueller R (2005). Expression of receptor kinase in roots is stimulated by the basidiomycete *Piriformospora indica* and the protein accumulates in Triton-X-100 insoluble plasma membrane microdomains. J Plant Physiol. 162: 945-958

Shearer JF (2001). Recovery of endophytic fungi from *Myriophyllum spicatum*. ERDC TN-APCRP-BC-03. 1-11

Smith SE, Smith FA (1990). Structure and function of the interfaces in biotrophic symbiosis as they relate to nutrient transport. New Phytol 114: 1-38

Smith SE, Read DJ (1997). Mycorrhizal symbiosis. San Diego. CA, USA: Academic Press

Smith FA, Smith SE (1997). Structural diversity in vesicular - arbuscular mycorrhizal symbiosis. New Phytol 137: 373-388

Smith SE, Smith FA, Jakobsen I (2003). Mycorrhizal fungi can dominate phosphate supply to plants irrespective of growth responses. Plant Physiol 133:16-20

Smith SE, Smith FA, Jakobsen I (2004). Functional diversity in arbuscular mycorrhizal (AM) symbioses: the contribution of the mycorrhizal P uptake path way is not correlated with mycorrhizal responses in growth or total P uptake. New Phytol 162: 511-524

Solaiman MZ, Abbott LK (2004). Functional diversity of arbuscular mycorrhizal fungi on root surface. In: Varma A, Abbott L, Werner D, Hampp R (Eds.). In: Plant surface microbiology. Springer-Verlag, Berlin Heidelberg Germany. pp 331-349

Subramanian KS, Charest C, Dwyer LM, Hamilton RI (1995). Arbuscular mycorrhizae and water relations in maize under drought stress at tasseling. New Phytol 129: 643-650

Sylvia DM, Williams SE (1992). Vesicular-arbuscular mycorrhizae and environmental stress. In: Bethlenfalvay GJ, Linderman RG (Eds) Mycorrhizae in Sustainable Agriculture. ASA Special Publication Number 54, Madison Wisconsin pp 101-124

Thingstrup I, Rosendahl S (1994). Quantification of fungal activity in arbuscular mycorrhizal symbiosis by polyacrilamide gel electrophoresis and densitometry of malate dehydroganate. Soil Biol Biochem 26: 1483-1489

Tisserant B, Gianinazzi-Pearson V, Gianinazzi S, Gollotte A (1993). *In planta* histochemical staining of fungal alkaline phosphatase activity for analysis of efficient arbuscular mycorrhizal infections. Mycol Res 97:245-250

Tonin C, Vandenkoornhuyse, Joner EJ, Straczek J, Leyval C (2001). Assessment of arbuscular mycorrhizal fungi diversity in the rhizosphere of *Viola calaminaria* and the effect of these fungi on heavy metal uptake by clover. Mycorrhiza 10: 161-168

Turnau K, Ryszka P, Gianinazzi-Pearson V (2001). Identification of arbuscular mycorrhizal fungi in soils and roots of plants colonizing zinc wastes in southern Poland. Mycorrhiza 10: 169-174

Varma A (1999). Functions and application of arbuscular mycorrhizal fungi in arid and semi-arid soils. In: Varma A, Hock B (Eds) Mycorrhiza: Structure, function and molecular biology, 2nd Edition, Springer-Verlag, pp 521-556

Varma A, Singh A, Sudha, Sahay NS, Sharma J, Roy A, Kumari M, Rana D, Thakran S, Deka D, Bharti K, Hurek T, Blechert O, Rexer K-H, Kost G, Hahn A, Maier W, Walter M, Strack D, Kranner I (2001). *Piriformospora indica*: an axenically culturable mycorrhiza-like endosymbiotic fungus. In: Hock B (Ed) The Mycota IX, Springer-Verlag, Berlin, Heidelberg. 125-150

Varma A, Abbott LK, Werner D, Hampp R (2004). The state of art. Plant surface microbiology. In: Varma A, Abbott L, Werner D, Hampp R (Eds) In Plant surface microbiology. Springer-Verlag. 1-11

Verma S, Varma A, Rexer KH, Hassel A, Kost G, Sarabhoy A, Bisen P, Bütenhorn B, Franken P (1998). *Piriformospora indica*, a new root colonizing fungus. Mycologia 90: 896-903

Weib M, Selosse MA, Rexer KH, Urban A, Oberwinkler F (2004). *Sebacinales*: a hitherto overlooked cosm of heterobasidiomycetes with abroad mycorrhizal potential. Mycol Res 108: 1-8

Weish J, McClelland M (1990). Fingerprinting genomes using PCR with arbitrary primers. Nucl Acids Res 18: 7213-7218

Williams JGK, Kubelik AR, Livak KJ, Rafalski JA, Tingey SV (1990). DNA polymorphism amplified by arbitrary primers are useful as genetic markers. Nucl Acids Res 18: 6531-6535

Wright SF, Upadhyaya A (1996). Extraction of an abundant and unusual protein from soil and comparison with hyphal protein of arbuscular mycorrhizal fungi. Soil Sci 161: 575-586

Wu T, Sharda JN, Koide RT (2003). Exploring interactions between saprotrophic microbes and ectomycorrhizal fungi using a protein-tannin complex as a N source by red pine. New Phytol 159:131-139

Wurst S, Dugassa-Gobena D, Langel R, Bonkowski M, Scheu S (2004). Combined effects of earthworms and vesicular arbuscular mycorrhizas on plant and aphid performance. New Phytol 163:169-176

Zak B (1971). Characterization and identification of Douglas fir mycorrhizae. In: Hecskaylo E (ed) Mycorrhiza.1189:38-53

Zhang Y, Zeng M, Xiong B, Yang X (2003). Ecological significance of arbuscular mycorrhiza biotechnology in modern agricultural system. Ying Yong Sheng Tai Xue Bao. 14: 613-617

Zinno TD, Longree H, Maraite H (1998). Diversity of *Pyrenophora tritici-repentis* isolates from warm wheat growing areas; Pathogenicity, toxin production and RAPD analysis. In: Duveiller E, Dubbin H.J, Reeves J, A. Mc Nab (eds) *Helminthosporium* blights of wheat, Spot blotch and Tan Spot. Mexico, D.F. CIMMYT, pp 302-311

Omega-3/6 Polyunsaturated Fatty Acids, Essential to Human Health, from Thraustochytrids

OWEN P. WARD and AJAY SINGH[*]

1. INTRODUCTION

The omega-3 polyunsaturated fatty acids (PUFA), docosahexaenoic (DHA) and eicosapentaenoic (EPA) acids, and the omega-6 fatty acid, arachidonic acid (ARA) are essential components of higher eukaryotes and hence they are important nutritional, neutraceutical and pharmaceutical targets (Colquhoun 2001). These polyunsaturated fatty acids, contained in membrane phospholipids, confer flexibility, fluidity and selective permeability properties to membranes. These phospholipids are also precursors for synthesis of prostaglandins, leukotrienes and thromboxanes which mediate cellular physiological responses to inflammation, vasodilation, blood pressure, pain and fever (Funk 2001). While fish oil has been used as a source of DHA and EPA (Uauy and Valenzuela 2000), the fishy odor has been considered a disadvantage. Moreover, for pregnant and nursing mothers and for infant formula, EPA content should be minimal because of its capacity to cause bleeding. Certain microbial species belonging to lower fungi and microalgae produce omega-3 fatty acids in large amount of and hence microbes as an alterative source of PUFA have been exploited (Yongmanitchai and Ward 1989; Ward 1995; Sijtsma and deSwaaf 2004).

Department of Biology, University of Waterloo, Waterloo, Ontario N2L 3G1, Canada.
* **Correspondence:** Ajay Singh, E-mail: ajasingh@sciborg.uwaterloo.ca

Microbiological and biotechnological aspects of thraustochytrids, hyperproducers of omega-3 polyunsaturated fatty acids, are discussed in this chapter. Lipids of thraustochytrids are rich in docosahexaenoic acid (DHA) (Ellenbogen et al. 1969; Singh and Ward 1997a; Ratledge 2004; Ward and Singh 2005). Our laboratory was first to describe fermentation conditions for cultivation of thraustochytrids and methods for production of DHA by different species of *Thraustochytrium* (Bajpai et al. 1991 a,b; Singh and Ward 1996). These organisms may also be potentially exploited as producers of other PUFAs such as eicosapentaenoic acid (EPA), arachidonic acid (ARA) and docosapentaenoic acid (DPA).

2. IMPORTANCE OF OMEGA-3/6 FATTY ACIDS IN HUMAN HEALTH

Omega-3 fatty acids, especially DHA and EPA support good cardiovascular health. However, the impact of EPA is specific to the individual and can result in thinning of artery walls which may cause bleeding. The mechanisms by which EPA and DHA reduce risk of cardiovaslcular disease are well established, ever since it was noted that populations deriving a substantial proportion of their food from fish had a much lower incidence of heart disease. Typically these fish-diet dependent populations consume 0.5-0.7 g/d DHA, whereas DHA and EPA account for only 0.1-0.2 g/d in the American diet.

The typical 'Western' diet contains low levels of omega-3 PUFAs. The main source of DHA in the diet is cold-water fish (about 0.2-1.1 g DHA/100 g fish). DHA intake in the USA in 1994 was reported at 92 mg/day among fish eaters, and 34 mg/day among non-fish eaters (Becker and Kyle 1998). Greenland Eskimos living on their traditional marine diet ingest 5-10 g/day of polyunsaturated fatty acids of the n-3 family, mainly as EPA and DHA. In Europe, the intake of long chain n-3 fatty acids (including DHA) is estimated to be approximately 0.1-0.5 g/day. For Australian and New Zealand breast-fed infants, the estimated exposure to DHA based on the DHA levels in breast milk is 1.5 g per day (FSANZ 2002).

DHA is a major structural component of the gray matter of the brain, an important component of heart tissue and of the eye retina where it is a ligand for the retinoid X-receptor (deUequiza 2000). The brain is also rich in arachidonic acid (ARA), which is the most abundant PUFA in humans, present in organs, muscle and blood tissue and has a major role as a structural lipid associated predominantly with PLs. As a result dietary DHA and ARA are important for proper development of the brain and eyes of infants. Omega-3 PUFAs have other potential applications in treatment of certain mental disorders (Fenton et al. 2000; Peet 2004) and carcinomas (Tisdale 1999).

Fish oil, being a rich source of PUFAs, has been used for commercial production of omega-3 fatty acids (Agren et al. 1996). However, fish and also plant oils can accumulate hazardous hydrophobic environmental contaminants from the environment and indeed these oils may be used for extraction of hydrophobic contaminants from soil (Pannu et al. 2003, 2004). Some microbial species are currently being used for commercial production of PUFAs. Industrial microbial sources of DHA and ARA for human consumption must pass stringent safety tests and in particular must be non-pathogenic and non-toxigenic organisms (Hammond et al. 2001 a,b). Fish oils for direct consumption are dispensed in the form of capsules. Application of microencapsulation methods has addressed concerns regarding the oxidative instability of PUFA-containing oils destined for incorporation of oils into dried foods.

Human breast milk is rich in ARA and DHA, and as far back as 1975 the Food and Agriculture Organization (FAO) and the World Health Organization (WHO) recommended that the composition of infant formula should mimic that of human milk. From 1990 onwards a number of health and nutrition organizations and the Food and Drug Administration (FDA) began making specific recommendations that ARA and DHA be included in pre-term and term infant formula (FDA 2001). Certain microbial ARA/DHA-containing oils were included in pre-term and term infant formula, microbial DHA-containing oil were recommended for consumption by pregnant and nursing women. The world wholesale market for infant formula is substantial, amounting to greater than $10 billion per annum and in many countries ARA and DHA are used for fortification of infant formulae (Arterburn et al. 2000). Inclusion of microbial single cell oils in infant formula can add 10-20% to the retail price, which establishes the market value of oil product for infant formula at $1 billion.

3. BIOLOGY AND ECOLOGY OF THRAUSTOCHYTRIDS

Thraustochytrids are the marine fungoid protists. Labyrinthulomycota which includes the thraustochytrids, the labyrinthulids and the aplanochytrids, represent a phylum of marine saprobes, and are classified as straemenopiles in the kingdom Protista. Leander and Porter (2001) used DNA sequence analysis to differentiate between the three groups. Apart from a few instances, these organisms were considered to have limited environmental impact other than having an important role in breakdown of organic matter and recycling of nutrients in marine habitats. A strain of *Aplanochytrium haliotidis* caused a major infestation of Abalone mariculture facilities. Deleterious impacts were also noted for *Labyrinthula zosterae*, responsible for wasting disease of eelgrass (*Zostera marina*), which caused major destruction of eelgrass along the North American and European Atlantic coast lines in the 1930s. This, in turn, led to extinction of the limpet, *Lottia alveus*, which required eelgrass for its habitat.

The thraustochytrids are primarily consumed in the marine ecosystem by filter feeding marine invertebrates, such as mussels and clams, and by fish. It is now well-established that the best microbial sources of DHA are from the order of Thraustochytriales, within the phylum of Labyrinthulomycota (Hammond et al. 2001a,b). Within the order Thraustochytriales, these polyunsaturated fatty acid (PUFA)-producing thraustochytrids are unicellular heterotrophic algal or algal-like protists from the family Thraustochytriaceae; genus *Thraustochytrium, Schizochytrium* or *Ulkenia*. *Schizochytrium* replicates by successive bipartition and by release of zoospores from sporangia, whereas *Thraustochytrium* strains only replicate by formation of sporangia/zoospores. Individual *Thraustochytrium* cells are larger and cell aggregation and clumping is often a problem (Barclay 1997). Molecular phylogenetic studies and zoospore ultrastructure, particularly related to the flagellar apparatus, indicated that the thraustochytrids have a close relationship with the heterokont algae (Cavelier-Smith et al. 1994; Huang et al. 2003). *Schizochytrium* comprising *S. aggregatum, S. mangrovei, S. minutum, S. octosporum* and *S. limacinum*, consists of fewer species than *Thraustochytrium* (Kumar et al. 2002). *Ulkenia* species form rhizoids, lack apophyses and form amoeba-like cells in specific media (Tanaka et al. 2003). *Ulkenia* was characterized by the release of amoeboid cells before the formation of zoospores (Gaertner 1977). Many of the early strains of the Thraustochytriales were isolated by so-called pollen baiting techniques, which are very specific for selection of these organisms.

The capacity of thraustochytrids to produce high levels of DHA appears to be related to the ecological habitat from which the strains were isolated. Bowles et al. (1999) isolated 57 thraustochytrids strains from a range of habitats and ecological niches. Three different ecological and climatic environments were chosen: a cold temperate littoral, a cool temperate littoral and subtropical mangroves. Thraustochytrids were isolated from water and sediment samples, mangrove leaf tissue, algae and marine angiosperms. Isolates from the subtropical location exhibited higher levels of biomass in culture, containing upto 37%, w/w oil, but the DHA fraction of fatty acids was low. In contrast, in some of the isolates from the cold temperate climate, DHA accounted for almost 50% of fatty acids. Isolates from the cool temperate location exhibited intermediate values.

Barclay (1997) obtained more than 150 thraustochytrid-like organisms from shallow inland saline ponds. In obtaining samples, special attention was paid to include living plant material or decaying plant and animal matter with the recovered water. Examples of the collection sites included: the saline warm springs in Glenwood Springs along the Colorado river, or in Utah at the western edge of Stansbury Mountains; in playas such as Goshen playa, Utah; in marine tide pools as are located in the Bird Rocks area of LaJolla, and in estuaries, such as the Tiajuana estuary, San Diego County, both in California. The collection and screening process was designed to recover high omega-3 producers having a low content of saturated and omega-6 PUFAs which could grow heterotrophically. Additional preferential characteristics sought were that the strains should be non-pigmented, white or colorless, that the cells could grow at low salt concentrations and at temperatures above 30°C. The isolation process involved water filtration with polycarbonate filter followed by plating on a glucose/protein hydrolyzate/yeast extract vitamin, mineral salts-enriched sea water containing agar.

Thraustochytrids are typically isolated from estuarine, intertidal and marine habitats, including sandy sea bottoms, surf areas and sea water, in muds or on decaying plant material or leaf tissue in the latter saline environments (Bowles et al. 1999; Fan et al. 2002; Bongiorni and Dini 2002). Mangroves, which are coastal forests of the tropical and subtropical intertidal zones represented by estuaries, backwaters, deltas, creeks, lagoons, marshes and mudflata, provides the most diverse ecosystem for investigation of these marine protists (Sridhar 2005). A quarter of the world's coastline is estimated to be impacted upon by mangroves, estimated to comprise 181,000 square kilometers (Spalding et al. 1997). Ninety percent of the world's mangroves are reported to be located in the Americas, South East Asia and Africa. Mangroves contain about eighty species of trees and shrubs, some other minor plant species and salt tolerant plants which grow in the vicinity of mangroves (Field 1995). These plants have adapted biologically to the harsh conditions found in these ecosystems, namely, high salinity, high winds, tidal variations, high temperatures and anaerobic clay-like soils.

Because of the extent of the tree canopy covering mangrove waters, these ecosystems are detritus-based, arising from leaves, barks, woods and other organic matter. These materials support a hugely diverse fungal population involved in the processing of decaying plant material (Manoharachary et al. 2005). Indeed mangrove fungi constitute the second group of organisms among the marine fungi. Microbial studies have been conducted in many ecological niches associated with mangroves: in the water and mud, in the rhizosphere, in/on leaves, stems, wood, seeds, marine algae, animal remnants, in sea grasses and in decomposing plant matter (Sridhar 2005). Other parameters which influence the nature of the mangrove microbial population include the amount of dissolved organic nutrients, pH, dissolved oxygen concentration, hydrostatic pressure, tidal amplitude and availability of light (Hyde et al. 2000).

Mangroves have been found to represent a rich source of thraustochytrid organisms (Schmidt and Schearer 2003). Thraustochytrids can also be isolated from marine bivalve mollusks and they also are found in marine microbial films on immersed surfaces in the sea (Kumar et al. 2000).

4. MICROBIAL PRODUCERS OF OMEGA 3/6 FATTY ACIDS

A number of fungal and algal groups produce omega 3 and 6 PUFAs including ARA, EPA and DHA (Cohen et al. 1995; Behrens and Kyle 1996; Singh and Ward 1997b; Ratledge 2004). The commercial arachidonic acid single cell oil (ARASCO) from *Mortierella alpina*, produced by Martek Biosciences, is approved for inclusion in infant formulae in many jurisdictions. This product contains 40-50% ARA and minimal amounts of other long chain PUFAs. Wuhan-Alking Engineering Co. of China also produces ARA-based single cell oil using *M. alpina* (http://www.alking.com.cn). Most algae are disadvantageous in that large amounts of PUFAs are not accumulated as triglycerides (TGs) and those that do accumulate TGs are obligate phototrophs (Apt and Behrens 1999). However, heterotrophic algae, which can be grown in conventional fermenters, especially the marine dinoflagellate *Crypthecodinium* species, produce PUFAs predominantly as TGs and PLs. With heterotrophs, photosynthesis for carbon and energy generation is replaced by supply of glucose or other utilizable carbon source to the medium.

The commercial docosahexaenoic acid single cell oil product (DHASCO) from *Crypthecodinium cohnii*, produced by Martek Biosciences is also approved for inclusion in infant formulae. This product contains 40-50% DHA but no EPA or other long chain PUFAs. In the patented 3-5 day *C. cohnii* process, for production of commercial DHASCO, biomass was produced in the range 20-40 g/L, and 4-12 g/L oil, 35% of which was DHA (Kyle 1994). Much research has focused on the importance of PUFAs, particularly ARA, DHA and EPA, in larval growth and attention was first specifically directed towards *Crypthecodinium* species. High biomass densities (upto 109 g/L) and DHA concentrations of ~20g/L have been achieved in carbon fed batch cultures of marine species such as *C. cohnii*, although prolonged culture times (400 h) were required.

Photoautotrophic algae, with a capacity to produce EPA, such as *P. tricornutum* and *Monodus,* can grow to high cell densities of about 10 g/L. Heterotrophic EPA-producing algae such as *Nitschia*, which do not require light, can grow to higher cell densities (50 g/L). However, in both cases EPA contents of biomass remained relatively low at 1-3%, w/w and their EPA productivities, amounting only to 0.1-0.3 g/L/d are not commercially viable. A few higher fungi, notably of *Mortierella* species, isolated from soils for production of omega-6 PUFAs can also produce omega-3 fatty acids, but concentrations of the latter represent only a small proportion of total fatty acids. *Mortierella alpina*, grew at similar cell densities to *Nitschia laeva*, and exhibited similar EPA productivities.

Some of the key specific PUFA-fermentation characteristics of *Schizochytrium* production species are listed in Table 1. Even though PUFA-producing protists are isolated from salt-containing marine ecosystems, industrial processes involve growing these strains at low salinities and at higher temperatures, such that high productivities are achievable. The organism should not be heavily pigmented, given the pigment would likely be present in the oil. The oils have to contain low concentrations of PUFAs other than the principal component since closely related molecular species are not easily separated from the desired product. The desired PUFAs are in the glyceride form, appropriate for human consumption. For mariculture applications, the harvested oil-containing biomass is small enough for small fish/larval consumption.

Commercial PUFA producing strains/processes must comply with general industrial process engineering, safety and regulatory requirements (Ward 1989,1991; Ratledge and Wynn 2001). Strains must be genetically stable, ensuring retention of desirable oil-producing characteristics and process consistency. Regulatory bodies demand that the product, process and the strain do not substantially change from that which received regulatory approval. Strains must be non-pathogenic and non-toxin forming, so that they are neither health hazards to production employees nor consumers. The strain has to be suitable for use in large aerated fermenters, being able to withstand shear due to impeller mixing and aeration. Strains must exhibit high rates of growth and product formation as well as high yields based on substrate conversion.

Table 1 Characteristics for DHA production by *Schizochytrium*.

Characteristics	Specific properties	References
Growth and DHA production	• Capable of heterotrophic growth. • Very high percentage of oil in biomass. • The oil is easily extracted from the biomass. • Glucose concentration >1% inhibits growth, importance of feeding and controlling glucose concentration. • pH controlled in range optimizes growth rate. • O$_2$ controlled at 40% saturation to maintain high growth rated. • Nitrogen deficient conditions favor DHA production. • Feeding of N as NH$_4$OH during pH control, maintains N-deficient conditions. • Sodium sulphate reduces biomass aggregate size as required for mariculture applications in small fish feeder consumption. • Achievement of very high density cultures (200 g/L) in 4d. • Achievement of very high DHA yields and productivities. • Produces a high proportion of total lipid as DHA. • Produces a low content of other PUFAs not desired in the product. • The organism is non-pigmented, white or colorless cells. • PUFA present as triglyceride which is suitable for human consumption.	Barclay and Zeller 1996; Kyle 1997; Yokochi et al. 1998; Simopoulos et al. 1999ab; FSANZ 2002; Baily et al. 2003
Food safety	• No reports that *Schizochytrium* is pathogenic. • No reports that *Schizochytrium* produces toxic chemicals and common algal toxins such as domoic acid and prynesium toxin are absent. • *Schizochytrium* spray dried microalgae used in aquaculture for 9 years as excellent stable dietary source of DHA for shrimp larva culture and finfish. • Dried *Schizochytrium* microalgae has GRAS status for use as DHA-rich ingredient in broiler chickens and laying hen feed. • Thraustochytrids are consumed directly through consumption of seafood.	Innis and Hansen 1999; Frankel et al. 2002; ISSFAL 2004; FSANZ, 2005

The Omega-Tech DHASCO-S process (now part of Martek) uses *Schizochytrium* sp. while Nutrinova GmbH uses an *Ulkenia* species (FSANZ 2005). The high DHA productivities which are achievable with *Schizochytrium* and *Ulkenia* species have facilitated production of low cost oil suitable for use as a dietary supplement in cheeses, yogurts, spreads and dressings, and breakfast cereal and in beverages, in health foods, in animal feeds and in mariculture. Other markets include foods for pregnant and nursing women and applications in cardiovascular health. DHA-rich *Schizochytrium* preparations are being marketed as a poultry feed (Simopoulos et al. 1999 a,b). In mariculture applications biomass aggregates of less than 50-150 µ containing microbial oils are fed to larval shrimp, brine shrimp, rotifers and mollusks. In aquaculture applications these species are enriched with a DHA spray dried feed, prepared from *Schizochytrium* (Barclay and Zeller 1996). *Schizochytrium* species are consumed directly by humans in meals containing mussels and clams. Taken together, these markets are estimated to have much greater growth potential than infant formulae.

5. PRODUCTION OF DHA BY THRAUSTOCHYTRIDS

Appropriate fermentation conditions produce *Schizochytrium* biomass with very high percentages of lipids in the form of PUFAs. Product (DHA) formation and high cell density development requires careful control of glucose, nitrogen, sodium and oxygen concentrations as well as temperature and pH. There is potential to produce *Schizochytrium*-based oil, low in EPA, for the infant formula market. With additional strain selection, manipulation and culture development there is potential to develop *Schizochytrium*-based fermentations for production of ARA and EPA.

A summary of selected investigations on production of DHA by thrausochydrids is presented in Table 2. Conditions for production of DHA by *T. aureum, T roseum and S. aggregatum* were described by Bajpai et al. (1991a,b), Li and Ward (1994a) and Iida et al. (1996). *T. aureum*, *T. roseum* and *S. aggregatum* grew well on maltose and starch. Growth of *T. aureum* was inhibited at glucose concentrations of >10g/L.

With the variety of thraustochytrid strains isolated by Bowles et al. (1999) DHA production was stimulated by a high C: N ratio. Maxima for biomass, lipid content of biomass, DHA content of lipid and DHA yield were 14g/L, 78% w/w, 25% w/w and 2.17g /L, respectively, after 107h. Fan et al. (2001) described production of DHA by nine thraustochytrid strains from

Table 2 Overview of process investigations on production of DHA by Thraustochytrids.

Organism	Culture time (h)	Biomass g/L	DHA g/L	DHA % of Biomass	DHA % of Oil	Productivity g/L/d	Comments	References
T. aureum ATCC 34304	144	4.9	0.51	10.4	57	0.09	Initial identification of thrausto-chytrids as DHA producers; culture development.	Bajpai et al. 1991ab
T. roseum ATCC 28210	Batch / Fed-batch	10.0 / 17.1	0.10 / 0.12	10.2 / 11.7	53 / 48	0.21 / 0.17	Improved biomass and DHA production using fed-batch.	Singh and Ward 1996, 1997a
S. limacinum SR21	120	-	>4	-	-	>0.8	Increased yields due to strain tolerance to high C.	Yokochi et al. 1998
Thraustochytrids	107	14	2.17	78	25	0.48	Screening large collection; DHA production stimulated at high C:N, >1% glucose inhibits growth.	Bowles et al. 1999
S. mangrovei Thraustochytrium and	52	-	-	-	-	1.27	Identified high producing strain.	Fan et al. 2001
Schizochytrium sp.	-	-	-	-	-	1.08	Screening of 151 newly isolated strains; isolation of high growth rate strains and high DHA producers.	Kyle 1997
Schizochytrium strains	90-100	200	40-45	20-22	40	~10.00	High biomass /DHA production by controlling O_2 and glucose at <7g/L; pH and N-feed.	Baily et al. 2003

subtropical mangroves. A strain of *S. mangrovei* produced 2.8 g/L DHA in 52h at 25°C on glucose yeast extract medium (Table 2).

Tanaka et al. (2003) described a process for production of DHA using *Ulkenia sp.* In a 3-day fermentation DHA volumetric yield was 5.5 g/L, representing a productivity of 1.8g/L/d. DHA constituted 90-95% of omega-3 fatty acids.

Shake flask and fermenter optimization studies were implemented with a high DHA-producing thraustochytrid, *Schizochytrium limacinum* SR21. In shake flask cultures, DHA yields of >4g/L were obtained in glucose (12%, w/v) or glycerol (9%, w/v) and corn steep liquor-salt-supplemented media in a 5-day incubation (Yokochi et al. 1998). This strain also grew at zero salinity, although at a lower growth rate (50% of maximum). TFA content increased with decreasing amount of nitrogen source to more than 50% of biomass weight. The relatively high yields of DHA obtained were attributed to the strains ability to grow in the presence of 9-12%, w/v carbon substrate. When growth and DHA production by this strain was investigated and optimized in fermenters, final yields of biomass and DHA in media containing 12%, w/v glucose and corn steep liquor of 48.1 g/L and 13.3 g/L, respectively, were achieved. DHA productivity increased with glucose concentrations upto 12% w/v and maximum DHA productivity was 3.3 g/L/d (Yaguchi et al. 1997). This strain also produced 1,2-DHA-phosphatidyl choline and 1-palmitoyl-2-DHA-phosphatidyl choline.

One hundred and fifty one strains were isolated and fermentation conditions were developed to optimize DHA production. A number of important observations contributed to the development of industrial process for production of DHA by Omega-Tech, USA:

- High DHA producing thraustochytrids could be cultured in media where most of the sodium requirement was supplied by sodium sulphate.
- Use of sodium sulphate also had the beneficial effect of reducing cell aggregate size in fermentation media typically to less than 50 or 100μ, making the product suitable for use as aquaculture feed material.
- Reduction of chloride content was additionally important since chloride forms have a deleterious corrosive impact on stainless steel vessels.
- High biomass yields (50% w/w of sugar substrate) could be achieved with chloride concentrations as low as 60-120mg/L.
- Lowering the oxygen concentration promote lipid production.

The distribution of key PUFAs, among selected isolates is presented in Table 3 (Kyle 1997). *Schizochytrium* S9 produced 94% of its omega-3 fatty acids as DHA and strain ATCC 20892 produced almost 80% of its fatty acids as omega-3. With regard to the growth rate, doubling time per day for the thraustochytrids is reported to be about 4.7 - 8.4 at 30°C (Barclay 1997).

Table 3 Distribution of PUFA in the lipid of selected isolates.

Strain no.	Distribution of key PUFAs					
	ARA C20:4 ω-6	DPA C22:5 ω-6	Total ω-6 fatty acids	EPA C20:5 ω-3	DHA C22:6 ω-3	Total ω-3 fatty acids
56B	16.7	18.4	35.1	8.6	22.5	31.1
43B	18.5	20.5	39.0	8.1	22.1	30.2
46B	18.6	19.5	38.1	9.0	21.7	30.7
55B	16.2	17.7	33.8	6.3	18.1	24.4
Schizochytrium 31 ATCC 20888	5.0	15	20	6.9	47.4	55.8
Schizochytrium S9 ATCC20889	0.4	15.5	15.9	2.3	47.0	49.7
Thraustochytrium ATCC 20890	1.4	0	1.4	18.9	43.5	67.3
Thraustochytrium ATCC 20891	4.2	9.5	13.7	7.3	51.2	59.6
Thraustochytrium ATCC 20892	1.5	0	1.5	13.5	60.0	79.5
Thraustochytrium S42	3.3	23.8	27.1	13.5	38.7	52.2
Thraustochytrium U30	2.3	18.4	20.7	15.0	43.4	59.4
T. aureum ATCC 28211	2.9	9.8	12.7	7.7	54.6	62.9
S. aggregatum ATCC 28209	10.7	12.6	23.4	4.3	20.6	26.4
T. aureum ATCC 34304	3.9	8.1	12.0	3.7	55.1	58.8
T. striatum ATCC 34473	1.6	11.4	12.9	6.9	17.8	24.7
T. roseum ATCC 28210	6.3	4.3	10.6	6.9	60.1	67.0

Adapted from Kyle (1994, 1996, 1997).

Strategies for development of high density cultures were exploited to further enhance DHA productivity (Baily et al. 2003). The fermentation development approach involved a first stage focusing on increasing biomass density, with an oxygen concentration of greater than 4% saturation and a second stage focused on high lipid production with a lower oxygen concentration of less than 3% saturation. The process was operated in carbohydrate fed-batch mode to maintain a sugar concentration of about 7 g/L. The major source of sodium was sodium sulphate. Ammonium hydroxide was used both for pH control and as a nitrogen source for growth. Fermentation duration was typically 90-100 h. Biomass productivities were higher in the first half of the fermentation (4-5 g/L/h), and later ranged from 2-3 g/L/h, with final biomass content reaching upto 200 g/L. Final total fatty acid content of biomass was about 40%w/w. DHA productivity and volumetric yield were 0.5 g/L/h and 40-45g/L, respectively. DHA contents of biomass and of total fatty acids were 20 g/L and 50%, w/w, respectively.

6. POTENTIAL PROCESSES FOR OTHER PUFAS BY THRAUSTOCHYTRIDS

The high DHA-producing strain, *Schizochytrium limacinum* SR21, described by Yaguchi et al. (1997) produced 95% of its lipid in the neutral lipid form but the remaining polar lipid fraction contained a substantial amount of a potentially interesting DHA-containing products, 1,2-DHA-phosphatidyl choline and 1-palmitoyl-2-DHA-phosphatidyl choline. There may be potential to manipulate these strains and to develop the fermentation process to enhance production of these phospholipids.

The thraustochytrids have also been shown to produce significant quantities of docosapentaenoic acid. In the 3-day process for production of DHA using an *Ulkenia* species, described by Tanaka et al. (2003), volumetric yield of DPA was 1.3g/L, representing a productivity of 0.43 g/L/d, respectively. DPA present in typically constitutes as much as 10% of the fatty acid content in these thraustochytrid oils.

Two of the thraustochytrid species (Strains 43B and 46B) isolated by Barclay (1997) produced 18.5-18.6% and 38.1-39% as ARA in biomass and as omega-6 fatty acids as % of TFAs, respectively, which may represent interesting sources of omega-6 fatty acids. Levels of omega-6 fatty acids produced by prior *Thraustochytrium* and *Schizochytrium* strains were much less. If research on fermentation development aspects of these strains can achieve 50% of the biomass productivities observed for DHA production in

Schizochytrium, i.e., 100 g/L in 4d, while retaining the above percentage of proportion of ARA in biomass, ARA productivities of greater than 4 - 5 g/L/ d may be achieved with these strains. Potential also exists to carry out strain development programs to reduce or completely eliminate the non-ARA PUFAs which would otherwise make ARA purification complicated and costly. The above high ARA-producing thraustochytrids also produce DHA, so it may be possible to manipulate these strains and fermentation conditions to the mix of PUFAs present in human mother's milk using just one strain.

Schizochytrium strains BRBG and BRBG1, described by Barclay (1997), produced 87-92% of their omega-3 fatty acids as EPA. These strains may have potential as producing organisms in commercial EPA fermentations. Again based on culture development studies with *Schizochytrium*, for DHA production, if cell densities of 100 g/L in 100 h, with the cells constituting 40% of their biomass as lipids, and if omega fatty acids accounted for 35% of total fatty acids, these strains have the potential to produce 2.5 – 3.0 g EPA /L/d.

6.1 Purification and Processing of PUFAs

With microbial oil-containing biomass, the cells or mycelium are typically harvested by centrifugation or filtration to remove the majority of the aqueous material after which the cells may be dried and solvent extracted. Homogenization of an aqueous suspension to reduce biomass particle size to less than 10 microns followed by extraction with a non-water miscible organic solvent has been found to be an effective extraction method (Hoeksema 2000). Alkali pretreatment of the microbial biomass can also facilitate solvent extraction of the single cell oil. Alternatively moisture content of the harvested cells may be reduced by spray drying or freeze drying prior to application of a solvent extraction step. Recovered oils can be further processes if desired to concentrate the PUFA fraction by the process of winterization, whereby the temperature of the oil is reduced to effect precipitation of the more saturated lipids which may then be separated out. A variation of the winterization process incorporates a polar solvent in the oil to reduce the possible development of haze in PUFA oil products (Dueppen et al. 2005). A solventless method for recovery of microbial oil has also been described which involved a cell lysis step followed by separation of the aqueous and oil phases (Ruecker et al. 2004). Barclay (1992) described a combined fermentation/extraction process for production of DHA from *Schizochytrium* to produce oil containing at least 94% by weight of total omega-3 fatty acids as DHA.

Applications of enzymes in bioprocessing are especially advantageous because they act under mild reaction conditions (Li and Ward 1994b; 1996). Non-positionally specific lipases can be used to hydrolyze glycerides into free fatty acids. Where the desired PUFA is concentrated at either the 2- or 1,3-positions of the glyceride, positional-specific lipases (with respect to the glycerol component) may be used to concentrate the PUFA in the free fatty acids or residual glyceride fraction. Lipases may also be exploited in the esterification directions, to selectively enrich PUFAs, based on their differential rates of transformation for PUFAs versus more saturated fatty acids. Ethyl esters of PUFAs are of great interest to the pharmaceutical industry and free fatty acid PUFAs can be are reacted with ethanol in a lipase-catalyzed system to produce ethyl PUFA of a purity of >90% (Shimada et al. 2001). A transesterification strategy can be used to concentrate PUFAs in the glyceride fraction. By incubating PUFA-containing oil with isopropanol in the presence of lipase under defined conditions, the lipase preferentially transesterifies more unsaturated fatty acids from the oil triglycerides to isopropanol, thereby concentrating the PUFAs in the residual glycerides. By substitution of glycerol with 1,2 isopropylidene glycerol, lipase can mediate selective esterification of PUFA free fatty acids to form PUFA monoglycerides. Interesterification processes may also be exploited to introduce, by substitution, new desired fatty acids into high volume oils.

For production of bulk PUFAs for dietary applications, it appears that the microbial strains are already available to generate high concentrations of the desired PUFA in the right form (usually triglycerides), such that appending a lipase conversion step is not necessary. Hence, the main application of lipases with processing of PUFAs relates to generation of non-natural esters of these products for use as pharmaceuticals or other synthetic bioactive compounds or their precursors.

6.2 Biochemical Pathways for Production of PUFAs

In order to understand why the thraustochytrid marine protists are exceptional producers of DHA it is necessary to discuss the metabolic pathways for fatty acid synthesis. PUFA biosynthesis pathways in microbes and mammals are shown in Figure 1. PUFAs are generally synthesized from saturated fatty acid precursors through insertion of double bonds at specific carbon locations in the fatty acid, mediated by desaturases and through extension of the chain in two-carbon increments, mediated by fatty acid elongases (Shanklin and Cahoon 1998; Metz et al. 2001; Cunnane 2003).

PUFA synthesis from acetyl CoA, involving these desaturation/elongation processes can involve as many as 30 enzymes (Sprecher 2000). A novel PUFA biosynthetic pathway has recently been characterized in some prokaryotic and eukaryotic organisms that does not depend on the fatty acid desaturase/elongase system but rather exploits a polyketide synthase mechanism to produce 20:5n3 and 22:6n3 (Fang et al. 2004).

With respect to the desaturation and elongation processes, the hydrophobic/membrane-bound fatty acid desaturases contain at least three

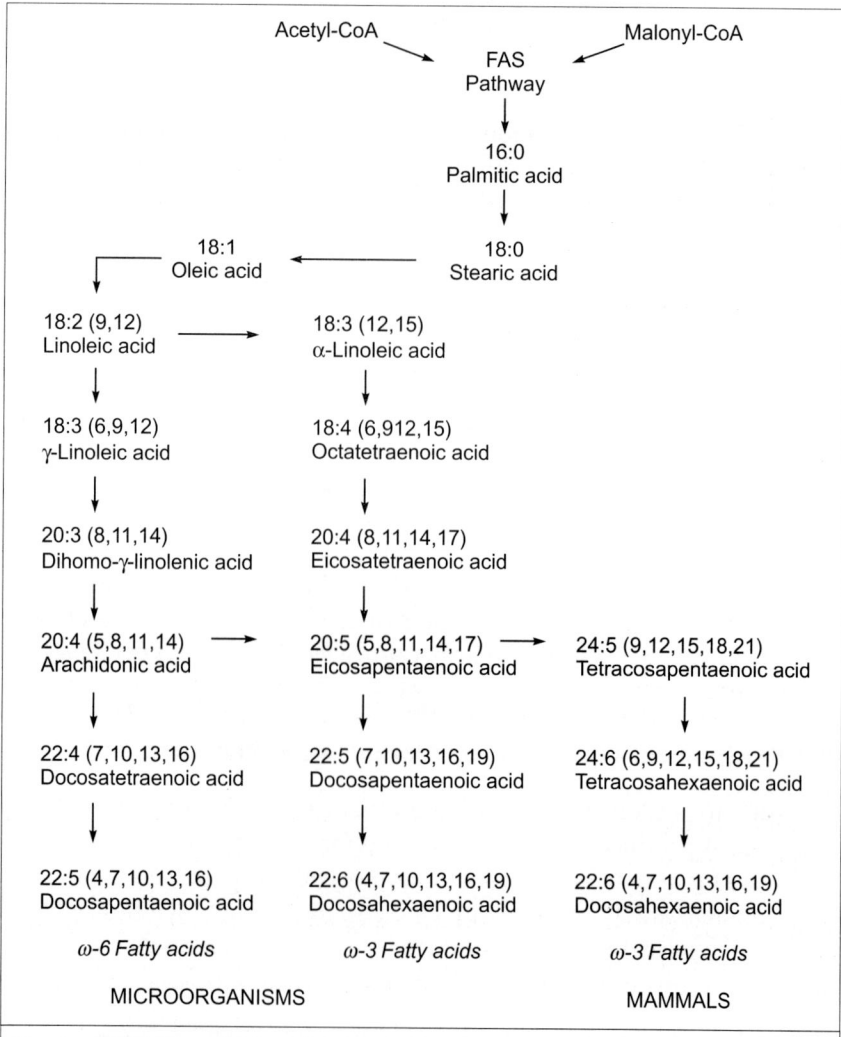

Fig. 1   PUFA biosynthesis using the conventional fatty acid synthase (FAS) pathways (aerobic) in microorganisms and mammals.

functional components: Cyt b5 reductase, Cyt b5 and a terminal reductase. Desaturases exhibit fairly stringent substrate specificities with respect to the nature of the fatty acid and regiospecificity with respect to the location of the bond to be transformed (Sakurdani et al. 1999 a,b):

- The Δ-5 desaturase can convert di-homo-GLA(20:3n-6) to ARA (20:4n-6) [n-6 pathway] or 20:4n-3 to EPA (20:5n-3) [n-3 pathway]
- The Δ-6 desaturases convert linoleic (18:2n-6) to GLA (18:3n-3) [n-6 pathway] and GLA(18:3n-3) to 18:4n-3 [n-3 pathway]
- The Δ-9 desaturase, which catalyses insertion of the first double bond into the fatty acid (C9-C10), preferentially uses palmitate (16:0) but can also use stearate (18:0) as an alternative substrate. In contrast to other desaturases which use phospholipids bound acyl groups, Δ-9 desaturase uses acylCoA as substrate
- The Δ-12 desaturases, which convert oleic (18:1n-9) to linoleic (18:2n-6), have been characterized from a variety of microbial species.

In *Schizochytrium* species, EPA biosynthesis is not affected by anaerobic growth suggesting a fatty acid oxygen-requiring desaturation route to EPA is not necessary (Lewis et al. 1999). Elongases participate in a reaction sequence for fatty acid elongation similar to that of fatty acid synthesis, involving 4 linked steps:

- Condensation of the acyl group to a malonyl CoA producing a ketoacyl product and CO_2
- Reduction to the γ-hydroxyl moiety
- Its dehydration to an enoyl group, followed by
- A second reduction, to give the elongated FA.

The key difference between these systems is that elongases are multiunit membrane-associated proteins, whereas *de novo* fatty acid synthesis occurs in the cytosol. It has been suggested that the unique part of the elongase system is the condensing enzyme, which also defines its specificity and rate limiting properties and that the other three components may be supplied by the existing fatty acid complex.

A much simpler pathway involving a specialized polyketide synthase system has been characterized for PUFA production in marine prokaryotic and eukaryotic microorganisms (Wallis et al. 2002; Fang et al. 2004). Synthesis of polyketide secondary metabolites including antibiotics involves a set of enzymatic reactions analogous to those of fatty acid synthesis (Hopwood and Sherman 1990; Katz and Donadio 1993; Hutchinson and Fujii 1995). Novel approaches to PUFA production involved use of a polyketide-like system, which utilize the same four basic reactions of FAS, but the cycle is often shortened to produce a carbon chain with many keto, hydroxyl and C=C double bonds. Polyketide synthesis (PKS) proteins have sequence similarities with FAS enzymes. *Shewanella putrefaciens* contains this kind of PKS-like gene cluster, which has a minimum of 5 open reading frames (ORFS), expressing the EPA synthesis pathway (Russel and Nichols 1999) and containing 11 putative enzyme domains. Eight of these domains, including malonyl CoA : ACP acetyltransferase, 3-ketoacyltransferase and chain length factor and a cluster of 6 putative ACP domains, were more strongly related to PKS rather than FAS proteins. Three other regions appeared to be homologues of the bacterial FAS proteins, enoyl reductase and dehydrase from bacteria (Bauer 1997). *Photobacterium profundum* and *Vibrio marinus* possess similar PKS-gene clusters (Feller and Gerday 1997).

The membrane-bound desaturase and elongase enzymes from *Schizochytrium* appear not to be involved in PUFA synthesis. Genetic analysis of fatty acid biosynthesis in *Schizochytrium* and *Ulkenia* species has indicated the presence of the PKS-like system (Ratledge 2004. Unlike the desaturation-elongation FA synthesis system, oxygen is not required for synthesis of the double bonds in the manner required by conventional desaturases and hence overall NADPH requirement for PUFA synthesis is greatly reduced. The 8 PKS domains found in *Shewanella* are also present in *Schizochytrium* although the genetic sequences are different. The domains encode:

- condensing enzyme
- 3-ketoacyl synthase (KS)
- malonyl-CoA:ACP acyltransferase (MAT)
- acyl carrier protein (ACP)
- 3-ketoacyl-ACP-reductase (KR)
- acyltransferase (AT)
- chain length factor (CLF)
- enoyl reductase (ER) and
- dehydrase/isomerase (DH)

EPA and DHA double bonds are located at every third carbon, while the PKS cycle adds 2-C units to the chain. Enzyme mechanisms, involving cis and trans isomerizations possibly encoded by the unassigned PKS protein domains, which would achieve biosynthesis of the correctly configured double bonds in DHA and EPA, have been postulated. A feature of the fatty

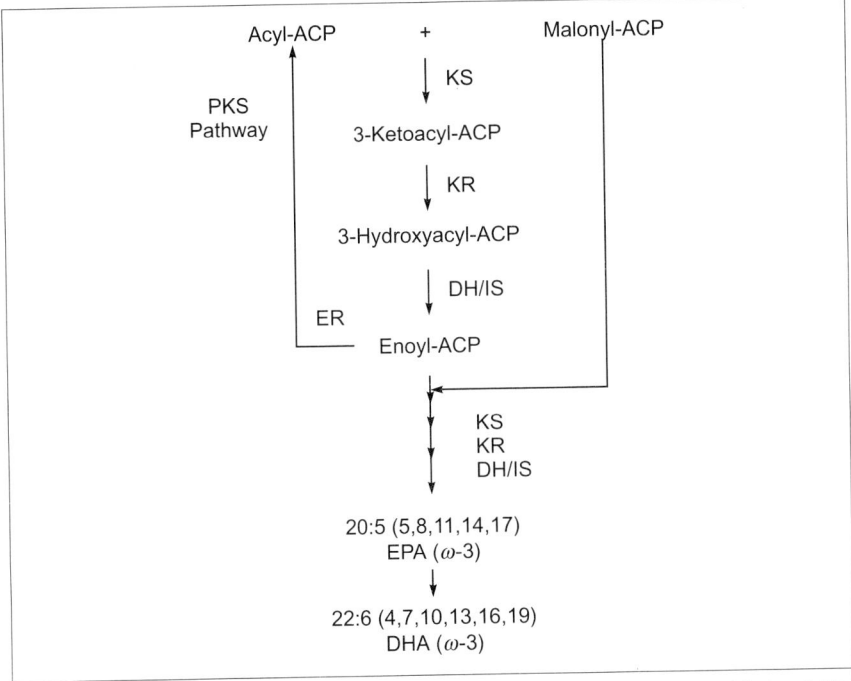

Fig. 2   The suggested polyketide synthase (PKS) pathway (anaerobic) for DHA biosynthesis in *Schizochytrium sp.* and the marine bacteria *Shewanella* sp. (Wallis et al. 2002; Qiu, 2003; Ratledge 2004). ACP, acyl carrier protein; KS, 3-ketoacyl-ACP synthase; KR, 3-ketoacyl-ACP reductase; DH/IS, dehydrase/isomerase; ER, enoyl reductase.

acid composition of these thraustochytrid oils is the presence of DPA and the mechanism of DPA synthesis is not yet clear.

A Δ-4 desaturase can catalyze the conversion of docosapentaenoic acid (22:5n-3) to form DHA (22:6n-3), and a gene encoding a protein with Δ-4 desaturase activity in yeast and plants has been cloned from *Thraustochytrium sp.* (Qiu 2003). This observation was used to suggest that while the PKS pathway predominates in *Schizochytrium*, the desaturation/ elongation pathway may dominate in *Thraustochytrium*. Given the similarities in fatty acid profiles between oils of *Schizochytrium*, *Ulkenia* and *Thraustochytrium* species and also their close taxonomic characteristics, Ratledge (2004) has cautioned against this conclusion. Indeed the use of a PKS-system by thraustochytrids in general in single cell oil production would provide an explanation as to why intermediate chain length fatty acids are not observed in the oils from these species.

CONCLUSION

The beneficial health effects of PUFAs have been recognized and generated a substantial public awareness of the importance of these essential fatty acids to general physical and mental well-being. These microbial supplements are now widely accepted by regulatory agencies and by the public. Future market for PUFAs appears to be related to increasing PUFA content of the human diet through dietary supplements using transgenic plant and animal products (Fraser et al. 2004). It is likely that designer plant oils containing PUFAs will emerge from transgenic plants for addition to animal milks, infant formulae and other foods and feeds (Lopez and Garcia 2000; Galili et al. 2002).

Microorganisms are capable of producing individual PUFAs as the dominant fatty acid in their lipid, which greatly simplifies the recovery and purification. Another advantage of microbial system is an ease of creating recombinant strains tailored to produce PUFA-oils of defined composition. The fermentation-based technology with high oil production rates is being exploited commercially for ARA and DHA production. Areas for future research in single cell oil production include increasing yields of existing products in current industrial strains of *Schizochytrium*, *Ulkenia* and *Thraustochytrium*, and investigate potential applications of genetically engineered heterotrophic algae. Opportunities exist in developing new strains capable of producing novel PUFA intermediates with novel bioactivity properties and develop biotransformation systems for bioorganic synthesis involving PUFAs.

REFERENCES

Agren JJ, Hanninen O, Julkenen A, Fogelholm L, Maliranta H, SchwabU, Pynnonen O, Uusitupa M (1996). Fish diet, fish oil and docosahexaenoic acid triglyceride (DHASCO oil). Food Chem Toxicol 38: 35-49.

Apt KE, Behrens PW (1999). Commercial developments in microalgal biotechnology. J Phycol 35: 215-226.

Arterburn LM, Boswell KD, Henwood SM, Kyle DJ (2000). A developmental safety study in rats using DHA- and ARA-rich single cell oils. Food Chem Toxicol 38: 763-771.

Baily RB, DiMasi D, Hansen JMM, Mirrasoul PJ, Ruecker CM, Veeder III GT, Kaneko T, Barclay WR (2003). Enhanced production of lipids containing polyenoic fatty acid by very high density cultures of eukaryotic microbes in fermenters. United States Patent 6,607,900.

Bajpai P, Bajpai P, Ward OP (1991a). Production of docosahexaenoic acid (DHA) by *Thraustochytrium aureum*. Appl Microbiol Biotechnol 35: 706-710.

Bajpai P, Bajpai P, Ward OP (1991b). Optimisation of production of docosahexaenoic acid (DHA) by *Thraustochytrium aureum* ATCC 34304. J Am Oil Chem Soc 68: 509-514.

Barclay WR, Zeller S (1996). Nutritional enhancement of n-3 and n-6 fatty acids in rotifers and *Artemia napuli* by feeding spray-dried *Schizochytrium* sp. J World Aquacult Soc 27: 314-322.

Barclay WR (1992). Process for heterotrophic production of microbial products with high concentrations of omega-3 highly unsaturated fatty acids. United States Patent 5,130,242.

Barclay WR (1997). Method of aquaculture feeding microflora having a small cell aggregate size. United States Patent 5,688,500.

Bauer CC (1997). Suppression of heterocyst differentiation in *Anabaena* PCC 7120 by a cosmid carrying wild-type genes encoding enzymes for fatty acid synthesis. FEMS Microbiol Lett 151: 23-30.

Becker CC, Kyle DJ (1998). Developing functional foods containing algal docosahexaenoic acid. Food Technology 52: 68-71.

Behrens PW, Kyle DJ (1996). Microalgae as a source of fatty acids. J Food Lipids 3: 259-272.

Bongiorni L, ad Dini F (2002). Distribution and abundance of thraustochytrids in different Mediterranean coastal habitats. Aquat Microb Ecol 30: 49-57.

Bowles RD, Hunt AE, Bremer GB, Duchars MG, Eaton RA (1999). Long-chain n-3 polyunsaturated fatty acid production by members of the marine protistan group the thraustochytrids: screening of isolates and optimization of docosahexaenoic acid production. J Biotechnol 70: 193-202.

Cavalier-Smith T, Allsop MTEP, Chao EE (1994). Thraustochytrids are chromists, not Fungi: 18s rRNA signatures of heterokonta. Phil Trans R Soc Lond 346: 387-397.

Cohen Z, Norman HA, Heimer YM (1995). Microalgae as a source of omega-3 fatty acids. World Rev Nutr Diet 77: 1-31.

Colquhoun DM (2001). Nutraceuticals: vitamins and other nutrients in coronary heart disease. Curr Opin Lipidol 12: 639-646.

Cunnane SC (2003). Enduring issues about essential fatty acids: time for a new paradigm Prog Lipid Res 48: 544-568.

deUequiza AM (2000). Docosahexaenoic acid, a ligand for the retinoid X receptor in mouse brain. Science 294: 1871-1875.

Dueppen DG, Zeller SG, Diltz SI, Driver R (2005). Extraction and winterization of lipids from oilseeds and microbial sources. US Patent Appl. 2005115897.

Ellenbogen BB, Aaronson S, Goldstein S, Belsky M (1969). Polyunsaturated fatty acids of aquatic fungi: possible phylogenetic significance. Comp Biochem Physiol 29: 805-811.

Fan KW, Chen F, Jones EBG, Vrijmoed LLP (2001). Eicosapentaenoic and docosahexaenoic acids production by and okara-utilizing potential of thraustochytrids. J Indust Microbiol Biotechnol 27: 199-202.

Fan KW, Vrijmoed LLP, Jones EBG (2002). Physiological studies of subtropical mangrove thraustochydrids. Botan. Mar 45: 50-57.

Fang J, Kato C, Sato T, Chan O, McKay D (2004). Biosynthesis and dietary uptake of polyunsaturated fatty acids by piezophilic bacteria. Comp Biochem Physiol Part B 137: 455–461.

FDA (2001). Consumer advisory: an important message for pregnant women and women of childbearing age who may become pregnant about the risks of mercury in fish. Center for Food Safety and Applied Nutrition (http://www.cfsan.fda.gov/~dms/admehg.html).

Feller G, Gerday C (1997). Psychrophilic enzymes - molecular basis of cold adaptation. Cell Mol Life Sci 53: 830-841.

Fenton WS, Hibbeln J, Knable M (2000). Essential fatty acids, lipid membrane abnormalities, and the diagnosis and treatment of schizophrenia. Biol Psych 47 : 8-21.

Field C (1995). Journey amongst mangroves. International Society for Mangrove Ecosystems, Okinawa, Japan. South China Printing Co., Hong Kong.

Frankel EN, Statue-Gracia T, Meyer AS, German JB (2002). Oxidative stability of fishb and algae oils containing long-chain polyunsaturated fatty acids in bulk and in oil-in-water emulsions. J Agric Food Chem 50: 2094-2099.

Fraser TCM, Qi B, Elhussein S, Chatrattanakunchai S, Stobart AK, Lazarus CM (2004). Expression of the *Isochrysis* C18-δ polyunsaturated fatty acid specific elongase component alters arabidopsis glycerolipid Profiles. Plant Physiol 135: 859-866.

Funk CD (2001). Prostaglandins and leukotrienes: advances in eicosanoid biology. Science 294: 1871-1875.

FSANZ. DHA-rich dried marine micro-algae (*Schizochytrium sp.*) and DHA-rich oil derived from *Schizochytrium* sp. as novel food ingredients. Final Report, Application no. A428, Food Standards Australia New Zealand (2002).

FSANZ. DHA-rich micro-algal oil from *Ulkenia* sp. as a novel food. Final Report, Application no. A522, Food Standards Australia New Zealand (2005).

Gaertner S (1977). Revision of the Thraustochytriaceae (lower marine fungi). I. *Ulkenia* nov. gen., with description of three new species. Veroffentlichungen des Instituts fur Meeresforschung in Bremerhaven 16: 139-157.

Galili G, Galili S, Lewinsohn E, Tadmor Y (2002). Genetic, molecular and genomic approaches to improve the value of plant foods and feeds. Crit Rev Plant Sci 21: 167–204.

Hammond BG, Mayhew DA, Naylor MW, Ruecker FA, Mast RW, Sander WJ (2001a). Safety assessment of DHA-rich microalgae from *Schizochytrium* sp. 1. Subchronic rat feeding study. Regul Toxicol Pharmacol 33: 192-204.

Hammond BG, Mayhew DA, Holson JF, Nemec MD, Mast RW, Sander WJ (2001 b). Safety assessment of DHA-rich microalgae from *Schizochytrium* sp. II. Developmental toxicity evaluation in rats and rabbits. Regul Toxicol Pharmacol 33: 205-217.

Hoeksema SD (2000). Two-phase extraction of oil from biomass. United States Patent 6: 166 -231.

Hopwood DA, Sherman DH (1990). Molecular genetics of polyketides and its comparison to fatty acid biosynthesis. Ann Rev Genet 24: 2437-2466.

Huang J, Aki T, Yokochi T, Nakahara T, Honda D, Kawamoto S (2003). Grouping newly isolated docosahexaenoic acid-producing thraustochytrids based on their polyunsaturated fatty acid profiles and comparative analysis of 18S rRNA genes. Mar Biotechnol 5: 450-457.

Hutchinson CR, Fujii I (1995). Polyketide synthase gene manipulation – a structure-function approach in engineering novel antibiotics. Ann Rev Microbiol 49: 201-238.

Hyde KD, Sarma VV, Jones EBG (2000). Morphology and taxonomy of higher marine fungi. In: Marine mycology - A practical approach. (Eds. K.D. Hyde and S.B. Pointing) pp. 172-204. Fungal Diversity Press, Hong Kong.

Iida I, Nakahara T, Yokochi T, Kamisaka Y, Yagi H, Yamaoka M, Suzuki O (1996). improvement of docosahexaenoic acid production in a culture of *Thraustochytrium aureum* by medium optimization. J Ferm Bioeng 81: 76-78.

ISSFAL. Recommendations for intake of polyunsaturated fatty acids in healthy adults.

Report of the Sub-Committee of International Society for the Study of Fatty Acids and Lipids, UK (2004).

Inniss SM, Hansen JW (1999). Plasma fatty acid responses, metabolic effects and safety of microalgal and fungal oils rich in arachidonic and docosahexaenoic acids in healthy adults. Am J Clin Nutr 64: 159-167.

Katz L, Donadio S (1993). Polyketide synthesis – prospects for hybrid antibiotics. Ann Rev Microbiol 47: 875-912.

Kumar SR, Anil AC, Khandeparker L, Patil JS (2000). Thraustochytrid protists as a component of marine microfilms. Mar Biol 136: 603-609.

Kumar SR, Chandramohan DR, Desa E (2002). Method for enhancing levels of polyunsaturated fatty acids in thraustochytrid fungi. US Patent 6,410,282.

Kyle DJ (1994). Microbial oil mixtures and uses thereof. US Patent 5,374,657.

Kyle DJ (1996). Arachidonic acid and methods for the production and use thereof. PCT Patent WO96/21037.

Kyle DJ (1997). Arachidonic acid and methods for the production and use thereof. United States Patent 5,658,767.

Leander CA (2002). Phylogeny of the labyrinthulomycota. Dissertation Abstracts International Part B: Science and Engineering 63, 25.

Lewis TE, Nichols PD, McMeekin TA (1999). The biotechnological potential of Thraustochytrids. Mar Biotechnol 1: 580-587.

Li Z-Y, Ward OP (1994 a). Production of docosahexaenoic acid by *Thraustochytrium roseum*. J Indust Microbiol 13: 238-241.

Li Z-Y, Ward OP (1994 b). Synthesis of monoglyceride containing omega-3 fatty acids by microbial lipase in organic solvent. J Indust Microbiol 13: 49-52.

Li Z-Y, Ward OP (1996). Enzymatically solvent-free synthesis of sugar ester containing ω-3 polyunsaturated fatty acid. Chin Chem Lett 7: 611-614.

Lopez AD, Garcia MF (2000). Plants as chemical factories for the production of polyunsaturated fatty acids. Biotechnol Adv 18: 481–497.

Manoharachary C, Sridhar K, Singh R, Adholeya A, Suryanarayanan TS, Rawat S, Johri BN (2005). Fungal biodiversity: distribution, conservation and prospecting of fungi from India. Curr Sci 89: 58-71.

Metz JG, Roessler P, Facciotti D, Levering C, Dittrich F, Lassner M, Valentine R, Lardizabal K, Domergue F, Yamada A, Yazawa K, Knauf V, Browse J (2001). Polyketide synthases produce polyunsaturated fatty acids in both prokaryotes and eukaryotes. Science 293: 290-293.

Pannu J, Singh A, Ward OP (2003). Influence of vegetable oil on microbial degradation of polycyclic aromatic hydrocarbons. Can J Microbiol 49: 508-513.

Pannu J, Singh A, Ward OP (2004). Vegetable oil as a contaminated soil remediation amendment: application of peanut oil for extraction of polycyclic aromatic hydrocarbons from soil. Proc Biochem 39: 1211-1216.

Peet M (2004). Nutrition and schizophrenia: beyond omega-3 fatty acids. Prostagl. Leukotr. Essen Fat Acid 70: 417-422.

Qiu X (2003). Biosynthesis of docosahexaenoic acid (DHA, 22:6-4, 7,10,13,16,19): two distinct pathways. Prostagl Leukotr Essent Fat Acid 68: 181-186.

Qiu X, Hong H, MacKenzie SL (2001). Identification of a delta-4 fatty acid desaturase from *Thraustochytrium* sp. involved in the biosynthesis of docosahexanoic acid by heterologous expression in *Saccharomyces cerevisiae* and *Brassica juncea*. J Biol Chem 276: 31561-31566.

Ratledge C, Wynn JP (2001). The biochemistry and molecular biology of lipid accumulation in oleaginous microorganisms. Adv Appl Microbiol 51: 1-51.

Ratledge C (2004). Fatty acid biosynthesis in microorganisms being used for single cell oil production. Biochemie 86: 807-815.

Ruecker CM, Adu-peasah SP, Engelhardt BS, Veeder III GT (2004). Solventless extraction process. USP 6,750,048.

Russell NJ, Nichols DS (1999). Polyunsaturated fatty acids in marine bacteria – a dogma rewritten. Microbiology 145: 767-779.

Sakurdani E, Kobayashi M, Shimizu S (1999a). ?(9) fatty acid desaturase from arachidonic acid-producing fungus - Unique gene sequence and its heterologous expression in a fungus, *Aspergillus*. Eur J Biochem 260: 208-216.

Sakurdani E, Kobayashi M, Ashikara T, Shimizu S (1999b). Identification of Delta 12- fatty acid desaturase from arachidonic acid-producing *Mortierella* fungus by heterologous expression in the yeast *Saccharomyces cerevisiae* and the fungus *Aspergillus oryzae*. Eur J Biochem 261: 812-820.

Schmidt JO, Schearer CA (2003). A checklist of mangrove associated fungi, their geographical distribution and known host plants. Mycotaxon 70: 423-477.

Shanklin J, Cahoon EB (1998). Desaturation and related modifications of fatty acids. Ann Rev Plant Physiol Plant Mol Biol 49: 611-641.

Shimada Y, Sugihara A, Tominaga Y (2001). Enzymatic purification of polyunsaturated fatty acids. J Biosci Bioeng 91: 529-538.

Sijtsma L, deSwaaf ME (2004). Biotechnological production and applications of ω-3 polyunsaturated fatty acid docosahexaenoic acid. Appl Microbiol Biotechnol 64: 146-153.

Simopoulos AP, Leaf A, Salem N Jr (1999). Essentiality of and recommended dietary intakes for omega-6 and omega-3 fatty acids. Ann Nutr Metab 43: 127-130.

Simopoulos AP, Leaf A, Salem N Jr (1999). Workshop on the essentiality of and recommended dietary intakes for omega-6 and omega-3 fatty acids. Food Australia 5: 332-333.

Singh A, Ward OP (1996). Production of high yields of docosahexaenoic acid by *Thraustochytrium roseum* ATCC 28210. J Indust Microbiol 16: 370-373.

Singh A, Ward OP (1997 a). Microbial Production of docosahexaenoic acid (DHA, C22:6). Adv Appl Microbiol 45: 271-312.

Singh A, Ward OP (1997b). Production of high yields of arachidonic acid in a fed-batch system by *Mortierella alpina* ATCC 32222. Appl Microbiol Biotechnol 48: 1-5.

Spalding MD, Blasco F, Field CD (1997). World Mangrove Atlas. The International Society for Mangrove Ecosystems, Okinawa, Japan.

Sprecher H (2000). Metabolism of highly unsaturated n-3 and n-6 fatty acids. Biochim Biophys Acta 1486: 219-231.

Sridhar KR (2005). Diversity of fungi in mangrove ecosystems. In: (Eds. T. Satyanarayana and B.N. Johri) Microbial diversity: Current perspectives and potential applications. pp. 129-148. I.K. International, New Delhi.

Tanaka S, Yaguchi T, Shimizy S, Sogo T, Fujizawa S (2003). Process for preparing docosahexaenoic and docosapentaenoic acid with *Ulkenia*. USP 6509178.

Tisdale MJ (1999). Wasting in cancer. J Nutr 129: 243S-246S.

Uauy R, Valenzuela A (2000). Marine oils: the health benefits of n-3 fatty acids. Nutrition 16: 680-684.

Wallis JG, Watts JL, Browse J (2002). Polyunsaturated fatty acid synthesis: what will they think of next? Trend Biochem Sci 27: 467-473.

Ward OP. Fermentation Biotechnology. Open University Press, UK (1989).

Ward OP. Bioprocessing. Open University Press, UK (1991).

Ward OP. Microbial production of long-chain polyunsaturated fatty acids. INFORM 6, 683-688 (1995).

Ward OP. Singh A (2005). Omega-3/6 fatty acids: alternative sources of production. Proc Biochem (in press).

Yaguchi T, Yokochi T, Nakahara T, Higashihara T (1997). Production of high yields of docosahexaenoic acid by *Schizochytrium* sp. strain SR21. J Am Oil Chem Soc 74: 1431-1434.

Yokochi T, Honda D, Higashihara T, Nakahara T (1998). Optimization of docosahexaenoic acid production by *Schozochytrium limacinum* SR21. Appl Microbiol Biotechnol. 49: 72-76.

Yongmanitchai W, Ward OP (1989). Omega-3 fatty acids: alternative sources of production. Proc Biochem. 24: 117-125.

Index